Bird Conservation

Global evidence for the effects of interventions

David R. Williams, Robert G. Pople,
David A. Showler, Lynn V. Dicks, Matthew F. Child,
Erasmus K. H. J. zu Ermgassen and
William J. Sutherland

Synopses of Conservation Evidence, Volume 2

Pelagic Publishing | www.pelagicpublishing.com

Published by **Pelagic Publishing**
www.pelagicpublishing.com
PO Box 725, Exeter EX1 9QU

Bird Conservation
Global evidence for the effects of interventions
Synopses of Conservation Evidence, Volume 2

ISBN 978-1-907807-19-0 (Pbk)
ISBN 978-1-907807-20-6 (Hbk)

This book should be quoted as Williams, D. R., Pople, R. G., Showler, D. A., Dicks, L. V., Child, M. F., zu Ermgassen, E. K. H. J. and Sutherland, W. J. (2013) *Bird Conservation: Global evidence for the effects of interventions.* Exeter, Pelagic Publishing.

British Library Cataloguing in Publication Data
A catalogue record for this book is available from the British Library.

Contents

Advisory board

We thank the following people for advising on the scope and content of this synopsis:

Dr Andrew Brown, Natural England
Dr David Gibbons, Royal Society for the Protection of Birds
Dr Roger Mitchell, Natura International
Dr Jörn Scharlemann, UNEP-World Conservation Monitoring Centre
Dr Gavin Siriwardena, British Trust for Ornithology
Dr Alison Stattersfield, BirdLife International
Dr David Stroud, Joint Nature Conservation Committee
Professor Des Thompson, Scottish Natural Heritage

About the authors

David Williams is a Doctoral Student in the Department of Zoology, University of Cambridge.

Robert G. Pople is a former Research Assistant in the Department of Zoology, University of Cambridge.

David A. Showler is a Research Associate in the School of Biological Sciences, University of East Anglia and the Department of Zoology, University of Cambridge.

Lynn Dicks is a Research Associate in the Department of Zoology, University of Cambridge.

Matthew F. Child is a Research Assistant in the Department of Zoology, University of Cambridge.

Erasmus K. H. J. zu Ermgassen is a Student in the Department of Veterinary Medicine, University of Cambridge.

William J. Sutherland is the Miriam Rothschild Professor of Conservation Biology at the University of Cambridge.

Acknowledgements

This synopsis was funded by the Natural Environment Research Council and Arcadia.

We would also like to thank Dr Stephanie Prior for providing support throughout the project and all the people who gave help and advice and allowed us access to their research: Helen Baker (Joint Nature Conservation Committee), Leon Bennun (BirdLife International), Mark Bolton (Royal Society for the Protection of Birds, RSPB), Chris Bowden (RSPB), Jöel Bried (University of the Azores), Rhys Bullman (Scottish Nature Conservancy), Bob Carling (Pelagic Publishing), Motti Charter (Tel Aviv University), Mariano Codesido (University of Buenos Aires), James Dwyer (Virginia Tech), Rob Field (RSPB), Graham Fulton (Edith Cowan University), Emma Gyuris (James Cook University), Wendy Johnson (Hawaii Audubon Society), Tim Johnson (UNEP – World Conservation Monitoring Centre), Naoki Katayama (University of Tokyo), Andrew Kelly (Royal Society for the Prevention of Cruelty to Animals), Will Kirby (RSPB), Raivo Mänd (University of Tartu), Nigel Massen (Pelagic Publishing), Marcos Moleón Páiz (University of Granada), Roberto Muriel (Consejo Superior de Investigaciones Cientificas), Itziar Olmedo (University of Cambridge), Ray Poulin (Royal Saskatchewan Museum), David Priddel (Department of Environment, Climate Change and Water, New South Wales), Casey Primacio (Hawaii Audubon Society), C. John Ralph (US Department of Agriculture Forest Service), Michel Robert (Canadian Wildlife Service), Maurizio Sarà (University of Palermo), Debbie Saunders (Australian National University), Phil Seddon (University of Otago), Danaë Sheehan (RSPB), Jennifer Smart (RSPB), Ian Smales (Biosis Research), John Smallwood (Montclair State University), Juliet Vickery (British Trust for Ornithology) and Matt Wagner (Texas Parks and Wildlife Department).

1. About this book

The purpose of Conservation Evidence synopses

Conservation Evidence synopses **do**	Conservation Evidence synopses **do not**
• Bring together scientific evidence captured by the Conservation Evidence project (over 3,000 studies so far) on the effects of interventions to conserve biodiversity	• Include evidence on the basic ecology of species or habitats, or threats to them
• List all realistic interventions for the species group or habitat in question, regardless of how much evidence for their effects is available	• Make any attempt to weight or prioritise interventions according to their importance or the size of their effects
• Describe each piece of evidence, including methods, as clearly as possible, allowing readers to assess the quality of evidence	• Weight or numerically evaluate the evidence according to its quality
• Work in partnership with conservation practitioners, policymakers and scientists to develop the list of interventions and ensure we have covered the most important literature	• Provide answers to conservation problems. We provide scientific information to help with decision-making

Who is this synopsis for?

If you are reading this, we hope you are someone who has to make decisions about how best to support or conserve biodiversity. You might be a land manager, a conservationist in the public or private sector, a farmer, a campaigner, an advisor or consultant, a policymaker, a researcher or someone taking action to protect your own local wildlife. Our synopses summarise scientific evidence relevant to your conservation objectives and the actions you could take to achieve them.

We do not aim to make your decisions for you, but to support your decision-making by telling you what evidence there is (or isn't) about the effects that your planned actions could have.

When decisions have to be made with particularly important consequences, we recommend carrying out a systematic review, as the latter is likely to be more comprehensive than the summary of evidence presented here. Guidance on how to carry out systematic reviews can be found from the Centre for Evidence-Based Conservation at the University of Bangor (www.cebc.bangor.ac.uk).

The Conservation Evidence project

The Conservation Evidence project has three parts:

An online, **open access journal** *Conservation Evidence* publishes new pieces of research on the effects of conservation management interventions. All our papers are written by, or in conjunction with, those who carried out the conservation work and include some monitoring of its effects.

An ever-expanding **database of summaries** of previously published scientific papers, reports, reviews or systematic reviews that document the effects of interventions.

Synopses of the evidence captured in parts one and two on particular species groups or habitats. Synopses bring together the evidence for each possible intervention. They are freely available online and available to purchase in printed book form.

These resources currently comprise over 3,000 pieces of evidence, all available in a searchable database on the website www.conservationevidence.com.

Alongside this project, the Centre for Evidence-Based Conservation (www.cebc.bangor.ac.uk) and the Collaboration for Environmental Evidence (www.environmentalevidence.org) carry out and compile systematic reviews of evidence on the effectiveness of particular conservation interventions. These systematic reviews are included on the Conservation Evidence database.

Of the 322 bird conservation interventions identified in this synopsis, five are the subjects of current systematic reviews:

- How does the impact of grazing on heathland compare with the impact of burning, cutting or no management? http://www.environmentalevidence.org/SR14.html
- Is predator control an effective strategy for enhancing bird populations? http://www.environmentalevidence.org/SR38.html.
- Do matrix features affect species movement? http://www.environmentalevidence.org/SR43.html
- Does structural connectivity facilitate dispersal of native species in Australia's fragmented terrestrial landscape? http://www.environmentalevidence.org/SR44.html
- How do thinning and burning treatments in south-western conifer forests in the United States affect wildlife distribution, abundance and population performance? http://www.environmentalevidence.org/SR66.html

In addition, three systematic reviews provide important information on the impacts of threats on bird populations:

- Effects of wind turbines on bird abundance. http://www.environmentalevidence.org/SR4.html
- What is the impact of public access on the breeding success of ground-nesting and cliff-nesting birds? http://www.environmentalevidence.org/SR16.html
- What are the impacts of human recreational activity on the distribution, nest-occupancy rates and reproductive success of breeding raptors? http://www.environmentalevidence.org/SR27.html

Another provides evidence for how to apply an intervention:

- Do trapping interventions effectively reduce or eradicate populations of the American mink (*Mustela vison*)? http://www.environmentalevidence.org/SR7.html.

There are several interventions which we feel would benefit significantly from systematic reviews:

- Interventions to reduce the impact of electricity pylons and power lines
- Interventions to reduce seabird bycatch
- The provision of artificial nest sites
- The provision of supplementary food

Scope of the Bird Conservation synopsis

This synopsis covers evidence for the effects of conservation interventions for native (see below), wild birds.

It is restricted to evidence captured on the website www.conservationevidence.com. It includes papers published in the journal *Conservation Evidence*, evidence summarised on our database and systematic reviews collated by the Collaboration for Environmental Evidence.

We have gathered evidence from all around the world, and the apparent over (or under-representation) of some regions reflects the current biases in published research papers available to Conservation Evidence.

Native vs. non-native species

This synopsis does not include evidence from the substantial literature on husbandry of domestic birds, or non-native gamebirds (e.g. common pheasants *Phasianus colchicus* in Europe and North America). However, where these interventions affect native species, or are relevant to the conservation of native, wild species, they are included (e.g. management of farmland for common pheasants has a significant impact on several declining native songbirds in the UK, see Stoate (2002) in 'Manage hedges to benefit wildlife', 'Plant nectar flower mixture/wildflower strips', 'Plant wild bird seed cover strips', 'Provide supplementary food for birds', 'Create beetle banks', 'Control predators not on islands – songbirds', 'Reduce pesticide or herbicide use generally'.

How we decided which conservation interventions to include

Our list of interventions has been agreed in partnership with an Advisory Board made up of international conservationists and academics with expertise in bird conservation. Although the list of interventions may not be exhaustive, we have tried to include all actions that have been carried out or advised to support populations or communities of wild birds.

How we reviewed the literature

In addition to evidence already captured by the Conservation Evidence project, we have searched the following sources for evidence relating to bird conservation:

- Fifteen specialist bird conservation journals, from their first publication to the end of 2010 (*African Bird Club Bulletin, The Auk, Bird Conservation International, Bird Study, BTO Research Reports, Emu, Ibis, Journal of Avian Biology* – formerly *Ornis Scandinavica, Journal of Field Ornithology, Journal Raptor Research* – formerly *Raptor Research, Ornitologia Neotropical, RSPB Research Reports, The Condor, Waterbirds* – formerly *Colonial Waterbirds, Wilson Journal of Ornithology* – formerly *Wilson Bulletin*)
- Twenty general conservation journals over the same time period.
- Where we knew of an intervention which we had not captured evidence for, we performed keyword searches on ISI Web of Science and www.scholar.google.com for this intervention.

Individual studies covered in this synopsis are all included in full or in summary on the Conservation Evidence website.

The criteria for inclusion of studies in the Conservation Evidence database are as follows:

- There must have been an intervention that conservationists would do
- Its effects must have been monitored quantitatively

In some cases, where a body of literature has strong implications for conservation of a particular species group or habitat, although it does not directly test interventions for their effects, we refer the reader to this literature, but present no evidence.

How the evidence is summarised

Conservation interventions are grouped primarily according to the relevant direct threats, as defined in the International Union for the Conservation of Nature (IUCN)'s Unified

Classification of Direct Threats (www.iucnredlist.org/technical-documents/classification-schemes/habitats-classification-scheme-ver3). In most cases, it is clear which main threat a particular intervention is meant to alleviate or counteract.

Not all IUCN threat types are included, only those that threaten birds, and for which realistic conservation interventions have been suggested.

Some important interventions can be used in response to many different threats, and it would not make sense to split studies up depending on the specific threat they were studying. We have therefore separated out these interventions, following the IUCN's Classification of Conservation Actions (http://www.iucnredlist.org/technical-documents/classification-schemes/conservation-actions-classification-scheme-ver2). The actions we have separated out are: 'Habitat protection', 'Education and community development', 'Habitat restoration and creation', 'General responses to small/declining populations' and 'Captive breeding, rearing and releases (*ex situ* conservation)'. These respectively match the following IUCN categories: 'Land/water protection', 'Education and awareness' and 'Livelihood, economic and other incentives', 'Land/water management – Habitat and natural process restoration', and 'Species Management'.

Normally, no intervention is listed in more than one place, and when there is ambiguity about where a particular intervention should fall there is clear cross-referencing. Some studies describe the effects of multiple interventions. When this is the case, cross-referencing is again used to direct readers to the other interventions investigated. Where a study has not separated out the effects of different interventions, the study is only described once, but readers are directed to it from the other interventions.

In the text of each section, studies are presented in chronological order, so the most recent evidence is presented at the end. The summary text at the start of each section groups studies according to their findings.

At the start of each chapter, a series of **key messages** provides a rapid overview of the evidence. These messages are condensed from the summary text for each intervention.

Background information is provided where we feel recent knowledge is required to interpret the evidence. This is presented separately and relevant references included in the reference list at the end of each intervention section.

References containing evidence for the effects of interventions are summarised in more detail on the Conservation Evidence website. In electronic versions of the synopsis, they are hyperlinked directly to the summary. If you do not have access to the electronic version of the synopsis, searching for the reference details or the species name on www.conservationevidence.com is the quickest way to locate summaries.

The information in this synopsis is available in three ways:

- As a book, printed by Pelagic Publishing and for sale from www.nhbs.com
- As a pdf to download from www.conservationevidence.com
- As text for individual interventions on the searchable database at www.conservationevidence.com.

Terminology used to describe evidence

Unlike systematic reviews of particular conservation questions, we do not quantitatively assess the evidence, or weight it according to quality. However, to allow you to interpret evidence, we make the size and design of each trial we report clear. The table below defines the terms that we have used to do this.

The strongest evidence comes from randomised, replicated, controlled trials with paired-sites and before and after monitoring.

Term	Meaning
Site comparison	A study that considers the effects of interventions by comparing sites that have historically had different interventions or levels of intervention.
Replicated	The intervention was repeated on more than one individual or site. In conservation and ecology, the number of replicates is much smaller than it would be for medical trials (when thousands of individuals are often tested). If the replicates are sites, pragmatism dictates that between five and ten replicates is a reasonable amount of replication, although more would be preferable. We provide the number of replicates wherever possible, and describe a replicated trial as 'small' if the number of replicates is small relative to similar studies of its kind.
Controlled	Individuals or sites treated with the intervention are compared with control individuals or sites not treated with the intervention.
Paired sites	Sites are considered in pairs, within which one was treated with the intervention and the other was not. Pairs of sites are selected with similar environmental conditions, such as soil type or surrounding landscape. This approach aims to reduce environmental variation and make it easier to detect a true effect of the intervention.
Randomised	The intervention was allocated randomly to individuals or sites. This means that the initial condition of those given the intervention is less likely to bias the outcome.
Before-and-after trial	Monitoring of effects was carried out before and after the intervention was imposed.
Review	A conventional review of literature. Generally, these have not used an agreed search protocol or quantitative assessments of the evidence.
Systematic review	A systematic review follows an agreed set of methods for identifying studies and carrying out a formal 'meta-analysis'. It will weight or evaluate studies according to the strength of evidence they offer, based on the size of each study and the rigour of its design. All environmental systematic reviews are available at: www.environmentalevidence.org/index.htm

Taxonomy

We have followed the taxonomy used in BirdLife International's 2011 checklist (http://www.birdlife.org/datazone/info/taxonomy), updating the names used in original papers where necessary. We have always referred to the species name used in the original paper as well. Where possible, common names and Latin names are both given the first time each species is mentioned within each intervention.

Where interventions have a large literature associated with them we have sometimes divided studies along taxonomic or functional lines. These do not follow strict taxonomic divisions, but instead are designed to maximise their utility. For example, storks, herons and ibises are often included together as both groups are large wading birds and may respond to interventions in similar ways.

Habitats

Where interventions have a large literature associated with them and effects could vary between habitats, we have divided the literature using the IUCN Habitat Classification Scheme (Version 3.0), available from www.iucnredlist.org.

Significant results

Throughout the synopsis we have quoted results from papers. Unless specifically stated, these results reflect statistical tests performed on the results.

Multiple interventions

Many studies investigate several interventions at once. When the effects of different interventions are separated, then the results are discussed separately in the relevant sections. However, often the effects of multiple interventions cannot be separated. When this is the case, the study is included in the section on each intervention, but the fact that several interventions were used is highlighted.

How you can help to change conservation practice

If you know of evidence relating to bird conservation that is not included in this synopsis, we invite you to contact us, via the www.conservationevidence.com website. Following guidelines provided on the site, you can submit a summary of a previously published study, or submit a paper describing new evidence to the Conservation Evidence journal. We particularly welcome summaries written by the authors of papers published elsewhere, and papers submitted by conservation practitioners.

2. Habitat protection

Habitat destruction is the largest single threat to biodiversity, and the spread of agriculture into natural habitats alone threatens 1,065 species of birds (87% of all threatened species) (BirdLife International 2008). Habitat protection is therefore one of the most frequently used conservation interventions, particularly in the tropics and in other areas with large areas of surviving natural vegetation.

Habitat protection can be through the designation of legally protected areas (PAs), using national or local laws; through the designation of Important Bird Areas (IBAs) or similar schemes, which, whilst providing no formal protection, may increase the profile of a site and make its conversion more difficult; or through the protection of entire habitat types, for example through the EU's Habitats Directive.

However, it can be difficult to measure the effectiveness of such areas: there may be no suitable controls; monitoring often only begins with the designation of the PA and PAs tend to be located in areas that would be less likely to be cleared even if it was not protected, making the prevention of agricultural expansion less politically difficult (Joppa & Pfaff 2011). Analysis of PAs often, therefore, requires large datasets, and this means that most studies investigating them use either satellite imagery (e.g. Joppa *et al.* 2008) or overall 'condition' scores (e.g. Mwangi *et al.* 2010), rather than data on bird populations, which are much harder to collect.

It is worth noting that the designation of a habitat or area as 'protected' (also known as *de jure* protection) does not necessarily mean protection in practical terms (*de facto* protection). The chapter on 'Threat: Biological resource use' contains several studies that examine the effect of greater *de facto* protection on bird populations.

Birdlife International (2008) *State of the world's birds: Indicators for our changing world.* Birdlife International.

Joppa, L. N., Loarie, S. R. & Pimm, S. L. (2008) On the protection of 'protected areas.' *Proceedings of the National Academy of Sciences,* 105, 6673–6678.

Joppa, L.N. & Pfaff, A. (2011) Global protected area impacts. Proceedings of the Royal Society B: *Biological Sciences,* 278, 1633–1638.

Mwangi, M. A. K., Butchart, S. H. M., Munyekenye, F. B., Bennun, L. A., Evans, M. I., Fishpool, L. D. C., Kanyanya, E., Madindou, I., Machekele, J., Matiku, P., Mulwa, R., Ngari, A., Siele, J. & Stattersfield, A. J. (2010) Tracking trends in key sites for biodiversity: a case study using Important Bird Areas in Kenya. *Bird Conservation International,* 20, 215–230.

Key messages

Legally protect habitats

Four studies from Europe found that populations increased after habitat protection and a review from China found high use of protected habitats by cranes. A replicated, randomised and controlled study from Argentina found that some, but not all bird groups had higher species richness or were at higher densities in protected habitats.

Ensure connectivity between habitat patches

Two studies of a replicated, controlled experiment in Canadian forests found that some species (not forest specialists) were found at higher densities in forest patches connected to continuous forest, compared to isolated patches and that some species used corridors more than clearcuts between patches.

Provide or retain un-harvested buffer strips

Three replicated studies from the USA found that species richness or abundances were higher in narrow (<100 m) strips of forest, but five replicated studies from North America found that wider strips retained a community more similar to that of uncut forest than narrow strips. Two replicated studies from the USA found no differences in productivity between wide and narrow buffers, but that predation of artificial nests was higher in buffers than in continuous forest.

2.1. Legally protect habitats

- Four studies (two replicated) from Europe (1,4–6) found population increases following habitat protection, more positive population trends in protected habitats, compared with outside, or with increases amounts of protected habitats.
- A literature review (2) reported that a large number of cranes (Gruidae) of seven species used nature reserves in China, whilst a replicated, randomised and controlled study from Argentina (3) found that some guilds of birds were found at higher species richnesses in protected forests, some at higher densities, and that some showed no differences.

A before-and-after study in the western Pyrenees, Spain (1), found that the population of griffon vultures *Gyps fulvus* increased from 282 pairs (in 23 colonies) in 1969–75 to 1,097 pairs (46 colonies) in 1989 following the initiation of multiple conservation interventions including the creation of reserves at nine breeding colonies (one in 1976, eight in 1987). This study is also discussed in more detail in 'Use legislative regulation to protect wild populations', 'Restrict certain pesticides or other agricultural chemicals' and 'Provide supplementary food to increase adult survival'.

A 1998 literature review (2) found that 25,500–31,800 cranes (Gruidae) of seven species used 32 nature reserves, established mainly for crane conservation, in China in 1994. This review is also discussed in 'Use education programmes and local engagement to help reduce pressures on species', 'Use legislative regulation to protect wild populations', 'Mark power lines to reduce incidental mortality', 'Provide supplementary food to increase adult survival' and 'Release captive-bred individuals'.

A replicated, randomised, controlled study in 1992–4 in Buenos Aires Province, Argentina (3), found that total bird abundance and species richness was significantly higher in September, December and March in a protected old-growth tala-coronillo woodland (free of human disturbance for >100 years) than in a woodland selectively logged for *Celtis tala* until 1960. Insect-eating bird density and species richness was higher in the protected woodland than in the exploited woodland and fruit-eating birds showed higher total density in the protected woodland in spring and summer but species richness was similar between both woodland types. There were no differences between sites for seed-eating birds.

A replicated study in 1997 in 19 nature reserves across England (4) found that they held consistently higher densities of northern lapwing *Vanellus vanellus*, common redshank *Tringa totanus* and common snipe *Gallinago gallinago* than the non-reserve Environmentally Sensitive Areas (an agri-environment scheme designation) surrounding them (densities approximately 730% higher for lapwing, 520% higher for redshank and 1600% higher for snipe). In addition, population trends were generally positive in reserves (except for snipe), but negative outside them (lapwing: 0.9–7.4% annual increase inside reserves vs. 0.7–13% decline outside; redshank: 3.9–8.6% increase vs. 1.8–18.6% decrease; snipe: 6.1–16.8% decrease in reserves vs. 7.3–29.7% decrease outside). The authors note that snipe have declined by approximately 70% across all reserves, due mainly to declines at a single reserve with a large population and on reserves with mineral soils (i.e. those with little organic matter). However, declines outside reserves were considerably higher than those on reserves (20% vs. 10% annually). This study is discussed in detail in 'Pay farmers to cover the costs of conservation measures' and 'Maintain traditional water meadows'.

A controlled, before-and-after study from 1970–2000 across Europe (5) found that targeted species in European Union (EU) countries, which were legally obliged to increase coverage of special protected areas (SPAs), had significantly more positive population trends after implementation of the directive and compared to non-EU countries (no implementation). Statistical models suggested that for every additional 1% increase in SPA area, the chances of all species experiencing positive population growth increased by 4%, with a 7% increase for target species. The authors argue that the stronger response of the target species provides direct evidence for the effectiveness of the EU Bird (79/409/EEC, est. 1979) and Habitats (92/43/EEC, est. 1992) Directives. Although non-target species' trends did not differ between EU and non-EU countries there was some evidence that these populations were more positive in EU countries with more extensive SPA networks.

A 2007 site comparison study of 677 plots covering 38,705 ha across southern England (6) found that for three wader species, population trends were most favourable in nature reserves, compared with farmland under the Environmentally Sensitive Areas scheme. Between 1982 and 2002, common redshank *Tringa totanus* declined by 70% in the wider countryside but increased by 160% in nature reserves outside Environmentally Sensitive Areas. Northern lapwing *Vanellus vanellus* showed a 48% decline in the wider countryside, and increased only in nature reserves outside Environmentally Sensitive Areas (by 55%) and reserves with Environmentally Sensitive Area enhancement (121%). Common snipe *Gallinago gallinago* breeding numbers decreased everywhere (commonly with declines of 90% or more), although declines were smaller in nature reserves outside Environmentally Sensitive Areas (–66%) and reserves in Environmentally Sensitive Area enhancement (–24%).

(1) Donazar, J. A. (1990) Population trends of the griffon vulture *Gyps fulvus* in northern Spain between 1969 and 1989 in relation to conservation measures. *Biological Conservation*, 53, 83–91.
(2) Davis, C. (1998) A review of the success of major crane conservation techniques. *Bird Conservation International*, 8, 19–30.
(3) Cueto, V. R. & Casenave, J. L. D. (2000) Bird assemblages of protected and exploited coastal woodlands in east-central Argentina. *The Wilson Bulletin*, 112, 395–402.
(4) Ausden, M. & Hirons, G. J. M. (2002) Grassland nature reserves for breeding wading birds in England and the implications for the ESA agri-environment scheme. *Biological Conservation*, 106, 279–291.
(5) Donald, P. F., Sanderson, F. J., Burfield, I. J., Bierman, S. M., Gregory, R. D. & Waliczky, Z. (2007) International conservation policy delivers benefits for birds in Europe. *Science*, 317, 810–813.
(6) Wilson, A., Vickery, J. & Pendlebury, C. (2007) Agri-environment schemes as a tool for reversing declining populations of grassland waders: mixed benefits from Environmentally Sensitive Areas in England. *Biological Conservation*, 136, 128–135.

2.2. Ensure connectivity between habitat patches

- A replicated, controlled study in Canada (2) found significantly higher abundances of some birds, but not forest specialists, in forest patches connected to a continuous area of forest, than in isolated patches.
- Another study of the same system (1) found evidence that corridors were used by some bird species more than clearcuts between patches, although corridors near cut forest were not used more than those near uncut stands.

Background

Habitat fragmentation, as well as destruction, may be an important driver of population declines. Small areas hold fewer species than large ones and if individuals are unable to cross areas of converted habitat then populations in separate habitat patches will become isolated. This potentially makes them more vulnerable to extinction, from natural variations in birth and death rates or sex ratios; from inbreeding depression and from outside pressures, both natural (such as storms or wildfires) and man-made (such as hunting or continued habitat loss), although the precise effects of habitat fragmentation, as opposed to loss, are debated (e.g. Fahrig 1997).

Theoretically, the number of species surviving in a habitat fragment is determined by its size and its effective distance to other habitat patches (MacArthur & Wilson 1967). Connecting remaining areas of habitat is therefore often seen as a way to increase the viability of populations, but there is considerable debate as to the effectiveness of such 'wildlife corridors' (e.g. Beier & Noss 1998).

Studies describing the effects of *creating* habitat corridors, rather than retaining them are described in 'Habitat restoration and creation'.

Beier, P. & Noss, R. F. (1998) Do habitat corridors provide connectivity? *Conservation Biology*, 12, 1241–1252.
Fahrig, L. (1997) Relative effects of habitat loss and fragmentation on population extinction. *The Journal of Wildlife Management*, 61, 603–610.
MacArthur, R. H. & Wilson, E. O. (1967) *The Theory of Island Biogeography*. Princeton University Press, Princeton, N.J.

A replicated, controlled study in boreal forests in 1993–5 in Alberta, Canada (1), found that significantly higher abundances of the ten most common under-storey birds were found in three riparian corridors between forest patches than in three clearcuts between patches. Only two of the ten were found nesting or foraging in clearcuts. In addition, significantly more juveniles used corridors following logging, than before, but only in one site. No more birds used the buffer strips near logged forest than similar strips near un-logged forest, when controlling for local abundances. Corridors consisted of 1–5 m of riparian vegetation and 90–110 m of forest. Visual surveys were used in clearcuts and mist nets in corridors.

A replicated, controlled study as part of the same study as (1) in mixed boreal forests in northern Alberta, Canada (2), found significantly higher abundances of resident songbirds and woodpeckers, but not of forest specialists, in forest plots connected to a continuous block when compared to isolated fragments. Resident species were found at similar abundances in connected fragments and unfragmented forests, whilst habitat generalists were found at similar abundances across all forest types. None of the individual species analysed appeared to benefit from connectivity. Forest fragments were 10 or 40 ha, either in continuous forest, isolated by a 200 m strip of clearcut on all sides or isolated on three of four sides for connected fragments. Three replicates of each treatment were established in the winter of 1993–4 and monitored until 1998.

(1) Machtans, C. S., Villard, M.-A. & Hannon, S. J. (1996) Use of riparian buffer strips as movement corridors by forest birds. *Conservation Biology*, 10, 1366–1379.
(2) Hannon, S. J. & Schmiegelow, F. K. A. (2002) Corridors may not improve the conservation value of small reserves for most boreal birds. *Ecological Applications*, 12, 1457–1468.

2.3. Provide or retain un-harvested buffer strips

- Four replicated studies from Canada and the USA (1,3,5,7) found that wider buffer strips retained a bird community more similar to that of uncut forest than narrower strips. Two replicated and controlled studies from the USA (3,4) found that several forest-specialist species were absent from buffers up to 70 m wide.
- Two replicated and controlled studies from the USA (3,7) found that richness was higher in buffers <100 m wide, compared to wider strips or forest. A replicated, controlled study in the USA (6) found that thinned buffer strips had lower abundances of forest species than unthinned strips, but higher abundances of early successional species. A replicated study from the USA (4) found that species richness was similar between 20–50 m buffers and original forest.
- A replicated study from the USA (4) found that bird abundances were higher in 20–50 m wide buffer strips than in original forest.
- A replicated study in the USA (8) found no differences in productivity of birds nests between buffer strips wider than 350 m, compared to those thinner than 250 m. Whilst a replicated, controlled study from the USA (2) found that predation of artificial nests was significantly higher in buffer strips compared with continuous forest, but that there was no difference between narrow and wide buffers.

Background

Provision or retention of forest strips in areas subject to timber extraction may be undertaken with the purported objectives of helping to mitigate the effects of loss of forest cover for woodland flora and fauna, as well as reducing potential problems such as soil erosion. Nature conservation aims include retaining valuable old forest features such as older trees with cavities and dead wood affording nest site and foraging opportunities for woodland birds. A similar intervention is 'Ensure connectivity between habitat patches', discussed in 'Habitat protection'.

A replicated study in balsam fir *Abies balsamea* stands in Quebec, Canada, (1), in 1989–92 found that 60 m-wide riparian forest buffer strips retained forest-dwelling breeding bird abundances and a species composition more similar to uncut areas than that of narrower strips. For one year before and three years following clearcutting, birds were surveyed in five buffer strips: 20 m-, 40 m-, 60 m-, and more than 300 m-wide (i.e. undisturbed control) strips, and a 20 m-wide thinned (33% of trees removed) strip. After initial increases in bird densities (30–70%) in all strips in the year after cutting, the 20 m- and 40 m-wide strips exhibited greatest decreases. Three years after cutting, forest species were less abundant (four songbirds becoming virtually absent) than habitat generalists in the 20 m strips (the thinned 20 m strip had densities around 20% less than the unthinned 20 m-wide strip).

A replicated, controlled study from June–July in 1994 in five mainstem buffer strips (60–80 m wide), five tributary buffer strips (20–40 m wide) and five continuously forested control sites within commercial forests in Maine, USA (2) found that avian nest predation rates were significantly higher in the buffer strips than in control sites (31% predation in tributary buffer strips, 23% in mainstem buffers vs. 15% in control sites). Red squirrels *Tamiasciurus hudsonicus* and blue jays *Cyanocitta cristata* were responsible for >50% of depredations. The authors suggest leaving wide (>150 m) buffer strips along riparian zones to reduce edge-related nest predation. Artificial nests were placed at five points (100 m apart) along transects that ran parallel to the stream.

A replicated, controlled study from May–June in 1994 in 12 riparian buffer-strip (18–70 m wide) sites and four unlogged riparian sites of old-growth forest in Oregon, USA (3) found that 27 species were recorded at unlogged and 42 species at buffer-strip sites: eight species were more abundant in unlogged and 12 species more abundant in buffer-strip sites. Four species that were more abundant in unlogged stands increased with increasing width of riparian buffers. However, four other species were rarely observed in even the widest buffers sampled (40–70 m). Overall, bird species richness and abundance were not related to buffer-strip width, but the author recommends buffer widths >40 m and maintaining a high density of trees within the buffer.

A replicated, controlled paired sites study from June–July in 1994–5 in 16 pairs of forested buffer strips (20–50 m) and undisturbed riparian coastal forest in Newfoundland, Canada (4), found that bird abundance was higher in the buffer strips (average of 10.5 individuals/transect for buffer strips vs. 7.9 for control sites), total species richness was similar (7.2 species/transect in buffers vs. 6.2 in controls) but that three of six specialist forest species were absent. Abundance of forest generalist, interior forest, and riparian species were similar between buffers and controls and did not increase in wider buffers. Buffer strips were adjacent to 3–5 year-old clearcuts (>10 ha) and were typically >300 m long.

A replicated controlled before-and-after study in a managed Douglas-fir *Pseudotsuga menziesii* forest in Washington, USA (5), found that 31 m wide riparian buffer strips contained bird communities that were more similar to control (unharvested) forests than 14 m strips. Forest species (black-throated grey warbler *Dendroica nigrescens*, golden-crowned kinglet *Regulus satrapa* and brown creeper *Certhia americana*) decreased in buffer treatments (especially the narrow buffer) relative to controls. Species of shrubby habitats (dark-eyed junco *Junco hyemalis*, cedar waxwing *Bombycilla cedrorum* and song sparrow *Melospiza melodia*) increased in one or both buffer treatments. Birds were surveyed in 18 sites (six of each treatment) in both pre-harvest (spring 1993) and post-harvest (1995 and 1996) years.

A replicated controlled trial along a stream in Minnesota, USA (6), found that bird species responded differently to timber harvest in riparian buffers, and that any amount of harvest affected breeding bird communities. Along the stream, 30 m wide forest buffers were established within plots with four treatments (3 plots/treatment): 1) no harvest in buffer; 2) reduction of tree basal area to 7–10 m^2/ha; 3) reduction to 2 m^2/ha (i.e. clearcut); and 4) no harvest in buffer or adjacent upland forest. Bird surveys were conducted 1 year prior to and for 4 years after, harvest. In the first year after harvest, bird community composition changed in all buffer treatments relative to control plots, and diverged over time. More individuals and species (primarily associated with edge or early-successional habitats) colonised harvested buffers; abundances and species richness of interior forest species declined.

A replicated, randomised, controlled study from May–June in 2001 and May–July in 2002 in 24 buffer-strip blocks and 18 continuous, old-growth forest blocks in coastal, temperate forests in Alaska, USA (7) found that species richness was similarly distributed across treatments and controls (both averaged 15 species/100 detections). Species richness and diversity were greatest in the narrowest buffers (<100 m) but species composition in the largest buffers (≥400 m) was most similar to that in control blocks. Only 3 of 10 common species differed in abundance across buffer treatments and controls (2 were positively and 1 was negatively related to buffer width). The authors conclude that forested beach buffers ≥250 m wide can support densities of forest-associated birds similar to that of natural stands but rare or uncommon species will benefit most from buffers ≥400 m in width.

A replicated study in 2003–4 in old-growth forest on Prince of Wales Island, Alaska, USA (8), there were no significant differences in average clutch size or number of young fledged across six species between nests in narrow (<250 m) buffers at four sites, compared to wide (>350 m) buffers at three sites. The buffers surrounded areas of 8–18 ha of forest and 76 nests of six species (Pacific-slope flycatcher *Empidonax difficilis*, chestnut-backed chickadee *Poecile rufescens*, winter wren *Troglodytes troglodytes*, Swainson's thrush *Catharus ustulatus*, hermit thrush *C. guttatus* and varied thrush *Ixoreus naevius*) were monitored. Of the 25 (18%) of nests that did not fledge young, 23 failed due to predation. Daily survival rates were slightly higher (0.2 to 2.5%) in wide buffers.

(1) Darveau, M., Beauchesne, P., Bélanger, L., Huot, J. & Larue, P. (1995) Riparian forest strips as habitat for breeding birds in boreal forest. *Journal of Wildlife Management*, 59, 67–78.
(2) Vander Haegen, W. M. & Degraaf, R. M. (1996) Predation on artificial nests in forested riparian buffer strips. *The Journal of Wildlife Management*, 60, 542–550.
(3) Hagar, J. C. (1999) Influence of riparian buffer width on bird assemblages in western Oregon. *Journal of Wildlife Management*, 63, 484–496.
(4) Whitaker, D. M. & Montevecchi, W. A. (1999) Breeding bird assemblages inhabiting riparian buffer strips in Newfoundland, Canada. *Journal of Wildlife Management*, 63, 167–179.
(5) Pearson, S. F. & Manuwal, D. A. (2001) Breeding bird response to riparian buffer width in managed Pacific northwest Douglas-fir forests. *Ecological Applications*, 11, 840–853.
(6) Hanowski, J., Danz, N., Lind, J. & Niemi, G. (2005) Breeding bird response to varying amounts of basal area retention in riparian buffers. *The Journal of Wildlife Management*, 69, 689–698.
(7) Kissling, M. L. & Garton, E. O. (2008) Forested buffer strips and breeding bird communities in southeast Alaska. *Journal of Wildlife Management*, 72, 674–681.
(8) Sperry, D. M., Kissling, M. & George, T. L. (2008) Avian nest survival in coastal forested buffer strips on Prince of Wales Island, Alaska. *The Condor*, 110, 740–746.

3. Education and awareness raising

Key messages

Raise awareness amongst the general public through campaigns and public information

A literature review from North America found that education was not sufficient to change behaviour, but that it was necessary for the success of economic incentives and law enforcement.

Provide bird feeding materials to families with young children

A single replicated, paired study from the USA found that most children involved in a programme providing families with bird food increased their knowledge of birds, but did not significantly change their environmental attitudes.

Enhance bird taxonomy skills through higher education and training

We captured no published evidence for the effects of enhancing bird taxonomy skills on bird populations.

Provide training to conservationists and land managers on bird ecology and conservation

We captured no published evidence on the effects of general awareness campaigns and public information on the state of bird populations.

3.1. Raise awareness amongst the general public through campaigns and public information

- A review of programmes in the USA and Canada (1) argues that education was not sufficient to change behaviour, although it was necessary as a catalytic factor for economic incentives and law enforcement.

Background

This intervention involves general information and awareness campaigns in response to a range of threats. Studies describing educational campaigns in response to specific threats are described in the chapter on that threat category (e.g. 'Threat: Biological Resource Use – Use education programmes and local engagement to help reduce persecution or exploitation of species').

A review of 16 case studies (six of which were directly related to birds) using before-and-after analyses in the USA and Canada (1) found that education and awareness initiatives were necessary but insufficient in effective conservation projects. Of the six case studies concerning birds, education and awareness decreased the hunting of American black duck *Anas rubripes* (USA and Canada) and threatened geese through more stringent regulations; did not decrease lead poisoning of common loons *Gavia immer* in New England, three years after pamphlet distribution; decreased oil contamination in Colorado and Wyoming pits (USA) by 67%; increased hatching rates of snowy plovers *Charadrius nivosus* in California (USA) by 18% in 5 years; and doubled seabird populations in a region in Quebec (Canada). Overall, education and awareness was almost never a sufficient factor in changing behaviour, although it was necessary as a catalytic factor for economic incentives and law enforcement.

(1) Byers, B. A. (2003) *Education, Communication and Outreach (ECO) success stories: Solving conservation problems by changing behavior*. U.S. Fish and Wildlife Service National Conservation Training Center Division of Education Outreach report.

3.2. Provide bird feeding materials to families with young children

- A single replicated before-and-after study from the USA (1) found that most children involved in a programme providing families with bird food increased their knowledge of birds, but there was no significant change in environmental attitudes.

Background

Feeding birds in gardens is a popular past time in many parts of the world, and there is the possibility that encouraging young children to feed birds may increase their knowledge of local species and their desire to conserve them. Studies describing the effects of feeding on bird populations and reproduction are described in 'General responses to small/declining populations – Provide supplementary food'.

A replicated before-and-after study in 65 families containing at least 1 child provisioned with bird feeding and educational materials for use in urban gardens in the USA (1) found that younger children showed significant gains in bird knowledge but there was no systematic change in environmental attitudes. Forty-nine (75%) children improved in bird knowledge, six (9%) showed no change and ten (15%) declined. Post-programme scores were significantly higher than pre-programme scores for both younger boys and girls (7–9 years old) but not older children (10–12 years old). Positive change was correlated with higher education levels of parents. Environmental attitudes, however, did not change and declined for one subgroup of children (younger boys). Over 80% of parents felt the programme increased family interaction and 80% reported they were still watching and feeding birds a year later. Of the children, 44% were boys and 56% girls.

(1) Beck, A. M., Melson, G. F., da Costa, P. L. & Liu, T. (2001) The educational benefits of a ten-week home-based wild bird feeding program for children. *Anthrozoos*, 14, 19–28.

3.3. Enhance bird taxonomy skills through higher education and training

- We captured no published evidence for the effects of enhancing bird taxonomy skills on bird populations.

3.4. Provide training to conservationists and land managers on bird ecology and conservation

- We captured no published evidence on the effects of general awareness campaigns and public information on the state of bird populations.

4. Threat: Residential and commercial development

Probably the biggest threats from development are from the destruction of habitat, pollution and 'transportation and service corridors'. Interventions in response to these threats are described in 'Habitat restoration and creation', 'Threat: Pollution' and 'Threat: Transportation and service corridors'.

The two interventions described in this section are designed to reduce collisions between birds and windows, which kill many birds each year. Approximately 25% of bird species in the USA have been recorded colliding with windows, with no environmental conditions apparently reducing this risk (Klem 1989). Studies that examine placing bird feeders in such a way as to minimise collision risk are described in 'Provide supplementary food'.

Klem, D. (1989) Bird-window collisions. *The Wilson Bulletin*, 101, 606–620.

Key messages

Angle windows to reduce bird collisions

A randomised, replicated and controlled experiment in the USA found that fewer birds collided with windows angled away from the vertical.

Mark or tint windows to reduce bird collisions

Two randomised, replicated and controlled studies found that marking windows did not appear to reduce bird collisions. However, when windows were largely covered with white cloth, or tinted, fewer birds flew towards or collided with them. A third randomised, replicated and controlled study found that fewer birds collided with tinted windows than with un-tinted ones, although the authors noted that the poor reflective quality of the glass could have influenced the results.

4.1. Angle windows to reduce collisions

- A single randomised, replicated and controlled experiment in the USA (1) found fewer birds collided with windows angled away from the vertical.

A randomised, replicated and controlled experiment in 1991 in Pennsylvania, USA (1), found that fewer birds collided with windows angled at 20° or 40° from the vertical (28% and 15% of 53 recorded collisions respectively) than with vertical windows (57% of collisions). Six plate glass, wooden framed windows (1.4 × 1.2 m, 1.2 m off the ground, 15–43 m apart) were used, between January and May, on the edge of deciduous woodland and farmland.

(1) Klem Jr, D., Keck, D. C., Marty, K. L., Ball, A. J., Niciu, E. E. & Platt, C. T. (2004) Effects of window angling, feeder placement, and scavengers on avian mortality at plate glass. *The Wilson Bulletin*, 116, 69–73.

4.2. Mark or tint windows to reduce collision mortality

- Two randomised, replicated and controlled studies (one *ex situ*, (1)) found that marking windows did not appear to reduce bird collisions. However, when windows were largely covered with white cloth, fewer birds flew towards them.
- A randomised, replicated and controlled study (2) found that fewer birds collided with tinted windows than with un-tinted ones, although the authors noted that the poor reflective quality of the glass could have influenced the results.

A randomised, replicated and controlled study over 52 days in Illinois, USA (1), found that marking windows in various ways did not reduce the number of birds colliding with the windows, compared to an unmarked control window. Similarly, a randomised, repeated and controlled choice experiment in a flight cage found that dark-eyed juncos *Junco hyemalis* did not consistently avoid windows marked with wind chimes, silhouettes of falcons, plants, stickers of eyes or model owls. However, birds tended to avoid windows that were completely covered by white cloth, or covered by closely spaced cloth strips and meshes. Widely spaced cloth strips and flashing lights partially increased avoidance.

A randomised, replicated and controlled experiment between January and May 1991 in Pennsylvania, USA (2), found that a smaller proportion of collisions were with tinted windows (32% of 53 recorded collisions) than with clear windows (68% of collisions). The same study found that, when platform feeders were placed at varying distances in front of the windows (see 'Provide supplementary food – Place feeders close to windows to reduce collisions'), only four of 52 fatal collisions (8%) occurred with tinted windows, the rest with clear glass windows. However, the authors note that the tinted glass was of a poor reflective quality and they believe this may have resulted in fewer fatalities than a highly reflective tinted glass. Experiments used six plate glass, wooden framed windows (1.4 × 1.2 m, 1.2 m off the ground, 15–43 m apart) on the edge of deciduous woodland and farmland.

(1) Klem Jr, D. (1990) Collisions between birds and windows: mortality and prevention. *Journal of Field Ornithology*, 120–128.

(2) Klem Jr, D., Keck, D. C., Marty, K. L., Ball, A. J., Niciu, E. E. & Platt, C. T. (2004) Effects of window angling, feeder placement, and scavengers on avian mortality at plate glass. *The Wilson Bulletin*, 116, 69–73.

5. Threat: Agriculture

In Europe, much of the conservation effort is directed at reducing the impacts of agricultural intensification on biodiversity on farmland and in the wider countryside, and the majority of the interventions we have captured reflect this. However, there is some debate as to whether 'wildlife-friendly farming' is the best overall strategy to conserve biodiversity across the world. Wildlife-friendly agriculture *may* be lower yielding than intensive agriculture, in which case a larger area of land will be required to produce the same amount of food. If this leads to increased habitat conversion, then intensifying production on current agricultural land and 'sparing' wild habitats for conservation may be preferable.

Whilst there has been considerable debate over the validity of the land-sparing approach, and Ewers *et al.* (2009) found some weak evidence that increased crop yields are associated with land sparing, we have captured no studies examining whether land-sparing benefits bird populations. To be successful, land-sparing may well require effective habitat protection (Ewers *et al.* 2009), studies on which are discussed in a separate chapter.

Ewers, R. M., Scharlemann, J. P. W., Balmford, A., Green, R. E. (2009) Do increases in agricultural yield spare land for nature? *Global Change Biology*, 15, 1716–1726.

Key messages – All farming systems

Support or maintain low-intensity agricultural systems

We captured no intervention-based evidence for the effects of supporting low-intensity agricultural systems on bird populations.

Food labelling schemes relating to biodiversity-friendly farming

We captured no evidence for the effects of food labelling schemes on bird populations.

Increase the proportion of natural/semi-natural habitat in the farmed landscape

Two studies from Switzerland and Australia, of the five we captured, found that areas with plantings of native species, or areas under a scheme designed to increase semi-natural habitats (the Swiss Ecological Compensation Areas scheme), held more bird species than other areas. One study from Switzerland found that populations of three bird species increased in areas under the Ecological Compensation Areas scheme. A third Swiss study found that some habitats near Ecological Compensation Areas held more birds than habitats further away, but the overall amount of Ecological Compensation Area had no effect on bird populations. A study from the UK found no effect of habitat-creation on grey partridge populations.

Pay farmers to cover the costs of conservation measures

Three out of 31 studies found national population increases in three species after payment schemes targeted at their conservation. One found that many other species continued declining. Twenty-two studies found that at least some species were found at higher densities on sites with agri-environment schemes; some differences were present only in summer or only in winter. Fifteen studies found some species at similar densities on agri-environment schemes and non-agri-environment scheme sites or appeared to respond negatively to agri-environment schemes. One study found that grey partridge survival was higher in some years on agri-environment scheme sites. Two studies found higher productivity on agri-environment scheme sites for some species, one found no effect of agri-environment schemes. A review found that some agri-environment schemes options were not being used enough to benefit many species of bird. A study from the UK found that there was no difference in the densities of seed-eating birds in winter between two agri-environment scheme designations.

Cross compliance standards for all subsidy payments

Apart from the Swiss Ecological Compensation Areas scheme (considered in another section), we found no studies investigating the effects of cross compliance standards on birds.

Reduce field size (or maintain small fields)

We found no intervention-based evidence on the effects of reducing field sizes on birds.

Provide (or retain) set-aside areas in farmland

Four out of 23 studies from Europe and North America found more species on set-aside than on crops. One study found fewer. Twenty-one studies found that some species were at higher densities on set-aside than other habitats, or that they used set-aside more often. Four found that some species were found at lower densities on set-aside than other habitats. Three studies found that waders and Eurasian skylarks had higher productivities on set-aside than other crops. One study found that skylarks on set-aside had lower similar or lower productivities than on crops. One study from the UK found that rotational set-aside was used more than non-rotational set-aside, another found no difference. A review from North America and Europe found that naturally regenerated set-aside held more birds and more species than sown set-aside.

Manage hedges to benefit wildlife

One of seven studies found no differences in the number of species in a UK site with wildlife-friendly hedge management and sites without. Seven studies found that some species increased in managed hedges or were more likely to be found in them than other habitats. One investigated several interventions at the same time. Four studies found that some species responded negatively or not at all to hedge management or that effects varied across regions of the UK.

Plant new hedges

A study from the USA found that populations of northern bobwhites increased following several interventions including the planting of new hedges.

Manage stone-faced hedge banks to benefit birds

We captured no evidence for the effects of managing stone-faced hedge banks on birds.

Manage ditches to benefit wildlife

One study of four from the UK found that bunded ditches were visitied more often by birds than non-bunded ditches. Three studies found that some birds responded positively to ditches managed for wildlife, but that other species did not respond to management, or responded negatively.

Protect in-field trees

We found no evidence for the effects of protecting in-field trees on birds.

Plant in-field trees

We found no evidence for the effects of planting in-field trees on birds.

Tree pollarding and tree surgery

We found no evidence for the effects of tree pollarding and tree surgery on bird populations.

Plant wild bird seed or cover mixture

Seven of 41 studies found that fields or farms with wild bird cover had higher diversity than other sites, or that wild bird cover held more species than other habitats. Thirty-two studies found that populations, or abundances of some or all species were higher on wild bird cover than other habitats, or that wild bird cover was used more than other habitats. Four of these studies investigated several interventions at once. Thirteen studies found that bird populations or densities were similar on wild bird cover and other habitats that some species were not associated with wild bird cover, or that birds rarely used wild bird cover. Three studies found higher productivities of birds on wild bird cover than other habitats. Two found no differences for some or all species studied. Two studies found that survival of grey partridge or artificial nests increased on wild bird cover; one found lower partridge survival in farms with wild bird cover than other farms. Five studies from the UK found that some wild bird cover crops were used more than others. A study and a review found that the arrangement of wild bird cover in the landscape affected its use by birds.

Plant nectar flower mixture/wildflower strips

Three of seven studies found that birds used wildflower strips more than other habitats; two found strips were not used more than other habitats. A study from Switzerland found that Eurasian skylarks were more likely to nest in patches sown with annual weeds than in crops and were less likely to abandon nests. A study from the UK found that management of field margins affected their use more than the seed mix used.

Create uncultivated margins around intensive arable or pasture fields

One of eight studies found that three sparrow species found on uncultivated margins on a site in the USA were not found on mown field edges. A replicated study from Canada found fewer species in uncultivated margins than in hedges or trees. Three studies found that some bird species were associated with uncultivated margins, or that birds were more abundant on margins than other habitats. One study found that these effects were very weak and four studies of three experiments found that uncultivated margins contained similar numbers of birds as other habitats in winter, or that several species studied did not show associations with margins. A study from the UK found that yellowhammers used uncultivated margins more than crops in early summer. Use fell in uncut (but not cut) margins later in the year. A study from the UK found that grey partridge released on uncultivated margins had high survival.

Plant grass buffer strips/margins around arable or pasture fields

One of 15 studies found more bird species in fields in the USA that were bordered by grass margins than in unbordered fields. Two studies from the UK found no effect of margins on species richness. One study found that more birds used grass strips in fields than used crops. Even more used grass margins. Nine studies from the USA and UK found that sites with grass margins had more positive population trends or higher populations for some birds, or that some species showed strong habitat associations with grass margins. Three studies found no such effect for some or all species. Two studies found that species used margins more than other habitats and one found that birds used cut margins more than uncut during winter, but less than other habitats during summer. A study from the UK found that grey partridge broods were smaller on grass margins than other habitat types.

Use mowing techniques to reduce chick mortality

One of three studies from the UK found a large increase in the national population of corncrakes after a scheme to delay mowing and promote corncrake-friendly mowing techniques. Two studies found lower levels of corncrake and Eurasian skylark mortality when wildlife-friendly mowing techniques were used.

Provide refuges in fields during harvest or mowing

One study found that fewer gamebirds came into contact with mowing machinery when refuges were left in fields. A study from the UK found that Eurasian skylarks did not nest at higher densities in uncut refuges than in the rest of the field.

Mark bird nests during harvest or mowing

A study from the Netherlands found that fewer northern lapwing nests were destroyed when they were marked with bamboo poles than when they were unmarked.

Relocate nests at harvest time to reduce nestling mortality

A study from Spain found that Montagu's harrier clutches had higher hatching and fledging rates when they were temporarily moved during harvest than control nests that were not moved.

Make direct payments per clutch for farmland birds

One of two studies from the Netherlands found slightly higher breeding densities of waders on farms with per clutch payment schemes but this and another study found no higher numbers overall. One study found higher hatching success on farms with payment schemes.

Control scrub on farmland

A study from the UK found farms with a combined intervention that included scrub control had lower numbers of young grey partridge per adult.

Take field corners out of management

A study from the UK found that overwinter survival of grey partridge was positively correlated with taking field corners out of management, but this relationship was only significant in one of three winters. There was no relationship with measures of productivity (brood size, young:adult).

Reduce conflict by deterring birds from taking crops

Three studies have found evidence that deterrents are or could be effective. One found less crop damage in almond orchards in the USA when crow distress calls were broadcast, compared to when they were not. A study from Pakistan found that pest species were less abundant when reflector ribbons were hung over crops. An *ex situ* study from the USA found that dickcissels consumed less rice if it was treated with repellent, compared to untreated rice.

Key messages – Arable farming

Increase crop diversity

A study from the UK found that more barnacle geese used a site after the amount of land under cereals was decreased and several other interventions were used.

Implement mosaic management

One of two studies from the Netherlands found that northern lapwing population trends, but not those of three other waders, became more positive following the introduction of mosaic management. The other found that black-tailed godwit productivity was higher under mosaic management than other management types.

Leave overwinter stubbles

Three of fourteen studies report positive population-level changes in two species after winter stubble provision. All investigated several interventions at once. Eight studies found that some farmland birds were found on stubbles or were positively associated with them, three investigated several interventions and one found no more positive associations than expected by chance. A study from the UK found that most species did not preferentially use stubble, compared to cover crops and another found that a greater area of stubble in a site meant lower grey partridge brood size. Five studies from the UK found that management of stubbles influenced their use by birds. One study found that only one species was more common on stubbles under agri-environment schemes.

Plant nettle strips

We found no evidence for the effects of planting nettle strips on bird populations.

Leave unharvested cereal headlands within arable fields

We found no evidence for the effects of leaving unharvested cereal headlands within arable fields on bird populations.

Plant crops in spring rather than autumn

One study from Sweden, of three examining the effects of spring-sown crops, found that more birds were found on areas with spring, rather than autumn-sown crops. A study from the UK found that several species used the study site for the first time after spring-sowing was started. All three studies found that some populations increased after the start of spring sowing. A study from the UK found that some species declined as well. A study from Sweden found that hatching success of songbirds and northern lapwing was lower on spring-sown, compared with autumn-sown crops.

Undersow spring cereals, with clover for example

Four of five studies from the UK found that bird densities were higher on undersown fields or margins than other fields, or that use of fields increased if they were undersown. Two studies of the same experiment found that not all species nested at higher densities in undersown habitats. A study from the UK found that grey partridge populations were lower on sites with large amounts of undersown cereal.

Plant more than one crop per field (intercropping)

A study from the USA found that 35 species of bird used fields with intercropping, with four nesting, but that productivity from the fields was very low.

Revert arable land to permanent grassland

All five studies looking at the effects of reverting arable land to grassland found no clear benefit to birds. The studies monitored birds in winter or grey partridges in the UK and wading birds in Denmark. They included three replicated controlled trials.

Reduce tillage

Six of ten studies found that some or all bird groups had higher species richness or diversity on reduced-tillage fields, compared to conventional fields in some areas. Two studies found that some groups had lower diversity on reduced-tillage sites, or that there was no difference between treatments. Nine studies found that some species were found at higher densities on reduced tillage fields, six found that some species were at similar or lower densities. Three studies found evidence for higher productivities on reduced-tillage fields. One found that not all measures of productivity were higher.

Add 1% barley into wheat crop for corn buntings

We have found no studies investigating the impact of adding barley to wheat on bird populations.

Leave uncropped cultivated margins or plots (includes lapwing and stone curlew plots)

Three of nine studies report that the UK population of Eurasian thick-knees increased following a scheme to promote lapwing plots (and other interventions). A study from the UK found that plots did not appear to influence grey partridge populations. Four studies from the UK found that at least one species was associated with lapwing plots, or used them for foraging or nesting. One study found that 11 species were not associated with plots, another that fewer used plots than used crops in two regions of the UK. Two studies found that nesting success was higher on lapwing plots and fallow than in crops. A third found fewer grey partridge chicks per adult on sites with lots of lapwing plots.

Create skylark plots

One study of seven found that the Eurasian skylark population on a farm increased after skylark plots were provided. Another found higher skylark densities on fields with plots in. Two studies from the UK found that skylark productivity was higher for birds with skylark plots in their territories, a study from Switzerland found no differences. Two studies from Denmark and Switzerland found that skylarks used plots more than expected, but a study from the UK found that seed-eating songbirds did not.

Create corn bunting plots

We have found no studies investigating the impact of corn bunting plots on bird populations.

Plant cereals in wide-spaced rows

One of three studies from the UK found that fields with wide-spaced rows held more Eurasian skylark nests than control fields. One study found that fields with wide-spaced rows held fewer nests. Both found that fields with wide-spaced rows held fewer nests than fields with skylark plots. A study from the UK found that skylark chicks in fields with wide-spaced rows had similar diets to those in control fields.

Create beetle banks

Two of six studies from the UK found that some bird populations were higher on sites with beetle banks. Both investigated several interventions at once. Two studies found no relationships between bird species abundances or populations and beetle banks. Two studies (including a review) from the UK found that three bird species used beetle banks more than expected, one used them less than expected.

Key messages – Livestock farming

Maintain species-rich, semi-natural grassland

One of two studies found that the populations of five species increased in an area of the UK after the start of management designed to maintain unimproved grasslands. A study from Switzerland found that wetland birds nested at greater densities on managed hay meadows than expected, but birds of open farmland used hay meadows less.

Reduce management intensity of permanent grasslands

Seven of eight European studies found that some or all birds studied were more abundant on grasslands with reduced management intensity, or used them more than other habitats for foraging. Five studies of four experiments found that some or all species were found at lower or similar abundances on reduced-management grasslands, compared to intensively-managed grasslands.

Reduce grazing intensity

Nine of eleven studies from the UK and USA found that the populations of some species were higher on fields with reduced grazing intensity, compared to conventionally-grazed fields, or found that birds used these fields more. Three studies investigated several interventions at once. Five studies from Europe found that some or all species were no more numerous, or were less abundant on fields with reduced grazing. A study from the UK found that black grouse populations increased at reduced grazing sites (whilst they declined elsewhere). However, large areas with reduced grazing had low female densities. A study from the USA found that the number of species on plots with reduced grazing increased over time. A study from four European countries found no differences in the number of species on sites with low- or high-intensity grazing.

Provide short grass for waders

A study from the UK found that common starlings and northern lapwing spent more time foraging on areas with short swards, compared to longer swards.

Raise mowing height on grasslands

One of two studies from the UK found that no more foraging birds were attracted to plots with raised mowing heights, compared to plots with shorter grass. A review from the UK found that Eurasian skylarks had higher productivity on sites with raised mowing heights, but this increase was not enough to maintain local populations.

Delay mowing date or first grazing date grasslands

Two of five studies (both reviews) found that the UK corncrake populations increased following two schemes to encourage farmers to delay mowing. A study from the Netherlands found no evidence that waders and other birds were more abundant in fields with delayed mowing. Another study from the Netherlands found that fields with delayed mowing held more birds than other fields, but differences were present before the scheme began and population trends did not differ between treatments. A study from the USA found that fewer nests were destroyed by machinery in late-cut fields, compared with early-cut fields.

Leave uncut rye grass in silage fields

All four studies from the UK (including two reviews) found that seed-eating birds were benefited by leaving uncut (or once-cut) rye grass in fields, or that seed-eating species were more abundant on uncut plots. Three studies found that seed-eating birds were more abundant on uncut and ungrazed plots than on uncut and grazed plots. A study from the UK found that the responses of non-seed-eating birds were less certain than seed-eating species, with some species avoiding uncut rye grass.

Plant cereals for wholecrop silage

Three studies of one experiment found that seed-eating birds used cereal-based wholecrop silage crops more than other crops in summer and winter. Insect-eating species used other crops and grassland more often.

Maintain lowland heathland

We found no intervention-based evidence on the effects of maintaining lowland heath on bird populations.

Maintain rush pastures

We found no intervention-based evidence on the effects of maintaining rush pastures on bird populations.

Maintain traditional water meadows

One of four studies (from the UK) found that the populations of two waders increased on reserves managed as water meadows. Two studies from the Netherlands found that there were more waders or birds overall on specially managed meadows or 12.5 ha plots, but one found that these differences were present before management began, the other found no differences between individual fields under different management. Two studies from the UK and Netherlands found that wader populations were no different between specially and conventionally managed meadows, or that wader populations decreased on specially-managed meadows. A study from the UK found that northern lapwing productivity was not high enough to maintain populations on three of four sites managed for waders.

Maintain upland heath/moor

A study from the UK found that bird populations in one region were increasing with agri-environment guidelines on moor management. There were some problems with overgrazing, burning and scrub encroachment.

Plant Brassica fodder crops

We found no evidence on the effects of planting Brassicas on bird populations.

Use mixed stocking

We found no evidence on the effects of mixed stocking on bird populations.

Use traditional breeds of livestock

A study from four countries in Europe found no differences in bird abundances in areas grazed with traditional or commercial breeds.

Maintain wood pasture and parkland

We found no intervention-based evidence on the effects of maintaining wood pasture and parkland on bird populations.

Exclude livestock from semi-natural habitat (including woodland)

Two studies from the USA, out of 11 overall, found higher species richness on sites with grazers excluded. A study from Argentina found lower species richness and one from the USA found no difference. Seven studies from the USA found that overall bird abundance, or the abundances of some species were higher in sites with grazers excluded. Seven studies from the USA and Argentina found that overall abundance or the abundance of some species were lower on sites without grazers, or did not differ. Three studies found that productivities were higher on sites with grazers excluded. In one, the difference was only found consistently in comparison with improved pastures, not unimproved.

Protect nests from livestock to reduce trampling

One of two studies found that a population of Chatham Island oystercatchers increased following several interventions including the erection of fencing around individual nests. A study from Sweden found that no southern dunlin nests were trampled when protected by cages; some unprotected nests were destroyed.

Mark fences to reduce bird collision mortality

A study from the UK found that fewer birds collided with marked sections of deer fences, compared to unmarked sections.

Create open patches or strips in permanent grassland

A study from the UK found that Eurasian skylarks used fields with open strips in, but that variations in skylark numbers were too great to draw conclusions from this finding.

Key messages – Perennial, non-timber crops

Maintain traditional orchards

Two site comparison studies from the UK and Switzerland found that traditional orchards offer little benefit to birds. In Switzerland only one breeding bird species was associated with traditional orchards. In the UK, the population density of cirl bunting was negatively related to the presence of orchards.

Manage perennial bioenergy crops to benefit wildlife

We captured no evidence for the effects of managing bioenergy crops for wildlife on bird populations.

Key messages – Aquaculture

Scare birds from fish farms

One study from Israel found a population increase in fish-eating birds after efforts to scare them from fish farms, possibly due to lower persecution. One of two studies found evidence for reduced loss of fish when birds were scared from farms. Two studies from Australia and Belgium found that disturbing birds using foot patrols was not effective. Ten of 11 studies from across the world found some effects for acoustic deterrents, five of seven found that visual deterrents were effective. In both cases some studies found that results were temporary, birds became habituated or that some deterrents were effective, whilst others were not. One study found that trained raptors were effective, one found little evidence for the effectiveness of helicopters or light aircraft.

Disturb birds at roosts

One study from the USA found reduced fish predation after fish-eating birds were disturbed at roosts. Five studies from the USA and Israel found that birds foraged less near disturbed roosts, or left the area after being disturbed. One found the effects were only temporary.

Use electric fencing to exclude fish-eating birds

Two before-and-after trials from the USA found lower use of fish ponds by herons after electric fencing was installed.

Use netting to exclude fish-eating birds

Two studies from Germany and the USA, and a review, found that netting over ponds reduced the loss of fish to predatory birds. Two studies from the USA and the Netherlands found that birds still landed on ponds with netting, but that they altered their behaviour, compared to open ponds. Two studies from Germany and Israel found that some birds became entangled in netting over ponds.

Disturb birds using foot patrols

Two replicated studies from Belgium and Australia found that using foot patrols to disturb birds from fish farms did not reduce the number of birds present or fish consumption.

Use 'mussel socks' to prevent birds from attacking shellfish

A study from Canada found that mussel socks with protective sleeves lost fewer medium-sized mussels (but not small or large mussels), compared to unprotected mussel socks.

Translocate birds away from fish farms

A study from the USA found that translocating birds appeared to reduce bird numbers at a fish farm. A study from Belgium found that it did not.

Increase water turbidity to reduce fish predation by birds

An *ex situ* study from France found that egret foraging efficiency was reduced in more turbid water.

Provide refuges for fish within ponds

A study from the UK found that cormorants caught fewer fish in a pond with fish refuges in, compared to a control pond.

Use in-water devices to reduce fish loss from ponds

A study from the USA found that fewer cormorants used two ponds after underwater ropes were installed; a study from Australia found that no fewer cormorants used ponds with gillnets in.

Spray water to deter birds from ponds

A study from Sweden found that spraying water deterred birds from fish ponds, but that some birds became habituated to the spray.

Deter birds from landing on shellfish culture gear

A study from Canada found that fewer birds landed on oyster cages fitted with spikes than control cages. The same study found that fewer birds landed on oyster bags suspended 6 cm, but not 3 cm, underwater, compared to bags on the surface.

All farming systems

5.1. Support or maintain low-intensity agricultural systems

- We captured no evidence for the effects of supporting low-intensity agricultural systems on bird populations.

Background

Low-intensity agricultural systems have consistently been shown to have higher biodiversity than more intensive systems, both in temperate regions and the tropics. Supporting such systems may therefore help declining farmland bird populations. However, whilst we captured many studies describing the distribution of birds across high- and low-intensity agricultural systems, we found no intervention-based evidence for the effects of legislation aimed at supporting and maintaining low-intensity agricultural systems on bird populations.

5.2. Practise integrated farm management

Background

Integrated Farm Management is a whole farm system that aims to provide profitable production whilst being environmentally responsible. It focuses on integrating beneficial natural processes, by using efficient soil management and crop rotations for example, into modern farming techniques. Practitioners of Integrated Farm Management need to be able to clearly demonstrate improvement to the quality of soil, water, air, wildlife habitat and the landscape.

We have not included studies describing the effects of Integrated Farm Management because farms are able to use a variety of different management interventions and which were used in any particular case is not always recorded. Where individual interventions are recorded, studies are described in the appropriate section.

5.3. Food labelling schemes relating to biodiversity-friendly farming

- We captured no evidence for the effects of food labelling schemes on bird populations.

Background

Food from many parts of the world now carries certification labels such as the LEAF Marque (Integrated Farm Management) or Rainforest Alliance, or labelling for shade-grown coffee or chocolate. These schemes are designed to allow biodiversity-friendly farming to attract a price premium, become more profitable and therefore spread, potentially benefiting biodiversity.

5.4. Increase the proportion of natural/semi-natural vegetation in the farmed landscape

- Of four studies captured, one, a replicated and controlled paired sites study from Australia (4), found that farms with plantings of native vegetation held more species than those without. The effect was smaller than that explained by variation in the amount of natural habitat remaining on farms. A replicated study from Switzerland (5) found more species in areas under the Ecological Compensation Area scheme than areas not under it.

- A before-and-after study from Switzerland (1) found that the populations of three bird species increased after an increase in the amount of land under the Ecological Compensation Scheme. This study found that three species were more found more than expected on Ecological Compensation Scheme land. Another replicated study from Switzerland (3) found that some habitats held more birds if they were close to ECA habitat but that the amount of Ecological Compensation Scheme in an area had no impact on population densities.
- A small study from the UK (2) found no effect of habitat creation on grey partridge populations.

Background

This intervention is concerned with general increases in the proportion of natural or semi-natural habitat in a landscape. Studies describing the effects of the creation of specific habitat types and the use of individual restored sites are discussed in 'Habitat restoration and creation'.

A before-and-after study in 6 km^2 of mixed farmland in Switzerland (1) found that the populations of corn buntings *Miliaria calandra*, whitethroat *Sylvia communis* and common stonechat *Saxicola torquata* all increased following an increase in the proportion of land under the Ecological Compensation Scheme from 0.7% to 8.2% between 1992 and 1996 (corn buntings: six pairs in 1992 vs. 26 in 1996; whitethroat: 15 vs. 44; stonechat: 14 vs. 35). In addition, across 23 study areas in Switzerland, Ecological Compensation Scheme land and a 25 m buffer around it occupied only 17% of farmland but contained more (37–38% of 68) red-backed shrike *Lanius collurio* territories. Only 6% of Eurasian skylark *Alauda arvensis* territories were found on Ecological Compensation Scheme land.

A small 2003 site comparison study of 20 farms in East Anglia and the West Midlands, UK (2), found that the intentional creation of wildlife habitat had no discernible effect on autumn grey partridge *Perdix perdix* densities. The change in partridge densities from 1998 to 2002 on farms with habitat creation (–32% and –1%, respectively) was not statistically different from farms without habitat creation (–51% and –28%, respectively). Surveys of grey partridge were made once each autumn in 1998 and 2002 on 20 farms: 12 farms that created wildlife habitat and 8 farms which did not.

A 2007 site comparison study of 23 sites in the lowlands north of the Alps, Switzerland (3), found that the percentage of farmland designated as an ecological compensated area had no effect on the population density of farmland bird species or bird species with territories incorporating several habitat types. Ecological compensated areas are areas managed for the primary function of providing plant and animal habitat – these include meadows farmed at a low intensity. For 37 species surveyed in 1998/1999 and again in 2003/2004, population densities in wetlands and rivers were not affected by proximity to ecological compensated areas, although hedges and traditional orchards close to ECAs did have higher bird population densities than those further away. Twenty-three out of one hundred hedges within ecological compensated areas had at least one of the 37 surveyed species present compared to 13 of 100 hedges outside the agri-environment scheme. The 23 selected sites (covering up to 3 km^2 each) were randomly selected and surveyed three times each between April and June in both years of study.

A replicated and controlled paired sites study in the springs of 2002, 2004 and 2006 and winter 2004 on 46 wheat and livestock farms across New South Wales, Australia (4), found that 23 farms with plantings of native vegetation had, on average, 3.4 more bird species than farms without plantings. If farms had more than 20 ha of plantings then this increased to 4.4 more species. In addition, 12 native species responded positively to planting, and six responded negatively. However, three times more variation in bird community assemblage was explained by the presence or absence of remnant natural vegetation and the size of remnant patches than by plantings. Plantings were of both locally endemic and non-local (but native) species and were at least seven years old.

A 2007 site comparison study of 181 plots in the canton of Aagau, Switzerland (5), found that, on average, two more bird species were identified in ecological compensated areas (10 species on average) than in non-ecological compensated areas (9 species). Although on average two more bird species were found in the second set of surveys (carried out from 2001–2005) than in the first set (1996–2000), this increase was uniform in both ecological compensated areas and non-ecological compensated areas. One hundred and twenty 100 m radius circle plots that contained some land designated as an ecological compensated area were compared with 61 plots not containing any ecological compensated areas. The authors note that ecological compensated areas were typically established on promising farmland with the potential for 'maximum biodiversity gain', which may have affected the relative species richness of ecological compensated areas and non-ecological compensated areas.

(1) Spiess, M., Marfurt, C. & Birrer, S. (2000) Ecological compensation – a chance for farmland birds? 441 in: T. Alfoldi, W. Lockeretz, U. Niggli (eds) *IFOAM 2000: the world grows organic*. vdf Hochschulverlag AG an der ETH Zurich, Basel, Switzerland, 28–31 August 2000.
(2) Browne, S. & Aebischer, N. (2003) *Arable stewardship: impact of the pilot scheme on grey partridge and brown hare after five years*. DEFRA Final Report RMP1870vs3. Department for Environment, Food and Rural Affairs, London, UK.
(3) Birrer, S., Spiess, M., Herzog, F., Jenny, M., Kohli, L. & Lugrin, B. (2007) The Swiss agri-environment scheme promotes farmland birds: but only moderately. *Journal of Ornithology*, 148, S295–303.
(4) Cunningham, R. B., Lindenmayer, D. B., Crane, M., Michael, D., MacGregor, C., Montague-Drake, R. & Fischer, J. (2008) The combined effects of remnant vegetation and tree planting on farmland birds. *Conservation Biology*, 22, 742–752.
(5) Roth, T., Amrhein, V., Peter, B. & Weber, D. (2008) A Swiss agri-environment scheme effectively enhances species richness for some taxa over time. *Agriculture, Ecosystems & Environment*, 125, 167–172.

5.5. Pay farmers to cover the costs of conservation measures

- Three reviews from the UK (1,10,13) of three studies captured reported population increases of three species after the introduction of specially-designed agri-environment schemes. These species were cirl buntings, corncrakes and Eurasian thick-knees. One of these found that many other species continued to decline (13).
- Twenty-two of 25 studies all from Europe, including a systematic review (2–9,12,14–18,21,23,24, 26–29,31), examining local population levels or densities found that at least some birds studied were at higher densities, had higher population levels or more positive population trends on sites with agri-environment schemes, compared to non-agri-environment scheme sites. Some studies found that differences were present in all seasons, others in either summer or winter. Fifteen studies from Europe, including a systematic review (4,5,7–9,11,14,15,17,19,24,25,27,28,31), found that some or all species were not found at higher densities, had similar or lower population levels, showed similar population trends on sites with agri-environment schemes, compared with non-agri-environment scheme sites, or showed negative population trends. A study from the Netherlands (20) found that many agri-environment scheme farms were sited in areas where they were unlikely to be effective.
- One small study from the UK (30) found no differences between winter densities of seed-eating birds on UK Higher Levels Stewardship sites, compared with those under Entry Level Stewardship.
- A replicated study from the UK (29) found that grey partridge survival was higher on agri-environment scheme sites than non-scheme sites. This difference was not significant every year.
- Two of three studies investigating reproductive productivity (8,24), including one replicated study, found that productivity was higher on farms under agri-environment schemes. One replicated study from the UK (29) found no effect of agri-environment schemes on productivity.
- A review (22) found that the amount of land entering an agri-environment scheme was on target, but that some options were not being used at high enough rates to help many species.

Background

Agri-environment schemes are government or inter-governmental schemes designed to compensate farmers financially for changing agricultural practice to be more favourable to biodiversity and landscape. In Europe, agri-environment schemes are an integral part of the European Common Agricultural Policy (CAP) and Member States devise their own agri-environment prescriptions to suit their agricultural economies and environmental contexts.

Agri-environment schemes represent many different specific interventions, and where a study's results can be clearly assigned to a specific intervention, they appear in the appropriate section. This section, meanwhile, includes evidence about the success of agri-environment policies overall.

Evidence relating to the Swiss Ecological Compensation Areas is placed under 'Increase the proportion of natural habitat in the landscape', if it involves monitoring biodiversity effects on a landscape scale, rather than focusing on specific aspects of habitat management.

In the USA and Canada, schemes such as the Conservation Reserve Program (USA) and the Permanent Cover Program (Canada) are aimed primarily at creating semi-natural and natural vegetation and are mainly discussed in 'Habitat creation and restoration'.

A 1998 literature review (1) found that cirl buntings *Emberiza cirlus* in Britain responded positively to Countryside Stewardship Schemes, reaching population levels of 360–388 occupied territories in 1995–1997, compared with 118 or so in the mid-1980s. Some of the interventions used include reducing grassland management intensity; sowing arable field margins; managing hedgerows for wildlife; growing spring barley; reducing herbicide use and maintaining overwinter stubbles. More studies describing the effects of these interventions are discussed in the relevant sections.

A 2000 literature review from the UK (2) found that the populations of four farmland birds (grey partridge *Perdix perdix*, cirl buntings *Emberiza cirlus*, corncrake *Crex crex* and Eurasian thick-knee *Burhinus oedicnemus*) increased following agri-environment schemes targeted at them. The individual schemes are discussed in the relevant interventions.

A 2001 replicated paired site comparison study in south Devon, England (3) found that the number of cirl bunting *Emberiza cirlus* increased significantly more (up 72%, from 54 to 93 breeding territories) in areas participating in the Countryside Stewardship Scheme, than on adjacent land not participating in the Countryside Stewardship Scheme (down 20%, from 124 to 96 territories) between 1992 and 1999. Countryside Stewardship Scheme land that was near to known bunting breeding territories saw greater increases in bunting numbers than Countryside Stewardship Scheme areas further away: of the nine agreements further than 2 km from the nearest known breeding site in 1992, seven remained un-colonised in 1999, one lost its only pair and one gained a pair. Forty-one 4 km^2 squares containing both land within the Countryside Stewardship Scheme and non-Countryside Stewardship Scheme land were surveyed in 1992, 1998 and 1999. In each year each tetrad was surveyed at least twice, the first time during mid-April to late May, and the second time between early June and the end of August.

A replicated 2002 study from nine areas of the UK under Environmentally Sensitive Areas schemes (4) found that the impacts of Environmentally Sensitive Area designation on farmland birds were mixed. There was evidence for population increases or high numbers of some species of birds on Environmentally Sensitive Areas-managed land for four Environmentally Sensitive Areas. Populations of some species were stable in six Environmentally Sensitive Areas, often in contrast to national trends, but four Environmentally Sensitive Areas saw falls in the populations of at least one target species. The authors also note that in five regions there were not adequate data for all target species. The Environmentally Sensitive Areas scheme was introduced in 1987 and offered payments for either maintaining or enhancing landscape quality and biodiversity.

A study in 1997 in two Environmentally Sensitive Areas in eastern England (5) found that higher tier options (i.e. those with more demanding prescriptions but higher financial compensation) held significantly higher densities of wading birds (northern lapwing *Vanellus vanellus*, common redshank *Tringa totanus* and common snipe *Gallinago gallinago*) than lower tiers (Tier 1: 0.02–0.04 pairs/ha; Tier 2: 0.07–0.22; Tier 3: 0.40). In addition, they held more waders for each unit of money spent on the Environmentally Sensitive Area (Tier 1: 18–46 pairs/£100,000; Tier 2: 29–114; Tier 3: 167). However, when examining 1988–1997 population trends in four Environmentally Sensitive Areas, the authors found all three species investigated declined significantly (lapwing: 0.7–13% decline each year; redshank: 1.8–18.6%; snipe: 7.3–29.7%). The impact of wetland protection and management on waders is discussed in 'Maintain traditional water meadows' and 'Legally protect habitats'.

A review of research on agri-environment schemes in the UK (6) summarised two reports (Wilson *et al.* 2000, ADAS 2001) evaluating the effects of the Arable Stewardship Pilot Scheme (ASPS) in two regions (East Anglia and the West Midlands) from 1998–2003. At the whole farm scale in winter, seed-eating songbirds, thrushes and wagtails showed some increase on agreement farms relative to control farms (numbers not given). In summer, numbers of breeding northern lapwing *Vanellus vanellus*, reed bunting *Emberiza schoeniclus*, greenfinch *Carduelis chloris*, house sparrow *Passer domesticus*, common starling *Sturnus vulgaris* and yellow wagtail *Motacilla flava* were higher on agreement farms. Agreement farms had some of the following options: overwinter stubbles (sometimes preceded by reduced herbicide, followed by fallow or a spring crop), undersown spring cereals (sometimes followed by a grass or grass/clover ley), arable crop margins with reduced spraying (conservation headlands), grass margins or beetle banks and sown wildlife seed mixtures (pollen and nectar or wild bird seed mix). Over-winter stubble (974 and 2200 ha in East Anglia and West Midlands respectively) and conservation headlands (605 and 1085 ha in East Anglia and West Midlands respectively) were the most widely implemented options. The effects of the pilot scheme on birds were monitored at the farm scale over three years, relative to control areas, or control farms.

A 2003 replicated site comparison study of 102 fields across East Anglia and the West Midlands in the UK (7) found that two years after the introduction of the Arable Stewardship Scheme there was no difference in the number of farmland bird species observed in winter on Arable Stewardship Scheme and non-Arable Stewardship Scheme fields. There were, however, significantly more seed-eating songbirds, wagtails, and pipits on fields participating in the scheme than on farms not participating in the scheme. A further survey of 98 fields in summer found that although there were significantly more northern lapwings, starlings, greenfinches and reed buntings on Arable Stewardship Scheme fields, there were also fewer woodpigeons *Columba palumbus*, sedge warblers *Acrocephalus schoenobaenus* and rooks *Corvus frugilegus* than on the non-Arable Stewardship Scheme fields. Fifty-four Arable Stewardship Schemes and 48 comparable non-Arable Stewardship Scheme fields were surveyed for farmland birds in both the winters of 1998/1999 and 1999/2000; 50 Arable Stewardship Schemes and 48 non-Arable Stewardship Scheme fields were surveyed in the summer months of 1999 and 2000. The seed-eating songbirds identified included 13 species of finches, buntings and sparrows; wagtails and pipits comprised three species.

A 2003 replicated site comparison study of 76 farms in East Anglia, UK, and the West Midlands (8) found that autumn densities of grey partridges fell across both Arable Stewardship Pilot Scheme and non-Arable Stewardship Pilot Scheme (ASPS) farms from 1998 (when Arable Stewardship Pilot Scheme was introduced) to 2002. In East Anglia densities fell 68% on non-ASPS farms (5.5 to 1.8 birds/km^2) and 21% on Arable Stewardship Pilot Scheme farms (9.6 to 7.6 birds/km^2); in the West Midlands densities fell 40% on non-ASPS farms (1.4 to 0.8 birds/km^2) and 78% on Arable Stewardship Pilot Scheme farms (3.0 to 0.8 birds/km^2). In East Anglia, however, the young-to-old ratio doubled on Arable Stewardship Pilot Scheme plots from 1998 to 2002 (1 to 2 young:adult birds), whereas on non-Arable Stewardship Pilot Scheme farms the ratio fell by more than 50% (1.2 to 0.5 young:adult birds), indicating that the change in productivity on Arable Stewardship Pilot Scheme farms was twice that on non-Arable Stewardship Pilot Scheme farms. Surveys of grey partridge were made once each autumn in 1998 and 2002 on 76 farms: 20 ASPS and 19 non-ASPS farms in East Anglia and 20 Arable Stewardship Pilot Schemes and 17 non-Arable Stewardship Pilot Scheme farms in the West Midlands.

A 2003 review of 29 studies from six European countries (9) found that agri-environment schemes had no consistent effect on bird species. While there were individual successes, such as the 83% increase in cirl bunting between 1992 and 1998 on land within the Countryside Stewardship Scheme compared with the 2% increase on adjacent land not in the scheme, only 13/29 studies found agri-environment schemes increased bird species richness or abundance. Two studies reported negative effects and nine reported both positive and negative effects. Of the 19 studies that involved statistical tests, only four found positive effects, 2 of 19 reported negative effects and 9 of 19 reported both positive and negative effects.

A 2004 review of agri-environment scheme uptake and effectiveness in Europe (10) found that an average of 9% of agricultural land in EU countries was under agri-environment scheme designation, but that this ranged from 7% or less in some countries (e.g. The Netherlands, Spain, Greece) to 78, 77 and 64% in Austria, Finland and Sweden, respectively. In the UK, four rare species (grey partridge, corncrake, stone curlew or Eurasian thick-knee and cirl bunting) benefited from agri-environment schemes, although the authors note that densities of some species were higher on agri-environment scheme farms *before* they were designated. Similar methodological issues were found with studies in the Netherlands, where studies found that, at both field and larger scales, there were no population-level benefits of agri-environment scheme designation, although hatching and fledging rates of some species were higher on agri-environment scheme farms.

A 2004 replicated site comparison study of 74 farms in East Anglia and the West Midlands (11) found few differences in the density of farmland birds on farms participating in the Arable Stewardship Pilot Scheme and non-Arable Stewardship Pilot Scheme and, five years after the introduction of the scheme. In the West Midlands, although seed-eating songbirds, wagtails and pipits, insectivores, and raptors were found at higher densities on Arable Stewardship Pilot Scheme land than non-Arable Stewardship Pilot Scheme land, these higher densities were already present when measured within one year of the introduction of the scheme. Moreover, in East Anglia there were no differences in the bird densities found on Arable Stewardship Pilot Schemes and non-Arable Stewardship Pilot Scheme fields. Surveys of grey partridge populations on 76 farms in 1998 and 2002 found that adult densities decreased uniformly on both Arable Stewardship Pilot Schemes and non-Arable Stewardship Pilot Scheme farms over the five-year period. Bird surveys were carried out twice each during the winters of 1998/1999 and 2002/1903 on 18 Arable Stewardship Pilot Scheme and 19 non-Arable Stewardship Pilot Scheme farms in East Anglia and 19 Arable Stewardship Pilot Schemes and 18 non-Arable Stewardship Pilot Scheme farms in the West Midlands.

A 2004 literature review of farmland bird declines in Britain (12) found that 12 of 30 declining species have shown local population density increases after the implementation of agri-environment scheme options. Five out of ten seed-eating birds responded positively to agri-environment schemes, one (cirl bunting) showing large increases. Three other songbirds as well as corncrake, grey partridge and two waders responded to agri-environment scheme options. A further seven species responded to local conservation measures and eleven species were not studied sufficiently, were found not to respond to conservation measures or were recovering following national legislation (i.e. the prohibition of organochlorine pesticides).

A 2004 literature review (13) describes how ten years of agri-environment schemes in the UK have failed to halt the decline of many formerly common farmland species. However, it also points out that specially-designed agri-environment scheme options have led to local-scale population increases of three rare and range-restricted species (corncrake, Eurasian thick-knee and cirl bunting).

A 2006 replicated site comparison study in Spain and the Netherlands (14) found that birds bred more often, or were more numerous in areas participating in two agri-environment schemes, than on conventionally-farmed fields. In Spain, birds bred more often, and rare species bred and foraged more often in areas under a scheme designed to promote the conservation of steppe-associated birds than on paired sites without the scheme. In the Netherlands, more birds bred on 12.5 ha plots consisting of a mixture of fields with postponed agricultural activities and fields with a per-clutch payment scheme. However, the number of bird species on each type of farmland also did not differ between agri-environment schemes and non-agri-environment scheme plots, and there was no difference in bird abundance and breeding on those fields with only postponed agricultural activities compared with conventionally farmed fields. In Spain, the agri-environment scheme included limits on annual fertiliser and pesticide application; a month of restricted agricultural activity between April and July; mandatory unploughed strips covering three percent of fields; ploughing restrictions and a ban on burning fallow vegetation. In the Netherlands, the scheme prohibited changes in field drainage, pesticide application (except for patch-wise control of problem weeds) and any agricultural activity between 1 April and early June. Additionally, farmers of surrounding fields were paid for each meadow bird clutch laid on their land (though no agricultural restrictions were in place on these fields). In both countries, seven pairs of fields were surveyed in three parts of the country, four times over the breeding season.

A replicated study in 1999 and 2003 on 84 farms in East Anglia and the West Midlands, England (15), found that only three species (two in East Anglia, one in the West Midlands) showed a significant positive response to the introduction of agri-environment schemes, whilst one showed a significant negative effect. Meadow pipits *Anthus pratensis*, carrion crows *Corvus corone* and reed buntings either declined less or increased on farms under agri-environment schemes, compared to conventionally managed farms. Corn buntings *Miliaria calandra* declined significantly faster on agri-environment scheme farms. Overall, only six species showed any positive response (significant or not) in both regions, ten showed negative responses in both and 12 showed a positive response in one region and a negative response in the other. The impacts of individual management options are discussed in the relevant interventions.

A single farm, Rawcliffe Bridge, East Yorkshire, UK (16), with a combination of conservation measures prescribed under the English Entry Level Stewardship Scheme had higher densities of some bird species than the average for UK lowland farms. Meadow pipit, reed bunting, Eurasian skylark, grey partridge, corn bunting and yellow wagtail occurred in higher numbers in each monitoring year than the average lowland farm density (provided by the British Trust for Ornithology). For example, there were between 12 and 22 meadow pipit pairs/100 ha at Rawbridge, compared to a national average of <3. Birds on the farm were monitored five times each year from 2003 to 2005, by walking the field boundaries. The number of breeding pairs/ha was estimated from clusters of sightings.

A 2007 systematic review of 29 studies incorporating data for 15 farmland bird species in the UK (17) found that there were significantly higher winter densities of farmland birds on fields under agri-environment schemes than on conventionally managed fields. Considering each scheme individually, there was greater winter densities of birds on fields within the Arable Stewardship Pilot Scheme, Countryside Stewardship Scheme, fields with set-aside, overwinter stubble and wild bird cover than on conventionally farmed fields. Overall, eight species (53%) had significantly higher winter densities on agri-environment fields compared to conventional cropping (corn bunting, greenfinch, grey partridge, northern lapwing, linnet, rook, Eurasian skylark and song thrush *Turdus philomelos*) and no species were found to have higher densities on conventional agricultural fields compared to those fields entered under agri-environment scheme agreements. Although set-aside fields in summer had significantly higher densities of farmland birds, there was no difference between the number of birds on conventionally farmed fields and Arable Stewardship Pilot Schemes fields in summer. Six (35%; grey partridge, northern lapwing, woodpigeon, Eurasian skylark, rook and cirl bunting) of the 17 species for which summer data were available were found at significantly higher densities on agri-environment scheme fields compared with fields under conventional systems. The migratory yellow wagtail *Motacilla flava* was found at lower densities on scheme fields than on conventionally managed fields. In total 29 papers describing experiments conducted between 1985 and 2005 on a total of 12,653 fields (5,381 fields under agri-environment schemes and 7,272 fields farmed conventionally) were used for the meta-analysis. The meta-analysis included seven site comparison studies, five randomised control trials and 17 controlled trials.

A 2007 site comparison study of 677 plots covering 38,705 ha across southern England (18) found that for three wader species, population trends were more favourable (increasing or declining less rapidly) in areas under the Environmentally Sensitive Areas scheme options aimed at enhancing habitat than in the less expensive Environmentally Sensitive Areas habitat maintenance options and in parts of the surrounding countryside not participating in the scheme. However, population trends were most favourable on nature reserves. Between 1982 and 2002, common redshank declined by 70% in the wider countryside but increased overall from 646 to 755 pairs (up 17%) on Environmentally Sensitive Areas designated land (compared with 160% increases on non-Environmentally Sensitive Areas reserves). Northern lapwing showed a 48% decline in the wider countryside, but increased in reserves with Environmentally Sensitive Areas enhancement by 121% (compared with a 55% increase in non-Environmentally Sensitive Areas reserves). Common snipe breeding numbers decreased everywhere, but declines were smaller in reserves in Environmentally Sensitive Areas (24% decline) compared with reserves outside Environmentally Sensitive Areas (66% decline) or the wider countryside (up to 90% declines). Breeding waders were surveyed in 1982 and 2002 at lowland wet grassland sites covering ten counties in England. In both years, three censuses were carried out at each site between mid-April and mid-June.

A before-and-after study, examining data from 1976–2003 from farms across southern Sweden (19) found that four locally migrant farmland birds (northern lapwing, Eurasian skylark, common starling and linnet) showed less negative (or positive) population trends during 1987–1995, a period of agricultural extensification which included the introduction of agri-environment schemes, compared to the preceding period of intensification (1976–1987). However, following the adoption of the Common Agricultural Policy in 1995–2003, the species showed more negative population trends again, despite the widespread adoption of agri-environment scheme options. Three non-migrant species (house sparrow *Passer domesticus*, tree sparrow *P. montanus* and yellowhammer *Emberiza citrinella*) showed more diverse population trends and responses to agricultural changes were largely non-significant.

A study of the locations of Meadow Bird Agreements in the Netherlands (20) found that 43% of the 71,982 ha of Meadow Bird Agreements area in 2004 was located on sites where meadow bird populations are constrained for reasons other than those addressed by the agri-environment management. Twenty-two percent (15,798 ha) were outside the area of known black-tailed godwit *Limosa limosa* occurrence (more than five breeding pairs/100 ha in a 1998–2000 survey; 90–95% of other specialist meadow bird species breed in suitable black-tailed godwit habitat). Within the black-tailed godwit area, 11% (6,166 ha) of the Meadow Bird Agreements area was on heavily drained land, 4% (2,500 ha) was in landscapes not considered open enough for meadow birds, 10% (5,400 ha) was in areas of high traffic disturbance and an estimated 8% (2,834 of the 35,000 ha for which data were available) was on sites with high predation. The authors advocated targeting Meadow Bird Agreements to the 285,000 ha of land in the Netherlands with more than five breeding pairs of black-tailed godwit/100 ha, but none of the other identified constraints.

A replicated 2008 site comparison study of 53 2 km^2 plots on 14 farms in south-east Scotland (21) observed that between 2002 and 2004, the number of territorial male corn buntings fell by only 5% on plots that managed land according to the Farmland Bird Lifeline Scheme, whereas numbers declined by 43% in non-Farmland Bird Lifeline plots in the same area. Between 2000 and 2002, before the 2002 introduction of the Farmland Bird Lifeline management practices, there was no observed change in the number of corn buntings on either group of plots – although plots destined to participate in the Farmland Bird Lifeline Scheme did already have 33% higher densities of corn bunting than comparison plots. The Farmland Bird Lifeline Scheme intended to reverse the declining numbers of corn bunting, a priority species in the UK Biodiversity Action Plan. Farmers were paid for a number of interventions, including providing grass margins to arable fields, farming spring cereals and turnips at low intensity, spring cropping, leaving unharvested crop and supplementary feeding. Fourteen farms, nine in Aberdeenshire and five in Fife, were surveyed every breeding season (late April to August) from 2000 to 2004.

A 2008 literature review of the Environmental Stewardship programme, particularly Entry Level Stewardship in the UK (22) found that the amount of land entering the scheme was on target, but that several classes of options were not being taken up at a high enough rate to maintain some farmland birds. The authors argue that 'in-field' options such as skylark plots, conservation headlands and stubbles (all are discussed in their own sections) need to be promoted, as do complex field-edge options such as 'enhanced hedgerow management'. The rate of Entry Level Stewardship uptake in 2008 was estimated to be sufficient to promote population growth in only two of 12 species studied, and close in another. Even with a 70% uptake rate, the scheme was not predicted to promote population growth in five species (northern lapwing, European turtle dove *Streptopelia turtur*, yellow wagtail, Eurasian linnet and yellowhammer). The authors warn, however, that their analysis may have underestimated the effectiveness of Entry Level Stewardship.

A 2008 site comparison study of ten 3 km^2 plots in Austria (23) showed that, compared to conventionally managed arable land, land farmed less intensively (under agri-environment schemes) had larger numbers of ground breeding birds (16 vs. 13 individuals/10 ha), red listed birds (3 vs. 2 individuals/10 ha) and Species of European Conservation Concern (14 vs. 10 individuals/10 ha). Arable land managed for the conservation of particular species had 27 Species of European Conservation Concern individuals/10 ha and 29 ground breeding individuals/10 ha compared with the 11 and 14, respectively, on conventionally managed farmland. Reed-breeding birds on grassland benefited from similar initiatives (11 vs. 3 individuals/10 ha of farmland). Habitat conservation measures appeared to benefit ground breeders on arable farmland (17 vs. 10 individuals/10 ha). Breeding birds were surveyed during three visits between April and June 2003.

A 2009 literature review of agri-environment schemes in England (24) found that options and schemes varied in effectiveness. Breeding populations of some nationally rare birds increased after the implementation of options on arable farms (cirl bunting pairs increased by 130%, Eurasian thick-knee pairs by 87%) and a case study from a single farm found that grey partridge numbers increased by more than 250%/year; corn buntings by over 100%/year and Eurasian skylarks by 71%/year following the implementation of a number of different options. Productivity of some species was found to be higher on agri-environment scheme farms, which also provided key habitats. However, there was little evidence for any population-level beneficial effects of Entry Level Stewardship designation on widespread birds such as skylarks or yellowhammers *E. citrinella*. Several studies reviewed argued that most agri-environment scheme schemes were not well targeted to provide habitat for waders, although other studies argued that wader populations had declined less in regions designated as agri-environment schemes than in the country overall. The effects of individual options on birds are discussed in the relevant sections.

A replicated paired sites study on farms across Scotland under two agri-environment scheme prescriptions (Countryside Premium Scheme and Rural Stewardship Scheme) in spring-summer 2004–2008 (25) concluded that the schemes had little impact on farmland biodiversity. Whilst 280 agri-environment scheme farms had more birds of more species than 193 non-scheme paired farms (averages of 140 birds of 23 species on 105 Countryside Premium Scheme farms vs. 108 of 20 on paired non-scheme farms; 108 birds of 19 species on 88 Rural Stewardship Scheme farms vs. 86 of 17 on paired farms), trends did not vary between scheme and non-scheme farms, and scheme farms had higher species richness and abundances before entering schemes. Differences held for all species and for nationally threatened species. Time since entry into the Countryside Premium Scheme did not appear to affect the number of species or bird abundance, except for a small decline in the abundance of tits *Parus* spp. In addition, no evidence was found for differing effects of schemes in different regions of Scotland, or on different farm types.

A controlled study in 2002–9 on mixed farmland in Hertfordshire, England (26), found that the estimated population density of grey partridges was significantly higher on land under agri-environment schemes than on conventional arable crops. This study also examined the densities found on set-aside (which were similar to those on land under other agri-environment schemes, see 'Provide or retain set-aside'), wild bird cover (which were considerably higher than on other land uses, see 'Plant wild bird seed or cover mixture') and the impact of predator control and supplementary food provision (see 'Provide supplementary food to increase adult survival' and 'Control predators not on islands').

A large 2010 site comparison study of 2,046, 1 km^2 plots of lowland farmland in England (27) found that the Countryside Stewardship Scheme and Entry Level Stewardship schemes had no consistent effect on farmland bird numbers three years after their introduction in 2005. Between 2005 and 2008 eight Farmland Bird Index species showed significant declines on arable plots, nine species declined significantly on pastoral plots and six species declined on mixed farmland squares (farmland plots covered with less than 50% arable and less than 50% pastoral farming). Only goldfinch *Carduelis carduelis*, jackdaw *Corvus mondedula* and woodpigeon showed population increases between 2005 and 2008. Although certain farmland bird species did show landscape-specific effects, there were no consistent relationships between farmland bird numbers and whether or not the plots contained Entry Level Stewardship and Countryside Stewardship Scheme land, or the financial cost of the agri-environment interventions, or the length of hedgerows or ditches under an agri-environment scheme, or the availability of wild bird seed mix and over-winter stubbles (i.e. some species showed increases in response to a particular intervention on a particular landscape-type but not on other landscape-types, and these changes were not consistent between species). The 2,046 1 km^2 lowland plots were surveyed in both 2005 and 2008 and classified as arable, pastoral or mixed farmland. Eighty-four percent of plots included some area managed according to the Entry Level Stewardship or Countryside Stewardship Scheme. In both survey years, two surveys were conducted along a 2 km pre-selected transect route through each 1 km^2 square.

A replicated site comparison of the same 2,046, 1 km squares of agricultural land across England as in (27) in April–June 2005 and 2008 (28) found that farmland bird population responses to Entry Level Stewardship schemes varied regionally. The authors suggest that detailed, regional prescriptions may be more effective in stimulating population growth than uniform agri-environment schemes. Field margin management took place in 36% of squares and did not have clear impacts on 'field margin' species: two responded positively in at least one region, three species showed positive and negative responses in different regions, one only negative responses and the other six showed no significant responses.

A replicated site comparison study on 1,031 agricultural sites across England in 2004–2008 (29) found that in three out of four year-on-year comparisons, grey partridge *Perdix perdix* density changes and overwinter survival were higher on sites under agri-environment schemes, than on sites not under schemes (density changes were more positive on agri-environment scheme sites than non-agri-environment scheme sites in all comparisons except 2007–2008; overwinter survival was higher for all except 2006–2007). However, these differences were only significant in 2005–6 for density changes (6% increase on agri-environment scheme sites vs. 11% decrease on non-agri-environment schemes sites) and 2006–2007 for overwinter survival. There were no consistent differences between agri-environment schemes and non-agri-environment scheme sites with respect to brood size. When schemes were investigated individually, only Countryside Stewardship Scheme sites and Environmentally Sensitive Areas sites had significantly more positive density trends than non-scheme sites, and only in 2005–2006 (6% increase on Countryside Stewardship Scheme and Environmentally Sensitive Areas sites vs. 12% decline on non-agri-environment scheme sites), although other years and schemes showed a similar pattern. Overwinter survival, brood size and the ratio of chicks to adults did not show consistent effects across different schemes. These individual options are discussed in the relevant sections. Various methods of succession management (rough grazing, scrub creation, scrub control, grassland creation) were negatively associated with the ratio of young to old partridges in 2008.

A small 2010 site comparison study of 75 fields in East Anglia and the West Midlands, UK, (30) found no difference between the numbers of seed-eating birds in fields managed under the Higher Level of the Environmental Stewardship scheme and numbers in fields managed under the Entry Level of the scheme. Entry Level Stewardship fields had stubbles and were prohibited from post-harvest herbicide and cultivation until mid-February, and were planted overwinter with wild bird seed mix. Higher Level Environmental Stewardship fields were planted with enhanced wild bird seed mix and the stubbles had the basic Entry Level Stewardship requirements plus reduced herbicide use. These interventions are discussed in more detail in 'Plant wild bird seed or cover mixture' and 'Leave overwinter stubbles'.

A 2010 before-and-after trial of the Entry Level Stewardship on a 1,000 ha lowland arable farm in central England (31) observed that the number of seed-eating birds was higher on both Entry Level Stewardship and conventionally farmed fields in the winter of 2006/2007 than during the previous winter – when the Entry Level Stewardship was first introduced. This increase was greater on Entry Level Stewardship plots setting aside five percent of farmland to provide winter bird food (with an average of 70 birds/km of transect in 2007 versus five birds/km of transect in 2006) than on conventionally farmed fields (25 birds/km of transect in 2007 versus ten birds/km of transect in 2006). Although there were also more summer breeding territories of seed-eating species, chaffinch *Fringilla coelebs*, dunnock *Prunella modularis*, and robin *Erithacus rubecula* on the farm as a whole in 2007 than in the previous breeding season, there was no difference in this increase between Entry Level Stewardship and conventional fields. Land managed according to the minimal environmental requirements was compared both with fields where five percent of land was removed from production and replaced with patches of winter bird food and field margins (6–8 m). Winter birds were surveyed from transects on three visits (November, December and January) in both the winters of 2005/2006 and 2006/2007 – i.e. before and after bird food patch establishment. Breeding territories were surveyed during four visits (April, May, June and July) in 2006 and 2007.

(1) Ovenden, G. N., Swash, A. R. H. & Smallshire, D. (1998) Agri-environment schemes and their contribution to the conservation of biodiversity in England. *Journal of Applied Ecology*, 35, 955–960.

(2) Aebischer, N. J., Green, R. E. & Evans, A. D. (2000) From science to recovery: four case studies of how research has been translated into conservation action in the UK. 43–54 in: N. J. Aebischer, A. D. Evans, P. V. Grice, J. A. Vickery (eds) *Ecology and Conservation of Lowland Farmland Birds*. British Ornithologists Union, Tring.

(3) Peach, W., Lovett, L., Wotton, S. & Jeffs, C. (2001) Countryside stewardship delivers cirl buntings (*Emberiza cirlus*) in Devon, UK. *Biological Conservation*, 101, 361–373.

(4) DEFRA (2003) Review of agri-environment scheme monitoring results and R&D. (RMP/1596). Final Report – Part A (V45). Ecoscope/CPM/CJC Consulting 15/04/03.

(5) Ausden, M. & Hirons, G. J. M. (2002) Grassland nature reserves for breeding wading birds in England and the implications for the ESA agri-environment scheme. *Biological Conservation*, 106, 279–291.

(6) Evans, A. D., Armstrong-Brown, S. & Grice, P. V. (2002) The role of research and development in the evolution of a 'smart' agri-environment scheme. *Aspects of Applied Biology*, 67, 253–264.

(7) Bradbury, R. & Allen, D. (2003) Evaluation of the impact of the pilot UK Arable Stewardship Scheme on breeding and wintering birds. *Bird Study*, 50, 131–141.

(8) Browne, S. & Aebischer, N. (2003) Arable stewardship: impact of the pilot scheme on grey partridge and brown hare after five years. DEFRA Final Report RMP1870vs3. Department for Environment, Food and Rural Affairs, London, UK.

(9) Kleijn, D. & Sutherland, W. J. (2003) How effective are European agri-environment schemes in conserving and promoting biodiversity? *Journal of Applied Ecology*, 40, 947–969.

(10) Berendse, F., Chamberlain, D., Kleijn, D. & Schekkerman, H. (2004) Declining biodiversity in agricultural landscapes and the effectiveness of agri-environment schemes. *Ambio*, 33, 499–502.

(11) Bradbury, R. B., Browne, S. J., Stevens, D. K. & Aebischer, N. J. (2004) Five-year evaluation of the impact of the Arable Stewardship Pilot Scheme on birds. *Ibis*, 146, 171–180.

(12) Newton, I. (2004) The recent declines of farmland bird populations in Britain: an appraisal of causal factors and conservation actions. *Ibis*, 146, 579–600.

(13) Vickery, J. A., Bradbury, R. B., Henderson, I. G., Eaton, M. A. & Grice, P. V. (2004) The role of agri-environment schemes and farm management practices in reversing the decline of farmland birds in England. *Biological Conservation*, 119, 19–39.

(14) Kleijn, D., Baquero, R. A., Clough, Y., Diaz, M., Esteban, J., Fernández, F., Gabriel, D., Herzog, F., Holzschuh, A., Jöhl, R., Knop, E. Kruess, A., Marshall, E. J. P., Steffan-Dewenter, I., Tscharntke, T., Verhulst, J., West, T. M. & Yela J. L. (2006) Mixed biodiversity benefits of agri-environment schemes in five European countries. *Ecology Letters*, 9, 243–254.

(15) Stevens, D. K. & Bradbury, R. B. (2006) Effects of the Arable Stewardship Pilot Scheme on breeding birds at field and farm-scales. *Agriculture, Ecosystems & Environment*, 112, 283–290.
(16) Bryson, R. J., Hartwell, G. & Gladwin, R. (2007) Rawcliffe Bridge, arable production and biodiversity, hand in hand. *Aspects of Applied Biology*, 81, 155.
(17) Roberts, P. D. & Pullin, A. S. (2007) *The effectiveness of land-based schemes (including agri-environment) at conserving farmland bird densities within the UK.* Systematic Review No. 11. Collaboration for Environmental Evidence/Centre for Evidence-Based Conservation, Birmingham, UK.
(18) Wilson, A., Vickery, J. & Pendlebury, C. (2007) Agri-environment schemes as a tool for reversing declining populations of grassland waders: mixed benefits from Environmentally Sensitive Areas in England. *Biological Conservation*, 136, 128–135.
(19) Wretenberg, J., Lindstrom, A., Svensson, S. & Part, T. (2007) Linking agricultural policies to population trends of Swedish farmland birds in different agricultural regions. *Journal of Applied Ecology*, 44, 933–941.
(20) Melman, T. C. P., Schotman, A. G. M., Hunink, S. & de Snoo, G. R. (2008) Evaluation of meadow bird management, especially black-tailed godwit (*Limosa limosa* L.), in the Netherlands. *Journal for Nature Conservation*, 16, 88–95.
(21) Perkins, A., Maggs, H., Wilson, J., Watson, A. & Smout, C. (2008) Targeted management intervention reduces rate of population decline of Corn Buntings *Emberiza calandra* in eastern Scotland. *Bird Study*, 55, 52–58.
(22) Vickery, J., Chamberlain, D., Evans, A., Ewing, S., Boatman, N., Pietravalle, S., Norris, K. & Butler, S. (2008) Predicting the impact of future agricultural change and uptake of Entry Level Stewardship on farmland birds. British Trust for Ornithology.
(23) Wrbka, T., Schindler, S., Pollheimer, M., Schmitzberger, I. & Peterseil, J. (2008) Impact of the Austrian agri-environmental scheme on diversity of landscapes, plants and birds. *Community Ecology*, 9, 217–227.
(24) Natural England (2009) *Agri-environment Schemes in England 2009. A Review of Results and Effectiveness.* Natural England, Peterborough.
(25) Parish, D., Hirst, D., Dadds, N., Brian, S., Manley, W., Smith, G. and Glendinning, B. (2009) Monitoring and Evaluation of Agri-environment Schemes. Final Report, May 2009. Scottish Government.
(26) Aebischer, N. J. & Ewald, J. A. (2010) Grey partridge *Perdix perdix* in the UK: recovery status, set-aside and shooting. *Ibis*, 152, 530–542.
(27) Davey, C. M., Vickery, J. A., Boatman, N. D., Chamberlain, D. E., Parry, H. R. & Siriwardena, G. M. (2010) Assessing the impact of Entry Level Stewardship on lowland farmland birds in England. *Ibis*, 152, 459–474.
(28) Davey, C. M., Vickery, J. A., Boatman, N. D., Chamberlain, D. E., Parry, H. R. & Siriwardena, G. M. (2010) Regional variation in the efficacy of Entry Level Stewardship in England. *Agriculture, Ecosystems & Environment*, 139, 121–128.
(29) Ewald, J. A., Aebischer, N. J., Richardson, S. M., Grice, P. V. & Cooke, A. I. (2010) The effect of agri-environment schemes on grey partridges at the farm level in England. *Agriculture, Ecosystems & Environment*, 138, 55–63.
(30) Field, R. H., Morris, A. J., Grice, P. V. & Cooke, A. (2010) The provision of winter bird food by the English Environmental Stewardship scheme. *Ibis*, 153, 14–26.
(31) Hinsley, S. A., Redhead, J. W., Bellamy, P. E., Broughton, R. K., Hill, R. A., Heard, M. S. & Pywell, R. F. (2010) Testing agri-environment delivery for farmland birds at the farm scale: the Hillesden experiment. *Ibis*, 152, 500–514.

5.6. Cross compliance standards for all subsidy payments

- Apart from the Swiss Ecological Compensation Areas scheme (considered in another section), we found no studies comparing the effects of cross compliance standards with other means of implementing agri-environmental measures, or that considered the effects of cross compliance by monitoring farmland bird populations before and after it was implemented.

Background

Cross compliance is when farmers have to meet certain statutory standards to qualify for direct support payments such as those under the first pillar of the current Common Agricultural Policy. The standards could include, for example, keeping the land in 'good agricultural condition' or managing soil to avoid erosion. The Swiss Ecological Compensation Areas scheme, under which farmers have to manage 7% of their land to qualify for area-based payments, was made obligatory in Switzerland under cross compliance in 1998. Studies examining the effects of this scheme are included in a different section: 'Increase the proportion of natural/semi-natural habitat in the farmed landscape'.

5.7. Reduce field size (or maintain small fields)

- We found no intervention-based evidence on the effects of reducing field sizes on bird populations.

Background

Reducing field size means having a greater number of smaller fields, with boundaries between them. One reason this approach is expected to enhance biodiversity is that field boundaries of any type provide heterogeneity, with heterogeneity thought to be a strong factor determining biodiversity on farmland.

5.8. Provide or retain set-aside areas in farmland

- Three replicated studies and a review of five studies from Europe and North America (1,8,13,15) examining species richness or diversity found that more species were found on set-aside than on crops. One (14) found fewer species on set-aside than other agricultural habitats.
- All 21 studies, including a systematic review, 12 replicated experiments and two reviews, from Europe and North America that investigated population trends or habitat associations found that some species were found at higher densities or used set-aside more than other habitats (1–8,11–14,16–22), or were found on set-aside (9,10,23). Four studies (three replicated) from the UK (4,5,11,14) found that some species were found at lower densities on set-aside compared to other habitats.
- Three of four replicated studies from the UK (1–3) found that waders and Eurasian skylarks had higher productivities on set-aside, compared to other habitats. One study (10) found that skylarks nesting on set-aside had lower productivity compared to those on cereal crops, and similar productivities to those on other crops.
- One replicated paired study from the UK (7) found that rotational set-aside was used more than non-rotational set-aside, a replicated paired study (8) found no differences between rotational and non-rotational set-aside. A review from Europe and North America (13) found that naturally regenerated set-aside held more birds and more species than sown set-aside.

Background

Allocation of some farmland to 'set-aside' (fields taken out of production) was compulsory under European agricultural policy from 1992 until 2008. Originally intended as a method of reducing production, set-aside has also been promoted as a way of protecting on-field biodiversity. Set-aside fields can be sown with fallow crops or left to naturally regenerate. Set-aside can be rotational (in a different place every year) or long term (retained for 5–20 years).

A 2008 literature review of the Environmental Stewardship programme, particularly Entry Level Stewardship (ELS) in the UK (Vickery *et al.* 2008) found that the population trends of all Farmland Bird Index species were positively correlated with the availability of set-aside in that year and that Entry Level Stewardship may not be able to effectively replace set-aside.

Vickery, J., Chamberlain, D., Evans, A., Ewing, S., Boatman, N., Pietravalle, S., Norris, K. & Butler, S. (2008) *Predicting the Impact of Future Agricultural Change and Uptake of Entry Level Stewardship on Farmland Birds*. British Trust for Ornithology, The Nunnery, Thetford.

A replicated, paired sites study on seven pairs of fields in north-east Scotland in 1989–91 (1) found that one-year-old set-aside fields held significantly more species of bird than similar, non-set-aside fields (average of 12 species/10 ha for first year set-aside vs. 5 species/10 ha for 'control' fields). There were no differences in the years before or after set-aside. In addition, there were higher breeding densities of grey partridge *Perdix perdix*, Eurasian skylark *Alauda arvensis* and Eurasian curlew *Numenius arquata* in set-aside compared with 'control' fields. Densities of curlew, partridge, northern lapwing *Vanellus vanellus* and Eurasian oystercatcher *Haematopus ostralegus* were higher in set-aside years than before set-aside (songbird densities were not recorded before set-aside was used). Wader breeding success appeared higher on set-aside, but numbers were too small for statistical tests. The densities and number of species declined over time in set-aside fields. Set-aside fields were previously arable fields but were not cropped for at least one year.

A replicated study in summers of 1993–95 on seven farms in southern England (2) found that there were significantly higher densities of Eurasian skylark *Alauda arvensis* nests on set-aside fields than on conventionally or organically managed crop fields (0.3–0.5 territories/ha for set-aside fields vs. a maximum of 0.4 territories/ha for cropped fields). Estimated nest survival was significantly higher on set-aside fields than conventionally managed cereal fields (44% survival to fledgling on set-aside vs. 11% for conventional cereals). Set-aside was either naturally regenerated from crop stubble or sown with grass.

A site comparison in April to August 1992 on three farms in south England (3) found that skylarks had significantly higher productivity in set-aside fields, compared to spring-sown cereals or grass (0.5 fledglings/ha in set-aside vs. 0.21 fledglings/ha in spring cereals and 0.1 fledglings/ha in silage grass). This difference was largely due to higher densities of territories (2–3 times higher in set-aside and grass, compared to cereals) and more successful nests (highest on grass, but twice as high in set-aside as in cereal crops) and larger clutches in set-aside (3.9 eggs/clutch for nests in set-aside vs. 3.3 eggs/clutch for spring cereals and 3.4 eggs/clutch in grass, eleven nests in each habitat type). Fledging success did not vary between habitats. No nests with chicks were found in winter-sown cereals. Set-aside consisted of four year-old permanent fallow sown with red fescue *Festuca rubra*, perenial rye-grass *Lolium perenne* and white clover *Trifolium pratense*.

A replicated study in summer 1995 on 89 fields in the South Downs, southern England (4), found that the density of singing Eurasian skylarks was higher on set-aside fields than on any other field type, except undersown spring barley fields (approximately 15 birds/km^2 on six set-aside fields vs. 22 birds/km^2 on four spring barley fields and 2–12 birds/km^2 on 79 other fields). Other field types were arable fields reverted to species-rich grassland ('Habitat restoration and creation') or permanent grassland ('Revert arable land to permanent grassland'); downland turf (close-cropped, nutrient-poor grassland); permanent grasslands; and winter wheat, barley and oil seed rape. This study is also described in 'Reduce grazing intensity on permanent grasslands' and 'Undersow spring cereals'.

A randomised and replicated site comparison in the winters of 1992–1993 and 1993–1994 on 40 farmland sites in Devon and East Anglia, UK (5) found that only one taxonomic group (finches, sparrows and buntings, seven species) showed a significant preference for set-aside habitats in both years, preferentially using sown set-aside less than one year old. Conversely, thrushes (four species) and hedge-dwelling species (European robin *Erithacus rubecula*, wren *Troglodytes troglodytes* and dunnock *Prunella modularis*) avoided regenerating set-aside less than one year old in Devon. At a species level, a preference for set-aside was seen in both winters by one species in Devon (cirl buntings *Emberiza cirlus* selecting sown set-aside more than one year-old) and two species (plus one introduced species not considered here) in East Anglia (grey partridge preferred older sown set-aside and yellowhammer *Emberiza citrinella* selected one year-old sown cover). A further 13 species in both East Anglia and Devon preferentially selected a set-aside habitat in one winter. Blackbirds *Turdus merula* and five other species avoided some set-aside in at least one year in Devon; no native species did so in East Anglia. The same 40 plots (50–100 ha) were surveyed each winter, although the amount of set-aside they contained varied due to rotation schemes.

A 2000 literature review from the UK (6) found that the populations of grey partridge, Eurasian thick-knee *Burhinus oedicnemus* and cirl buntings all increased following multiple measures including the provision of set-aside. Partridge numbers were 600% higher on farms with conservation measures aimed at partridges (including conservation headlands, planting cover crops, using set-aside and creating beetle banks) in place, compared to farms without these measures; the UK thick-knee population increased from 150 to 233 pairs from 1991 to 1999 (interventions were set-aside provision and uncultivated plots in fields); the UK cirl bunting population increased from 118–132 pairs in 1989 to 453 pairs in 1998, with a 70% increase on fields under schemes (with overwinter stubbles, grass margins and beneficially managed hedges and set-aside), compared to a 2% increase elsewhere.

A replicated paired sites study in 1996–7 across 92 arable farms in England (7) found that five of six bird functional groups examined were at higher densities on set-aside fields, compared to winter cereals or grassland (although thrushes only showed this preference in one year). On ten farms with rotational and non-rotational set-aside, all groups except crows were found at higher densities on rotational set-aside fields. All groups except gamebirds (which showed no significant field preferences) were also more likely to be found on set-aside than on other field types. Functional groups of birds were gamebirds, pigeons, crows, skylarks, thrushes and seed-eating songbirds (sparrows, buntings and finches).

A replicated paired sites study in 1996–7 on 11 farms in east and west England (8), found that set-aside fields supported more species and higher densities of birds than adjacent crop fields (1–7 birds/ha and 7–21 species for 11 set-aside fields vs. 0.2–0.8 birds/ha and 2–5 species on 11 crop fields). Between 78% and 100% of species found on both field types were more abundant on set-aside. These preferences were stronger (although not significantly so) for rotational set-aside, compared to non-rotational.

Another analysis (9) as part of the same study as in (7) found that skylark densities on set-aside fields ranged from zero to approximately three birds/ha. A total of 74 set-aside fields (36 rotational and 38 non-rotational) were examined, each from a different farm. The authors' note that fields with approximately 30% bare earth, straw and litter had the highest densities of skylarks.

A replicated study in 1996–98 on 22 farms in southern England (10) found that skylark nests had significantly lower survival in set-aside, compared to in cereals (22% overall survival for 525 nests in set-aside vs. 38% survival for 183 nests in cereal fields). There were no differences between set-aside and other crop types (19% survival for 173 nests in grass fields, 29% survival for 60 nests in other field types) or between rotational and non-rotational set-aside. On one intensively-studied farm, over 90% of 422 skylark nests were found on ten fields of well-established, non-rotational set-aside. This study also describes the impact of predator control on skylark nest survival, discussed in 'Control predators not on islands'.

A study of different set-aside crops at Allerton Research and Educational Trust Loddington farm, UK (11) found that Eurasian skylark, but not yellowhammer *Emberiza citrinella*, used unmanaged set-aside more than expected compared to availability. Skylarks used unmanaged set-aside more than expected compared to availability, but significantly less than kale set-aside 'wild bird cover', 'wild bird cover' strips and beetle banks. Cereal (wheat, barley) and broad-leaved crops (beans, rape) were used less than expected. Yellowhammer used unmanaged set-aside as expected compared to availability and used it significantly less than cereal and cereal set-aside 'wild bird cover' and 'wild bird cover' strips. Set-aside strips (field margin and midfield) were sown with kale-based and cereal-based mixtures for 'wild bird cover' and 'beetle banks'. Other habitat types were: unmanaged set-aside, cereal (wheat, barley), broad-leaved crop (beans, rape) and 'other' habitats (including permanent pasture, woodland, hedgerows, tracks and riparian areas). Thirteen skylark and 15 yellowhammer nests with chicks between 3–10 days old were observed. Foraging habitat used by the adults was recorded for 90 minutes during three periods of the day. This study is also discussed in 'Create beetle banks' and 'Plant wild bird seed/cover'.

A replicated, randomised study of 200 farms in England with set-aside (12) found that an increase in bird numbers was reported by 47% of farmers with rotational set-aside and 69% of farmers with non-rotational set-aside. Bird density in rotational set-aside was nine times, and in non-rotational sown grassland set-aside seven times, that in crops. Management of set-aside had minimal effect on bird abundance. Breeding bird territories were mapped on 63–92 farms (1996–1997). More intensive surveys were undertaken for habitat use by birds on 11 farms (1996–1997).

A meta-analysis of 127 studies comparing set-aside and conventional land (13) found that species richness and population densities of birds were significantly higher on set-aside land than on nearby conventional fields in Europe and North America. Positive effects were greatest on larger and older areas of set-aside, when the comparison conventional field was crops rather than grasses and in countries with more arable land under agri-environment schemes and with less intensive agriculture. Overall, variation in establishment methods and types of set-aside made little difference to the positive effect on biodiversity, although species richness was increased more when set-aside was naturally regenerated rather than sown.

A replicated, randomised, controlled study from November–February in 2000/2001 and 2001/2002 in 20 arable farms in eastern Scotland (14) found that, of 23 species recorded, only Eurasian skylarks were found at higher densities in fields with set-aside than fields with wild bird cover crops or conventional crops. Bird density was up to 100 times greater in wild bird cover crops than on set-aside fields. The wild bird cover crops attracted 50% more species than set-aside fields. Of eight species with sufficient data for individual analysis, seven were consistently significantly more abundant in wild bird cover than in set-aside fields. Set-aside fields were those in which cereal stubble was left to regenerate naturally. Between 6 and 28 ha were sampled on each farm annually.

A replicated paired sites comparison in summer 2003 in County Laois and County Kildare, Ireland (15), found that 18 set-aside fields had significantly higher avian species diversity and richness than 18 adjacent agricultural fields (an average of 13 species on set-aside vs. 9 species on farmed fields). Three species were significantly more abundant on set-aside and whilst six species showed a preference for non-set-aside fields, these preferences were not significant and the species (whitethroat *Sylvia communis*, goldcrest *Regulus regulus*, blackcap *Sylvia atricapilla*, stonechat *Saxicola torquata*, tree sparrow *Passer montanus* and treecreeper *Certhia familiaris*) were more likely to be selecting habitats based on field margins, rather than field management. Six species were associated with non-rotational set-aside; two with rotational set-aside; one with long-term grazed pasture set-aside and three with first-year pasture set-aside.

A replicated study in 1999 and 2003 on 256 arable and pastoral fields across 84 farms in East Anglia and the West Midlands, England (16), found that only two of twelve farmland bird species analysed were positively associated with the provision of set-aside, wildlife seed mixtures (see 'Plant wild bird seed or cover mix') or overwinter stubble (see 'Leave overwinter stubbles'). These were skylarks *Alauda arvensis* (a field-nesting species) and linnets *Carduelis cannabina* (a boundary-nesting species). The study did not distinguish between set-aside, wildlife seed mixtures or overwinter stubble, classing all as interventions to provide seeds for farmland birds. This study describes several other interventions, discussed in the relevant section.

A 2007 systematic review identified 11 papers investigating the effect of set-aside provision on farmland bird densities in the UK (17). In both winter and summer surveys there were significantly higher densities of farmland birds on fields removed from production and under set-aside designation than on conventionally farmed fields. The meta-analysis included experiments conducted between 1988 and 2002 from eight controlled trials and three site comparison studies.

A before-and-after study, examining data from 1976–2003 from farms across southern Sweden (18) found that four locally migrant farmland birds showed less negative (or positive) population trends during a period of agricultural extensification, which included an increase in the area of set-aside. The authors suggest that the two could be causally linked. This study is discussed in 'Pay farmers to cover the costs of conservation measures'.

A before-and-after site comparison study in 2000–2005 in Bedfordshire, England (19), found that set-aside fields sprayed in May or June supported higher densities of grey partridge, seed-eating songbirds and skylarks *Alauda arvensis*, compared to set-aside sprayed in April or crop fields (although seed-eating passerines were equally numerous on oilseed rape *Brassica napus* fields). Early-sprayed set-aside had consistently lower densities of all species, compared to all land uses except winter-sown wheat. The site-level effects of set-aside and sowing crops in spring are discussed in 'Plant crops in spring rather than autumn'. This study also investigated the impact of reducing pesticide and fertiliser inputs (see 'Threat: Pollution').

A controlled study in 2002–2009 on mixed farmland in Hertfordshire, England (20), found that the estimated population density of grey partridges *Perdix perdix* was significantly higher on set-aside land, than on conventional arable crops. The difference was strongest for rotational set-aside, with non-rotational set-aside not having a significant positive impact on partridge densities. This study also examined the densities found on land under various agri-environment schemes (which were similar to those on set-aside, see 'Pay farmers to cover the costs of conservation measures'), wild bird cover (which were higher than those on set-aside, see 'Plant wild bird seed or cover mixture') and the impact of predator control and supplementary food provision (see 'Provide supplementary food to increase adult survival' and 'Control predators not on islands').

A small study on four farms in Aberdeenshire, north-east Scotland, in summer 2005 (21) found that yellowhammers from ten nests preferentially foraged on set-aside land, compared to cereal fields, but that this preference was not significant (set-aside comprising 23% of available habitat but used for 42% of foraging flights vs. cereals comprising 42% of habitat and being used 25% of the time).

A study in April–May 2004 and 2005 (22) found that four birds of conservation concern were all found on set-aside on 210 fields in pseudo-steppe farmland in Catalonia, Spain. Little bustards *Tetrax tetrax* were found on 23–50% of fields within their range at densities of 0.3–0.8 birds/ha (68 fields surveyed in 2004, 86 in 2005), Eurasian thick-knee on 43–52% at 0.4–0.6 birds/ha (93 fields in 2004, 117 in 2005), short-toed larks *Calandrella brachydactyla* on 28–32% at 0.2–0.4 birds/ha (50 fields in 2004, 64 in 2005) and calandra larks *Melanocorphya calandra* on 27–34% at 0.4–0.7 birds/ha (93 fields in 2004, 117 in 2005). Only male bustards were recorded, due to problems surveying cryptic females.

(1) Watson, A. & Rae, R. (1997) Some effects of set-aside on breeding birds in northeast Scotland. *Bird Study*, 44, 245.

(2) Wilson, J. D., Evans, J., Browne, S. J. & King, J. R. (1997) Territory distribution and breeding success of skylarks *Alauda arvensis* on organic and intensive farmland in southern England. *Journal of Applied Ecology*, 34, 1462–1478.

(3) Poulsen, J. G., Sotherton, N. W. & Aebischer, N. J. (1998) Comparative nesting and feeding ecology of skylarks *Alauda arvensis* on arable farmland in southern England with special reference to set-aside. *Journal of Applied Ecology*, 35, 131–147.

(4) Wakeham-Dawson, A., Szoszkiewicz, K., Stern, K. & Aebischer, N. J. (1998) Breeding skylarks *Alauda arvensis* on Environmentally Sensitive Area arable reversion grass in southern England: survey-based and experimental determination of density. *Journal of Applied Ecology*, 35, 635–648.

(5) Buckingham, D. L., Evans, A. D., Morris, A. J., Orsman, C. J. & Yaxley, R. (1999) Use of set-aside land in winter by declining farmland bird species in the UK. *Bird Study*, 46, 157–169.

(6) Aebischer, N. J., Green, R. E. & Evans, A. D. (2000) From science to recovery: four case studies of how research has been translated into conservation action in the UK. 43–54 in: N. J. Aebischer, A. D. Evans, P. V. Grice, J. A. Vickery (eds) *Ecology and Conservation of Lowland Farmland Birds* British Ornithologists Union, Tring.

(7) Henderson, I. G., Cooper, J., Fuller, R. J. & Vickery, J. (2000) The relative abundance of birds on set-aside and neighbouring fields in summer. *Journal of Applied Ecology*, 37, 335–347.

(8) Henderson, I. G., Vickery, J. A. & Fuller, R. J. (2000) Summer bird abundance and distribution on set-aside fields on intensive arable farms in England. *Ecography*, 23, 50–59.

(9) Henderson, I. G., Critchley, N. R., Cooper, J. & Fowbert, J. A. (2001) Breeding season responses of skylarks *Alauda arvensis* to vegetation structure in set-aside (fallow arable land). *Ibis*, 143, 317–321.

(10) Donald, P. F., Evans, A. D., Muirhead, L. B., Buckingham, D. L., Kirby, W. B. & Schmitt, S. I. A. (2002) Survival rates, causes of failure and productivity of skylark *Alauda arvensis* nests on lowland farmland. *Ibis*, 144, 652–664.

(11) Murray, K. A., Wilcox, A. & Stoate, C. (2002) A simultaneous assessment of farmland habitat use by breeding skylarks and yellowhammers. *Aspects of Applied Biology*, 67, 121–127.

(12) Firbank, L. G., Smart, S. M., Crabb, J., Critchley, C. N. R., Fowbert, J. W., Fuller, R. J., Gladders, P., Green, D. B., Henderson, I. & Hill, M. O. (2003) Agronomic and ecological costs and benefits of set-aside in England. *Agriculture, Ecosystems & Environment*, 95, 73–85.

(13) Van Buskirk, J. & Willi, Y. (2004) Enhancement of farmland biodiversity within set-aside land. *Conservation Biology*, 18, 987–994.

(14) Parish, D. M. B. & Sotherton, N. W. (2004) Game crops and threatened farmland songbirds in Scotland: a step towards halting population declines? *Bird Study*, 51, 107.
(15) Bracken, F. & Bolger, T. (2006) Effects of set-aside management on birds breeding in lowland Ireland. *Agriculture, Ecosystems & Environment*, 117, 178–184.
(16) Stevens, D. K. & Bradbury, R. B. (2006) Effects of the Arable Stewardship Pilot Scheme on breeding birds at field and farm-scales. *Agriculture, Ecosystems & Environment*, 112, 283–290.
(17) Roberts, P. D. & Pullin, A. S. (2007) *The effectiveness of land-based schemes (including agri-environment) at conserving farmland bird densities within the UK.* Systematic Review No. 11. Collaboration for Environmental Evidence/Centre for Evidence-Based Conservation, Birmingham, UK.
(18) Wretenberg, J., Lindstrom, A., Svensson, S. & Part, T. (2007) Linking agricultural policies to population trends of Swedish farmland birds in different agricultural regions. *Journal of Applied Ecology*, 44, 933–941.
(19) Henderson, I. G., Ravenscroft, N., Smith, G. & Holloway, S. (2009) Effects of crop diversification and low pesticide inputs on bird populations on arable land. *Agriculture, Ecosystems & Environment*, 129, 149–156.
(20) Aebischer, N. J. & Ewald, J. A. (2010) Grey partridge *Perdix perdix* in the UK: recovery status, set-aside and shooting. *Ibis*, 152, 530–542.
(21) Douglas, D. J. T., Benton, T. G. & Vickery, J. A. (2010) Contrasting patch selection of breeding yellowhammers *Emberiza citrinella* in set-aside and cereal crops. *Bird Study*, 57, 69–74.
(22) Mcmahon, B. J., Giralt, D., Raurell, M., Brotons, L. & Bota, G. (2010) Identifying set-aside features for bird conservation and management in northeast Iberian pseudo-steppes. *Bird Study*, 57, 289.

5.9. Manage hedges to benefit wildlife

- The one study of six that investigated species richness (2) found no difference in species richness between a UK site with wildlife-friendly hedge management and three control sites.
- Seven studies from the UK (1,2,4–7) and Switzerland (3), five replicated, found that some species studied increased in relation to managed hedges or were more likely to be found in managed hedges, compared to other habitats. Two (1,2) investigated several interventions at once.
- One replicated study (5) found that species that showed positive responses to hedge management in some regions showed weak or negative responses in other parts of the UK. Four studies from the UK (2,4–6) found that some species declined or showed no response to wildlife-friendly management of hedges.

Background

Hedges can be key habitats for farmland biodiversity, but they may need managing to maximise their value. Managing hedges to benefit wildlife involves one or more of the following management changes: reduce cutting frequency; reduce or avoid spraying; mowing vegetation beneath hedgerows or filling gaps in hedges.

A 2000 literature review (1) found that the UK population of cirl buntings *Emberiza cirlus* increased from between 118 and 132 pairs in 1989 to 453 pairs in 1998 following a series of schemes designed to provide overwinter stubbles, grass margins and beneficially managed hedges and set-aside. Numbers on fields under the specific agri-environmental scheme increased by 70%, compared with a 2% increase elsewhere.

A small replicated controlled study from May–June in 1992–8 in Leicestershire, England (2), found that the abundance of nationally declining songbirds and species of conservation concern significantly increased on a 3 km^2 site where hedges were managed to benefit wildlife (alongside several other interventions), although there was no overall difference in bird abundance, species richness or diversity between the experimental and three control sites. Numbers of nationally declining species rose by 102% (except for Eurasian skylark *Alauda arvensis* and yellowhammer *Emberiza citrinella*). Nationally stable species rose (insignificantly) by 47% (eight species increased, four decreased). The other interventions employed were: 'Create beetle banks', 'Plant nectar flower mixture/wildflower strips', 'Plant wild bird seed cover strips', 'Provide supplementary food', 'Control predators' and 'Reduce pesticide or herbicide use generally'.

A replicated site comparison study across eleven areas in the Swiss plateau between 1998 and 2001 (3) found that the centres of territories of hedgerow birds were significantly more frequent in or near Ecological Compensation Areas than expected by an even distribution across the landscape (293 territories found in ECA hedgerows), suggesting that hedgerow birds were attracted to or favoured by these areas. Territories of breeding birds were mapped in 23 study areas, based on three visits between mid-April and mid-June.

A replicated study in February 2008 across 97, 1 km^2 plots in East Anglia, England (4), found that four farmland birds showed strong positive responses to field boundaries (hedges and ditches) managed under agri-environment schemes. These were blue tits *Parus caeruleus* (also called *Cyanistes caeruleus*), dunnock *Prunella modularis*, common whitethroat *Sylvia communis* and yellowhammer. A further five (Eurasian blackbird *Turdus merula*, song thrush *T. philomelos*, Eurasian bullfinch *Pyrrhula pyrrhula*, long-tailed tit *Aegithalos caudatus* and winter wren *Troglodytes troglodytes*) showed weak positive responses and Eurasian reed bunting *Acrocephalus scirpaceus* showed a weak negative response. The boundaries were classed as either hedges, ditches or hedges and ditches and most were managed under the Entry Level Stewardship scheme.

A replicated site comparison of 2,046, 1 km squares of agricultural land across England in 2005 and 2008 (5) found that management of hedges and ditches (see 'Manage ditches to benefit wildlife') under Entry Level Stewardship did not have clear impacts on farmland bird species. Management had significant positive impacts on five species in at least one region of England, but these effects were often very weak and four of the same species showed negative responses in other regions. The other five 'hedgerow species' investigated were never positively associated with boundary management. Generally, effects appeared to be more positive in the north of England.

A replicated 2010 site comparison study of 2,046 1 km^2 plots of lowland farmland in England (6) found that three years after the 2005 introduction of the Countryside Stewardship Scheme and Entry Level Stewardship schemes, there was no association between the length of hedgerow managed according to the agri-environment scheme and farmland bird numbers. Hedgerow specialist species, including the yellowhammer *Emberiza citrinella* and common whitethroat, showed no significant population response, whereas there were greater numbers of common starling *Sturnus vulgaris* on arable, pastoral and mixed farmland with hedgerow management. For example, in mixed farmland plots starling populations increased by 0.2 individuals for each 1 km of hedgerow. On the other hand, the grey partridge *Perdix perdix* appeared to be detrimentally affected, with an apparent decline of 0.3 individuals for every 1.1 km of hedgerow managed according to the agri-environment schemes. The 2,046 1 km^2 lowland plots were surveyed in both 2005 and 2008 and classified as arable, pastoral or mixed farmland. Eighty-four percent of plots included some area managed according to the schemes. In both survey years, two surveys were conducted along a 2 km pre-selected transect route through each 1 km^2 plot.

A replicated site comparison study on farms in two English regions (7) found that summer yellowhammer numbers were significantly higher in hedges under environmental stewardship management than in conventionally managed hedges. On East Anglian farms, this was true for both Entry Level Stewardship and Higher Level Stewardship hedge management options (estimated >1.5 yellowhammers/m in Higher Level Stewardship hedges compared to <0.5 yellowhammers/m in conventional hedges). On farms in the Cotswolds, UK, it was only true for hedges managed as 'high environmental value hedges' under Higher Level Stewardship (estimated 0.5 yellowhammers/m), while hedges managed under Entry Level Stewardship did not have more yellowhammers than conventional hedges (estimated <0.2 yellowhammers/m). Hedgerows managed under Entry Level Stewardship are cut every two or three years in winter only. Surveys were carried out in the summers of 2008 and 2009, on up to 30 Higher Level Stewardship farms and 15 non-stewardship farms in East Anglia, and up to 19 Higher Level Stewardship and 8 non-stewardship farms in the Cotswolds. This study also discusses several other interventions.

(1) Aebischer, N. J., Green, R. E. & Evans, A. D. (2000) From science to recovery: four case studies of how research has been translated into conservation action in the UK. 43–54 in: N. J. Aebischer, A. D. Evans, P. V. Grice, J. A. Vickery (eds) *Ecology and Conservation of Lowland Farmland Birds* British Ornithologists Union, Tring.
(2) Stoate, C. (2002) Multifunctional use of a natural resource on farmland: wild pheasant (*Phasianus colchicus*) management and the conservation of farmland passerines. *Biodiversity and Conservation*, 11, 561–573.
(3) Herzog, F., Dreier, S., Hofer, G., Marfurt, C., Schupbach, B., Spiess, M. & Walter, T. (2005) Effect of ecological compensation areas on floristic and breeding bird diversity in Swiss agricultural landscapes. *Agriculture, Ecosystems & Environment*, 108, 189–204.
(4) Davey, C. M., Vickery, J. A., Boatman, N. D., Chamberlain, D. E. & Siriwardena, G. M. (2010) Entry Level Stewardship may enhance bird numbers in boundary habitats. *Bird Study*, 57, 415–420.
(5) Davey, C. M., Vickery, J. A., Boatman, N. D., Chamberlain, D. E., Parry, H. R. & Siriwardena, G. M. (2010) Regional variation in the efficacy of Entry Level Stewardship in England. *Agriculture, Ecosystems & Environment*, 139, 121–128.
(6) Davey, C. M., Vickery, J. A., Boatman, N. D., Chamberlain, D. E., Parry, H. R. & Siriwardena, G. M. (2010) Assessing the impact of Entry Level Stewardship on lowland farmland birds in England. *Ibis*, 152, 459–474.
(7) Field, R. H., Morris, A. J., Grice, P. V. & Cooke, A. I. (2010) Evaluating the English Higher Level Stewardship scheme for farmland birds. *Aspects of Applied Biology*, 100, 59–68.

5.10. Plant new hedges

- A small study from the USA (1) found that the population of northern bobwhites increased following several interventions including the planting of new hedges.

Background

Hedges are used to separate fields but are also extremely important habitats on many farms, providing heterogeneity in the landscape and resources not found elsewhere. In much of Europe, hedges are being removed as field sizes are increased, potentially reducing the biodiversity value of farmland. Planting new hedges may mitigate this change, but may be both costly and unattractive to farmers, as they can reduce the efficiency of farming.

A small 1967 before-and-after study on a 1,214 ha farm in Maryland, USA (1), found that after the introduction in 1957 of a number of management interventions, including planting 11.4 miles of new hedges, the number of coveys of northern bobwhites *Colinus virginianus* increased from five coveys identified in the winter of 1956/1957 to 38 in the winter of 1964/1965. Although this study does not isolate the effect of the individual interventions made, it is noted that 14 of 33 new coveys were located in multi-flora hedges planted during the eight years of management interventions. Interventions included planting shrub lespedeza *Lespedeza thunbergii* and sericea lespedeza *Lespedeza cuneata* strips, seeding 20 ha of grassland and limiting livestock grazing. Sightings of coveys were reported by farm employees and hunting parties during each winter from 1956 to 1965.

(1) Burger, G. V. & Linduska, J. P. (1967) Habitat management related to bobwhite populations at Remington farms. *The Journal of Wildlife Management*, 31, 1–12.

5.11. Manage stone-faced hedge banks to benefit birds

- We captured no evidence for the effects of managing stone-faced hedge banks on bird populations.

Background

Stone-faced hedge banks are traditional boundary features in some agricultural landscapes, such as in the south-west of England. Management for biodiversity involves maintaining the wall with traditional materials.

5.12. Manage ditches to benefit wildlife

- Three out of four replicated studies from the UK (2–4) found that some farmland birds responded positively to the presence of ditches managed for wildlife. All three also found that some species did not respond positively or responded negatively to management.
- A replicated, controlled and paired sites study from the UK (1) found that bunded ditches were visited by more birds than non-bunded ditches.

Background

Managing ditches to benefit wildlife can involve reduced or delayed cutting of vegetation on ditch banks and restricted fertiliser, herbicide or pesticide use on ditch banks or in fields adjoining ditches. 'Bunded' ditches are blocked to allow them to fill with water.

A replicated, controlled and paired sites study of bunded and non-bunded drainage ditches in arable and pastoral areas of Leicestershire, UK (1), found that bird visit rates were significantly higher in bunded compared to non-bunded ditches (1.0 vs. 0.5 visits/month). Sampling involved bird observations (45 minutes, 1–2/month between April 2005 and March 2007.

A replicated study in February 2008 across 97 1 km^2 plots in East Anglia, England (2), found that four farmland birds showed strong positive responses to field boundaries (hedges and ditches) managed under agri-environment schemes. Six others showed weak or negative responses. This study is discussed in detail in 'Manage hedges to benefit wildlife'.

A replicated site comparison of 2,046, 1 km squares of agricultural land across England in 2005 and 2008 (3) found that management of hedges (see 'Manage hedges to benefit wildlife') and ditches under Entry Level Stewardship did not have clear impacts on farmland bird species. Management had significant positive impacts on five species in at least region of England, but these effects were often very weak and four of the same species showed negative responses in other regions. The other five 'hedgerow' species investigated were never positively associated with boundary management. Generally, effects appeared to be more positive in the north of England.

A replicated 2010 site comparison study (4) of the same 2,046, 1 km^2 plots of lowland farmland in England as in (3) found that three years after the 2005 introduction of the Entry Level Stewardship and Countryside Stewardship Schemes, there was no consistent association between the length of ditches managed according to the agri-environment scheme on a plot and farmland bird numbers. Although there were higher numbers of linnet *Carduelis cannabina* and reed bunting *Emberiza schoeniclus* (two species known to nest in vegetation at the side of ditches) in plots with ditches managed according to the Countryside Stewardship Scheme and Entry Level Stewardship than in other plots, this difference was not observed for other species also expected to benefit from ditch management, including the yellowhammer *Emberiza citrinella* and yellow wagtail *Motacilla flava*. Between 2005 and 2008, skylark *Alauda arvensis* and grey partridge *Perdix perdix* declines were greater in plots with lengths of ditch management than other plots. For example, grey partridges showed decreases of 1.3 birds for each 0.08 km of ditch on pastoral farmland. The 2,046 1 km^2 lowland plots were surveyed in both 2005 and 2008 and classified as arable, pastoral or mixed farmland. Eighty-four percent of plots included some area managed according to the Entry Level Stewardship or Countryside Stewardship Schemes. In both survey years, two surveys were conducted along a 2 km pre-selected transect route through each 1 km^2 square.

(1) Anon (2007) Wetting up farmland for birds and other biodiversity, Defra Report BD1323.
(2) Davey, C. M., Vickery, J. A., Boatman, N. D., Chamberlain, D. E. & Siriwardena, G. M. (2010) Entry Level Stewardship may enhance bird numbers in boundary habitats. *Bird Study*, 57, 415–420.
(3) Davey, C. M., Vickery, J. A., Boatman, N. D., Chamberlain, D. E., Parry, H. R. & Siriwardena, G. M. (2010) Regional variation in the efficacy of Entry Level Stewardship in England. *Agriculture, Ecosystems & Environment*, 139, 121–128.

(4) Davey, C. M., Vickery, J. A., Boatman, N. D., Chamberlain, D. E., Parry, H. R. & Siriwardena, G. M. (2010) Assessing the impact of Entry Level Stewardship on lowland farmland birds in England. *Ibis*, 152, 459–474.

5.13. Protect in-field trees

- We found no evidence for the effects of protecting in-field trees on bird populations.

Background
Retaining in-field trees and developing agro-forestry systems has the potential to retain on-farm biodiversity. There has been considerable work on the importance of agro-forestry systems for biodiversity in general and especially birds, particularly in the tropics, where traditional farming practices often use such systems. Agroforestry is discussed in more detail in the text at the beginning of the chapter.

5.14. Plant in-field trees

- We found no evidence for the effects of planting in-field trees on bird populations.

5.15. Tree pollarding and tree surgery

- We found no evidence for the effects of tree pollarding and tree surgery on bird populations.

5.16. Plant wild bird seed or cover mixture

- All seven studies (based on five replicated experiments and a review) that investigated species richness or diversity were from the UK and found that fields or farms with wild bird cover had higher bird diversity than those without, or that more species were found in wild bird cover than in surrounding habitats (17,19,20,24–26,30).
- Thirty-two studies out of 33 from the UK and North America that examined abundance and population data (4–10,12,13,15–36,40), found that bird densities, abundances, nesting densities or use of wild bird cover was higher than in other habitats or management regimes, or that sites with wild bird cover had higher populations than those without. These studies included a systematic review (27) and seven randomised, replicated and controlled studies (10,16–18,23,24,28). Some studies found that this was the case across all species or all species studied, while others found that only a subset showed a preference. Four studies investigated other interventions at the same time. Thirteen of the 33 studies (all replicated and from Europe and the USA), found that bird populations or densities were similar on wild bird cover and other habitats, that some species were not associated with wild bird cover or that birds rarely used wild bird cover (7,10,11,13,14,17,21–23,25,35,36,40).
- Three studies from the UK and Canada (3,6,37), two replicated, found higher productivities for some or all species monitored on wild bird cover, compared to other habitats. Two replicated and controlled studies from Canada and France (3,14) found no differences in reproductive success between wild bird cover and other habitats for some or all species studied.
- Three studies from Europe and the USA investigated survival, with two finding higher survival of grey partridge *Perdix perdix* released on wild bird cover (41) or of artificial nests in some cover crops (ref no. 1 needs to go here). The third (14) found that survival of grey partridge was lower on farms with wild bird cover, possibly due to high predation.
- Five studies from the UK (8,10,16,38,39), three replicated, found that some wild bird cover crops were preferred to others. A randomised, replicated and controlled study and a review from the UK (19,28) found that the landscape surrounding wild bird cover and their configuration within it affected use by birds.

Background

The loss of food supplies, especially seeds, is thought to be a key driver of farmland bird declines. Plants that provide seed food for wild birds include maize, sunflower and cereals. They can be planted in blocks or 6 m wide strips and are left unharvested. These plants can also provide cover for nesting birds or juveniles and are sometimes called wild bird cover, 'game crops' or 'game cover crops'.

A replicated, controlled study from May–June in 1955–1958 in three treatment cover types and six natural (control) cover types in Idaho, USA (1), found that artificial nests in some cover crops were less likely to be predated than those in other crops. Over ten days, 30% of nests in cereal crops or cattail *Typha angustifolia*, bulrush *Scirpuss acutus* or *S. validus* margins were predated, compared with 40% in alfalfa *Medicago sativa* and 80% in tall weeds, willows *Salix* spp., sagebrush *Artemisia tridentate* or downy chess *Bromus tectorum*. Overall, 52% of nests were destroyed within 10 days. Grain fields provided significantly greater protection (average ten 'safe' days and only 3% nests destroyed) compared to alfalfa and irrigation ditches (average of seven and five 'safe' days) or any control cover types. A total of 529 nests, each containing four eggs were placed randomly in cover types (32–68 nests/cover type).

A study of habitat use by yellowhammers *Emberiza citrinella* on a mixed farm in Leicestershire, UK (2) found that in summer yellowhammers used both cropped and uncropped habitats including Wild Bird Cover, whereas in winter Wild Bird Cover was used more than all other habitats relative to its availability. In summer, Wild Bird Cover strips (8 m wide) were used significantly more than wheat or field boundaries (2 m wide), but less than barley. In winter, cereal-based Wild Bird Cover was used significantly more than all other habitats and kale-based Bird Cover was used significantly more than cereal and rape crops. A 15% area of the arable land was managed for gamebirds. Yellowhammer nests were observed for 1.5–2 hours when nestlings were 4–10 days old and 5–15 foraging sorties per nest were plotted during May–June 1993 and 1995. A 60 ha area of the farm was also walked seven times in November–December and February–March 1997 and habitat use was recorded.

A replicated, controlled study in May–July 1992–94 of 31 wild bird cover and 31 control prairie-parkland plots in Saskatchewan, Canada (3) found that mallard *Anas platyrhyncos* and gadwall *A. strepera* displayed higher nest survival rates in wild bird cover than in unmanaged plots (14–16% vs. 4%). There was no difference in nest survival for blue-winged teal *A. discors* and northern shoveler *A. clypeata* nests (10–15% vs. 10–14%). Nest survival rates differed significantly between years (8–26% in wild bird cover and 4–16% in control plots) and overall nesting density in wild bird cover plots was low (1.1–1.4 nests/ha). Consequently, the authors suggest that wild bird cover plots would need prohibitively large areas of establishment to be effective. The wild bird cover plots were planted on previously cultivated land with a grass-legume mix (average 37 ha); unmanaged plots were cropland (average 40.4 ha).

A 2000 literature review from the UK (4) found that the populations of grey partridge *Perdix perdix* was 600% higher on farms with conservation measures aimed at partridges in place, compared to farms without these measures. Measures included the provision of conservation headlands, planting cover crops, using set-aside and creating beetle banks.

A small study of set-aside strips over five years at Loddington, Leicestershire, UK (5), found that set-aside sown with wild bird cover was used by nesting Eurasian skylarks *Alauda arvensis* significantly more than other habitats. The majority of skylark territories found were within set-aside strips (margins or midfield) sown with wild bird cover (55–76% each year), although the habitat covered only 8–10% of the area. The habitat was also used more for foraging than all others, except linseed. Wild bird cover was sown with either cereal-based or kale-based mixtures. Skylark territories were recorded in 1995–1997 and 1999. Nests were located in 1999 and foraging trips observed for two one-and-a-half-hour periods.

A small before-and-after study from May–July in 1992–1994 in river islands in Quebec, Canada (6), found that the number of dabbling ducks *Anas* spp. nesting in the study area increased from 143 to 263 nests, following the establishment of dense nesting cover and rotational grazing (see 'Graze semi-natural habitats'). Density of nests on fields seeded with dense nesting cover in 1993 as higher than other habitats in 1994 (7 nests/ha vs. 1.1–2.8 nests/ha for other habitats). Nesting success in seeded fields was also higher (82% success for 64 nests) than in improved pastures (15% for 39 nests).

A replicated, randomised study of annual and biennial crops over three years in Norfolk, Hertfordshire and Leicestershire, UK (7) found that bird species tended to use a variety of cover crops, but whereas yellowhammers *Emberiza citrinella* used mainly cereals, greenfinches *Carduelis chloris* tended to use borage, sunflowers and mustard. Crops used by several species included kale, quinoa, fat-hen and linseed. Buckwheat was used a small amount, and apart from greenfinch, few others used sunflower or borage. Crops were sown in a randomised block design with three replicates at each of the three farms. Plots sizes were 20 or 50 m × 12 or 16 m. Numbers of birds feeding in, or flushed from each plot were recorded before 11:00 at weekly intervals from October–March 1998–2000.

A study of different set-aside crops at Allerton Research and Educational Trust Loddington farm, Leicestershire, UK (8), found that Eurasian skylark and yellowhammer used wild bird cover set-aside (kale set-aside, cereal set-aside, annual/biennial crop strips) more than expected compared to availability. Skylarks also used wild bird cover more than unmanaged set-aside, broad-leaved crops and other habitats. Yellowhammer used wild bird cover strips more than expected. Cereal set-aside wild bird cover was used significantly more than beetle banks, kale set-aside wild bird cover, unmanaged set-aside and 'other' habitats. Wild bird cover strips were used significantly more than kale set-aside, unmanaged set-aside and other habitats. Field margin and midfield set-aside strips were sown with kale-based and cereal-based mixtures for wild bird cover and 'beetle banks'. Other habitat types were: unmanaged set-aside, cereal (wheat, barley), broad-leaved crop (beans, rape) and 'other' habitats. Thirteen skylark and 15 yellowhammer nests with chicks between 3–10 days old were observed. Foraging habitat used by the adults was recorded for 90 minutes during three periods of the day.

A small replicated controlled study from May–June in 1992–98 in Leicestershire, England (9), found that the abundance of nationally declining songbirds and species of conservation concern significantly increased on a 3 km^2 site where 20 m wide mid-field and field-edge strips were planted with game cover crops (alongside several other interventions), although there was no overall difference in bird abundance, species richness or diversity between the experimental and three control sites. Numbers of nationally declining species rose by 102% (except for Eurasian skylark and yellowhammer). Nationally stable species rose (insignificantly) by 47% (eight species increased, four decreased). The other interventions employed were: 'Manage hedges to benefit wildlife', 'Create beetle banks', 'Provide supplementary food', 'Control predators' and 'Reduce pesticide or herbicide use generally'.

A replicated, randomised, controlled study over the winters of 1998–2001 in 192 sites on 161 arable farms across England (10) found that, of all the wild bird cover crops trialled, kale (*Brassica* spp.) was used by the widest range of species. Overall, all species analysed exhibited higher densities on wild bird cover crops over conventional crops except Eurasian skylarks, which preferred cereal stubbles. Although all species showed non-random and different wild bird cover crop preferences, kale was preferred by the greatest number of species. Additionally, bird abundance was significantly greater on wild bird cover crops located adjacent to hedgerows than those located midfield. Ten annual crops and four biennial crops were planted each year at each site with three replicates/crop. At 11 and 13 sites for 1999–2000 and 2000–2001 respectively strips containing the same crop were grown in pairs, one against a hedgerow and one infield, to determine location preference.

A replicated 2003 site comparison study of 88 farms in East Anglia and the West Midlands (11) found that between 1998 and 2002 there was no difference in the decrease in autumn densities of grey partridge on farms that planted wild bird cover mixtures and farms that did not. Surveys for grey partridge were made once each autumn in 1998 and 2002 on 88 farms: 38 farms that planted wild bird cover and 50 farms that did not.

A replicated, controlled study over the winters of 1997–1998, 1998–1999 and 2000/01 in approximately 15 experimental and 15 control fields on one arable, autumn-sown crop farm in County Durham, England (12) found that farmland bird abundance was significantly higher in wild bird cover crops than commercial crops (420 birds/km^2 in wild bird cover vs. 30–40/km^2 for commercial crops). Of 11 species with sufficient data for analysis, exhibited significant preference for wild bird cover crops in all species-year combinations birds. Of the wild bird cover crops, kale *Brassica napus* crops were preferred by nine species and quinoa *Chenopodium quinoa* crops by six species, although cereals and linseed were also used. The wild bird cover crops were planted in approximately 20 cm wide strips along one edge of arable wheat, barley or oil-seed rape fields. Bird counts were conducted twice monthly from October–March in 1997–1998; and three times per month from October–December as well as twice monthly from January–March in 1998–1999 and 2000–2001.

A replicated, randomised study between November 2003 and March 2004 in 205 cereal stubble fields under a range of management intensities in arable farmland in south Devon, UK (13) found no clear changes in habitat use by seed-eating birds after the establishment of wild bird cover crops on some stubble fields. The target species, cirl bunting *Emberiza cirlus*, made insignificant use of wild bird cover crops (average of two individuals/plot). Only two plots contained >5 individuals and use of the habitat dropped drastically in March, which the authors suggest makes the habitat a poor alternative to stubbles. High numbers of other seed-eating species were recorded on the wild bird cover crops, especially those containing a mixture of rape, millet, linseed, kale and quinoa (maximum seed-eating bird count = 491 vs. 191 on barley fields). Only song thrush *Turdus philomelos* abundance was significantly positively related to wild bird cover presence. However, few stubble fields contained wild bird cover crops (13 fields with 24 wild bird cover strips) and the results may have been confounded by low sample size.

A replicated, controlled, before-and-after study from 1998–2003 (three years habitat manipulation and three years monitoring) in four cereal farms (12–20 km^2) in the Beauce, Grande Beauce and Champagne Berrichonne regions, France (14) found that grey partridge populations were unaffected by cover strips. Neither breeding density nor the reproductive success of breeding pairs increased in managed compared to control areas. The survival rate was significantly lower in managed areas for all winters except for one winter in one site. Observations suggested that cover strips attracted predators, such as foxes *Vulpes vulpes* and hen harriers *Circus cyaneus*, causing the managed land to become 'ecological traps'. Cover strips (500–1,000 ha/farm) were either set-asides or, typically, a maize-sorghum mixture.

A review of experiments on the effects of agri-environment measures on livestock farms in the UK (15) found that in one experiment in south-west England (the PEBIL project, also reported in (23), birds preferred grass margins sown with plants providing seed food and cover over plots of grassland subject to various managements. The review assessed results from seven experiments (some incomplete at the time of the review) in Europe.

A replicated, randomised, controlled study over the winters of 1998–2001 in 192 plots of arable fields in lowland England (16) found that farmland birds were significantly greater in density and diversity on wild bird cover crops than on conventional crops. Although there were no significant differences between wild bird covers containing a single plant species and conventional crops, bird density was 50 times higher on 'preferred' wild bird covers. Kale *Brassica oleracae* viridus-dominated wild bird cover supported the widest range of species (especially insectivores and seed-eaters), quinoa *Chenopodium quinoa* dominated wild bird cover were mainly used by finches and tree sparrows *Passer montanus* and (unharvested) seeding cereals were mainly used by buntings. Sunflowers, phacelia and buckwheat were the least preferred wild bird cover. All bird species, besides Eurasian skylarks, corn buntings *Miliaria calandra* and rooks *Corvus frugilegus*, were significantly denser on wild bird cover. The differences between wild bird cover were more marked in late-winter as kale and quinoa retained seeds for longer periods. Within each plot, one wild bird cover and up to four conventional crops were surveyed at least once.

A replicated, randomised, controlled study from November–February in 2000–2001 and 2001–2002 in 20 arable farms in eastern Scotland (17) found that farmland bird abundance and diversity were significantly higher in fields containing wild bird cover crops (0.6–4.2 ha sampled annually) than fields with set-aside, fields with overwinter stubble or fields with conventional crops. Bird density was up to 100 times higher/ha in wild bird cover crops than on control fields. The wild bird cover crops attracted 50% more species than set-aside and stubble fields; and 91% more than the conventional fields. Of eight species with sufficient data for individual analysis, seven were consistently significantly more abundant in wild bird cover than in control crops. However, skylarks were significantly more abundant in set-aside and stubble fields. The authors point out that many of the species that favour wild bird cover crops are those currently causing concern because of their declining populations.

A replicated, randomised, controlled study from June–September in 2001–02 of 21 cereal farms in eastern Scotland (18) found that farmland birds were significantly more abundant on fields containing wild bird cover crops than on fields with conventional crops. A total of 25 species were recorded, with up to 80 times more birds seen in wild bird cover than in conventional crops. Over all month-crop combinations bird density was significantly higher on wild bird cover crops for all groups except finches in July. Bird density increased steadily over all months of the study on wild bird cover crops but remained relatively constant on conventional crops. Wild bird cover crops contained up to 90% more weed species and 280% more important bird-food weeds, than conventional crops. The wild bird cover crops were composed mainly of kale *Brassica* spp., quinoa *Chenopodium quinoa* and triticale *Triticosecale* spp. and were sown in strips (20 × 650 m). A random sample of 4.9 ha of conventional crops was made on each farm.

A review of the results of four projects conducted from 1998–2004 of wild bird cover crops planted in arable farms in England (19) found that the density and diversity of bird species increased significantly when wild bird cover crops were included in the farm. Four studies reported greater use of wild bird cover crops than of commercial crops during winter (October–March). One study reported an increase in bird abundance when wild bird cover crops were introduced into areas that previously lacked them. Kale *Brassica napus* and quinoa *Chenopodium quinoa* were used by the most species. Buckwheat was rarely used by species in any of the studies. Millet was used by more species than any other cereal. Three other studies also found that the location of wild bird covers within the whole-farm configuration had an effect on bird densities. Wild bird covers located close to hedges were favoured. Four studies found that a mixture of wild bird cover crops will produce the highest bird density and diversity.

A replicated, controlled, paired site study over winter (1997–1998) and summer (1999–2000) in arable farmlands in southern England and the Scottish lowlands (20) found that songbird density and species richness was higher in wild bird cover crops in both seasons. In total, more species were recorded in wild bird cover winter crops than control plots (26 vs. 10 species). Similarly, summer wild bird cover crops contained more species (14 vs. 10 species). Songbird abundance was significantly higher on wild bird cover winter (10–50 individuals/ha vs. 1) and summer (3 individuals/ha vs. 0.4) crops. There was significantly higher abundance of declining songbird species in the kale *Brassica oleracea* and quinoa *Chenopodium quinoa* but not cereal wild bird cover crops. Winter wild bird cover plots were sown with kale, quinoa or cereal while summer wild bird cover plots were predominantly triticale. Thirty experimental and 30 control plots were used in winter, with six experimental and six control plots in summer.

A replicated, controlled study in February–March 2002–03 on three arable farms in Mississippi, USA (21), found that densities of song sparrow *Melospiza melodia* were significantly higher in field margins seeded with Kobe lespedeza *Lespedeza striata* and partridge pea *Chamaecrista fasciculata*, compared to control field margins, when fields bordered blocks (>30 m) of herbaceous vegetation (31 birds/ha vs. 8 birds/ha) or strips (<30 m) of woodland (38 birds/ha vs. 10 birds/ha), but not when fields bordered herbaceous strips (96 birds/ha vs. 70 birds/ha) or blocks of woodland (25 birds/ha vs. 28 birds/ha). Savanna sparrows

Passerculus sandwichensis did not show any such variation, whilst other sparrow species (notably swamp sparrow *M. georgiana*) were significantly higher in uncultivated margins adjacent to herbaceous blocks (78 birds/ha vs. 19 birds/ha), herbaceous strips (139 birds/ha vs. 30 birds/ha) and wooded blocks (51 birds/ha vs. 12.6 birds/ha). Borders were established in 2000 and were seeded in 2000 and early 2001.

A replicated study in 1999 and 2003 on 256 arable and pastoral fields across 84 farms in East Anglia and the West Midlands, England (22), found that only two of twelve farmland bird species analysed were positively associated with the provision of wildlife seed mixtures, overwinter stubble (see 'Leave overwinter stubbles') or set-aside (see 'Provide or retain set-aside areas in farmland'). These were Eurasian skylarks (a field-nesting species) and Eurasian linnets *Carduelis cannabina* (a boundary-nesting species). The study did not distinguish between set-aside, wildlife seed mixtures or overwinter stubble, classing all as interventions to provide seeds for farmland birds.

A randomised, replicated, controlled trial on four farms in south-west England in 2003–2006 (23) found that 12, 50 × 10 m plots of permanent pasture sown with a wild bird seed attracted more foraging songbirds (dunnock *Prunella modularis*, winter wren *Troglodytes troglodytes*, European robin *Erithacus rubecula*, seed-eating finches and buntings) than 12 control plots managed as silage (cut twice in May and July, and grazed in autumn/winter). Dunnocks, but not chaffinches *Fringilla coelebs* or blackbirds *Turdus merula*, nested in hedgerows next to the sown plots more than expected, with 2.5 nests/km, compared to less than 0.5 nests/km in hedges next to experimental grass plots. Experimental plots were sown with a mix of crops including linseed and legumes. There were twelve replicates of each management type, monitored over the four years (2003–2006).

A randomised, replicated, controlled trial on four farms in south-west England (24) (same study as Defra 2007) found that 50 × 10 m plots of permanent pasture sown with a mix of crops including linseed and legumes attracted more birds, and more bird species than control treatments, in both summer and winter. Plots were established in 2002, re-sown in new plots each year and monitored annually from 2003 to 2006. Legumes sown included white clover, red clover, common vetch and bird's-foot trefoil. There were twelve replicates of each treatment.

A replicated trial on four farms in England (25) found that the numbers of birds and bird species were higher in sown wild bird mix than crops in December and January (around 100 birds of over three species per count on average in the wild bird mix, compared to fewer than 10 birds or <1 species in the crop), but not in February and March. Eurasian linnet *Carduelis cannabina* (at three sites) and reed bunting *Emberiza schoeniclus* (at one site) were the most abundant bird species recorded in the wild bird mix. A seed mix containing white millet *Echinochloa esculenta*, linseed *Linum usitatissimum*, radish *Raphanus sativus* and quinoa *Chenopodium quinoa* was sown in a 150 × 30 m patch in the centre of an arable field (winter wheat) on each of four farms in Cambridgeshire, Bedfordshire, Oxfordshire and Buckinghamshire, in April 2004 and 2005. Birds were counted once a month between December 2004 and March 2005.

A replicated controlled trial on one farm in Warwickshire, UK, in 2005–2006 (26) found that field corners or margins sown with a wild bird seed mix had more birds and bird species in winter than all other treatments. Fifty-five birds/plot from four species on average were recorded on the wild bird seed plots, compared to 0.1–1 bird/plot, or 0.1–0.7 species on average on control crop plots, plots sown with wildflower seed mix or left to naturally regenerate. The wild bird seed mix (five species) was sown in April 2006 and fertilised in late May 2006. The crop, oats, was sown in October 2005. Each treatment was tested in one section of margin and one corner in each of four fields. Farmland birds were counted on each plot on seven counts between December 2006 and March 2007.

A 2007 systematic review identified five papers investigating the effect of winter bird cover on farmland bird densities in the UK (27). There were significantly higher densities of farmland birds in winter on fields with winter bird cover than on adjacent conventionally managed fields. The meta-analysis included experiments conducted between 1998 and 2001 from two controlled trials and one randomised control trial.

A replicated, randomised, controlled study in September, November, December and February in 2004–2005 in seven grassland farms (87–96% grass) in western Scotland (28) found that songbirds responded significantly more positively to wild bird cover crops in grassland compared to arable regions. Average songbird densities were two orders of magnitude greater in wild bird cover crops than conventional crops (average 51 birds/ha vs. 0.2). The average density of songbirds in wild bird cover in the grassland region was more than double that in wild bird cover in the arable region at the same time of year (average 61.3 and 29.0 birds/ha respectively). Average densities in grassland conventional crops were just 14% of that in the arable region. On each site, an average of 1.2 ha of wild bird cover and 10.3 ha of conventional crops was randomly sampled. Arable farm data from a previous study was used for comparison.

A replicated experiment in north-east Scotland over three winters (2002–2005) (29), found that unharvested seed-bearing crops were most frequently selected by birds (28% of all birds despite these patches occupying less than 5% of the area surveyed). For nine species, seed-bearing crops were used more than expected (based on available crop area) in at least one winter. Outside agri-environment schemes (the Rural Stewardship Scheme and Farmland Bird Lifeline), cereal stubble was the most selected habitat. In total, 53 lowland farms (23 in Rural Stewardship Scheme, 14 in Farmland Bird Lifeline and 16 not in a scheme were assessed. Over 36,000 birds of 10 species were recorded.

The second monitoring year of the same study as (26) in 2005–2007 (30) found that wild bird cover plots had more birds of more species in winter (86 birds/plot, of six species on average) than control cereal plots, plots sown with wildflower seed mix or left to naturally regenerate (2 birds/plot or less, and 0.4–1.6 species/plot on average). Farmland birds were counted on each plot on four counts between December 2007 and March 2008. The crop control in year two was winter wheat.

A 2009 literature review of agri-environment schemes in England (31) found that high densities of seed-eating songbirds and Eurasian skylarks were found on land planted with wild bird seed or cover mix and on stubble fields (see 'Leave overwinter stubbles'). A survey in 2007–2008 found that densities of seed-eating songbirds were highest on wild bird seed or cover mix, compared to other agri-environment schemes options. This review also examines several other interventions, discussed in the relevant sections.

A 2009 literature review of European farmland conservation practices (32) found that margins sown with wild bird cover crops such as quinoa *Chenopodium quinoa* and kale provided more food for seed-eating birds in late winter than other field margin types and supported large numbers of some songbird species.

A controlled study in 2002–2009 on mixed farmland in Hertfordshire, England (33), found that the estimated population density of grey partridges was significantly higher on land sown with wild bird cover than on conventional arable crops. This study also examined the densities found on land under various agri-environment schemes and set-aside (which were higher than those on wild bird cover, see 'Pay farmers to cover the costs of conservation measures' and 'Provide or retain set-aside') and the impact of predator control and supplementary food provision (see see 'Provide supplementary food to increase adult survival' and 'Control predators not on islands').

A follow-up review of experiments on the effects of agri-environment measures on livestock farms in the UK (34), found that in one experiment in south-west England (the PEBIL project, also reported (23), small insect-eating birds preferred grassland margins sown with plants providing seed food over plots of grassland subject to various managements, despite there being no difference in insect numbers between the two sets of treatments. The preference for wild bird cover was attributed to easier accessibility (less dense ground cover). The review assessed results from four experimental projects (one incomplete at the time of the review) in the UK.

A replicated 2010 site comparison study of 2,046 1 km² plots of lowland farmland in England (35) found that three years after the 2005 introduction of the two agri-environment schemes, Countryside Stewardship Scheme and Environmental Stewardship, there was no consistent association between the provision of wild bird cover and farmland bird numbers. European greenfinch, stock dove *Columba oenas*, starling *Sturnus vulgaris* and woodpigeon *Columba palumbus* showed more positive population change (population increases or smaller decreases relative to other plots) in the 9 km^2 and 25 km^2 areas immediately surrounding plots planted with wild bird cover mix than in the area surrounding plots not planted with wildlife seed mixture. Although Eurasian linnet and rook also showed positive associations with wild bird cover mix at the 25 km^2 scale, plots with wild bird cover were associated with a greater decline in grey partridge populations at both scales between 2005 and 2008. The 2,046 1 km^2 lowland plots were surveyed in both 2005 and 2008 and classified as arable, pastoral or mixed farmland. Eighty-four percent of plots included some area managed according to the Entry Level Stewardship or CSS. In both survey years, two surveys were conducted along a 2 km pre-selected transect route through each 1 km^2 square.

A replicated site comparison of 2,046, 1 km squares of agricultural land across England in 2005 and 2008 (36) found that four of eight regions of England had at least two farmland birds that showed positive responses to wild bird cover and overwinter stubble fields (see 'Leave overwinter stubbles'). Across all 15 species thought to benefit from these interventions, only one region (the North West) showed significantly more positive responses than would be expected by chance. Some species responded positively in some regions and negatively in others.

A replicated site comparison study on 1,031 agricultural sites across England in 2004–2008 (37) found that the proportion of young grey partridges in the population was higher in 2007 and 2008 on sites with higher proportions of wild bird cover. Brood sizes were also related to wild bird cover in 2008 only. Overwinter survival was positively related to wild bird cover in 2004–2005 but negatively in 2007–2008. There were no relationships between wild bird cover and year-on-year density trends. This study describes the effects of several other interventions, discussed in the relevant sections.

A replicated 2010 site comparison study of 52 fields in East Anglia and the West Midlands, UK (38), found no difference between the number of seed-eating birds in fields managed under the Higher Level Strata of the Environmental Stewardship scheme (i.e. on fields planted with Enhanced Wild Bird Seed Mix) than in fields managed under the Entry Level Strata of the Environmental Stewardship scheme (i.e. fields planted with wild bird cover mix). In East Anglia, but not the West Midlands, there were significantly more seed-eating birds on fields planted with wild bird cover under the Environmental Stewardship scheme (59 birds/ha) than non-Environmental Stewardship fields planted with a game cover (2 birds/ha). Seed-eating birds were surveyed on two visits to each site between 1 November 2007 and 29 February 2008.

A replicated site comparison study on farms in two English regions (39) found that more seed-eating farmland songbirds (including tree sparrow and corn bunting) were found on Higher Level Stewardship wild bird seed mix sites than on non-stewardship game cover crops in East Anglia (6–11 birds/ha on wild bird seed mix, compared to <0.5 birds/ha on game cover), but not in the West Midlands (2–4 birds/ha on both types). The survey was carried out in winter 2007–2008 on 27 farms with Higher Level Stewardship, 13 farms with Entry Level Stewardship and 14 with no environmental stewardship, in East Anglia or the West Midlands.

A replicated study from April–July in 2006 on four livestock farms (3 replicates/farm) in south-west England (40) found that dunnock *Prunella modularis*, but not Eurasian blackbird *Turdus merula* or chaffinch, nested at higher densities in hedges alongside field margins sown with wild bird seed crops, or barley undersown with grass and clover, compared to those next to grassy field edges under various management options (dunnocks: approximately 2.5 nests/km for seed crops vs. 0.3/km for grass margins; blackbirds: 1.0 vs. 1.3; chaffinch: 1.5 vs. 1.4). Margins were 10 m wide, 50 m long and located adjacent to existing hedgerows. Seed crop margins were sown with barley (undersown with grass/legumes) or a kale/quinoa mix. There were 12 replicates of each treatment. This study reports on results from the same experiment as (23).

A replicated study on four farms in Gloucestershire and Oxfordshire, England, in 2007 (41) found that grey partridge released in coveys in the autumn used cover crops more frequently than birds released in pairs in the spring. This study is discussed in 'Captive breeding, rearing and releases (*ex situ* conservation)'.

(1) Jones, R. E. & Hungerford, K. E. (1972) Evaluation of nesting cover as protection from magpie predation. *The Journal of Wildlife Management*, 36, 727–732.
(2) Stoate, C. & Szczur, J. (1997) Seasonal changes in habitat use by yellowhammers (*Emberiza citrinella*). 1167–1172 *1997 Brighton Crop Protection Conference – Weeds, Conference Proceedings Vols 1–3* British Crop Protection Council, Farnham.
(3) McKinnon, D. T. & Duncan, D. C. (1999) Effectiveness of dense nesting cover for increasing duck production in Saskatchewan. *The Journal of Wildlife Management*, 63, 382–389.
(4) Aebischer, N. J., Green, R. E. & Evans, A. D. (2000) From science to recovery: four case studies of how research has been translated into conservation action in the UK. 43–54 in: N. J. Aebischer, A. D. Evans, P. V. Grice, J. A. Vickery (eds) *Ecology and Conservation of Lowland Farmland Birds* British Ornithologists Union, Tring.
(5) Boatman, N. D. & Bence, S. L. (2000) Management of set-aside to enhance biodiversity: the wild bird cover option. *Aspects of Applied Biology*, 73–78.
(6) Lapointe, S., Giroux, J. F., Belanger, L. & Filion, B. (2000) Benefits of rotational grazing and dense nesting cover for island-nesting waterfowl in southern Quebec. *Agriculture, Ecosystems & Environment*, 78, 261–272.
(7) Boatman, N. D. & Stoate, C. (2002) Growing crops to provide food for seed-eating birds in winter. *Aspects of Applied Biology*, 67, 229–235.
(8) Murray, K. A., Wilcox, A. & Stoate, C. (2002) A simultaneous assessment of farmland habitat use by breeding skylarks and yellowhammers. *Aspects of Applied Biology*, 67, 121–127.
(9) Stoate, C. (2002) Multifunctional use of a natural resource on farmland: wild pheasant (*Phasianus colchicus*) management and the conservation of farmland passerines. *Biodiversity and Conservation*, 11, 561–573.
(10) Boatman, N. D., Stoate, C., Henderson, I. G., Vickery, J. A., Thompson, P. G. & Bence, S. L. (2003) *Designing crop/plant mixtures to provide food for seed-eating farmland birds in winter*. British Trust for Ornithology.
(11) Browne, S. & Aebischer, N. (2003) *Arable stewardship: impact of the pilot scheme on grey partridge and brown hare after five years*. DEFRA Final Report RMP1870vs3. Department for Environment, Food and Rural Affairs, London, UK.
(12) Stoate, C., Szczur, J. & Aebische, N. (2003) Winter use of wild bird cover crops by passerines on farmland in northeast England: Declining farmland species were more abundant in those crops which can be matched to the birds' requirements. *Bird Study*, 50, 15–21.
(13) Defra (2004) *Comparative quality of winter food sources for cirl bunting delivered through countryside stewardship special project and CS arable options*. RSPB/Defra Report BD1626.
(14) Bro, E., Mayot, P., Corda, E. & Reitz, F. (2004) Impact of habitat management on grey partridge populations: assessing wildlife cover using a multisite BACI experiment. *Journal of Applied Ecology*, 41, 846–857.
(15) Buckingham, D. L., Atkinson, P. W. & Rook, A. J. (2004) Testing solutions in grass-dominated landscapes: a review of current research. *Ibis*, 146, 163–170.
(16) Henderson, I. G., Vickery, J. A. & Carter, N. (2004) The use of winter bird crops by farmland birds in lowland England. *Biological Conservation*, 118, 21–32.
(17) Parish, D. M. B. & Sotherton, N. W. (2004) Game crops and threatened farmland songbirds in Scotland: a step towards halting population declines? *Bird Study*, 51, 107.
(18) Parish, D. M. B. & Sotherton, N. W. (2004) Game crops as summer habitat for farmland songbirds in Scotland. *Agriculture, Ecosystems & Environment*, 104, 429–438.
(19) Stoate, C., Henderson, I. G. & Parish, D. M. B. (2004) Development of an agri-environment scheme option: seed-bearing crops for farmland birds. *Ibis*, 146, 203–209.
(20) Sage, R. B., Parish, D. M. B., Woodburn, M. I. A. & Thompson, P. G. L. (2005) Songbirds using crops planted on farmland as cover for game birds. *European Journal of Wildlife Research*, 51, 248–253.
(21) Smith, M. D., Barbour, P. J., Burger Jr, L. W. & Dinsmore, S. J. (2005) Density and diversity of overwintering birds in managed field borders in Mississippi. *The Wilson Bulletin*, 117, 258–269.
(22) Stevens, D. K. & Bradbury, R. B. (2006) Effects of the Arable Stewardship Pilot Scheme on breeding birds at field and farm-scales. *Agriculture, Ecosystems & Environment*, 112, 283–290.
(23) Defra (2007) *Potential for enhancing biodiversity on intensive livestock farms* (PEBIL). Defra Report BD1444.

(24) Pilgrim, E. S., Potts, S. G., Vickery, J., Parkinson, A. E., Woodcock, B. A., Holt, C., Gundrey, A. L., Ramsay, A. J., Fuller, R. & Tallowin, J. R. B. (2007) Enhancing wildlife in the margins of intensively managed grass fields. 293–296 in: J. J. Hopkins, A. J. Duncan, D. I. McCracken, S. Peel, J. R. B. Tallowin (eds) *High Value Grassland: Providing Biodiversity, a Clean Environment and Premium Products. British Grassland Society Occasional Symposium No.38* British Grassland Society (BGS), Reading.
(25) Pywell, R. F., Meek, W., Carvell, C. & Hulmes, L. (2007) The SAFFIE project: enhancing the value of arable field margins for pollinating insects. *Aspects of Applied Biology*, 81, 239–245.
(26) Pywell, R. & Noweakowski, M. (2007) Farming for Wildlife Project: Annual Report 2006/7.
(27) Roberts, P. D. & Pullin, A. S. (2007) *The effectiveness of land-based schemes (including agri-environment) at conserving farmland bird densities within the UK*. Systematic Review No. 11. Collaboration for Environmental Evidence/Centre for Evidence-Based Conservation, Birmingham, UK.
(28) Parish, D. M. B. & Sotherton, N. W. (2008) Landscape-dependent use of a seed-rich habitat by farmland passerines: relative importance of game cover crops in a grassland versus an arable region of Scotland. *Bird Study*, 55, 118.
(29) Perkins, A. J., Maggs, H. E. & Wilson, J. D. (2008) Winter bird use of seed-rich habitats in agri-environment schemes. *Agriculture, Ecosystems & Environment*, 126, 189–194.
(30) Pywell, R. & Noweakowski, M. (2008) Farming for Wildlife Project: Annual Report 2007/8. NERC/Centre for Ecology and Hydrology, 19pp. (CEH Project Number: C03242).
(31) Natural England (2009) *Agri-environment schemes in England 2009. A review of results and effectiveness.* Natural England report.
(32) Vickery, J. A., Feber, R. E. & Fuller, R. J. (2009) Arable field margins managed for biodiversity conservation: a review of food resource provision for farmland birds. *Agriculture, Ecosystems & Environment*, 133, 1–13.
(33) Aebischer, N. J. & Ewald, J. A. (2010) Grey partridge *Perdix perdix* in the UK: recovery status, set-aside and shooting. *Ibis*, 152, 530–542.
(34) Buckingham, D. L., Atkinson, P. W., Peel, S. & Peach, W. (2010) New conservation measures for birds on grasslands and livestock farms. *BOU Proceedings – Lowland Farmland Birds III: delivering solutions in an uncertain world*. British Ornithologists Union.
(35) Davey, C. M., Vickery, J. A., Boatman, N. D., Chamberlain, D. E., Parry, H. R. & Siriwardena, G. M. (2010) Assessing the impact of Entry Level Stewardship on lowland farmland birds in England. *Ibis*, 152, 459–474.
(36) Davey, C. M., Vickery, J. A., Boatman, N. D., Chamberlain, D. E., Parry, H. R. & Siriwardena, G. M. (2010) Regional variation in the efficacy of Entry Level Stewardship in England. *Agriculture, Ecosystems & Environment*, 139, 121–128.
(37) Ewald, J. A., Aebischer, N. J., Richardson, S. M., Grice, P. V. & Cooke, A. I. (2010) The effect of agri-environment schemes on grey partridges at the farm level in England. *Agriculture, Ecosystems & Environment*, 138, 55–63.
(38) Field, R. H., Morris, A. J., Grice, P. V. & Cooke, A. (2010) The provision of winter bird food by the English Environmental Stewardship scheme. *Ibis*, 153, 14–26.
(39) Field, R. H., Morris, A. J., Grice, P. V. & Cooke, A. I. (2010) Evaluating the English Higher Level Stewardship scheme for farmland birds. *Aspects of Applied Biology*, 100, 59–68.
(40) Holt, C. A., Atkinson, P. W., Vickery, J. A. & Fuller, R. J. (2010) Do field margin characteristics influence songbird nest-site selection in adjacent hedgerows? *Bird Study*, 57, 392.
(41) Rantanen, E. M., Buner, F., Riordan, P., Sotherton, N. & Macdonald, D. W. (2010) Habitat preferences and survival in wildlife reintroductions: an ecological trap in reintroduced grey partridges. *Journal of Applied Ecology*, 47, 1357–1364.

5.17. Plant nectar flower mixture/wildflower strips

- Two replicated and controlled studies from the UK (one randomised) and a European review (1,3,7) out of seven studies captured found that more birds used nectar/wildflower strips than crops or land under other management. Two studies of a replicated and controlled experiment in the UK (4,5) found that no more birds used nectar/wildflower strips in winter than used land under other management.

- A replicated, controlled study from Switzerland (6) found that Eurasian skylarks *Alauda arvensis* were more likely to nest in patches of fields sown with annual weeds than in crops, and were less likely to abandon nests in these patches.

- A randomised, replicated and controlled study from the UK (2) found that field margin management affected their use by birds more than the seed mix used on them.

Background

Flowering plants are sown in strips or blocks, for bees and other flower-visiting insects. Nectar flower mixture can include agricultural varieties of flowering plants such as clovers. Increased numbers of insects may then provide food for more birds.

A replicated, controlled study in summer and autumn of 1995 and 1996 on 15 sown set-aside strips on a farm in Cambridgeshire, UK (1), found that more bird individuals (average 20% of the total) and species (average 56%) used the strips than the adjacent crop area (average 7% of individuals and 33% of species) in both years. However, the highest proportions of both individuals and species were recorded in the field boundaries (average 68% individuals and 80% of species). This study is discussed in detail in 'Plant grass buffer strips/margins around arable or pasture fields'.

A randomised, replicated, controlled trial of sown grassy field margins from 2002 to 2006 in eastern England (2) found that the management of margins affected bird use more than the seed mix used. The number of birds using the margins in summer increased by 29% between 2003 and 2006. Bird densities were higher on disturbed and graminicide-treated plots than on cut plots (no actual bird densities given, only model results). Bird densities were linked to densities of diurnal ground beetles (Carabidae), especially in disturbed and graminicide-treated plots. In winter, there were twice as many birds on cut margins as uncut margins, and twice as many birds in the second year than the first. Field margin plots (6 × 30 m) were established using one of three seed mixes: 1) Countryside Stewardship mix, 2) tussock grass mix and 3) a mixture of grasses and forbs designed for pollinating insects. The margins were managed in spring from 2003 to 2005 with one of three treatments: 1) cut to 15 cm, 2) soil disturbed by scarification until 60% of the area was bare ground, 3) treated with graminicide at half the recommended rate. There were five replicates of each treatment combination, at two farms – one in Boxworth, Cambridgeshire, England, and one in High Mowthorpe, Yorkshire, England. Birds were surveyed five to eight times between April and July from 2002 to 2006. In winters of 2004–2005 and 2005–2006, birds were also surveyed on 6 m margins on 10 farms in eastern England with two seed mixes (tussocky grass and fine grass). Margins were either cut in autumn or uncut. There were four replicates of each treatment combination per farm.

A randomised, replicated, controlled trial on four farms in south-west England (3) found that 50 × 10 m plots of permanent pasture sown with a grass and legume seed mix attracted more birds, and more bird species than control treatments in both summer and winter. Plots were established in 2002, re-sown in new plots each year and monitored annually from 2003 to 2006. Legumes sown included white clover *Trifolium repens*, red clover *T. pratense*, common vetch *Vicia sativa* and bird's-foot trefoil *Lotus corniculatus*. There were twelve replicates of each management type.

A replicated controlled trial on one farm in Warwickshire, UK in 2005–2006 (4) found that field corners or margins sown with a wildflower mix did not have more birds in winter (species or individuals) than control crop plots. Average counts were close to zero birds/plot for both. The wildflower mix (25 broadleaved non-grass species, making up 10% by weight, with 90% grass from four species) was sown in August 2005 and treated with graminicide in November 2005. Plots were cut three times in 2006, and cuttings removed. The crop, oats, was sown in October 2005. Each treatment was tested in one section of margin and one corner in each of four fields. Farmland birds were counted on each plot on seven counts between December 2006 and March 2007.

The second monitoring year of the same study as (4), from 2005–2007 (5) found that wildflower plots did not have more birds in winter than control cereal plots. There were two birds/plot or fewer, and 0.4–1.6 bird species/plot on average on all treatments except those sown with wild bird seed mix. Farmland birds were counted on each plot on four counts between December 2007 and March 2008. The crop control in year two was winter wheat.

A replicated, controlled study from March–July in 2006 in winter wheat fields in mixed farming lands near Berne, Switzerland (6), found that Eurasian skylarks *Alauda arvensis* with territories that included undrilled patches were significantly less likely to abandon their territory than birds without patches, and more likely to use the undrilled patches as nesting and foraging sites than expected by chance. The strips were sown with six annual weed species but otherwise resembled skylark plots and this study is discussed in detail in 'Create skylark plots'.

A 2009 literature review of European farmland conservation practices (7) found that the availability of bird food-species was higher in nectar-rich field margins than in crops, and several species used margins planted with wildflower mixes more than grass-only strips (see 'Plant grass buffer strips/margins around arable or pasture fields'). This study discusses several other field-margin agri-environment options, which are described in the relevant sections.

(1) Clarke, J. H., Jones, N. E., Hill, D. A. & Tucker, G. M. (1997) The management of set-aside within a farm and its impact on birds. *Proceeding of the 1997 Brighton Crop Protection Conference*, 1–3, 1179–1184.

(2) Henderson, I. G., Morris, A. J., Westbury, D. B., Woodcock, B. A., Potts, S. G., Ramsay, A. & Coombes, R. (2007) Effects of field margin management on bird distributions around cereal fields. *Aspects of Applied Biology*, 81, 53–60.

(3) Pilgrim, E. S., Potts, S. G., Vickery, J., Parkinson, A. E., Woodcock, B. A., Holt, C., Gundrey, A. L., Ramsay, A. J., Fuller, R. & Tallowin, J. R. B. (2007) Enhancing wildlife in the margins of intensively managed grass fields. 293–296 in: J. J. Hopkins, A. J. Duncan, D. I. McCracken, S. Peel, J. R. B. Tallowin (eds) *High Value Grassland: Providing Biodiversity, a Clean Environment and Premium Products. British Grassland Society Occasional Symposium No.38* British Grassland Society (BGS), Reading.

(4) Pywell, R. & Noweakowski, M. (2007) Farming for Wildlife Project: Annual Report 2006/7. Centre for Ecology and Hydrology.

(5) Pywell, R. & Noweakowski, M. (2008) Farming for Wildlife Project: Annual Report 2007/8. NERC/ Centre for Ecology and Hydrology, 19pp. (CEH Project Number: C03242).

(6) Fischer, J., Jenny, M. & Jenni, L. (2009) Suitability of patches and in-field strips for skylarks *Alauda arvensis* in a small-parcelled mixed farming area. *Bird Study*, 56, 34–42.

(7) Vickery, J. A., Feber, R. E. & Fuller, R. J. (2009) Arable field margins managed for biodiversity conservation: a review of food resource provision for farmland birds. *Agriculture, Ecosystems & Environment*, 133, 1–13.

5.18. Create uncultivated margins around intensive arable or pasture fields

- A replicated, controlled study from the USA (1) found that three sparrow species found on uncultivated margins were not found on mown field edges. A replicated study from Canada (2) found fewer species in uncultivated margins than in hedges or in trees planted as windbreaks.
- Three replicated studies from the USA and UK (1,3,7), one controlled, found that some birds were associated with uncultivated margins, or that birds were more abundant on margins than on other habitats. One study found that these effects were very weak. Four replicated studies (two of the same experiment) from the UK (3–5,7), two controlled, found that uncultivated margins contained similar numbers of birds in winter, or that several species studied did not show associations with margins.
- A replicated, controlled study from the UK (6) found that yellowhammers *Emberiza citrinella* used uncultivated margins more than crops in early summer, but use fell in uncut margins in late summer. Cut margins however, were used more than other habitat types late in summer.
- A replicated study from the UK (8) found high rates of survival for grey partridge *Perdix perdix* released in margins.

Background

This intervention allows vegetation in field margins to regenerate naturally, without planting, although it can involve subsequent mowing. The margins are not fertilised and only spot-treated with herbicides if necessary.

A replicated, controlled study in February 1997 and 1998 on eight arable farms in North Carolina, USA (1), found that sparrow species were significantly more abundant on farms with uncultivated field margins (set up in 1996) than on those with mown field edges (34–36 sparrows/ha for uncultivated margins vs. 6–21 sparrows/ha for mowed edges). In addition, uncultivated field margins contained three species (white-throated sparrow *Zonotrichia albicollis*, field sparrow *Spizella pusilla* and chipping sparrow *S. passerina*) not found in mowed edges; all four species found in mowed edges (savanna sparrow *Passerculus sandwichensis*, song sparrow *Melospiza melodia*, swamp sparrow *M. georgiana* and dark-eyed junco *Junco hyemalis*) were also found in uncultivated field margins. In total, 93% of birds detected in field edges were sparrows.

A replicated study in southern Quebec, Canada, in July 1995 (2), found that herbaceous borders around arable fields held significantly fewer individuals and species than either hedges or trees planted as windbreaks (19 species found in 17 herbaceous borders, at 19 birds/ha vs. 25 species at 51 birds/ha for 17 windbreaks and 39 species at 57 birds/ha for 27 hedges). Differences were significant, even when adjusting for different sample sizes.

A replicated study in 1999 and 2003 on 256 arable and pastoral fields across 84 farms in East Anglia and the West Midlands, England (3), found that a combination of creating planted (see 'Plant grass buffer strips/margins around arable or pasture fields') and uncultivated margins around fields was strongly positively associated with four out of twelve farmland bird species analysed. These were skylark *Alauda arvensis* (a field-nesting species) and chaffinch *Fringilla coelebs*, whitethroat *Sylvia communis* and yellowhammers *Emberiza citrinella* (all boundary-nesting species). The study did not distinguish between uncultivated and planted margins. This study describes several other interventions, discussed in the relevant sections.

A replicated controlled trial on one farm in Warwickshire, UK, in 2005–2006 (4) found that field corners or margins left to naturally regenerate for one year did not have more birds in winter (species or individuals) than control crop plots. Average counts were one bird/plot or fewer for both treatments. The plots were left as unmanaged wheat stubble for all of 2006. The crop, oats, was sown in October 2005. Each treatment was tested in one section of margin and one corner in each of four fields. Farmland birds were counted on each plot on seven counts between December 2006 and March 2007.

The second monitoring year of the same study as (4), from 2005–2007 (5) found that naturally regenerated plots did not have more birds in winter than control cereal plots. There were two birds/plot or fewer, and 0.4–1.6 bird species/plot on average on all treatments except wild bird seed mix (see 'Plant wild bird seed or cover mixture'). Farmland birds were counted on each plot on four counts between December 2007 and March 2008. The crop control in the second year was winter wheat.

A replicated, controlled study in May–August 2005–6 on five farms in Aberdeenshire, Scotland (6), found that a larger proportion of early-summer yellowhammer *Emberiza citrinella* foraging flights were in field margins (32% of 233 flights from ten nests), compared to cereal crops (8%). However, in late summer, cereal fields were used more (up to 56% of 506 flights) and field margins less (down to 15%). In 2006, sections of margins around some nests were cut down to the soil. These patches comprised 2.3–2.4% of margin area, and were used for 2.9% of 172 foraging flights in early summer and 34% of 77 foraging flights in late summer. The authors suggest that yellowhammers used cut patches disproportionately as the uncut sections grew taller and so reduced the access to invertebrates.

A large 2010 site comparison study of 2,046 1 km^2 plots of lowland farmland in England (7) found that three years after the 2005 introduction of the Countryside Stewardship Scheme and Entry Level Stewardship schemes, there was no consistent association between the provision of uncultivated field margins on arable or pastoral farmland and farmland bird numbers. Although plots with field margins did see more positive population changes (increases or smaller decreases relative to other plots) of rook *Corvus frugilegus*, starling *Sturnus vulgaris* and woodpigeon *Columba palumbus*, the effect was small, with starlings, for example, showing increases of only 0.0002 individuals for every 0.001 km^2 of margin in mixed farmland plots. Other species expected to benefit from margin provision including corn bunting *Emberiza calandra*, grey partridge *Perdix perdix*, kestrel *Falco tinnunculus*, jackdaw *Corvus monedula*, reed bunting *Emberiza schoeniclus* and common whitethroat *Sylvia communis* all showed no effect of margin management. Yellowhammer *Emberiza citrinella*, also expected to benefit from margin creation, showed a positive association in mixed landscapes and a negative association on grassland plots. The 2,046 1 km^2 lowland plots were surveyed in both 2005 and 2008 and classified as arable, pastoral or mixed farmland. Eighty-four percent of plots included some area managed according to the Entry Level Stewardship or the Countryside Stewardship Scheme. In both survey years, two surveys were conducted along a 2 km pre-selected transect route through each 1 km^2 plot.

A replicated study on four farms in Gloucestershire and Oxfordshire, England, in 2007 (8) found that grey partridge *Perdix perdix* released in pairs in the spring used field margins more frequently than birds released as family groups in the autumn. This study is discussed in detail in 'Captive breeding, rearing and releases (*ex situ* conservation)'.

(1) Marcus, J. F., Palmer, W. E. & Bromley, P. T. (2000) The effects of farm field borders on overwintering sparrow densities. *The Wilson Bulletin*, 112, 517–523.
(2) Jobin, B., Choiniere, L. & Belanger, L. (2001) Bird use of three types of field margins in relation to intensive agriculture in Quebec, Canada. *Agriculture, Ecosystems & Environment*, 84, 131–143.
(3) Stevens, D. K. & Bradbury, R. B. (2006) Effects of the Arable Stewardship Pilot Scheme on breeding birds at field and farm-scales. *Agriculture, Ecosystems & Environment*, 112, 283–290.
(4) Pywell, R. & Noweakowski, M. (2007) Farming for Wildlife Project: Annual Report 2006/7.
(5) Pywell, R. & Noweakowski, M. (2008) Farming for Wildlife Project: Annual Report 2007/8.
(6) Douglas, D. J. T., Vickery, J. A. & Benton, T. G. (2009) Improving the value of field margins as foraging habitat for farmland birds. *Journal of Applied Ecology*, 46, 353–362.
(7) Davey, C. M., Vickery, J. A., Boatman, N. D., Chamberlain, D. E., Parry, H. R. & Siriwardena, G. M. (2010) Assessing the impact of Entry Level Stewardship on lowland farmland birds in England. *Ibis*, 152, 459–474.
(8) Rantanen, E. M., Buner, F., Riordan, P., Sotherton, N. & Macdonald, D. W. (2010) Habitat preferences and survival in wildlife reintroductions: an ecological trap in reintroduced grey partridges. *Journal of Applied Ecology*, 47, 1357–1364.

5.19. Plant grass buffer strips/margins around arable or pasture fields

- One replicated controlled study from the USA (11) found that there were more species in fields bordered by margins than unbordered fields. Two replicated studies from the UK (6,7), one with paired sites, found no effect of field margins on species richness. A replicated, controlled study from the UK (1) found that more birds and more species used sown strips in fields than the fields themselves, but even more used field margins.
- Nine studies from the UK and USA, seven replicated, two controlled, found more positive population trends, higher populations or strong habitat associations for some or all species for sites with grass margins to fields (2–4,8,10,11,13–15). One study investigated multiple interventions. Three replicated studies from the UK (6,8,13) found that grass field margins did not have a positive effect on populations of some or all bird species investigated.
- Both studies that examined habitat use (one replicated, both from the UK) found that species used margins more than other habitats (5,12). A randomised, replicated and controlled study from the UK (9) found that birds used cut margins more than uncut margins during winter but less than other management regimes during summer. The authors argue that management type is more important than the seed mix used to sow the margins.
- A replicated study from the UK (14) found that grey partridge *Perdix perdix* had smaller broods in grass margins than other habitat types.

Background

This intervention involves planting field margins with a grass-rich seed mixture. It includes 'floristically-enhanced' grass margins available under the English Higher Level Stewardship scheme. The margins are not fertilised and only spot-treated with herbicides if necessary.

A replicated, controlled study in summer and autumn of 1995 and 1996 on 15 sown set-aside strips on a farm in Cambridgeshire, UK (1) found that more bird individuals (average 20%) and species (average 56%) used the strips than the adjacent crop area (average 7% individuals and 33% species) in both years. However, the highest proportions of both individuals and species were recorded in the field boundaries (average 68% ind. and 80% spp.). The highest species richness was found in the most diverse grass mix. The seed mixture 'Tübinger Mischung' with only wildflowers attracted most individuals, but the lowest species numbers. Note that no statistical analyses were performed on these data. Five seed mixtures were sown on set-aside areas (minimum 20 m wide and 100 m long) in the autumns of 1993 and 1994. Seed mixtures contained either only grass species (three mixes including three to six species, cost £15–£70/ha), a mix of grasses and herbs (six grass and eight herb species, cost £300/ha) or only herbs 11 species, £35/ha). Birds were recorded during 15 min point counts on 10 occasions between June and September 1995 and July and October 1996. Each bird's location was recorded in three categories: field boundary, set-aside strip and crop. After each count, the strips were walked to flush any birds present but not visible during the count.

A 2000 literature review (2) found that the UK population of cirl bunting *Emberiza cirlus* increased from between 118 and 132 pairs in 1989 to 453 pairs in 1998 following a series of schemes designed to provide overwinter stubbles, grass margins and beneficially managed hedges and set-aside. Numbers on fields under the specific agri-environment scheme increased by 70%, compared with a 2% increase elsewhere.

A 2001 replicated paired site comparison study in south Devon (3) found that fields with 6 m grass margin were associated with increases in cirl bunting *Emberiza cirlus* numbers. Six of 7 Countryside Stewardship Scheme (CSS) plots that had 6 m grass margins and were within 2.5 km of former bunting territories gained birds, whereas more generally there were declines of 20% in bunting numbers on land not-participating within the CSS. Forty-one 2×2 km^2 squares containing both land within CSS and non-CSS land were surveyed in 1992, 1998 and 1999. In each year each square was surveyed at least twice, the first time during mid-April to late May, and the second time between early June and the end of August.

A replicated, controlled study in winter 1999/2000 and summer 2000 on 23 pastoral farms in the West Midlands, UK (4), found 16 times higher winter densities of seed-eating birds within 6 m of boundaries on fields with CSS grass margins than on fields without (1.1 vs. 0.1 birds/ha), and twice as many Eurasian blackbirds *Turdus merula* near the boundaries on fields without CSS grass margins than with (1.8 vs. 0.9 birds/ha). A total of 388 grass fields were surveyed four times each in winter and in summer. No statistical analysis was performed.

A controlled study from May–August in 1995–7 and 1999 on a mixed arable and pastoral farm in Oxfordshire, UK (5), found that yellowhammers *Emberiza citrinella* spent significantly greater time foraging in grass margins and field boundaries than in other habitats. There was no difference between margins and boundaries, or between cut and uncut grass margins. However, greater use was made of both cut and uncut grass margins combined than field boundaries. Habitats surveyed were cut (1.8 km) or uncut (1.6 km) grass margins (2 or 10 m wide, at edges of arable field), field boundaries, arable fields (winter-sown cereals) and grass fields (pasture, silage and hay) found. Total area surveyed was 143 ha in 1995–7 and 107 ha in 1999.

A 2006 replicated site comparison study of 42 fields in the UK (6) found that installing 6 m-wide grass field margin strips along arable fields had no effect on the number of birds or bird species found to breed or forage on farmland. Under the CSS, these 6-m-wide grass field margin strips were grown through natural regeneration, the sowing of grass, or grass/forbs mixture. Pesticides applications were prohibited – except for the patch-wise control of problem weeds. The margin, which could not be used for regular access by farm vehicles, may have been mown once a year after mid-July, and dense cuttings must be removed. The study surveyed seven pairs of fields (one with field margins managed under the CSS, one conventionally farmed) and the 12.5 ha area surrounding each field, from three different regions of the UK four times during the breeding season.

A replicated, paired sites comparison in mid-summer 2003 on 42 arable fields in southern England (7) found that there were no more farmland bird species and birds were no more abundant on fields with 6 m wide grassy margins, compared to control fields without margins (11–18 species/site for 21 fields with margins vs. 11–15 species/site for 21 without).

A replicated study in 1999 and 2003 on 256 arable and pastoral fields across 84 farms in East Anglia and the West Midlands, England (8), found that a combination of creating uncultivated (see 'Create uncultivated margins around intensive arable or pasture fields') and planted margins around fields was strongly positively associated with four out of 12 farmland bird species analysed. These were skylarks *Alauda arvensis* (a field-nesting species) and chaffinch *Fringilla coelebs*, whitethroat *Sylvia communis* and yellowhammer *Emberiza citrinella* (all boundary-nesting species). The study did not distinguish between uncultivated and planted margins. This study describes several other interventions, discussed in 'Leave headlands in fields unsprayed (conservation headlands)'; 'Create beetle banks'; 'Leave uncropped, cultivated margins or plots, including lapwing and stone curlew plots'; 'Leave overwinter stubbles'; 'Plant wild bird seed or cover mixture'; 'Provide or retain set-aside areas in farmland'; 'Pay farmers to cover the costs of conservation measures'; 'Reduce pesticide or herbicide use generally'.

A randomised, replicated, controlled trial of sown grassy field margins from 2002 to 2006 in eastern England (9) found that the management of margins affect bird use more than the seed mix used. The number of birds using margins on two farms in summer increased by 29% between 2003 and 2006 and bird densities were higher on disturbed and plots treated with grass-killing herbicides (graminicides) than on cut plots (no actual bird densities given, only model results). Bird densities were linked to densities of diurnal ground beetles (Carabidae), especially in disturbed and graminicide-treated plots. In winter, there were twice as many birds on cut margins on 10 farms as on uncut margins, and twice as many birds in the second year than the first. Field margin plots (6 × 30 m) were established using one of three seed mixes: 1) Countryside Stewardship mix, 2) tussock grass mix and 3) a mixture of grasses and forbs designed for pollinating insects. The margins were managed in spring from 2003 to 2005 with one of three treatments: 1) cut to 15 cm, 2) soil disturbed by scarification until 60% of the area was bare ground, 3) treated with graminicide at half the recommended rate. There were five replicates of each treatment combination, at two farms – one in Boxworth, Cambridgeshire, England, and High Mowthorpe, Yorkshire, England. Birds were surveyed five to eight times between April and July from 2002 to 2006. In winters of 2004/5 and 2005/6, birds were also surveyed on 6 m margins on 10 farms in eastern England with two seed mixes (tussocky grass and fine grass). Margins were either cut in autumn or uncut. There were four replicates of each treatment combination per farm.

A 2007 literature review discussing research on a farm in Leicestershire, UK (10), found that grass margins around fields contained high numbers of yellowhammer *Emberiza citrinella* and whitethroat *Sylvia communis* nests, the former of which had higher survival than in adjacent hedgerows. This study is also discussed in 'Leave uncropped, cultivated margins or plots, including lapwing and stone curlew plots', 'Create skylark plots' and 'Create beetle banks'.

A replicated controlled study in May and June 2003–4 on six arable farms in Mississippi, USA (11), found that there were significantly more farmland bird species in bordered field margins, compared to unbordered margins (approximately 5 species/ha for 35 bordered margins vs. 0.5 species/ha for 21 unbordered margins). There were higher densities of farmland birds on margins and crops for fields with wide borders (35 birds/ha for 7–11 wide borders and 27–29 birds/ha for adjacent cropland), compared with narrow margins (18 birds/ha for 24–27 narrow borders and 13–15 birds/ha for cropland) or fields without borders (3 birds/ha for 21 unbordered margins and 1–9 birds/ha for cropland). Four species (red-winged blackbird *Agelaius phoeniceus*, dickcissel *Spiza americana*, northern cardinal *Cardinalis cardinalis*, indigo bunting *Passerina cyanea*) were significantly more abundant on bordered margins. Borders consisted of strips either 6–12 m (narrow) or 20–56 m (wide) around arable fields and planted in spring 2002 with grasses and legumes. If non-native species were dominant, the borders were also treated with selective herbicide.

A replicated study in Aberdeenshire, Scotland, in May–August 2004–6 (12), investigated the impact of cutting sown and naturally regenerated field margins, with yellowhammers *Emberiza citrinella* appearing to use cut patches of margins for 3% (of 172) in early summer, compared to 34% (of 77) foraging flights in late summer. This study is discussed in 'Create uncultivated margins around intensive arable or pasture fields'.

A replicated study in February 2008 across 97 1 km^2 plots in East Anglia, England (13), found that 19 of 24 farmland bird species responded positively to field margins managed under agri-environment schemes, but only yellowhammer *Emberiza citrinella* and possibly blackcaps *Sylvia altricapilla* showed strong responses. Great tits *Parus major* and common starlings *Sturnus vulgaris* showed weak positive responses. Field margins were categorised as grassy/weedy, bare/fallow or wild-bird cover (although very few fields had wild bird cover) and most were managed under the Entry Level Stewardship scheme. This study also investigated the effects of field boundary management; see 'Manage hedges to benefit wildlife'.

A replicated site comparison study on 1,031 agricultural sites across England in 2004–8 (14) found that grey partridge *Perdix perdix* brood size was negatively associated with the proportion of a site under planted grass buffer strips, with a significant relationship in 2008. The ratio of young partridges to old was negatively related to the proportion of grass strips in 2005 and 2008. However, year-on-year changes in partridge density and overwinter survival were positively correlated with the proportion of grass buffer strips on a site in some years – 2006 to 2007 (year-on-year changes) and 2005–6 (overwinter survival). This study describes the effects of several other interventions, discussed in the relevant sections.

A replicated site comparison study on farms in three English regions (15) found that hedges alongside grass field margins 'floristically enhanced' under Higher Level Stewardship had more yellowhammers (estimate of 0.4 birds/m) compared to hedges without a grass margin (estimated 0.2 birds/m). Hedges alongside unenhanced grass margins, either conventionally managed or managed under Entry Level Stewardship, did not have more yellowhammers. Surveys were carried out on 69 farms with Higher Level Stewardship in East Anglia, the West Midlands and the Cotswolds and on 31 farms across all three regions with no environmental stewardship.

(1) Clarke, J. H., Jones, N. E., Hill, D. A. & Tucker, G. M. (1997) The management of set-aside within a farm and its impact on birds. *Proceeding of the 1997 Brighton Crop Protection Conference*, 1–3, 1179–1184.

(2) Aebischer, N. J., Green, R. E. & Evans, A. D. (2000) From science to recovery: four case studies of how research has been translated into conservation action in the UK. 43–54 in: N. J. Aebischer, A. D. Evans, P. V. Grice, J. A. Vickery (eds) *Ecology and Conservation of Lowland Farmland Birds*. British Ornithologists Union, Tring.

(3) Peach, W., Lovett, L., Wotton, S. & Jeffs, C. (2001) Countryside stewardship delivers cirl buntings (*Emberiza cirlus*) in Devon, UK. *Biological Conservation*, 101, 361–373.

(4) Buckingham, D. L., Peach, W. J. and Fox, D. (2002) Factors influencing bird use in different pastoral systems. In: J. Frame (ed) *Conservation pays? Reconciling environmental benefits with profitable grassland systems*. British Grassland Society occasional symposium no. 36, pp. 55–58. British Grassland Society, Reading, UK.

(5) Perkins, A. J., Whittingham, M. J., Morris, A. J. & Bradbury, R. B. (2002) Use of field margins by foraging yellowhammers *Emberiza citrinella*. *Agriculture, Ecosystems & Environment*, 93, 413–420.

(6) Kleijn, D., Baquero, R. A., Clough, Y., Diaz, M., Esteban, J., Fernández, F., Gabriel, D., Herzog, F., Holzschuh, A., Jöhl, R., Knop, E. Kruess, A., Marshall, E. J. P., Steffan-Dewenter, I., Tscharntke, T., Verhulst, J., West, T. M. & Yela J. L. (2006) Mixed biodiversity benefits of agri-environment schemes in five European countries. *Ecology Letters*, 9, 243–254.

(7) Marshall, E. J. P., West, T. M. & Kleijn, D. (2006) Impacts of an agri-environment field margin prescription on the flora and fauna of arable farmland in different landscapes. *Agriculture, Ecosystems & Environment*, 113, 36–44.

(8) Stevens, D. K. & Bradbury, R. B. (2006) Effects of the Arable Stewardship Pilot Scheme on breeding birds at field and farm-scales. *Agriculture, Ecosystems & Environment*, 112, 283–290.

(9) Henderson, I. G., Morris, A. J., Westbury, D. B., Woodcock, B. A., Potts, S. G., Ramsay, A. & Coombes, R. (2007) Effects of field margin management on bird distributions around cereal fields. *Aspects of Applied Biology*, 81, 53–60.

(10) Stoate, C. & Moorcroft, D. (2007) Research-based conservation at the farm scale: Development and assessment of agri-environment scheme options. *Aspects of Applied Biology*, 81, 161–168.
(11) Conover, R. R., Burger, L. W. & Linder, E. T. (2009) Breeding bird response to field border presence and width. *Wilson Journal of Ornithology*, 121, 548–555.
(12) Douglas, D. J. T., Vickery, J. A. & Benton, T. G. (2009) Improving the value of field margins as foraging habitat for farmland birds. *Journal of Applied Ecology*, 46, 353–362.
(13) Davey, C. M., Vickery, J. A., Boatman, N. D., Chamberlain, D. E. & Siriwardena, G. M. (2010) Entry Level Stewardship may enhance bird numbers in boundary habitats. *Bird Study*, 57, 415–420.
(14) Ewald, J. A., Aebischer, N. J., Richardson, S. M., Grice, P. V. & Cooke, A. I. (2010) The effect of agri-environment schemes on grey partridges at the farm level in England. *Agriculture, Ecosystems & Environment*, 138, 55–63.
(15) Field, R. H., Morris, A. J., Grice, P. V. & Cooke, A. I. (2010) Evaluating the English Higher Level Stewardship scheme for farmland birds. *Aspects of Applied Biology*, 100, 59–68.

5.20. Use mowing techniques to reduce chick mortality

- A review from the UK (2) found a large increase in corncrake *Crex crex* populations in the UK following a scheme to delay mowing and promote corncrake-friendly mowing techniques.
- One replicated controlled study from the UK (1) and a review from the UK (3) found lower levels of mortality of corncrakes and Eurasian skylark *Alauda arvensis* when wildlife-friendly mowing techniques were used, compared to other techniques.

Background

During mowing and harvesting operations, ground-nesting birds frequently remain in long grass or crops for as long as possible. If mowing/harvest occurs from the outside of the field inwards, this behaviour can leave the birds trapped in the centre of the field and killed as the last patch is harvested. Adjusting mowing techniques, for example starting from the inside of the field, can therefore allow chicks to escape into field margins.

A replicated controlled study in three areas in Ireland between 1992 and 1995 (1) found that corncrake *Crex crex* chicks were more likely to survive in hay and silage meadows when they were mown from the inside-out (I-O), compared to the traditional outside-in (O-I) mowing pattern (68% survival for 76 chicks in I-O fields vs. 45% survival for 31 chicks in O-I fields). Most chicks (80%) were killed during the last eight sweeps of the harvester for O-I and the last five for I-O, and mortality was zero for both methods when the nearest tall vegetation was within 5 m of the edge of the field. Chicks that were more than one day old were able to move fast enough away from the mower to escape, so long as a route to unmown cover was available.

A 2000 literature review (2) found that the UK population of corncrakes *Crex crex* increased from 480 to 589 males between 1993 and 1998 (an average rise of 3.5%/year) following schemes to get farmers to delay mowing dates and to leave leaving unmown 'corridors' to allow chicks to escape to field edges.

A review of four experiments on the effects of agri-environment measures on livestock farms in the UK (3) found one trial from 2006 to 2008 that tested the effect of mowing techniques to reduce mortality of Eurasian skylarks *Alauda arvensis* nesting in silage fields. Preliminary results showed that chick survival was strongly affected by the type of machinery used. Survival was four times higher using wider machinery and reducing the number of machinery passes than without these changes. However, the number of new birds produced each year (productivity) was more sensitive to re-nesting rates than chick survival. This study formed part of a Department for Environment, Food and Rural Affairs (Defra) funded project (BD1454) for which no reference is given in the review.

(1) Tyler, G. A., Green, R. E. & Casey, C. (1998) Survival and behaviour of corncrake *Crex crex* chicks during the mowing of agricultural grassland. *Bird Study*, 45, 35–50.
(2) Aebischer, N. J., Green, R. E. & Evans, A. D. (2000) From science to recovery: four case studies of how research has been translated into conservation action in the UK. 43–54 in: N. J. Aebischer, A. D. Evans,

P. V. Grice, J. A. Vickery (eds) *Ecology and Conservation of Lowland Farmland Birds* British Ornithologists Union, Tring.

(3) Buckingham, D. L., Atkinson, P. W., Peel, S. & Peach, W. (2010) New conservation measures for birds on grasslands and livestock farms. *BOU Proceedings – Lowland Farmland Birds III: delivering solutions in an uncertain world*. British Ornithologists Union.

5.21. Provide refuges in fields during harvest or mowing

- A replicated study in France (1) found that fewer gamebirds came into contact with mowing machinery when refuges were left in fields than when they were not left.
- A review from the UK (2) found that Eurasian skylarks *Alauda arvensis* did not nest at higher densities in uncut refuges than in the rest of the fields.

Background

During mowing and harvesting operations, ground-nesting birds frequently remain in long grass or crops for as long as possible. If mowing/harvest occurs from the outside of the field inwards, this behaviour can leave the birds trapped in the centre of the field and killed as the last patch is harvested. However, if unharvested refuges are left in fields then it is possible that chicks and adults will remain in them and survive.

A replicated study in 1996–7 in 62 hay fields in Bourgogne, France (1), found that contact between mowing machinery and unfledged common quail *Coturnix coturnix* and corncrake *Crex crex* was reduced by approximately 50% and 33% respectively, by leaving 10 m wide, uncut strips in the centre of fields. In addition, unmown strips held the highest concentrations of corncrakes, quails and passerines (7.7 birds/ha, 3.8 birds/ha and 10.8 birds/ha respectively in 1996).

A review of four experiments on the effects of agri-environment measures on livestock farms in the UK (2) found one trial from 2006 to 2008 demonstrating that uncut nesting refuges for skylarks *Alauda arvensis* in silage fields were not used more than other areas. Refuge plots of 1 ha were cut with raised mowing height in the first silage cut, then left uncut for the rest of the season. The plots were preferred for re-nesting for two weeks following the first cut, but subsequently did not have higher nest densities than other areas. Skylarks continually re-nest rather than re-nesting in a batch after each cut. After the second cut, safe areas were completely avoided by skylarks. This study formed part of a Defra-funded project (BD1454) for which no reference is given in the review.

(1) Broyer, J. (2003) Unmown refuge areas and their influence on the survival of grassland birds in the Saône valley (France). *Biodiversity and Conservation*, 12, 1219–1237.

(2) Buckingham, D. L., Atkinson, P. W., Peel, S. & Peach, W. (2010) New conservation measures for birds on grasslands and livestock farms. *BOU Proceedings – Lowland Farmland Birds III: delivering solutions in an uncertain world*. British Ornithologists Union.

5.22. Mark bird nests during harvest or mowing

- A replicated study from the Netherlands (1) found that northern lapwing *Vanellus vanellus* nests were less likely to be destroyed when they were marked, compared to when they were not.

Background

Marking the nests of ground-nesting birds may reduce the accidental destruction by farmers during harvest or mowing.

A replicated study in 2005–6 on arable farms in Noordoostpolder and Oostelijk Flevoland, the Netherlands (1), found that marked northern lapwing *Vanellus vanellus* nests were significantly less likely to fail as a result of farming operations than unmarked nests (0–9% of 1,644 marked nests destroyed vs. 15–42% of 229 unmarked nests). However, overall survival rates did not differ significantly (37–73% success for marked nests vs. 38–66% for unmarked), with some evidence that marked nests were deserted or predated more often. Nests on the marked farms (121 in 2005, 113 in 2006) were marked with two bamboo poles (1 m high) by 151–171 volunteers, and farmers told of their presence. On the control farms, no markers were put in place and farmers were not informed of the nests.

(1) Kragten, S., Nagel, J. A. N. & De Snoo, G. R. (2008) The effectiveness of volunteer nest protection on the nest success of northern lapwings *Vanellus vanellus* on Dutch arable farms. *Ibis*, 150, 667–673.

5.23. Relocate nests at harvest time to reduce nestling mortality

- A replicated controlled study from Spain (1) found that clutches that were temporarily removed from fields during harvest and then replaced had higher hatching and fledging rates than control clutches. Effects were greater on clutches that were older when moved.

Background

If nests are likely to be destroyed by machinery during harvest or mowing, it may be possible to move them and then return them after the danger has passed. If nests are extremely likely to be destroyed during harvest or mowing then it may be best to remove the chicks and hand-rear them. Studies on the effects of this intervention are found in the chapter on captive breeding and hand-rearing.

A replicated, controlled study from 1987–91 in five areas of cereal fields in south-west Spain (1) found that nestling mortality of Montagu's harriers *Circus pygarus* was significantly lower, and fledging success significantly higher, for clutches that were removed from fields before harvesting, and returned within an hour, compared to control (unmoved) clutches (28% mortality and 75% of nests fledging at least one chick in 72 managed clutches vs. 67% mortality and 29% fledging success in 39 controls). Outcome was highly dependent on clutch age at time of harvest: no clutches less than ten days old at harvest fledged young, whilst nest management increased the proportion of successful clutches aged 11–20 days at harvesting from 14% to 75%. The average harvest date of barley fields was later than for wheat or oat fields, but the small number of clutches (13) in barley fields made it impossible to assess the influence of nesting habitat on unmanaged clutch success. The nature of the crop (wheat and/or oat vs. barley) did not influence breeding success in managed clutches.

(1) Corbacho, C., Sánchez, J. M. & Sánchez, A. (1999) Effectiveness of conservation measures on Montagu's harriers in agricultural areas of Spain. *Journal of Raptor Research*, 33, 117–122.

5.24. Make direct payments per clutch for farmland birds

- One of two replicated and controlled study from the Netherlands (2) found that farms with per clutch payments held slightly higher breeding densities of waders, but not higher overall numbers than control farms. One study found no population effects over three years (1).
- A replicated and controlled study (1) found higher hatching success on farms with payment schemes than control farms.

Background

Most agri-environment schemes aim to compensate farmers for the cost of conservation management on their land, irrespective of the outcomes. The Netherlands, however, also has a scheme where farmers are paid directly, based on the number of breeding bird pairs on their land.

A replicated and controlled study on intensive dairy grassland in the western Netherlands between 1993 and 1996 (1) found that northern lapwing *Vanellus vanellus* and black-tailed godwit *Limosa limosa* showed higher hatching success on 15 farms offered per-clutch payments for farmland birds than on nine control farms (65% vs. 48% for lapwing, 63% vs. 39% for godwits). A non-significant difference was also seen for common redshank *Tringa totanus* (39% vs. 21%). There were no differences in treatment during 1993–4, before payments. The number of control farms was reduced to three in 1995–6, because the farmers on other farms had become too involved in conservation for their farms still to be considered true controls. No other bird conservation measures were in place and the cost was estimated at €40/clutch. Population-level impacts were not observed, possibly due to the relatively short time-scale and small number of farms.

A replicated and controlled paired sites study in the western Netherlands in 2003 (2) found slightly higher breeding densities of birds on 19 grassland plots with per-clutch payments for wader clutches, compared to 19 paired, control plots, both when delayed mowing was also used and when per-clutch payment was the only scheme used (13 territories/plot for combined schemes; 13 territories/plot for per-clutch payment and 11 territories/plot for controls). However, birds were not more abundant under either scheme, compared with controls (approximately 125 birds/plot for combined schemes; 125 birds/plot for per-clutch payment and 110 birds/plot for controls). Wader breeding densities were higher (but not significantly so) on combined and per-clutch payment plots (approximately 7 territories/plot for combined schemes; 7 territories/plot for per-clutch payment and 5 territories/plot for controls). When individual wader species were analysed, there were higher numbers of redshank *Tringa totanus* on combined or per-clutch payment plots (approximately 5 birds/plot for combined schemes; 5 birds/plot for per-clutch payment and 3 birds/plot for controls), but there were no significant differences in breeding densities for redshank, northern lapwing *Vanellus vanellus*, Eurasian oystercatcher *Haematopus ostralegus* or black-tailed godwit *Limosa limosa*. The authors suggest that groundwater depth, soil hardness and prey density drove these patterns. All farms had been operating the schemes for at least three (and an average of four) years before the study. This study is also discussed in 'Delay haying/mowing'.

(1) Musters, C. J. M., Kruk, M., De Graaf, H. J. & Keurs, W. J. T. (2001) Breeding birds as a farm product. *Conservation Biology*, 15, 363–369.

(2) Verhulst, J., Kleijn, D. & Berendse, F. (2006) Direct and indirect effects of the most widely implemented Dutch agri-environment schemes on breeding waders. *Journal of Applied Ecology*, 44, 70–80.

5.25. Control scrub on farmland

- A replicated study from the UK (1) found a negative relationship between the number of young grey partridge *Perdix perdix* per adult and a combined intervention of scrub control, rough grazing and the restoration of various semi-natural habitats.

Background

Scrub on farmland can add complexity and heterogeneity to farmland. However, if scrub dominates non-productive land on farms it may lead to declines in species that require grassland and other farmland habitats.

A replicated site comparison study on 1,031 agricultural sites across England in 2004–8 (1) investigated the impact of scrub control on grey partridge *Perdix perdix*. However, the study does not distinguish between the impacts of scrub control, rough grazing and the restoration of various semi-natural habitats. There was a negative relationship between the combined intervention and the ratio of young to old partridges in 2008. This study investigated several other interventions, discussed in the relevant sections.

(1) Ewald, J. A., Aebischer, N. J., Richardson, S. M., Grice, P. V. & Cooke, A. I. (2010) The effect of agri-environment schemes on grey partridges at the farm level in England. *Agriculture, Ecosystems & Environment*, 138, 55–63.

5.26. Take field corners out of management

- A replicated study in the UK (1) found that overwinter survival of grey partridge *Perdix perdix* was higher where field corners were taken out of management than on other sites for one of three winters. There was no relationship with the intervention and brood size, the ratio of young to old birds or density changes.

Background

Field corners can be taken out of management on both arable and livestock farms. This can either involve simply not managing or planting corners with grass (see also 'Plant grass buffer strips').

A replicated site comparison study on 1,031 agricultural sites across England in 2004–8 (1) found that grey partridge *Perdix perdix* overwinter survival was positively correlated with taking field corners out of management, significantly so in 2007–8. There were no relationships with brood size, the ratio of young to old birds or year-on-year density changes. This study describes the effects of several other interventions, discussed in the relevant sections.

(1) Ewald, J. A., Aebischer, N. J., Richardson, S. M., Grice, P. V. & Cooke, A. I. (2010) The effect of agri-environment schemes on grey partridges at the farm level in England. *Agriculture, Ecosystems & Environment*, 138, 55–63.

5.27. Reduce conflict by deterring birds from taking crops

Background

In some parts of the world, the persecution of birds that take crops can be a serious threat to the survival of populations. Methods to reduce the damage done by birds can therefore be important in reducing the pressure on populations.

5.27.1. Use bird scarers

- A controlled paired study in the USA (1) found reduced levels of damage to almond orchards when American crow *Corvus brachyrhynchos* distress calls were broadcast, compared to the previous year. There were no decreases in control orchards.
- A replicated study in Pakistan (2) found that four pest species were less abundant when reflector ribbons were hung above crops, compared to when ribbons were not used.

A controlled, paired study in central California, USA (1), found that two of three almond orchards with crow distress calls broadcast had reduced damage from American crows *Corvus brachyrhynchos* in 2003, compared to 2002, when broadcasts were not used. Damage reduced from 6.0 kg/ha to 1.1 kg/ha, and 18.2 kg/ha to 4.8 kg/ha. There was no change in three paired sites without broadcasts. Orchards were 16–30 ha in area and monitored in June–August. Broadcasting units were deployed at onset of crow damage until almond harvest (1 unit/1.6 ha; hung in trees at 1–2 m) throughout the orchard, moved to a new tree every two weeks, switched to a different call every 3–4 days, broadcast dawn to dusk, each 25 seconds long with approximately 12 min between calls.

A replicated two-month study (November–December) on agricultural land in Punjab, Pakistan (2), found that hanging reflector ribbons 65–100 cm above crops was a low cost technique that significantly decreased abundances of four main bird pest species (house crow *Corvus splendens*, ring-necked parakeet *Psittacula krameri*, common myna *Acridotheres tristis* and bank myna *A. ginginianus*) that heavily damage young wheat *Triticum aestivum* and maize *Zea mays* crops.

(1) Houk, A., Delwiche, M., Gorenzel, P. & Salmon, T. (2004) Electronic repeller and field protocol for control of crows in almonds in California. 130 in: *Proceedings-Vertebrate Pest Conference*, 21, 130–135.
(2) Hafeez, S., Khan, T. H., Khan, T. A. J., Shabaz, M. & Ahmed, M. (2008) Use of reflector ribbon as a pest birds repellent in wheat and maize crop. *Journal of Agriculture and Social Sciences*, 4, 92–94.

5.27.2. Use repellents

- A replicated, randomised and controlled *ex situ* study in the USA (1) found that dickcissels *Spiza americana* consumed less rice if it was treated with two repellents, compared to controls. Two other repellents did not reduce consumption as effectively.

A replicated, randomised and controlled *ex situ* study in the USA (1) found that dickcissels *Spiza americana* captured in Venezuela consumed 70% less rice if it was treated with methiocarb (at 0.05 g/g rice) or anthraquinone (0.5 g/g), compared to control (untreated) rice offered previously. Methyl anthranilate and lower doses of anthraquinone did not reduce consumption of rice when treated rice was offered after untreated rice. However, when a choice of rice treated with 0.05% or 0.1% anthraquinone or untreated millet was offered at the same time, birds significantly reduced their consumption of rice, with the preference growing over eight days of testing. Rice was offered over five days (control rice on the first, followed by treated rice), with rice and millet being offered over eight days. The number of birds tested is not provided.

(1) Avery, M. L., Tillman, E. A. & Laukert, C. C. (2001) Evaluation of chemical repellents for reducing crop damage by dickcissels in Venezuela. *International Journal of Pest Management*, 47, 311–314.

Arable farming

5.28. Increase crop diversity

- A before-and-after study in the UK (1) found that more barnacle geese *Branta leucopsis* used a site after the amount of land used to grow cereals was reduced and other interventions were used.

Background

Farmland heterogeneity is thought to be key in determining on-farm biodiversity (Benton *et al.* 2003). Therefore, increasing the range of different crops grown in a given year may increase the biological value of a farm.

Benton, T. G., Vickery, J. A. & Wilson, J. D. (2003) Farmland biodiversity: is habitat heterogeneity the key? *Trends in Ecology & Evolution*, 18, 182–188.

A before-and-after study in Dumfries, southern Scotland (1), found that the number of barnacle geese *Branta leucopsis* on a mixed agricultural site and nature reserve increased from 3,200 in 1970 to 6,000 in 1975 following a reduction in the amount of cereals grown on arable land. From 1970 onwards, only 16.7% of the 50 ha of arable land was used for cereals. In addition, all cereals were undersown (see 'Undersow spring cereals') and no stock were allowed to graze on the arable land after November (see 'Reduce grazing intensity on permanent grassland').

(1) Owen, M. (1977) The role of wildfowl refuges on agricultural land in lessening the conflict between farmers and geese in Britain. *Biological Conservation*, 11, 209–222.

5.29. Implement 'mosaic management'

- A replicated, controlled before-and-after study from the Netherlands (2) found that northern lapwing *Vanellus vanellus* population trends changed from decreases to increases following the introduction of mosaic management. Three other waders did not show such a response.
- A replicated, paired sites study in the Netherlands (1) found that black-tailed godwits *Limosa limosa* had higher productivity under mosaic management than other management types, and nests were less likely to be trampled by livestock.

Background

Mosaic management is a Dutch agri-environment scheme that, rather than concentrating on individual farms, attempts to coordinate management across groups of farms. Interventions include delayed and staggered mowing, refuge strips and nest protection and aim to provide suitable foraging habitat for wader chicks throughout the year.

A replicated paired sites comparison in 2004–5 on six wet grassland sites in the Netherlands (1) found that the reproductive productivity of black-tailed godwits *Limosa limosa* was significantly higher on sites managed under a 'mosaic management' agri-environment scheme, compared to on non-scheme sites (average of 0.28 chicks fledged/breeding pair for scheme sites vs. 0.16 chicks/pair on non-scheme sites). Differences were due to higher nest survival on mosaic management sites (50% vs. 33%), as there were no differences in the number of chicks hatching in successful nests (3.4 chicks/successful nest vs. 3.2 chicks/successful nest), or the fledging rate of chicks (11% fledging success on all sites). Nests were equally likely to be predated on scheme and non-scheme sites (32% predated vs. 37%), but were more likely to be trampled or destroyed by mowing on non-scheme sites (6% vs. 29%). Most fields in five scheme sites and about 50% in the sixth, had nests marked (to reduce losses due to farming activities); at non-scheme sites almost 100% of nest were marked in three, some in two, and none in one. The number of nests on different sites was not provided.

A replicated, controlled before-and-after study in 1996–2008 in eight wet grassland areas in Friesland and Groningen, the Netherlands (2), found that northern lapwing *Vanellus vanellus* population trends moved from a 7% annual decrease to a 4% annual increase following the introduction of mosaic management in 2000–1. Three other species (black-tailed godwit *Limosa limosa*, common redshank *Tringa totanus* and Eurasian oystercatcher *Haematopus ostralegus*) did not show any change in trend after the introduction. When comparing trends on the mosaic management sites with 29 farms using individual conservation management, 46 farms with standard management and 42 nature reserves, only lapwing populations increased significantly more on mosaic management sites, compared to the others. Oystercatcher populations did significantly less well on mosaic management sites, compared to nature reserves.

(1) Schekkerman, H., Teunissen, W. & Oosterveld, E. (2008) The effect of 'mosaic management' on the demography of black-tailed godwit *Limosa limosa* on farmland. *Journal of Applied Ecology*, 45, 1067–1075.

(2) Oosterveld, E. B., Nijland, F., Musters, C. J. M. & Snoo, G. R. (2010) Effectiveness of spatial mosaic management for grassland breeding shorebirds. *Journal of Ornithology*, 152, 161–170.

5.30. Leave overwinter stubbles

- The three studies from the UK (one replicated) that report population-level changes found positive effects of over-winter stubble provision (1,9,12), but all investigated multiple interventions at once.
- Eight studies from the UK (2,4,6,8–12), including a systematic review, found that at least some species or groups of farmland birds were positively associated with over-winter stubbles, or were found on stubbles. Three studies (6,9,10) investigated multiple interventions without separating

the effects of each. Two studies (8,9) reported that seed-eating birds in particular were more abundant on stubbles.

- One of the eight studies (10) found that no more positive responses to stubbles were found than would be expected by chance. A replicated, randomised and controlled study from the UK (4) found that 22 of 23 species did not preferentially use stubbles compared to cover crops. A replicated study from the UK (12) found that the area of stubble in a site was negatively related to grey partridge *Perdix perdix* brood size.
- Five studies from the UK (3,5,7,13,14), four replicated, found that stubble management affected use by birds. Some species or groups were more common on cut stubbles, some on uncut and some showed preferences for barley over wheat. One study (13) found that only Eurasian skylarks *Alauda arvensis* were more common on stubbles under agri-environment schemes, and only on highly prescriptive schemes. One study (14) found that all seed-eating species were more abundant on stubbles under agri-environment schemes in one of two regions studied.

Background

A 2008 literature review and analysis of the Environmental Stewardship programme, particularly Entry Level Stewardship in the UK (Vickery *et al.* 2008), suggested that, for Eurasian skylarks *Alauda arvensis*, approximately 0.1 km^2 of stubble/km^2 would be needed to prevent population declines. The authors also suggest that having these patches over 1 km apart would maximise winter use.

Vickery, J., Chamberlain, D., Evans, A., Ewing, S., Boatman, N., Pietravalle, S., Norris, K. & Butler, S. (2008) *Predicting the impact of future agricultural change and uptake of Entry Level Stewardship on farmland birds*. British Trust for Ornithology, The Nunnery, Thetford.

A 2000 literature review (1) found that the UK population of cirl buntings *Emberiza cirlus* increased from between 118 and 132 pairs in 1989 to 453 pairs in 1998 following a series of schemes designed to provide overwinter stubbles, grass margins and beneficially managed hedges and set-aside areas. Abundance on fields under the specific agri-environment schemes increased by 70%, compared with a 2% increase elsewhere.

A replicated study in the winters of 1997–8 and 1998–9 on 122 stubble fields on 32 farms in central England (2) found 13 bird species using stubble fields. Four species (Eurasian linnet *Carduelis cannabina*, Eurasian skylark *Alauda arvensis*, reed bunting *E. schoeniclus* and corn bunting *Miliaria calandria*) were found more frequently on intensively-farmed barley stubbles than intensive or organic wheat, whilst woodpigeons *Columba palumbus* were found most frequently on organic wheat.

A replicated, randomised study from November 2003 to March 2004 in 205 cereal stubble fields under a range of management intensities in arable farmland in south Devon, UK (3), found that barley stubbles following low-input herbicide were more beneficial for cirl buntings *Emberiza cirlus* than wheat or conventionally managed stubbles. Higher population sizes were also associated with the number of breeding bunting territories the previous season, and with small field size. The effect of small field size may be because cirl buntings prefer to forage near hedgerows and because smaller fields are less intensively managed. The authors argue for strategic spatial targeting of stubble prescriptions. Overall, barley fields were generally preferred by seed-eating species. Low-input barley stubbles had significantly higher seed abundance and broad-leaved weed cover (approximately four times greater). Fields where stubbles were grazed over winter had significantly lower densities of seed-eating birds in general. The authors point out that seed-eating species' preference for barley stubbles was independent from the positive correlation with broad-leaved weed density and should be taken into account when planning prescriptions.

A replicated, randomised, controlled study from November–February in 2000–2001 and 2001–2002 in 20 arable farms in eastern Scotland (4) found that, of 23 species recorded, only skylarks *Alauda arvensis* were significantly denser in fields with stubble left over winter than fields with wild bird cover crops or conventional crops. Stubble fields were those in which cereal and oilseed rape stubbles were left over winter. Between 6.2 and 28.3 ha were sampled on each farm annually. This study is discussed in more detail in 'Plant wild bird cover crops' and 'Provide set-aside areas'.

A replicated controlled study in winter 2003–2004 on 20 wheat fields from 12 lowland farms in central England (5) found that seed-eating songbirds and invertebrate-feeding birds were more abundant on stubble fields cut to 6 cm, whereas skylark *Alauda arvensis* and partridge *Perdix perdix* were more abundant on fields with uncut stubble, approximately 14 cm tall (fields were visited six times each for a total of 120 visits. Seed-eaters: 343 individuals were seen on approximately 25 visits to cut fields vs. 89 individuals on 15 visits to control fields; invertebrate-eaters: 623 birds on 17 visits vs. 34 on five visits; skylarks: 557 on 50 visits vs. 814 on 80 visits; partridges: five on two visits vs. 235 on 27 visits). Crows and pigeons showed no response to stubble cutting. Each field was split so that half was cut to approximately 6 cm tall, with the other half left as a control.

A replicated study in 1999 and 2003 on 256 arable and pastoral fields across 84 farms in East Anglia and the West Midlands, England (6), found that only two of 12 farmland bird species analysed were positively associated with the provision of overwinter stubble, set-aside areas (see 'Provide or retain set-aside areas in farmland') or wildlife seed mixtures (see 'Plant wild bird seed or cover mixture'). These were Eurasian skylarks *Alauda arvensis* (a field-nesting species) and Eurasian linnets *Carduelis cannabina* (a boundary-nesting species). The study did not distinguish between set-aside, wildlife seed mixtures or overwinter stubble, classing all as interventions to provide seeds for farmland birds. This study describes several other interventions, discussed in the relevant sections.

A small randomised site comparison study in winter 2004–5 in central England (7) found that seed-eating songbirds and invertebrate-feeding birds were found at higher density on sections of fields where stubble had been cut short (642 seed-eaters and 1,207 invertebrate-feeders recorded on cut stubble plots vs. 364 and 415 on cut stubble). Eurasian skylarks *Alavda arvenis*, partridges, pigeons *Columba* spp., and meadow pipits *Anthus pratensis* were at higher densities in areas of uncut stubble (241 skylarks, 100 partridges, 37 pigeons and 81 meadow pipits on uncut plots vs. 27, 7, 12 and 9 on cut plots). In addition, skylarks and invertebrate feeders were found at higher densities on scarified (i.e. lightly tilled) sections of fields than control (unscarified) sections (339 skylarks and 1371 invertebrate feeders on scarified plots vs. 241 and 251 on controls). The stubble on one half of each field was cut in the winter of 2004–2005 before the fields were surveyed between December 2004 and March 2005.

A 2007 systematic review identified five papers investigating the effect of overwinter stubble provision on farmland bird densities in the UK (8). There were significantly higher densities of farmland birds in winter on fields with stubbles than on conventionally managed fields. In particular, there were greater densities of seed-eating songbirds and crows on fields with stubbles than on control fields. The meta-analysis included experiments conducted between 1992 and 2002 from three controlled trials, before-and-after study and one site comparison study.

A 2009 literature review of agri-environment schemes in England (9) found that there was a 146% increase in cirl bunting *Emberiza cirlus* territory density on land under a Countryside Stewardship Scheme 'special project', which (amongst other interventions) increased the amount of weedy overwinter stubbles in the target area between 1992 and 2003. In addition, the national population increased from 319 to nearly 700 pairs over the same period. Generally, the review found high densities of seed-eating songbirds and Eurasian skylarks *Alauda arvensis* on stubbles and wild bird seed or cover mix (see 'Plant wild bird seed or cover mixture'), compared to other land uses, and a survey in the winter of 2007–8 found the highest densities of skylarks on stubble fields, compared with other agri-environment schemes options. This review also examines several other interventions, discussed in the relevant sections.

A replicated site comparison of 2,046 1 km squares of agricultural land across England in 2005 and 2008 (10) found that four of eight regions of England had at least two farmland birds that showed positive responses to wild bird cover (see 'Plant wild bird seed or cover mixture') and overwinter stubble fields. Across all 15 species thought to benefit from these interventions, only one region (the north-west) showed significantly more positive responses than would be expected by chance. Some species responded positively in some regions and negatively in others. This study is also discussed in 'Pay farmers to cover the costs of conservation measures', 'Manage ditches to benefit wildlife' and 'Manage hedges to benefit wildlife'.

A large 2010 site comparison study of the same 2,046 1 km^2 plots of lowland farmland in England (11), as in (10), found that three years after the 2005 introduction of the Countryside Stewardship Scheme and Entry Level Stewardship schemes, there was no consistent association between the provision of stubbles and farmland bird numbers. Grey partridge *Perdix perdix* and tree sparrow *Passer montanus* were the only two species that showed more positive population change (population increases or smaller decreases relative to other plots) from 2005 to 2008 in the 9 km^2 and 25 km^2 areas immediately surrounding plots planted with stubble than in the area surrounding plots without stubbles. The effect of stubbles was small, however, with tree sparrow numbers increasing by 0.05 at the 9 km^2 scale for each 0.07 km^2 of stubble and by 0.07 at the 25 km scale for each 0.14 km^2 of stubble. The 2,046 1 km^2 lowland plots were surveyed in both 2005 and 2008 and classified as arable, pastoral or mixed farmland. Eighty-four percent of plots included some area managed according to the Entry Level Stewardship or Countryside Stewardship Scheme. In both survey years, two surveys were conducted along a 2 km pre-selected transect route through each 1 km^2 square.

A replicated site comparison study on 1,031 agricultural sites across England in 2004–2008 (12) found that the ratio of young to old grey partridges *Perdix perdix* on sites was positively related to the proportion of sites left as overwinter stubble. However, when stubbles were used in conjunction with other interventions, the results were mixed. In conjunction with small field sizes and reduced chemical inputs, stubbles were weakly positively correlated with year-on-year changes in partridge density but negatively related to brood size. In conjunction with undersowing spring cereals, stubbles were negatively associated with year on year changes (2006–2007) and overwinter survival (2004–2005, 2005–2006 and generally). This study describes the effects of several other interventions, discussed in the relevant sections.

A replicated site comparison study of 75 fields in East Anglia and the West Midlands (13) found no difference in the number of seed-eating birds or Eurasian skylarks *Alauda arvensis* on Environmental Stewardship stubbles and non-Environmental Stewardship stubbles. There was also no significant difference in the number of seed-eating birds on stubbles managed under the Higher Level Stewardship (18.0 birds/ha) than in fields managed under the Entry Level Stewardship (8.5 birds/ha). Skylarks, however, were found to be more numerous on Higher Level Stewardship fields (9.3 birds/ha) than ELS fields (1.2 birds/ha). Entry Level Stewardship stubbles prohibited post-harvest herbicide and cultivation until mid-February; Higher Level Stewardship stubbles had the basic Entry Level Stewardship requirements plus reduced herbicide use. Non-ES stubbles were rotational stubbles without restrictions on herbicide or cultivation practices. Seed-eating birds were surveyed on two visits to each site between 1 November 2007 and 29 February 2008.

A replicated site comparison study on farms in two English regions (14) found more seed-eating farmland songbirds on overwinter stubbles managed under Entry Level Stewardship than on non-stewardship stubbles in the West Midlands (average 6 birds/ha on Entry Level Stewardship compared with 2.5 bird/ha on conventionally managed stubble). This difference was not significant for farms in East Anglia (3.5 birds/ha on Entry Level Stewardship stubble compared with 0.7 birds/ha on conventionally managed stubble fields). Overwinter stubble fields in stewardship schemes have restrictions on herbicide use and cultivation times. The survey was carried out in winter 2007–2008 on 27 farms with Higher Level Stewardship, 13 farms with Entry Level Stewardship and 14 with no environmental stewardship, in East Anglia or the West Midlands. The group of birds analysed included tree sparrow *Passer montanus* and corn bunting *Emberiza calandra*, but not grey partridge *Perdix perdix*. More of these birds used overwinter stubbles on Higher Level Stewardship farms than on Entry Level

Stewardship farms. There were 5 birds/ha compared to 2 birds/ha on average, on stubble fields in Higher Level Stewardship and Entry Level Stewardship farms respectively.

(1) Aebischer, N. J., Green, R. E. & Evans, A. D. (2000) From science to recovery: four case studies of how research has been translated into conservation action in the UK. 43–54 in: N. J. Aebischer, A. D. Evans, P. V. Grice, J. A. Vickery (eds) *Ecology and Conservation of Lowland Farmland Birds.* British Ornithologists Union, Tring.
(2) Moorcroft, D., Whittingham, M. J., Bradbury, R. B. & Wilson, J. D. (2002) The selection of stubble fields by wintering granivorous birds reflects vegetation cover and food abundance. *Journal of Applied Ecology*, 535–547.
(3) Defra (2004) Comparative quality of winter food sources for cirl bunting delivered through countryside stewardship special project and CS arable options. RSPB/Defra Report BD1626.
(4) Parish, D. M. B. & Sotherton, N. W. (2004) Game crops and threatened farmland songbirds in Scotland: a step towards halting population declines? *Bird Study*, 51, 107.
(5) Butler, S. J., Bradbury, R. B. & Whittingham, M. J. (2005) Stubble height affects the use of stubble fields by farmland birds. *Journal of Applied Ecology*, 42, 469–476.
(6) Stevens, D. K. & Bradbury, R. B. (2006) Effects of the Arable Stewardship Pilot Scheme on breeding birds at field and farm-scales. *Agriculture, Ecosystems & Environment*, 112, 283–290.
(7) Whttingham, M. J., Devereux, C. L., Evans, A. D. & Bradbury, R. B. (2006) Altering perceived predation risk and food availability: management prescriptions to benefit farmland birds on stubble fields. *Journal of Applied Ecology*, 43, 640–650.
(8) Roberts, P. D. & Pullin, A. S. (2007) *The effectiveness of land-based schemes (including agri-environment) at conserving farmland bird densities within the UK.* Systematic Review No. 11. Collaboration for Environmental Evidence/Centre for Evidence-Based Conservation, Birmingham, UK.
(9) Natural England (2009) *Agri-environment schemes in England 2009: A review of results and effectiveness.* Natural England, Peterborough.
(10) Davey, C. M., Vickery, J. A., Boatman, N. D., Chamberlain, D. E., Parry, H. R. & Siriwardena, G. M. (2010) Regional variation in the efficacy of Entry Level Stewardship in England. *Agriculture, Ecosystems & Environment*, 139, 121–128.
(11) Davey, C. M., Vickery, J. A., Boatman, N. D., Chamberlain, D. E., Parry, H. R. & Siriwardena, G. M. (2010) Assessing the impact of Entry Level Stewardship on lowland farmland birds in England. *Ibis*, 152, 459–474.
(12) Ewald, J. A., Aebischer, N. J., Richardson, S. M., Grice, P. V. & Cooke, A. I. (2010) The effect of agri-environment schemes on grey partridges at the farm level in England. *Agriculture, Ecosystems & Environment*, 138, 55–63.
(13) Field, R. H., Morris, A. J., Grice, P. V. & Cooke, A. (2010) The provision of winter bird food by the English Environmental Stewardship scheme. *Ibis*, 153, 14–26.
(14) Field, R. H., Morris, A. J., Grice, P. V. & Cooke, A. I. (2010) Evaluating the English Higher Level Stewardship scheme for farmland birds. *Aspects of Applied Biology*, 100, 59–68.

5.31. Plant nettle strips

- We found no evidence for the effects of planting nettle strips on bird populations.

5.32. Leave unharvested cereal headlands within arable fields

- We found no evidence for the effects of leaving unharvested cereal headlands within arable fields on bird populations.

Background

Unharvested cereal headlands are strips of cereal crop around the edge of arable fields that are left throughout the winter. In addition, they are often treated less intensively with few fertilisers and no broad-leaved herbicides.

5.33. Plant crops in spring rather than autumn

- A replicated, controlled, paired sites study from Sweden (3) found more bird species on areas with spring sown cereals, compared with winter cereals. A before-and-after study from the UK (2) found that several species bred in the study site for the first time after the start of spring sowing.

- Three studies from Sweden and the UK (1–3), two replicated and controlled, found population increases after the start of spring sowing, or higher populations on sites with spring-sown cereals, compared to sites with winter cereals. A before-and-after study from the UK (2) found that ten species did not increase after spring sowing began. No species decreased. A replicated, controlled paired sites study from Sweden (3) found that the benefits of spring-sowing decreased with the proportion of autumn-sown crops in the surrounding area.
- A replicated, controlled study from Sweden (1) found that hatching success was lower on spring-sown crops than autumn sown.

Background

Changes in farming practice in northern Europe have included a shift from sowing crops in spring to sowing them the preceding autumn/winter. This change is considered to have adversely affected farmland biodiversity including invertebrates and farmland birds (see for example, Donald & Vickery, 2000).

Donald, P. F. & Vickery, J. A. (2000) The importance of cereal fields to breeding and wintering Skylarks *Alauda arvensis* in the UK. 140–150 in: N. J. Aebischer, A. D. Evans, P. V. Grice and J. A. Vickery (eds) Ecology and Conservation of Lowland Farmland Birds, British Ornithologists Union, Tring.

A replicated, controlled study between 1984 and 1994 in Västmanland, Sweden (1), found that northern lapwings *Vanellus vanellus* nested on spring-sown crops more than expected based on their availability, and on autumn-sown crops less than expected. However, hatching success on spring crops was lower than on autumn crops (29–50% for 1,236 nests on spring crops vs. approximately 85% for 27 nests on autumn crops). This study is discussed in more detail in 'Restore or create traditional water meadows'.

A before-and-after site comparison study in 2000–2005 in Bedfordshire, England (2), found that fields sown with wheat in spring held significantly more skylarks *Alauda arvensis*, seed-eating songbirds and insect-eating birds than winter-sown wheat. In addition, 20 bird species showed significant population increases on a 61 ha site where the area of spring-sown wheat and naturally regenerated set-aside was increased over the study period. Increases were lower or absent on an 80 ha area of farmland adjacent to the experimental area and without the land use change. Five species were recorded breeding for the first time after management started. Ten species showed no significant increase on the study site, whilst none decreased significantly. The biggest increases occurred in the first three years of management and were higher for farmland birds than for woodland birds. This study also investigated the impact of reducing pesticide and fertiliser inputs (see 'Reduce pesticide or herbicide use generally') and of set-aside (see 'Provide or maintain set-aside').

A replicated, controlled paired sites study in 2004 in Uppsala, Sweden (3), found that there were significantly greater numbers of ground-foraging breeding birds and more species in spring-sown barley than in autumn-sown wheat (0.8 species/ha in spring-sown vs. 0.5 species/ha in autumn-sown plots). Territory densities of lapwing *Vanellus vanellus* and wheatear *Oenanthe oenanthe* were also higher in spring-sown (lapwing: 0.08 territories/ha; wheatear: 0.12) compared to autumn-sown cereal plots (lapwing: 0.02; wheatear: 0.05). There was no effect of sowing time on skylark *Alauda arvensis* or yellowhammer *Emberiza citrinella* breeding density. In spring-sown plots, numbers of species decreased significantly as the proportion of autumn-sown cereals in the surrounding landscape increased. Forty-one independent pairs of autumn-sown wheat and spring-sown barley plots were selected, each centred on an infield non-crop island. Non-crop islands were surveyed for cover of trees, shrubs and weeds and cereal height was measured on five occasions in each field. All birds were recorded within a radius of 100 m from the centre of each plot during five point counts of seven minutes (mid-May – end of June 2004).

(1) Berg, A., Jonsson, M., Lindberg, T. & Källebrink, K. G. (2002) Population dynamics and reproduction of northern lapwings *Vanellus vanellus* in a meadow restoration area in central Sweden. *Ibis*, 144, E131–E140.
(2) Henderson, I. G., Ravenscroft, N., Smith, G. & Holloway, S. (2009) Effects of crop diversification and low pesticide inputs on bird populations on arable land. *Agriculture, Ecosystems & Environment*, 129, 149–156.
(3) Eggers, S., Unell, M. & Pärt, T. (2011) Autumn-sowing of cereals reduces breeding bird numbers in a heterogeneous agricultural landscape. *Biological Conservation*, 144, 1137–1144.

5.34. Undersow spring cereals, with clover for example

- Three studies from the UK (1–3,5), two replicated, found that there were higher densities of some study species on undersown fields or margins, compared with other fields, or that use of fields increased after they were undersown. One of these (reported in two places) found that not all species nested at higher densities (3,5). One replicated study from the UK (4) found that various measures of grey partridge population health declined as the amount of undersown cereal on sites increased.
- A replicated study from the UK (4) found no relationship between the amount of undersown cereals on a site and the productivity of grey partridge on that site.

A before-and-after study in Dumfries, southern Scotland (1), found that the number of barnacle geese *Branta leucopsis* on a mixed agricultural site and nature reserve increased from 3,200 in 1970 to 6,000 in 1975 after all cereals sown on the site were undersown from 1970 onwards. The nature reserve consisted of 220 ha of salt pasture, whilst the agricultural land was 50 ha of arable fields. Most of the extra geese fed on the arable land. In addition to undersowing, the proportion of cereals grown on the arable land decreased (see 'Increase crop diversity' for details) and no stock were allowed to graze on the arable land after November. The paper also discussed the impact of reducing grazing intensity, see 'Reduce grazing intensity on permanent grassland'.

A replicated, controlled study in summer 1995 on 89 fields in the South Downs, southern England (2), found that the density of singing Eurasian skylarks *Alauda arvensis* was higher on undersown spring barley fields than on any other field type (approximately 22 birds/km^2 on four spring barley fields vs. 2–15 birds/km^2 on 85 other fields). Other field types were arable fields reverted to species-rich grassland (see 'Habitat restoration and creation: Grasslands') or permanent grassland ('Revert arable land to permanent grassland'); downland turf (close-cropped, nutrient-poor grassland); permanent grasslands; winter wheat, barley and oil seed rape and set-aside ('Provide or maintain set aide areas in farmland').

A randomised, replicated, controlled trial on four farms in south-west England, in 2003–2006 (3), found that 12, 50 × 10 m plots of undersown spring barley attracted more small passerines (dunnock *Prunella modularis*, wren *Troglodytes* troglodytes, European robin *Erithacus rubecula*, seed-eating finches and buntings) than 12 control (not-undersown) plots. In addition, dunnocks, but not chaffinches or blackbirds, nested in hedgerows next to the sown plots more than expected, with 2.5 nests/km, compared to less than 0.5 nests/km in hedges next to experimental grass plots. Experimental plots were sown with spring barley *Hordeum vulgare* and a grass and legume mix, whereas control plots were managed as silage – cut twice in May and July, and grazed in autumn/winter. This study is also discussed in 'Reduce management intensity on permanent grassland', 'Reduce pesticide or herbicide use generally', 'Raise mowing height on grasslands', 'Reduce grazing intensity on permanent grasslands' and 'Plant wild bird seed or cover mixture'.

A replicated site comparison study on 1,031 agricultural sites across England in 2004–8 (4) found that various measures of grey partridge *Perdix perdix* population health declined as the amount of undersown cereal on sites increased. There were significant changes for year-on-year density changes in 2006–2007. When undersown cereals were combined with overwinter stubbles, overwinter survival rates were lower in sites with higher amounts of undersown cereals. There were no changes in brood size or the ratio of young to old birds.

A replicated study from April–July in 2006 on four livestock farms (3 replicates/farm) in south-west England (5) – the same study as (3) – found that dunnock *Prunella modularis*, but not Eurasian blackbird *Turdus merula* or chaffinch *Fringilla coelebs*, nested at higher densities in hedges alongside field margins sown with either wild bird seed crops or barley undersown with grass and clover, compared to those next to grassy field edges under various management options (dunnocks: approximately 2.5 nests/km for seed crops vs. 0.3/km for grass margins; blackbirds: 1.0 vs. 1.3; chaffinch: 1.5 vs. 1.4). Margins were 10 m wide, 50 m long and located adjacent to existing hedgerows. Seed crop margins were sown with barley (undersown with grass/legumes) or a kale/quinoa mix. There were 12 replicates of each treatment.

(1) Owen, M. (1977) The role of wildfowl refuges on agricultural land in lessening the conflict between farmers and geese in Britain. *Biological Conservation*, 11, 209–222.
(2) Wakeham-Dawson, A., Szoszkiewicz, K., Stern, K. & Aebischer, N. J. (1998) Breeding skylarks *Alauda arvensis* on Environmentally Sensitive Area arable reversion grass in southern England: survey-based and experimental determination of density. *Journal of Applied Ecology*, 35, 635–648.
(3) Defra (2007) *Potential for enhancing biodiversity on intensive livestock farms (PEBIL)*. Defra Report BD1444.
(4) Ewald, J. A., Aebischer, N. J., Richardson, S. M., Grice, P. V. & Cooke, A. I. (2010) The effect of agri-environment schemes on grey partridges at the farm level in England. *Agriculture, Ecosystems & Environment*, 138, 55–63.
(5) Holt, C. A., Atkinson, P. W., Vickery, J. A. & Fuller, R. J. (2010) Do field margin characteristics influence songbird nest-site selection in adjacent hedgerows? *Bird Study*, 57, 392.

5.35. Plant more than one crop per field (intercropping)

- A study from the USA (1) found that 35 species of bird used fields with intercropping, with four nesting, but that productivity from the fields was very low.

Background

Planting more than one crop in each field increases habitat heterogeneity at a smaller scale than increasing crop diversity at a landscape scale (see 'Increase crop diversity'). Heterogeneity is thought to be key for increasing farmland biodiversity and so planting multiple crops may help birds of different functional groups to use a single field.

A study on two arable farms in Iowa, USA, in May–August 1992–3 (1), found that 35 bird species used fields under an experimental intercropping system, with an average of 108 birds/count/100 ha. Three native species (red-winged blackbird *Agelaius phoeniceus*, common grackle *Quiscalus quiscula* and vesper sparrow *Pooecetes gramineus*) nested in the fields, but that only one nest of forty (2.5%) successfully fledged young. Destruction by farming activities was the largest cause of nest mortality (39%) followed by predation (29%). Desertion only occurred at 5% of nests. Strips were 4.6 m wide and contained corn, soybeans and oats as well as mammoth red clover *Trifolium pratense*.

(1) Stallman, H. R. & Best, L. B. (1996) Bird use of an experimental strip intercropping system in northeast Iowa. *The Journal of Wildlife Management*, 60, 354–362.

5.36. Revert arable land to permanent grassland

- All five studies looking at the effects of reverting arable land to grassland found no clear benefit to birds. The studies monitored birds (2,3) or grey partridges (1,5) in the UK and wading birds in Denmark (4). They included three replicated controlled trials (2–4).
- One of the studies, a controlled before-and-after study from the UK, showed that grey partridge numbers fell significantly following the reversion of arable fields to grassland (1).

Background

This intervention involves changing from an arable crop to sown agricultural grassland, to be used for grazing or silage. It is not the same as the creation of species-rich or other semi-natural grasslands, which is discussed in the chapter on habitat restoration and creation.

See also 'Provide or retain set-aside areas' for some studies where non-rotational set-aside land was sown with grass, but managed as set-aside rather than as permanent agricultural grassland.

A controlled before-and-after study in 1970–94 in a 28 km^2 area of arable farmland in Sussex, England (1), found that grey partridge *Perdix perdix* numbers declined rapidly on arable fields in 1987–94, following their reversion to grassland, beginning in 1987 (average of 6.5 coveys/km^2 in 1970–86 vs. 1.1 coveys/km^2 in 1987–94). There was a considerably smaller decline on arable fields that were not reverted to grassland (average of 4.9 coveys/km^2 in 1970–86 vs. 2.5 coveys/km^2 in 1987–94). The reversed fields went from being more favoured by partridges before reversion to less favoured afterwards, equating to a 23% per year decrease in relative habitat quality on the reversion fields.

A replicated, controlled study in the winters of 1994–7 on farmland in southern England (2) found that Eurasian skylark *Alauda arvensis*, corn bunting *Miliaria calandra* and meadow pipit *Anthus pratensis* were not consistently more abundant on arable land reverted to grassland than on intensively managed permanent grassland or winter wheat fields (4–11 birds/km^2 for skylarks on reverted fields vs. 0–10 and 1–8 birds/km^2 for permanent grassland and winter wheat; values were 0.1–0.2, 0 and 0–1 birds/km^2 for buntings and 0–1.1 0 and 0–4 birds/km^2 for pipits). Densities of rooks *Corvus frugilegus* did not differ across field types. Reverted arable fields were sown with agricultural grass mixtures and managed under specific guidelines, whilst the permanent grassland fields were mown frequently and fertilised. This study also describes the effects of 'Habitat restoration/creation' and is discussed in 'Create open patches or strips in permanent grassland'.

A replicated, controlled study in the spring and summer 1994–6 on 40 farms in southern England (3) found that arable fields reverted to permanent grassland held similar densities of Eurasian skylarks *Alauda arvensis* as winter wheat and intensively managed permanent grassland, except in summer 1994, when they held significantly higher densities, and summer 1995, when they held lower densities than winter wheat (summer 1994: 11.9 birds/km^2 on 65 reverted fields vs. 2.6 and 4.4 birds/km^2 for 29 and 47 fields of permanent grassland and winter wheat, respectively; summer 1995: 2.1 birds/km^2 for 15 reverted fields vs. 3.0 and 11.0 birds/km^2 for seven and 26 fields of permanent grassland and winter wheat; other seasons: 5.7–9.1 birds/km^2 vs. 3.6–4.0 and 8.5–13.0 birds/km^2). Densities of carrion crows *Corvus corone* tended to be higher on reverted land, significantly so in some seasons (1.8–4.8 birds/km^2 on reverted fields vs. 0–3.0 and 0–1.1 birds/km^2 for grassland and wheat) and rooks *C. frugilegus* were never found on winter wheat. Between 65 and 82 reverted arable fields were surveyed, each sown with agricultural grass mixtures and managed under specific guidelines, whilst the 15–29 permanent grassland fields were frequently mown and fertilised. Between 38 and 47 winter wheat fields were surveyed. This study is also described in 'Undersow spring cereals', 'Reduce grazing intensity on permanent grasslands', 'Habitat restoration and creation: Grasslands' and 'Provide or maintain set-aside areas in farmland'.

A replicated, controlled study in 615 grassland fields in Jutland, Denmark (4), found that the populations of four waders (northern lapwing *Vanellus vanellus*, black-tailed godwit *Limosa limosa*, common redshank *Tringa totanus* and Eurasian oystercatcher *Haematopus ostrolagus*) did not increase on restored grasslands (formerly croplands), whether or not they were under a scheme designed to increase water levels in fields. There were increases on some other field types. This study is discussed in detail in 'Raise water levels in ditches or grassland'.

A replicated site comparison study on 1,031 agricultural sites across England in 2004–8 (5) investigated the impact of the restoration of different grasslands on grey partridge *Perdix perdix*. However, the study does not distinguish between the impacts of grassland restoration, scrub restoration and control and rough grazing. Sites with more of the combined intervention had a lower proportion of young partridges in the population in 2008. This study describes the effects of several other interventions, discussed in the relevant sections.

(1) Aebischer, N. J. & Potts, G. R. (1998) Spatial changes in grey partridge (*Perdix perdix*) distribution in relation to 25 years of changing agriculture in Sussex, UK. *Gibier faune sauvage, Game Wildlife*, 15, 293–308.
(2) Wakeham-Dawson, A. & Aebischer, N. J. (1998) Factors determining winter densities of birds on environmentally sensitive area arable reversion grassland in southern England, with special reference to skylarks (*Alauda arvensis*). *Agriculture, Ecosystems & Environment*, 70, 189–201.
(3) Wakeham-Dawson, A., Szoszkiewicz, K., Stern, K. & Aebischer, N. J. (1998) Breeding skylarks *Alauda arvensis* on Environmentally Sensitive Area arable reversion grass in southern England: survey-based and experimental determination of density. *Journal of Applied Ecology*, 35, 635–648.
(4) Kahlert, J., Clausen, P., Hounisen, J. P. & Petersen, I. K. (2007) Response of breeding waders to agri-environmental schemes may be obscured by effects of existing hydrology and farming history. *Journal of Ornithology*, 148, 287–293.
(5) Ewald, J. A., Aebischer, N. J., Richardson, S. M., Grice, P. V. & Cooke, A. I. (2010) The effect of agri-environment schemes on grey partridges at the farm level in England. *Agriculture, Ecosystems & Environment*, 138, 55–63.

5.37. Reduce tillage

- Four replicated and controlled studies from North America and Canada (1–3) and the UK (6) and two literature reviews (4,5) found that some or all bird groups had higher species richness or diversity on reduced-tillage fields, compared to conventional field in some areas. Two replicated and controlled studies from Canada (3) and the UK (9) and a review (4) found that some measures of diversity were lower, or no different, on reduced-tillage fields, compared to conventional fields.
- Five replicated and controlled studies from the USA and Europe (1,2,6,7,10), a small study (8) and two reviews (4,5) all found that some bird species are found at higher densities on fields with reduced tillage than conventional fields. Five replicated and controlled studies from the USA, Canada and Europe (2,3,6,9,10), and a review (5) found that some or all species were found at similar or lower densities on reduced-tillage fields compared to conventional fields, with one (7) finding that preferences decreased over time (possibly due to extreme weather) and another (2) finding that preferences were only found in spring.
- Two controlled studies (one replicated) and a review (1,4,8) found evidence for higher productivity, nesting success or earlier laying on reduced tillage fields, compared to conventional fields. One controlled study found no evidence for greater success or larger chicks on reduced-tillage fields (8).

Background

Conventional ploughing uses a mould-board plough, cultivating to a depth of around 20 cm. This can damage soil structure and potentially reduce the abundance of soil invertebrates, a food source for many farmland birds. This intervention includes various methods to reduce the depth or intensity of ploughing, such as layered cultivation, non-inversion tillage and conservation tillage. It also includes stopping tillage altogether, sometimes called 'no till'.

A replicated, controlled study from May–July in 1982–1984 in nine experimental sites and three control sites in cropland in Iowa, USA (1), found that farmland bird species richness and nesting density and success were higher in fields without tillage. In total, 12 species were found nesting in the non-tillage fields with an average density of 36 nests/100 ha whereas only three species with an average of 4 nests/100 ha nested in tilled fields. Nest success was greatest in fields with corn residue (48% nestling survival rate). Three no-tillage treatments (corn planted in corn residue: 125 ha); corn planted in sod residue: 117 ha); and soybeans planted in corn residue: 113) and one control treatment (corn planted in tilled fields: 129 ha) were surveyed each year. Discovered nests were monitored every 2–3 days.

A replicated, controlled, site comparison study from 1991–1993 in ten reduced tillage, ten organic and ten conventional agricultural fields in North Dakota, USA (2), found that more farmland birds nested on reduced-tillage than conventional fields (1 nest/10 ha vs. 0.5 nests/10 ha). Minimum tillage fields also possessed a significantly greater diversity of nesting species (2 species/field vs. 1). In spring, bird densities in minimum tillage fallow fields were higher than those in organic fallow, minimum tillage sunflower and wheat fields and all conventional fields. There were no differences in bird abundance between treatments in other seasons but fallow fields (across treatments) exhibited the highest densities in summer (1–2 individuals/ha). There were no significant differences in nest loss or daily survival rate between treatments.

A replicated, controlled study from June–July in 1996–1997 in 37 conservation tillage, 40 organic, 38 conventional and 31 wild (control) sites in both upland and wetland areas of crop farms (75% wheat) in Saskatchewan, Canada (3), found that bird diversity and abundance were highest overall in wild areas compared to farmed areas, highest in conservation tillage farms in upland areas and in organic farms in wetland areas. In upland areas, of 37 species recorded, one was more abundant on farms, four more abundant in wild areas while the rest showed no distinct preference. Conservation tillage wetlands had similar bird communities to conventional wetland farms. Clusters of four treatments were located within a 25 km radius of one another. Fixed-radius (100 m) point-count surveys were used to survey twice per year.

A 2004 review of the effects of non-inversion tillage (NIT) on farmland bird abundance across the world, with special reference to the UK and Europe (4) found that the evidence for positive bird responses to NIT is inconclusive. Four studies in North America found higher bird density, diversity and nest productivity on NIT fields and another found greater bird diversity in summer on NIT fields (but not in autumn, winter or spring). Three studies found that Eurasian skylarks *Alauda arvensis*, gamebirds and seed-eating songbirds are more abundant on NIT fields. However, one study found that NIT fields act as ecological 'traps' when nests are destroyed by mechanical weeding. The authors point out that this type of weed control is not common in Europe.

A review of the effects of conservation tillage relative to conventional ploughing (5) found mixed effects for birds. One study showed no effect on five bird species in the context of organic farming. Another showed a higher number and diversity of birds on conservation tillage fields in Spain.

A replicated, controlled study in the winters of 2000–2003 in 63 experimental and 58 control winter wheat and barley fields in Oxfordshire, Leicestershire and Shropshire, UK (6), found that Eurasian skylarks *Alauda arvensis*, seed-eating songbirds and gamebirds occupied a significantly higher proportion of fields managed through non-inversion tillage than conventionally ploughed fields in late winter (January–March). Species richness of seed-eating songbirds was also higher on non-inversion tillage fields (five species vs. one on conservation tillage fields). No birds showed any preference for field type in early winter (October to December), and crows, pigeons and insect-eaters showed no preference across the study period. Field size ranged from 1.6 to 22.3 ha, with similar numbers of non-inversion tillage and conventionally ploughed farms censused each year.

A replicated controlled paired site study from October to March 2003–6 in 12 pairs of winter wheat fields in Dióskál, Hungary (7), found that the preference of some farmland birds for conservation tillage fields over adjacent ploughed fields decreased over the study period. In one farm (with eight field pairs), Eurasian skylarks *Alauda arvensis* and seed-eating songbirds (mostly European goldfinches *Carduelis carduelis*) were more abundant on conservation tillage fields in the first winter (2003–4), whilst starlings *Sturna vulgaris* and skylarks were more abundant on conservation tillage fields over the second and third winters respectively. In a second farm, with four fields, skylarks and crows were more abundant on conservation tillage fields in the first winter only. The number of days with ground snow cover increased over the three years. The authors suggest such abnormal weather may have confounded the results.

A small replicated, randomised, controlled study from April–July 2005 in two experimental and two control fields of winter wheat in Rutland, England (8), found that Eurasian skylark *Alauda arvensis* nest density was higher in fields managed through conservation tillage than control fields that were ploughed, with 24 of 32 nests found in conservation tillage fields. Average laying date was also significantly earlier on conservation tillage fields by 25 days. The authors suggest the effect was due to conservation tillage fields containing more crop residue than control fields (32% residue compared to none). Foraging distance of provisioning adult skylarks was 50% lower on conservation tillage fields (48 m vs. 93 m). However, nest success and nestling size were similar in both field types. Control fields were sown with winter wheat after mould-board ploughing, while conservation tillage fields were direct drilled into oil-seed rape residue after light rotary harrow.

A replicated, controlled study in the winters of 2006–8 in four (2006–7) and two (2007–8) fields of winter oilseed rape on a single farm in Cambridgeshire, UK (9), found that bird densities were similar between non-inversion tillage and control fields. Neither individual species nor groups of species responded to differences in crop establishment. However, the Farmland Bird Index (which included omnivores, carnivores, insect-eating birds and seed-eating species) was significantly higher on control fields. The authors point out that the overall densities on both treatments were still relatively low compared to other interventions (such as wild bird seed and over-winter cereal stubble). Two surveys were made in each field each month between September–March across the whole field area.

A replicated, controlled study from April–June in 2006–2007 in 48 conservation tillage, 31 organic and 63 conventional winter barley and wheat fields in Seine-et-Marne, France (10), found that that species differed in their responses to management. Two species were more abundant in conservation tillage fields than conventional fields, whilst seven were more abundant on conservation tillage fields than on organic. One species was more abundant on conventional fields and five on organic, compared to conservation tillage. Specialist species were least abundant on conservation tillage fields, whilst insect-eating birds were more abundant. The authors point out that conservation tillage fields were more intensely managed than conventional fields and experienced much disturbance.

(1) Basore, N. S., Best, L. B. & Wooley, J. B. (1986) Bird nesting in Iowa no-tillage and tilled cropland. *The Journal of Wildlife Management*, 50, 19–28.

(2) Lokemoen, J. T. & Beiser, J. A. (1997) Bird use and nesting in conventional, minimum-tillage and organic cropland. *The Journal of Wildlife Management*, 61, 644–655.

(3) Shutler, D., Mullie, A. & Clark, R. G. (2000) Bird communities of prairie uplands and wetlands in relation to farming practices in Saskatchewan. *Conservation Biology*, 14, 1441–1451.

(4) Cunningham, H. M., Chaney, K., Bradbury, R. B. & Wilcox, A. (2004) Non-inversion tillage and farmland birds: a review with special reference to the UK and Europe. *Ibis*, 146, 192–202.

(5) Holland, J. (2004) The environmental consequences of adopting conservation tillage in Europe: reviewing the evidence. *Agriculture, Ecosystems & Environment*, 103, 1–25.

(6) Cunningham, H. M., Bradbury, R. B., Chaney, K. & Wilcox, A. (2005) Effect of non-inversion tillage on field usage by UK – farmland birds in winter. *Bird Study*, 52, 173.

(7) Field, R. H., Benke, S., Bádonyi, K. & Bradbury, R. B. (2007) Influence of conservation tillage on winter bird use of arable fields in Hungary. *Agriculture, Ecosystems & Environment*, 120, 399–404.

(8) Field, R., Kirby, W. & Bradbury, R. (2007) Conservation tillage encourages early breeding by Skylarks *Alauda arvensis*. *Bird Study*, 54, 137–141.

(9) Dillon, I. A., Morris, A. J. & Bailey, C. M. (2009) Comparing the benefits to wintering birds of oil-seed rape establishment by broadcast and non-inversion tillage at Grange Farm, Cambridgeshire, England. *Conservation Evidence*, 6, 18–25.

(10) Ondine, F. C., Jean, C. & Romain, J. (2009) Effects of organic and soil conservation management on specialist bird species. *Agriculture, Ecosystems & Environment*, 129, 140–143.

5.38. Add 1% barley into wheat crop for corn buntings

- We have found no studies investigating the impact of adding barley to wheat on corn bunting *Miliaria calandra* populations.

Background
This is a suggested way of providing the preferred food source of corn buntings within a wheat crop.

5.39. Leave uncropped, cultivated margins or plots (includes lapwing and stone curlew plots)

- Two studies and two reviews examined population-level effects of uncropped margins or plots. A before-and-after study from the UK (3) and two reviews (1,7) found an increase in Eurasian thick-knee *Burhinus oedicnemus* numbers following a scheme that promoted plots (amongst other interventions); a replicated study from the UK (8) found no effect of plots on grey partridge density changes.
- Four studies (three replicated) and a review from the UK (2,4,6–8) found that at least one species was associated with lapwing plots or used them for foraging or nesting. One replicated study from the UK (2) found that 11 species were not associated with plots; another (9) found that fewer birds used the plots than cropland in two out of three UK regions.
- Two of the three studies that examined productivity (one replicated) (4,5) found that nesting success of birds was higher in fallow fields or lapwing plots than in crops. A replicated study from the UK (8) found that grey partridge *Perdix perdix* productivity was not related to the amount of lapwing plots on a site and that the proportion of young partridges in the population was lower on sites with lots of cultivated fallow plots.

Background
Lapwing and stone curlew plots are cultivated plots or strips that are left undrilled to encourage northern lapwings *Vanellus vanellus* and stone curlews (Eurasian thick-knees) *Burhinus oedicnemus* to nest successfully. They are normally >2 ha in size and different from 'skylark plots' (see separate section), which are much smaller and usually created in groups. Similar interventions include set-aside, which involves fields that are not cultivated at all.

A 2000 literature review (1) found that the UK population of Eurasian thick-knees *Burhinus oedicnemus* increased from 150 pairs in 1991 to 233 in 1999, following an agri-environment scheme designed to provide uncultivated plots in fields and set-aside.

A replicated study in 1999 and 2003 on 256 arable and pastoral fields across 84 farms in East Anglia and the West Midlands, England (2), found that only reed buntings *Emberiza schoeniclus* (out of 12 farmland birds analysed) were strongly and positively associated with uncropped, cultivated strips. No other species showed a strong association (positive or negative) with the strips. This study describes several other interventions, discussed in the relevant sections.

A before-and-after study of a Countryside Stewardship Scheme in southern England (3) found that the population of stone curlews (Eurasian thick-knees) *Burhinus oedicnemus* increased from 71 breeding pairs in 2000 to 103 in 2005, following the creation of 156 stone curlew plots over the study period. A further 51 were created in 2006 and the UK population of stone curlews increased from 160 pairs in the 1980s to 300 pairs in 2005. Stone curlew plots consisted of 1–2 ha of arable or set-aside land cultivated to create a 'rough fallow' in spring.

Preferably they should be located close (<1 km) to pasture, pig farms or other food sources and away from edges of fields.

A replicated, controlled study in the breeding seasons of 1999–2000 on 28 farms in western England (4) found that 85% of 34 northern lapwing *Vanellus vanellus* nests successfully hatched at least one chick on fields with cultivated 'lapwing plots', compared to 64% of 154 nests on all other fields types. Nest survival estimates were also significantly higher (99% daily survival vs. 96–96%), and no nests were lost to agricultural operations, compared to over 50% in other fields.

A study in 2003–5 in Cambridgeshire, UK (5), found that the nesting success of Eurasian skylarks *Alauda arvensis* was significantly higher in a field that was fallowed after harvest, compared to in cereal crop fields (84% success in the fallow field vs. 35%), whilst the number of nests in the field increased from two to eight following the fallow. Overwinter counts of yellowhammers *Emberiza citrinella*, reed buntings *E. schoeniclus*, linnets *Carduelis cannabina* and skylarks on the fallow field were also far higher than in previous years. This study is also discussed in 'Create skylark plots', 'Plant grass buffer strips/margins around arable or pasture fields' and 'Create beetle banks'.

A replicated study in 2007 (6) found that northern lapwings *Vanellus vanellus* used 39% of 212 lapwing plots on 180 farms across England, with breeding suspected on 25% of plots. In addition, Eurasian skylark *Alauda arvensis*, grey partridge *Perdix perdix* and yellow wagtail *Motacilla flava* were recorded breeding in 73%, 17% and 6% of plots respectively. There were no significant differences in lapwing occurrence or breeding in plots managed under Higher Level Stewardship compared with those under the Countryside Stewardship Scheme. Lapwing occurrence decreased if there was woodland adjacent, and the probability of breeding increased with the proportion of bare ground present on plots. Skylarks were less likely to be found on plots near hedgerows.

A 2009 literature review of agri-environment schemes in England (7) found that spring and summer fallows provided nesting habitats for northern lapwings *Vanellus vanellus*, with 40% of fallow plots used by lapwings and breeding suspected on 25%. In addition, the number of breeding pairs of Eurasian thick-knees (stone curlews) *Burhinus oedicnemus* in southern England increased from 63 in 1997 to 103 in 2005 following the implementation of a Country Stewardship Scheme 'special project', which included the provision of fallow plots. This review also examines several other interventions, discussed in the relevant sections.

A replicated site comparison study on 1,031 agricultural sites across England in 2004–8 (8) found a lower proportion of young grey partridges *Perdix perdix* in the population in 2007 on sites with a large area of uncropped but cultivated margins and plots. There were no significant relationships with changes in partridge density, brood size or overwinter survival. This study describes the effects of several other interventions, discussed in the relevant sections.

A replicated site comparison study on farms in three English regions (9) found that in two of the three regions Higher Level Stewardship fallow plots for ground-nesting birds had significantly fewer seed-eating farmland songbirds than conventional crop fields during summer. On farms in East Anglia and the Cotswolds, there were approximately 2.5 birds/ha on crops compared to 1 bird/ha on fallow plots. However, in a third region, the West Midlands, more seed-eating farmland birds were recorded on fallow plots than in crop fields (1.5 birds/ha on fallow plots compared to <0.5 birds/ha on crops). The group of birds analysed included tree sparrow *Passer montanus* and corn bunting *Emberiza calandra*, but not grey partridge *Perdix perdix*. Surveys were carried out in the summers of 2008 and 2009, on 69 farms with Higher Level Stewardship in East Anglia, the West Midlands or the Cotswolds and on 31 farms across all three regions with no environmental stewardship.

(1) Aebischer, N. J., Green, R. E. & Evans, A. D. (2000) From science to recovery: four case studies of how research has been translated into conservation action in the UK. 43–54 in: N. J. Aebischer, A. D. Evans, P. V. Grice, J. A. Vickery (eds) *Ecology and Conservation of Lowland Farmland Birds* British Ornithologists Union, Tring.

(2) Stevens, D. K. & Bradbury, R. B. (2006) Effects of the Arable Stewardship Pilot Scheme on breeding birds at field and farm-scales. *Agriculture, Ecosystems & Environment*, 112, 283–290.

(3) Evans, A. D. & Green, R. E. (2007) An example of a two-tiered agri-environment scheme designed to deliver effectively the ecological requirements of both localised and widespread bird species in England. *Journal of Ornithology*, 148, S279–S286.
(4) Sheldon, R., Chaney, K. & Tyler, G. (2007) Factors affecting nest survival of northern lapwings *Vanellus vanellus* in arable farmland: an agri-environment scheme prescription can enhance nest survival. *Bird Study*, 54, 168–175.
(5) Stoate, C. & Moorcroft, D. (2007) Research-based conservation at the farm scale: development and assessment of agri-environment scheme options. *Aspects of Applied Biology*, 81, 161–168.
(6) Chamberlain, D., Gough, S., Anderson, G., Macdonald, M., Grice, P. & Vickery, J. (2009) Bird use of cultivated fallow 'lapwing plots' within English agri-environment schemes. *Bird Study*, 56, 289–297.
(7) Natural England (2009) *Agri-environment schemes in England 2009. A review of results and effectiveness.* Natural England, Peterborough.
(8) Ewald, J. A., Aebischer, N. J., Richardson, S. M., Grice, P. V. & Cooke, A. I. (2010) The effect of agri-environment schemes on grey partridges at the farm level in England. *Agriculture, Ecosystems & Environment*, 138, 55–63.
(9) Field, R. H., Morris, A. J., Grice, P. V. & Cooke, A. I. (2010) Evaluating the English Higher Level Stewardship scheme for farmland birds. *Aspects of Applied Biology*, 100, 59–68.

5.40. Create skylark plots

- A before-and-after study found an increase in Eurasian skylark *Alauda arvensis* population on a farm after the creation of skylark plots (3); a replicated, controlled study from the UK found higher densities of skylarks on fields with plots, compared to those without (2). No other studies investigated population-level effects.
- Two UK studies (2,5), one replicated and controlled, found that skylark productivity was higher in plots or in fields with plots than in controls. One replicated and controlled study from Switzerland (6) found no differences in productivity between territories that included plots and those that did not.
- Two replicated studies (one controlled) from Denmark (1) and Switzerland (6) found that skylark plots were used by skylarks more than expected. A replicated and controlled study from the UK (7) found that seed-eating songbirds did not use skylark plots more than surrounding crops.

Background

Eurasian skylarks *Alauda arvensis* require short vegetation to nest in and skylark plots are small (usually 4–16 m^2) undrilled patches within cereal fields that provide this, with little impact on overall yield. They are similar to lapwing plots (see above) but much smaller.

A replicated study from April–May in 1990–3 in five spring-sown barley fields in eastern Jutland, Denmark (1) found that Eurasian skylarks *Alauda arvensis* used unsown plots in the fields significantly more than expected by an even distribution across the landscape. Radio-tracked birds were observed more in tramlines and unsown plots and mean faecal density was significantly higher in unsown areas than in crops (1.4 droppings/ha vs. 0.1). One 22 ha field with 100, 40 m^2 plots had higher densities of skylarks than four fields with an average of seven plots/ha, each of 7 m^2.

A replicated, controlled study from April–August in 2002–3 in 15 sites in northern and eastern England (2) found that Eurasian skylark breeding density, duration and success were higher in winter wheat fields with undrilled patches (4 × 4 m) than in fields with widely-spaced (25 cm apart) rows or under conventional management (0.3 nests/ha in fields with undrilled plots vs. 0.2 for the other treatments). Fields with undrilled patches also lost fewer territorial and nesting birds over the breeding season and by the end of the breeding season nests in these fields produced on average one more chick than control nests. Body condition of nestlings decreased in control nests over the breeding season but increased in experimental fields. The proportion of within-treatment foraging flights remained constant in fields with undrilled patches but decreased over time in other treatments.

A before-and-after study in Cambridgeshire, England (3), found that the population of Eurasian skylarks on an arable farm increased from 10 territorial males in 2000 to 34 in 2005, following the use of skylark plots from 2001 (in addition to 6 m margins around fields and set-aside). Nests were also aggregated in fields with skylark plots. The study also reports that fields on 15 experimental farms with skylark plots held 30% more skylarks than control fields. In addition, nests in fields with plots produced 0.5 more chicks/breeding attempt.

A replicated, controlled study in 2002–3 on ten farms in England (4) found that 45% of 159 Eurasian skylark nests monitored were found in fields with skylark plots. By June, fields with plots held 30% more skylarks and 100% more nests than control fields. At the start of the breeding season there was little difference in success between treatments, but by June fields with plots in had more nests (1 nest/ha vs. 0.4) and more chicks/nest than controls (1.75 chicks/nest vs. 0.9). Over the whole season nests in experimental fields raised 0.5 more chicks per breeding attempt (and 1.5 more late in the season) than controls.

A 2007 literature review (5) reports that on two experimental farms in the UK Eurasian skylarks were able to raise 49% more young in fields with skylark plots, compared to fields without plots, by prolonging the length of the breeding season. This study is also discussed in 'Leave uncropped, cultivated margins or plots, including lapwing and stone curlew plots', 'Plant grass buffer strips/margins around arable or pasture fields' and 'Create beetle banks'.

A replicated, controlled study near Berne, Switzerland (6) found that skylarks *Alauda arvensis* with territories that included undrilled patches were significantly less likely to abandon their territory than birds without patches, and more likely to use the undrilled patches as nesting and foraging sites than expected by chance. The study was from March–July in 2006 in 21 experimental sites and 16 control sites of winter wheat fields in mixed farming lands From June to July, the percentage of control fields in skylark territories decreased from 60% to 38%, whilst 55% of fields with undrilled patches remained in territories. Nest productivity was identical between control and fields with undrilled patches (1.4 chicks/territory) and there was no difference in chick body mass or tarsus length. Undrilled patches were composed of either 4 patches/ha (each 3 × 12 m, in seven fields) or a single strip (2.5 × 80 m, in 14 fields) sown with a mixture of six annual weed species. This study is also discussed in 'Plant nectar flower mixture/wildflower strips'.

A replicated site comparison study on farms in three English regions (7) found that skylark plots were well used (1–3 seed-eating farmland songbirds/ha) but did not have significantly more birds in than crop fields or fallow plots. Surveys were carried out on 69 farms with Higher Level Stewardship in East Anglia, the West Midlands or the Cotswolds and 31 farms across all three regions with no environmental stewardship.

(1) Odderskær, P., Prang, A., Poulsen, J. G., Andersen, P. N. & Elmegaard, N. (1997) Skylark (*Alauda arvensis*) utilisation of micro-habitats in spring barley fields. *Agriculture, Ecosystems & Environment*, 62, 21–29.

(2) Morris, A. J., Holland, J. M., Smith, B. & Jones, N. E. (2004) Sustainable Arable Farming For an Improved Environment (SAFFIE): managing winter wheat sward structure for skylarks *Alauda arvensis*. *Ibis*, 146 Supplement 155–162.

(3) Donald, P. F. & Morris, T. J. (2005) Saving the skylark: new solutions for a declining farmland bird. *British Birds*, 98, 570–578.

(4) Ogilvy, S. E., Clarke, J. H., Wiltshire, J. J. J., Harris, D., Morris, A., Jones, N., Smith, B., Henderson, I., Westbury, D. B., Potts, S. G., Woodcock, B. A. & Pywell, R. G. (2006) SAFFIE – research into practice and policy. *HGCA Conference: Arable crop protection in the balance: Profit and the environment.*

(5) Stoate, C. & Moorcroft, D. (2007) Research-based conservation at the farm scale: Development and assessment of agri-environment scheme options. *Aspects of Applied Biology*, 81, 161–168.

(6) Fischer, J., Jenny, M. & Jenni, L. (2009) Suitability of patches and in-field strips for skylarks *Alauda arvensis* in a small-parcelled mixed farming area. *Bird Study*, 56, 34–42.

(7) Field, R. H., Morris, A. J., Grice, P. V. & Cooke, A. I. (2010) Evaluating the English Higher Level Stewardship scheme for farmland birds. *Aspects of Applied Biology*, 100, 59–68.

5.41. Create corn bunting plots

- We have found no studies investigating the impact of corn bunting plots on corn bunting *Miliaria calandra* or other bird populations.

Background
Corn bunting plots are sown patches (normally 0.15 or 0.6 ha in size) of either grass or a cereal mix designed to provide nesting habitat for corn buntings *Miliaria calandra*.

5.42. Plant cereals in wide-spaced rows

- A replicated and controlled study from the UK (2) found that planting cereals in wide-spaced rows 'offered benefits over conventional wheat for Eurasian skylarks, but details were not given. Another replicated and controlled study from the UK (1) found that fields with wide-spaced rows had fewer skylark nests than control or skylark plot fields.
- A replicated and controlled study from the UK (3) found that the faecal content (and therefore diet) of skylark nestlings was similar between control fields and those with wide-spaced rows.

Background
Planting cereals in widely spaced rows can increase the proportion of habitat in the farmland that can be used by birds, as spaces between rows can be left fallow or planted with grass or legumes.

A replicated, controlled study from April–August in 2002–2003 in 15 winter wheat fields in northern and eastern England (1) found that Eurasian skylark *Alauda arvensis* nests were significantly less abundant on fields with wide-spaced rows than on control fields or those with undrilled patches (see 'Create skylark plots') (0.16 nests/ha vs. 0.18 for controls and 0.31 for those with undrilled patches). The proportion of within-treatment foraging flights decreased over time in control and wide-spaced fields but remained constant in fields with undrilled patches. Body condition of nestlings, however, decreased in control nests but increased in the other treatments over the breeding season.

A replicated, controlled study in 2002–2003 on ten farms in England (2) found that wide-spaced rows offered 'significant benefits' to Eurasian skylarks, but details were not given. The authors note that skylark plots (see 'Create skylark plots') were more consistently beneficial.

A replicated, controlled study from April–August in 2002–3 in 30 treatment and 30 control fields of winter wheat in northern and eastern England, UK (3) found no difference in faecal content of Eurasian skylark *Alauda arvensis* nestlings in fields with wide-spaced rows, compared to control fields.

(1) Morris, A. J., Holland, J. M., Smith, B. & Jones, N. E. (2004) Sustainable Arable Farming For an Improved Environment (SAFFIE): managing winter wheat sward structure for skylarks *Alauda arvensis*. *Ibis*, 146, 155–162.
(2) Ogilvy, S. E., Clarke, J. H., Wiltshire, J. J. J., Harris, D., Morris, A., Jones, N., Smith, B., Henderson, I., Westbury, D. B., Potts, S. G., Woodcock, B. A. & Pywell, R. G. (2006) SAFFIE – research into practice and policy. *HGCA Conference: Arable crop protection in the balance: Profit and the environment*.
(3) Smith, B., Holland, J., Jones, N., Moreby, S., Morris, A. J. & Southway, S. (2009) Enhancing invertebrate food resources for skylarks in cereal ecosystems: how useful are in-crop agri-environment scheme management options? *Journal of Applied Ecology*, 46, 692–702.

5.43. Create beetle banks

- A small UK study (3) found that a site with beetle banks had increasing populations of rare or declining species, although several other interventions were used on this site. A literature review from the UK (1) found that grey partridge *Perdix perdix* populations were far larger on sites with beetle banks and other interventions than on other farms. Two replicated studies from the UK also investigated population-level effects: one (4) found that no bird species were strongly associated with beetle banks; the second (6) found no relationship between beetle banks and grey partridge population density trends.

- A UK literature review (5) found that two bird species nested in beetle banks and that some species were more likely to forage in them than others. A study in the UK (2) found that one of two species used beetle banks more than expected. The other used them less than other agri-environment options.

Background

Beetle banks are grassy mounds, about 2 m wide, that run across the middle of large arable fields. They are intended to provide habitat, especially during winter, for predatory insects such as beetles and spiders and therefore could also provide foraging habitats for birds.

A 2000 literature review from the UK (1) found that the populations of grey partridge *Perdix perdix* was 600% higher on farms with conservation measures aimed at partridges in place, compared to farms without these measures. Measures included the provision of conservation headlands, planting cover crops, using set-aside and creating beetle banks.

A study of different set-aside crops at Allerton Research and Educational Trust Loddington farm, Leicestershire, (2) found that Eurasian skylarks *Alauda arvensis*, but not yellowhammer *Emberiza citrinella*, used beetle banks more than expected compared to availability. Skylarks used them significantly more than unmanaged set-aside, broad-leaved crops and other habitats, while yellowhammers used them significantly less than cereal and set-aside with 'wild bird cover'. Field margin and midfield set-aside strips were sown with kale-based and cereal-based mixtures for 'wild bird cover', and 'beetle banks'. Other habitat types were: unmanaged set-aside, cereal (wheat, barley), broad-leaved crop (beans, rape) and 'other' habitats. Thirteen skylark and 15 yellowhammer nests with chicks between 3–10 days old were observed. Foraging habitat used by the adults was recorded for 90 minutes during three periods of the day. This study is also discussed in 'Plant wild bird seed/cover and Provide (or retain) set-aside areas in farmland'.

A small replicated study from May–June in 1992–8 in Leicestershire, England (3), found that the abundance of nationally declining songbirds and species of conservation concern significantly increased on a 3 km^2 site where beetle banks were created (alongside several other interventions), although there was no overall difference in bird abundance, species richness or diversity between the experimental and three control sites. Numbers of nationally declining species rose by 102% (except for Eurasian skylark *Alauda arvensis* and yellowhammer *Emberiza citrinella*). Nationally stable species rose (insignificantly) by 47% (eight species increased, four decreased). The other interventions employed were: 'Manage hedges to benefit wildlife', 'Plant nectar flower mixture/wildflower strips', 'Plant wild bird seed cover strips', 'Provide supplementary food', 'Control predators' and 'Reduce pesticide or herbicide use generally'.

A replicated study in 1999 and 2003 on 256 arable and pastoral fields across 84 farms in East Anglia and the West Midlands, England (4), found that none of 12 species of farmland bird were strongly associated (either positively or negatively) with beetle banks. The species analysed were skylark *Alauda arvensis*, corn bunting *Miliaria calandra*, lapwing *Vanellus vanellus*, yellow wagtail *Motacilla flava*, chaffinch *Fringilla coelebs*, dunnock *Prunella modularis*, greenfinch *Carduelis chloris*, linnet *C. cannabina*, reed bunting *Emberiza schoeniclus*, tree sparrow *Passer montanus*, whitethroat *Sylvia communis* and yellowhammer *E. citrinella*. This study describes several other interventions, discussed in 'Leave headlands in fields unsprayed (conservation headlands)'; 'Leave uncropped, cultivated margins or plots, including lapwing and stone curlew plots'; 'Leave overwinter stubbles'; 'Plant grass buffer strips/margins around arable or pasture fields'; 'Create uncultivated margins around intensive arable or pasture fields'; 'Plant wild bird seed or cover mixture'; 'Provide or retain set-aside areas in farmland'; 'Pay farmers to cover the costs of conservation measures' and 'Reduce pesticide or herbicide use generally'.

A 2007 UK literature review (5) describes studies that found grey partridge *Perdix perdix* and Eurasian skylarks *Alauda arvensis* nesting in beetle banks. One study also found that skylarks were more likely than yellowhammers *Emberiza citrinella* to forage in beetle banks. This study is also discussed in 'Leave uncropped, cultivated margins or plots, including lapwing and stone curlew plots', 'Plant grass buffer strips/margins around arable or pasture fields' and 'Create skylark plots'.

A replicated site comparison study on 1,031 agricultural sites across England in 2004–8 (6) found that grey partridge *Perdix perdix* overwinter survival was significantly and positively correlated with the presence of beetle banks in 2007–8. Across all years there was a positive relationship with the ratio of young to old birds. There were no relationships with brood size or year-on-year density changes. This study describes the effects of several other interventions, discussed in the relevant sections.

(1) Aebischer, N. J., Green, R. E. & Evans, A. D. (2000) From science to recovery: four case studies of how research has been translated into conservation action in the UK. 43–54 in: N. J. Aebischer, A. D. Evans, P. V. Grice, J. A. Vickery (eds) *Ecology and Conservation of Lowland Farmland Birds* British Ornithologists Union, Tring.
(2) Murray, K. A., Wilcox, A. & Stoate, C. (2002) A simultaneous assessment of farmland habitat use by breeding skylarks and yellowhammers. *Aspects of Applied Biology*, 67, 121–127.
(3) Stoate, C. (2002) Multifunctional use of a natural resource on farmland: wild pheasant (*Phasianus colchicus*) management and the conservation of farmland passerines. *Biodiversity and Conservation*, 11, 561–573.
(4) Stevens, D. K. & Bradbury, R. B. (2006) Effects of the Arable Stewardship Pilot Scheme on breeding birds at field and farm-scales. *Agriculture, Ecosystems & Environment*, 112, 283–290.
(5) Stoate, C. & Moorcroft, D. (2007) Research-based conservation at the farm scale: Development and assessment of agri-environment scheme options. *Aspects of Applied Biology*, 81, 161.
(6) Ewald, J. A., Aebischer, N. J., Richardson, S. M., Grice, P. V. & Cooke, A. I. (2010) The effect of agri-environment schemes on grey partridges at the farm level in England. *Agriculture, Ecosystems & Environment*, 138, 55–63.

Livestock farming

5.44. Maintain species-rich, semi-natural grassland

- A before-and-after study from the UK (1) found five species of conservation concern increased after the implementation of management designed to maintain unimproved grasslands.
- A replicated study from Switzerland (2) found that wetland birds appeared to preferentially choose managed hay meadows; birds of open farmland avoided it.

Background

Low-intensity management of grasslands has produced some of the most species-rich habitats in Europe and there are several agri-environment schemes designed to maintain these grassland. Such schemes may include several different interventions, attempting to replicate traditional management.

A before and after trial in England (1) concluded that management prescriptions in the Exmoor Environmentally Sensitive Area are maintaining the condition of unimproved grassland, based on trends in bird populations in parts of the Environmentally Sensitive Areas under long term management agreements. The study found that five red/amber-listed species of conservation concern (linnet *Carduelis cannabina*, bullfinch *Pyrrhula pyrrhula*, grey partridge *Perdix perdix*, house sparrow *Passer domesticus* and garden warbler *Sylvia borin*) appeared to be increasing in density within the Cotswolds Environmentally Sensitive Areas while declining nationally, suggesting that they benefit from some aspect of Environmentally Sensitive Areas management. In each Environmentally Sensitive Areas, breeding birds were surveyed in May–August 2002, and results were compared with baseline survey information from 1992/3 (Exmoor) and 1997 (Cotswolds). In the Cotswolds Environmentally Sensitive

Area, birds were surveyed in 96 randomly-selected 1 km squares, while the majority (153 km^2) of the Exmoor Environmentally Sensitive Area was surveyed.

In a replicated site comparison study, Herzog *et al.* (2005) (2) found that on average 86% of litter meadows in Ecological Compensation Areas on the Swiss plateau were of 'good ecological quality' (based on national guidelines for Ecological Compensation Areas target vegetation), compared to only 20% of hay meadows. While wetland birds appeared to benefit from litter meadow Ecological Compensation Areas, with significantly more territories (52) than expected (31) in these areas, birds of open cultivated land had fewer territories (68) than expected (151) on hay meadow Ecological Compensation Areas. For hay meadow Ecological Compensation Areas, ecological quality was significantly lower in the more intensively farmed 'lowland' zone of the Swiss plateau, compared to 'pre-alpine hills' zone. Territories of breeding birds were mapped in 23 study areas, based on 3 visits between mid-April and mid-June. This study is also discussed in 'Maintain traditional orchards' and 'Manage hedges to benefit wildlife (includes no spray, gap-filling and laying)'.

(1) Defra (2002) Breeding bird survey of the Cotswold Hills ESA and Exmoor ESA. Defra, UK.
(2) Herzog, F., Dreier, S., Hofer, G., Marfurt, C., Schupbach, B., Spiess, M. & Walter, T. (2005) Effect of ecological compensation areas on floristic and breeding bird diversity in Swiss agricultural landscapes. *Agriculture, Ecosystems & Environment*, 108, 189–204.

5.45. Reduce management intensity on permanent grasslands

- Four replicated trials and a review (2–6), of seven studies in total, found that some or all birds monitored were more abundant or foraged more on grasslands with lower management intensity than on conventionally managed agricultural grasslands.
- Four analyses from three replicated trials (1–3,7), of seven studies in total, found that some or all birds monitored were less or similarly abundant on grasslands with lower management intensity than on conventionally managed agricultural grasslands.

Background

Reducing the intensity of grassland management involves one or more of: reducing or stopping the use of fertilisers, herbicides and pesticides; delaying the mowing date; reducing the number of cuttings taken.

A replicated controlled, paired site study in the Netherlands (1) found that the density of breeding bird territories was not significantly different between 20 fields with meadow bird agreements and 20 control fields, both for all bird species and just for waders. Oystercatcher *Haematopus ostralegus*, black-tailed godwit *Limosa limosa*, common redshank *Tringa totanus* and lapwing *Vanellus vanellus* were all significantly less abundant on management agreement fields than on control fields. There was no significant difference in the number of territories between field types for three of these species, but oystercatchers had significantly fewer territories on management agreement fields than on control fields (0.13 vs. 0.52). Paired fields were within 1 km of each other, similar in size and soil type. Fertiliser inputs were significantly lower and mowing dates later on fields with management agreements than on conventionally managed fields. Birds were surveyed five times between March and June.

Further analysis of the same data used in Kleijn *et al.* 2001 (2), found that wading birds were less abundant on fields under meadow bird agreements (average of seven birds and 1.3 territories on agreement fields vs. 12 and 2.1 on conventional fields), whilst meadow songbirds were more abundant on meadow bird agreement fields, when analysed as a 12.5 ha scale (9.9 birds/plot on agreement fields vs. 7.7 on conventional fields). Duck and non-meadow bird breeding densities did not differ between management types at either the field, or 12.5 ha scale.

A 2006 replicated site comparison study of 42 fields in Switzerland (3) found that more birds, but not more bird breeding territories, were found in fields participating in the ecological compensation area scheme than in conventionally farmed fields. There was no difference in the numbers of bird species on each type of farmland. Ecological Compensation Areas are typically hay meadows farmed at low intensity: no fertilisers or pesticides (except for patch-wise control of problem weeds) are permitted, and vegetation must be cut and removed at least once a year – but not before 15 June (lowlands) or early July (mountains). The study surveyed seven pairs of fields (one within an Ecological Compensation Area, one conventionally farmed) and a 1-ha area surrounding each field, from each of three different parts of Switzerland four times during the breeding season.

A randomised, replicated, controlled trial on four farms in south-west England, in 2003–2006 (4), found that 50 × 10 m plots of permanent pasture cut just once in May or July or not at all during the summer and left unfertilised attracted more insectivorous and seed-eating songbirds than control plots (fertilised plots cut in May and July, as in conventional silage management). The preference was shown by dunnocks *Prunella modularis*, winter wren *Troglodytes troglodytes*, European robin *Erithacus rubecula*, seed-eating finches and buntings, and was particularly strong for plots left uncut in summer. There were twelve replicates of each management type. This study is also discussed in 'Reduce pesticide or herbicide use generally', 'Undersow spring cereals', 'Raise mowing height on grasslands', 'Reduce grazing intensity on permanent grasslands' and 'Plant wild bird seed or cover mixture'.

A replicated, controlled before-and-after study in 615 grassland fields in Jutland, Denmark (5), found that permanent grasslands fields under an agri-environment scheme designed to increase water levels had significantly higher numbers of three species of wader (northern lapwing *Vanellus vanellus*, black-tailed godwit *Limosa limosa*, common redshank *Tringa totanus*) in 2004–2005 after the scheme was implemented, compared to in 1999–2001, before the scheme. Eurasian oystercatchers *Haematopus ostrolagus* did not increase and effects varied between restored and permanent grasslands, and between wet and dry fields. The scheme involved promoting wet grasslands (see 'Raise water levels in ditches or grassland') as well as reducing fertiliser inputs, grazing pressure and the period of mowing.

A review of four experiments on the effects of agri-environment measures on livestock farms in the UK (6) found two replicated trials in south-west England showing that reduced management intensity on permanent grasslands benefits foraging birds. Both found higher numbers of invertebrates, seed heads and foraging birds at lower management intensity (less fertiliser, less cutting, less grazing or a combination of these). One study was the PEBIL project, also reported in (4). The other was part of a Defra-funded project focussed largely on the effects of reduced grazing pressure (Defra report BD1454) for which no reference is given in the review. See 'Reduce grazing intensity on permanent grasslands' for more information.

A replicated site comparison study on farms in three English regions (7) found that grassland managed under Higher or Entry Level Stewardship Schemes with low or very low inputs was not used significantly more by seed-eating farmland songbirds than improved grassland or open rough grassland. Between 0.5 and 2 birds/ha were recorded on average on the different types of grassland. The stewardship grassland category also included land being maintained as semi-natural grassland under the schemes. It is not clear how many sites of the different management types were used in the analysis. Surveys were done in the summers of 2008 and 2009 on 69 farms with Higher Level Stewardship in East Anglia, the West Midlands or the Cotswolds and on 31 farms across all three regions with no environmental stewardship.

(1) Kleijn, D., Berendse, F., Smit, R. & Gilissen, N. (2001) Agri-environment schemes do not effectively protect biodiversity in Dutch agricultural landscapes. *Nature*, 413, 723–725.

(2) Kleijn, D., Berendse, F., Smit, R., Gilissen, N., Smit, J., Brak, B. & Groeneveld, R. (2004) Ecological effectiveness of agri-environment schemes in different agricultural landscapes in the Netherlands. *Conservation Biology*, 18, 775–786.

(3) Kleijn, D., Baquero, R. A., Clough, Y., Diaz, M., Esteban, J., Fernández, F., Gabriel, D., Herzog, F., Holzschuh, A., Jöhl, R., Knop, E. Kruess, A., Marshall, E. J. P., Steffan-Dewenter, I., Tscharntke, T., Verhulst, J., West, T. M. & Yela J. L. (2006) Mixed biodiversity benefits of agri-environment schemes in five European countries. *Ecology Letters*, 9, 243–254.

(4) Defra (2007) Potential for enhancing biodiversity on intensive livestock farms (PEBIL). Defra Report BD1444.
(5) Kahlert, J., Clausen, P., Hounisen, J. P. & Petersen, I. K. (2007) Response of breeding waders to agri-environmental schemes may be obscured by effects of existing hydrology and farming history. *Journal of Ornithology*, 148, 287–293.
(6) Buckingham, D. L., Atkinson, P. W., Peel, S. & Peach, W. (2010) New conservation measures for birds on grasslands and livestock farms. *BOU Proceedings – Lowland Farmland Birds III: delivering solutions in an uncertain world*. British Ornithologists Union.
(7) Field, R. H., Morris, A. J., Grice, P. V. & Cooke, A. I. (2010) Evaluating the English Higher Level Stewardship scheme for farmland birds. *Aspects of Applied Biology*, 100, 59–68.

5.46. Reduce grazing intensity

- Nine studies from the USA (1,2) and the UK (3–6,8,10,11), one replicated and controlled, found increases in populations of some species on fields with reduced grazing, or increased use of such fields by birds. Three of the studies used multiple interventions at once. Five studies from Europe (4,7–10), four replicated and controlled, found that some or all species were no more numerous on fields with reduced grazing, compared to intensively-grazed fields. One paired sites study from the UK (5) found that black grouse *Tetrao tetrix* populations increased at reduced grazing sites (and declined elsewhere), but that large areas of reduced grazing had lower densities of female grouse.
- A before-and-after study from the USA (2) found that the number of species on plots with reduced grazing increased over time. A replicated, controlled study from four countries in Europe (9) found no differences in the number of species on sites with low-intensity or high-intensity grazing.
- One replicated trial in the UK (10) found that some bird groups preferred grassland short in winter (grazing effect simulated by mowing), and others preferred it long (unmown to simulate removal of livestock). Frequency and timing of the simulated grazing did not alter this preference.

Background

Overgrazing is responsible for the degradation of habitats across the world, being especially damaging in arid environments, where the removal of vegetation can quickly lead to soil erosion. Reducing grazing intensity may reduce the damage to vegetation and can also help reduce disturbance to birds and accidental loss of nests.

A small 1967 study in Maryland, USA (1), investigated the impact of limiting livestock grazing, as well as other interventions, on northern bobwhites *Colinus virginianus* and found that the population on the farm increased from five to 38 coveys in eight years. This study is described in 'Threat: Agriculture – Plant new hedges'.

A before-and-after study in an 8,357 ha grassland site under rest-rotation grazing since 1967 in Montana, USA (2), found that the number of wildfowl nesting on the site, the species richness and the number of broods produced all increased between 1970 and 1973–4 (190 pairs of seven species producing 127 broods in 1970 vs. 270 pairs of 12 species producing 191 broods in 1974). The grazing regime involved five areas of the site being grazed at different times each year to allow the vegetation to recover. The highest densities of wildfowl were found in areas that had been rested in the previous year.

A before-and-after study in Gloucestershire, England, (3), found that the proportion of geese on a grassland site using a specifically managed 130 ha area increased from 33% in the winter of 1970–1971 to 87% by 1975–1976, following a reduction in grazing intensity over this period. Starting in 1970, stock were sequentially removed from three sections of the area: the first was ungrazed from the 30th September, the second from the 31st October and the third from the 30th November. A fourth area was not grazed at all. Other interventions are discussed in 'Increase crop diversity' and 'Undersow spring cereals'.

A randomised, replicated and controlled study in spring and summer 1995–6 on 12 fields in Sussex, England (4), found that Eurasian skylark *Alauda arvensis* densities were significantly higher on fields grazed at lower intensities (4.4–14.3 birds/km^2 for six lightly-grazed fields vs. 1.3–2.4 birds/km^2 on six intensely-grazed fields). The density of carrion crows *Corvus corone* and rooks *C. frugilegus* did not vary between treatments. Intensively-grazed fields were managed to keep the sward under 10 cm long, less intensively managed fields had a 15–25 cm sward. This study is also described in 'Undersow spring cereals', 'Revert arable land to permanent grassland', 'Habitat restoration and creation' and 'Provide or maintain set aide areas in farmland'.

A paired sites study on moorland in 1996–2000 in northern England (5) found that the number of displaying black grouse *Tetrao tetrix* males increased by an average of 5% each year at 10 sites where levels of sheep grazing were reduced, compared with average declines of 2% each year at ten control sites. Changes were most positive in the first years after grazing reduction. The proportion of females with chicks was also significantly higher at treatment sites (average of 54%) than at control sites (32%). However, there were declines in female densities at sites where restricted grazing areas exceeded approximately 1 km^2. Grazing was reduced to below 1.1 sheep/ha in summer and 0.5 sheep/ha in winter for at between one and five years on treatment sites. Densities were two or three times higher on control sites.

A before-and-after study of grazing marshes in east England from 1993–2003 (6) found that the number of northern lapwing *Vanellus vanellus* and wildfowl increased and vegetation communities changed following a reduction in grazing intensity and improved footdrain management in 1996. This study is discussed in 'Raise water levels in ditches or grassland'

A randomised, replicated, controlled trial on four farms in south-west England in 2003–6 (7) found that 12, 50 × 10 m plots of permanent pasture managed as conventional silage but without autumn/winter grazing did not attract more foraging birds than 12 control plots, managed identically but with autumn and winter grazing. Plots were fertilised and cut twice in May and July. This study is also discussed in 'Reduce management intensity on permanent grassland', 'Reduce pesticide or herbicide use generally', 'Raise mowing height on grasslands', 'Undersow spring cereals' and 'Plant wild bird seed or cover mixture' .

A controlled replicated trial in the UK (8) found that the response of bird populations to the removal of grazing from upland improved grassland between late May and July varied between functional groups of birds and depended on the time of year. Plots with seasonal removal of grazing had the greatest number of birds of songbird species between May and July (126 birds compared to 60 in control plots), and between July and September (312 birds compared to 169 in control plots), but numbers were similar to those in control plots between October and January (13 and 11, respectively). Between July and September, there were more birds of invertebrate-feeding species on plots with seasonal removal of grazing (105 birds, compared to 41 on control plots), but between October and January there were more birds on continuously grazed plots (5,833 birds, compared to 1,458 on plots with seasonal removal of grazing). At all times of year, crows were more abundant on continuously grazed plots. Bird numbers and species were recorded in plots with and without seasonal removal of grazing for silage making (10 replicates).

A replicated, controlled trial in four European countries (UK, Germany, France and Italy) from 2002–4 (9) found that numbers of birds and bird species were not different between fields under low-intensity grazing, compared to intensively-grazed fields. Birds were counted every two weeks in early morning, from May to October in 2002–4, with a 7 minute observation period and a walking transect. Exact grazing regimes differed between countries.

A randomised, replicated trial of different winter cutting regimes, designed to simulate grazing intensity on grasslands in Oxfordshire, England (10), found that different groups of birds prefer different treatments. Foraging song thrushes *Turdus philomenus* and common starlings *Sturnus vulgaris*, crows and Eurasian kestrels *Falco tinnunculus* preferred mown (grazed) plots to unmown (ungrazed) plots. Grey herons *Ardea cinerea* and meadow pipits *Anthus pratensis* preferred unmown plots to plots that were mown once or twice. For gamebirds, woodpigeons and hedgerow species, there was no significant difference in numbers between the different mowing regimes. Seventeen grass fields (average size 5 ha) were used in the experiment, with two treatments (mown once vs. unmown) or four treatments (unmown, mown once at two different times or mown twice) in each. Winter mowing simulates the effects of grazing or cutting for silage. Grass height did not differ between the 14 replicate plots mown once in November/December, once in January or twice during winter, so one winter cut or grazing period was sufficient to create the habitat advantage for bird groups that prefer short grass.

A 2009 literature review of agri-environment schemes in England (11) describes a case study of a farm on Exmoor, Devon, which found that three species increased on the farm from 1993–2003, following a reduction in grazing intensity on moorland areas (Eurasian skylark *Alauda arvensis* increased from none to 13 birds; Eurasian linnet *Carduelis cannabina* from none to nine birds; common stonechat *Saxicola torquata* from none to one territory). One species (meadow pipit *Anthus pratensis*) showed little change (nine birds vs. eight) and another (northern wheatear *Oenanthe oenanthe*) declined slightly, from one territory to none. This review also examines several other interventions, discussed in the relevant sections.

A review of UK experiments on the effects of agri-environment measures on livestock farms in the UK (12) found two replicated controlled trials that reduced grazing pressure (fewer cattle, cattle removed from July onwards, or both) over two to four years. One also reduced fertiliser input from 150 to 50 kg N/ha. Reduced grazing significantly increased the number of foraging skylarks *Alauda arvensis* on the trial fields in both studies. Birds that eat only seeds – European goldfinch *Carduelis carduelis* and linnet *Carduelis cannabina* – preferred plots with cattle removed in July. These studies formed part of a Defra-funded project (BD1454) for which no reference is given in the review. The study including low fertiliser input used eight replicates, the other used 14. The review assessed results from four experimental projects (one incomplete at the time of the review) in the UK. This study also discusses other interventions, described in the relevant sections.

(1) Burger, G. V. & Linduska, J. P. (1967) Habitat management related to bobwhite populations at Remington farms. *The Journal of Wildlife Management*, 31, 1–12.
(2) Mundinger, J. G. (1976) Waterfowl response to rest-rotation grazing. *The Journal of Wildlife Management*, 40, 60–68.
(3) Owen, M. (1977) The role of wildfowl refuges on agricultural land in lessening the conflict between farmers and geese in Britain. *Biological Conservation*, 11, 209–222.
(4) Wakeham-Dawson, A., Szoszkiewicz, K., Stern, K. & Aebischer, N. J. (1998) Breeding skylarks *Alauda arvensis* on Environmentally Sensitive Area arable reversion grass in southern England: survey-based and experimental determination of density. *Journal of Applied Ecology*, 35, 635–648.
(5) Calladine, J., Baines, D. & Warren, P. (2002) Effects of reduced grazing on population density and breeding success of black grouse in northern England. *Journal of Applied Ecology*, 39, 772–780.
(6) Smart, M. & Coutts, K. (2004) Footdrain management to enhance habitat for breeding waders on lowland wet grassland at Buckenham and Cantley Marshes, Mid-Yare RSPB Reserve, Norfolk, England. *Conservation Evidence*, 1, 16–19.
(7) Defra (2007) Potential for enhancing biodiversity on intensive livestock farms (PEBIL). Defra Report BD1444.
(8) Vale, J. E. & Fraser, M. D. (2007) Effect of sward type and management on diversity of upland birds. 333–336 in: J. J. Hopkins, A. J. Duncan, D. I. McCracken, S. Peel, J. R. B. Tallowin (eds) *British Grassland Society Occasional Symposium No.38* British Grassland Society, Reading.
(9) Wallis De Vries, M., Parkinson, A., Dulphy, J., Sayer, M. & Diana, E. (2007) Effects of livestock breed and grazing intensity on biodiversity and production in grazing systems. 4. Effects on animal diversity. *Grass and Forage Science*, 62, 185–197.
(10) Whittingham, M. J. & Devereux, C. L. (2008) Changing grass height alters foraging site selection by wintering farmland birds. *Basic and Applied Ecology*, 9, 779–788.
(11) Natural England (2009) *Agri-environment schemes in England 2009. A review of results and effectiveness.* Natural England, Peterborough.
(12) Buckingham, D. L., Atkinson, P. W., Peel, S. & Peach, W. (2010) New conservation measures for birds on grasslands and livestock farms. *BOU Proceedings – Lowland Farmland Birds III: delivering solutions in an uncertain world.* British Ornithologists Union.

5.47. Provide short grass for waders

- A replicated UK study (1) found that common starlings and northern lapwings spent more time foraging on short swards, compared to longer grass, and that starlings captured more prey in short grass.

Background

Vegetation height is important in determining the value of a grassland to wildlife, with short vegetation allowing birds access to the ground for foraging and potentially reducing predation risk. However, high vegetation can provide more complex environments and more habitats (see Raise mowing height on permanent grassland).

A replicated study from January-May in 2002 that observed 15 northern lapwing *Vanellus vanellus* chicks on the Isle of Islay, UK, and 20 common starlings *Sturnus vulgaris* in Oxfordshire, UK (1) found that both species experienced significantly greater foraging success in shorter grass swards. For lapwing chicks, foraging rate declined as sward height increased. In short swards, starlings spent 30% more time actively foraging and captured 33% more prey, although intake rate (captures per second of active foraging) did not differ between swards. Invertebrate abundance did not differ between long and short swards. Fertiliser application and water level was manipulated to provide a range of sward heights on the lapwing site. Starlings were observed in enclosures placed within intensively managed permanent pasture that was mown to either 3 cm (short sward) or 13 cm (tall sward).

(1) Devereux, C. L., Mckeever, C. U., Benton, T. G. & Whittingham, M. J. (2004) The effect of sward height and drainage on common starlings *Sturnus vulgaris* and northern lapwings *Vanellus vanellus* foraging in grassland habitats. *Ibis*, 146, 115–122.

5.48. Raise mowing height on grasslands

- A review from the UK (2) found that raising mowing height may have increased productivity of Eurasian skylarks, but not sufficiently to maintain the local population.
- A randomised, replicated and controlled study from the UK (1) found that no more foraging birds were attracted to plots with raised mowing heights, compared to plots with shorter grass.

Background

Vegetation height is important in determining the value of a grassland to wildlife. High vegetation can provide more complex environments and more habitats, but short vegetation can allow birds access to the ground which can help foraging, and can reduce the risk of predation (see Provide short grass for waders).

A randomised, replicated, controlled trial on four farms in south-west England in 2003–6 (1) found that 12, 50 × 10 m plots of permanent pasture cut to 10 cm in May and July did not attract more foraging birds than 12 control plots cut to 5 cm. Plots were cut twice in May and July, and grazed in autumn/winter. This study is also discussed in 'Reduce management intensity on permanent grassland', 'Reduce pesticide or herbicide use generally', 'Undersow spring cereals', 'Reduce grazing intensity on permanent grasslands' and 'Plant wild bird seed or cover mixture'.

A review of four experiments on the effects of agri-environment measures on livestock farms in the UK (2) found one trial from 2006 to 2008 that tested the effect of mowing height on skylarks *Alauda arvensis* nesting in silage fields. Preliminary results showed that chick survival was not affected by raised cutting height. However, the number of new birds produced each year (productivity) was more sensitive to re-nesting rates than chick survival. Raised cutting height slightly increased productivity, because skylarks re-nested sooner after cutting, but this was not enough to maintain a local population given survival rates. This study formed part of a Defra-funded project (BD1454) for which no reference is given in the review.

(1) Defra (2007) Potential for enhancing biodiversity on intensive livestock farms (PEBIL), Defra Report BD1444.

(2) Buckingham, D. L., Atkinson, P. W., Peel, S. & Peach, W. (2010) New conservation measures for birds on grasslands and livestock farms. *BOU Proceedings – Lowland Farmland Birds III: delivering solutions in an uncertain world*. British Ornithologists Union.

5.49. Delay mowing date or first grazing date on grasslands

- Two reviews from the UK (1,2) found that the population of corncrakes *Crex crex* increased following the implementation of two initiatives to encourage farmers to delay mowing (and provide cover and use corncrake-friendly techniques).
- A replicated and controlled paired sites studies from the Netherlands (3) found no evidence that waders and other birds were more abundant in fields with delayed mowing, compared to paired controls. A replicated and controlled before-and-after study from the Netherlands (4) found that fields with delayed mowing held more birds than controls, but did so before the start of the scheme. Population trends did not differ between treatments.
- A replicated, controlled study from the USA (5) found that destruction of nests by machinery was lower and late-season nesting higher in late-cut fields, compared with early-cut fields.

Background

Early-season, mechanised mowing is thought to be responsible for declines in the UK and elsewhere of species such as the corncrake *Crex crex*, with chicks killed and nests destroyed by mowing machinery. Delaying mowing until after chicks can escape is therefore a part of many agri-environment schemes.

A 2000 literature review (1) found that the UK population of corncrakes *Crex crex* increased from 480 to 589 males between 1993 and 1998 (an average rise of 3.5%/year) following schemes to get farmers to delay mowing dates and to leave unmown 'corridors' to allow chicks to escape to field edges which are thought to increase chick survival.

A 2002 review (2) states that the British population of corncrakes *Crex crex* increased by 34% between 1993 and 2001, following the implementation of the 'Corncrake Initiative' which financially compensates farmers who agree to delay mowing until after chicks can escape machinery. A second programme, begun in 1999, also included the provision of suitable cover. Both were based in western Scotland, where the remaining British population was found.

A replicated and controlled paired sites study in the western Netherlands in 2003 (3) found that 19 grassland plots with delayed mowing had significantly higher breeding densities of waders, compared to 19 paired, control plots (approximately 8 territories/plot for delayed-mowing plots vs. approximately 3 territories/plot for controls). This difference was not apparent when delayed mowing was combined with per-clutch payment, and there were no differences in abundances of waders or all bird species. However, when delayed mowing was combined with per-clutch payment, breeding densities of all bird species was significantly higher (13 territories/plot for combined schemes; 11 territories/plot for controls). There were higher numbers of redshank *Tringa tetanus* on combined plots (approximately 5 birds/plot for combined schemes; 5 birds/plot for per-clutch payment and 3 birds/plot for controls), but not on delayed-mowing plots. There were higher abundances of northern lapwing *Vanellus vanellus* on control plots, compared to delayed-mowing plots, but this difference was not significant (approximately 18 birds/plot for controls vs. 13 birds/plot for delayed-mowing plots). There were no significant differences in breeding densities for redshank, northern lapwing, Eurasian oystercatcher *Haematopus ostralegus* or black-tailed godwit *Limosa limosa*. The authors suggest that groundwater depth, soil hardness and prey density were drove these patterns. All farms had been operating the schemes for an average of four years before the study. This study is also discussed in 'Offer per-clutch payment for farmland birds'.

A replicated, controlled, before-and-after study in 1,040 grassland areas in the Netherlands, between 1990 and 2002 (4), found that nesting densities of black-tailed godwit *Limosa limosa* and redshank *Tringa totanus* were higher in areas with management agreements with postponed mowing, but these differences were present before the agreements came into effect. Population trends were similar between management and control areas for godwits and Eurasian oystercatchers *Haematopus ostralegus*, but northern lapwing *Vanellus vanellus* and redshank declined on management areas, relative to controls. Mowing was postponed on management areas to the end of May or beginning of June.

A replicated, controlled study in Arkensas, USA, in 2003 (5) found that a far higher percentage of grassland bird nests were destroyed by haying operations in two early-cut fields (cut from 26–31 May), compared to four late-cut fields (cut 17–26 June) (88% of 17 nests destroyed in early-cut fields vs. 4% of 52 nests destroyed in late-cut fields). The two surviving nests in early-cut fields did not fledge any chicks. Following early cutting, only one nest was started in early cut fields (0.03 nests/ha) compared with 0.13 nests/ha in uncut fields (seven nests) and 0.13 nests/ha in late-cut fields (11 nests). Nests were of dickcissel *Spiza americana* (32 nests), red-winged blackbird *Agelaius phoeniceus* (30 nests), field sparrow *Spizella pusilla* (14 nests) and eastern meadowlark *Sturnella magna* (13 nests) and nest densities were similar across field types before haying (0.3–0.5 nests/ha).

(1) Aebischer, N. J., Green, R. E. & Evans, A. D. (2000) From science to recovery: four case studies of how research has been translated into conservation action in the UK. 43–54 in: N. J. Aebischer, A. D. Evans, P. V. Grice, J. A. Vickery (eds) *Ecology and Conservation of Lowland Farmland Birds* British Ornithologists Union, Tring.
(2) Green, R. E. (2002) Corncrakes, conservation management and agri-environment schemes. *Aspects of Applied Biology*, 67, 189.
(3) Verhulst, J., Kleijn, D. & Berendse, F. (2006) Direct and indirect effects of the most widely implemented Dutch agri-environment schemes on breeding waders. *Journal of Applied Ecology*, 44, 70–80.
(4) Breeuwer, A., Berendse, F., Willems, F., Foppen, R., Teunissen, W., Schekkerman, H. & Goedhart, P. (2009) Do meadow birds profit from agri-environment schemes in Dutch agricultural landscapes? *Biological Conservation*, 142, 2949–2953.
(5) Luscier, J. D. & Thompson, W. L. (2009) Short-term responses of breeding birds of grassland and early successional habitat to timing of haying in northwestern Arkansas. *The Condor*, 111, 538–544.

5.50. Leave uncut rye grass in silage fields

- Two reviews from the UK (1,3) found that leaving rye grass uncut, or with only a single cut, benefited seed-eating birds and two replicated, controlled studies from the UK (2,4) found that seed-eating birds were more abundant on uncut plots.
- Two replicated and controlled studies (2,4) and a review (1), all from the UK, found that seed-eating birds were more abundant on uncut and ungrazed plots than on uncut and grazed plots.
- A replicated, controlled study from the UK (4) found that the responses of non-seed-eating birds were less certain than seed-eaters, with some species avoiding uncut rye grass.

Background

In the UK, seed-eating songbirds have declined across farmland, probably in part because of a lack of winter food. Rye grass *Lolium perenne* seeds are a potential food source, but cutting rye grass fields multiple times a year for silage removes seed heads before they can ripen and so reduces the food available to birds the following winter. Leaving fields or plots uncut may therefore provide valuable overwinter food.

A review of experiments on the effects of agri-environment measures on livestock farms in the UK (1) found that leaving rye grass silage uncut was shown to benefit seed-eating birds in winter in one experiment. No reference was given in the review for these results. The birds were only found in any numbers on plots left unmown, and were more abundant on plots left ungrazed rather than being grazed from September. Yellowhammers *Emberiza citrinella* and reed buntings *Emberiza schoeniclus* reached densities of 132 and 52 birds/ha respectively on unmown, ungrazed plots.

A replicated, controlled study of four silage fields on separate dairy farms in England (2) found that numbers of yellowhammer *Emberiza citrinella*, reed bunting *Emberiza schoeniclus*, wren *Troglodytes troglodytes*, song thrush *Turdus philomelos* and skylark *Alauda arvensis* were higher in plots left to set seed compared to mown plots, and in ungrazed seeded plots compared to grazed seeded plots. Significantly higher numbers of yellowhammer were observed in seeded plots (458 birds seen) compared to mown (one bird) and in ungrazed seeded plots (423) than grazed seeded plots (35). Reed buntings showed a similar response (seeded ungrazed: 160; grazed: 29; mown ungrazed: 3; grazed: 0). As did wren (seeded ungrazed: 22, grazed: 1; mown ungrazed: 2, grazed: 0) and song thrush (seeded ungrazed: 7, grazed: 3; mown ungrazed: 4, grazed: 0). There were more skylark in seeded than mown plots (18 vs. 0), and more in grazed (17) than ungrazed seeded plots (1). Two of four plots (0.5 ha) in each field were left uncut when the third silage cut was taken in July–August 2002 so that the grass set seed. One mown and one seeded plot was grazed by cattle until October, cattle were excluded from the other two plots. Numbers and species of birds using each plot were recorded over eight one hour periods between November 2002 and February 2003.

A review of four experiments on the effects of agri-environment measures on livestock farms in the UK (3) found that leaving rye grass *Lolium perenne* silage uncut was shown to benefit seed-eating birds in winter in one experiment. These are further results from a study discussed in Buckingham *et al.* (2004), with no reference given (Defra project BD1455). Only plots cut once during previous season produced large seed crops and attracted yellowhammers *Emberiza citrinella* (0.5 birds per visit on average) and reed buntings *E. schoeniclus*, (approximately 2 birds/visit on average) but not finches. Plots cut twice or three times (control) did not attract these birds. Birds were observed over two winters.

A replicated, controlled study on 12 farms in the West Midlands, UK (4), in the winters of 2007–9, found that seed-eating birds (yellowhammer *Emberiza citrinella* and reed bunting *E. schoeniclus*) preferentially foraged in rye grass fields that were only one cut once for silage and ungrazed, compared to twice cut (ungrazed) or control (two or more cuts and grazed) plots. Meadow pipits *Anthus pratensis* (which eat seeds and insects) did not show a preference for perennial rye grass fields under different treatments and showed a weak preference for other rye grasses that were only cut once. Insect-eating winter wrens *Troglodytes troglodytes* preferentially foraged in all treatments except controls. Insect-eating European robins *Erithacus rubecula* preferentially foraged on control plots.

(1) Buckingham, D. L., Atkinson, P. W. & Rook, A. J. (2004) Testing solutions in grass-dominated landscapes: a review of current research. *Ibis*, 146, 163–170.
(2) Buckingham, D. L. & Peach, W. J. (2006) Leaving final-cut grass silage *in situ* overwinter as a seed resource for declining farmland birds. *Biodiversity and Conservation*, 15, 3827–3845.
(3) Buckingham, D. L., Atkinson, P. W., Peel, S. & Peach, W. (2010) New conservation measures for birds on grasslands and livestock farms. *BOU Proceedings – Lowland Farmland Birds III: delivering solutions in an uncertain world*. British Ornithologists Union.
(4) Defra (2011) *Grass silage as a new source of winter food for declining farmland birds*. Defra, UK.

5.51. Plant cereals for wholecrop silage

- A replicated, controlled trial in the UK (1–3) found that seed-eating birds used CBWCS fields, especially those planted with barley, more than other crops in both summer and winter. Insect-eating species used other crops and grassland more.

Background

Cereal-based wholecrop silage (CBWCS) is an intervention that involves growing crops, not grass, to turn into silage. This may provide seed resources for grain-eating farmland birds throughout the year.

A replicated, controlled trial in 2004–2006 in north-west England (1) found that seed-eating songbirds and swallows and martins were more abundant on cereal (wheat and barley) fields planted in livestock areas than in grass and maize fields. In winter 2005/6, 1,390–1,564 seed-eaters were recorded on barley stubbles compared to 48 on grass fields and 406 on maize. Large insect-eating birds (thrushes) were far more abundant on grass fields in winter (2,272 birds in total, compared to 28–789 on other field types. Winter wheat and spring barley were sown in 16 trial fields, each on a separate farm in Cheshire, Staffordshire and north Shropshire. Neighbouring maize or short-term grass silage fields were monitored for comparison. Plants, invertebrates and birds were monitored on each field, in summer 2005 and winter 2005/06.

A review of four experiments on the effects of agri-environment measures on livestock farms in the UK (2) found one study of CBWCS in which winter wheat planted for silage was avoided by seed-eating birds during winter, but used as much as a control spring barley crop during summer. Maize planted for silage was little used by birds in summer or winter. These results are reported in more detail by Peach *et al.* (2011). This study also describes the results of several other interventions, discussed in the relevant sections.

An update of Mortimer *et al.* 2007 included data from winter 2004/5 (3) and found that CBWCS fields were used significantly more by farmland birds than other crop types. Each farm contained two CBWCS fields (autumn-sown wheat, 5.3 ha, and spring-sown barley, 4.4 ha), one maize field (6.1 ha) and one grass field (2.1 ha). During summer, a total of 1,535 seed-eaters and 1,901 swallows and martins were found on barley CBWCS fields, compared with 847 and 197 for wheat CBWCS fields; 441 and 95 for maize fields; and 41 and 480 for grass fields. Northern lapwing *Vanellus vanellus*, insect-eating species, and crows did not use CBWCS fields more than other types in summer. In winter, seed-eating species (seed-eating songbirds, Eurasian skylark *Alauda arvensis*, meadow pipit *Anthus pratensis*) used barley stubbles extensively, whilst insect-eating species used other crop stubbles more. The authors argue that CBWCS (with selectively applied herbicide, retention of over-winter stubbles and delayed harvesting) offer a practical conservation measure for seed-eating farmland birds. This study uses data from Defra report number BD1448 (Defra 2007).

(1) Mortimer, S., Westbury, D., Dodd, S., Brook, A., Harris, S., Kessock-Philip, R., Chaney, K., Lewis, P., Buckingham, D. & Peach, W. (2007) Cereal-based whole crop silages: potential biodiversity benefits of cereal production in pastoral landscapes. *Aspects of Applied Biology*, 81, 77–86.

(2) Buckingham, D. L., Atkinson, P. W., Peel, S. & Peach, W. (2010) New conservation measures for birds on grasslands and livestock farms. *BOU Proceedings – Lowland Farmland Birds III: delivering solutions in an uncertain world*. British Ornithologists Union.

(3) Peach, W. J., Dodd, S., Westbury, D. B., Mortimer, S. R., Lewis, P., Brook, A. J., Harris, S. J., Kessock-Philip, R., Buckingham, D. L. & Chaney, K. (2011) Cereal-based wholecrop silages: a potential conservation measure for farmland birds in pastoral landscapes. *Biological Conservation*, 144, 836–850.

5.52. Maintain lowland heathland

- We found no intervention-based evidence on the effects of maintaining lowland heath on bird populations.

5.53. Maintain rush pastures

- We found no intervention-based evidence on the effects of maintaining rush pastures on bird populations.

5.54. Maintain traditional water meadows

- A replicated study from the UK (1) found that northern lapwing and common redshank populations increased on nature reserves managed to maintain water meadows. Two replicated studies from the Netherlands (2,3) found that there were more waders or birds overall on specially managed meadows or 12.5 ha plots including several management interventions than on conventional fields, but one study (2) found that these differences were present before the management scheme was introduced and the other (3) found no differences between individual fields under different management.
- A replicated study from the UK (1) found that common snipe decreased on nature reserves managed to maintain water meadows and a replicated before-and-after study from the Netherlands (2) found that wader population trends on specially managed meadows were no different to those on conventionally-managed meadows.
- A replicated study from the UK (4) found that lapwing populations on three of four water meadow sites managed for conservation did not have high enough productivity to maintain population levels. All three sites were judged deficient in at least one management category.

Background

Water meadows are areas of grazing land or hay meadow that have carefully controlled water levels to keep the soil damp. In Europe they provide valuable breeding habitats for waders and other biodiversity. The studies below describe instances where multiple interventions have been used to maintain meadows. When the effects of multiple interventions, such as raising water levels and adding foot drains, can be separated, they are discussed under the relevant interventions in 'Threat: Natural system modifications'. The creation of new water meadows and the restoration of degraded ones are discussed in 'Habitat restoration and creation'.

A replicated study in 19 nature reserves established across England between 1983 and 1999 (1) found that the number of northern lapwing *Vanellus vanellus* and common redshank *Tringa totanus* on 13 nature reserves increased by 300% and 500% respectively in the first seven years following the initiation of management aimed at wading birds. Numbers then declined but were still higher than before the initiation of management. However, across all reserves, common snipe *Gallinago gallinago* declined, largely due to population collapses on reserves with mineral soils. Management included immediate changes to grazing (reduced during breeding seasons and adjusted to produce a favourable sward) and mowing (delayed until after nesting) and hydrological changes (raising water levels, surface flooding) introduced over two or more years. This study is also discussed in 'Pay farmers to cover the costs of conservation measures' and 'Legally protect habitats'.

A replicated, before-and-after site comparison study of 34 fields in Zeeland, the Netherlands (2), found no conclusive evidence that meadow bird conservation efforts resulted in higher territory numbers. Although there were significantly more meadow birds and territories of lapwing and black-tailed godwit *Limosa limosa* on fields managed for meadow bird conservation than on conventionally farmed fields in 1995, these differences were at least partly because those meadows in the bird agreements scheme also had higher groundwater levels. Moreover, population trends between 1989 and 1995 were similar for fields with and without meadow bird agreements, and the observed difference in settlement density in 1995 was also already present in 1989. 17 pairs of fields were matched for landscape structure and were surveyed in 1989, 1992 and 1995.

A 2006 replicated site comparison study of 42 fields in the Netherlands in 2006 (3) found that more birds bred on 12.5-ha scheme plots consisting of a mixture of fields with postponed agricultural activities and fields with a per-clutch payment scheme than on conventionally farmed plots. A survey of individual fields found there was no difference in bird abundance and breeding on those fields with postponed agricultural activities only and on conventionally farmed fields. The number of bird species on each type of farmland also did not differ between agri-environment schemes and non-agri-environment scheme plots. The agri-environment scheme, which intended to promote the conservation of Dutch meadow birds, prohibited changes in field drainage, pesticide application (except for patch-wise control of problem weeds) and any agricultural activity between 1 April and early June. Additionally, farmers of surrounding fields were paid for each meadow bird clutch laid on their land (though no agricultural restrictions were in place on these fields). The study surveyed seven pairs of fields (one within the agri-environment scheme, one conventionally farmed) and the 12.5-ha area surrounding each field, from each of three different parts of the Netherlands four times during the breeding season.

A replicated study in 2010 on four areas of wet grassland managed for wildlife in Kent, England (4), found that productivity of northern lapwings *Vanellus vanellus* was too low to sustain populations on three of the four (i.e. below 0.7 chicks/pair/year, which is thought to be the level necessary to maintain populations). The author identifies five management practices thought to be important for lapwing success: grazing regime; water availability; 'micro-topography' (changes in ground level to provide a range of habitats); reduced fertiliser inputs and predator control. At least one of these was rated as 'fair' or 'poor' in all three sites with low productivity.

(1) Ausden, M. & Hirons, G. J. M. (2002) Grassland nature reserves for breeding wading birds in England and the implications for the ESA agri-environment scheme. *Biological Conservation*, 106, 279–291.
(2) Kleijn, D. & van Zuijlen, G. J. C. (2004) The conservation effects of meadow bird agreements on farmland in Zeeland, The Netherlands, in the period 1989–1995. *Biological Conservation*, 117, 443–451.
(3) Kleijn, D., Baquero, R. A., Clough, Y., Diaz, M., Esteban, J., Fernández, F., Gabriel, D., Herzog, F., Holzschuh, A., Jöhl, R., Knop, E. Kruess, A., Marshall, E. J. P., Steffan-Dewenter, I., Tscharntke, T., Verhulst, J., West, T. M. & Yela, J. L. (2006) Mixed biodiversity benefits of agri-environment schemes in five European countries. *Ecology Letters*, 9, 243–254.
(4) Merricks, P. (2010) Lapwings, farming and environmental stewardship. *British Wildlife*, 22, 10–13.

5.55. Maintain upland heath/moor

- A literature review from the UK (1) found that agri-environment guidelines on moorland grazing were leading to increased bird populations in one region. There were localised problems with overgrazing, burning and scrub encroachment.

Background

Unpland heath and moorland is maintained through unenclosed upland grazing. This intervention includes grazing on acid grassland, dry and wet upland heath.

A 2009 literature review of agri-environment schemes in England (1) found studies that concluded that Environmentally Sensitive Area management prescriptions were having positive effects on moorland bird populations in Dartmoor Environmentally Sensitive Areas, UK. However, a study warned that localised problems such as overgrazing, burning or scrub encroachment were negatively affecting species such as tree pipit *Anthus trivialis*, whinchat *Saxicola rubetra* and ring ouzel *Turdus torquatus*. This review also examines several other interventions, discussed in the relevant sections.

(1) Natural England (2009) *Agri-environment schemes in England 2009. A review of results and effectiveness.* Natural England, Peterborough.

5.56. Plant Brassica fodder crops

- We found no evidence on the effects of planting brassicas on bird populations.

5.57. Use mixed stocking

- We found no evidence on the effects of mixed stocking on bird populations.

Background
Different livestock forage differently and so stocking multiple species in one area may help create a more diverse habitat.

5.58. Use traditional breeds of livestock

- A replicated controlled study in four European countries (1) found no differences in bird abundances between areas grazed with traditional or commercial breeds of livestock.

A replicated and controlled trial in four European countries (UK, Germany, France and Italy) from 2002–4 (1) found no differences in bird numbers between areas grazed with traditional breeds of livestock and those grazed by commercial breeds. Birds were counted every two weeks in early morning, from May to October, with a 7 minute observation period and a walking transect. The traditional breeds were Devon, German Angus and Salers, compared with commercial Charolais x Fresian, Simmental and Charolais, in the UK, Germany and France respectively. In Italy traditional Karst sheep were compared with commercial Finnish Romanovs. Animals were monitored in 2002, 2003 and 2004.

(1) Wallis De Vries, M., Parkinson, A., Dulphy, J., Sayer, M. & Diana, E. (2007) Effects of livestock breed and grazing intensity on biodiversity and production in grazing systems. 4. Effects on animal diversity. *Grass and Forage Science*, 62, 185–197.

5.59. Maintain wood pasture and parkland

- We found no intervention-based evidence on the effects of maintaining wood pasture and parkland on bird populations.

5.60. Exclude grazers from semi-natural habitats (including woodland)

- Two replicated (one controlled) studies from the USA (3,6) found higher species richnesses on sites with grazers excluded; a replicated and controlled study from Argentina (10) found lower species richness in ungrazed sites and a study from the USA (4) found no difference.

- Seven studies from the USA (three controlled, two replicated) found that overall bird abundance, or the abundances of some species were higher in sites with grazers excluded (1,3,5–8,11); seven studies from the USA (1,3,4,7,8,10,11) and Argentina found that overall abundance or the abundances of some species were lower on sites without grazers, or did not differ between treatments.

- Three studies from the USA investigated productivity (2,7,9) and found it higher in sites with grazers excluded. In one study (7) this difference was only found on improved, not unimproved pastures.

Background

Whilst grazing can be used to maintain early-successional habitats (see 'Natural System Modifications'), this may not always be desirable. In addition, over-grazing can be a severe problem and whilst a reduction in stocking densities (see 'Reduce grazing intensity') can reduce damage, sometimes livestock need to be excluded to allow vegetation to recover.

Excluding wild grazers and browsers is discussed in 'Threat: Invasive alien and other problematic species'.

A controlled study in 1981–1983 at a semi-desert grassland site in Arizona, USA (1), found that bird communities differed between an area from which cattle had been excluded since 1968 and one that had been continuously grazed. Total bird numbers were higher on grazed plots than ungrazed in summer, with no difference in winter (summer: 193 birds counted in ungrazed sites vs. 270 in grazed; winter: 242 birds in grazed vs. 247 in ungrazed). Open-ground foraging species were significantly more abundant in the grazed area, whilst species that prefer grass and shrub cover were the most abundant birds in protected sites, but absent on grazed pasture. The authors argue that the bird communities prevalent in grazed areas were more typical of lower elevations and dry habitats, and may be an indication of desertification of intensively grazed semi-desert and plains grasslands.

In a 1994 site comparison study in Little Valley, Nevada, USA (2), the nesting success rates of riparian bird species were found to be lower in an area grazed by cattle than an adjacent area rested from grazing for 30 years (grazed area: 83% of six above-ground nests successful and 67% of 12 ground nests predated; rested area 36% of 14 above-ground nests successful and 43% of seven ground nests predated). Experimental data from placing artificial nests baited with a Japanese quail *Coturnix japonica* egg and one painted plasticine egg in both areas showed a similar trend (daily survival rates of 55–95% of 120 eggs in grazed area vs. 77–98% of 120 in rested area). The authors suggest that grazing may facilitate nesting predation through changes in predator assemblage or increasing nest detectability.

A replicated, controlled study in 1991–1994 in semi-arid riparian habitats in Oregon, USA (3), found that bird species richness and relative abundance were significantly higher on three ungrazed 1.5 ha plots, compared to three grazed 1.5 ha plots (approximately 10–12 species/plot for ungrazed plots vs. 7–10 species/plot for grazed plots). In addition, ten species associated with riparian and wetlands habitats were found only on exclosure plots, and five species associated with uplands habitats only on open plots. Ungrazed plots had not been grazed for 30 years, whilst grazed plots were grazed until 1990. In the final year of study, four years after grazing had been stopped, key wet-meadow species (sora *Prozana carolina*, Wilson's phalarope *Phalaropus tricolor*, green-winged teal *Anas crecca*, and gadwall *A. strepera*) were found on open plots. Throughout the study, sedge cover, forb cover and foliage height diversity of herbs were greater within the exclosure; bare ground, litter cover, shrub cover and shrub foliage height diversity were greater on open plots.

A study from 1992–5 in New Mexico, USA (4), found no significant differences in songbird abundance or species richness between pinyon-juniper woodland sites that were actively grazed and sites from which livestock grazing had been excluded for 20 years (39 species on ungrazed sites, 36 on grazed). However, the authors argue that the slow growing woodland may not have had time to recover over the study period. One species, the western scrub-jay *Aphelocoma californica*, was more common on ungrazed sites. The authors note that over 75% of blue-gray gnatcatcher *Polioptila caerulea*, solitary vireo *Vireo solitarius* and western tanager *Piranger ludoviciana* nests were parasitised by brown-headed cowbirds *Molothrus ater*, raising concern that pinyon-juniper woodland habitat close to grazed areas could act as a population sink for songbirds due to cowbird parasitism.

A site comparison study from December–March in 1996–8 in oak savanna in Arizona, USA (5), found that 19 seed-eating birds were 270% more abundant in a livestock exclosure (former cattle ranch, ungrazed since 1968) than on a 'holistically managed' ranch, where 60 paddocks (covering 3,238 ha) were grazed intensively on a short rotation. Twenty-four other species (predators, fruit-eaters and insect-eaters) made up a smaller proportion of total bird abundance and did not differ in abundance between grazed and ungrazed sites. Grasses in the ungrazed area were significantly taller (4.4 times) and had higher basal-area ground cover (2.5 times) and higher overall canopy (2.2 times). The study sites were separated by a 7 km boundary fence, which was divided into 1 km sampling transects.

A replicated study in 1994–1995 in the Mojave Desert, California, USA (6), found that bird abundance and species richness were higher inside two 2.25 ha sites protected from sheep grazing and off-highway vehicles (OHV) since 1978, compared to adjacent sites that were grazed and driven over by OHVs. Significant differences were observed for sage sparrow *Amphispizia belli*, Le Conte's thrasher *Toxostoma lecontei*, loggerhead shrike *Lanius ludovicianus*, verdin *Auriparus flaviceps* and ash-throated flycatcher *Myiarchus cinerascens*. The authors suggest the increased abundance of bird species within the protected area is linked to a greater food supply.

A small controlled study from May–July in 1992–4 in river islands in Quebec, Canada (7), found that, in 1993, more duck nests than expected by an even distribution were found in idle fields, from which cattle were excluded, whilst fewer than expected were found on improved or unimproved pasture. In 1994, unimproved pasture held more than expected as well, but improved pasture held fewer. Nests on improved pasture had significantly lower success than those in other habitats (15% success of 39 nests vs. 47–82% elsewhere), with 33% being trampled. Nesting densities were no higher on idle areas than other habitat types.

A before-and-after study from 1986–1990 (8) found that more birds were detected in an area of riparian, mesquite and Chihuahuan desert-scrub in Arizona, USA, after the removal of cattle and the onset of a grazing moratorium in 1988 (average of 221 birds detected/km of transect in 1990 vs. 103 birds/km for 1986). Detections increased for 42 species, 26 significantly, and decreased for 19 species, eight significantly. Only four species in the study showed similar trends in regional Breeding Bird Surveys. Insectivores, granivores, midstory species, upperstory species and riparian species were most likely to increase, and migrants tended to show greater increases than residents. Chihuahuan desert-scrub species showed the smallest increases and were most likely to decline, possibly because the Chihuahuan scrub changed the least with the grazing moratorium. Surveys were conducted three times a month, every month over the study period.

A study in May–July of 2000 and 2001 in Kaibab National Forest, Arizona, USA (9), observed significantly higher fledging success rates of ground-nesting dark-eyed Juncos *Junco hyemalis* breeding in areas not grazed by cattle (48% of 21 nests) than in immediately adjacent, grazed areas (12% of 17 nests). The authors suggest that reduced nest cover may expose nests to more extreme climatic conditions and make them more conspicuous to predators.

A replicated, controlled study from December 2002 to March 2003 in 46 sampling transects (300 m long, 60 m wide, 1.8 ha, 2–40 km apart) across eight vegetation units and two grazing regimes (6 transects/vegetation unit; 3/grazing regime) in woodland, grassland and rocky habitats in the Córdoba Mountains, Argentina (10) found that bird species richness and abundance was significantly lower in livestock-excluded areas. Livestock exclusion reduced bird density and species richness across all vegetation units for all species and for endemic subspecies alone. Similarly, species richness was higher in grazed sites than in livestock-excluded areas for both insect-eating birds (5.0 compared to 3.8) and seed-eating birds (1.8 compared to 1.6 species/1.8 ha). Community composition was different between vegetation units, but not between grazing regimes. Traditional livestock management stocking rates ranged from 0.4–1.5 cattle equivalents/ha. Livestock exclusion areas were without cattle since 1998.

A study in northern Hawaii, USA (11), found that seven species in an open koa *Acacia koa* forest from which feral grazers were excluded showed long-term population stability or growth, but only two were increasing in a closed forest with grazers excluded. This study is discussed in 'Threat: Invasive and other problematic species – Reduce adverse habitat alterations by excluding problematic species' and 'Habitat restoration and creation – Forest restoration'.

(1) Bock, C. E., Bock, J. H., Kenney, W. R. & Hawthorne, V. M. (1984) Responses of birds, rodents, and vegetation to livestock exclosure in a semidesert grassland site. *Journal of Range Management*, 37, 239–242.
(2) Ammon, E. M. & Stacey, P. B. (1997) Avian nest success in relation to past grazing regimes in a montane riparian system. *The Condor*, 99, 7–13.
(3) Dobkin, D. S., Rich, A. C. & Pyle, W. H. (1998) Habitat and avifaunal recovery from livestock grazing in a riparian meadow system of the northwestern Great Basin. *Conservation Biology*, 12, 209–221.
(4) Goguen, C. B. & Mathews, N. E. (1998) Songbird Community composition and nesting success in grazed and ungrazed pinyon-Juniper woodlands. *The Journal of Wildlife Management*, 62, 474–484.
(5) Bock, C. E. & Bock, J. H. (1999) Response of winter birds to drought and short-duration grazing in southeastern Arizona. *Conservation Biology*, 13, 1117–1123.
(6) Brooks, M. (1999) Effects of protective fencing on birds, lizards, and black-tailed hares in the western Mojave Desert. *Environmental Management*, 23, 387–400.
(7) Lapointe, S., Giroux, J. F., Belanger, L. & Filion, B. (2000) Benefits of rotational grazing and dense nesting cover for island-nesting waterfowl in southern Quebec. *Agriculture, Ecosystems & Environment*, 78, 261–272.
(8) Krueper, D., Bart, J. & Rich, T. D. (2003) Response of vegetation and breeding birds to the removal of cattle on the San Pedro River, Arizona (USA). *Conservation Biology*, 17, 607–615.
(9) Walsberg, G. E. (2005) Cattle grazing in a national forest greatly reduces nesting success in a ground-nesting sparrow. *The Condor*, 107, 714–716.
(10) Garcia, C., Renison, D., Cingolani, A. M. & Fernandez-Juricic, E. (2008) Avifaunal changes as a consequence of large-scale livestock exclusion in the mountains of Central Argentina. *Journal of Applied Ecology*, 45, 351–360.
(11) Camp, R. J., Pratt, T. K., Gorresen, P. M., Jeffrey, J. J. & Woodworth, B. L. (2010) Population trends of forest birds at Hakalau Forest National Wildlife Refuge, Hawai'i. *The Condor*, 112, 196–212.

5.61. Protect nests from livestock to reduce trampling

- A before-and-after study from the Chatham Islands, New Zealand (1) found that the population of Chatham Island oystercatcher increased following several interventions including the erection of fencing around individual nests.
- A replicated, controlled study in Sweden (2) found that no southern dunlin nests were trampled when protected by cages; some unprotected nests were destroyed.

Background

As well as altering vegetation (see previous intervention), livestock can also reduce the breeding success of ground-nesting birds by trampling nests.

A study in the Chatham Islands from 1999 to 2005 (1) found that the number of Chatham Island oystercatcher *Haematopus chathamensis* pairs in a 14 km stretch of coastal land increased from 16 to 35 within six years, following several interventions including erecting 10 × 10 m enclosures of 1 m high electric fencing around individual nests to reduce disturbance and trampling by livestock. Other interventions used are discussed in the relevant sections.

A replicated, controlled study between 1999 and 2004 on pastures in south-west Sweden (2) found that none of 77 southern dunlin *Calidris alpina schinzii* nests protected with cages were trampled by cattle, whereas 31 of 291 unprotected nests (11%) failed because of grazing livestock. Cages were 20 cm high truncated cones with 7.5 cm gaps between vertical bars and 4 × 4 cm steel mesh covering the top. The effect of cages on predation of nests and adults is discussed in 'Threat: Invasive alien and other problematic species'.

(1) Moore, P. (2005) Stock fencing and electric fence exclosures to prevent trampling of Chatham Island oystercatcher *Haematopus chathamensis* eggs, Chatham Island, New Zealand. *Conservation Evidence*, 2, 76–77.
(2) Pauliny, A., Larsson, M. & Bloqvist, D. (2008) Nest predation management: effects on reproductive success in endangered shorebirds. *Journal of Wildlife Management*, 72, 1579–1583.

5.62. Mark fences to reduce bird collision mortality

- A randomised, replicated and controlled study from the UK (1) found that fewer birds collided with deer fence marked with orange netting than with unmarked sections.

Background

Fences erected around young plantations (to exclude deer and other browsers) or to delineate property can be hard to see and low-flying birds such as grouse can be killed by flying into them. See 'Mark power lines to reduce incidental mortality' for similar interventions designed to reduce collision mortality with power lines.

In a randomised, replicated and controlled study at thirteen sites in the Scottish Highlands from April 1995 to May 1997 (1), significantly fewer birds collided with sections of deer fence marked with orange netting (0.35 collisions/km/month) than with unmarked control sections (1.13 collisions/km/month). A total of 437 birds collided with the fences, 92% of which were gamebirds (red grouse *Lagopus lagopus scoticus* accounted for 42%, black grouse *Tetrao tetrix* 29% and western capercaillie *Tetrao urogallus* 20%). Collision rates in marked sections were 91% lower for black grouse and 64% lower for capercaillie than in control sections. A total of 20 km of ten different fences was tested, with two 1 km stretches of each fence being randomly assigned to treatments.

(1) Baines, D. & Andrew, M. (2003) Marking of deer fences to reduce frequency of collisions by woodland grouse. *Biological Conservation*, 110, 169–176.

5.63. Create open patches or strips in permanent grassland

- A randomised, replicated and controlled study from the UK (1) found that more Eurasian skylarks used fields with open strips in, but that variations in skylark numbers were too great to draw conclusions from this finding.

Background

Open patches and strips in permanent grassland can be used, in a similar way to skylark plots (see 'Create skylark plots') to provide short, open swards for ground-nesting birds.

A randomised, replicated and controlled trial on 14 fields in southern England in winter 1995–6 (1), found more Eurasian skylarks *Alauda arvensis* on seven fields that had open strips created in them, than in seven control fields, but the variation in numbers was so great that these differences were not significant (2–55 skylarks/km^2 on treated fields vs. 0 on controls). Open strips were created in a grid pattern, 25 m apart, using a tine-cultivator in November 1995. Experimental fields were still significantly more open in May 1996, but the swards had closed entirely by February 1997. This study is also described in 'Revert arable land to permanent grassland' and 'Habitat restoration and creation'.

(1) Wakeham-Dawson, A. & Aebischer, N. J. (1998) Factors determining winter densities of birds on environmentally sensitive area arable reversion grassland in southern England, with special reference to skylarks (*Alauda arvensis*). *Agriculture, Ecosystems & Environment*, 70, 189–201.

Perennial, non-timber crops

Agroforestry and silvopastoralism are the practices of growing crops or raising livestock under shade trees, which can be native and remnants of cleared vegetation, leguminous species that increase soil fertility, or fruit and other crop trees. These practices are often traditional forms of agriculture and are still widely practiced in many parts of the world, including much of the tropics.

Agroforestry provides a more complex habitat than many forms of farming and has been shown to support higher levels of biodiversity than monocultures. However, all the studies we found examined correlations between land-uses and biodiversity patterns, rather than studying the effects of converting agricultural land to agroforestry systems.

5.64. Maintain traditional orchards

- Two site comparison studies from the UK (1) and Switzerland (2) found that traditional orchards did not benefit many birds. In Switzerland, one species was associated with orchards; in the UK, the focal species was negatively related to the presence of orchards.

Background

Traditionally-managed orchards are low-intensity systems that have the potential to provide unique habitats for wildlife and that tend to hold older and rarer varieties of fruit. However, they are threatened in many countries, with 60% of traditional orchards in Britain having been lost and another 30% converted to intensive production since the 1950s. Studies investigating the impact of restoring or creating traditional orchards are described in 'Habitat restoration and creation'.

A 2001 paired site comparison study in south Devon (1) found that the presence of traditional orchards was associated with reduced cirl bunting *Emberiza cirlus* numbers. This effect, however, may have been at least partly because orchards typically had few areas of spring-sown barley and scrub clearance – both practices identified as benefiting cirl bunting. Traditional orchard management was encouraged as a prescription within the Countryside Stewardship Scheme (CSS). Forty-one 2×2 km^2 squares containing both land within the CSS and non-CSS land were surveyed in 1992, 1998 and 1999. In each year each tetrad was surveyed for cirl bunting at least twice, the first time during mid-April to late May, and the second time between early June and the end of August.

A replicated site comparison (2) found that on average only 12% of traditional orchards in Ecological Compensation Areas on the Swiss plateau were of 'good ecological quality' (based on national guidelines for Ecological Compensation Area target vegetation). Orchard Ecological Compensation Areas appeared to offer little benefit to orchard birds, with territories of only one species (green woodpecker *Picus viridis*) found more frequently in or near Ecological Compensation Area orchards (11 territories) than expected. Plant species and orchard characteristics were recorded for 187 Ecological Compensation Area orchards (total area 108 ha) between 1998 and 2001. Territories of breeding birds were mapped in 23 study areas, based on 3 visits between mid-April and mid-June. This study is also discussed in 'Manage hedges to benefit wildlife (includes no spray, gap-filling and laying)' and 'Maintain species-rich, semi-natural grassland'.

(1) Peach, W., Lovett, L., Wotton, S. & Jeffs, C. (2001) Countryside stewardship delivers cirl buntings (*Emberiza cirlus*) in Devon, UK. *Biological Conservation*, 101, 361–373.
(2) Herzog, F., Dreier, S., Hofer, G., Marfurt, C., Schupbach, B., Spiess, M. & Walter, T. (2005) Effect of ecological compensation areas on floristic and breeding bird diversity in Swiss agricultural landscapes. *Agriculture, Ecosystems & Environment*, 108, 189–204.

5.65. Manage perennial bioenergy crops to benefit wildlife

- We captured no evidence for the effects of managing bioenergy crops for wildlife on bird populations.

Background

Several perennial crops are grown solely for biomass, or conversion to fuel. They include elephant grass (Miscanthus) and willow (*Salix* sp.) grown as short rotation coppice.

Aquaculture

5.66. Reduce conflict with humans to reduce persecution

Background

Predation by birds at aquaculture facilities (e.g. fish ponds, raceways and shellfish farms) can cause significant commercial loss (Draulans 1987) and all the interventions discussed below are designed to reduce conflict between aquaculturists and birds, and therefore reduce persecution. With increasing protective wildlife legislation, demand for non-lethal, environmentally safe methods of bird exclusion and scaring have increased. Most fish farmers now rely primarily on non-lethal techniques to accomplish control (i.e. to reduce abundance or exclude fish-eating birds in and around the vicinity of fish farms). Control efforts may be optimised by compiling evidence relating to deterrent or exclusion device efficacy, taking into account costs, practicality of use and the possibility of developing integrated strategies (i.e. combining more than one deterrent method). Avian deterrents can be categorised as auditory, visual, chemical, exclusion, habitat modification and lethal (Bishop *et al.* 2003). Lethal deterrents are not considered here (except where, very rarely, used as part of an integrated approach). Presented here are synopses of numerous published studies and reviews; it is acknowledged that much information exists as grey literature and is not presented here.

Bishop, J., McKay, H., Parrott, D. & Allan, J. (2003) *Review of international research literature regarding the effectiveness of auditory bird scaring techniques and potential alternatives.* Department for Environment Food and Rural Affairs.

Draulans, D. (1987) The effectiveness of attempts to reduce predation by fish-eating birds: a review. *Biological Conservation*, 41, 219–232.

5.67. Scare birds from fish farms

- One before-and-after study from Israel (15) found that the population of pygmy cormorants in the area increased after birds were scared away from fish farms, possibly due to lower persecution.
- One of two studies that examined fish stocks found that fewer fish were taken from a farm when heron distress calls were played (2). The other study, a literature review (5), found no evidence for the effects of scaring birds on fish stocks.
- Two replicated studies from Belgium (4) and Australia (9) found that using foot patrols to disturb birds from fish farms did not reduce the number of birds present or fish consumption.
- Ten of eleven studies from across the world (1,2,5,6,8,11–13,15,16), three controlled, found evidence that playing distress calls or using other acoustic deterrents (some with flashes of light) reduced the number of birds at fish farms, or changed bird behaviours. One of these involved underwater broadcasting (13). One study found effects were only temporary (16) and five (2,3,6,8,11) found that birds became habituated to noises. Four studies (1,7,11,16), one replicated and controlled, two before-and-after, found that acoustic deterrents were not effective in scaring birds.
- Five of seven studies (7,8,10,12,14), one controlled, found evidence that visual deterrents (including inflatable 'Scarey Man' scarecrows) reduced the number of birds at fish farms. Three found evidence for habituation to deterrents (8,10,12) and three studies (3,5,12) found no evidence that visual deterrents were effective.
- Two studies examined other deterrents, finding that trained raptors were effective (5) but that the effects of helicopters and ultra-light aircraft were either inconclusive or very temporary (6).

A trial at a fish pond in western Germany in 1976 (1) found that broadcasting the flight call of a grey heron *Ardea cinerea* during daylight caused herons standing near the pond to take off and deterred flying herons from landing. However, at dawn trials, the majority (ten of 12) of herons were not deterred from landing. Preliminary trials had revealed that distress calls were more effective at scaring herons than heron alarm calls, combined heron and other bird alarm calls or a white-tailed sea eagle *Haliaeetus albicilla* call.

A controlled trial in northern Israel in September and October 1978 (2) found that 88% (1,122 of 1,265) of black-crowned night-herons *Nycticorax nycticorax* feeding at fishpond were scared off when heron distress calls (both adult and juvenile) were played on 12 observation nights. Over the study period there was no apparent habituation to the distress calls, in contrast to when recordings of a gas gun were used, when herons became habituated after only one night (60% of birds remained at ponds after 12 nights). Distress calls also reduced the number of herons perched in nearby trees by approximately 50% and, despite less than 5% of scared herons leaving the area, the scaring significantly reduced fish losses over the study period.

A replicated trial in the winter of 1962–3 in North Rhine-Westphalia, Germany (3), tested the effectiveness of visual and acoustic deterrents on deterring grey herons *Ardea cinerea* from fish ponds and found that a 'Flash-Harry' (wind-powered rotating orange cross on a pole) had no effect, whilst birds quickly became used to streamer bands, broadcasting of bird distress/ alarm calls and scarecrows. Shooting guns close to birds proved impractical. This study also investigated the use of netting on ponds, discussed in 'Use netting to reduce fish loss to birds'.

A replicated study from Limburg, Belgium, over 49 nights in 1982–3 (4), found that using foot patrols to disturb grey herons *Ardea cinerea* from 12 fish ponds did not necessarily reduce fish consumption. Low frequency disturbance (e.g. 3–5 farmer visits/night) caused a significant decrease in heron numbers but became less effective as heron numbers increased. Reduced numbers did not necessarily reduce fish consumption, as maximum predation occurred soon after bird arrival and disturbance mostly discouraged only well-fed birds from returning.

A 1987 literature review (5) found that there was little evidence that scaring devices at fish farms succeeded in increasing fish stocks/reducing losses. Devices could be classified as visual (scarecrows, flags, reflectors, lights, model etc), acoustic (gun shots, firecrackers and gas cannons) or biological (recordings of distress calls etc). Black-crowned night herons *Nycticorax nycticorax* were deterred by distress calls but only close to the speaker and no data are presented to support any positive effects on fish stocks. The use of dead birds, model predators and dogs is reported by several authors as almost completely ineffectual. Trained raptors seemed effective but expensive.

A series of experiments in Flevoland, the Netherlands, in 1981–3 (6) found that pistol-fired flash cartridges (detonation after a light flash, or a flash only) appeared the most effective method of deterring great cormorants *Phalacrocorax carbo* from ponds and scared most birds away (although some alighted on nearby ponds). Gas cannons (producing a bang at regular or irregular intervals) had little effect as birds soon habituated to the noise; an overflying helicopter scared cormorants from ponds but they soon returned (on the day following the 2-day trial large numbers were present). An ultra-light aircraft proved inconclusive.

A before-and-after trial at 25 fish ponds in a catfish farm over 45 days in February and March 1992 in the Mississippi delta region, USA (7), found that using a 'Scarey Man' resulted in a rapid decrease in double-crested cormorant *Phalacrocorax auritus* numbers (320 birds/ patrol before treatment vs. 8/patrol for the first seven days after erection and 16/patrol for the whole 46 day experiment). Clothing the devices to resemble marksmen, replacing them with actual marksmen and using propane gas exploders (at up to six Scarey Man positions for 23 days) did not further reduce cormorant numbers.

A 1995 review assessed effectiveness of techniques used to prevent double-crested cormorant *Phalacrocorax auritus* predation at aquaculture facilities in the Mississippi delta region, USA (8), and concluded that there was little good evidence for what worked and what did not. Pyrotechnics, human effigies, gas cannons, and live ammunition have been used with varying degrees of success in frightening cormorants, but the authors warn that birds can become habituated to them.

A replicated trial in New South Wales, Australia (9), found that hanging gillnets in fish ponds and using harassment patrols to deter cormorants *Phalacrocorax* spp. from fish farms was not effective. This study is discussed in 'Use in-water devices to reduce fish loss from ponds'.

A replicated, controlled, paired sites study in winter-spring 1991 at four pairs of catfish farms in the Mississippi delta region, USA (10), found a 71–99% reduction in double-crested cormorant *Phalacrocorax auritus* numbers following the deployment of six 'Scarey Man' devices for 10–19 days. However, signs of habituation became apparent (reduced flush success) within the trial period at three sites.

A before-and-study in May–June 1993 at a trout-rearing farm in Colorado, USA (11), found a 48% reduction in black-crowned night heron *Nycticorax nycticorax* numbers following the broadcasting heron alarm/distress calls for 11 days (pre-treatment average 77 birds; treatment 40; post-treatment 69). However, numbers and the proportion remaining increased from nights over the treatment phase, indicating habituation. Great blue heron *Ardea herodias* numbers were unaffected (pre-treatment 15; treatment 13; post-treatment 16. Herons were counted during six pre-treatment (12–13 to 18–19 May 1993), five treatment (21–22 May to 31 May-1 June 1993), and five post-treatment (1–2 to 11–12 June 1993) nights. Calls were broadcast through each night of the 11 day treatment period: 15-sec sequences of night heron calls followed by 14 min without calls, then a similar sequence of great blue heron calls.

A series of before-and-after trials in trout farms in Colorado, USA, in 1990–2 (12), found that pyrotechnics were effective at decreasing the number of black-crowned night-herons *Nycticorax nycticorax* and great blue herons *Ardea herodia* at farms. Firing pyrotechnics for 14 consecutive nights was more successful than doing so for seven nights. Frightening every fifth night was unsuccessful. Rotating lights did not reduce the number of attempted or successful fish captures. 'Scarey Man' reduced heron numbers during the first four nights but numbers of both species subsequently increased substantially to night 18, indicating habituation.

A controlled, replicated before-and-after experiment in January–April and October 1998, and March 1999, in Argyll, Scotland (13), found that an underwater playback system (UPS) was effective in deterring common eider *Somateria mollissima* from feeding on mussels *Mytilus edulis* at farms on two sea lochs (47–80% fewer birds feeding after use of the UPS; 2–37 birds feeding before use). Underwater recordings of an approaching 'scare boat' (scaring by boat being a conventional deterrent method) were played via an underwater loudspeaker, and also a 'control' i.e. playback of an unassociated sound. Average return time of eiders after chasing by boat also increased significantly, suggesting that effectiveness was strengthened by UPS.

Replicated *ex situ* experiments in Ohio, USA (14), found that µ10 mW, 633 nm laser did not repel brown-headed cowbirds *Molothrus ater* or European starlings *Sturnus vulgaris* from a perch over three trials with stationary and moving laser beams treating a randomly selected perch. Effectiveness of a 68 mW, 650 nm laser in dispersing starlings and rock doves *Columba livia* from perches, and Canada geese *Branta canadensis* and mallard *Anas platyrhynchos* from grass plots was also tested. Starlings did not disperse when targeted with the beam, doves dispersed only in the first 5 min of six 80 min treatment periods. An average, 96% of individual geese in six groups of four birds, dispersed from laser-treated plots during 20-min periods (23 replicates). Mallard dispersed (average 57% of individuals) during 20-min treatment periods, but habituated to the beam after about 20 min.

A before-and-after study in northern Israel (15) found that pygmy cormorants *Phalacrocorax pygmeus* relocated away from colonies near fish farms during 1999–2002, following the use of gas cannons and pyrotechnics to scare birds before the start of nesting in winters between 1999–2000 and 2002–3. Between 1998 and 2004, the overall number of cormorant nests in the area increased from 60 to approximately 110 (reaching a high of approximately 155 in 2001), possibly due to greater reproductive success with lower levels of persecution following relocation.

At four sites around Lake Como, Italy, a controlled replicated experiment (16) found that none of three deterrents (gas cannon detonations, fire crackers and shooting near birds) were effective or practical in deterring great crested grebes *Podiceps cristatus* from areas with commercially important common bleak *Alburnus alburnus* shoals. Grebe behavioural response was recorded during 3-h observation periods when deterrents were or were not, in use. Cannons had little effect. Crackers and shooting caused significant behavioural changes (less time feeding, resting and preening, and more time swimming) compared to control periods, but grebe numbers were only temporarily reduced in the vicinity.

(1) Von R. Behlert, H. (1977) Phonoakustische methode zur vergrämung von Graureihern (*Ardea cinerea*) in Fischzuchtanlagen. *Zeitschrift für Jagdwissenschaft*, 23, 144–152.
(2) Spanier, E. (1980) The use of distress calls to repel night herons (*Nycticorax nycticorax*) from fish ponds. *Journal of Applied Ecology*, 287–294.
(3) Ueckermann, E., Spittler, H. & Graumann, F. (1981) Technische Maßnahmen zur Abwehr des Graureihers (*Ardea cinerea*) von Fischteichen und Fischzuchtanlagen. *Zeitschrift für Jagdwissenschaft*, 27, 271–282.
(4) Draulans, D. & Van Vessem, J. (1985) The effect of disturbance on nocturnal abundance and behaviour of grey herons (*Ardea cinerea*) at a fish-farm in winter. *Journal of Applied Ecology*, 22, 19–27.
(5) Draulans, D. (1987) The effectiveness of attempts to reduce predation by fish-eating birds: a review. *Biological Conservation*, 41, 219–232.
(6) Moerbeek, D. J., Van Dobben, W. H., Osieck, E. R., Boere, G. C. & Bungenberg de Jong, C. M. (1987) Cormorant damage prevention at a fish farm in the Netherlands. *Biological Conservation*, 39, 23–38.
(7) Stickley Jr, A. R. & King, J. O. (1993) Long-term trial of an inflatable effigy scare device or repelling cormorants from catfish ponds. 89–92 *Proceedings of the Eastern Wildlife Damage Control Conference*, 6, 89–92.
(8) Mott, D. F. & Boyd, F. L. (1995) A review of techniques for preventing cormorant depredations at aquaculture facilities in the southeastern United States. *Colonial Waterbirds*, 18, 176–180.
(9) Rowland, S. J. (1995) Predation of *Bidyanus bidyanus* (Teraponidae) in ponds by cormorants. *The Progressive Fish-Culturist*, 57, 248–249.
(10) Stickley, A. R., Mott, D. F. & King, J. O. (1995) Short-term effects of an inflatable effigy on cormorants at catfish farms. *Wildlife Society Bulletin*, 23, 73–77.
(11) Andelt, W. F. & Hopper, S. N. (1996) Effectiveness of alarm–distress calls for frightening herons from a fish rearing facility. *The Progressive fish-culturist*, 58, 258–262.
(12) Andelt, W. F., Woolley, T. P. & Hopper, S. N. (1997) Effectiveness of barriers, pyrotechnics, flashing lights, and Scarey Man® for deterring heron predation on fish. *Wildlife Society Bulletin*, 25, 686–694.
(13) Ross, B. P., Lien, J. & Furness, R. W. (2001) Use of underwater playback to reduce the impact of eiders on mussel farms. *ICES Journal of Marine Science: Journal du Conseil*, 58, 517.
(14) Blackwell, B. F., Bernhardt, G. E. & Dolbeer, R. A. (2002) Lasers as nonlethal avian repellents. *The Journal of Wildlife Management*, 66, 250–258.
(15) Nemtzov, S. C. (2005) Relocation of pygmy cormorants *Phalacrocorax pygmeus* using scare tactics to reduce conflict with fish farmers in the Bet She'an Valley, Israel. *Conservation Evidence*, 2, 3–5.
(16) Gagliardi, A., Martinoli, A., Preatoni, D., Wauters, L. A. & Tosi, G. (2006) Behavioral responses of wintering great crested grebes to dissuasion experiments: Implications for management. *Waterbirds*, 29, 105–114.

5.68. Disturb birds at roosts

- One controlled study from the USA (2) investigated the effects of harassment on fish predation, and found there were fewer double-crested cormorants *Phalacrocorax auritus* on, and fewer fish were taken from, fish ponds near roosts which were harassed, compared with undisturbed roosts.

- A review (1) found that there was a reduction in the number of cormorants foraging near roosts after night-time disturbance. Four studies, two replicated, from the USA and Israel (2–5), found that cormorants moved away from roosts where they were disturbed at night. One study (4) found that this effect was only temporary.

A 1995 review assessed effectiveness of techniques used to prevent double-crested cormorant *Phalacrocorax auritus* predation at aquaculture facilities in the Mississippi delta region, USA (1), and concluded that disturbing birds at their roosts was more effective than scaring birds from fish farms during the day, with one study finding a 75–90% reduction in cormorant numbers foraging in the area after disturbance.

A controlled experiment over three winters in the vicinity of channel catfish *Ictalurus punctatus* rearing ponds in the Mississippi delta region, USA (2), found that the number of double-crested cormorants *Phalacrocorax auritus* on or near fish ponds was reduced by approximately 70% following a winter of extensive harassment at roosts. Catfish farmers within an area of intensive roost harassment also reported a reduction in fish predation by cormorants. Significantly fewer cormorants used intensely harassed night roosts than less intensely harassed or roosts that were not harassed. Pyrotechnics (designed to scare birds) were fired at roosting cormorants and those flying towards the roost during the two hours before sunset. Cormorants were counted on or near ponds and comparisons made between intensely harassed, less intensely harassed and non-harassed roost sites.

Replicated before-and-after trials in January–March 2003 at double-crested cormorant *Phalacrocorax auritus* night roosts (with 2,500 to 34,000 individuals) near catfish farms in Mississippi and Alabama, USA (3), found that hand-held lasers reduced cormorant numbers by 94% to 100%. The time required to achieve success varied (the most effective was 16 min to achieve 100% success at one roost; the least effective 113 min to achieve 94% success at another), but cormorants typically abandoned roosts after three nights of harassment. Six trials (at six sites) were conducted using a Desman™ Laser and five using a Laser Dissuader™. From sunset to 1 hour after sunset (on one to three consecutive evenings), a laser beam was directed at roosting cormorants, from 100–1,000 m distant. Birds were counted before and after treatment. (Note: Laboratory trials found no ocular damage to cormorants exposed to the Desman Laser at distances to the minimum of 1 m tested).

A replicated, controlled study in 1997 near channel catfish *Ictalurus punctatus* ponds in the Mississippi delta, USA (4), found that harassed double-crested cormorants *Phalacrocorax auritus* flew farther to their next roost than birds not harassed the previous night. Only 11% of harassed birds returned to the same roost within 48 hours, compared with 81% return to non-harassed roosts. Harassment shifted birds away from areas of catfish farm concentrations, but effects were temporary. Farmers undertook co-ordinated night-roost harassment patrols and prior to patrols, 50 cormorants were radio-tagged and their movements monitored from January through March 1997.

A before-and-after study in Bet She'an Valley, Israel (5), found that pygmy cormorants *Phalacrocorax pygmeus* relocated away from fish farms and bred successfully in nearby wetlands after harassment at all roosting sites in the valley using gas cannons and pyrotechnics in the winters of 1999/2000 and 2000/01. Between 1998 and 2004, number of cormorant nests in the area increased from 60 to 110 (peaking at around 155 in 2001).

(1) Mott, D. F. & Boyd, F. L. (1995) A review of techniques for preventing cormorant depredations at aquaculture facilities in the southeastern United States. *Colonial Waterbirds*, 18, 176–180.

(2) Mott, D. F., Glahn, J. F., Smith, P. L., Reinhold, D. S., Bruce, K. J. & Sloan, C. A. (1998) An evaluation of winter roost harassment for dispersing double-crested cormorants away from catfish production areas in Mississippi. *Wildlife Society Bulletin*, 26, 584–591.

(3) Glahn, J. F., Ellis, G., Fioranelli, P. & Dorr, B. S. (2000) Evaluation of moderate and low-powered lasers for dispersing double-crested cormorants from their night roosts. *Wildlife Damage Management Conferences–Proceedings*, 11, 89–92.

(4) Tobin, M. E., King, D. T., Dorr, B. S., Werner, S. J. & Reinhold, D. S. (2002) Effect of roost harassment on cormorant movements and roosting in the delta region of Mississippi. *Waterbirds*, 25, 44–51.

(5) Nemtzov, S. C. (2005) Relocation of pygmy cormorants *Phalacrocorax pygmeus* using scare tactics to reduce conflict with fish farmers in the Bet She'an Valley, Israel. *Conservation Evidence*, 2, 3–5.

5.69. Use electric fencing to exclude fish-eating birds

- **Two before-and-after studies from the USA (1, 2) found that electric fencing reduced the use of fish ponds by great blue herons *Ardea herodias* and great egrets *Casmerodius albus*.**

Before-and-after trials in Mississippi, USA (1), found that a two-strand electric fence reduced pond use by great blue herons *Ardea herodias* and great egrets *Casmerodius albus* by 91%. Five ponds (0.3–2.2 ha in area) containing channel catfish *Ictalurus punctatus* were tested.

A before-and-after trial in August–November 1996 in Pennsylvania, USA (2), found that electric fencing was fairly effective in deterring great blue herons *Ardea herodias* from raceways (long, 3–6 m wide fish ponds) at two trout hatcheries (declines from 6–14 birds/h/day and 76–159/h/day to <3 and <58 after electric fencing was erected). Fences comprised two strands of polyethylene tape (1.6 cm wide, 15–30 cm apart). Herons were counted (4 counts/week before and after installation; additional counts to 62 days after installation). Reductions in fish predation were not assessed.

(1) Mott, D. E. & Flynt, R. D. (1995) Evaluation of an electric fence system for excluding wading birds at catfish ponds. *The Progressive Fish-Culturist*, 57, 88–90.
(2) Tobin, M. E., Glahn, J. F. & Rasmussen, E. S. (1997) Electric fencing reduces heron predation at north-eastern trout hatcheries. *Proceedings of the Eighth Eastern Wildlife Damage Management Conference (1997)*, 8, 16–19.

5.70. Use netting to exclude fish-eating birds

- Two replicated studies from Germany (1) and the USA (5) found that netting or closely-spaced string barriers reduced losses of fish or deterred fish-eating birds from fish ponds. A review (3) concluded that excluding birds was an effective way to reduce damage.
- A series of tests in the Netherlands (2) found that netting or nylon lines over ponds did not prevent birds from landing, but did alter behaviour, whilst a before-and-after study from the USA (4) found that fewer great blue herons *Ardea Herodias* landed at fish ponds with netting, but that they stayed longer.
- Two replicated studies from Germany (1) and Israel (6) found that some birds became entangled in netting or closely-spaced string barriers over fish ponds. The Israeli study found that dark, small meshed netting entangled fewer birds than other netting types.

A replicated trial in the winter of 1962–3 in North Rhine-Westphalia, Germany (1), found that using string barriers and netting were effective at deterring grey herons *Ardea cinerea* from fish farms, if strings were 30 cm apart (larger distances being significantly less effective), but that occasional fatalities (entanglement) occurred. String barriers were ropes laid parallel across the water, 0.3 to 2 m apart; netting had 10 cm mesh size; and stumble ropes were also tested. This study also investigated the effects of scaring birds from fish farms, discussed in 'Scare birds from fish farms'.

A series of experiments in Flevoland, the Netherlands, in 1981–3 (2) found that nylon lines and ropes did not prevent cormorants from landing on fish ponds, but did change behaviour (from quick incursions by many birds to smaller numbers staying in ponds for longer). Effects on fish stocks are not reported. In 1981 nylon lines (forming 20 × 20 m squares) were placed over a 5 ha pond; in winter 1981–1982, 10 × 10 m squares were tested, and four ponds fitted with 20 × 20 m squares, with an irregular pattern over another; in 1983 over a 5 ha pond, ropes were stretched downwards from 10 m towers in the pond centre to the sides, with spacing between lines of 14–15 m). The effect of scaring birds from fish ponds is discussed in 'Scare birds from fish farms'.

A 1995 review assessed effectiveness of techniques used to prevent double-crested cormorant *Phalacrocorax auritus* predation at aquaculture facilities in the Mississippi delta region, USA (3), and concluded that excluding birds using netting or wires was an effective way to reduce damage.

A before-and-after trial in trout farms in Colorado, USA, in 1990–2 (4), found that 7 cm nylon netting significantly reduced great blue heron *Ardea Herodias* numbers next to ponds, but they stayed longer to capture fish after netting installation. Netting was 2.3 m long, 0.5 m high and slanted toward the water. The effects of various scaring devices is discussed in 'Scare birds from fish farms'.

A controlled, replicated experiment in 1995–6 at 18 ponds in seven aquaculture facilities in Florida, USA (5), found that fish losses from ponds covered in netting were significantly lower (11%) than for uncovered ponds (38%). Main species predating fish were snowy egret *Egretta thula*, green-backed heron *Butorides striatus*, tricoloured heron *E. tricolor* and little blue heron *E. caerulea*.

A replicated study in 2000–1 at 101 fish ponds, covered in 20 net types, in two fish farms in Israel (6) found that few or no dead birds were found in small mesh sizes with thick or dark-coloured netting. However, the largest numbers of entangled (dead) birds were found in two, thin monofilament net types, even though they had small mesh sizes (35 birds equivalent to 34/ha and 30 birds, equivalent to 25/ha). Ponds with slack, horizontal nets also had fewer dead birds, probably because they were more visible (flapping in the wind) than tight-strung nets. Entrapment within some ponds was attributed to poor net maintenance (e.g. holes/tears). A total of 327 dead (entangled) and 4,575 live birds (under nets) of 31 species were recorded.

(1) Ueckermann, E., Spittler, H. & Graumann, F. (1981) Technische Maßnahmen zur Abwehr des Graureihers (*Ardea cinerea*) von Fischteichen und Fischzuchtanlagen. *Zeitschrift für Jagdwissenschaft*, 27, 271–282.
(2) Moerbeek, D. J., Van Dobben, W. H., Osieck, E. R., Boere, G. C. & Bungenberg de Jong, C. M. (1987) Cormorant damage prevention at a fish farm in the Netherlands. *Biological Conservation*, 39, 23–38.
(3) Mott, D. F. & Boyd, F. L. (1995) A review of techniques for preventing cormorant depredations at aquaculture facilities in the southeastern United States. *Colonial Waterbirds*, 18, 176–180.
(4) Andelt, W. F., Woolley, T. P. & Hopper, S. N. (1997) Effectiveness of barriers, pyrotechnics, flashing lights, and Scarey Man® for deterring heron predation on fish. *Wildlife Society Bulletin*, 25, 686–694.
(5) Avery, M. L., Eiselman, D. S., Young, M. K., Humphrey, J. S. & Decker, D. G. (1999) Wading bird predation at tropical aquaculture facilities in central Florida. *North American Journal of aquaculture*, 61, 64–69.
(6) Nemtzov, S. C. & Olsvig-Whittaker, L. (2003) The use of netting over fishponds as a hazard to waterbirds. *Waterbirds*, 26, 416–423.

5.71. Disturb birds using foot patrols

- Two replicated studies from Belgium (1) and Australia (2) found that using foot patrols to disturb birds from fish farms did not reduce the number of birds present or fish consumption.

A replicated study from Limburg, Belgium, over 49 nights in 1982–3 (1), found that using foot patrols to disturb grey herons *Ardea cinerea* from 12 fish ponds did not necessarily reduce fish consumption. Low frequency disturbance (e.g. 3–5 farmer visits/night) caused a significant decrease in heron numbers but became less effective as heron numbers increased. Reduced numbers did not necessarily reduce fish consumption, as maximum predation occurred soon after bird arrival and disturbance mostly discouraged only well-fed birds from returning.

A replicated trial in New South Wales, Australia (2), found that hanging gillnets in fish ponds and using harassment patrols to deter cormorants *Phalacrocorax* spp. from fish farms was not effective. This study is discussed in 'Use in-water devices to reduce fish loss from ponds'.

(1) Draulans, D. & Van Vessem, J. (1985) The effect of disturbance on nocturnal abundance and behaviour of grey herons (*Ardea cinerea*) at a fish-farm in winter. *Journal of Applied Ecology*, 22, 19–27.
(2) Rowland, S. J. (1995) Predation of *Bidyanus bidyanus* (Teraponidae) in ponds by cormorants. *The Progressive Fish-Culturist*, 57, 248–249.

5.72. Use 'mussel socks' to prevent birds from attacking shellfish

- A randomised, replicated controlled experiment in Canada (1) found that fewer medium-sized mussels were taken from mussel socks with a protective 'sleeve', compared to un-sleeved socks. There were no differences for small or large mussels.

Background
'Mussel socks' are protective netting materials designed to reduce predation by birds on shellfish. The socks are made from polypropylene material with a biodegradable layer sown around it. The second layer (which has smaller mesh openings) prevents mussels from migrating towards the outside of the sock in order to filter feed and so reduces the threat from predation. The mussels are buffered between these layers over winter until the following spring when the protective layer decomposes and allows mussel growth to continue unhindered.

A randomised, replicated controlled experiment in October 2002 in three bays on Prince Edward Island, Canada (1), found that mussel socks with a 'sleeve' of a biodegradable cotton-polyester mesh lost fewer medium-sized (20 mm) mussels to greater scaup *Aythya marila* predation than un-sleeved socks. Losses were similar for small (14 mm) and large (26 mm) mussels, but more small mussels migrated through sleeved socks (thus more vulnerable to predation).

(1) Dionne, M., Lauzon-Guay, J. S., Hamilton, D. J. & Barbeau, M. A. (2006) Protective socking material for cultivated mussels: a potential non-disruptive deterrent to reduce losses to diving ducks. *Aquaculture International*, 14, 595–613.

5.73. Translocate birds away from fish farms

- A study from the USA cited in a review (1) found that translocating birds away from a fish farm appeared to reduce the number of birds at the farm. A study from Belgium (2) found that translocating herons did not seem to be an effective way to reduce bird numbers at fish farms.

Background
Studies detailing the effects of translocating birds (for many reasons) are discussed in a separate chapter. The two studies described below specifically deal with catching and releasing birds away from fish farms.

A study cited in (1), effectively reduced a population of green herons *Butorides virescens* feeding at a fish hatchery in USA, by catching and releasing them 40 km away. No details are given regards the time period over which the reduced population level was maintained or the effort needed to catch the birds. The effects on fish farm productivity were not measured.

A study in Belgium found that capturing grey herons *Ardea cinerea* (using clap nets) at a fish farm and releasing them at various distances away (2) did not appear to reduce heron abundance. Catching herons was very time-consuming (a maximum of four, but usually 1–2 caught each day). Some birds released 30 to 80 km returned within a month; one released 150 km away reappeared the following winter. Numbers relocated were too low to reduce heron abundance and relocation was considered ineffective.

(1) Mott, D. F. (1978) Control of wading bird predation at fish-rearing facilities. 131–132 in: A Sprunt IV, J. C. Ogden, S. Winckler (eds) *Wading Birds. National Audubon Society, New York, NY* Aududon Society, New York.
(2) Van Vessem, J., Draulans, D. & De Bont, A. F. (1985) The effects of killing and removal on the abundance of grey herons at fish farms. 337–343 *Proceedings of the XVIIth Congress of the International Union of Game Biologists* Brussels, Belgium.

5.74. Increase water turbidity to reduce fish predation by birds

- A randomised trial in France (1) found that little egret *Egretta garzetta* foraging efficiency was lower in turbid water than clear.

Background

Birds are visual predators and rely on detecting prey within the water in order to hunt. By obscuring predators' search images overall predation rates might decrease and reduce bird-human conflict in aquaculture operations.

A randomised trial in southern France found that little egret *Egretta garzetta* foraging efficiency (capture rate) declined significantly in turbid water, under captive conditions (1). A 3 × 6 m pool (water depth 10 cm) was used. Nine adult egrets either foraged alone or in threes. Three trials were undertaken using clear or turbid water, with three densities of prey i.e. mosquito fish *Gambusia affinis* (1, 2, and 4/m^2). Treatment order was randomised and trials lasted 5 min. Turbid water was created by adding clay (Secchi disc lost from view at 10 cm depth).

(1) Cezilly, F. (1992) Turbidity as an ecological solution to reduce the impact of fish-eating colonial waterbirds on fish farms. *Colonial Waterbirds*, 15, 249–252.

5.75. Provide refuges for fish within ponds

- A controlled cross-over trial in the UK (1) found that great cormorant *Phalacrocorax carbo* foraging success was lower in a pond with artificial refuges, compared to a control pond.

In Berkshire, England, a controlled cross-over trial (in the winters of 2003 and 2004) found that great cormorant *Phalacrocorax carbo* foraging success was lower in a pond with artificial refuges, compared to a control pond (on average, 67% less fish weight consumed/cormorant visit); fish loss was significantly lower (79% less fish weight lost); and there were 77% fewer cormorant visits. Two adjacent ponds (45.0 × 27.2 m; 1.35 m depth) were used, with provided in one pond each year. Refuges were 12 mesh covered cages (2 × 2 × 1.2 m high) in two blocks of six, with shade cloth and containing 4–6 tree saplings to provide cover. At commencement of each trial, ponds were similarly stocked with freshwater fish. Fish weight was recorded at the beginning and end of each trial.

(1) Russell, I., Parrott, D., Ives, M., Goldsmith, D., Fox, S., Clifton-Dey, D., Prickett, A. & Drew, T. (2008) Reducing fish losses to cormorants using artificial fish refuges: an experimental study. *Fisheries Management and Ecology*, 15, 189–198.

5.76. Use in-water devices to reduce fish loss from ponds

- A before-and-after study from the USA (1) found a 95% reduction in the number of double-crested cormorant *Phalacrocorax auritus* at two ponds in a fish farm following the installation of underwater ropes.
- A replicated study at a fish farm in Australia (2) found that hanging gillnets in ponds did not decrease the number of cormorants swimming in ponds.

Background

Cormorants *Phalacrocorax* spp. hunt by pursuing fish underwater. Putting nets or other obstacles underwater may reduce hunting effectiveness and so reduce aquaculture losses.

A before-and-after study at a channel catfish *Ictalurus punctatus* farm in Mississippi, USA, in January–April 1992 (1) found a 95% reduction in double-crested cormorant *Phalacrocorax auritus* on two ponds following the installation of parallel lengths of 9.5 mm yellow polyethylene rope with foam floats, 6.1 m apart (0.8–2.2 birds/min/day before installation vs. 0.03–0.08 afterwards). Eleven helium balloons also appeared useful in frightening cormorants habituated to ropes. During the week prior to addition 0.29 cormorants/min/day entered, whereas in the week after 0.02 entered. A 4.6 ha pond was monitored for 1,019 minutes before rope installation and 2,418 minutes after.

A replicated study at four silver perch *Bidyanus bidyanus* rearing ponds in New South Wales, Australia, found that hanging of gillnets and harassment patrols proved ineffective in deterring cormorants (great *Phalacrocorax carbo*, little black *P. sulcirostris* and little pied *P. melanoleucos*) from fishing in the ponds (2). Gillnets (large diameter mesh enabling young fish to swim through) were hung vertically in the water and harassment patrols (people walking around the ponds) were undertaken. Three ponds of 0.1 ha were stocked with fry (50,000–60,000 fish/ha) and one of 0.3 ha stocked with fingerlings (8,000/ha). Survival rates of fry in the three smaller ponds was 0.3%, 0.8% and 3.5% (average 1.5%), and survival of fingerlings 0.9%.

(1) Mott, D. F., Flynt, R. D. & King, J. O. (1993) An evaluation of floating ropes for reducing cormorant damage at catfish ponds. 24 *Sixth Eastern Wildlife Damage Control Conference*, 6, 93–97.
(2) Rowland, S. J. (1995) Predation of *Bidyanus bidyanus* (Teraponidae) in ponds by cormorants. *The Progressive Fish-Culturist*, 57, 248–249.

5.77. Spray water to deter birds from ponds

- A replicated study from Sweden (1) found that a rotating water spray deterred birds from fish ponds, but that birds often became used to the spray.

A replicated study from Älvkarleby municipality, Sweden (1), found that a rotator device for protecting small, circular fish ponds from gulls and terns deterred birds. Increased water pressure and intermittent spraying (instead of continuous) increased effectiveness but birds often became accustomed to the spray.

(1) Svensson, K. M. (1976) Rotator for protecting circular fish ponds against predatory birds. *The Progressive Fish-Culturist*, 38, 152–154.

5.78. Deter birds from landing on shellfish culture gear

Background

Oyster bags are mesh bags, attached to floats and anchored on the seabed into which young 'seed' oysters are placed. They then grow and are harvested when large enough. Oyster cages are similar, but are made of rigid wire mesh.

5.78.1. Use spikes on oyster cages

- A replicated and controlled study from Canada (1) found that significantly fewer birds landed on oyster cages with spikes attached, compared to control cages.

A replicated and controlled experiment in summer and autumn 2006 and 2007 in oyster *Crassostrea virginica* farms off the coast of New Brunswick, Canada (1), found that 'AntiCormo' devices (spikes attached to oyster cage floats) significantly reduced the number of birds roosting on oyster cages (0–1.3 birds/100 cages at one site with AnitCormo cages; 0–42 birds at a second vs. 100–499 birds/100 cages at a control farm). Five species were seen on oyster cages (double-crested cormorants *Phalacrocorax auritus*, herring gull *Larus argentatus*, great black-backed gull *L. marinus*, common tern *Sterna hirundo* and great blue heron *Ardea herodias*).

(1) Comeau, L. A., St-Onge, P., Pernet, F. & Lanteigne, L. (2009) Deterring coastal birds from roosting on oyster culture gear in eastern New Brunswick, Canada. *Aquacultural Engineering*, 40, 87–94.

5.78.2. Suspend oyster bags under water

- A replicated and controlled study from Canada (1) found that significantly fewer birds roosted on oyster bags suspended 6 cm below the water, compared with non-submerged bags. Birds roosted on bags suspended 3 cm below the water as frequently as control bags.

A replicated and controlled experiment in summer and autumn 2006 and 2007 in oyster *Crassostrea virginica* farms off the coast of New Brunswick, Canada (1), found that suspending oyster bags 6 cm under the water significantly reduced the number of birds roosting on them (1 bird/100 bags on submerged bags vs. 37 on controls). However, suspending bags 3 cm below the water did not significantly reduce the number of birds roosting (15 birds roosting/100 bags).

(1) Comeau, L. A., St-Onge, P., Pernet, F. & Lanteigne, L. (2009) Deterring coastal birds from roosting on oyster culture gear in eastern New Brunswick, Canada. *Aquacultural Engineering*, 40, 87–94.

6. Threat: Energy production and mining

Mining can have immense impacts on natural environments through the destruction and pollution of habitats. For example, coal mining in the eastern USA has been implicated in the catastrophic decline of cerulean warbler *Dendroica cerulea* in the past 40 years (Weakland & Wood 2005). Similarly, energy production, both renewable and non-renewable can impact on bird populations, again through pollution, but also through direct collisions with wind farms. Most of the interventions in response to these threats are discussed elsewhere, in 'Threat: Pollution' and 'Habitat restoration and creation'.

Weakland, C. A. & Wood, P. B. (2005) Cerulean Warbler Dendroica cerulea microhabitat and landscape-level habitat characteristics in southern West Virginia. *Auk*, 122, 497–508.

Key messages

Paint wind turbines to increase their visibility

A single, controlled *ex situ* experiment found that thick black stripes running across a wind turbine's blades made them more conspicuous to an American kestrel *Falco sparverius* than control (unpatterned) blades. Other designs were less visible or indistinguishable from controls.

6.1. Paint wind turbines to increase their visibility

- A single *ex situ* experiment (1) found that thick black stripes running across a wind turbine's blades made them more conspicuous to an American kestrel than control (unpatterned) blades, but that other designs were less visible, or indistinguishable from controls.

Background

Wind turbines can have a detrimental impact on birds both through direct collisions with the blades or because the vortices created force birds to the ground. Direct collisions may be reduced if birds can clearly see and avoid blades.

A randomised, controlled *ex situ* experiment on an American kestrel *Falco sparverius* (1) found that it could discriminate between a control stimulus and an image of rotating wind turbine blades better if the blades were painted with two thick black bands running across the width of the blade (a visibility ratio of 2.4 in bright light, decreasing to 1.5 in low light, no difference in very low light). Blades with narrow black bands running across the width of the blade were less conspicuous in bright light (ratio of 0.1) and were indistinguishable in low light. A pattern of three stripes running the length of the blade were not significantly more or less conspicuous than plain white blades. Both the control (grey blade rotating in front of a grey background) and experimental stimuli rotated at 30 rpm.

(1) McIsaac, H. P. (2001) Raptor acuity and wind turbine blade conspicuity. *Proceedings of National Avian-Wind Power Planning Meeting IV*, 59–87.

7. Threat: Transportation and service corridors

Key messages – Verges and airports

Mow roadside verges

A single replicated, controlled trial in the USA found that mowed roadside verges were less attractive to ducks as nesting sites, but had higher nesting success after four years.

Sow roadside verges

We captured no intervention-based evidence for the effects of sowing roadside verges on bird populations.

Scare or otherwise deter birds from airports

Two replicated studies in the UK and USA found that fewer birds used areas of long grass at airports, but no data were provided on the effect of long grass on strike rates or bird mortality.

Key messages – Power lines and electricity pylons

Bury or isolate power lines

A single before-and-after study in Spain found a dramatic increase in juvenile eagle survival following the burial or isolation of dangerous power lines.

Remove earth wires from power lines

Two before-and-after studies from Norway and the USA describe significant reductions in bird collision mortalities after earth wires were removed from sections of power lines.

Thicken earth wires

A single paired sites trial in the USA found no reduction in crane *Grus* spp. collision rates in a wire span with an earth wire three times thicker than normal.

Mark power lines

A total of eight studies and two literature reviews from across the world found that marking power lines led to significant reductions in bird collision mortalities. Different markers had different impacts.

Use raptor models to deter birds from power lines

A single paired sites trial in Spain found that installing raptor models near power lines had no impact on bird collision mortalities.

Add perches to electricity pylons

A single before-and-after study in Spain found that adding perches to electricity pylons did not reduce electrocutions of Spanish imperial eagles *Aquila adalberti*.

Insulate electricity pylons

A single before-and-after study in the USA found that insulating power pylons significantly reduced the number of Harris's hawks *Parabuteo unicinctus* electrocuted.

Use perch-deterrent lines

A single controlled study in the USA found that significantly fewer raptors were found near perch-deterrent lines, compared to controls, but no information on electrocutions was provided.

Reduce electrocutions by using plastic, not metal, leg rings to mark birds

A single replicated and controlled study in the USA found no evidence that using plastic leg rings resulted in fewer raptors being electrocuted.

Verges and airports

7.1. Mow roadside verges

- A single replicated, controlled study in the USA (1) found that more ducks nested on unmown roadside verges, but that over four years, nesting success on unmown verges fell to below that on mown verges.

Background

Roadside verges could potentially act as a valuable nesting or foraging habitat for many species of birds, but can become overgrown with rank vegetation. Mowing, therefore, may increase the value of this habitat.

Alternatively, conservationists may want to limit the number of birds using roads verges, to reduce the risk of collisions with vehicles.

A replicated, controlled trial in 1969–72 in North Dakota, USA (1), found that more ducks nested in unmown than mown road verges (although this difference was not significant in 1972). However, nesting success remained between 40–60% in mowed strips, whereas it fell from >70% to <30% in unmown strips due to an increase in mammalian nest predation. Alternate mowed and unmown 1.6 km strips of roadside vegetation were compared along 37 km of Interstate 94. The 23 mowed strips (totalling 123 ha) were mown once in autumn. Six duck species nested over the four years: blue-winged teal *Anas discors*, mallard *A. platyrhynchos*, gadwall *A. strepera*, northern shoveler *A. clypeata*, pintail *A. acuta* and lesser scaup *Aythya affinis*.

(1) Voorhees, L. D. & Cassel, J. F. (1980) Highway right-of-way: mowing versus succession as related to duck nesting. *Journal of Wildlife Management*, 44, 155–163.

7.2. Sow roadside verges

- We captured no evidence for the effects of sowing roadside verges on bird populations.

Background

Sowing roadside verges with natural vegetation could provide valuable habitats for birds, for nesting, foraging and escaping predators.

7.3. Scare or otherwise deter birds from airports

- Two replicated studies in the UK and USA (1,2) found that fewer birds (mainly gulls *Larus* spp.) used areas of long grass at airports.
- However, no data were provided on the effect of long grass on strike rates or mortality of birds.

Background

Collisions between birds and aircraft ('birdstrikes') at airports are potentially dangerous to planes, whilst also harming bird populations. Therefore making airports less attractive to birds, or scaring them off, has the potential for multiple benefits.

A replicated study in the UK (1) found that fewer birds used grass on Royal Air Force airfields when it was allowed to grow long, compared to when it was kept short. In 1967–1968, ten English airfields were included, with data from 1972–1973 available for three more airfields (including one in Scotland and one in Wales). Grass was kept 15–20 cm high in some areas whilst others were maintained at 5–10 cm. The repellent effect of long grass was almost 100% for gulls *Larus* spp. and golden plover *Pluvialis apricaria*, and very good for northern lapwing *Vanellus vanellus*, Eurasian oystercatcher *Haematopus ostralegus* and crows (rook *Corvus frugilegus*, carrion crow *C. corone*, Eurasian jackdaw *C. monedula*).

A replicated study in June–August 1985–6 at Kennedy International Airport, New York City, USA (2), found that fewer laughing gulls *Larus atricilla* were found on areas of grass allowed to grow long, than on short-cropped areas. Thirty-six plots across three experimental blocks were used, with long grass being grown to 45 cm in length and short-cropped areas kept at 5 cm.

(1) Brough, T. & Bridgman, C. J. (1980) An evaluation of long grass as a bird deterrent on British airfields. *Journal of Applied Ecology*, 17, 243–253.
(2) Buckley, P. A. & McCarthy, M. G. (1994) Insects, vegetation, and the control of laughing gulls (*Larus atricilla*) at Kennedy International Airport, New York City. *Journal of Applied Ecology*, 31, 291–302.

Power lines and electricity pylons

In 2008, there were an estimated 65 million km of medium- and high-voltage power lines in use across the world, with the network growing at perhaps 5% a year (Jenkins, Smallie, & Diamond 2010). These power lines are a significant threat to many species of birds, with both direct collision mortality and electrocution (from touching multiple wires at once) killing birds. As well as causing potentially unsustainable mortality, these collisions can be economically costly, especially when large-bodied species are involved.

Species are most at risk if they are large and heavy, with poor eyesight and limited manoeuvrability (Jenkins, Smallie, & Diamond 2010), but some relatively small species, such as gamebirds, are also very vulnerable (Bevanger & Brøseth 2001). Appropriate planning of power lines to avoid high-risk areas (such as those between feeding and roosting grounds) and to follow features such as roads may well reduce bird mortalities (Bevanger 1994), but in many cases wires are already in potentially dangerous locations and techniques are needed to mitigate the damage they cause.

Bevanger, K. (1994) Bird interactions with utility structures: collision and electrocution, causes and mitigating measures. *Ibis*, 136, 412–425.
Bevanger, K. & Brųseth, H. (2001) Bird collisions with power lines–an experiment with ptarmigan (*Lagopus* spp.). *Biological Conservation*, 99, 341–346.
Jenkins, A. R., Smallie, J. J. & Diamond, M. (2010) Avian collisions with power lines: a global review of causes and mitigation with a South African perspective. *Bird Conservation International*, 20, 263–278.

7.4. Bury or isolate power lines to reduce incidental mortality

- A single before-and-after trial in Spain (1) showed a dramatic increase in the survival of juvenile Spanish imperial eagles *Aquila adalberti* following the burial or isolation of power lines.

A before-and-after trial in the wetlands of the Doñana National Park (1), Andalusia, Spain, found that six-month survival for radio-marked juvenile Spanish imperial eagles *Aquila adalberti* increased from 18% of 17 individuals in 1986–7 to 80% of 15 in 1988–9 following the isolation or burial of previously identified dangerous power lines. This study discusses other eagle management techniques, described in 'Add perches to electricity pylons to reduce electrocution', 'Use signs and access restrictions to reduce disturbance at nest sites', 'Foster eggs or chicks with wild conspecifics' and 'Remove/treat endoparasites'.

(1) Ferrer, M. & Hiraldo, F. (1991) Evaluation of management techniques for the Spanish imperial eagle. *Wildlife Society Bulletin*, 19, 436–442.

7.5. Remove earth wires to reduce incidental mortality

- A before-and-after study and a literature review (2,3) describe significant reductions in collision mortalities of cranes *Grus* spp. and grouse *Lagopus* spp. following the removal of earth wires.

Background

Earth wires act to protect transmission wires against lightning, by dispersing excessive electricity. They are normally positioned either above or below the transmission wires, increasing the vertical height of wires and therefore the chances of collisions. In addition, earth wires tend to be thinner than transmission wires and so less visible, which, when they are positioned above transmission wires, increases the chances of collisions as birds climb to avoid the transmission wires.

A controlled before-and-after trial between 1989 and 1995 in boreal and subalpine forests in southern Norway (2) found a 51% decrease in collision mortality of willow grouse *Lagopus lagopus* and rock ptarmigan *L. mutus* by a 2.5 km section of 22 kV power line from which the earth wire was removed (49 fatalities before removal, 24 afterwards). There was no corresponding decrease in two control sections (61 vs. 50 fatalities and 20 vs. 27 fatalities). The earth wire was located 1.5 m below the phase conductor lines.

A literature review (3) described a before-and-after trial (1) that found an 80% reduction in collision mortality of sandhill cranes *Grus canadensis* and whooping cranes *G. americana* following removal of the earth wire from a 3.2 km span of 116 kV line in Colorado, USA.

(1) Brown, W. M., Drewien, R. C. & Bizeau, E. G. (1987) Mortality of cranes and waterfowl from power-line collisions in the San Luis Valley, Colorado. 128–136 *Proceedings of the crane workshop, 1985* Platte River Whooping Crane Maintenance Trust, Grand Island, Nebraska.
(2) Bevanger, K. & Brųseth, H. (2001) Bird collisions with power lines–an experiment with ptarmigan (*Lagopus* spp.). *Biological Conservation*, 99, 341–346.
(3) Jenkins, A. R., Smallie, J. J. & Diamond, M. (2010) Avian collisions with power lines: a global review of causes and mitigation with a South African perspective. *Bird Conservation International*, 20, 263–278.

7.6. Thicken earth wire to reduce incidental mortality

- A literature review (2) found no evidence that thickening the earth wire had any impact on collision mortality of cranes *Grus* spp.

Background

Removing earth wires is not always practical, given their protective nature. However, they can be rendered more visible by thickening the wire.

A literature review (2) describes how a paired site study in Colorado, USA (1), found that replacing 50% of the earth wire from 3.2 km of 115 kV wire with an earth wire three times thicker than normal had no effect on crane *Grus* spp. collision mortality compared to the span with a normal thickness earth wire.

(1) Brown, W. M., Drewien, R. C. & Bizeau, E. G. (1987) Mortality of cranes and waterfowl from power-line collisions in the San Luis Valley, Colorado. 128–136 *Proceedings of the crane workshop, 1985* Platte River Whooping Crane Maintenance Trust, Grand Island, Nebraska.
(2) Jenkins, A. R., Smallie, J. J. & Diamond, M. (2010) Avian collisions with power lines: a global review of causes and mitigation with a South African perspective. *Bird Conservation International*, 20, 263–278.

7.7. Mark power lines to reduce incidental mortality

- A total of studies (1–4,6,9–11) and two literature review (5,12) from across the world found that marking power lines led to significant reductions in collision rates or dangerous flight behaviour (i.e. approaching close to power lines) in cranes *Grus* spp., mute swans *Cygnus olor* and other bird species.

- All markers except thin, black plastic strips or neoprene crosses (6) were effective, with no differences in effectiveness between Bird Flight Diverters (BFDs: brightly coloured plastic spirals) and static fibreglass plates (3) and only a small possible difference between BFDs and 'flappers' (moving markers).

Background

Birds may fly into power lines accidentally, either because they do not see them at all, or because they see them too late to react; this is a particular issue for large species that cannot change direction quickly. Increasing power line visibility by marking them is therefore frequently proposed as a way of reducing collision-induced mortalities without the need to remove earth wires or bury transmission wires.

A replicated, paired sites study (1) in south-central Nebraska, USA, between 1988–1990, found that marking the static wire of nine spans of high-voltage transmission wire reduced collision mortality in sandhill cranes *Grus canadensis* by 66% compared with unmarked spans (11 fatalities vs. 25). Crane flocks were also more likely to increase altitude (454 flocks vs. 397) or change direction gradually (114 vs. 92) when flying close to marked spans, but were less likely to react quickly (and potentially dangerously) to marked spans (19 vs. 36) or show no reaction (1,200 vs. 768). Experimental spans were 1–2.5 km in length and marked with yellow aviation balls (30 cm in diameter with a vertical black stripe) at 100 m intervals, staggered to appear more closely spaced.

A controlled before-and-after trial in the winters of 1989–90 and 1990–1 in mixed woodland and farmland in Extremadura, Spain (2), found there were fewer collisions and fewer birds flying between lines following the marking of four sections of line (60% reduction in fatalities: 45 vs. 18; 61% decrease in birds flying between lines: 357 birds/day vs. 124 birds/day). There were no such decreases at unmarked spans of wire (19 vs. 15 fatalities). Markers were coloured PVC spirals (1 m long, 30 cm in diameter) attached every 10 m.

A paired sites study in autumn and spring 1988–1991 in mixed wetlands, croplands and uplands in south-central Colorado, USA (3), found that collision mortality was 61% lower in four spans marked by 'dampers' and 63% lower in four spans marked by 'plates' compared with eight unmarked spans. Birds also reacted to marked lines earlier and flew over them at a greater height. 'Dampers' were yellow, spiral vibration dampers, 112–125 cm long, placed at 3.3 m intervals; 'plates' were yellow fibreglass plates, 30.5 × 30.5 cm with a 5 cm diagonal black strip, placed at 23–32 m intervals. Both markers were placed on two spans of 7.2 kV distribution lines and two of 69–115 kV transmission lines, with all spans totalling 13.2 km. Most birds killed were sandhill cranes *Grus canadensis* and wildfowl.

A paired sites study in 1991–4 in coastal wetlands in South Carolina, USA (4), found a 53% reduction in collision mortalities at a 3.9 km span of 115 kV transmission lines where the static wires were marked, compared to a 1.2 km unmarked span. A higher proportion of birds approaching marked wires at the most dangerous height (between transmission wires and earth wire) reacted to them, compared to unmarked wires (98% of 9,819 flocks vs. 89% of 4,209 flocks respectively) and fewer crossed the wires at this height (4% vs. 24%). However, overall, a higher percentage of birds reacted to lines at unmarked spans (40% of 17,391 flocks vs. 34% of 64,512 flocks). The experimental span was marked with yellow aviation balls (30 cm in diameter with a vertical black stripe) at 61 m intervals, staggered to give the appearance of a 30.5 m spacing.

A controlled before-and-after study of three different markers at three sites in Spain in 1991–5 (6) found that markers differed in their effects. In grassland and arable land in Badajoz, collision mortality across all species was 81% lower following the installation of white plastic spirals (expected mortality of 47 birds, actual mortality of nine). In mixed cattle grazing and cereal cropland in Cáceres, the total number of collisions (72) remained constant following the installation of neoprene strips, but there was a 76% reduction in collision mortality when great bustards *Otis tarda* were excluded from the analysis. In a wetland in Huelva there was no difference in collision mortality after the installation of black plastic strips (6 collisions in marked vs. 12 in unmarked spans). Plastic spirals were 1 m long, 30 cm maximum diameter and placed every 10 m on static wires, staggered to give the appearance of a 5 m placement. Neoprene bands were crossed black strips, 35 × 5 cm with a 5 × 4 cm phosphorescent stripe, installed every 20 m on conductor wires and staggered to give the appearance of a 10 m placement. Black plastic stripes were 70 × 0.8 cm and hung every 12 m on a distribution line.

A literature review (5) found three studies, all of which show a reduction in crane *Grus* spp. mortality, following the marking of power lines. (1) and (3) are discussed elsewhere. The third study found that marking, burying or removing power lines in Japan reduced the percentage of red-crowned crane *Grus japonicus* mortalities attributed to power line collisions from 70.9% (1970–4) to 26.8% (1980–4). The rate of population increase rose from 5.9% (1970–4) to 18.9% (1980–4). There was also an increase in the number of breeding pairs and the percentage of juveniles in the population.

A controlled before-and-after study in 1997–2000 in northern Colombia (9) found that the average number of bird deaths due to collision with power lines was significantly lower (5.3 deaths/ha) in a site where the ground wires were marked with yellow plastic spirals than for an unmarked circuit (13.6 deaths/ha); no significant difference in collision mortality between the circuits was detected prior to marking. A total of 810 birds of 47 species were found.

A before-and-after trial at a wetland site in Essex, England (10), installed 500 red, spiral-shaped 'flight diverters' (32 cm long, 17.5 cm in diameter) at 5 m intervals along 1.5 km of power lines. In two springs preceding installation (2004 and 2006), 28 mute swans *Cygnus olor* were killed through collisions with the wires. Following installation, one swan was killed in the springs of 2007 and 2008 combined.

A replicated, controlled before-and-after trial over the winters of 2003–4, 2004–5 and 2005–6, in mixed wetland and crops on Staten Island, California, USA (11), found a 60% reduction in collision mortality across all species following the installation of 'Firefly bird flappers' on 20 spans of a 5.6 km section of 12 kV wire. There was also a smaller (approximately 10%) reduction in mortality in nine spans adjacent to marked spans. There was no corresponding decrease in collision mortality in 20 unmarked spans. However, markers did not appear to have an effect on bird flight behaviour. Markers were 15 × 9 cm acrylic sheets of two contrasting colours with a luminescent strip and placed at 15 m intervals, staggered to appear 5 m apart. A total of 65 fatalities were recorded over the three winters.

A 2010 literature review (12) found significant reductions in bird mortality following the marking of power lines in the USA and South Africa. A before-and-after trial over two years in Indiana (7), USA, found significantly reduced mortality with both Bird Flight Diverters (BFDs, 73% fewer fatalities) and larger 'Swan Flight Diverters' (50% fewer). The reviewers noted that there was considerable variability in collision rates across all sites. A before-and-after trial over three years in Karoo, South Africa (8) found a 67% reduction in collision mortality following the marking of both earth wires of 10 km of 132 kV line with BFDs (30 cm long) every 10 m. The same study also found that spans marked with BFDs and 'flappers' (loosely suspended polycarbon discs) had 52% lower collision mortality than spans marked just with BFDs, and 80% lower collision mortality than the same spans prior to marking. Spans with just flappers had 60% lower mortality than those with BFDs but the results are inconclusive. The reviewers noted a considerable decrease in blue crane *Anthropoides paradiseus* and Ludwig's bustard *Neotis ludwigii* (the main species colliding with the line) populations in the area following marking.

(1) Morkill, A. E. & Anderson, S. H. (1991) Effectiveness of marking powerlines to reduce sandhill crane collisions. *Wildlife Society Bulletin*, 19, 442–449.

(2) Alonso, J. C., Alonso, J. A. & Munoz-Pulido, R. (1994) Mitigation of bird collisions with transmission lines through groundwire marking. *Biological Conservation*, 67, 129–134.

(3) Brown, W. M. & Drewien, R. C. (1995) Evaluation of two power line markers to reduce crane and waterfowl collision mortality. *Wildlife Society Bulletin*, 23, 217–227.
(4) Savereno, A. J., Savereno, L. A., Boettcher, R. & Haig, S. M. (1996) Avian behavior and mortality at power lines in coastal South Carolina. *Wildlife Society Bulletin*, 24, 636–648.
(5) Davis, C. (1998) A review of the success of major crane conservation techniques. *Bird Conservation International*, 8, 19–30.
(6) Janss, G. F. & Ferrer, M. (1998) Rate of bird collision with power lines: effects of conductor-marking and static wire-marking. *Journal of Field Ornithology*, 69, 8–17.
(7) Crowder, M. R. (2000) Assessment of devices designed to lower the incidence of avian power line strikes. Masters of Science Thesis, Purdue University, East LaFayette, IN. 91p.
(8) Anderson, M. D. (2002) Large terrestrial bird powerline project. *Unpublished report* Eskom, Johannesburg.
(9) De La Zerda, S. & Roselli, L. (2003) Mitigation of collisions of birds with high-tension electric power lines by marking the guard wire. *Ornitologia Colombiana*, 1, 42–63.
(10) Frost, D. (2008) The use of 'flight diverters' reduces mute swan *Cygnus olor* collision with power lines at Abberton Reservoir, Essex, England. *Conservation Evidence*, 5, 83–91.
(11) Yee, M. L. (2008) Testing the effectiveness of an avian flight diverter for reducing avian collisions with distribution power lines in the Sacramento Valley, California: PIER final project report. California Energy Commission.
(12) Jenkins, A. R., Smallie, J. J. & Diamond, M. (2010) Avian collisions with power lines: a global review of causes and mitigation with a South African perspective. *Bird Conservation International*, 20, 263–278.

7.8. Use raptor models to deter birds and so reduce incidental mortality

- A single paired sites study in Spain (1) found no evidence that raptor models were effective in deterring birds from crossing power lines and may even have attracted some species to the area.

Background

Models of raptors are used in a variety of situations in attempts to keep smaller birds away from areas, on the basis that they will avoid predators.

A controlled paired sites study in two sites in Andalusia, Spain (1), found no effect of raptor models (one golden eagle *Aquila chrysaetos* and two *Accipiter* spp.) on the number of flocks crossing power lines in either migration or non-migration areas. In the migration area south of Cadiz, there was no difference in the number of flocks of birds coming within 100 m of the power lines, although there was a higher proportion of raptors in the section with the eagle model (42% of 119 records were raptors vs. 21% of 43 records) and flocks flew above 20 m more frequently in the section with the eagle model. In the non-migration area (the Coto Doñana National Park), more flocks and a higher proportion of raptors, waterfowl and corvids were seen in sections near the eagle model, compared with control sections (the *Accipiter* models were not tested in the park).

(1) Janss, G. F., Lazo, A. & Ferrer, M. (1999) Use of raptor models to reduce avian collisions with powerlines. *Journal of Raptor Research*, 33, 154–159.

7.9. Add perches to electricity pylons to reduce electrocution

- A single before-and-after study in Spain (1) found that adding perches did not reduce electrocutions of Spanish imperial eagles *Aquila adalberti*.

Background

Raptors and other large birds can become electrocuted when they perch on pylons and touch multiple wires with their wings. Adding perches to pylons may therefore reduce mortality by allowing birds to land safely away from the wires.

A before-and-after study in a wetland national park in Andalusia, Spain (1), found that adding perches to 80 electricity pylons did not reduce electrocution rates of Spanish imperial eagle *Aquila adalberti*. This study discusses other eagle management techniques, described in 'Bury or isolate power lines', 'Use signs and access restrictions to reduce disturbance at nest sites', 'Foster eggs or chicks with wild conspecifics' and 'Remove/treat endoparasites'.

(1) Ferrer, M. & Hiraldo, F. (1991) Evaluation of management techniques for the Spanish imperial eagle. *Wildlife Society Bulletin*, 19, 436–442.

7.10. Insulate power pylons to prevent electrocution

- A single before-and-after study in the USA (1) found the insulating power pylons significantly reduced the number of Harris's hawks *Parabuteo unicinctus* electrocuted.

Background

Another method of preventing electrocution is to insulate parts of pylons that large birds are likely to touch at the same time.

A before-and-after study in 2003 in urban environments in Tucson, Arizona, USA (1), found that there were significantly fewer electrocutions following the insulation of potentially dangerous elements of power pylons (1.4 electrocutions/nest before insulation vs. 0.2 electrocutions/nest afterwards). Any differentially energised hardware components, closer than 60 cm apart, on half the power poles within 300 m of 58 Harris's hawk *Parabuteo unicinctus* nests were insulated.

(1) Dwyer, J. F. & Mannan, R. (2007) Preventing raptor electrocutions in an urban environment. *Journal of Raptor Research*, 41, 259–267.

7.11. Use perch-deterrents to stop raptors perching on pylons

- A single controlled study from the USA (1) found significantly lower raptor activity close to perch-deterrent power lines, compared to control lines. No data were provided on electrocution rates.

Background

Preventing raptors from landing on pylons at all may reduce electrocution, for example, perch-deterrent power lines have spikes and pole caps on the pylons to deter birds.

A controlled study from September–August in 2006–2007 in shrubland in Wyoming, USA (1), found that raptor activity and predation rates were significantly lower near a 24.9 km perch-deterrent line, compared to a 16.4 km section of control line (42 sightings vs. 551 and 69 prey items found vs. 277). Golden eagles *Aquila chrysaetos* and common ravens *Corvus corax* were the species most commonly observed successfully overcoming deterrent devices (76% of deterrent-line sightings). More raptors perched on wires (rather than pylons) on perch-deterrent lines (228 compared to 11 sightings on control lines; 68% of sightings were rough-legged hawks *Buteo lagopus*).

(1) Slater, S. J. & Smith, J. P. (2010) Effectiveness of raptor perch deterrents on an electrical transmission line in southwestern Wyoming. *Journal of Wildlife Management*, 74, 1080–1088.

7.12. Reduce electrocutions by using plastic, not aluminium, leg rings to mark birds

- A replicated and controlled study in the USA (1) found no evidence for lower electrocution rates for raptors marked with plastic leg rings, compared to metal ones.

Background

Leg rings, used to identify birds, could increase electrocution rates if made of metal. Therefore, plastic bands are sometimes recommended.

A replicated, controlled study in Tucson, Arizona, USA (1), found that electrocution rates for 38 Harris's hawks *Parabuteo unicinctus* marked with plastic leg bands were no lower than for 41 birds marked with aluminium bands. In addition, there were no signs of burning/ electrocution close to the rings on dead birds' legs.

(1) Dwyer, J. F. & Mannan, R. W. (2009) Return rates of aluminum versus plastic leg bands from electrocuted Harris's hawks (*Parabuteo unicinctus*). *Journal of Raptor Research*, 43, 152–154.

8. Threat: Biological resource use

Key messages – reducing exploitation and conflict

Use legislative regulation to protect wild populations

Five out of six studies from Europe, Asia, North America and across the world, found evidence that stricter legislative protection was correlated with increased survival, lower harvests or increased populations. The sixth, a before-and-after study from Australia, found that legislative protection did not reduce harvest rates.

Increase 'on-the-ground' protection to reduce unsustainable levels of exploitation

Two before-and-after studies from Europe and Central America found increases in bird populations and recruitment following stricter anti-poaching methods or the stationing of a warden on the island in question. However, the increases in Central America were only short-term, and were lost when the intensive effort was reduced.

Promote sustainable alternative livelihoods

A single before-and-after study in Costa Rica found that a scarlet macaw *Ara macao* population increased following several interventions including the promotion of sustainable, macaw-based livelihoods.

Use education programmes and local engagement to help reduce persecution or exploitation of species

Six out of seven studies from across the world found increases in bird populations or decreases in mortality following education programmes, whilst one study from Venezuela found no evidence that poaching decreased following an educational programme. In all but one study reporting successes, other interventions were also used, and a literature review from the USA and Canada argues that education was not sufficient to change behaviour, although a Canadian study found that there was a significant shift in local peoples' attitudes to conservation and exploited species following educational programmes.

Employ local people as 'biomonitors'

A single replicated study in Venezuela found that poaching of parrot nestlings was significantly lower in years following the employment of five local people as 'biomonitors'.

Mark eggs to reduce their appeal to collectors

A single before-and-after study in Australia found increased fledging success of raptor eggs in a year they were marked with a permanent pen.

Relocate nestlings to reduce poaching

A single replicated study in Venezuela found a significant reduction in poaching rates and an increase in fledging rates of yellow-shouldered amazons *Amazona barbadensis* when nestlings were moved into police premises overnight.

Use wildlife refuges to reduce hunting disturbance

Three studies from the USA and Europe found that more birds used refuges where hunting was not allowed, compared to areas with hunting, and more used the refuges during the open season. However, no studies examined the population-level effects of refuges.

Introduce voluntary 'maximum shoot distances'

A single study from Denmark found a significant reduction in the injury rates of pink-footed geese *Anser brachyrhynchus* following the implementation of a voluntary maximum shooting distance.

Use alerts during shoots to reduce mortality of non-target species

We found no evidence as to the effectiveness of using alerts during shoots.

Provide 'sacrificial' grasslands to reduce the impact of wild geese on crops

Two UK studies found that more geese used areas of grassland managed for them, but that this did not appear to attract geese from outside the study site and therefore was unlikely to reduce conflict with farmers.

Move fish-eating birds to reduce conflict with fishermen

A single before-and-after study in the USA found that Caspian tern *Sterna caspia* chicks had a lower proportion of commercial fish in their diet following the movement of the colony away from an important fishery.

Scare fish-eating birds from areas to reduce conflict

Studies investigating scaring birds from fishing areas are discussed in 'Threat: Agriculture – Aquaculture'.

Key messages – reducing fisheries bycatch

Set longlines at night to reduce seabird bycatch

Six out of eight studies from around the world found lower bycatch rates when longlines were set at night, but the remaining two found higher bycatch rates (of northern fulmar *Fulmarus glacialis* in the North Pacific and white-chinned petrels *Procellaria aequinoctialis* in the South Atlantic, respectively). Knowing whether bycatch species are night- or day-feeding is therefore important in reducing bycatch rates.

Turn deck lights off during night-time setting of longlines to reduce bycatch

A single replicated and controlled study in the South Atlantic found lower seabird bycatch rates on night-set longlines when deck lights were turned off.

Use streamer lines to reduce seabird bycatch on longlines

Ten studies from coastal and pelagic fisheries across the globe found strong evidence for reductions in bycatch when streamer lines were used. Five studies from the South Atlantic, New Zealand and Australia were inconclusive, uncontrolled or had weak evidence for reductions. One study from the sub-Antarctic Indian Ocean found no evidence for reductions. Three studies from around the world found that bycatch rates were lower when two streamers were used compared to one, and one study found rates were lower still with three streamers.

Use larger hooks to reduce seabird bycatch on longlines

We captured no intervention-based evidence on the impact of large hooks on seabird bycatch. However, one study found a negative correlation between hook size and bycatch rate.

Use a water cannon when setting longlines to reduce seabird bycatch

We found no evidence for the effects on seabird bycatch rates when using a water cannon during the setting longlines.

Set lines underwater to reduce seabird bycatch

Five studies in Norway, South Africa and the North Pacific found lower seabird bycatch rates on longlines set underwater. However, results were species-specific, with shearwaters *Puffinus* spp. and possibly albatrosses continuing to take baits set underwater.

Set longlines at the side of the boat to reduce seabird bycatch

We found no evidence for the effects on seabird bycatch rates of setting longlines from the side of the boat.

Use a line shooter to reduce seabird bycatch

Two randomised, replicated and controlled trials found that seabird bycatch rates were higher (in the North Pacific) or the same (in Norway) on longlines set using line shooters, compared to those set without a shooter.

Use bait throwers to reduce seabird bycatch

A single analysis found significantly lower seabird bycatch on Australian longliners when a bait thrower was used to set lines.

Tow buoys behind longlining boats to reduce seabird bycatch

We found no evidence for the effects on seabird bycatch rates of towing buoys behind longlining boats.

Dye baits to reduce seabird bycatch

A single randomised, replicated and controlled trial in Hawaii, USA, found that albatross *Phoebastria* spp. attacked baits at significantly lower rates when baits were dyed blue.

Use high-visibility longlines to reduce seabird bycatch

We captured no intervention-based evidence on the impact of high-visibility longlines on seabird bycatch.

Use a sonic scarer when setting longlines to reduce seabird bycatch

A single study from the South Atlantic found that seabirds only temporarily changed behaviour when a sonic scarer was used, and seabird bycatch rates did not appear to be lower on lines set with a scarer.

Weight baits or lines to reduce longline bycatch of seabirds

Three replicated and controlled studies from the Pacific found lower bycatch rates of some seabird species on weighted longlines. An uncontrolled study found low bycatch rates with weighted lines but that weights only increased sink rates in small sections of the line. Some species were found to attack weighted lines more than control lines.

Use shark liver oil to deter birds when setting lines

Two out of three replicated and controlled trials in New Zealand found that fewer birds followed boats or dived for baits when non-commercial shark oil was dripped off the back of the boat.

Thaw bait before setting lines to reduce seabird bycatch

A study from Australia found that longlines set using thawed baits caught significantly fewer seabirds than controls.

Reduce seabird bycatch by releasing offal overboard when setting longlines

Two replicated and controlled studies in the South Atlantic and sub-Antarctic Indian Ocean found significantly lower seabird bycatch rates when offal was released overboard as lines were being set.

Use bird exclusion devices such as 'Brickle curtains' to reduce seabird mortality when hauling longlines

A single replicated study found that Brickle curtains reduced the number of seabirds caught, when compared to an exclusion device using only a single boom. Using purse seine buoys as well as the curtain appeared to be even more effective, but sample sizes did not allow useful comparisons.

Use acoustic alerts on gillnets to reduce seabird bycatch

A randomised, replicated and controlled trial in a coastal fishery in the USA found that fewer guillemots (common murres) *Uria aalge* but not rhinoceros auklets *Cerorhinca monocerata* were caught in gillnets fitted with sonic alerts.

Use high visibility mesh on gillnets to reduce seabird bycatch

A single randomised, replicated and controlled trial in a coastal fishery in the USA found that fewer guillemots (common murres) *Uria aalge* and rhinoceros auklets *Cerorhinca monocerata* were caught in gillnets with higher percentages of brightly coloured netting. However, such netting also reduced the catch of the target salmon.

Reduce gillnet deployment time to reduce seabird bycatch

We found no evidence for the effects of reducing gillnet deployment time on seabird bycatch rates.

Mark trawler warp cables to reduce seabird collisions

A single replicated and controlled study in Argentina found lower seabird mortality (from colliding with warp cables) when warp cables were marked with orange traffic cones.

Reduce 'ghost fishing' by lost/discarded gear

We captured no evidence for the effects of reducing ghost fishing on seabird bycatch rates or populations.

Reduce bycatch through seasonal or area closures

We captured no evidence for the effects of seasonal or area closures on seabird populations or bycatch rates.

Reducing exploitation and conflict

8.1. Use legislative regulation to protect wild populations

- Six out of seven (1,3,5,6,8,9) before-and-after studies and two literature reviews/meta-analyses (4,7) found evidence that legislation protects bird populations.
- Five studies in Europe (3,8,9), Indonesia (6) and across the world (4) found increased population levels of vultures, raptors, cranes and cockatoos following protective legislation (amongst other interventions). However, one (9) found populations of raptors declined soon after. The literature review (4) also found two cases of limited or no impact of legislation.
- Two before-and-after studies from Denmark (1) and the USA and Canada (5) and the meta-analysis (7) found increased estimated survival of falcons, ducks and parrots with stricter protection, but not necessarily population-level responses.
- A meta-analysis (7) found decreased harvest of parrots in areas with stricter protection regimes, but a before-and-after study (2) found no evidence for reduced shearwater harvest with legislation.

Background

Perhaps the most commonly used intervention in response to declining species is to provide legal protection for the species. This alone may be enough to protect a species or population, but if not, de facto or 'on-the-ground' protection may also be required to ensure people abide by the law.

A before-and-after study examining 524 kestrels *Falco tinnunculus* recovered during 1917–80 in Denmark (1) found that estimated survival rates of birds ringed as chicks increased during 1967–72 (66% annual survival) compared to 1945–66 (50%), following the introduction of legal protection for all birds of prey in 1967. However, the increase in survival rate following kestrel-specific legislation in 1926 was insignificant (45% for 1917–25 vs. 55% for 1926–39) and there was a significant fall in 1973–80 (to 53%). There were similar (although insignificant) patterns for birds ringed as juveniles or adults. There were significant decreases in the proportion of recoveries that were shot following each piece of legislation, from 1917–25 (59% of 29) to 1926–39 (14% of 35) and again from 1945–66 (17% of 76) to 1976–80 (2% of 192).

A before-and-after study at non-commercially exploited short-tailed shearwater *Puffinus tenuirostris* colonies in Tasmania, Australia (2), found that attempts to reduce non-commercial harvests of shearwater chicks in 1979 did not reduce the percentage of chicks taken, which remained at over 90% until 1985. Legal measures included shortening the non-commercial harvest season (so that it ended in mid- not late-April), reducing the daily bag limit, closing colonies on rotation, protecting colonies, education and stricter regulation enforcement. Takes at the time of the study were considered to be well in excessive of the maximum sustainable yield at the colonies.

A before-and-after study in the western Pyrenees, Spain (3), found that the population of griffon vultures *Gyps fulvus* increased from 282 pairs (in 23 colonies) in 1969–75 to 1,097 pairs (46 colonies) in 1989 following the initiation of multiple conservation interventions including: legal protection (in 1973); the creation of a reserve at nine breeding colonies (one in 1976, eight in 1987); the banning of strychnine (in 1984); and the installation of feeding stations between 1969 and 1979. Surveys in 1979 and 1984 found 364 pairs (in 26 colonies), and 589 pairs (32 colonies) respectively. This study is also discussed in 'Habitat protection', 'Restrict certain pesticides or other agricultural chemicals' and 'Provide supplementary food to increase adult survival'.

A 1998 literature review of crane *Grus* spp. conservation (4) described how two of four case studies found a positive response of crane populations to legal protection, with one showing partial success and the final study finding that legal protection had no impact. A small population of red-crowned cranes *G. japonensis* remained stable and increased slightly (from 20 to 35 birds between 1925–52) following legal protection at Kushiro Marsh, Hokkaido, Japan; whilst the American population of migratory sandhill cranes *G. canadensis* increased dramatically following a ban on hunting. Whooping cranes *G. americana* continued to decline following protection in 1916, but after public education, only four birds are thought to have been shot (see 'Use education programmes and local engagement to help reduce pressures on species'). The central Asian population of Siberian cranes *G. leucogeranus* continued to decline in India, from 200 in 1964–5 to three in 1996–7 despite protection in flyways in Pakistan. This study is also discussed in 'Habitat protection', 'Use education programmes and local engagement to help reduce pressures on species', 'Mark power lines to reduce incidental mortality', 'Provide supplementary food to increase adult survival' and 'Release captive-bred individuals'.

A before-and-after study in six areas in Canada and the USA (5) found that (for areas with more than 100 ducks recovered), the introduction of progressively more restrictive legislation on harvesting American black ducks *Anas rubripes* in 1967 and 1983 (USA) and 1984 (Canada) appeared to reduce hunting mortality: recovery rates of ringed ducks fell by 14.5% (adults) and 7% (immatures) from 1955–66 to 1967–82 and by 37% (adults) and 27% (immatures) from 1967–82 to 1983–93. Models calculating survival rates, however, estimated that these changes would not necessarily see a corresponding increase in survival.

A before-and-after study on Sumba, Indonesia (6), found that estimated population densities of citron-crested cockatoos *Cacatua sulphurea citrinocristata* increased between 1992 and 2002, following the imposition of a ban on trade in wild-caught birds in 1993. Increases were seen over the entire study and at two out of four forest sites (by 130–700%). One further site showed no change in density and the final site a possible decrease. No evidence was found for forest contraction (i.e. increased densities are thought to reflect an increase in total population sizes), total recorded birds increased by 56% from 1992–2002; significantly more birds were in groups of two or more and the estimated population size was 90% larger in 2002. The authors note that the trade ban has not been enforced perfectly, but that it has significantly reduced the number of wild-caught birds being traded.

A literature review and meta-analysis (7) found that protective legislation reduced nest-take (the percentage of nests from which chicks were removed by people) in wild parrots and may increase nest success. Across 20 species-country combinations, medium (involving either national or local protection) and high (involving both) levels of protection significantly reduced nest-take between 4.5 and 50 times compared to low levels of protection (unenforced, ambiguous or absent local or national protection). These results excluded a Nigerian study based on trapper surveys and with 100% nest-take, but included Australian data – Australia being the most developed country in the analysis and with a disproportionate number of studies. If both countries were excluded then medium and high protection resulted in a significant (170%) increase in nest success. However, if all data were included then there was a non-significant increase in nest success.

A controlled, before-and-after study from 1970–2000 across Europe (8) found that target bird species appeared to benefit from the passing of the EU Birds (79/409/EEC, est. 1979) and Habitats (92/43/EEC, est. 1992) Directives. This study is discussed in 'Habitat protection – Legally protect habitats'.

A before-and-after study on a grouse moor in Dumfries and Galloway, south Scotland (9), found that the numbers of hen harriers *Circus cyaneus* and peregrine falcons *Falco peregrinus* increased after birds were given full protection from persecution in 1990 (harriers increased from two pairs in 1992 to 20 pairs in 1997, whilst peregrines increased from two to six pairs). However, following the discontinuation of moor management in 2000, harriers declined again to two to four pairs in 2003–6 (see 'Shrubland modifications – Use fire suppression or control'). Both species were legally protected since 1961, but until 1990 many were still killed illegally on the moor. Three wader species and red grouse *Lagopus lagopus* all declined following harrier protection and the cessation of management. Meadow pipits *Anthus pratensis* and stonechats *Saxicola rubicola* both declined as harriers increased but increased again after 2000. Carrion crows *Corvus corone* increased from 2000, after they were no longer shot by gamekeepers.

(1) Noer, H. & Secher, H. (1983) Survival of Danish kestrels *Falco tinnunculus* in relation to protection of birds of prey. *Ornis Scandinavica*, 14, 104–114.
(2) Skira, I. J., Wapstra, J. E., Towney, G. N. & Naarding, J. A. (1986) Conservation of the short-tailed shearwater *Puffinus tenuirostris* in Tasmania, Australia. *Biological Conservation*, 37, 225–236.
(3) Donazar, J. A. (1990) Population trends of the griffon vulture *Gyps fulvus* in northern Spain between 1969 and 1989 in relation to conservation measures. *Biological Conservation*, 53, 83–91.
(4) Davis, C. (1998) A review of the success of major crane conservation techniques. *Bird Conservation International*, 8, 19–30.
(5) Francis, C. M., Sauer, J. R. & Serie, J. R. (1998) Effect of restrictive harvest regulations on survival and recovery rates of American black ducks. *The Journal of Wildlife Management*, 62, 1544–1557.
(6) Cahill, A. J., Walker, J. S. & Marsden, S. J. (2006) Recovery within a population of the critically endangered citron-crested cockatoo *Cacatua sulphurea citrinocristata* in Indonesia after 10 years of international trade control. *Oryx*, 40, 161–167.
(7) Pain, D. J., Martins, T. L. F., Boussekey, M., Diaz, S. H., Downs, C. T., Ekstrom, J. M. M., Garnett, S., Gilardi, J. D., McNiven, D., Primot, P., Rouys, S., Saoumoe, M., Symes, C. T., Tamungang, S. A., Theuerkauf, J., Villafuerte, D., Verfailles, L., Widmann, P. & Widmann, I. D. (2006) Impact of protection on nest take and nesting success of parrots in Africa, Asia and Australasia. *Animal Conservation*, 9, 322–330.
(8) Donald, P. F., Sanderson, F. J., Burfield, I. J., Bierman, S. M., Gregory, R. D. & Waliczky, Z. (2007) International conservation policy delivers benefits for birds in Europe. *Science*, 317, 810–813.
(9) Baines, D., Redpath, S., Richardson, M. & Thirgood, S. (2008) The direct and indirect effects of predation by hen harriers *Circus cyaneus* on trends in breeding birds on a Scottish grouse moor. *Ibis*, 150, 27–36.

8.2. Increase 'on-the-ground' protection to reduce unsustainable levels of exploitation

- Two before-and-after studies from Central America (1) and Europe (2) found increases in recruitment and population levels following either stricter anti-poaching measures (1) or stricter protection and the stationing a warden on an island (2).
- However, the Central American study (1) found that recruitment increases were only maintained for as long as the intensive effort was continued.

Background

Whilst providing legal protection for species may be effective on its own, it is possible that those exploiting a species or population will not stop unless there is 'on-the-ground' or de facto protection – wardens or guards at a site preventing exploitation. However, increasing such protection is expensive and may be unpopular with local people if it is done in an insensitive way.

A before-and-after study in western Costa Rica (1) found an increase in a scarlet macaw *Ara macao* population from 185–225 individuals in 1990–4 to 225–265 in 1997–2003, following an increase in anti-poaching patrols and the confiscation of ladders and tree-climbing equipment (used to remove nestlings from nests) and several other interventions (see 'Use education programmes and local engagement to reduce pressures on species', 'Promote sustainable alternative livelihoods based on species', 'Provide artificial nesting sites' and 'Guard nests to increase nest success'). In 1990–4 the population had been showing a 4%/year decline. In addition, following the start of intensive anti-poaching activities, the young-to-adult ratio (which is related to recruitment rate) was 9% in 1995–6 (compared to an average of 6% for

1990–2003). However, the intensity of the anti-poaching effort could not be maintained and when it was reduced the ratio fell back to 6%.

A before-and-after study on Selvagem Grande, Madeira, Portugal (2), found that the population of Cory's shearwaters *Calonectris diomedea borealis* increased from approximately 7,000 pairs in 1980 and only 64 chicks on the island in 1976, to 18,100 breeding pairs in 1995 (a 5% annual increase) following the installation of a permanent warden and stricter *de facto* protection on the island. Before this, a series of severe harvesting events by Portuguese and Spanish fishermen in 1975–6 had reduced the population from 130,000–150,000 in the early 1900s, despite the *de jure* protection of the island from 1971. Despite a 13% decrease over 1995–8, the population was estimated at 29,540 pairs in 2005.

(1) Vaughan, C., Nemeth, N. M., Cary, J. & Temple, S. (2005) Response of a scarlet macaw *Ara macao* population to conservation practices in Costa Rica. *Bird Conservation International*, 15, 119–130.
(2) Granadeiro, J. P., Dias, M. P., Rebelo, R., Santos, C. D. & Catry, P. (2006) Numbers and population trends of Cory's shearwater *Calonectris diomedea* at Selvagem Grande, northeast Atlantic. *Waterbirds*, 29, 56–60.

8.3. Promote sustainable alternative livelihoods

- A single before-and-after study in Costa Rica (1) found an increase in a scarlet macaw *Ara macao* population following several interventions including the promotion of sustainable, macaw-based livelihoods.

Background

Conserving biodiversity and eliminating poverty are linked global challenges. The poor, particularly the rural poor, depend on nature for many elements of their livelihoods, including food, fuel, shelter and medicines. Working alongside people who will ultimately benefit from conservation can build social capital, improve accountability and reduce poverty.

A before-and-after study in western Costa Rica (1) found an increase in a scarlet macaw *Ara macao* population from 185–225 individuals in 1990–4 to 225–265 in 1997–2003, following the promotion of economic incentives relating to macaws and several other interventions ('Use education programmes and local engagement to help reduce persecution or exploitation of species', 'Provide artificial nesting sites' and 'Guard nests to increase nest success'). In 1990–4 the population had been showing a 4%/year decline. This study is discussed in more detail in 'Increase 'on-the-ground' protection to reduce unsustainable levels of exploitation'.

(1) Vaughan, C., Nemeth, N. M., Cary, J. & Temple, S. (2005) Response of a scarlet macaw *Ara macao* population to conservation practices in Costa Rica. *Bird Conservation International*, 15, 119–130.

8.4. Use education programmes and local engagement to help reduce persecution or exploitation of species

- Five out of six studies (1–5) from across the world found increases in bird populations or decreases in mortality following education programmes. In all but one case *(1)*, education was one of several interventions employed.
- A replicated before-and-after study from Canada (1) also found that there was a significant shift in local peoples' attitudes to conservation and exploited species following educational programmes.
- One study from Venezuela (6) found no evidence for decreases in yellow-shouldered parrot *Amazona barbadensis* poaching following an educational programme in local schools. The authors argue that the benefits would probably be seen later in the project.

Background

If a community relies on exploiting bird populations then they may well want to exploit the population sustainably, to ensure it persists and can be exploited in the future. However, exactly how to exploit a population sustainably may not be clear, and so education programmes may be useful in reducing over-exploitation.

Other, more general education programmes are discussed in 'Education and community development'.

A replicated, before-and-after study from 1978–1988 using 141 heads-of-households interviewed before (1981–1982) and after (1988) the wide-scale implementation of the Marine Bird Conservation Project in the coastal North Shore region of Quebec, Canada (1) found that conservation behaviour and seabird populations significantly increased after educational campaigns (including hands-on lessons, field trips, local volunteering, academic materials and creative productions). All seabird populations, especially the alcids (average increase >50%), significantly increased from 1978–1988 following a decline from 1955–1978. There were significant reductions in respondents who believed that it should be legal to hunt Atlantic puffins *Fratercula arctica* (54% to 30%), razorbills *Alca torda* (59% to 38%), and common guillemot *Uria aalge* (76% to 65%) but no change in perception of common eider *Somateria mollissima* or herring gull *Larus argentatus* hunting. The percentage of family members involved in hunting dropped significantly from 76% to 48% and the average number of birds reported as needed per year dropped from 44 to 24.

A ten-year study of a griffon vulture *Gyps fulvus* reintroduction programme in river gorges in Aveyron, southern France (2) found that an education programme run at the same time appeared to reduce persecution of vultures. No shooting or poisoning was recorded in the study area and only a single nest was disturbed by a climber. In addition, farmers were reported to be leaving carcasses in fields more frequently, thus providing a source of food for the vultures. No details are provided on the education programme. The reintroduction programme itself is discussed in 'Release captive-bred individuals' and 'Release birds as adult or subadults, not juveniles'.

A 1998 literature review of crane *Grus* spp. conservation (3) describes how whooping cranes *G. americana* continued to decline in the USA following legal protection, until intensive public education programmes. Before education, fewer than half the recorded whooping crane mortalities were due to natural causes, but between 1968 and 1998, only four whooping cranes are known to have been shot. This review is discussed in more detail in 'Habitat protection', 'Use legislative regulation to protect wild populations', 'Mark power lines to reduce incidental mortality', 'Provide supplementary food to increase adult survival' and 'Release captive-bred individuals'.

A 1998 review of a yellow-shouldered amazon *Amazona barbadensis* release programme on Margarita Island, Venezuela (4), found that the population on the island increased from 750 to approximately 1,900 individuals between 1989 and 1996. The authors argue that the success was dependent to a large degree on a five-year public education and awareness programme, with promotions such as making the parrot the state bird, visiting local schools, involving youth conservation brigades and local news reports. The details of the reintroduction programme are in 'Release captive-bred individuals', 'Artificially incubate and hand-rear birds in captivity' and 'Foster eggs or chicks with wild conspecifics'.

A before-and-after study in western Costa Rica (5) found an increase in a scarlet macaw *Ara macao* population from 185–225 individuals in 1990–4 to 225–265 in 1997–2003, following the formation of a local conservation organisation; environmental education programmes; meetings with local stakeholders and several other interventions ('Promote sustainable alternative livelihoods based on species', 'Provide artificial nesting sites' and 'Guard nests to increase nest success'). In 1990–4 the population had been showing a 4%/year decline. This study is discussed in more detail in 'Increase "on-the-ground" protection to reduce unsustainable levels of exploitation'.

A replicated study from 2000–2003 (part of a longer study from 2000–2009) of 10 monitored yellow-shouldered parrot *Amazona barbadensis* nests in tropical forest habitat on Margarita Island, Venezuela (6), found that there was no short-term decrease in poaching rates after raising environmental awareness at schools. Poaching increased from 25% (during 1990–1999) to 100% in 2000–2003 of monitored fledglings. The environmental education programme, focused on older elementary schoolchildren (8–13 years old), was established by providing information, training, resources and support to local elementary school teachers. Schools hosted environmental days and started environmental brigades. At the end of each breeding season, a 'parrot festival' was organised by the people of one of the towns. The authors point out that the benefits of this programme are likely to be detected at later stages of the project. This study is also discussed in 'Relocate nestlings to reduce poaching', 'Provide artificial nest sites', 'Employ locals as biomonitors' and 'Foster eggs or chicks with wild conspecifics'.

(1) Blanchard, K. A. & Monroe, M. C. (1990) Culture and conservation: strategies for reversing population declines in seabirds. *Endangered Species Update*, 7, 1–5.
(2) Sarrazin, F., Bagnolini, C., Pinna, J. L., Danchin, E. & Clobert, J. (1994) High survival estimates of griffon vultures (*Gyps fulvus fulvus*) in a reintroduced population. *The Auk*, 111, 853–862.
(3) Davis, C. (1998) A review of the success of major crane conservation techniques. *Bird Conservation International*, 8, 19–30.
(4) Sanz, V. & Grajal, A. (1998) Successful reintroduction of captive-raised yellow-shouldered Amazon parrots on Margarita Island, Venezuela. *Conservation Biology*, 12, 430–441.
(5) Vaughan, C., Nemeth, N. M., Cary, J. & Temple, S. (2005) Response of a scarlet macaw *Ara macao* population to conservation practices in Costa Rica. *Bird Conservation International*, 15, 119–130.
(6) Briceño-Linares, J. M., Rodríguez, J. P., Rodríguez-Clark, K. M., Rojas-Suárez, F., Millán, P. A., Vittori, E. G. & Carrasco-Muñoz, M. (2011) Adapting to changing poaching intensity of yellow-shouldered parrot (*Amazona barbadensis*) nestlings in Margarita Island, Venezuela. *Biological Conservation*, 144, 1188–1193.

8.5. Employ local people as 'biomonitors'

- A single replicated study in Venezuela (1) found that poaching of parrot nestlings was significantly lower following the employment of five young men as 'biomonitors'.

Background

Employing local people in conservation can give them an active stake in ensuring that species or habitats survive. In addition, if locals are used to reduce threats such as poaching then those responsible may respond more positively than to police or conservationists.

A replicated study in 2004 (part of a longer study from 2000–2009) of 10 monitored yellow-shouldered parrot *Amazona barbadensis* nests in tropical forest habitat on Margarita Island, Venezuela (1) found that the recruitment of young people as nest biomonitors significantly decreased poaching rate. Implementation of 24 h surveillance by the biomonitors resulted in a decrease in poaching from nearly 100% between 2000 and 2003, to 56% in 2004. A team of five young people of a similar age, background and social context as poachers were recruited from local communities to monitor nests. The authors point out that it was their hope that the biomonitors would be well-placed to encourage positive conservation action by peers. This study is also discussed in 'Relocate nestlings to reduce poaching', 'Use education programmes and local engagement to help reduce pressures on species', 'Provide artificial nest sites' and 'Foster eggs or chicks with wild conspecifics'.

(1) Briceño-Linares, J. M., Rodríguez, J. P., Rodríguez-Clark, K. M., Rojas-Suárez, F., Millán, P. A., Vittori, E. G. & Carrasco-Muñoz, M. (2011) Adapting to changing poaching intensity of yellow-shouldered parrot (*Amazona barbadensis*) nestlings in Margarita Island, Venezuela. *Biological Conservation*, 144, 1188–1193.

8.6. Mark eggs to reduce their appeal to egg collectors

- A single before-and-after (1) study found that marking eggs greatly increased the number of chicks fledging from six raptor nests in Australia in 1979 and 1980.

Background

Egg collecting involves removing eggs from nests, blowing them to remove the contents and presenting the eggs in a collection. This practice can be extremely damaging for some slow-reproducing species (Ewins 1997) and is made more so by the greater value of, and demand for, rarer species' eggs. This has led to the prohibition of egg collecting in many countries, but the practice still continues.

Marking eggs in a non-damaging way may reduce the desirability of eggs because collectors are interested in eggs for their aesthetic appeal.

Ewins, P. J. (1997) Osprey (*Pandion haliaetus*) populations in forested areas of North America: changes, their causes and management recommendations. *Journal of Raptor Research*, 31, 138–150.

A small before-and-after study in the Flinders Ranges, South Australia (1), found that twice as many young fledged from five peregrine falcon *Falco peregrinus* nests (12 young fledged from 16 eggs) and one wedge-tailed eagle *Aquila audax* nest (one chick from two eggs) in 1980, when eggs were marked with a single line drawn in black, waterproof ink, as in 1979, when no eggs were marked (6/15 and 0/2 fledged respectively).

(1) Olsen, J., Billet, T. & Olsen, P. (1982) A method for reducing illegal removal of eggs from raptor nests. *Emu*, 82, 225.

8.7. Relocate nestlings to reduce poaching

- A replicated before-and-after study in Venezuela (1) found significant decreases in poaching rate and increased fledging rates of parrots after wild chicks were moved into police premises each night.

Background

Nestlings are particularly vulnerable to exploitation, as they cannot fly and are confined to nests. This is especially problematic with species such as parrots that are exploited for the pet trade, as nestlings can be easily raised in captivity and sold on. Moving nestlings into safer areas may reduce the threat of exploitation, but is likely to be expensive or time-consuming. In the study below, chicks were taken from nests overnight and then returned in the morning, something that is only likely to be possible for species under very intense management.

A replicated before-and-after study in 2008–2009 (part of a longer study from 2000–2009) in 15 monitored yellow-shouldered parrot *Amazona barbadensis* nests in tropical forest habitat on Margarita Island, Venezuela (1) found that moving nestlings into municipal police premises overnight significantly decreased poaching rates. In 2008, the municipal police received the birds nightly during the breeding season, which brought poaching rates down from 60% at the end of 2007 to 16% in 2008 and 1% in 2009 (with the help of the National Guard). The Macanao municipal police helped with surveillance in the field and escorted the fledglings every night to police headquarters instead of the local field base. In 2009, birds were taken nightly to the National Guard headquarters. Overall, the fledging rate doubled while the poaching rate was halved from 2000–2009 compared to pre-intervention period of 1990–1999 (3.8 and 1.6 birds/nest; 25% and 49% respectively). This study is also discussed in 'Use education programmes and local engagement to help reduce pressures on species',

'Provide artificial nesting sites', 'Employ locals as biomonitors' and 'Foster eggs or chicks with wild conspecifics'.

(1) Briceño-Linares, J. M., Rodríguez, J. P., Rodríguez-Clark, K. M., Rojas-Suárez, F., Millán, P. A., Vittori, E. G. & Carrasco-Muñoz, M. (2011) Adapting to changing poaching intensity of yellow-shouldered parrot (*Amazona barbadensis*) nestlings in Margarita Island, Venezuela. *Biological Conservation*, 144, 1188–1193.

8.8. Use wildlife refuges to reduce hunting disturbance

- Three studies from the USA (1) and Europe (2,3) found that bird densities were higher in refuges where hunting was prohibited, compared to areas with hunting. In addition, two studies (2,3) found that more birds used hunting-free areas during the open season and on hunting days.
- No studies investigated the population-level impacts of these refuges.

Background

Wildlife refuges are a type of protected area where hunting is prohibited. Often situated near hunting areas, they can be effective both by reducing the number of birds shot, but also in reducing the disturbance to non-target species. Other interventions designed to reduce disturbance are discussed in 'Threat: Disturbance'.

A site comparison study from 1940–1951 in two natural, 400 private and 13 public waterfowl refuges of wetland habitat in Illinois, USA (1), found that waterfowl refuges should cover at least 400 ha if shooting is permitted. Refuges where hunting is prohibited exhibit higher waterfowl abundance (for example, duck populations increased on average by 37,075 ducks/refuge over seven years in sites where hunting became closed). Similarly, hunting-restricted refuges exhibit greater duck usage (4,010, 911 and 56 duck-days/ha over 50 days for a non-hunting refuge, a hunting refuge and a non-refuge respectively). Refuge size affects hunting impact: one smaller refuge containing higher concentrations of duck food than a larger, nearby refuge exhibits significantly lower average duck density (1,504 compared to 4,327 ducks/ha), but significantly higher hunting pressure (15 compared to 8% of the population hunted). In total, refuges cover 23,209 ha, of which 2,023 ha are open to hunting.

A study on a lake in Northern Ireland, UK, in the boreal winter of 1997–8 (2), found that significantly more wildfowl were found on a bay used as a wildlife refuge (i.e. closed to hunting) during the hunting season, compared with the closed season (average of 1,027 individuals on the lake during open season vs. 597 during the closed season). A significant increase in usage was also observed within the open season at weekends, when hunting intensity was highest, a pattern most noticeable for mallard *Anas platyrhynchos* and common coot *Fulica atra*. There was a corresponding decrease in wildfowl numbers in an area of the lake used for shooting. A total of 20 waterfowl species were recorded at the refuge, the most common being mallard, common goldeneye *Bucephala clangula*, tufted duck *Aythya fuligula* and common coot.

A controlled study from October–December in 2003 on one 10 km^2 site within intensively cultivated farmland in Tauché and Sainte Blandine villages, France (3), found that northern lapwings *Vanellus vanellus*, golden plovers *Pluvialis apricaria* and little bustards *Tetrax tetrax* were affected by hunting activities and used hunting-free areas in response. Hunting activity increased flight probability and time spent vigilant (higher on hunting days than just before and after a hunting day), to the detriment of resting. Foraging was unaffected by hunting. The hunting-free reserve was used significantly more frequently during hunting days. Little bustards used the hunting-free reserve almost exclusively (96% of observations within hunting-free reserve). The authors suggest that reserves can mitigate the disturbance caused by hunting.

(1) Bellrose, F. C. (1954) The value of waterfowl refuges in Illinois. *The Journal of Wildlife Management*, 18, 160–169.
(2) Evans, D. M. & Day, K. R. (2002) Hunting disturbance on a large shallow lake: the effectiveness of waterfowl refuges. *Ibis*, 144, 2–8.

(3) Casas, F., Mougeot, F., Viñuela, J. & Bretagnolle, V. (2009) Effects of hunting on the behaviour and spatial distribution of farmland birds: importance of hunting-free refuges in agricultural areas. *Animal Conservation*, 12, 346–354.

8.9. Introduce voluntary 'maximum shoot distances'

- A replicated, randomised before-and-after study from Denmark (1) found that significantly fewer pink-footed geese *Anser brachyrhynchus* were wounded but not killed, following the implementation of a voluntary maximum shooting distance.

Background

The probability of successfully killing a bird when shooting decreases with distance from the bird. If hunting is regulated by 'bag limits' (i.e. the number of birds individuals can harvest), then shooting at birds from a long distance may be detrimental, as many birds are hit but do not die instantly. These escape and die later but are not collected and so are not included in a bag limit. This means that actual mortality from hunting is considerably higher than the set bag limits.

A replicated, randomised before-and-after study in March from 1998–2005 in one wetland area in which 150–500 pink-footed geese *Anser brachyrhynchus* were monitored annually in Jutland, Denmark (1) found that the implementation of a voluntary restriction on maximum shooting distance (25 m) in 1997 significantly reduced injury rate during the hunting season. The proportion of wounded first-year and older geese significantly decreased over the study period (7–11% and 18% decrease by 2005 respectively). A simple population dynamic model predicted these decreases to be consistent with a *c.* 60% reduction of numbers wounded for both age classes. Since 1997, the total annual number of harvested geese in Denmark increased from 15,000 to 30,000. Thus, the authors point out, reductions in numbers wounded did not appear to have had any negative impact on harvest size. A mobile surgical X-ray unit was used to screen for shot. Recaptures accounted for just 1% of the sample.

(1) Noer, H., Madsen, J. & Hartmann, P. (2007) Reducing wounding of game by shotgun hunting: effects of a Danish action plan on pink-footed geese. *Journal of Applied Ecology*, 44, 653–662.

8.10. Provide 'sacrificial' grasslands to reduce the impact of wild geese on crops

- Two studies in the UK (1,2) found that managing grasslands for geese increased the number grazing there. However, both found that the birds were moving within a relatively small area (i.e. within the study sites) and therefore the grasslands may not reduce conflict with farmers.

Background

There have been dramatic increases in many species of goose in recent decades (Madsen *et al.* 1999) and this has led to increasing conflict with farmers, as many species graze on arable land, potentially ruining crops. One potential solution, to reduce conflict whilst maintaining the populations of geese is to provide 'sacrificial grasslands' – areas set-aside for geese to feed on, which keeps them away from agricultural fields.

To be useful, such areas need to be more attractive than neighbouring fields. Interventions detailing specific management practices are mainly described in 'Threat: Natural system modifications', whilst the studies below describe whether the provision of sacrificial grasslands actually affects the distribution of geese in an area.

Madsen, J., Cracknell, G. and Fox, A. D. (1999) *Goose Populations of the Western Palearctic. A Review of Status and Distribution*, Wetlands International Publication No. 48. Wetlands International, Wageningen, The Netherlands.

A before-and-after study in Gloucestershire, England (1) found that up to 87% of geese on a grassland site used a 130 ha area managed for them in 1975–6. The interventions used are discussed in 'Reduce grazing intensity', 'Increase crop diversity' and 'Undersow spring cereals'.

A replicated, controlled trial in 1984–7 on a reserve on the island of Islay, west Scotland (2), found more barnacle geese *Branta leucopsis* used wet pasture fields if they were re-seeded or fertilised than if they were unmanaged. However, increases were due to a redistribution of local birds, rather than new birds visiting the reserve. The author therefore suggests that improving the reserve grasslands will only minimally reduce conflict with farmers elsewhere on the island. The details of management interventions are discussed in 'Re-seed grasslands' and 'Fertilise grasslands'.

(1) Owen, M. (1977) The role of wildfowl refuges on agricultural land in lessening the conflict between farmers and geese in Britain. *Biological Conservation*, 11, 209–222.
(2) Percival, S. M. (1993) The effects of reseeding, fertiliser application and disturbance on the use of grasslands by barnacle geese, and the implications for refuge management. *Journal of Applied Ecology*, 30, 437–443.

8.11. Move fish-eating birds to reduce conflict with fishermen

- A single before-and-after study in the USA (1) found that Caspian tern *Sterna caspia* chicks had a lower proportion of commercial fish in their diet following the movement of the colony away from an important fishery.

A before-and-after study in 1999–2001 on two small islands in the Columbia River Estuary, Oregon, USA (1), found that various interventions led to the relocation of a Caspian tern *Sterna caspia* colony (8,900 pairs in total) away from an important fishery. Tern chicks had a significantly lower proportion of juvenile commercial fish (such as coho salmon *Oncorhynchus kisutch*, steelhead *O. mykiss* and chinook salmon *O. tshawytscha*) in their diet following the movement of the colony to East Sand Island, compared with its original position on Rice Island (commercial fish making up 42% of the diet on East Sand Island vs. 83% on Rice Island, approximately 120 bill-loads sampled at each site each year). The predation of commercial fish by terns was a significant source of conflict with local fishermen, and translocation may well have reduced this conflict. Tern productivity was significantly higher at East Island than Rice Island in every year of the study. Individual interventions are discussed in 'Use decoys to attract birds to new nesting areas', 'Use vocalisations to attract birds to new nesting areas', 'Control avian predators on islands', 'Habitat restoration/creation – intertidal habitats' and 'Translocations – Alter habitat to encourage birds to leave an area'.

(1) Roby, D. D., Collis, K., Lyons, D. E., Craig, D. P., Adkins, J. Y., Myers, A. M. & Suryan, R. M. (2002) Effects of colony relocation on diet and productivity of Caspian terns. *The Journal of Wildlife Management*, 66, 662–673.

8.12. Scare fish-eating birds from areas to reduce conflict

Studies investigating scaring birds from fishing areas are discussed in 'Threat: Agriculture – Aquaculture'.

Reduce fisheries bycatch

Commercial fisheries have expanded dramatically across the world since the 1960s, and are having increasing direct impacts on seabirds worldwide. Commercial longline, trawl and gillnet fisheries are responsible for the deaths of hundreds of thousands of seabirds each year, threatening the survival of a number of species, especially albatrosses.

Longline fleets set more than one billion hooks each year and up to 300,000 seabirds, including 100,000 albatrosses, are killed on longlines as they scavenge for bait and offal and are accidentally caught on the fishhooks and drowned. This form of bycatch is now the single greatest threat to albatrosses. All 22 albatross species are classed as globally threatened or Near Threatened.

Trawl fisheries are also a threat, with birds killed when they are hit by the powerful cables that attach the trawl net to the vessel, and by being entangled by the net as they scavenge for fish.

Gillnets – static curtains of netting designed to entangle fish by their gills – can be responsible for the accidental entanglement of large numbers of pelagic birds. Despite a ban on their use in the high seas, commercial gillnet fisheries continue to operate in territorial and coastal waters around the world, where they pose a significant threat to numerous seabird populations, especially 'pursuit-diving' species, such as divers (loons), grebes, sea ducks, auks and cormorants.

This bycatch is extremely detrimental to bird populations, but is also expensive for fishermen, as each bait taken by a bird is one lost for catching fish. Methods to prevent bycatch are therefore also likely to increase profits. Both conservation-based and commercial fisheries organisations are now working together to try and uncover which methods are the most effective and then to ensure that every boat employs them.

A literature review in 2000 (Melvin & Robertson 2000) argued that much of the research into bycatch is confused and non-standardised, with different methodologies being used and experiments often uncontrolled. Bycatch can often vary hugely from one voyage to the next and from year to year, due to movements of birds and changes in fish stocks, but also because of factors, such as the phase of the moon, which alter the effectiveness of, for example, setting lines at night. Evidence from single-year comparisons or uncontrolled tests therefore need to be treated with caution.

Melvin, E. F. and Robertson, G. (2000) Appendix 3. Seabird mitigation research in longline fisheries: status and priorities for future research and actions. In: Cooper, J. (Ed.). Albatross and Petrel Mortality from Longline Fishing International Workshop, Honolulu, Hawaii, USA, 11–12 May 2000. Report and presented papers. *Marine Ornithology*, 28, 179–182.

8.13. Set longlines at night to reduce seabird bycatch

- Six out of eight studies from fisheries around the world (1–3, 5–7) found lower rates of seabird bycatch on longlines set at night, compared with during the day, or with previously collected data (5). However, effects seemed to depend on the species caught.
- Two studies (4,8) found higher rates of bycatch on night-set longlines, due to high numbers of white-chinned petrels *Procellaria aequinoctialis* or northern fulmars *Fulmarus glacialis* being caught at night.

Background

Most seabirds feed during the day and so longlines set at night may catch fewer birds. Some species, however, do hunt at night and so night-setting is unlikely to be an effective mitigation measure. An adequate knowledge of bycatch species' ecology is therefore important in determining whether to promote night-setting.

Many of the studies in this section do not explicitly say whether night-setting was done for conservation purposes or not. However, given the valuable information they contain, we have included them below.

A study using data from bluefin tuna *Thunnus thynnus* boats operating off New Zealand between 1988 and 1992 (1) found that the effects of night-setting on seabird bycatch rates varied between fishing areas, probably due to different species making up the majority of bycatch. In southern areas, 87% of 47 identified birds caught were albatrosses and 73% of 88 birds were caught between 0600 and 1400 (when 41% of 1,009 lines were set). In contrast, in northern fishing grounds, where 59% of 75 identified birds were petrels, 44% of the 181 birds caught were hooked within 90 minutes of dawn or dusk and 42% were caught at night (when 54% of 1,180 lines were set).

A small study on a commercial longlining boat in the South Atlantic in March–April 1993 (2) found that no birds were caught on seven lines, set predominantly at night. The number of birds following the boat increased during the day: seven birds were seen following the boat during setting before 05:00 hr (first light was at 04:00 hr), with 84 being seen between 05:00 hr and 07:30 hr and several hundred following the boat during daytime hauling operations. This study also investigated the impact of streamer lines (see 'Use streamer lines to reduce seabird bycatch on longlines').

A replicated, controlled study on a commercial fishing boat in the South Atlantic in April–May 1994 (3) found that longlines set at night appeared to catch fewer birds as bycatch, and of fewer species, compared to lines set during the day (15 birds caught on 16 lines set at night, all white-chinned petrels *Procellaria aequinoctialis* vs. 83 birds of six species on four lines set during the day). A further four day-set and three night-set lines caught 1–5 birds each but were not analysed separately. The vessel was fishing for Patagonian toothfish *Dissostichus eleginoides* off South Georgia, UK. This study is also discussed in 'Use streamer lines to reduce seabird bycatch on longlines' and 'Use a sonic scarer when setting longlines to reduce seabird bycatch'.

A replicated and controlled study (4) in the Patagonian toothfish *Dissostichus eleginoides* fishery in the South Atlantic found that 38 birds (two grey-headed albatross *Thalassarche chrysostoma*, formerly *Diomedea chrysostoma*, and 36 white-chinned petrels *Procellaria aequinoctialis*) caught on 72 longlines (174,000 hooks) set in February 1994, were caught at a much higher rate on lines set at night, than during the day (1.00 vs. 0.38 birds/1,000 hooks). The study took place around South Georgia (UK) and Kerguelen Islands (France) (sectors 332 and 333). This study is also discussed in 'Turn decklights off during night-time setting of longlines to reduce bycatch'.

A study in March–May 1997 near South Georgia, South Atlantic (5) found that 61 line sets, laid at night from a Patagonian toothfish *Dissostichus eleginoides* vessel, caught a total of 12 birds. This was the equivalent of 0.1 birds/1,000 hooks, which the authors' state is considerably lower than many boats fishing near South Georgia, possibly due to night-setting of lines. Nine birds were white-chinned petrels *Procellaria aequinoctialis*, two black-browed albatross *Thalassarche melanophrys* and one was unidentified.

A replicated, controlled study using data from 86 longlining vessels operating in Australian waters, between April 1992 and March 1995 (6) found that longlines set at night caught approximately five times fewer seabirds than those set during the day (1.0 birds/1,000 hooks for 924 line-sets set at night vs. 4.8 birds/1,000 hooks for 1,372 line-sets set during the day). The difference was greatest on nights close to a new moon (with 7% of the bycatch rates of day sets), but there were always significant reductions (sets on full moon nights had approximately one-third the bycatch rates of day sets). This study does not discuss which birds were caught, but previous studies have shown that this fishery catches mainly albatross. This study is also discussed in 'Use bait throwers to reduce seabird bycatch' and 'Thaw bait to reduce seabird bycatch'.

A replicated, controlled study in the sub-Antarctic Indian Ocean in 1994–7 (7) found that longlines set at night had significantly lower seabird bycatch, compared to those set during the day (0.91 birds caught/1,000 hooks for day-set lines vs. 0.17 birds/1,000 hooks for night-set lines, total of 524 lines studied). This result was consistent across all species, except for wandering albatross *Diomedea exulans*, which was only caught on 12 occasions. The authors note that whilst the number of white-chinned petrels *Procellaria aequinoctialis* caught was half that caught during the day, these levels may still be unsustainably high: they quote an unpublished figure of 340 petrels caught during a single line set off Kerguelen Island. The study took place on three Ukranian and one Japanese longliners fishing for Patagonian toothfish *Dissostichus eleginoides* off Kerguelen Island (France). Deck lights were not switched on during night setting. This study is also discussed in 'Use streamer lines to reduce seabird bycatch on longlines' and 'Reduce seabird bycatch by releasing offal overboard when setting longlines'.

A replicated and controlled study (8) in the North Pacific in late summer 1999 and 2000, found that lines set at night or during sunrise had higher rates of seabird bycatch (0.13 and 0.07 birds/1,000 hooks respectively) than those set during the day or at sunset (0.02 and 0.01 birds/1,000 hooks respectively). Differences were due to bycatch rates of northern fulmars *Fulmarus glacialis* (the most numerous species caught and the only species caught at night). Shearwaters *Puffinus* spp. constituted 67% of species caught during the day, but were never caught during the night. A total of 490 line sets were studied from the Pacific cod *Gadus macrocephalus* and walleye pollock *Theragra chalcogramma* fishery south-east of the Pribilof Islands, USA. This study is also discussed in 'Weight baits or lines to reduce longline bycatch of seabirds', 'Use streamer lines to reduce seabird bycatch on longlines', 'Use a line shooter to reduce seabird bycatch' and 'Set lines underwater to reduce seabird bycatch'.

(1) Murray, T. E., Bartle, J. A., Kalish, S. R. & Taylor, P. R. (1993) Incidental capture of seabirds by Japanese southern bluefin tuna longline vessels in New Zealand waters, 1988–1992. *Bird Conservation International*, 3, 181–210.
(2) Ashford, J. R., Croxall, J. P., Rubilar, P. S. & Moreno, C. A. (1994) Seabird interactions with longlining operations for *Dissostichus eleginoides* at the South Sandwich islands and South Georgia. *CCAMLR Science*, 1, 143–153.
(3) Ashford, J. R., Croxall, J. P., Rubilar, P. S. & Moreno, C. A. (1995) Seabird interactions with longlining operations for *Dissostichus eleginoides* around South Georgia, April to May 1994. *CCAMLR Science*, 2, 111–121.
(4) Cherel, Y., Weimerskirch, H. & Duhamel, G. (1996) Interactions between longline vessels and seabirds in Kerguelen waters and a method to reduce seabird mortality. *Biological Conservation*, 75, 63–70.
(5) Ashford, J. R. & Croxall, J. P. (1998) An assessment of CCAMLR measures employed to mitigate seabird mortality in longlining operations for *Dissostichus eleginoides* around South Georgia. *CCAMLR Science*, 5, 217–230.
(6) Klaer, N. & Polacheck, T. (1998) The influence of environmental factors and mitigation measures on bycatch rates of seabirds by Japanese longline fishing vessels in the Australian region. *Emu*, 98, 305–316.
(7) Weimerskirch, H., Capdeville, D. & Duhamel, G. (2000) Factors affecting the number and mortality of seabirds attending trawlers and long-liners in the Kerguelen area. *Polar Biology*, 23, 236–249.
(8) Melvin, E. F., Parrish, J. K., Dietrich, K. S. & Hamel, O. S. (2001) *Solutions to seabird bycatch in Alaska's demersal longline fisheries*. Washington Sea Grant Program, University of Washington.

8.14. Turn deck lights off during night-time setting of longlines to reduce bycatch

- A single replicated and controlled study in the South Atlantic (1) found significantly lower bycatch rates when deck lights were turned off during line setting at night.

Background
Birds might be attracted to deck lights on fishing boats either to use them to forage, or because birds are generally attracted to lights at night (see 'Threat: Pollution' for interventions designed to reduce the threat of light pollution). In either case, birds may be vulnerable, so turning off the lights may reduce mortality.

A replicated and controlled study (1) in February 1994 in the South Atlantic Patagonian toothfish *Dissostichus eleginoides* fishery found that seabird bycatch rates on longlines set at night were much lower for lines set when the decklights were turned off, compared to sets with the lights on (0.15 vs. 0.59 birds/1,000 hooks). A total of 21 white-chinned petrels *Procellaria aequinoctialis* were caught. This study is also discussed in 'Set longlines at night to reduce seabird bycatch'.

(1) Cherel, Y., Weimerskirch, H. & Duhamel, G. (1996) Interactions between longline vessels and seabirds in Kerguelen waters and a method to reduce seabird mortality. *Biological Conservation*, 75, 63–70.

8.15. Use streamer lines to reduce seabird bycatch on longlines

- A total of eight studies (1,2,5,6,10–12,14) and two literature reviews (15,16) from coastal and pelagic fisheries across the world found strong evidence for reduced seabird bycatch on longlines when streamer lines were used.
- A replicated, controlled trial from the sub-Antarctic Indian Ocean (9) found no reduction in bycatch rates when using streamer lines, whilst five studies (3,4,7,8,13) were inconclusive, uncontrolled or had weak evidence for reductions.
- The effect of streamer lines appears to vary between seabird species: northern fulmars *Fulmarus glacialis* were consistently caught at lower rates when streamers were used (6,10,12,14,15) but shearwaters *Puffinus* spp. and white-chinned petrels *Procellaria aequinoctialis* were caught at similar rates with and without streamers in one study each (9,12).
- The three studies that investigated the use of multiple streamer lines all found that fewer birds were caught when two streamer lines were used, compared to one (11,12,16), with even fewer caught when three were used (16).

Background

Streamer lines are long lines attached to a high point on the stern of a fishing boat so that they fall slowly to the surface of the water. On the 25 m or so of the lines, multiple 'streamers' are attached: strands of often brightly coloured line attached on a swivel to the main line. These move in the wind and interfere with birds attempting to reach the baited longline below. Streamer lines can vary in effectiveness with design, and the Commission for the Conservation of Antarctic Marine Living Resources (CCAMLR) and some other fisheries organisations have specific recommendations on their design. For example, boats in CCAMLR fishing areas must use a line at least 150 m long, attached 4.5 m above the water, with five sets of streamers attached at 5 m interval (Miller *et al.* 2003).

Multiple streamer lines may also be more effective than a single streamer, and how the streamer is deployed is thought to be important: if the wind moves the streamer so it is not directly over the longline then it is unlikely to have an effect.

Miller, D. G. M., Sabourenkov, E. N. and Ramm, D. C. (2003) CCAMLR's approach to managing Antarctic marine living resources. *Conference Reports of: Deep Sea 2003: Conference on the Governance of Deep-sea Fisheries*, Queenstown, New Zealand.

A controlled study on a fishing voyage off the south-west coast of Tasmania, Australia (1), found that using a streamer line reduced the number of baits taken by albatross from 5.8/1000 to 1.7/1,000. Fewer attempts at baits were also made within 50 m of the ship (from 12.8/1,000 baits and 63% of the total attempts to 0.2/1,000 baits and 2.3%). Streamer lines were 150 m long with seven 'double-line vertical streamers' attached at 4.1 m intervals.

A comparative study of ten bluefin tuna *Thunnus thynnus* longlining boats fishing off New Zealand in 1992 (2), found that none of five vessels that used a streamer line over 51% of the time caught any birds (over 100 line sets). The remaining five boats used streamer lines for less than 12% of the time and caught 14 birds over 157 sets. However, the authors cautioned that these results were preliminary, with limited observer coverage, no controls and no statistical tests.

A small study on a commercial longlining boat in the South Atlantic in March–April 1993 (3) found that no birds were caught on seven lines set for Patagonian toothfish *Dissostichus eleginoides*. The authors state that they could not draw conclusions on the effectiveness of the streamers because of the small number of repeats and the lack of birds interacting with line setting. This study also investigated the impact of setting lines at night (see 'Set longlines at night to reduce seabird bycatch').

A study of 20 longlines set in April–May 1994 in the South Atlantic (4) found that 98 birds were caught as bycatch on the lines, with 45 on two lines set without streamer lines, 38 on two set with a streamer line and 1–5 birds/line for seven set with a different streamer design. A further 15 birds (all white-chinned petrels *Procellaria aequinoctialis*) were caught on 15 sets, of which three used the first streamer design. The authors concluded that streamer lines 'were observed to interrupt birds' behaviour . . . and to reduce mortality'. A streamer line was also use during the hauling of seven line sets, but no birds were caught during hauling, either with or without streamer lines. This study is also discussed in 'Set longlines at night to reduce seabird bycatch' and 'Use a sonic scarer when setting longlines to reduce seabird bycatch'.

Analysis of data from 14 longlining trips by 12 vessels in the South Atlantic between March and May 1995 (5), found that seven vessels using streamer lines caught significantly fewer birds per unit fishing effort (BPUE <0.1 birds/1,000 hooks) than those without streamer lines (BPUE >0.3 birds/1,000 hooks). A total of 1,428 birds were caught. The authors note that there was considerable variation in BPUE due to confounding factors, but that line sets compared were similar apart from the use of streamers. Vessels were fishing for Patagonian toothfish *Dissostichus eleginoides* in Subarea 48.3 (South Georgia and the South Sandwich Islands).

A replicated, controlled study off the coast of Norway (6) found that seabird bycatch on 13 days in May 1996 was significantly lower for 13 daytime line sets when a streamer line was used (two birds caught, 0.04 birds/1,000 hooks), compared to control sets with no streamer line (99 birds, 1.75 birds/1,000 hooks). Bycatch was mainly northern fulmars *Fulmarus glacialis* and the streamer line was 8 mm nylon with 8 cm wide, 0.5–3 m long yellow tarpaulin streamers at 5.1 m intervals.

A randomised, replicated and controlled study on a boat in the South Atlantic on 27 nights between March and May 1997 (7), found that, when the vessel followed other guidelines from the CCAMLR streamer lines did not significantly reduce BPUE (a total of 12 birds were killed by lines). However, the authors predict that 125 replicates of each treatment would be needed to detect the effect of streamer lines, far more than in this study. The vessel was fishing for Patagonian toothfish *Dissostichus eleginoides* in Subarea 48.3 of the south-west Atlantic (South Georgia and the South Sandwich Islands), setting an average of 12,990 hooks each night. CCAMLR guidelines say that lines should be set at night and weighted, that deck lights should be extinguished and no offal discarded during line setting.

An analysis of data from tuna vessels fishing in Australian waters in 1991–5 (8) does not provide conclusive evidence for the effectiveness of streamer lines. Voyages with streamer lines caught more birds than those without, but when catch rates for individual seasons and areas were analysed, catch rates were lower with streamer lines, but not significantly so. The authors argue that the lack of conclusive evidence is due to the lack of a controlled analysis and the disproportionate use of streamer lines in areas with higher catch rates and during the day (see 'Set longlines at night to reduce seabird bycatch'). A total of 3,477 line sets were studied.

A replicated, controlled study in the sub-Antarctic Indian Ocean in 1994–7 (9) found that using a streamer line whilst setting longlines did not appear to reduce seabird bycatch for all species combined, or for white-chinned petrels *Procellaria aequinoctialis*, the most frequently caught birds (0.57 birds/1,000 hooks on sets with a streamer vs. 0.52 birds/1,000 hooks for sets without, total of 524 lines studied). Streamer lines were 150–175 m long propylene ribbons (2 m long) every 2–3 m. This study is also discussed in 'Set longlines at night to reduce seabird bycatch' and 'Reduce seabird bycatch by releasing offal overboard when setting longlines'.

A randomised, replicated and controlled experiment in February 1999, in the north-western Islands, Hawaii, USA (10), found that using a streamer line when setting hookless bait lines lowered attacks by black-footed *Phoebastria nigripes* and Laysan *P. immutabilis* albatrosses by 75% and 77% respectively, compared to controls. Streamer lines were 150 m long: a 10 m attachment section of 6.25 mm twisted yellow polypropylene; 40 m with seven forked 'aerial streamers'; 85 m of red 3 mm nylon with eight small streamers in the first 40 m, 15 m of 12 mm yellow polypropylene. Lines were attached 8 m above the stern so that the first streamer touched the water approximately 5 m behind the bait entry point. Twenty four repeats of each treatment were used, with lines set during the day, mimicking swordfish longline techniques.

A randomised, replicated and controlled experiment off the coast of mid-Norway in August 1998 (11), found that two streamer lines both significantly reduced the seabird bycatch on longlines compared with lines set with no streamer (no birds caught with the 11 repeats using the advanced streamer, two birds and 0.03 birds/1,000 hooks with the 11 repeats of the simple streamer vs. 74 birds and 1.06 birds/1,000 hooks for 11 sets with no streamer line). The majority of hooked birds were northern fulmars *Fulmarus glacialis*. Each set contained approximately 6,500 hooks and was set during daylight. Both streamers were 80 m long and hung 7–8 m above sea level; the advanced line had four gillnet float rings at the trailing end and twelve 8 cm wide yellow tarpaulin streamers, 5 m apart, 0.5–3 m long; the simple line had a punctured buoy at the trailing end and six, equally placed, 30 cm, red plastic streamers.

Two randomised, replicated and controlled trials in 1999 and 2000 (12) found that seabird bycatch was 88–100% lower on longlines set with paired streamer lines, compared to controls (0.00–0.04 birds/1,000 hooks with paired streamers vs. 0.22–0.37 birds/1,000 hooks for controls). Similarly, bycatch was lower with single streamers in the Pacific halibut *Hippoglossus stenolepis* and sablefish *Anoplopoma fimbria* fishery in the Gulf of Alaska and the Aleutian Islands, USA (0.01 birds/1,000 hooks), although reduction for paired streamers were higher (50% and 80% reductions in Laysan albatross *Phoebastria immutabilis* and northern fulmar *Fulmarus glacialis* bycatch, respectively). However, in the Pacific cod *Gadus macrocephalus* and walleye pollock *Theragra chalcogramma* fishery south-east of the Pribilof Islands, USA, lines set with single streamers caught as many birds as controls. This was due to similar numbers of shearwaters *Puffinus* spp. caught on lines with single streamers; no northern fulmars *Fulmarus glacialis* were caught on lines set with streamers. Streamer lines were 90 m of 21 mm blue polyester, with streamers of 6.4 mm orange tubing attached at 5 metre intervals for the first 50 m. This study is also discussed in 'Weight baits or lines to reduce longline bycatch of seabirds', 'Set longlines at night to reduce seabird bycatch', 'Use a line shooter to reduce seabird bycatch' and 'Set lines underwater to reduce seabird bycatch'.

A study on a longlining vessel on the Chatham Rise, New Zealand, in July–August 1998 (13) and using weighted lines (see 'Weight baits or lines to reduce longline bycatch of seabirds') and a streamer line caught an average of 0.0093 birds/1,000 hooks – far lower than many other studies. The streamer line extended 75–85 m behind the boat, covering the longline to a depth of 2–5 m. Many seabirds can dive up to 10 m (a depth not reached until 170 m behind the streamer), so the authors caution that the streamer may not offer as high protection as it appeared.

A randomised, replicated and controlled trial on a commercial longlining vessel off the coast of mid-Norway in August 1999 (14), found that bycatch of northern fulmar *Fulmarus glacialis* fell to zero when a streamer line was deployed during line setting and just one bird (0.02 birds/1,000 hooks) when both a streamer line and line shooter were used, compared with 32 fulmars (0.52 birds/1,000 hooks) during control line sets, and 13 fulmars (0.22 birds/1,000 hooks) when just a line shooter was used. Eleven repeats of each treatment were used, with lines set during daylight. Streamer lines were 90 m long, with a 69 m streamer section with twelve 8 cm wide yellow tarpaulin streamers, 5.4 m apart, 0.5–2 m long. This study is also discussed in 'Use a line shooter to reduce seabird bycatch'.

A literature review of three replicated and controlled studies off the coast of Norway (15), found that only two northern fulmars *Fulmarus glacialis* were caught on 185,000 longline hooks when a streamer line was deployed, compared with 205 birds (mostly fulmars) from a similar number of hooks without streamer lines. The three studies (6,11,14) are outlined in detail above.

A literature review (16) described evidence for the effectiveness of bycatch reduction methods in pelagic longline fisheries as 'inconclusive', highlighting the need for further research into the design and configuration of streamer lines. Two studies (one controlled) from Alaska and the Falkland Islands found that streamer lines reduced seabird mortality by 71% (single line), 75% (two lines) or 97% (three lines). An uncontrolled study near New Zealand suggests that using two streamer lines and an acoustic cannon almost eliminated white-chinned petrels *Procellaria aequinoctialis* diving within 50 m of the boat, but that the 'boom and bridle' system of attaching streamers to boats did not reduce diving within the aerial range of the streamer line.

(1) Brothers, N. (1991) Albatross mortality and associated bait loss in the Japanese longline fishery in the Southern Ocean. *Biological Conservation*, 55, 255–268.
(2) Murray, T. E., Bartle, J. A., Kalish, S. R. & Taylor, P. R. (1993) Incidental capture of seabirds by Japanese southern bluefin tuna longline vessels in New Zealand waters, 1988–1992. *Bird Conservation International*, 3, 181–210.
(3) Ashford, J. R., Croxall, J. P., Rubilar, P. S. & Moreno, C. A. (1994) Seabird interactions with longlining operations for *Dissostichus eleginoides* at the South Sandwich islands and South Georgia. *CCAMLR Science*, *1*, 143–153.
(4) Ashford, J. R., Croxall, J. P., Rubilar, P. S. & Moreno, C. A. (1995) Seabird interactions with longlining operations for *Dissostichus eleginoides* around South Georgia, April to May 1994. *CCAMLR Science*, 2, 111–121.
(5) Moreno, C. A., Rubilar, P. S., Marschoff, E. & Benzaquen, L. (1996) Factors affecting the incidental mortality of seabirds in the *Dissostichus eleginoides* fishery in the southwest Atlantic (subarea 48.3, 1995 season). *CCAMLR Science*, 3, 79–91.
(6) Løkkeborg, S. (1998) Seabird by-catch and bait loss in long-lining using different setting methods. *ICES Journal of Marine Science*, 55, 145–149.
(7) Ashford, J. R. & Croxall, J. P. (1998) An assessment of CCAMLR measures employed to mitigate seabird mortality in longlining operations for *Dissostichus eleginoides* around South Georgia. *CCAMLR Science*, 5, 217–230.
(8) Brothers, N., Gales, R. & Reid, T. (1999) The influence of environmental variables and mitigation measures on seabird catch rates in the Japanese tuna longline fishery within the Australian Fishing Zone, 1991–1995. *Biological Conservation*, 88, 85–101.
(9) Weimerskirch, H., Capdeville, D. & Duhamel, G. (2000) Factors affecting the number and mortality of seabirds attending trawlers and long-liners in the Kerguelen area. *Polar Biology*, *23*, 236–249.
(10) Boggs, C. H. (2001) Deterring albatrosses from contacting baits during swordfish longline sets. 79–94 In: Melvin, E. F. & Parrish, J. K. (eds) *Seabird Bycatch: trends, roadblocks and solutions*. Alaska Sea Grant, Fairbanks.
(11) Løkkeborg, S. (2001) Reducing seabird bycatch in longline fisheries by means of bird-scaring lines and underwater setting. 33–41 In: Melvin, E. F. & Parrish, J. K. (eds) *Seabird Bycatch: trends, roadblocks and solutions*. Alaska Sea Grant, Fairbanks.
(12) Melvin, E. F., Parrish, J. K., Dietrich, K. S. & Hamel, O. S. (2001) *Solutions to seabird bycatch in Alaska's demersal longline fisheries*. Washington Sea Grant Program, University of Washington.
(13) Smith, N. W. M. (2001) Longline sink rates of an autoline vessel, and notes on seabird interactions. *Science for Conservation*, 183, 5–32.
(14) Løkkeborg, S. & Robertson, G. (2002) Seabird and longline interactions: effects of a bird-scaring streamer line and line shooter on the incidental capture of northern fulmars *Fulmarus glacialis*. *Biological Conservation*, 106, 359–364.
(15) Løkkeborg, S. (2003) Review and evaluation of three mitigation measures – bird-scaring line, underwater setting and line shooter – to reduce seabird bycatch in the north Atlantic longline fishery. *Fisheries Research*, 60, 11–16.
(16) Melvin, E. F., Sullivan, B., Robertson, G. & Wienecke, B. (2004) A review of the effectiveness of streamer lines as a seabird by-catch mitigation technique in longline fisheries and CCAMLR streamer line requirements. *CCAMLR Science*, 11, 189–201.

8.16. Use larger hooks to reduce seabird bycatch

- We captured no intervention-based evidence on the impact of large hooks on seabird bycatch.

Background

Large hooks may be more difficult for birds to swallow and so may reduce bycatch rates. A correlative study found a negative relationship between hook size and bycatch rate (Moreno *et al.* 1996). However, large hooks may be more likely to hook birds through the body or wings ('foul hooking').

(1) Moreno, C. A., Rubilar, P. S., Marschoff, E. & Benzaquen, L. (1996) Factors affecting the incidental mortality of seabirds in the *Dissostichus eleginoides* fishery in the southwest Atlantic (subarea 48.3, 1995 season). *CCAMLR Science*, 3, 79–91.

8.17. Use a water cannon when setting longlines to reduce seabird bycatch

- We found no evidence for the effects on seabird bycatch rates of using water cannon when setting longlines.

Background

A water cannon used when longlines are set could unsettle birds and stop them from coming close enough to the boat to attack baits and get hooked.

8.18. Set lines underwater to reduce seabird bycatch

- Four replicated and controlled studies (1–4) and a literature review (5) in Norway, South Africa and the North Pacific found reductions in northern fulmar *Fulmarus glacialis*, albatross and petrel bycatch rates when using an underwater setting funnel. Although one (4) found a disproportionate number of albatross were caught during day line setting.
- A replicated and controlled study (3) found that underwater setting increased attack rates of shearwaters *Puffinus* spp. on longlines and did not reduce bycatch.

Background

Most birds caught on longlines are caught when the lines are being set, after the bait has been attached to hooks but before they have sunk out of the reach of surface-feeding birds. If lines are set underwater then it may reduce the time they are available to birds and therefore reduce bycatch.

A randomised, replicated and controlled study off the coast of Norway (1) found that seabird bycatch in May 1996 was significantly lower when an underwater setting funnel was used to set longlines (28 birds caught, 0.49 birds/1,000 hooks for 13 line sets) than in control sets without the funnel (99 birds caught, 1.75 birds/1,000 hooks for 13 line sets). Lines were set during daylight and bycatch was mainly northern fulmars *Fulmarus glacialis*.

A randomised, replicated and controlled experiment, with 11 repeats of each treatment (each approximately 6,500 hooks), off the coast of mid-Norway during daylight in August 1998 (2), found that longlines set using an underwater setting funnel significantly reduced the number of seabirds caught compared with control lines, set without a funnel (six birds and 0.08/1,000 hooks vs. 74 birds and 1.75/1,000 hooks). The majority of hooked birds were northern fulmars *Fulmarus glacialis*.

A randomised, replicated and controlled study (3) in a North Pacific fishery in August 1999 found that using a setting funnel ('lining tube') reduced the number of seabirds caught on longlines, compared to controls (0.045 birds/1000 hooks vs. 0.218 birds/1000 hooks). However, this decline was only fewer northern fulmars *Fulmarus glacialis* being caught, with the bycatch rate of shearwater *Puffinus* spp. remaining constant. Overall, the number of birds following and the rate of attacks on baits were consistent across treatments, although attack rates by shearwaters were twice as high when using a setting funnel. The study took place in the Pacific cod *Gadus macrocephalus* and walleye pollock *Theragra chalcogramma* fishery south-east of the Pribilof Islands, USA. This study is also discussed in 'Weight baits or lines to reduce longline bycatch of seabirds', 'Set longlines at night to reduce seabird bycatch', 'Use a line shooter to reduce seabird bycatch' and 'Use streamer lines to reduce seabird bycatch on longlines'.

A replicated, controlled study in a South African fishery, between 1998 and 2000 (4) found that seabird bycatch in longline sets that used an underwater setting funnel was significantly lower than sets that did not use a funnel, both during the day (funnel: 2,255,150 hooks set, 23 petrels, 10 albatross killed, 0.015 birds/1,000 hooks; no funnel: 434,598 hooks set, 20 petrels, 1 albatross killed, 0.048 birds/1,000 hooks) and night (funnel: 317,503 hooks set, 3 petrels and no albatross killed, 0.009 birds/1,000 hooks; no funnel: 2,045,912 hooks set, 56 petrels, one albatross killed, 0.028 birds/1,000 hooks). However, a disproportionate number of albatrosses were killed during day-time sets using the setting funnel. The study took place in the Patagonian toothfish *Dissostichus eleginoides* fishery off the coast of the Prince Edward Islands.

A review of two randomised, replicated and controlled studies off the coast of Norway (5) found that fewer northern fulmars *Fulmarus glacialis* were caught on longline hooks when an underwater setting funnel was used (34 birds and 0.08–0.49 birds/1,000 hooks) compared to control line sets (174 birds and 1.06–1.75 birds/1,000 hooks). Both papers are outlined above.

(1) Løkkeborg, S. (1998) Seabird by-catch and bait loss in long-lining using different setting methods. *ICES Journal of Marine Science*, *55*, 145–149.
(2) Løkkeborg, S. (2001) Reducing seabird bycatch in longline fisheries by means of bird-scaring lines and underwater setting. 33–41 In: Melvin, E. F. & Parrish, J. K. (eds) *Seabird Bycatch: trends, roadblocks and solutions*. Alaska Sea Grant, Fairbanks.
(3) Melvin, E. F., Parrish, J. K., Dietrich, K. S. & Hamel, O. S. (2001) *Solutions to seabird bycatch in Alaska's demersal longline fisheries*. Washington Sea Grant Program, University of Washington.
(4) Ryan, P. G. & Watkins, B. P. (2002) Reducing incidental mortality of seabirds with an underwater longline setting funnel. *Biological Conservation*, 104, 127–131.
(5) Løkkeborg, S. (2003) Review and evaluation of three mitigation measures – bird-scaring line, underwater setting and line shooter – to reduce seabird bycatch in the north Atlantic longline fishery. *Fisheries Research*, 60, 11–16.

8.19. Set longlines at the side of the boat to reduce seabird bycatch

- We found no evidence for the effects on seabird bycatch rates of setting longlines from the side of the boat.

Background

Setting longlines off the side of the boat allows baits to sink before they reach the stern, where most of the birds following boats will be flying. However, it may require refitting boats, which is potentially expensive.

8.20. Use a line shooter to reduce seabird bycatch

- A randomised, replicated and controlled trial from a pelagic fishery in the North Pacific (1) found significantly higher seabird bycatch when a line shooter was used to set longlines.
- A second randomised, replicated and controlled trial (from Norway, (2)), found no effect of a line shooter on bycatch rates.

Background

As a longline is deployed it will be put under tension as the movement of the boat stretches the line. This can delay the sinking of the line and baits, increasing the time available for birds to get hooked. A line shooter works by deploying the line faster than the boat is moving, removing tension from it and allowing the line to sink faster.

A randomised, replicated and controlled study (1) in the North Pacific in August 1999, found that using a line shooter significantly increased the number of birds caught, compared to controls (0.336 birds/1,000 hooks vs. 0.218 birds/1,000 hooks). Treatment had no effect on the number of birds following vessels, or the attack rate on baits. A total of 156 line sets were studied, set for Pacific cod *Gadus macrocephalus* and walleye pollock *Theragra chalcogramma* south-east of the Pribilof Islands, USA. This study is also discussed in 'Weight baits or lines to reduce longline bycatch of seabirds', 'Use streamer lines to reduce seabird bycatch on longlines', 'Set longlines at night to reduce seabird bycatch' and 'Set lines underwater to reduce seabird bycatch'.

A replicated, randomised and controlled trial on a commercial long-lining vessel off the coast of mid-Norway in August 1999 (2), found that by-catch of northern fulmar *Fulmarus glacialis* was not significantly lower when a line shooter was used during line setting (13 fulmars hooked during 11 sets, 0.22 birds/1,000 hooks), compared with either control sets (32 fulmars in 11 sets, 0.52 birds/1,000 hooks) or with lines set using a streamer line as well (no birds caught on 11 sets with just the streamer line vs. a single bird or 0.02 birds/1,000 hooks on 11 sets with the streamer and shooter). This study is also discussed in 'Use streamer lines to reduce seabird bycatch on longlines'.

(1) Melvin, E. F., Parrish, J. K., Dietrich, K. S. & Hamel, O. S. (2001) *Solutions to seabird bycatch in Alaska's demersal longline fisheries.* Washington Sea Grant Program, University of Washington.

(2) Løkkeborg, S. & Robertson, G. (2002) Seabird and longline interactions: effects of a bird-scaring streamer line and line shooter on the incidental capture of northern fulmars *Fulmarus glacialis*. *Biological Conservation*, 106, 359–364.

8.21. Use bait throwers to reduce seabird bycatch

- A study from Australia (1) found significantly lower seabird bycatch on longlines set with a bait thrower.

Background

The turbulence created by a boat's wake can stop bait from sinking quickly, allowing birds more time to become hooked. Bait throwers work by throwing bait clear of the wake, so that it sinks faster. Throwers also allow fishermen to direct lines, for example aiming them to lie under streamer lines (see 'Use streamer lines to reduce seabird bycatch on longlines').

A comparative analysis of data from 86 longlining vessels operating around Tasmania, Australia, between April 1992 and March 1995 (1) and studying a total of 141 line sets, found that sets that using a bait thrower had significantly lower seabird bycatch rates, compared to sets without a thrower. This study is also described in 'Set longlines at night to reduce seabird bycatch' and 'Thaw bait to reduce seabird bycatch'.

(1) Klaer, N. & Polacheck, T. (1998) The influence of environmental factors and mitigation measures on by-catch rates of seabirds by Japanese longline fishing vessels in the Australian region. *Emu*, 98, 305–316.

8.22. Tow buoys behind longlining boats to reduce seabird bycatch

- We found no evidence for the effects on seabird bycatch rates of towing buoys behind longlining boats.

Background

Buoys may interfere with birds landing on the water and attempting to take baits in a similar way to streamer lines interfering with birds flying behind boats.

8.23. Dye baits to reduce seabird bycatch

- A randomised replicated and controlled study in Hawaii (1) found that dying bait blue significantly reduced the number of attacks from albatross on baits being set.

Background

Most seabirds are visual predators, and so are less likely to take bait if it is difficult to see. Dyeing bait blue so that it is hard to see against the water may therefore reduce bait loss and bycatch rates.

A randomised, replicated and controlled experiment in February 1999, in the north-western Islands, Hawaii, USA (1), found that dyeing squid bait blue when setting hookless bait lines reduced attacks by black-footed *Phoebastria nigripes* and Laysan *P. immutabilis* albatrosses by 95% and 94% respectively, compared to lines set with un-dyed baits (measured as attacks/bird/100 branch lines). Twenty-four repeats of each treatment were used, set during the day to mimic longline setting for swordfish.

(1) Boggs, C. H. (2001) Deterring albatrosses from contacting baits during swordfish longline sets. 79–94 In: Melvin, E. F. & Parrish, J. K. (eds) *Seabird Bycatch: trends, roadblocks and solutions*. Alaska Sea Grant, Fairbanks.

8.24. Use high-visibility longlines to reduce seabird bycatch

- We captured no intervention-based evidence on the impact on seabird bycatch of high-visibility longlines.

Background

If birds can see longlines then they may avoid them as they forage for baits, potentially reducing the risk of 'foul hooking' (birds being hooked through the body or wing), if not the risk of being hooked whilst taking bait.

8.25. Use a sonic scarer when setting longlines to reduce seabird bycatch

- A single study from the South Atlantic (1) found that seabird bycatch rates did not appear to be lower on longlines set with a sonic scarer, and that changes in seabird behaviour due to the scarer were only temporary.

Background

It might be possible to stop birds from taking baits from longlines by scaring them away from boats using loud noises.

A study on a commercial fishing boat in the South Atlantic in April-May 1994 (1) found that longlines set with a sonic scarer did not appear to catch fewer seabirds than lines set without a scarer. Lines set with and without the scarer were not analysed separately (with 15 birds caught on six sets without a scarer and six with, as well as three with a streamer line), although the authors observed changes in behaviour when a scarer was first used, but that birds quickly got used to the noise. The scarer emitted periodic bursts of compressed gas. This study is also discussed in 'Set longlines at night to reduce seabird bycatch' and 'Use streamer lines to reduce seabird bycatch on longlines'.

(1) Ashford, J. R., Croxall, J. P., Rubilar, P. S. & Moreno, C. A. (1995) Seabird interactions with longlining operations for *Dissostichus eleginoides* around South Georgia, April to May 1994. *CCAMLR Science*, 2, 111–121.

8.26. Weight baits or lines to reduce longline bycatch of seabirds

- Three replicated and controlled studies (1,2,4) found evidence for reduced bycatch in some species when using weighted lines. One study (3) found low bycatch rates, but was uncontrolled.
- In Hawaii (1) and New Zealand (4), rates of bait loss and bycatch of albatrosses *Phoebastria* spp., white-chinned petrels *Procellaria aequinoctialis* and sooty shearwaters *Puffinus griseus* were much lower with weighted baits or integrated weight lines than with control lines.

- In the North Pacific, two trials (2) found that bycatch rates of some species was reduced when using weights, but that shearwaters *Puffinus* spp. attacked weighted lines more often.
- A study off New Zealand (3) found that attaching weights to lines had only localised effects on sink-rate.

Background

Most birds take baits when they are close to the surface. Reducing the time that baits are at the surface by weighting them may, therefore, reduce bycatch rates.

A randomised, replicated and controlled experiment in February 1999, in the north-western Islands, Hawaii, USA (1), found that weighing bait when setting hook-less bait lines, reduced attacks by black-footed *Phoebastria nigripes* and Laysan *P. immutabilis* albatrosses by 93% and 91% respectively, compared to lines set with un-dyed baits (measured as attacks/bird/100 branch lines). There were 24 replicates of each treatment, with squid bait weighed down with a 60 g swivel weight. Lines were set during the day and mimicked swordfish longline setting.

Two randomised, replicated and controlled studies in 1999 (2) found that seabird bycatch rates were significantly lower on weighted longlines, than on controls (0.052 and 0.234 birds/1,000 hooks for weighted lines vs. 0.218 and 0.371 birds/1,000 hooks for controls; totals of 156 and 121 line sets). However, results in one fishery (for Pacific cod *Gadus macrocephalus* and walleye pollock *Theragra chalcogramma* fishery, south-east of the Pribilof Islands, USA) were due purely to reductions in northern fulmars *Fulmarus glacialis* caught, with shearwater *Puffinus* spp. bycatch remaining constant. In the second fishery (for Pacific halibut *Hippoglossus stenolepis* and sablefish *Anoplopoma fimbria* in the Gulf of Alaska and Aleutian Islands, USA), removing the single line set with the highest bycatch (of 27 birds) made the reduction in bycatch non-significant. A further trial in 2000 found no birds were caught on weighted or unweighted lines set under a streamer line (see 'Use streamer lines to reduce seabird bycatch on longlines'. Weighted lines had 4.5 kg weights every 90 m. This study is also discussed in 'Use a line shooter to reduce seabird bycatch', 'Set longlines at night to reduce seabird bycatch' and 'Set lines underwater to reduce seabird bycatch'.

A study on a longlining vessel on the Chatham Rise, New Zealand, in July–August 1998 (3) and using weighted lines and a streamer line (see 'Use streamer lines to reduce seabird bycatch on longlines'), caught an average of 0.0093 birds/1,000 hooks – far lower than many other studies. Weights of 5 kg were attached every 400 m, but only caused faster sinking for approximately 40 m either side.

A replicated, controlled and paired study using longlines in November 2002 and 2003 in New Zealand (4), found that using 'integrated weight' longlines reduced mortality of white-chinned petrels *Procellaria aequinoctialis* by 94–99%, and of sooty shearwaters *Puffinus griseus* by 61%, compared to unweighted lines (petrels: one and four birds on weighted lines vs. 81 and 43 in 2002 and 2003 respectively; shearwaters: 15 birds on weighted lines in 2003 vs. 38 on unweighted lines; a single shearwater was killed on an unweighted line in 2002). A total of 213 paired sets of lines were set, with integrated lines containing 50 g/m of lead. All lines were set under a single streamer line (see 'Use streamer lines to reduce seabird bycatch on longlines' for studies testing this intervention).

(1) Løkkeborg, S. (2001) Reducing seabird bycatch in longline fisheries by means of bird-scaring lines and underwater setting. 33–41 In: Melvin, E. F. & Parrish, J. K. (eds) *Seabird Bycatch: trends, roadblocks and solutions*. Alaska Sea Grant, Fairbanks.
(2) Melvin, E. F., Parrish, J. K., Dietrich, K. S. & Hamel, O. S. (2001) *Solutions to seabird bycatch in Alaska's demersal longline fisheries.* Washington Sea Grant Program, University of Washington.
(3) Smith, N. W. M. (2001) Longline sink rates of an autoline vessel, and notes on seabird interactions. *Science for Conservation*, 183, 5–32.
(4) Robertson, G., McNeill, M., Smith, N., Wienecke, B., Candy, S. & Olivier, F. (2006) Fast sinking (integrated weight) longlines reduce mortality of white-chinned petrels (*Procellaria aequinoctialis*) and sooty shearwaters (*Puffinus griseus*) in demersal longline fisheries. *Biological Conservation*, 132, 458–471.

8.27. Use shark liver oil to deter birds when setting lines

- Two replicated and controlled trials (1, 2) found reductions in the number of seabirds following boats, or diving for baits, when shark liver oil was dripped behind the boats. Other oils had no effect.
- A third replicated and controlled trial in (2) found no differences in the number of seabirds following a bait-laying boat with shark liver oil.

Background

Seabirds may be predated by large fish such as sharks and so may show avoid the smell of them. Dripping shark liver oil whilst setting lines may, therefore, reduce bycatch.

A replicated, controlled experiment off the coast off north-east New Zealand (1) found that the number of dives made by seabirds in pursuit of pilchard baits behind a longline fishing vessel was dramatically lower (<5 birds/min) when small quantities of shark liver oil were dripped onto the water behind the vessel than during control trials using vegetable oil (always >30 birds/min) or sea water (20–40 birds/min). Diving birds were mainly flesh-footed shearwaters *Puffinus carneipes*, but also Buller's shearwaters *Puffinus bulleri* and white-faced storm petrels *Pelagodroma marina*.

One replicated, controlled experiment off Kaikoura, South Island, New Zealand, in 2005 (2) found no significant differences in the number of seabirds following a bait-laying boat when it was dripping shark liver oil (both commercially available and made by fishermen) behind the boat, compared to control conditions. However, a second trial in April 2006 off Hauraki Gulf, North Island, New Zealand found the number of seabirds following a bait-laying boat decreased significantly faster if fisherman-produced shark liver oil was dripped behind the boat, compared to controls dripping seawater. Other fish oils (anchovy, pollock and commercially available shark liver oil) did not have a significant impact on the number of following birds. However, all oils except for anchovy did significantly reduce the number of dives made by seabirds.

(1) Pierre, J. & Norden, W. (2005) Trials using shark liver oil to deter seabirds from eating bait during long-line fishing, Leigh, New Zealand. *Conservation Evidence*, 2, 99–100.
(2) Norden, W. S. & Pierre, J. P. (2007) Exploiting sensory ecology to reduce seabird by-catch. *Emu*, 107, 38–43.

8.28. Thaw bait before setting lines to reduce seabird bycatch

- A single study from Australia (1) found that lines set using thawed baits caught significantly fewer seabirds than controls.

Background

Thawed bait sinks more quickly than frozen bait and so using it may reduce the time that bait is available to birds and therefore reduce bycatch.

An analysis of data from 86 longlining vessels operating around Tasmania, Australia, between April 1992 and March 1995 (1) and studying a total of 141 line sets, found that sets that using partially or completely thawed bait had significantly lower bycatch rates, compared to sets using unthawed bait. This study is also described in 'Set longlines at night to reduce seabird bycatch' and 'Use bait throwers to reduce seabird bycatch'.

(1) Klaer, N. & Polacheck, T. (1998) The influence of environmental factors and mitigation measures on by-catch rates of seabirds by Japanese longline fishing vessels in the Australian region, *Emu*. 98, 305–316.

8.29. Reduce seabird bycatch by releasing offal overboard when setting longlines

- Two replicated and controlled studies in the South Atlantic and sub-Antarctic Indian Ocean (1,2) found significant reductions in the number of albatross and petrels attacking baits and being caught when offal was released overboard during line setting.

Background

Many fishing boats prepare fish onboard, after catching them, in order to maximise the catch that can be stored. The offal (waste) is then normally discarded overboard. Moreno *et al.* 1996 describe the highest bycatch rates of a series of voyages coming from a vessel that piped offal overboard on the same side that the line was being set. However, other studies suggest that piping offal overboard from a different point to where the lines are being set will reduce bycatch, as birds spend their time eating the offal, rather than attempting to take bait.

Moreno, C. A., Rubilar, P. S., Marschoff, E. & Benzaquen, L. (1996) Factors affecting the incidental mortality of seabirds in the *Dissostichus eleginoides* fishery in the southwest Atlantic (subarea 48.3, 1995 season). *CCAMLR Science*, 3, 79–91.

A replicated and controlled study in the South Atlantic in February 1994 (1) found that piping offal overboard whilst setting longlines greatly reduced the number of birds caught on 69 line sets, from 33 birds to three, equivalent to a decrease from 0.49 to 0.01 birds/1000 hooks. Caught birds were white-chinned petrels *Procellaria aequinoctialis*, and two grey-headed albatross *Thalassarche chrysostoma* (formerly *Diomedea chrysostoma*). There was a corresponding significant decrease in the rate of bird attacks on the bait for all species except wandering albatross *D. exulans* (92% decrease for black-browed albatross *T. melanophris* (formerly *D. melanophris*), 96% for grey-headed albatross and 96% for white-chinned petrel). Vessels were fishing for Patagonian toothfish *Dissostichus eleginoides* around South Georgia and Kerguelen Islands.

A replicated, controlled study in the sub-Antarctic Indian Ocean in 1994–7 (2) found that significantly fewer white-chinned petrels *Procellaria aequinoctialis* were caught on longlines when offal was released over the side of the boat as lines were being set (0.46 birds/1,000 hooks caught when offal was released vs. 1.00 birds/1,000 hooks with no release, total of 524 lines studied). There were no significant changes in other species caught, possibly due to smaller sample sizes. The authors caution that significantly more birds followed fishing boats when they released offal (significant increases for six of ten species) and the practice may therefore enforce long-term associations between fishing boats and food. This study is also discussed in 'Set longlines at night to reduce seabird bycatch' and 'Use streamer lines to reduce seabird bycatch on longlines'.

(1) Cherel, Y., Weimerskirch, H. & Duhamel, G. (1996) Interactions between longline vessels and seabirds in Kerguelen waters and a method to reduce seabird mortality. *Biological Conservation*, 75, 63–70.

(2) Weimerskirch, H., Capdeville, D. & Duhamel, G. (2000) Factors affecting the number and mortality of seabirds attending trawlers and long-liners in the Kerguelen area. *Polar Biology*, 23, 236–249.

8.30. Use bird exclusion devices such as 'Brickle curtains' to reduce seabird mortality when hauling longlines

- A study of longliners in the South Atlantic (1) found that fewer seabirds were caught on longlines hauled under BEDs with two booms, compared to those with a single boom.

Background

Although most birds caught as bycatch are caught as the lines are set, they may also be vulnerable when the lines are hauled in, with birds attempting to take the remaining baits or even caught fish. Bird exclusion devices such as 'Brickle curtains' hang around the hauling point and are designed to prevent birds getting to the line, which will only be visible for a short time as it reaches the boat.

A study in 2005–8 on longline vessels near South Georgia, South Atlantic (1), found that fewer birds were caught during longline hauling operations when 'Brickle curtains' were used (28 birds caught by 20 vessels, 19 in a single incident where the line broke), compared to a single boom exclusion device (43 birds caught by 20 vessels). Brickle curtains consisted of two booms extending over the hauling area with streamers attached, the single boom had one or more objects suspended from it. Boats using a third exclusion device consisting of two booms and a line of purse seine buoys (with or without streamers) caught no birds over three years, but the sample size was too small for meaningful comparisons to be made.

(1) Reid, E., Sullivan, B. & Clark, J. (2010) Mitigation of seabird captures during hauling in CCAMLR longline fisheries. *CCAMLR Science*, 17, 155–162.

8.31. Use acoustic alerts on gillnets to reduce seabird bycatch

- A repeated, randomised and controlled trial (1) in the USA found that sonic alerts reduced the number of common guillemots *Uria aalge* but not rhinoceros auklets *Cerorhinca monocerata* caught in gillnets.

Background

Using acoustic alerts on gillnets may deter birds from approaching them and becoming entangled.

A repeated, randomised and controlled trial in a drift gillnet fishery in North Puget Sound, Washington, USA, in July and August 1996 (1), found that nets fitted with acoustic alerts ('pingers') caught significantly fewer common guillemots (common murres) *Uria aalge* than control nets (0.31 vs. 0.60 entanglements/net). There was no significant change in the number of rhinoceros auklets *Cerorhinca monocerata* caught. A total of eight boats and 321 net sets were studied.

(1) Melvin, E. F., Parrish, J. K. & Conquest, L. L. (1999) Novel tools to reduce seabird bycatch in coastal gillnet fisheries. *Conservation Biology*, 13, 1386–1397.

8.32. Use high-visibility mesh on gillnets to reduce seabird bycatch

- A repeated, randomised and controlled trial (1) in the USA found that having gillnets made partially from high-visibility mesh was effective in reducing seabird bycatch.
- Having a greater percentage (25% vs. 10%) of the net made from high-visibility mesh was more effective, but also reduced catch of the target species.

Background

Pursuit-diving birds (those most vulnerable to gillnets) are visual hunters, and so high-visibility mesh may reduce the number caught as bycatch. There is also the possibility, however, that highly-visible nets will reduce the number of target species caught.

A repeated, randomised and controlled trial in a drift gillnet fishery in North Puget Sound, Washington, USA, in July–August 1996 (1), found that nets fitted with highly visible mesh in the top 25% caught significantly fewer common guillemots (common murres) *Uria aalge* and rhinoceros auklets *Cerorhinca monocerata* than control nets (guillemots: 0.37 vs. 0.6 entanglements/net; auklets: approximately 0.05 vs. 0.2 entanglements/net). Nets fitted with highly visible mesh in the top 10% caught significantly fewer guillemots than controls (0.32 vs. 0.6 entanglements/net), but there was no significant change in the number of auklets caught. Nets with 25% high visibility mesh also caught significantly fewer sockeye salmon *Oncorhynchus nerka*, the target species, compared to controls (10 vs. 36 entanglements/net). A total of eight boats and 482 net sets were studied.

(1) Melvin, E. F., Parrish, J. K. & Conquest, L. L. (1999) Novel tools to reduce seabird bycatch in coastal gillnet fisheries. *Conservation Biology*, 13, 1386–1397.

8.33. Reduce gillnet deployment time to reduce seabird bycatch

- We found no evidence for the effects on seabird bycatch rates of reducing gillnet deployment time.

Background

If birds are attracted to the fish caught in gillnets then removing the nets more quickly may reduce their attractiveness to birds and so reduce bycatch.

8.34. Mark trawler warp cables to reduce seabird collisions

- A replicated, controlled study in Argentina (1) found that seabird mortality from collisions with trawler warp cables was much lower when the cables were marked.

A replicated and controlled study in Golfo San Jorge, Argentina between December 2004 and April 2005 (1) found that there were significantly fewer seabird collisions with warp cables during net-hauling activities (and cable-related mortality was nil) when an orange-coloured plastic traffic cone was attached to each cable (average of 5.4 contacts/haul and no mortalities in 12 hauls) compared with when cones were not used (average of 58.5 contacts/haul and a total of 11 mortalities in ten hauls). Trials were on three commercial trawlers using cones 1 m long and 20–10 cm in diameter. Mortalities were eight kelp gulls *Larus dominicanus* and three black-browed albatross *Thalassarche melanophrys*.

(1) González-Zevallos, D., Yorio, P. & Caille, G. (2007) Seabird mortality at trawler warp cables and a proposed mitigation measure: A case of study in Golfo San Jorge, Patagonia, Argentina. *Biological Conservation*, 136, 108–116.

8.35. Reduce 'ghost fishing' by lost/discarded gear

- We captured no evidence for the effects on seabird bycatch rates or populations of reducing ghost fishing.

Background

Abandoned fishing gear or gear that is lost at sea can continue catching fish, birds and other animals for years, or even decades, as the strong monofilament or steel nets slowly break up. Such gear has no commercial purpose and can damage fish stocks or even equipment as well killing birds. Reducing this 'ghost fishing' is therefore likely to benefit both wildlife and fishermen.

8.36. Reduce bycatch through seasonal or area closures

- We captured no evidence for the effects on seabird populations or bycatch rates of seasonal or area closures.

Background

Birds show seasonal patterns of habitat use and therefore reducing fishing in areas where there are large populations of foraging birds may reduce bycatch. However, these areas may also be the most profitable to fish in, as seabirds are likely to follow fish stocks. Closures may therefore not be a viable option.

9. Threat: Human intrusions and disturbance

Key messages

Use signs and access restrictions to reduce disturbance at nest sites

Six studies from across the world found increased numbers of breeders, higher reproductive success or lower levels of disturbance in waders and terns following the start of access restrictions or the erection of signs near nesting areas. Two studies from Europe and Antarctica found no effect of access restrictions on reproductive success in eagles and penguins, respectively.

Set minimum distances for approaching birds (buffer zones)

We captured no intervention-based evidence for the effects on bird populations of setting minimum distances for approaching birds.

Provide paths to limit disturbance

A study from the UK found that two waders nested closer to a path, or at higher densities near the path, following resurfacing, which resulted in far fewer people leaving the path.

Reduce visitor group sizes

We found no intervention-based evidence for the effects of limiting visitor group sizes on bird populations.

Use voluntary agreements with local people to reduce disturbance

A before-and-after trial in the USA found significantly lower rates of waterfowl disturbance following the establishment of a voluntary waterfowl avoidance area, despite an overall increase in boat traffic.

Start educational programmes for personal watercraft owners

A before-and-after study in the USA found that common tern reproduction increased, and rates of disturbance decreased, following a series of educational programmes aimed at recreational boat users.

Habituate birds to visitors

A study from Australia found that bridled terns from heavily disturbed sites had similar or higher reproductive success compared with less-disturbed sites, possibly suggesting that habituation had occurred.

Use nest covers to reduce the impact of research on predation of ground-nesting seabirds

A before-and-after study from Canada found that hatching success of Caspian terns was significantly higher when researchers protected nests after disturbing adults from them.

9.1. Use wildlife refuges to reduce hunting disturbance

Background

Wildlife refuges are areas where hunting is prohibited. They reduce mortality from hunting and the disturbance it causes. Studies describing the effects of refuges are discussed in 'Threat: Biological resource use'.

9.2. Use signs and access restrictions to reduce disturbance at nest sites

- Six studies (two replicated and controlled, two before-and-after and two small studies) from across the world (2,3,5,7–10) found increased numbers of breeders (2,8), higher reproductive success (3,5,9,10) or lower levels of disturbance (5,7) in waders and terns following the start of access restrictions or the erection of signs near nesting areas. One Canadian study (2) involved the use of multiple interventions.

- A before-and-after study from the USA (6) found that a colony of black-crowned night herons *Nycticorax nycticorax* was successfully relocated to an area with no public access.
- One small study from Europe (1) and one replicated and controlled study from Antarctica (4) found no effect of access restrictions on the reproductive success of eagles or penguins, respectively.

Background

If most disturbance to nesting birds is accidental, then simply warning the public that there are birds present, or restricting access at certain times of the year may help reduce disturbance and the abandonment or destruction of nests. Conversely, if disturbance is coming from people attempting to see nests or birds, then alerting the public to their presence with signs could be counter-productive.

A small study in 1976–88 in the wetlands of the Doñana National Park, Andalucia, Spain (1), found that there were no differences in number of Spanish imperial eagle *Aquila adalberti* pairs that laid, clutch size, hatching size or nestling survival after trails near nests were temporarily closed. This study discusses other eagle management techniques, described in 'Add perches to electricity pylons to reduce electrocution', 'Bury or isolate power lines', 'Foster eggs or chicks with wild conspecifics' and 'Remove/treat endoparasites'.

Two before-and-after studies in 1977–89 at two common tern *Sterna hirundo* colonies in Lake Ontario, Canada (2), found that the nesting population increased at one colony but decreased at the second following the use of several interventions, including the erection of signs highlighting the presence of nesting birds. This study is discussed in 'Replace nesting substrate following severe weather'.

A replicated controlled study on a 28 km stretch of coast in a heavily-visited national park in Victoria, Australia (3), found that hooded plovers *Thinornis rubricollis* had significantly higher reproductive success in 1991–8 under three restricted-access regimes, compared to two regimes that allowed dogs on the beach (0.55 fledglings/clutch for 40 restricted access clutches vs. 0.10 fledglings/clutch for 131 open-access clutches). Hatching success was 31–40% and fledging success 31–68% for the 40 clutches in areas with no access for dogs; both dogs and people or under a 'Plover Watch' scheme, where volunteers ask people to avoid nests and control dogs. This compared with hatching and fledging success of 0–12% and 0–16% for 131 clutches in areas where dogs were prohibited from 0900–1700 each day or where there was unrestricted access to people and dogs. Overall, the average number of fledglings increased over the study period.

A replicated, controlled study on Goudier Island (2 ha), Antarctica (4), in the austral summer of 1996–7, found that gentoo penguins *Pygoscelis papua* did not have significantly higher reproductive success at colonies with no access by tourists, compared to six colonies that tourists could walk through. The percentage of birds laying and eggs hatching did not vary between colonies (82–98% of birds laying and 71–95% of eggs hatching for 556 nests in disturbed colonies vs. 89–100% laying and 88–90% hatching for 170 nests in control colonies), nor did the growth rates of chicks. Overall, 3,103 tourists visited the island making a total of 7,938 'man-hours' of visits over four months. Tourists could walk under supervision through six disturbed colonies but could not approach closer than 25 m to four protected colonies (with two 70 m away and partially concealed by rocks).

A before-and-after trial from July–August in 1997–1998 in the waterways surrounding a common tern *Sterna hirundo* nesting island in Barnegat Bay, USA (5), found that disturbance and reproductive costs caused by personal watercraft disturbance significantly decreased after the implementation of educational campaigns and increased signage in late 1997. This study is discussed in 'Start educational programs for personal watercraft owners'.

A before-and-after trial at a coastal site in Long Beach, California, USA (6), reported the successful translocation of a black-crowned night heron *Nycticorax nycticorax* colony to a site where public access was stopped. This study is discussed in 'Translocate individuals'.

A replicated before-and-after study in 1982, 1987, 1992 and 2002 at 17 local beaches within Delaware Bay, USA (7) found that disturbance to shorebirds decreased markedly following intensive management intervention to control birdwatchers and crab collectors. Both the mean disruption rate and the mean time that shorebirds were disturbed increased during the 1980s when there were no restrictions or viewing platforms and then declined by 2002 after viewing platforms were constructed and beach access restrictions were enforced (5.6 disruptions/hour and 53 minutes disturbed/hour in 1987 vs. 0.4 and 3.6 in 2002). Fewer people were observed on the beaches after restrictions were enforced and only one bird watcher disturbed the birds in 2002. However, the percentage of disturbed shorebirds that flew away (and did not return within 10 min) did not change during the 1980s and increased in 2002. Observations were made on 12–20 days each year for 6–10 h per day.

A small before-and-after study on a beach in California, USA (8), found that the number of breeding snowy plovers *Charadrius alexandrinus* increased from one pair in 2001 to 26 pairs (fledging 74 young) in 2004, following the installation of a simple rope fence in June 2001. The probability of eggs being trampled in 2002 was 8% outside the roped area, compared with 0% inside. The fence consisted of metal posts every 5 m and a single rope strung across the top. In 2001, 265 m of beach was roped off; this increased to 400 m in 2002 and further increased in 2003–4.

A replicated, controlled study at three sandy beaches in Algarve, Portugal (9), found that little tern *Sterna albifrons* breeding success in 2003–5 was significantly higher on two beaches with information and warning signs and weekend wardening, compared to a beach without protective measures (50–91% nesting success for 339 nests on the two protected beaches vs. 0–35% success for 153 nests on the unprotected beach). The presence/absence of protective measures was the most important predictor of nesting success. The main causes of nest failure were predation, destruction by humans and dogs and abandonment.

A small study in Victoria, Australia, between 2003 and 2007 (10) found that two hooded plover *Thinornis rubricollis* nests located on beaches that were being cleaned following an oil spill, both survived and fledged young, after they were marked using signs and rope fences. In addition, cleaning crews worked for 20 minutes and then stopped for 20 minutes to allow adults to incubate the eggs. This study is also discussed in 'Clean birds following oil spills'.

(1) Ferrer, M. & Hiraldo, F. (1991) Evaluation of management techniques for the Spanish imperial eagle. *Wildlife Society Bulletin*, 19, 436–442.
(2) Morris, H., Blokpoel, H. & Tessier, G. D. (1992) Management efforts for the conservation of common tern *Sterna hirundo* colonies in the Great Lakes: two case histories. *Biological Conservation*, 60, 7–14.
(3) Dowling, B. & Weston, M. A. (1999) Managing a breeding population of the hooded plover *Thinornis rubricollis* in a high-use recreational environment. *Bird Conservation International*, 9, 255–270.
(4) Cobley, N. D. & Shears, J. R. (1999) Breeding performance of gentoo penguins (*Pygoscelis papua*) at a colony exposed to high levels of human disturbance. *Polar Biology*, 21, 355–360.
(5) Burger, J. & Leonard, J. (2000) Conflict resolution in coastal waters: the case of personal watercraft. *Marine Policy*, 24, 61–67.
(6) Crouch, S., Paquette, C. & Vilas, D. (2002) Relocation of a large black-crowned night heron colony in southern California. *Waterbirds*, 25, 474–478.
(7) Burger, J., Jeitner, C., Clark, K. & N, L. J. (2004) The effect of human activities on migrant shorebirds: successful adaptive management. *Environmental Conservation*, 31, 283–288.
(8) Lafferty, K. D., Goodman, D. & Sandoval, C. P. (2006) Restoration of breeding by snowy plovers following protection from disturbance. *Biodiversity and Conservation*, 15, 2217–2230.
(9) Medeiros, R., Ramos, J. A., Paiva, V. H., Almeida, A., Pedro, P. & Antunes, S. (2007) Signage reduces the impact of human disturbance on little tern nesting success in Portugal. *Biological Conservation*, 135, 99–106.
(10) Weston, M. A., Dann, P., Jessop, R., Fallaw, J., Dakin, R. & Ball, D. (2008) Can oiled shorebirds and their nests and eggs be successfully rehabilitated? A case study involving the threatened hooded plover *Thinornis rubricollis* in south-eastern Australia. *Waterbirds*, 31, 127–132.

9.3. Set minimum distances for approaching birds (buffer zones)

- We captured no intervention-based evidence for the effects on bird populations of setting minimum distances for approaching birds.

Background

Disturbance by people on foot or in vehicles can reduce birds' use of habitats, or drive them into less favourable habitats. Preventing people from approaching birds too closely may help reduce disturbance. We found two studies that investigated the distances at which birds were disturbed by people: one found that people on foot disturbed birds at greater distances than people in boats (Rodgers & Smoth 1995); the second found that different species reacted differently to disturbance (Rodgers & Schwikert 2002). Both recommended buffer zones of between 100 and 180 m around breeding or feeding birds.

Rodgers, J. A. & Smith, H. T. (1995) Set-back distances to protect nesting bird colonies from human disturbance in Florida. *Conservation Biology*, 9, 89–99.

Rodgers, J. A. & Schwikert, S. T. (2002) Buffer-zone distances to protect foraging and loafing waterbirds from disturbance by personal watercraft and outboard-powered boats. *Conservation Biology*, 16, 216–224.

9.4. Provide paths to limit the extent of disturbance

- A before-and-after study from the UK (1,2) found that two species of wader nested closer to a path, or at higher densities near the path, following resurfacing, which resulted in far fewer people leaving the path.

Background

Studies have shown that visitors keep to paths when they are provided, thereby reducing disturbance to the wider habitat without the need for specific access restrictions (e.g. Pearce-Higgins & Yalden 1997).

Pearce-Higgins, J. W. & Yalden, D. W. (1997) The effect of resurfacing the Pennine Way on recreational use of blanket bog in the Peak District National Park, England. *Biological Conservation*, 82, 337–343.

A before-and-after study from March–July in 1986–1988 and 1996–1998 at a moor and bog site within the Peak District, England (1), found that Eurasian golden plovers *Pluvialis apricaria* avoided a significantly smaller area surrounding a path after it was re-surfaced, compared with before (birds avoided areas up to 200 m from the path before re-surfacing vs. areas 50 m from the path afterwards; birds showed no avoidance on weekdays after re-surfacing). Before resurfacing, up to 30% of walkers left the path, afterwards only 4% left it. The study found no evidence that plover reproduction was adversely affected by disturbance around footpaths.

A before-and-after study (2) using data from the same surveys as in (1) found that dunlin *Calidris alpine* occupancy within 200 m of the footpath increased by 50% following path resurfacing in 1994 (35 birds seen before resurfacing vs. 57 afterwards). However, the authors caution that this was not a significant increase, probably due to small sample sizes. The study found no evidence that dunlin reproduction was adversely affected by disturbance around footpaths.

(1) Finney, S. K., Pearce-Higgins, J. W. & Yalden, D. W. (2005) The effect of recreational disturbance on an upland breeding bird, the golden plover Pluvialis apricaria. *Biological Conservation*, 121, 53–63.

(2) Pearce-Higgins, J. W., Finney, S. K., Yalden, D. W. & Langston, R. H. W. (2007) Testing the effects of recreational disturbance on two upland breeding waders. *Ibis*, 149, 45–55.

9.5. Reduce visitor group size

- We captured no intervention-based studies examining the effects of reducing visitor numbers on bird populations. However, a single, replicated study in forests in Spain in 2004 (1) found that fewer birds, but not fewer species, were observed as visitor number increased. This effect was largely due to decreases in collared dove *Streptopelia decaocto* presence and serin *Serinus serinus* abundance.

Background

Larger groups of visitors to sites will probably cause more disturbances, so limiting the number of people visiting at once may reduce detrimental impacts on birds.

(1) Remacha, C., Perez-Tris, J. & Antonio Delgado, J. (2011) Reducing visitors' group size increases the number of birds during educational activities: Implications for management of nature-based recreation, *Journal of Environmental Management*, 92, 1564–1568.

9.6. Use voluntary agreements with local people to reduce disturbance

- A before-and-after trial in the USA (1) found significantly lower disturbance rates following the establishment of a voluntary waterfowl avoidance area (VWAA), despite an overall increase in boat traffic.

Background

Many people will be willing to avoid certain areas or activities to help local bird populations if they are consulted and kept informed about conservation issues. Under these circumstances, voluntary agreements to avoid birds may be at least as effective in reducing disturbance as restricting access.

A before-and-after study 1986 in Lake Onalaska, Wisconsin and Minnesota, USA (1), found that disturbances to waterfowl within a voluntary waterfowl avoidance area (VWAA) established in 1986 decreased significantly over time. Despite an increase in boating traffic (1.82 boating events/hour in 1986–8 vs. 2.58 in 1997), the 1997 disturbance rate were comparable to that in 1981. Rate of intrusion into the VWAA was lower in 1997 (0.11 intrusions/boating event) than in either 1986–8 (0.18) or 1993 (0.21). Boating disturbances to waterfowl within the VWAA occurred at about half the rate (0.24–0.28 disturbances/hour) observed prior to establishment of the programme (0.48 disturbances/hour). The total number of waterfowl displacements observed as a result of boating events was 435,770 in 1993 and 71,155 in 1997. More than 90% of all waterfowl were observed within the VWAA.

(1) Kenow, K. P., Korschgen, C. E., Nissen, J. M., Elfessi, A. & Steinbach, R. (2003) A voluntary program to curtail boat disturbance to waterfowl during migration. *Waterbirds*, 26, 77–87.

9.7. Start educational programmes for personal watercraft owners

- A before-and-after trial in the USA (1) found that rates of disturbance by personal watercraft decreased and reproductive success of common terns *Sterna hirundo* increased following a series of educational programmes aimed at recreational boat users.

Background

It is possible that people will respond more positively to access restrictions or signs warning them of disturbing birds if they are more aware of the issue. Educational programmes may be able to help with this, informing the public of why certain actions are being taken. More general education programmes are discussed in 'Use education programmes and local engagement to help reduce pressures on species'.

A before-and-after trial from July–August in 1997–8 in the waterways surrounding a common tern *Sterna hirundo* nesting island in Barnegat Bay, USA (1), found that disturbance and reproductive costs caused by personal watercraft (PWC) disturbance significantly decreased after the implementation of educational campaigns in late 1997. The proportion of PWCs moving past the nesting island decreased from 60% in 1997 to 30% in 1998 and PWCs moved at significantly slower speeds. No PWCs went all the way around the nesting island in 1998. The number of terns displaced by PWCs significantly declined (from an average of 40 birds flying over the nesting island in 1997 to 20 birds in 1998). Fledging rate in 1998 was almost 1 chick/nest, similar to before PWC influx (pre-1996). The educational campaign was aimed at local PWC rental businesses and owners and signs were posted around tern nesting islands.

(1) Burger, J. & Leonard, J. (2000) Conflict resolution in coastal waters: the case of personal watercraft. *Marine Policy*, 24, 61–67.

9.8. Habituate birds to human visitors

- A replicated, controlled study from Australia (1) found that bridled terns *Sterna anaethetus* in heavily disturbed had similar or higher reproductive success compared with less-disturbed sites, possibly suggesting that habituation had occurred.

Background

Disturbance may well be more damaging to birds that are not use to humans. Therefore, habituating birds to the presence of human visitors may decrease nest abandonment or mortality, or increase reproductive success.

A replicated, controlled study from December–February in 1995–8 on rocky islets in Queensland, Australia (1), found that bridled terns *Sterna anaethetus* on three high-disturbance sites had similar reproductive success to birds on two low-disturbance sites, but that intermediate-aged chicks from the disturbed sites were significantly heavier in one of two breeding seasons (average weight of 80 g for 12–13 day-old nestlings in the high-disturbance site vs. 80 g in the low). The author argues that this may be caused by birds habituating to humans faster at the heavily disturbed sites. High-disturbance sites were disturbed by 'visiting' (3–6 people, variable walking speeds and noise levels). Visitation was 3 continuous hours/week or 3 × 1 hours/week and disturbance regimes were rotated between plots. Low-disturbance sites experienced ambient disturbance during data collection and monitoring.

(1) Gyuris, E. (2003) An experimental investigation of the effects of human intrusion into breeding colonies of bridled terns *Sterna anaethetus* in the Great Barrier Reef. *Pacific Conservation Biology*, 9, 265–272.

9.9. Use nest covers to reduce the impact of research on predation of ground-nesting seabirds

- A before-and-after study in Canada (1) found that protecting Caspian tern *Sterna caspia* nests after researchers disturbed parents from them significantly increased hatching success. This was due to a reduction in predation by ring-billed gulls *Larus delawarensis*.

Background

Researchers arriving at nesting colonies may disturb birds and, particularly with ground-nesting species, may leave them vulnerable to opportunistic predators. This predation might be reduced by protecting the nests with cages or individual barriers.

Studies describing the more general use of nest cages and predator barriers are discussed in 'Reduce nest predation by excluding predators from nests or nesting areas'.

A before-and-after study at a site on the South Limestone Islands, Lake Huron, Canada (1), found that the hatching rate in a Caspian tern *Sterna caspia* colony was significantly higher in 1979 when researchers visiting the site added nest covers to nests when they arrived at the colony and removed them as they left, compared to in 1978, when covers were not used (77% of 156 eggs hatching in 1979 vs. 62% of 188 in 1978). This difference was mainly due to large numbers of eggs being eaten by ring-billed gulls *Larus delawarensis* in 1978. The author notes that two eggs in 1978 were crushed by poor placement of the covers. Covers were 38 cm diameter hemisphere made from wood, steel and chicken wire.

(1) Quinn, J. S. (1984) Egg predation reduced by nest covers during researcher activities in a Caspian tern colony. *Colonial Waterbirds*, 7, 149–151.

10. Threat: Natural system modifications

Key messages

Use prescribed burning

We captured 70 studies that investigated the effects of prescribed burning on birds. Nine investigated burning alongside other interventions. Three studies, one from savanna and two from pine forests, found that species richness or diversity was higher in burned areas. Eleven studies from pine forests, Australian sclerophyl forests and grasslands found no differences in species richness between burned and unburned areas. Thirty-two studies on savannas, shrublands, grasslands and pine forests, found that some or all species were found at higher densities on burned areas, or that community composition changed with burning. Thirty-three studies found that some or all birds were similarly or less abundant on burned areas, or that there was no change in community composition. Six studies found higher reproductive productivities in burned areas, or that reproduction was not disrupted by burning. Eight studies found no changes in productivity, survival or predation rates after burning. Three studies found that the effects of burning varied with geography, season or climate; three found no effect of burn season. Six studies found that species were lost temporarily from burned areas but returned, or that results varied depending on the length of time between burning and monitoring.

Use fire suppression/control

All three studies we captured, from the USA, UK and Australia, found that some bird species increased after fire suppression, and in one case that woodland species appeared in a site. Two studies (from the UK and USA) found that some species declined following fire suppression. The USA study identified open country species as being negatively affected.

Protect nest trees before burning

We found no studies describing the effects of protecting nest trees on bird populations.

Clear or open patches in forests

Seven out of nine studies from the UK and USA found that early-successional species increased in clearcut areas of forests, compared to other management. Two studies found that mature-forest species declined. One study found no differences in species richness between treatments, another found no consistent differences. A study from the USA found that a mosaic of cut and uncut areas supported a variety of species.

Clearcut and re-seed forests

One of two studies from the USA found that stands of pines replanted with native species held more species typical of scrub habitats than stands under different management. The other study found similar bird densities in clearcut and re-seeded sites and those under different management.

Thin trees within forests

One study of 14 (from the USA) found higher bird species richness in sites with tree thinning and several other interventions, compared to unmanaged sites. Three studies from the UK and USA found no such differences. Seven studies (four investigating multiple interventions) found that overall bird abundance or the abundance of some species was higher in thinned plots, compared to those under different management. Five studies found that found that abundances were similar, or that some species were less abundant in areas with thinning. Two studies from the USA found no effect of thinning on wood thrushes, a species thought to be sensitive to it. A study from the USA found that a higher proportion of nests were in nest boxes in a thinned site, compared to a control. A study from the USA found no differences in bird abundances between burned sites with high-retention thinning, compared to low-retention sites.

Coppice trees

One of three studies found a population increase in European nightjars on a UK site after the introduction of coppicing and other interventions. Two studies from the UK and USA found that the use of coppices by some bird species declined over time. A UK study found that species richness decreased with the age of a coppice, but that some species were more abundant in older stands.

Use patch retention harvesting instead of clearcutting

One of two studies (from the USA) found that areas under patch retention harvesting contained more birds of more species than clearcut areas, retaining similar numbers to unharvested areas. Two studies found that forest specialist species were found more frequently in patch retention plots than under other management. Habitat generalists declined on patch retention sites compared to other managements.

Use selective harvesting/logging instead of clearcutting

Six of seven studies from the USA and Canada found that some species were more, and other less, abundant in selectively logged forests compared to unlogged stands, or those under other management. One study found that differences between treatments were not consistent. A study from the USA found that species richness of cavity-nesting birds was lower in selectively logged forests than in clearcuts. One study from the USA found that brood parasitism was higher in selectively logged forests for two species and lower for two others, compared to control stands.

Use variable retention management during forestry operations

A study from the USA found that nine species were more abundant and five less so in stands under variable retention management, compared to unmanaged stands.

Use shelterwood cutting instead of clearcutting

A study from the USA found that bird community composition differed between shelterwood stands and those under other forestry practices: some species were more abundant, others less so.

Manage woodland edges for birds

One of three studies found that a local population of European nightjars increased at a UK site following the start of a management regime that included the management of woodland edges for birds. Two studies of an experiment in the USA found that bird abundance (but not species richness or nesting success) was higher in woodland edges managed for wildlife than unmanaged edges.

Manually control or remove mid-storey and ground-level vegetation (including mowing, chaining, cutting etc)

One of the 34 studies captured found higher species richness in forests in the USA with mid- and under-storey control. Four studies found similar, or lower, richness or diversity on sites with vegetation control. One study from the UK reported an increase in wader populations after artificial grasslands were cut. Sixteen studies from natural and artificial grasslands, forests, shrublands and savannas found that overall bird densities, or those of some species, were higher on sites with mechanical vegetation control. Five of these investigated several interventions at once. Fifteen studies from the same habitats and reedbeds found that overall densities, or those of some species, were similar or lower on sites with vegetation control, compared to those without, or under different, management. Two investigated several interventions at once. Four studies found that productivity in forests, grasslands and reedbeds was higher with vegetation control, two found no differences. One study found that northern bobwhite chicks had greater foraging success in sites with vegetation controls than in burned sites, but not compared to other managements. One study found that Henslow's sparrows were more likely to be recaptured on artificial grasslands with vegetation control and a study found that mid- and under-storey control in forests did not alter inter-specific competition. One study found that burn season did not alter the effects of burning in shrublands and another that geese grazed at the highest densities on reedbeds cut 5–12 years earlier.

Replace non-native species of tree/shrub

A study from the USA found that the number of black-chinned hummingbird nests increased after fuel reduction and the planting of native species, but that the increase was smaller than at sites without planting.

Provide deadwood/snags in forests

One of six studies found higher bird diversity in forests in the USA when snags were provided. Two studies from the USA and Australia found that total bird abundances or those of some species were higher where trees were ring-barked or debris was added. Three studies from the UK and the USA found that birds nested in and foraged on artificially created snags. One UK study found that the target species (crested tits) did not use snags created for them.

Remove coarse woody debris from forests

Two studies from the USA found that some species increased in sites with woody debris removal. One found that overall breeding bird abundance and diversity were lower in removal plots; the other that survival of black-chinned hummingbird nests was lower.

Apply herbicide to mid- and under-storey vegetation

One of seven studies from North America found that bird species richness in a forest declined after deciduous trees were treated with herbicide. Three studies found increases in total bird densities, or those of some species, after herbicide treatment, although one found no differences between treatment and control areas. One study found that densities of one species decreased and another remained steady after treatment. Three studies found that nest survival was lower in herbicide-treated areas and one found lower nesting densities. One study found that northern bobwhite chicks higher had foraging success in forest areas treated with herbicide compared to under other managements.

Treat wetlands with herbicides

All four studies from the USA found higher densities of birds in wetlands sprayed with herbicide, compared with unsprayed areas. Two found that some species were at lower densities compared to unsprayed areas or those under other management.

Employ grazing in natural and semi-natural habitats

One of 30 studies found more bird species on grazed than ungrazed savannas in Kenya and one from Canada found that duck populations increased on artificial grasslands after grazing was started. One study found lower species richness on grazed artificial grasslands and one found a decrease in species number after an increase in grazing intensity. Thirteen studies found that some species were at higher densities on grazed than ungrazed habitats, or increased after grazing started, although one found that there were no such differences in drought years. Twenty found that some species declined, or did not differ, between grazed and ungrazed areas. Two studies found higher nesting success for ducks on grazed grasslands, ten studies found lower productivities for a variety of species on grazed, compared to ungrazed sites, one investigated several interventions at once. One study found that two groups of birds in the UK were more abundant on grazed non-grasslands than on permanent pasture. Another group was less abundant.

Plant trees to act as windbreaks

One of two studies found that a population of European nightjars increased at a UK site after multiple interventions including the planting of windbreak trees. A study from the USA found that such trees appeared to disrupt lekking behaviour in greater prairie chickens.

Re-seed grasslands

One of two studies from the UK found that geese grazed at higher densities on re-seeded grasslands than on control or fertilised grasslands. Another study from the UK found that geese grazed at higher densities on areas sown with clover, rather than grass seed.

Fertilise grasslands

All four studies captured (all from the UK) found that more geese grazed on fertilised areas of grass more than control areas. Two investigated cutting and fertilising at the same time. One study found that fertilised areas were used less than re-seeded areas. One study found that fertilisation had an effect at applications of 50 kg N/ha, but not at 18 kg N/ha. Another found that the effects of fertilisation did not increase at applications over 80 kg N/ha.

Raise water levels in ditches or grassland

One of seven studies found that three waders were found to have recolonised a UK site or be found at very high densities after water levels were raised. Three studies from Europe found that raising water levels on grassland provided habitat for waders. A study from Denmark found that oystercatchers did not nest at higher densities on sites with raised water levels. A study from the UK found that birds visited sites with raised water levels more frequently than other fields, but another UK study found that feeding rates did not differ between sites with raised water levels and those without. A study from the USA found that predation rates on seaside sparrow nests increased as water levels were raised.

Manage water level in wetlands

Three studies (of six) from the USA, UK and Canada found that different species were more abundant at different water heights. One found that diversity levels also changed. One study found that great bitterns in the UK established territories earlier when deep water levels were maintained, but productivity did not vary. A study from Spain found that water management successfully retained water near a greater flamingo nesting area, but did not measure the effects on productivity or survival.

Use environmentally sensitive flood management

One of two studies found more bird territories on a stretch of river in the UK with flood beams, compared to a channelised river. The other found that 13 out of 20 species of bird increased at sites in the USA where a river's hydrological dynamics were restored.

Use greentree reservoir management

A study from the USA found that fewer mid- and under-storey birds were found at a greentree reservoir site than at a control site. Canopy-nesting species were not affected.

Plough habitats

One of four studies found that bird densities were higher on ploughed wetlands in the USA than unploughed ones. Three studies of one experiment in the UK found that few whimbrels nested on areas of heathland ploughed and re-seeded, but that they were used for foraging in early spring. There were no differences in chick survival between birds that used ploughed and re-seeded heathland and those that did not.

Create scrapes and pools in wetlands and wet grasslands

Four out of six studies from the UK and North America found that more bird used sites, or breeding populations on sites increased, after ponds or scrapes were created. A study from the USA found that some duck species used newly created ponds and others used older ponds. A study from the UK found that northern lapwing chicks foraged in newly created features and that chick condition was higher in sites with a large number of footdrains.

10.1. Use prescribed burning

Controlled burning of accumulated litter (fuel reduction) may reduce the risk of hotter, more extensive and potentially more damaging, wildfires in temperate woodland habitats (e.g. Wooller & Brooker 1980). Periodic prescribed burns may also be used to reinstate/restore ecosystem processes in forests historically subject to occasional wildfires but where active fire suppression has occurred, often over many decades. Changes likely to occur with burning include a reduction in hardwood under-storey vegetation and an increase in grasses and herbaceous vegetation. Other habitat modifications or interventions may be under-taken in combination with fire, such as thinning trees or the removal of mid- and under-storey vegetation. Sometimes conflicts arise as to conservation priorities and possible detrimental effects on non-target species or communities.

Wooller, R. D. & Brooker, K. S. (1980) The effects of controlled burning on some birds of the understorey in karri forest. *Emu*, 80, pp.165–166.

10.1.1. Deciduous forests

- Of four studies found, one paired sites study from the USA (1) found that bird species richness was similar in burned and unburned aspen forests, although there were significant changes in the relative abundances of some species. A replicated, controlled study in the USA (2) found no evidence for changes in community composition in oak and hickory forests following burning.

- A replicated controlled trial from the USA (3) found no differences in wood thrush nest survival in burned compared to unburned areas. Another replicated controlled trial from the USA (4) found a reduction in the number of black-chinned hummingbird nests following fuel reduction treatments that included burning.

A paired site comparison study in 1994–1995 in Bridger-Teton National Forest Wyoming, USA (1), found no difference in average bird species richness in burned compared to unburned forest sites. There were significantly higher relative abundances of mountain bluebird *Sialia currucoides* and pine siskin *Carduelis pinus* in burned than unburned sites. Six areas of trembling aspen *Populus tremuloides*-dominated forest (38–407 ha) were burned during 1988–1993 and paired with similar-sized unburned areas for comparison.

A replicated, controlled study in Wayne National Forest and Vinton Furnace Experimental Forest, Ohio, USA (2), found that overall, there were no differences in breeding bird community composition in areas of forest under early spring burning compared to unburned areas, although species responses varied. Four areas dominated by oak *Quercus* spp. and hickory *Carya* spp. were each divided into three treatment units of 20–30 ha: unburned; burned 4-years in a row (1996–1999); and burned twice (1996 and 1999). Burning reduced habitat suitability for ground- and low-shrub nesting birds: some species declined in response to repeated burning. Conditions for ground- and aerial-foraging species appeared improved by burning.

A replicated controlled study in 1995–1999 at four mixed-oak *Quecus* spp. forest sites in Ohio, USA (3), found there were no significant differences in wood thrush *Hylocichla mustelina* nest survival rates in burned plots (of 20–35 ha), compared to unburned ones. Within burn plots, nests were situated more frequently in areas subject to low or moderate burn intensity and less so high intensity areas. Nest concealment (i.e. percentage overhead and side cover) was similar in burned and unburned plots but nests were located significantly higher, and in taller and larger-stemmed trees and shrubs in burned than unburned areas.

A replicated, controlled study in riparian forest along the Middle Rio Grande, New Mexico, USA, in 2002–2004 (4), found a 62% reduction in the number of black-chinned hummingbird *Archilochus alexandri* nests (from 42 to 16) on two sites where exotic shrubs and woody debris were cut and burned before herbicide was applied to the root crowns of exotic species. This compared with 8–18% increases at sites with fuel reduction treatments that did not involve burning. These results are discussed in more detail in 'Control/remove under-storey and mid-storey vegetation' and 'Plant native shrubs following fuel reduction'.

(1) Dieni, J. S., and Anderson, S. H. (1999) Effects of recent burning on breeding bird community structure in aspen forests. *Journal of Field Ornithology*, 70, 491–503.
(2) Artman, V. L., Sutherland, E. K., and Downhower, J. F. (2001) Prescribed burning to restore mixed-oak communities in southern Ohio: effects on breeding-bird populations. *Conservation Biology*, 15, 1423–1434.
(3) Artman, V. L., and Downhower, J. F. (2003) Wood thrush (*Hylocichla mustelina*) nesting ecology in relation to prescribed burning of mixed-oak forest in Ohio. *The Auk*, 120, 874–882.
(4) Smith, D. M., Finch, D. M., and Hawksworth, D. L. (2009) Black-chinned hummingbird nest-site selection and nest survival in response to fuel reduction in a southwestern riparian forest. *The Condor*, 111, 641–652.

10.1.2. Pine forests

- Two studies of the 28 captured (all from the USA) found higher bird species richness in sites with prescribed burning, tree thinning and mid- or under-storey control (11) or just burning and tree thinning (22), compared to control sites. Five studies found no differences in species richness or community composition between sites with prescribed burning (2,5,8); prescribed burning, tree thinning and mid- or under-storey control (3); or prescribed burning and tree thinning only (24), compared to control sites, or those with other management.

- Eight studies (8,16–19,26,27) found that some species or guilds (such as open habitat species) were more abundant or more likely to be found in burned areas of pine forest than control areas. One study (28) found that the responses of Henslow's sparrows to burning varied considerably with geography and habitat.
- Three studies (12,22,24) found that some species were more abundant in thinned and burned stands, compared to controls or other management (12,21,24). Three studies found that overall bird densities (3,11) or abundances of red-cockaded woodpeckers (1) were higher in open pine forests with prescribed burning, tree thinning and mid- or under-storey control, compared with control areas (1,11) or those thinned but not burned (3). One (11) found differences were more marked in spring. A study found that a red-cockaded woodpecker population increased following the start of intensive management consisting of prescribed burning and other interventions.
- Ten studies found that total bird densities or those of some species was the same or lower in sites with prescribed burning (2,3,5,7–9,18,24,25,27) compared to control sites, or those with other management. Five studies (3,7,9,18,24) investigated several interventions at once. Generally, closed-forest species and ground nesters appeared to be adversely affected by burning.
- Three studies found higher productivities or survival of species in burned (10,20) or burned and thinned (13) areas, compared to control areas or those burned less recently. Seven studies found no differences in productivity, behaviour or survival (including of artificial nests) in burned areas (14,20–22,25) or burned and thinned areas (9,15), compared to controls. One study (23) found that northern bobwhite chicks had lower foraging success in burned areas, compared to other management regimes, whilst another (14) found that different predators were dominant under different management.
- The three studies that investigated it (5,16,20) found that burning season did not appear to affect the effects of burning.

Background

Fire is a fundamental part of many pine forest ecosystems, and is particularly well studied in the USA, where open pine forests and savannas are intensively managed for red-cockaded woodpeckers *Picoides borealis* and other species such as Bachman's sparrows *Peucaea aestivalis* (formerly *Aimophila aestivalis*).

A replicated before-and-after study in four National Forests in Texas, USA (1), found that red-cockaded woodpecker *Picoides borealis* populations increased at all four sites in the early 1990s after management for woodpeckers was intensified in 1989. This followed declines in the 1980s. Management included: prescribed burning and mechanical mid-storey vegetation removal and pine thinning; provision of 716 artificial cavities and the translocation of 19 woodpeckers. Status (i.e. active or inactive) of woodpecker cavity-tree clusters (groups of trees occupied by a breeding group of woodpeckers) was monitored (1983–1993) and by 1993, 98% of active clusters and 58% of inactive clusters had been burned at least once.

A replicated study in 1991 in Ocala National Forest, an area of sand pine scrub in Florida, USA (2), found similar bird densities and species richness in areas that were burned, compared to areas that were clearcut and 'brake-seeded'. This study is discussed in detail in 'Clearcut and re-seed forests'.

A replicated controlled study in 33 pine-grassland stands in Ouachita National Forest, Arkansas, USA (3), found that overall bird species richness was similar across a series of different managements aimed at red-cockaded woodpecker conservation. Bird densities were highest in the second growing season after both burning and mid-storey and tree thinning, compared with those subject to only mid-storey and tree thinning or untreated stands. Ground-nesting species were most abundant in untreated stands. Management appeared beneficial to several species of conservation concern e.g. Bachman's sparrow *Peucaea aestivalis* (formerly *Aimophila aestivalis*), as well as red-cockaded woodpeckers.

A study in mixed pine *Pinus* spp. forests in South Carolina, USA (4) found that a population of red-cockaded woodpeckers increased from four individuals and one breeding pair in 1985 to 99 and 19 pairs in 1996 following intensive management, including prescribed burning. The authors emphasise that hardwood midstory control using prescribed burning as well as cutting, herbicide applications and thinning of trees mimicked the natural fire regime and was essential to the success of the project. In addition, artificial cavities were installed; southern flying squirrels *Glaucomys volans* (a cavity competitor) were controlled and 54 woodpeckers were translocated into areas with artificial or natural cavities.

A replicated study in mid-December 1995 to mid-February 1996 at a mixed pine forest at Fort Benning Military Reservation, Georgia, USA (5), found no significant differences between wintering bird abundances or community composition in nine plots burned during the growing season (April–August) of 1994, compared to nine plots burned during the dormant season (January–March). A total of 48 species were recorded (during point count censuses): 41 in growing-season burn plots and 41 in dormant-season burn plots (34 common to both). On average, 27 species were recorded in a growing-season burn plot and 25 in a dormant-season burn plot. Abundance of individual species was similar between plot types.

A replicated study in 2004–2005 in open pine forests at Eglin Air Force Base, Florida, USA (6), found that eight of 21 plots managed intensively for red-cockaded woodpeckers contained Bachman's sparrows. Management was mainly prescribed burning but included some mid-storey clearing. Some areas with woody midstory vegetation and lacking dense ground cover were unsuitable for breeding sparrows. A burn rotation of 3–5 years (burning early in the growing season) appeared best to encourage the dense grassy understory and sparse midstory preferred by Bachman's sparrows for nesting, whilst not decreasing habitat suitability for red-cockaded woodpeckers.

A replicated study in 1995–1996 in pine savanna in South Carolina, USA (7), found that there were fewer scrub-successional species in stands managed for red-cockaded woodpeckers (including prescribed burning) than in stands which were clearcut to remove non-native pines and replanted with longleaf pines *Pinus palustris*. This study is discussed in detail in 'Clearcut and re-seed forests'.

A replicated, controlled study in 1993–1995 in loblolly pine *Pinus taeda*-dominated forest in Piedmont National Wildlife Refuge, Georgia, USA (8), found that red-cockaded woodpeckers were found in 18 rotationally burned plots (each >100 ha) but not in six unburned plots. Average species richness was similar for burned and unburned plots (42 vs. 41 species) and all other species were found in both plot types. Of 29 species that showed significant differences in abundance between burned and unburned areas, 22 were more abundant in burned plots and seven were more abundant in unburned plots. During 1994–1995, 224 nests of 20 species were found in burned plots; only nine nests (six species) were found in unburned plots.

At Piedmont National Wildlife Refuge, Georgia, USA, a four-year replicated controlled study found no evidence that winter burning and tree thinning (management primarily to improve red-cockaded woodpecker habitat) negatively affected wood thrushes *Hylocichla mustelina* in loblolly pine-dominated forest (9). Thrush density and adult or juvenile survival during the breeding season did not vary between compartments subject to thinning and prescribed burning and unmanaged ones. Overall, thrush numbers increased on treatment compartments (0.91–0.97 birds present before vs. 0.98–1.05 after management), and declined slightly in controls (0.98–1.05 vs. 0.87–0.92). Plots were surveyed eight times in 1993 and sixteen times a year in 1995–1996.

A replicated study in 1996–1998 in a longleaf pine forest in South Carolina, USA (10), found that great crested flycatchers *Myiarchus crinitus* had larger clutches and slightly higher productivity in nest boxes in plots burned during the warm season (April–June: 4.9 eggs/clutch and 2.7 fledglings/clutch for 24 clutches) compared to those burned during the cool season (December–March: 4.5 eggs/clutch and 2.5 fledglings/clutch for 14 clutches). There were no differences in overall occupancy rates or hatching rates (21% occupancy and 61% hatching success for 210 boxes in plots burned in the warm season vs. 19% and 61% for 120 boxes in those burned in the cool season). This study is also discussed in 'Provide artificial nest sites'.

A replicated, paired, controlled study in 1995–1997 in two pine *Pinus* spp. habitats at Angelina National Forest, Texas, USA (11), found that bird species richness and abundances in spring were significantly higher in plots managed for red-cockaded woodpecker, compared to unmanaged plots (longleaf pine forests: 7 species and 22 individuals in managed forests vs. 6 and 13 in controls; loblolly pine-shortleaf pine *P. echinata* forests: 10 species and 32 individuals vs. 7 and 20). Differences were also present in loblolly-shortleaf, but not longleaf pine forests during winter (loblolly-shortleaf forests: 8 species and 42 individuals vs. 5 and 28; longleaf: 7 and 30 vs. 7 and 26). Management consisted of prescribed burning, mechanical mid-storey vegetation removal and thinning of pine trees.

A replicated study in 1999–2000 across 40 shortleaf pine-hardwood stands in Ouachita National Forest, Arkansas, USA (12), found that northern bobwhite *Colinus virginianus* abundances were highest in stands thinned and burned three years previously (1.5 males heard/count), but were not significantly higher than in stands which were only thinned (1.1) and those thinned and burned two years previously (0.8). Control (unmanaged) stands and those burned a year previously had significantly lower abundances than those burned three years previously or thinned but not burned (control stands: 0.1 calls/count; 0.4 for stands one year after burning). The stands with the highest abundances also had greater understory shrub cover. The management was aimed at restoring pine-grassland habitats.

A controlled study in 1998–2001 on Mt. Trumbull, Arizona, USA (13), found that western bluebirds *Sialia mexicana* were more likely to successfully fledge young and fledged more chicks in restored ponderosa pine *Pinus ponderosa* forest, compare to unrestored forest (75% of 56 nests in restored forest fledging young, with an average of 3 chicks/nest vs. 39% and 2 chicks in control stands). Clutch size and number of nestlings per nest were similar between treatments but an average of 91% of nests were infested with parasitic blowfly *Protocalliphora* spp. larvae in treated forest compared with 46% in unmanaged forest; any effects on post-fledging survival are unknown. Restoration treatments comprised tree thinning, slash manipulation (e.g. chopping) and burning.

A replicated, randomised and controlled study in May–July 2000 in 28 longleaf pine forest plots in Georgia, USA (14) found that survival of 770 artificial nests was similar between burned (42%) and unburned plots (41%). Bird predation was greater in burned (14%) than unburned (10%) plots. Small mammal predation was greater in unburned (31%) than in burned (15%) plots. There was no interaction between the supplementary feeding of predators, burning and nest predation.

At Piedmont National Wildlife Refuge, Georgia, USA (15), a controlled before-and-after study found that wood thrush habitat use ands movements in loblolly pine stands were very similar in stands managed for red-cockaded woodpeckers and control stands. Management consisted of thinning forests and prescribed burning, mostly on small scales, in stands up to 50 ha. Juvenile and adult thrushes were monitored by radio-tracking in two breeding seasons before management (1993–1994) and two after (1995–1996) on an experimental compartment, and for four years on a control (1993–1996).

A replicated study in 1999–2000 in longleaf pine forest at Conecuh National Forest Alabama, and Blackwater River State Forest, Florida, USA (16), found that Bachman's sparrow density was greater in forest patches during the first three years after burning (compared to four or more years). Density did not differ between stands burned during the growing season (April–September) or dormant season (October–March). More sparrows occurred in areas of denser, taller grass, regardless of burn season. As tree canopy cover increased, grass cover decreased. Fire suppression or burning at intervals of more than 4–5 years, resulted in a greatly reduced grass/herbaceous under-storey.

A site comparison study in the winters of 2001–2002 at three pine savanna sites in southeast Louisiana, USA (17), found that wintering Henslow's sparrow *Ammodramus henslowii* abundance was significantly higher in areas burned the preceding growing season (average of 3 birds/ha) than in areas burned two or three years previously (1 bird/ha). Eight areas were burned once in May–August 1999–2001. There was a trend for decreasing sparrow abundance with time since burn: 1.0/ha in areas burned one year previously, around 0.7/ha in areas burned two years previously and 0.2/ha in those burned three years previously. Of vegetation characteristics measured, average seed abundance (which decreased with time since burn) was the best predictor of sparrow abundance.

A replicated, controlled study in 1994–1997 in open-longleaf pine and pocosin woodlands at Fort Bragg, North Carolina, USA (18), found that species associated with open longleaf habitats (e.g. red-cockaded woodpecker and Bachman's sparrow) were most common in burned areas of forest. Fire-suppression-associated species (e.g. wood thrush and ovenbird *Seiurus aurocapilla*) were confined to denser vegetation around pocosins (woodland with a dense under-storey around stream-heads) in burned areas, but were abundant in fire-suppressed areas with a dense under-storey. Overall bird abundance and diversity was greater closer to the pocosins.

A replicated study in 2000 in pine and mixed forests in eight 'Sky Island' mountain ranges in Arizona, USA (19), found that the distributions of 11 of 65 species of birds were affected by burning: 73% of species were positively associated with burned areas and showed stronger associations with severe rather than less severe burns. Strong positive associations with severe fires were apparent for western wood-pewee *Contopus sordidulus* and house wren *Troglodytes aedon*, but negative associations were found for warbling vireo *Vireo gilvus* and red-breasted nuthatch *Sitta canadensis*.

A replicated experiment in 1999–2001 in 13 longleaf pine forest plots within Conecuh National Forest, Alabama, USA (20), found that a significantly lower proportion of Bachman's sparrow territories in compartments burned four years previously (20% of 20) successfully produced young than those in compartments burned less than three years previously (52% of 50). In addition, a higher proportion of male sparrows remained unpaired in compartments burned four years previously compared to those in more recently burned plots (50% of 20 males in plots burned four years before remain unpaired vs. 28% of 50 in more recently burned areas). There was no significant difference between growing or dormant season burn plots. Daily survival rates were similar in compartments burned 4-years previously (93%) and those burned more recently (94%), and those burned during the growing or dormant season (89% and 95% respectively).

A controlled cross-over study over four breeding seasons (2003–2006) in old-growth longleaf pine forest at Wade Tract, Georgia, USA (21), found no significant effect of breeding season prescribed burns on the territory size or site fidelity of male Bachman's sparrows. Approximately half (40 ha) of the study area was burned each year, with the unburned half burned the following year. On average, sparrow territories that were burned (1.9 ha) were similar in size from those that were not (2.1 ha). The proportion of ringed males observed after a prescribed burn was similar in burned (73%) and unburned (77%) areas.

A replicated, randomised, controlled study in 1998–2005 in 12 ponderosa pine stands (15–20 ha) in the North Cascade Range, Washington, USA (22), found that there was a trend towards higher bird density in restored stands, compared to controls (13 birds/ha in eight restored stands vs. 10 in four controls). Management consisted of thinning and under-storey burning. Thinning took place in 1998–1999, with burns in spring 2000 and 2004. Breeding birds were censused in 2001 and 2005. White-headed woodpecker *Picoides albolarvatus* and western bluebird *Sialia mexicana*, Cassin's finch *Carpodacus casinii* and yellow-rumped warbler *Dendroica coronata* had higher densities in treated stands. Mountain chickadee *Poecile gambeli*, western tanager *Piranga ludoviciana* and red-breasted nuthatch *Sitta canadensis* were more common in control stands.

A controlled study within a loblolly pine plantation in Louisiana, USA, in 2003–2005 (23) found that northern bobwhite chicks were 50% less likely to successfully capture arthropods in burned areas of forest than in mown areas, areas burned and treated with imazapyr herbicide, or control areas.

A controlled before-and-after study in 2000–2006 at three ponderosa pine forest sites in northern Arizona, USA (24), found no differences in species richness or evenness between blocks (16–30 ha) which were thinned, those which were burned, those both thinned and burned and control blocks. Western bluebird, and pygmy nuthatch *Sitta pygmea* were significantly more abundant after both burning and thinning (with bluebirds also being more abundant in burned-only sites), compared to controls. Mountain chickadees *Poecile gambeli* were less abundant under burning and burning-and-thinning, than in controls, whilst yellow-rumped warblers *Dendroica coronata* were less abundant after thinning-and-burning, compared to control sites and dark-eyed juncos *Junco hyemalis* did show a response to treatments. Thinning was undertaken in autumn 2002 and burns in autumn 2003. Birds were surveyed (point sampling) in May–July 2000–2002 and 2003–2006.

A replicated controlled trial in 2001–2003 in pine-dominated forests in Klamath National Forest, California, USA (25), found that the average number of dark-eyed juncos, the number of territories, nest survival and nestling weights and sizes were all similar between five 40 ha plots which were burned in autumn 2001 and unburned plots (8–17 birds, 3–6 territories and 95–97% survival for burned plots vs. 8–18 birds, 3–5 territories and 95–96%). Juncos were surveyed in 2002 and 2003. All ten plots had small trees removed in 1998–2000.

A replicated controlled trial in November–January 2005–2008 in longleaf pine plantations in Pebble Hill and Arcadia, Georgia, USA (26), found that there were more Bachman's sparrows in burned plots (10–60 ha), compared to unburned plots (0.6 birds/count vs. 0.3). Approximately 50% of habitat was burned each May with individual blocks burned every second year. Sparrow numbers were positively correlated with percentage bare ground cover, and negatively so with numbers of low woody shrubs, grass cover and grass height.

A paired, controlled study in the winters of 2004–2006 in ponderosa pine woodlands in Coconino and Kaibab National Forests, Arizona, USA (27), found that hairy woodpecker *Picoides villosus* densities were higher in burned plots than controls (11 birds/100 ha in burned plots vs. 2 in controls). Pygmy nuthatch (45 in burn units; 40 in controls) and white-breasted nuthatch *Sitta carolinensis* (10 and 12) densities were similar between treatments. Activity of bark beetles (potential bird food) was greater in burn units (10% of trees having signs of beetles vs. 5% of controls). Burns took place in autumn 2003 (Coconino) and autumn 2003 and spring 2004 (Kaibab).

A replicated study in the winters of 2005–2007 in 19 longleaf pine and flatwoods pitcher plant savannas in Louisiana, USA (28), found that the response of over-wintering Henslow's sparrows to burning varied regionally: abundance in burn plots increased over the first three years after burning in western plots (from approximately 1 bird/ha to 2 birds/ha), but decreased in the east (from 2 to 1 birds/ha). Different habitats may explain the regional responses, as seven of ten eastern plots were in wet flatwoods pitcher plant bogs, five of six western plots were in (drier) longleaf savanna. Throughout Louisiana 19, 2.25 ha plots were established, with 18 burned every 2–3 years. Sparrows were surveyed by flush netting in November–April. Habitat generally became unsuitable for sparrows by about 5-years post-burn due to woody plant encroachment.

(1) Conner, R. N., Rudolph, D. C., and Bonner, L. H. (1995) Red-cockaded woodpecker population trends and management on Texas National Forests. *Journal of Field Ornithology*, 66, 140–151.
(2) Greenberg, C. H., Harris, L. D., and Neary, D. G. (1995) A comparison of bird communities in burned and salvage-logged, clearcut, and forested Florida sand pine scrub. *The Wilson Bulletin*, 107, 40–54.
(3) Wilson, C. W., Masters, R. E., and Bukenhofer, G. A. (1995) Breeding bird response to pine-grassland community restoration for red-cockaded woodpeckers. *The Journal of Wildlife Management*, 59, 56–67.
(4) Franzreb, K. E. (1997) Success of intensive management of a critically imperiled population of red-cockaded woodpeckers in South Carolina. *Journal of Field Ornithology*, 68, 458–470.
(5) King, T. G., Howell, M. A., Chapman, B. R., Miller, K. V., and Schorr, R. A. (1998) Comparisons of wintering bird communities in mature pine stands managed by prescribed burning. *The Wilson Bulletin*, 110, 570–574.
(6) Plentovich, S., Tucker, J. W., Holler, N. R., and Hill, G. E. (1998) Enhancing Bachman's sparrow habitat via management of red-cockaded woodpeckers. *The Journal of Wildlife Management*, 62, 347–354.
(7) Krementz, D. G., and Christie, J. S. (1999) Scrub-successional bird community dynamics in young and mature longleaf pine-wiregrass savannahs. *The Journal of Wildlife Management*, 63, 803–814.
(8) White, D. H., Chapman, B. R., Brunjes IV, J. H., Raftovich Jr, R. V., and Seginak, J. T. (1999) Abundance and reproduction of songbirds in burned and unburned pine forests of the Georgia Piedmont. *Journal of Field Ornithology*, 70, 414–424.

(9) Powell, L. A., Lang, J. D., Conroy, M. J., and Krementz, D. G. (2000) Effects of forest management on density, survival, and population growth of wood thrushes. *The Journal of Wildlife Management*, 64, 11–23.
(10) White, D. H., and Seginak, J. T. (2000) Nest box use and productivity of great crested flycatchers in prescribed-burned longleaf pine forests. *Journal of Field Ornithology*, 71, 147–152.
(11) Conner, R. N., Shackelford, C. E., Schaefer, R. R., Saenz, D., and Rudolph, D. C. (2002) Avian community response to southern pine ecosystem restoration for red-cockaded woodpeckers. *The Wilson Bulletin*, 114, 324–332.
(12) Cram, D. S., Masters, R. E., Guthery, F. S., Engle, D. M., and Montague, W. G. (2002) Northern bobwhite population and habitat response to pine-grassland restoration. *The Journal of Wildlife Management*, 66, 1031–1039.
(13) Germaine, H. L., and Germaine, S. S. (2002) Forest restoration treatment effects on the nesting success of western bluebirds (*Sialia mexicana*). *Restoration Ecology*, 10, 362–367.
(14) Jones, D. D., Conner, L. M., Warren, R. J., and Ware, G. O. (2002) The effect of supplemental prey and prescribed fire on success of artificial nests. *The Journal of Wildlife Management*, 66, 1112–1117.
(15) Lang, J., Powell, L. A., Krementz, D. G., and Conroy, M. J. (2002) Wood thrush movements and habitat use: effects of forest management for red-cockaded woodpeckers. *The Auk*, 119, 109–124.
(16) Tucker, J. W., Robinson, W. D., and Grand, J. B. (2004) Influence of fire on Bachman's sparrow, an endemic North American songbird. *The Journal of Wildlife Management*, 68, 1114–1123.
(17) Bechtoldt, C. L., and Stouffer, P. C. (2005) Home-range size, response to fire, and habitat preferences of wintering Henslow's sparrows. *The Wilson Bulletin*, 117, 211–225.
(18) Allen, J. C., Krieger, S. M., Walters, J. R., and Collazo, J. A. (2006) Associations of breeding birds with fire-influenced and riparian-upland gradients in a longleaf pine ecosystem. *The Auk*, 123, 1110–1128.
(19) Kirkpatrick, C., Conway, C. J., and Jones, P. B. (2006) Distribution and relative abundance of forest birds in relation to burn severity in southeastern Arizona. *The Journal of Wildlife Management*, 70, 1005–1012.
(20) Tucker Jr, J. W., Robinson, W. D., and Grand, J. B. (2006) Breeding productivity of Bachman's sparrows in fire-managed longleaf pine forests. *Wilson Journal of Ornithology*, 118, 131–137.
(21) Cox, J. A., and Jones, C. D. (2007) Home range and survival characteristics of male Bachman's sparrows in an old-growth forest managed with breeding season burns. *Journal of Field Ornithology*, 78, 263–269.
(22) Gaines, W. L., Haggard, M., Lehmkuhl, J. F., Lyons, A. L., and Harrod, R. J. (2007) Short-term response of land birds to ponderosa pine restoration. *Restoration Ecology*, 15, 670–678.
(23) Burke, J. D., Chamberlain, M. J., and Geaghan, J. P. (2008) Effects of understory vegetation management on brood habitat for northern bobwhites. *Journal of Wildlife Management*, 72, 1361–1368.
(24) Hurteau, S. R., Sisk, T. D., Block, W. M., and Dickson, B. G. (2008) Fuel-reduction treatment effects on avian community structure and diversity. *Journal of Wildlife Management*, 72, 1168–1174.
(25) Sperry, J. H., George, T. L., and Zach, S. (2008) Ecological factors affecting response of dark-eyed juncos to prescribed burning. *Wilson Journal of Ornithology*, 120, 131–138.
(26) Cox, J. A., and Jones, C. D. (2009) Influence of prescribed fire on winter abundance of Bachman's Sparrow. *The Wilson Journal of Ornithology*, 121, 359–365.
(27) Pope, T. L., Block, W. M., and Beier, P. (2009) Prescribed fire effects on wintering, bark-foraging birds in northern Arizona. *Journal of Wildlife Management*, 73, 695–700.
(28) Palasz, L. M., Brooks, M. E., and Stouffer, P. C. (2010) Regional variation in abundance and response to fire by Henslow's sparrows in Louisiana. *Journal of Field Ornithology*, 81, 139–150.

10.1.3. Australian sclerophyll forest

- Two of three studies from Australia (1,3) found no differences in bird species richness in burned sites compared to unburned areas.

- Three studies (1–3) found differences in species assemblages in burned and unburned areas, with some species lost and others gained from areas after fire.

A before-and-after study at Treen Brook State Forest, Western Australia, Australia (1), in 1978–1979 found no significant effect on overall bird abundance of under-storey burning of karri *Eucalyptus diversicolor* forest, but two species (of low conservation concern) present prior to the burn were absent afterwards and five species not caught pre-burn were captured post-burn. Capture data from pre-burn (May 1978; 31 net-days) and post-burn (May 1979; 37 net-days) mist-netting was used to monitor birds. In May 1979, 83 individuals of 15 species were captured, comparable to the 66 individuals of 12 species caught in May 1978 (allowing for difference in catch effort).

A replicated study in 1999 in 36 sclerophyll forest sites in south-east Queensland, Australia (2), at found that two of the eleven hollow-nesting species analysed (brown treecreeper *Climacteris picumnus*, little lorikeet *Climacteris picumnus*) were positively associated with more frequent prescribed burning. No species were negatively associated with more frequent prescribed burning.

A randomised, controlled study in January–March 2001 in eucalypt and riparian woodland at three sites along seasonally dry watercourses in northeast Queensland, Australia (3), found no significant differences in overall bird species richness between 10 ha plots burned during the dry season (August 2000), the wet season (December 1999) or unburned plots. Two species were more abundant in the dry season burned compared to unburned sites, while one was more abundant in the dry season burned areas than under other treatments and two species were less abundant after burns. Two other species' responses varied depending on which forest type they were in. Follow-up surveys in 2004 found no differences in species richness between fire treatments.

(1) Wooller, R. D., and Brooker, K. S. (1980) The effects of controlled burning on some birds of the understorey in karri forest. *Emu*, 80, 165–166.
(2) Smyth, A., Mac Nally, R., and Lamb, D. (2002) Comparative influence of forest management and habitat structural factors on the abundances of hollow-nesting bird species in subtropical Australian eucalypt forest. *Environmental Management*, 30, 547–559.
(3) Valentine, L. E., Schwarzkopf, L., Johnson, C. N., and Grice, A. C. (2007) Burning season influences the response of bird assemblages to fire in tropical savannas. *Biological Conservation*, 137, 90–101.

10.1.4. Savannas

- A replicated and controlled study from Kenya (5), of five studies captured, found that burned areas of savanna tended to have more birds and more species than control or grazed areas. However, the authors note that differences were not present during drought years and burned sites showed significant annual variation, unlike grazed sites. A replicated and controlled study from Australia (4) found that the effects of burning on bird abundances depended on burn season, and habitat type.
- Two replicated studies in the USA (2,3) found that some open country species were more common in burned areas than unburned, whilst other species were less so.
- A small study from the USA (1) found that two eastern bluebird *Sialia sialis* successfully raised chicks after the habitat around their nest boxes was subject to a prescribed burn.

Background

Periodic wildfires and grazing/browsing by wild animals maintained open savanna ecosystems in parts of the world, retarding woody plant encroachment and succession to woodland. In some regions where these natural processes have been lost, and where active fire suppression may have also occurred, rotational burning may be used to reinstate and maintain more open habitat conditions amenable. Other methods commonly used include thinning of trees and shrubs, timber harvesting, and livestock grazing.

There is a continuum between habitats classified as 'savannas' and those classified as 'forests' and 'grasslands'. Relevant information for management of habitats with both trees and grasses may, therefore, be found in all three sections.

A small study in Minnesota, USA (1), found that eastern bluebird *Sialia sialis* clutches in two nest boxes in an area that underwent prescribed burning both had 100% success (i.e. all eggs produced fledglings), compared with an average success of 93% for 23 nest boxes nearby. Flame height was 1 m or less and did not reach the boxes at 1.2 and 1.4 m above ground. Adult bluebirds left their boxes as the fire approached, returning when it had passed.

At Cedar Creek Natural History Area, Minnesota, USA (2), a replicated controlled study of oak savanna restoration by prescribed burning (initiated in 1964) found that 'open country' bird abundance increased as restoration progressed. Seven units (8–18 ha) were subject to one of seven burn frequencies, ranging from nearly every year to no burning over the previous 31 years. Burns were conducted in spring (except two in late summer). Bird species richness in the two unburned units (17 and 23 species) was lower than that of two frequently burned units (30 and 32) in June 1995 and similar in 1996. As woodland became more open, upper tree canopy insectivores declined, whilst omnivorous birds, particularly ground and lower canopy foragers increased. Woodpeckers increased as standing dead tree abundance increased.

A replicated, paired study in oak *Quercus* spp. savanna in Illinois, USA (3), found that bird community composition was significantly different in areas with prescribed burning compared to closed-canopy oak forest. Of the 31 bird species analysed, 12 were more common in burned savanna and five more common in unburned forest. Twelve savanna sites maintained by burns (spring, autumn or both, on a 3–5 year rotation, with periodic removal of *Acer* spp. and European buckthorn *Rhamnus catharticus*) were paired with 12 forest sites (no burning for over 50 years). Point counts were conducted for 3–5 years (between 25 May–10 July 1995–1999) to assess bird abundance. There was no effect of burning on brood parasitism by brown-headed cowbird *Molothrus ater*.

A replicated, controlled study in January–March 2001 and 2004 in open eucalypt and riparian woodland along three creeks in Queensland, Australia (4), found that plots burned in the dry season of 2000 had significantly fewer birds in 2004 than control (unburned) sites, whilst sites burned in the wet season of 2000 had higher abundances. Species richness did not vary in 2004. However, several species showed short-term changes in abundance after fire: three species were more abundant in dry-season burned sites in 2001; two were less abundant after burns. Pied butcherbirds *Cracticus nigrogularis* were more abundant in burned eucalypt sites and little friarbirds *Philemon citreogularis* were more abundant in dry season burnt sites and riparian habitat of wet season burnt sites.

In Laikipia District, Kenya, a replicated controlled study in 2005–2007 (5) found that five burned plots of savanna had, on average, but not consistently, higher densities of birds and more species than five grazed or four unmanaged control areas (3–17 birds and 3–8 species/ 100 m^2 for burned areas vs. 4–6 birds and 2.5–4.0 species for controls; 5–8 birds and 4–5.5 species for grazed areas). The authors note that there were no differences between treatments in drought years, and that the yearly variation in burned plots was greater than in grazed plots, suggesting that grazing may have longer term benefits. In addition, some species were only recorded in unmanaged areas. The impact of burning appeared to decrease over time.

(1) Cox, C. A. (1987) Nesting bluebirds tolerate prescribed burn (Minnesota). *Restoration and Management Notes*, 5, 48.
(2) Davis, M. A., Peterson, D. W., Reich, P. B., Crozier, M., Query, T., Mitchell, E., Huntington, J., and Bazakas, P. (2000) Restoring savanna using fire: impact on the breeding bird community. *Restoration Ecology*, 8, 30–40.
(3) Brawn, J. D. (2006) Effects of restoring oak savannas on bird communities and populations. *Conservation Biology*, 20, 460–469.
(4) Valentine, L. E., Schwarzkopf, L., Johnson, C. N., and Grice, A. C. (2007) Burning season influences the response of bird assemblages to fire in tropical savannas. *Biological Conservation*, 137, 90–101.
(5) Gregory, N. C., Sensenig, R. L., and Wilcove, D. S. (2010) Effects of controlled fire and livestock grazing on bird communities in east African savannas. *Conservation Biology*, 24, 1606–1616.

10.1.5. Shrublands

- One controlled study from the USA (2), of eight captured, found that overall bird densities were similar between burned and unburned areas, whilst a replicated and controlled study (4) found that species numbers and bird densities did not vary between areas burned in summer and those burned in winter.
- Three studies (1,2,8) found that some species were more abundant on areas that were burned, compared to those managed differently, or not at all. Four studies (2,5,7,8) found that the densities of individual species were similar or lower on burned areas compared to control areas or those under different management.

- A before-and-after study (3) found that sage sparrows chose different nest sites before burning compared to after. A controlled study (6) found no differences in greater sage grouse movement between burned and unburned areas.

Background

Fire can play an important part in shrublands by maintaining the balance between woody and herbaceous vegetation types. If the natural systems have been modified so that fire does not occur as regularly as it used to, for example to reduce risks to homes and property, then prescribed burning may be necessary to maintain the natural dynamics of the habitat.

A before-and-after study in shrubland in 1962–1963 in Wisconsin, USA (1), found that the number of male prairie chickens *Tympanuchus cupido* displaying at a site increased from seven to 13 following prescribed burning. However, the authors note that the number at other sites without burning also increased over the period, (by a single male each time).

A controlled study in 1980 in Utah, USA (2), found that response of breeding songbirds in sagebrush habitat chained or burned 3–4 years earlier varied between species. Total bird densities and diversity were similar between a chained site (i.e. vegetation knocked down by dragging a large chain), burned sites and sites without any intervention for 17 years. However, a burned site had 50–86% fewer Brewer's sparrow *Spizella breweri* (a sagebrush specialist) territories than chained or untreated sites. Horned lark *Eremophila alpestris* densities were 200–250% higher on the burned site compared to the untreated one. Vesper sparrow *Pooecetes gramineus* and western meadowlark *Sturnella neglecta* densities appeared unaffected by sagebrush control.

A before-and-after study in *Artemisia* spp. sagebrush habitat around the Idaho National Engineering Laboratory, Idaho, USA (3), found that in 1982 all 34 sage sparrow *Amphispiza belli* nests found were within sagebrush plants. In 1983, after a prescribed burn created a mosaic of burned and unburned patches, 23 of 29 nests (79%) were within sagebrush but six were in atypical locations: five on the ground under small plants and one within a grass clump.

A replicated and controlled study in shrub dominated by saw-palmetto *Serenoa repens* (a type of palm) in 1988–1989 in Myakka River State Park, Florida, USA (4), found that the total number of birds and the number of species found did not vary between two sites burned in winter (January 1988) and two burned in summer (June 1988) (1.7 species and 2.4 individuals/winter burned site vs. 1.5 and 1.9 for summer burned). There were no differences between winter-burned and control (unburned) sites, but summer-burned sites had significantly fewer species and individuals (2.0 species and 2.7 individuals/unburned site).

A controlled before-and-after study in May–August 1989 in Wyoming big sagebrush *Artemisia tridentata wyomingensis* and threetip sagebrush *A. tripartita* scrub in Big Desert, Idaho, USA (5), found that relative abundances of greater sage grouse *Centrocercus urophasianus* were similar between a burned and unburned area before and after burning. Abundances of Hymenoptera species (an important part in grouse diets) were significantly lower in the burned area two and three years after burning.

A controlled study in 1987–1992 in the Big Desert sagebrush ecosystem (dominated by Wyoming big sagebrush) of south-east Idaho, USA (6), found that there were no differences in timing, distance or direction of movement of 81 greater sage grouse between burned and unburned areas. A 5,800 ha area of sagebrush was burned in late summer 1989, removing vegetation from approximately 57% of the area.

A replicated, controlled study in montane shrubland in Rocky Mountain National Park, Colorado, USA, (7) found that green-tailed towhee *Pipilo chlorurus* occurrence and nesting density was significantly lower for 3–5 years after burning in four areas of subject to prescribed burning (0.05–0.30 birds/ha) compared with three unburned sites (0.60–1.95 birds/ha). Two sites, Deer Ridge Low (55 ha) and Deer Ridge High (80 ha), were burned in 1998 and

1999 respectively, with towhee density estimated in June 2002–2003. Of 179 nests found, only 14 (8%) were at burned sites, and were within remnant patches of live shrubs (in areas where burn severity had been lower).

A controlled study in 1999–2001 on Nantucket Island, Massachusetts, USA (8), found that eastern towhee *Pipilo erythrophthalmus* and common yellowthroat *Geothlypis trichas* were significantly more abundant in burned areas of shrubland than in mown areas. Towhees (but not yellowthroats) were also more common in burned areas than controls (towhees: 1.4 birds/ha for burned areas vs. 1.1 for control areas; yellowthroats: 0.4 for both control and burned areas). Song sparrows *Melospiza melodia* were not significantly more abundant on burned areas than on control or mown areas (0.3 birds/ha for mown areas vs. 0.4 for controls and burned areas).

(1) Anderson, R. K. (1969) Prairie chicken responses to changing booming-ground cover type and height. *The Journal of Wildlife Management*, 33, 636–643.
(2) Castrale, J. S. (1982) Effects of two sagebrush control methods on non gamebirds. *The Journal of Wildlife Management*, 46, 945–952.
(3) Winter, B. M., and Best, L. B. (1985) Effect of prescribed burning on placement of sage sparrow nests. *The Condor*, 87, 294–295.
(4) Fitzgerald, S. M., and Tanner, G. W. (1992) Avian community response to fire and mechanical shrub control in south Florida. *Journal of Range Management*, 45, 396–400.
(5) Fischer, R. A., Reese, K. P., and Connelly, J. W. (1996) An investigation on fire effects within xeric sage grouse brood habitat. *Journal of Range Management*, 49, 194–198.
(6) Fischer, R. A., Wakkinen, W. L., Reese, K. P., and Connelly, J. W. (1997) Effects of prescribed fire on movements of female sage grouse from breeding to summer ranges. *The Wilson Bulletin*, 109, 82–91.
(7) Jehle, G., Savidge, J. A., and Kotliar, N. B. (2006) Green-tailed towhee response to prescribed fire in montane shrubland. *The Condor*, 108, 634–646.
(8) Zuckerberg, B., and Vickery, P. D. (2006) Effects of mowing and burning on shrubland and grassland birds on Nantucket Island, Massachusetts. *Wilson Journal of Ornithology*, 118, 353–363.

10.1.6. Grasslands

- Four studies from the USA (6,13,16,17), Of 21 studies captured, found that overall species richness did not vary between burned areas, or areas burned recently, and unburned sites. One study (13) found that community composition was also similar whilst others found that species showed individual responses.

- Nine studies from across the world (2,3,5,7,8,12,14,19,20) found that at least some study species were found at higher densities or were more abundant in burned areas than in unburned areas or areas under different management. One study investigated multiple interventions at once (7). Fourteen studies (1,3–11,14,15,19,20) found that at least one study species was less abundant or found at similar abundances on burned areas of grassland, compared to unburned areas or those under different management. However, four studies (2,16,19,21) found that apparent responses varied depending on how soon after fires measurements were taken. Care should therefore be taken when interpreting the results of studies on prescribed burning.

- One study from the USA (11) found that Florida grasshopper sparrow had significantly higher reproductive success soon after plots were burned, whilst another American study (18) founds that dickcissel reproductive success was higher in patch-burned areas than burned and grazed areas.

Background

Suppression of fires may shift grassland communities from grasses and herbs towards woody-stemmed species. Alternatively, it may allow the fuel load to build up to such an extent that when fires do occur, they are hotter and burn for longer. This can mean that even fire-adapted plants (such as some pines and palms) are killed.

A possible way to reduce these effects is to ensure there are frequent, smaller fires, which prevent trees from establishing and constantly reduce the fuel load of a system.

A replicated and controlled study in meadows in North Dakota, USA (1), in 1961 found that there were 38% fewer pairs of ducks in burned areas of meadow, compared to unburned areas. This study is discussed in more detail in 'Mow natural grasslands'.

A controlled trial in upland grassland at Highmoor State Forest, KwaZulu-Natal, South Africa (2), in 1973–1977 found that grey-winged *Francolinus africanus* and red-winged *F. levaillantii* francolins were found at higher densities in unburned, rather than burned areas immediately after fire, but that preferences were reversed in the year after burning. At the start of the experimental period (spring 1975), francolin densities were higher in an area sunject to large scale burns than an area under patch burning, but by autumn 1977 there were significantly higher densities in the patch-burned area.

A replicated before-and-after study on Matagorda Island, Texas, USA (3), found that northern harrier *Circus cyaneus hudsonius* numbers decreased significantly on two winter-burned natural grassland plots (from 39 to 12 birds), whilst American kestrels *Falco sparverius* increased (from two to ten birds). Raptor use during the month before and after burning was assessed on two 140 ha plots (burned 4–5 January 1993), by one hour weekly counts (13 December 1992 to 14 February 1993). Surveys (2 December–7 February) across the whole island, found a non-significant increase in total raptor numbers, and of the two commonest species (northern harrier and American kestrel). Total raptor numbers using plots pre- and post-burn were similar but with slight (non-significant) decreases.

A randomised, controlled, replicated before-and-after study in 1980–1988 in mixed-grass prairie at Lostwood National Wildlife Refuge, North Dakota, USA (4), found that nest densities of gadwall *Anas strepera* (but not six other duck species) were lower in areas and years with summer burning (and for several years after) compared to control areas. Densities of gadwall and blue-winged teal *A. discors* were also lower in areas with a combination of burning and spring cattle grazing. Nest success was generally high (31–45%) and unaffected by treatment. The authors argue that grazing reduced brush cover that provided nesting habitat for ducks.

A replicated, controlled trial in March–May 1993–1994 on Matagorda Island, Texas, USA (5), found few differences in spring bird abundance in cordgrass *Spartina*-dune paspalum *Paspalum monostachyum* grassland subject to summer compared to winter burns. Six 122 ha plots were established each year: two unburned; two burned late August (1992–1993); and two in January (1993–1994). Wrens (Troglodytidae) were consistently most abundant in unburned plots (21–28 birds/40 ha in unburned areas vs. 2–3 for summer burn and 1 for winter burn); 18–22 months after burning wren abundance increased but was still less than within unburned plots. Sparrows (Emberizinae) were most abundant on burn plots both 6–10 months (unburned: 3–8 birds/40 ha vs. summer burn: 16–19; winter burn: 11–16) and 18–22 months post-burn.

A replicated, controlled study in 1992–1995 in native grass-sown Conservation Reserve Program fields in Riley County, Kansas, USA (6), found significantly lower bird nesting density in fields with spring (mid-April–May) burning in the year of the burn, compared to control fields (27 of 399 nests found were on burned fields vs. 372 on controls). Nest success was 22% on burned and 34% on unburned fields. Average bird abundance on burned fields (year of burn) was 6 birds/km of transect vs. 9/km on unburned fields. Species richness was similar (12–21 burned vs. 10–19 unburned).

A replicated and controlled study in 1990–1994 in two intensively managed grassland sanctuaries in south-east Illinois, USA (7), found that northern harriers tended to nest in fields not disturbed by grassland management (burning and mowing) within the last 12 months (a total of 22 nests in unmanaged fields vs. seven in burned and grazed fields). Short-eared owl *Asio flammeus* nest-site selection could only be assessed in 1990: all 13 nests were in fields subject to management within the last 12 months. One study site comprised 550 ha of grassland among 10 tracts, the second 308 ha among seven tracts. Each tract comprised 3–32 'sub-fields' (0.5–15 ha) usually subject to one management type (all burned or all mowed).

A replicated, controlled before-and-after study in 1997–1999 in six semi-desert grassland plots at Buenos Aires National Wildlife Refuge, Arizona, USA (8), found that the population responses of five wintering sparrow species to a prescribed spring burn varied between species. Over the three years, vesper *Pocecetes gramineus* and savannah *Passerculus sandwichensis* sparrow populations increased in three burned plots, whilst Cassin's sparrow *Aimophila cassinii* populations decreased. Grasshopper sparrow *A. savannarum* increased up to two years post-burn. Baird's sparrow *Ammodramus bairdii* had consistently low abundance (<0.7/plot).

A replicated, controlled trial in May–June 1995–1996 in grasslands in Prairie Ridge State Natural Area, Illinois, USA (9), found that burned, non-native 'cool season' grassland plots held lower average densities of five grasslands birds than native 'warm season' grasslands under any management (burning, grazing, mowing or no management) and non-native grassland under mowing, haying, grazing or no management. However, species showed individual responses to different managements. The species surveyed were eastern meadowlark *Sturnella magna* and dickcissel *Spiza americana*, Henslow's sparrow *Ammodramus henslowii*, field sparrow *Spizella pusilla* and grasshopper sparrow.

A randomised, controlled study in 1995–1997 in six tracts of tallgrass prairie in Melvern Wildlife Management Area, Kansas, USA (10), found that Henslow's sparrow were significantly less abundant in burned areas than in unburned tracts (1 bird/ha vs. 4); dickcissel abundance was similar in burned and unburned tracts (12 birds/ha). In spring 1995 four tracts were burned, in 1996 one was burned and in 1997 two were burned. In total, 22 Henslow's sparrows (overall relative abundance 0.2 birds/ha) and 200 dickcissel (1.1 birds/ha) were recorded. Abundance was not correlated to tract perimeter length, or different distances to each other.

In prairie grassland at Air Force Avon Park Range, USA, a replicated study in March-August 1997–1999 (11) found that Florida grasshopper sparrow *Ammodramus savannarum floridanus* reproductive success was significantly higher six months after burning (38% of 32 nests successful) than at 18 (14% of 35 nests successful) and 30 months (0% of 13 nests successful) after burning. Four prairie pastures (165–324 ha) were burned in December to mid-March on three year rotations. Sparrow territory density (up to 0.2 territories/ha) appeared unaffected by time elapsed since burning.

A replicated study in June 1998–1999 in grassland and shrubland on a subalpine hillside in the Pyrénées-Orientales, France (12) found that four bird species with an unfavourable conservation status in Europe (rock bunting *Emberiza cia*, woodlark *Lullula arborea*, stonechat *Saxicola torquata* and red-backed shrike *Lanius collurio*) were found at highest abundances on recently burned grassland with scattered shrubs. A wildfire burned about one half of the hillside in 1980. From 1990 onwards, grassland management comprised prescribed winter burns (on a one-to-seven year rotation) and summer cattle grazing.

A replicated, randomised and controlled study in DeSoto National Wildlife Refuge, Iowa, USA, in 1998–1999 (13), found that the average species richness of tallgrass prairie blocks (3–10 ha) was similar for four burned sites (10), mowed sites (12) and controls (11). Community composition was also similar. Burning and mowing took place from 22 April–11 May 1999.

A replicated study in tallgrass prairie in Kansas, USA (14), found that six of seven birds surveyed showed a significant response to burning: Henslow's sparrow, grasshopper sparrow, dickcissel, eastern meadowlark (grassland species) and Bell's vireo *Vireo bellii* (a shrub-dependent species) were least abundant in the breeding season following a burn (with Bell's vireo being absent from sites burned annually); upland sandpipers *Bartramia longicauda* were most abundant in the season following a burn. Brown-headed cowbird *Molothrus ater* did not show any significant response.

A controlled study in 1999–2001 on Nantucket Island, Massachusetts, USA (15), found that song sparrows *Melospiza melodia* were significantly less abundant on burned grasslands, compared to controls (0.1 birds/ha on burned grasslands vs. 0.6 on controls). There was no significant difference between burned and mown grasslands. Savanna sparrows were equally abundant (0.7–0.9 birds/ha) on all treatments.

A replicated study in the winters of 2002–2003 in coastal prairie at Brazoria National Wildlife Refuge, Texas, USA (16), found no significant difference in average number of bird species in five plots one year after burning, compared to plots two or three years after burning. Three rarer species (sandhill crane *Grus canadensis*, Sprague's pipit *Anthus spragueii* and grasshopper sparrow) were only observed in first-year burn plots. Of the four commonest species, Le Conte's sparrow *A. leconteii* was significantly more abundant in second-year than third-year burn plots, savanna sparrow more so in first- than second- or third-year burn plots, and sedge wren *Cistothorus platensis* more common in second- and third-year than first-year burn plots. No significant differences were found for swamp sparrow *Melospiza georgiana*.

A replicated, randomised, controlled study in May–July 1995–1997 in tallgrass prairie at DeSoto National Wildlife Refuge, Iowa, USA (17), found that overall, bird species richness and diversity did not vary between burned patches (3–9.3 ha) of prairie and unburned controls (9–12 species/plot in five burned plots vs. 7–10 species on five controls).

A study in Oklahoma, USA, in 2003–2004 (18), found that dickcissel reproductive success was lower in traditionally-managed pastures (annual burning followed by early-intensive grazing) compared to patch-burn management of tallgrass prairie. Dickcissels (296 nests monitored) tended to start nesting later, but nest densities were higher, in traditionally managed pasture. The average number of eggs per clutch and fledglings produced were similar between treatments. Predation was the main cause of nest failure and was higher in the traditionally managed pastures, as was parasitism by brown-headed cowbird *Molothrus ater*.

A replicated study in 2002–2003 (19) at the same tallgrass prairie site in Kansas, USA, as in (14), found that all seven bird species surveyed showed a significant response to burning: upland sandpipers were more abundant in the breeding season following a burn, whilst Henslow's sparrow, grasshopper sparrow, dickcissel, eastern meadowlark, brown-headed cowbird (all grassland species) and Bell's vireo (a shrub-dependent species) were less abundant or absent. Grasshopper and Henslow's sparrows, and meadowlark were more abundant in areas not burned the preceding spring, and less abundant at sites burned every four years. Bell's vireo was commonest at sites burned every four years.

A replicated study in the winters of 2006–2007 in four wiregrass *Aristida beyrichiana*-dominated prairie sites in south-central Florida, USA (20), found that grasshopper sparrow were six times more likely to be found if transects were burned within the previous 12 months. Sedge wrens *Cistothorus platensis* were more abundant in grassland with longer intervals between fires.

A replicated study in prairie in May–June 1998–2003 at J. Clark Salyer National Wildlife Refuge, North Dakota, USA (21), found that prescribed burning in seven 41–69 ha blocks (each burned in one year of the study) initially reduced densities of some grassland passerines but that total numbers soon increased. Twenty-two grassland bird species were recorded. Species richness and number of pairs was lowest in the first post-burn breeding season, increasing in the second and stabilising up to 4-years post-burn. Fire significantly affected five of eight species analysed: numbers of pairs of sedge wren, clay-coloured sparrow *Spizella pallida*, Le Conte's sparrow *Ammodramus leconteii*, savanna sparrow and bobolink *Dolichonyx oryzivorus* were lowest in the year following burning but then generally increased and stabilised within three years.

(1) Martz, G. F. (1967) Effects of nesting cover removal on breeding puddle ducks. *The Journal of Wildlife Management*, 31, 236–247.

(2) Mentis, R. C., and Bigalke, R. C. (1981) The effect of scale of burn on the densities of grassland francolins in the Natal Drakensberg. *Biological Conservation*, 21, 247–261.

(3) Chavez-Ramirez, F., and Prieto, F. G. (1994) Effects of prescribed fires on habitat use by wintering raptors on a Texas barrier island grassland. *Journal of Raptor Research*, 28, 262–265.

(4) Kruse, A. D., and Bowen, B. S. (1996) Effects of grazing and burning on densities and habitats of breeding ducks in North Dakota. *The Journal of Wildlife Management*, 60, 233–246.

(5) Hul, J. T. V., Lutz, R. S., and Mathews, N. E. (1997) Impact of prescribed burning on vegetation and bird abundance at Matagorda Island, Texas. *Journal of Range Management*, 50, 346–350.

(6) Robel, R. J., Hughes, J. P., Hull, S. D., Kemp, K. E., and Klute, D. S. (1998) Spring burning: resulting avian abundance and nesting in Kansas CRP. *Journal of Range Management*, 51, 132–138.

(7) Herkert, J. R., Simpson, S. A., Westemeier, R. L., Esker, T. L., and Walk, J. W. (1999) Response of northern harriers and short-eared owls to grassland management in Illinois. *The Journal of Wildlife Management*, 63, 517–523.

(8) Gordon, C. E. (2000) Fire and cattle grazing on wintering sparrows in Arizona grasslands. *Journal of Range Management*, 53, 384–389.
(9) Walk, J. W., and Warner, R. E. (2000) Grassland management for the conservation of songbirds in the midwestern USA. *Biological Conservation*, 94, 165–172.
(10) Applegate, R. D., Flock, B. E., and Horak, G. J. (2002) Spring burning and grassland area: effects on Henslow's sparrow (*Ammodramus henslowii*) and dickcissel (*Spiza americana*) in Kansas, USA. *Natural Areas Journal*, 22, 160–162.
(11) Delany, M. F., Linda, S. B., Pranty, B., and Perkins, D. W. (2002) Density and reproductive success of Florida grasshopper sparrows following fire. *Journal of Range Management*, 55, 336–340.
(12) Pons, P., Lambert, B., Rigolot, E., and Prodon, R. (2003) The effects of grassland management using fire on habitat occupancy and conservation of birds in a mosaic landscape. *Biodiversity and Conservation*, 12, 1843–1860.
(13) Van Dyke, F., Van Kley, S. E., Page, C. E., and Van Beek, J. G. (2004) Restoration efforts for plant and bird communities in tallgrass prairies using prescribed burning and mowing. *Restoration Ecology*, 12, 575–585.
(14) Powell, A. F. L. A. (2006) Effects of prescribed burns and bison (*Bos bison*) grazing on breeding bird abundances in tallgrass prairie. *The Auk*, 123, 183–197.
(15) Zuckerberg, B., and Vickery, P. D. (2006) Effects of mowing and burning on shrubland and grassland birds on Nantucket Island, Massachusetts. *Wilson Journal of Ornithology*, 118, 353–363.
(16) Baldwin, H. Q., Grace, J. B., Barrow JR, W. C., and Rohwer, F. C. (2007) Habitat relationships of birds overwintering in a managed coastal prairie. *Wilson Journal of Ornithology*, 119, 189–197.
(17) van Dyke, F., Schmeling, J. D., Starkenburg, S., Yoo, S. H., and Stewart, P. W. (2007) Responses of plant and bird communities to prescribed burning in tallgrass prairies. *Biodiversity and Conservation*, 16, 827–839.
(18) Churchwell, R. T., Davis, C. A., Fuhlendorf, S. D., and Engle, D. M. (2008) Effects of patch-burn management on dickcissel nest success in a tallgrass prairie. *Journal of Wildlife Management*, 72, 1596–1604.
(19) Powell, A. F. L. A. (2008) Responses of breeding birds in tallgrass prairie to fire and cattle grazing. *Journal of Field Ornithology*, 79, 41–52.
(20) Butler, A. B., Martin, J. A., Palmer, W. E., and Carroll, J. P. (2009) Winter use of south Florida dry prairie by two declining grassland passerines. *The Condor*, 111, 511–522.
(21) Grant, T. A., Madden, E. M., Shaffer, T. L., and Dockens, J. S. (2010) Effects of prescribed fire on vegetation and passerine birds in northern mixed-grass prairie. *Journal of Wildlife Management*, 74, 1841–1851.

10.1.7. Coastal habitats

- Of three studies captured, one replicated, controlled, paired sites study from the USA (1) found that there was a fall in breeding seaside sparrow numbers on a burned site in the year it was burned. The next year, numbers were higher than on an unburned site. A controlled study in Argentina (2) found that tall-grass specialist species were lost from burned areas in the year of burning, but that some habitats recovered by the following year.
- A replicated controlled study from the USA (3) found no differences in nest predation rates between burned and unburned areas for two years after burning.

A replicated, controlled paired sites trial in 1996–1998 on a saltmarsh in Rockefeller State Wildlife Refuge, Louisiana, USA (1), found that seaside sparrow *Ammodramus maritimus* numbers were lower on four burned plots than on unburned controls in the breeding season after the burn (0–3 males/survey for burned areas vs. 7–12 for unburned). However, numbers were higher on burned plots the following year (16 males/survey vs. 8). Sparrow numbers were similar across plots in 1996, before burning, which occurred in 9–18 January 1997. Territorial male sparrows were recorded in April–July in 1996–1998 in each 250 × 250 m plot.

A controlled study on saltmarsh at Mar Chiquita Biosphere Reserve, Chaco, Argentina (2), found that during 12 months after prescribed burns, specialist grassland birds reliant on taller grassland were absent, whilst common widespread species remained. A 200 ha spring burn in September 1995 encompassed *Spartina* spp. marsh and *Juncus* spp. marsh and although *Juncus* marsh recovered pre-burn vegetation structure within a year, *Spartina* marsh had not recovered to its original condition (the vegetation was still short). The bird community and relative abundances of bird species using *Juncus* marsh a year after burning were similar to that in unburned areas, but bay-capped wren-spinetail *Spartonoica maluroides* was present within burned areas at lower abundance than unburned habitat.

A replicated, controlled study on *Spartina* marshland at Blackwater National Wildlife Refuge and Fishing Bay Wildlife Management Area, Maryland, USA (3), found no significant difference in overall predation rates of seaside sparrow nests between three plots burned in January–March 2002–2003 and three plots last burned in 1994 (37–41% of 130 nests in burned areas predated vs. 27–43% of 112 in unburned areas). In 2002, but not 2003, predation rates in the incubation period were higher for burned areas (35% of 51 nests) than unburned areas (13% of 45).

(1) Gabrey, S. W., and Afton, A. D. (2000) Effects of winter marsh burning on abundance and nesting activity of Louisiana seaside sparrows in the Gulf Coast Chenier Plain. *The Wilson Bulletin*, 112, 365–372.
(2) Isacch, J. P., Holz, S., Ricci, L., and Martínez, M. M. (2004) Post-fire vegetation change and bird use of a salt marsh in coastal Argentina. *Wetlands*, 24, 235–243.
(3) Almario, B. S., Marra, P. P., Gates, J. E., and Mitchell, L. (2009) Effects of prescribed fire on depredation rates of natural and artificial seaside sparrow nests. *Wilson Journal of Ornithology*, 121, 770–777.

10.2. Use fire suppression/control

- Two out of three before-and-after studies, from Australia (2) and the UK (3), found that five species of bird (including noisy scrub-bird, the target species of one study) increased following fire suppression measures.
- A before-and-after study in the USA (1) found that open habitat species declined in a pine forest site after fire exclusion, whilst mesic woodland species appeared. A before-and-after study from the UK (3) found that five bird species declined following fire suppression.

Background

In some environments, fires can damage important habitats, particularly if habitat patches are small or fragmented, meaning that entire patches can be destroyed in fires. Under these circumstances it may be beneficial to reduce fire frequency or severity, but there may be long time issues due to the build-up of fuel (i.e. dead vegetation).

A before-and-after study in 1967–1981 in loblolly pine *Pinus taeda*-shortleaf pine *P. echinata* woodland at Tall Timbers Research Station, Florida, USA (1), found the breeding bird community changed dramatically in an 8.6 ha plot from which fire was was excluded for 15 years. The plot was burned in March 1967, after which fire excluded, with annual burns in the surrounding woodland. Species of more open habitat (e.g. blue grosbeak *Passerina caerulea* and Bachman's sparrow *Aimophila aestivalis*) disappeared within five years of fire exclusion although abundance of species peaked during the 'brushy' stage (years 3–7) and mesic woodland species (e.g. wood thrush *Hylocichla mustelina*) appeared following sub-canopy development. The total number of species recorded regularly in the plot fluctuated between 15 and 29 species. Numbers of red-cockaded woodpecker *Picoides borealis* declined (over the site as a whole) over the study period.

A before-and-after study in Two Peoples Bay Nature Reserve (4,637 ha), Western Australia, Australia (2), found that the local population of noisy scrub-bird *Atrichornis clamosus* increased from 45 to 189 singing males over a period of 25 years following the implementation of fire prevention measures from 1970 to 1994, which excluded wildfires. The population also expanded outwards from its initial stronghold to colonise new areas.

A before-and-after study in 2000–2006 on a grouse moor in Dumfries and Galloway, south Scotland (3), found that five bird species decreased following the discontinuation of moor management in 2000, whilst four increased. Before 2000, the moor underwent rotational burning and red foxes *Vulpes vulpes*, carrion crows *Corvus corone*, stoats *Mustela erminea* and weasels *M. nivalis* were controlled.

(1) Engstrom, R. T., Crawford, R. L., and Baker, W. W. (1984) Breeding bird populations in relation to changing forest structure following fire exclusion: a 15-year study. *The Wilson Bulletin*, 437–450.

(2) Smith, G. T. (1996) Habitat use and management for the noisy scrub-bird *Atrichornis clamosus*. *Bird Conservation International*, 6, 33–48.

(3) Baines, D., Redpath, S., Richardson, M., and Thirgood, S. (2008) The direct and indirect effects of predation by hen harriers *Circus cyaneus* on trends in breeding birds on a Scottish grouse moor. *Ibis*, 150, 27–36.

10.3. Protect nest trees before burning

- We captured no evidence for the effects of protecting nest trees of bird populations.

Background

Rare species, or those restricted to small areas, may suffer population declines if fires damage nests or kill chicks. Protecting nest trees may help reduce this negative effect. We found no studies describing the effects of nest tree protection on bird populations but (Williams *et al.* 2006) found that protected nest trees in a longleaf pine *Pinus palustris* forest in Florida, USA, had significantly lower mortality than unprotected trees.

Williams, B.W., Moser, E.B., Hiers, J.K., Gault, K. & Thurber, D.K. (2006) Protecting red-cockaded woodpecker cavity trees predisposed to fire-induced mortality. *The Journal of Wildlife Management*, **70**, 702–707.

10.4. Clear or open patches in forests

- Of nine studies, seven from the UK (2) and the USA (3–8) found that early-successional species increased in clearcut areas or opened forests, compared to control areas, areas before management, or other management techniques. One study (3) found that population increases only occurred in clearcuts up to 20 ha in size. Two studies (6,8) report that mature-forest species declined in cut/opened areas of forest.
- A replicated, randomised, controlled study from the USA (5) found no differences in species richness between clearcuts of different sizes, whilst another American study (1) found that a mosaic of cut and uncut areas supported a variety of species. A long-term study from the USA (9) of a landscape with opened patches found that there were no consistent differences between clearcut and controlled areas, although some species were only seen in clearcuts.

Background

Forests naturally undergo disturbances, from storms, lightning and even large animals. These disturbances can create a mix of different habitats, with open clearings allowing a greater range of species to survive in a forest. Deliberately creating open patches may, therefore, encourage woodland edge and 'early-successional' species.

A replicated study between December 1981 and June 1984 in a mosaic of aspen *Populus* spp. and oak *Quercus* spp. managed for ruffed grouse *Bonasa umbellus* at Barrens Grouse Management Area, Pennsylvania, USA (1), found that 13 species in winter and 69 species in spring were recorded. In winter, birds were significantly more abundant in the interior of mature aspen (>60 years old) stands than in young (1–3 years since clearcutting) aspen stands and the edge of intermediate (4–8 years since clearcutting) aspen stands. In spring, birds were significantly more abundant in intermediate aspen and oak stands (interior and edge) and the interior of mature aspen, than in the interior of mature oak stands and the edge of mature and young aspen stands.

A before-and-after study at Minsmere reserve (151 ha), Suffolk, UK, in 1978–1988 (2), found that the number of churring (calling) male European nightjars *Caprimulgus europaeus* increased significantly from eight to 23 following a series of management interventions, including the creation of woodland 'glades'. Other interventions included increasing the length of woodland edge habitat; creating potential nesting sites (10–50/ha), mainly by clearing 1 m square patches of heather *Calluna vulgaris* at the base of small (1–3 m tall) birch *Betula* spp. trees (previously shown to be the most frequently-used nest sites); planting windbreaks; coppicing birch trees and the opening of areas of heath.

A replicated study in 1989–1990 at three mixed forest sites in Maine, USA (3), found some evidence of increased bird species richness in clearcuts from 2 ha up to 20 ha in area: of the 15 most common clearcut species in both years, ten (in 1989) and 12 (in 1990) were more abundant in larger clearcuts, up to 20 ha, beyond which no preference for clearcut size was apparent, however, average species richness showed no trend amongst the range of clearcut sizes. Study sites comprised 45 clearcuts (2 to 112 ha in area) from 3–10 years post-cut age. These were surveyed in May–June 1989 and 1990; 69 bird species were recorded.

A replicated study in 1993–4 in mixed forests in the Missouri Ozarks, Missouri, USA (4), found that eight species (brown-headed cowbird *Molothrus ater*, blue-winged warbler *Vermivora pinus*, prairie warbler *Dendroica discolor*, rufous-sided towhee *Pipilo erythrophthalmus*, white-eyed vireo *Vireo griseus* and yellow-breasted chat *Icteria virens*) were more abundant in 12 clearcuts than in 12 shelterwood stands (see 'Use shelterwood cutting instead of clearcutting'), 22 stands under selective logging (see 'Use selective harvesting/logging instead of clearcutting') or 12 mature stands. Six species were more abundant in selectively-logged or mature forest than in clearcuts.

In oak-hickory forest in the Missouri Ozarks, USA, in 1991–2000, a replicated, randomised, controlled study (5) found that early successional species increased in response to even- (i.e. clearcutting) and uneven-aged (i.e. selection cutting) management, whereas mature forest species declined. Mature forest bird abundance declined as trees were removed, with harvest disturbance affecting densities of some species in adjacent forest for three years or more. Nest success (average of 29% for all species) did not change after treatment. Each of nine sites was randomly assigned even- or uneven-aged treatment (undertaken May 1996 to May 1997) with a patch of about 10% of each site left uncut. The two treatments are compared in 'Use selective harvesting/logging instead of clearcutting'.

A before-and-after study in mixed woodlands in Pennsylvania, USA (6), at the same study site as (1) found that three early successional species (indigo bunting *Passerina cyanea*, eastern towhee *Pipilo erythrophthalmus* and field sparrow *Spizella pusilla*) were more abundant and three woodland species (red-eyed vireo *Vireo olivaceus*, ovenbird *Seiurus aurocapilla* and American redstart *Setophaga ruticilla*) were less abundant on test plots in 2001–2002, compared with 1998–1999, following the completion of a cutting cycle. Across the entire site (both test and control plots) total bird abundance and species richness increased over the study period, with several species showing significant population increases. The authors suggest this is because the cutting management increased the heterogeneity of habitats across the site.

In Ouachita National Forest, Arkansas and Oklahoma, USA, a replicated study (7) found that three species of songbird known to favour early-successional habitats were all more abundant in three 'seed-tree' stands (10–25 mature trees left/ha), compared to in the openings made by group-selection harvesting (typically 10% of stand cut every 10 years in patches of 0.8 ha or less): indigo bunting (54 nests and 31% success in seed-tree stands vs. 28 and 42% in group-selection stands); yellow-breasted chat (50 nests and 31% success vs. two and 0%) and prairie warbler (14 nests with 45% success, all in seed-tree stands). The authors conclude that group-selection openings appeared too small to support nesting yellow-breasted chat and prairie warbler. Nests were monitored in May–August 2000–2001, within three-, six- and seven-year-old openings created by the two management techniques.

A replicated controlled before-and-after study in oak and hickory *Carya* spp. forests in the Missouri Ozarks, USA (8), found that densities of early-successional species (indigo bunting, prairie warbler and yellow-breasted chat) increased after even-aged forest management (clearcutting), compared to control (no harvest) stands, whilst some mature forest species (Acadian flycatcher *Empidonax virescens*, ovenbird, and worm-eating warbler *Helmitheros vermivorus*) declined. Bird territories were recorded during before (1991–1995) and after cutting (1997–2000) in six sites (312–512 ha), three randomly assigned to even-aged management.

Each even-aged site was partitioned into: clearcut (average 5.4 ha), buffer (0–100 m from clearcut), and interior (>100 m from clearcut) bird. No effects of cutting were found >100 m from clearcuts.

A study from mixed woods in Pennsylvania, USA (9), at the same site as (6), compared the results from (6) with bird surveys in 2005–7. Species composition and abundance differed but early successional species did not decline (despite forest maturation). Overall habitat management for ruffed grouse did not affect other bird populations since the last cutting cycle. During 2005–2007, 46 species were recorded. Of the 17 species recorded 10 or more times, six were observed only in managed area plots (grey catbird *Dumetella carolinensis*, chestnut-sided warbler *Dendroica pensylvanica*, common yellowthroat *Geothlypis trichas*, indigo bunting, field sparrow *Spizella pusilla* and chipping sparrow *S. passerina*). No species were observed only within unmanaged plots.

(1) Yahner, R. H. (1987) Use of even-aged stands by winter and spring bird communities. *The Wilson Bulletin*, 99, 218–232.
(2) Burgess, N. D., Evans, C. E., and Sorensen, J. (1990) The management of lowland heath for nightjars at Minsmere, Suffolk, Great-Britain. *Journal of Environmental Management*, 31, 351–359.
(3) Rudnicky, T. C., and Hunter, M. L. (1993) Reversing the fragmentation perspective: effects of clearcut size on bird species richness in Maine. *Ecological Applications*, 3, 357–366.
(4) Annand, E. M., and Thompson, F. R. (1997) Forest bird response to regeneration practices in central hardwood forests. *The Journal of Wildlife Management*, 61, 159–171.
(5) Gram, W. K., Porneluzi, P. A., Clawson, R. L., Faaborg, J., and Richter, S. C. (2003) Effects of experimental forest management on density and nesting success of bird species in Missouri ozark forests. *Conservation Biology*, 17, 1324–1337.
(6) Yahner, R. H. (2003) Responses of bird communities to early successional habitat in a managed landscape. *The Wilson Bulletin*, 115, 292–298.
(7) Alterman, L. E., Bednarz, J. C., and Thill, R. E. (2005) Use of group-selection and seed-tree cuts by three early-successional migratory species in Arkansas. *The Wilson Bulletin*, 117, 353–363.
(8) Wallendorf, M. J., Porneluzi, P. A., Gram, W. K., Clawson, R. L., and Faaborg, J. (2007) Bird response to clear cutting in Missouri Ozark forests. *The Journal of Wildlife Management*, 71, 1899–1905.
(9) Yahner, R. H. (2008) Bird responses to a managed forested landscape. *Wilson Journal of Ornithology*, 120, 897–900.

10.5. Clearcut and re-seed forests

- One of two replicated studies from the USA (1) found similar bird densities in clearcut and re-seeded sites as in sites under other managements.
- A replicated study from the USA (2) found that pine stands replanted with native pines held more scrub-successional species than stands managed with tree thinning, midstory tree removal and burning.

A replicated study in 1991 in Ocala National Forest, an area of sand pine *Pinus clausa* scrub in Florida, USA (1), found similar densities and species richness of birds in areas that were clearcut and 'brake-seeded' (i.e. direct seeding on to small, machine-made mounds), compared with areas that were burned, or were clearcut with the under-storey also mown. Results were similar for the breeding season (389 birds/km^2 and five species for clearcut and re seeded areas vs. 581 birds/km^2, six species for clearcut and mown; 389 birds/km^2, five species for burned) and winter (894 birds/km^2 and 11 species for clearcut and re seeded areas vs. 594 birds/km^2, ten species for clearcut and mown; 531 birds/km^2, 12 species for burned). Shrub-nesting species were most abundant in mown plots. In summer, the threatened Florida scrub-jay *Aphelocoma coerulescens* was evenly distributed across plots, in winter it was found only in re-seeded plots. All management occurred 5–7 years before the study in 1991.

A replicated study in 1995–1996 in pine *Pinus* spp. savanna in South Carolina, USA (2), found that stands managed for red-cockaded woodpeckers *Picoides borealis* held fewer scrub-successional species than stands where non-native pines were removed and replanted with longleaf pines *Pinus palustris* (31–36 species in managed stands vs. 54–55 in replanted stands). However, no differences in survival rates were apparent for Bachman's sparrow *Aimophila aestivalis* (a near-threatened species), indigo bunting *Passerina cyanea*, and combined scrub-successional birds between stand types. Management for woodpeckers involved tree thinning, midstory tree removal and burning.

(1) Greenberg, C. H., Harris, L. D., and Neary, D. G. (1995) A comparison of bird communities in burned and salvage-logged, clearcut, and forested Florida sand pine scrub. *The Wilson Bulletin*, 107, 40–54.
(2) Krementz, D. G., and Christie, J. S. (1999) Scrub-successional bird community dynamics in young and mature longleaf pine-wiregrass savannahs. *The Journal of Wildlife Management*, 63, 803–814.

10.6. Thin trees within forests

- Of 13 studies, one from the USA (9) which used several interventions found higher species richness in managed sites. Three studies from the USA and the UK (3,13,14) found no differences between thinned and control sites.
- Seven studies from the USA and Sweden found that total bird abundance, or that of some species, were higher in thinned plots than control plots or those under different management (2,4,5,8–10). Four of these used other interventions as well (2,4,8,10). Five (3,5,6,13,14) studies found that abundances were similar, or that some species were less abundant in areas with thinning.
- Two studies from the USA (7,11) found no effect of thinning on wood thrushes, a species thought to be sensitive to it. A controlled before-and-after study (1) found that more nests were in nest boxes in a thinned site, compared to a control site.
- A replicated randomised, controlled study in the USA (12) found no differences in bird abundances between burned sites with high-retention thinning, compared to low-retention.

Background

Thinning of trees (i.e. removal of, trees to reduce density) may be undertaken as a timber management practice e.g. within forestry plantations where saplings may be planted at unnaturally high densities, or as a deliberate conservation management practice to reinstate more natural open woodland conditions that have been lost due to active fire suppression and/or loss of populations large mammal grazers and browsers. Whilst mechanical thinning may be aimed at benefitting certain target species (flora or fauna), impacts on non-target species inevitably occur.

Many studies (especially in pine forests and savannas) simultaneously thin forests whilst introducing a prescribed fire regime. These studies are described in detail in 'Use prescribed burning'.

A controlled before-and-after study in 1973–1983 in pine- *Pinus* spp. oak *Quercus* spp. woodlands in Arizona, USA (1), found that over 90% of nests on two managed plots were in nest boxes, compared to 30% on an unmanaged plot. This study is discussed in more detail in 'Provide artificial nesting sites'.

A replicated before-and-after study in four National Forests in Texas, USA (2), found that red-cockaded woodpecker *Picoides borealis* populations increased at all four sites in the early 1990s after management, including reducing pine tree basal area to 14 m^2/ha, was intensified in 1989. This study is discussed in detail 'Provide artificial nesting sites', 'Translocate individuals' and 'Use prescribed burning'.

A replicated controlled study in 1992–1993 in 33 pine-grassland stands in Ouachita National Forest, Arkansas, USA (3), found that overall bird species richness and abundances were similar in stands with tree thinning, compared to control stands. This study is discussed in more detail in 'Use prescribed burning'.

A study in mixed pine forests in 1985–1996 in South Carolina, USA (4) found that a population of red-cockaded woodpeckers increased following the thinning of trees, reducing basal area to 14–18 m^2/ha, amongst other interventions. The results of this study are discussed in more detail in 'Use prescribed burning'.

A replicated, controlled study in 1992–1994 in oak-hickory *Carya* forests in the Ozark Mountains, Arkansas, USA (5), found that three of 14 species analysed were more abundant in plots with both thinning and under-storey control, compared to control plots or those with just under-storey control. This study is discussed in detail in 'Manually control/remove under-storey and mid-storey vegetation'.

A replicated study in 1995–1996 in pine savanna in South Carolina, USA (6), found that there were fewer scrub-successional species in stands managed for red-cockaded woodpeckers (including tree thinning) than in stands which were clearcut to remove non-native pines and replanted with longleaf pines *Pinus palustris*. This study is discussed in detail in 'Clearcut and re-seed forests'.

At Piedmont National Wildlife Refuge, Georgia, USA (7), a replicated controlled study in 1993–1996 found no impact of thinning and prescribed burning on wood thrushes *Hylocichla mustelina*. This study is discussed in detail in 'Use prescribed burning'.

A replicated study in four oak-hazel *Corylus avellana* woodlands (average size 5.3 ha) in 1996–1999 in Uppland and Åland, Sweden (8), found that sites that were subject to brush cutting and tree thinning had similar numbers of migrant and breeding birds as grazed sites, and more than some abandoned sites. Sites under traditional management (cleared in spring, mown in mid-late summer and grazed in autumn) had higher abundances of migrant birds. This study is discussed in detail in 'Employ grazing in natural and semi-natural habitats'.

A replicated, controlled paired sites study in 1995–1997 in pine forests in Angelina National Forest, Texas, USA (9), found that spring bird species richness and abundances were significantly higher in plots managed for red-cockaded woodpecker, compared to unmanaged plots. This study is discussed in detail in 'Use prescribed burning'.

A replicated study across 40 shortleaf pine *Pinus echinata*-hardwood stands in Ouachita National Forest, Arkansas, USA, in 1999–2000 (10) found that northern bobwhite *Colinus virginianus* abundances were higher in thinned stands, compared to controls. This study is discussed in more detail in 'Use prescribed burning'.

A controlled before-and-after study in 1993–1996 in loblolly pine *Pinus taeda* forests in Piedmont National Wildlife Refuge, Georgia, USA (11), found that habitat management for red-cockaded woodpecker (largely prescribed burning and thinning) had little effect on wood thrushes. This study is discussed in detail in 'Use prescribed burning – pine forests'.

A replicated, randomised, controlled study in 1998–2005 in 12 ponderosa pine stands (15–20 ha) in the North Cascade Range, Washington, USA (12), found that there were no differences in bird densities between four stands with low-retention thinning and prescribed burning and those with high retention thinning and burning (averages of 13 birds/ha for both). This study is described in detail in 'Use prescribed burning'.

A controlled before-and-after study in 2000–2006 at three ponderosa pine *Pinus ponderosa* forest sites in northern Arizona, USA (13), found that species richness and evenness did not differ between thinned forest blocks and controls. In addition, none of five common species were more abundant after thinning, but two (yellow-rumped warbler *Dendroica coronata* and mountain chickadees *Poecile gambeli* were less abundant in thinned plots. This study is discussed in more detail in 'Prescribed burning'.

A replicated, paired site study from April–June in 2006 in 20 conifer plantation sites in Moray Firth, Scotland (14), found that bird species richness and abundance was similar between thinned and unthinned plantations. Sites were in first-rotation and 18–32 years since planting and consisting of ten thinned sites paired with ten unthinned (average of 11 and 16 trees within a 5 m radius around count stations). Average species richness was 19 (range 13–24) at the thinned sites and 19 (range 15–22) at control sites. No significant differences between treatments were found in occurrence rates or abundance for any bird species. As the authors did not find any difference in species richness, they concluded that thinning within the study areas was also unlikely to have influenced the breeding populations of the scarcer species. No significant differences in ground cover, the presence of shrubs or stem diameter at breast height were found between treatments.

(1) Brawn, J. D., and Balda, R. P. (1988) Population biology of cavity nesters in northern Arizona: do nest sites limit breeding densities? *The Condor*, 90, 61–71.
(2) Conner, R. N., Rudolph, D. C., and Bonner, L. H. (1995) Red-cockaded woodpecker population trends and management on Texas National Forests. *Journal of Field Ornithology*, 66, 140–151.
(3) Wilson, C. W., Masters, R. E., and Bukenhofer, G. A. (1995) Breeding bird response to pine-grassland community restoration for red-cockaded woodpeckers. *The Journal of Wildlife Management*, 59, 56–67.
(4) Franzreb, K. E. (1997) Success of intensive management of a critically imperiled population of red-cockaded woodpeckers in South Carolina. *Journal of Field Ornithology*, 68, 458–470.
(5) Rodewald, P. G., and Smith, K. G. (1998) Short-term effects of understory and overstory management on breeding birds in Arkansas oak-hickory forests. *The Journal of Wildlife Management*, 62, 1411–1417.
(6) Krementz, D. G., and Christie, J. S. (1999) Scrub-successional bird community dynamics in young and mature longleaf pine-wiregrass savannahs. *The Journal of Wildlife Management*, 63, 803–814.
(7) Powell, L. A., Lang, J. D., Conroy, M. J., and Krementz, D. G. (2000) Effects of forest management on density, survival, and population growth of wood thrushes. *The Journal of Wildlife Management*, 64, 11–23.
(8) Hansson, L. (2001) Traditional management of forests: plant and bird community responses to alternative restoration of oak–hazel woodland in Sweden. *Biodiversity and Conservation*, 10, 1865–1873.
(9) Conner, R. N., Shackelford, C. E., Schaefer, R. R., Saenz, D., and Rudolph, D. C. (2002) Avian community response to southern pine ecosystem restoration for red-cockaded woodpeckers. *The Wilson Bulletin*, 114, 324–332.
(10) Cram, D. S., Masters, R. E., Guthery, F. S., Engle, D. M., and Montague, W. G. (2002) Northern bobwhite population and habitat response to pine-grassland restoration. *The Journal of Wildlife Management*, 66, 1031–1039.
(11) Lang, J., Powell, L. A., Krementz, D. G., and Conroy, M. J. (2002) Wood thrush movements and habitat use: effects of forest management for red-cockaded woodpeckers. *The Auk*, 119, 109–124.
(12) Gaines, W. L., Haggard, M., Lehmkuhl, J. F., Lyons, A. L., and Harrod, R. J. (2007) Short-term response of land birds to ponderosa pine restoration. *Restoration Ecology*, 15, 670–678.
(13) Hurteau, S. R., Sisk, T. D., Block, W. M., and Dickson, B. G. (2008) Fuel-reduction treatment effects on avian community structure and diversity. *Journal of Wildlife Management*, 72, 1168–1174.
(14) Calladine, J., Humphreys, E. M., Strachan, F., and Jardine, D. C. (2009) Forestry thinning in commercial conifer plantations has little effect on bird species richness and breeding abundance. *Bird Study*, 56, 137.

10.7. Coppice trees

- Of three studies, one, a before-and-after study in the UK (3) found that a population of European nightjars increased following a series of management interventions, including the coppicing of some birch trees.

- Two before-and-after studies from the UK and the USA (1,2) found that the use of coppices by some bird species declined over time. The UK study (2) also found that overall species richness decreased with age, but that some species were more abundant in older stands.

Background

Coppicing is a management practice typical of Eurasian northern temperate zone deciduous woodlands and wood pastures, in which stems of tree species, such as hazel *Corylus avellana* and sweet chestnut *Castanea sativa*, are cut near ground level once every few years, often in defined coppice compartments. These then regrow from the cut 'stool' giving a sustainable yield of woody material harvested on a rotational basis.

A before-and-after study between 1950 and 1962 in a pine-oak forest in Pennsylvania, USA (1), found that the local population of ruffed grouse *Bonasa umbellus* declined over time, as coppiced woodlands became more mature and developed thick ground cover and mid-storey canopy. Similarly, the use of coppiced woodland by grouse broods decreased over time.

A before-and-after study between 1975 and 1984 at Longbeech Wood (300 ha), Kent, England (2), found that overall bird diversity decreased with coppice age and declined markedly at canopy closure. Warblers, finches and buntings were most abundant in young coppice (0–3 years of growth), whilst thrushes and tits increased in abundance with age since coppicing.

A before-and-after study at Minsmere reserve (151 ha), Suffolk, UK, in 1978–1988 (3), found that the local population of European nightjars *Caprimulgus europaeus* increased following a series of management interventions, including the coppicing of some birch trees. This study is discussed in detail in 'Clear or open patches in forests'.

(1) Sharp, W. M. (1963) The effects of habitat manipulation and forest succession on ruffed grouse. *The Journal of Wildlife Management*, 27, 664–671.
(2) Fuller, R. J., and Moreton, B. D. (1987) Breeding bird populations of Kentish sweet chestnut (*Castanea sativa*) coppice in relation to age and structure of the coppice. *Journal of Applied Ecology*, 13–27.
(3) Burgess, N. D., Evans, C. E., and Sorensen, J. (1990) The management of lowland heath for nightjars at Minsmere, Suffolk, Great-Britain. *Journal of Environmental Management*, 31, 351–359.

10.8. Use patch retention harvesting instead of clearcutting

- One before-and-after study of two from the USA (2) found that areas under patch retention harvesting contained more birds of more species than clearcut areas, retaining similar numbers to unharvested areas.
- Two studies from the USA (1,2) found that forest specialist species were found with greater frequency in patch retention plots than other management types. One found that habitat generalists increased on other management types, relative to patch retention areas (1).

Background

In forests in which trees are commercially exploited for timber, a system known as patch retention harvesting may be used as an alternative to a total clearcut. Typically, around 10% of mature and/or immature trees are retained in patches within an otherwise, clearcut harvest compartment, with 'prompt reforestation' subsequent to timber extraction in the other 90%. These retained patches could help maintain characteristic forest species and act as reservoirs for re-colonisation by forest dependant species.

In McDonald-Dunn Forest, Oregon, USA, a replicated, controlled study (1) found that patch-group-harvested stands (33% of tree volume removed in 0.2 ha patches) retained an old forest-associated bird composition more similar to that of control (unharvested, old-growth Douglas-fir *Pseudotsuga menziesii*) stands, compared to two-story (66% of wood removed) and modified clearcut (1.2 trees retained/ha) stands. Of ten abundant forest species in patch group stands, five restricted-range species declined in modified clearcut and two-story harvested stands, whilst nine mostly habitat generalists species increased in these two treatments. Seven to 11 stands of each treatment were studied, with birds surveyed in the breeding season prior to, and in the two years after, timber harvest (1989–1993).

A controlled before-and-after study in May–June 1999–2001 in bottomland hardwood forest in South Carolina, USA (2), found that a small increase in species richness in the short-term in an area with patch-retention harvesting and a control area, whilst richness decreased in an area with clearcutting (patch retention area: 21 species in 1999, 15 in 2000, 25 in 2001; clearcut area: 25, nine, five; control area: 18 in 1999 and 30 in 2001). Species lost from the clearcut plot were mostly forest specialists. Estimated bird density in the patch-retention area fell from c.3.5 pairs/ha in 1999 to 17 in 2000, recovering to around 34 in 2001. In the clearcut area, it fell from 3.3 pairs/ha before harvest to around three in 2000 and 14 in 2001. Densities in the control remained relatively constant (c.3.2 pairs/ha). Estimated bird density in the patch-retention area fell from 3.5 pairs/ha in 1999 to 1.7 in 2000, recovering to around 3.4 in

2001. In the clearcut area, it fell from 3.3 pairs/ha before harvest to around 0.3 in 2000 and 1.4 in 2001. Densities in the control remained relatively constant (3.2 pairs/ha).

(1) Chambers, C. L., McComb, W. C., and II, J. C. T. (1999) Breeding bird responses to three silvicultural treatments in the Oregon coast range. *Ecological Applications*, 9, 171–185.
(2) Harrison, C. A., and Kilgo, J. C. (2004) Short-term breeding bird response to two harvest practices in a bottomland hardwood forest. *The Wilson Bulletin*, 116, 314–323.

10.9. Use selective harvesting/logging instead of clearcutting

- Six studies of seven from the USA and Canada (1,3–7) found that some species were more abundant in selective-logged forests, whilst others were less abundant, compared to both control stands and other managements. One study (3) found that there were no consistent differences between selectively harvested and clearcut stands.
- A replicated study from the USA (6) found a lower species richness of cavity-nesting birds in snags in selectively-logged stands, compared to clearcuts.
- A replicated study from the USA (2) found that brood parasitism of two species by brown-headed cowbirds was higher in harvested stands compared to controls, but it was lower for two others.

Background

It has been suggested that conservation aimed at maintaining bird populations and economic need for timber extraction can be compatible if extraction methods minimise creation of large clearcut areas. One such method is *selective logging* (removing one or two trees and leaving the rest intact) that maintains an uneven-age forest structure and creates openings typically smaller than 0.4 ha, and often considered a sustainable alternative to clearcutting. However, predators and avian nest parasites are often most common in edge habitat, thus potentially reducing reproductive success of birds breeding near forest edges.

A replicated study in 1993–1994 in mixed forests in the Missouri Ozarks, Missouri, USA (1), found that four species (hooded warbler *Wilsonia citrina*, northern parula *Parula americana*, Acadian flycatcher *Empidonax virescens* and red-eyed vireo *Vireo olivaceus*) were more abundant in 22 stands under selective logging or in 12 mature stands, than in 12 clearcuts or 12 shelterwood stands. Eight species were less abundant in selectively-logged forests than in clearcuts and two were less abundant than in mature forest.

In deciduous forest in Illinois, USA, a replicated study in 1990–1991 (2) found that parasitism by brown-headed cowbirds *Molothrus ater* was significantly higher on Acadian flycatchers and Kentucky warbler *Oporornis formosus* in recently selectively-harvested compartments, compared to compartments 10–15 years post-harvest; and compartments uncut for at least 40 years. Parasitism on wood thrush *Hylocichla mustelina* and northern cardinal *Cardinalis cardinalis* was unaffected by cutting and no species had consistently greater nest losses attributable to predation in cut than uncut forests.

In oak-hickory forest in the Missouri Ozarks, USA, in 1991–2000, a replicated, randomised, controlled study (3) found no consistent differences in bird community responses to even- (i.e. clearcutting) and uneven-aged (i.e. selection cutting) management. However, some mature-forest species, such as ovenbirds *Seiurus aurocapillus* were less common on even-aged sites, whilst some early-successional species were more common on these sites. This study is discussed in more detail in 'Clear or open patches in forests'.

A replicated study in deciduous forest in 1998 in Algonquin Provincial Park, Ontario, Canada (4), found that white-throated sparrow *Zonotrichia albicollis*, chestnut-sided warbler *Dendroica pensylvanica*, and mourning warbler *Oporornis philadelphia* were significantly more abundant in stands recently (1–5 years previously) subject to single-tree selection harvest than in other treatments (logging 15–20 years previously or controls). Ovenbird abundance was approximately 50% lower in stands logged either recently or 15–20 years previously than in controls. Management was designed to mimic natural small-scale disturbances that create forest gaps. In June–August 1998, birds were surveyed in: 24, 1–5 years post-harvest stands; 23, 15–20 years post-harvest stands; and 24 stands subject to no harvest for over 30 years. Shrub and slash cover was highest in recently logged stands and appeared important in influencing bird species composition.

A replicated study in 2000–2001 in Ouachita National Forest, Arkansas and Oklahoma, USA (5), found that three early-successional species were more abundant in three 'seed-tree' stands (10–25 mature trees left/ha), compared to in the openings made by group-selection harvesting (typically 10% of stand cut every 10 years in patches of 0.8 ha or less). This study is discussed in detail in 'Clear or open patches in forests'.

A replicated study in 2001 in 30 Douglas-fir *Pseudotsuga menziesii* stands in the Coast Range of Oregon, USA (6), found that stands managed with group selection cuts had lower species richness of cavity-nesting birds using artificially-created snags and fewer nesting birds than clearcut stands with trees retained. This study is discussed in detail in 'Provide deadwood/snags in forests'.

A replicated study in April–June 2003–2004 in three bottomland hardwood forest wildlife management areas in Louisiana, USA (7), found that 14 species were more abundant in 12 stands that had been subject to selective harvest either recently or 12–18 years previously, compared to 12 control stands (not harvested for at least 30 years). Three species were more abundant in control stands than in harvested ones. A further 18 species did not differ between stands.

(1) Annand, E. M., and Thompson, F. R. (1997) Forest bird response to regeneration practices in central hardwood forests. *The Journal of Wildlife Management*, 61, 159–171.
(2) Robinson, S. K., and Robinson, W. D. (2001) Avian nesting success in a selectively harvested north temperate deciduous forest. *Conservation Biology*, 15, 1763–1771.
(3) Gram, W. K., Porneluzi, P. A., Clawson, R. L., Faaborg, J., and Richter, S. C. (2003) Effects of experimental forest management on density and nesting success of bird species in Missouri ozark forests. *Conservation Biology*, 17, 1324–1337.
(4) Jobes, A. P., Nol, E., and Voigt, D. R. (2004) Effects of selection cutting on bird communities in contiguous eastern hardwood forests. *The Journal of Wildlife Management*, 68, 51–60.
(5) Alterman, L. E., Bednarz, J. C., and Thill, R. E. (2005) Use of group-selection and seed-tree cuts by three early-successional migratory species in Arkansas. *The Wilson Bulletin*, 117, 353–363.
(6) Walter, S. T., and Maguire, C. C. (2005) Snags, cavity-nesting birds, and silvicultural treatments in western Oregon. *The Journal of Wildlife Management*, 69, 1578–1591.
(7) Heltzel, J. M., and Leberg, P. L. (2006) Effects of selective logging on breeding bird communities in bottomland hardwood forests in Louisiana. *The Journal of Wildlife Management*, 70, 1416–1424.

10.10. Use variable retention management during forestry operations

- A replicated, controlled study from the USA (1) found that nine bird species occurred at higher densities in stands under variable retention management, compared to control stands. Five were found at lower densities.

Background

Variable retention timber management is a silvicutural technique designed to retain habitat features important for wildlife e.g. large trees, snags and woody debris. This method of harvesting does not seek to maximise timber production.

A replicated, controlled study in the summers of 2003–2004 in bottomland hardwood forest on Tensas River National Wildlife Refuge, Louisiana, USA (1), found that densities of nine species of birds (six of conservation concern) were higher in stands under variable-retention timber harvests, compared to control (untreated) stands. Densities of five species were greater in untreated stands. Conservation concern scores and detection rates of 30 species, suggest that the mosaic of treated stands afforded greater community-wide bird conservation value than untreated stands. Bird densities were estimated (distance sampling) within forest subject to variable-retention harvests within a 13-year chronosequence (i.e. a set of forested sites with similar attributes but of different ages), and untreated stands.

(1) Twedt, D. J., and Somershoe, S. G. (2009) Bird response to prescribed silvicultural treatments in bottomland hardwood forests. *Journal of Wildlife Management*, 73, 1140–1150.

10.11. Use shelterwood cutting instead of clearcutting

- A replicated study from the USA (1) found that community composition of birds in shelterwood stands differed from other forestry practices, with some species more abundant and others less so.

Background

Shelterwood cutting is a management technique designed to avoid clearcutting, but to provide even-aged timber. It involves cutting trees in a series of cuttings, allowing new seedlings to grow from the seeds of older trees.

A replicated study in 1993–1994 in mixed forests in the Missouri Ozarks, Missouri, USA (1), found that indigo bunting *Passerina cyanea* and field sparrow *Spizella pusilla* were more abundant in 12 shelterwood stands and in 12 clearcuts than in 22 stands under selective logging or 12 mature stands. Six species were more abundant in clearcuts than in shelterwood stands and six were more common in selectively-logged or mature stands.

(1) Annand, E. M., and Thompson, F. R. (1997) Forest bird response to regeneration practices in central hardwood forests. *The Journal of Wildlife Management*, 61, 159–171.

10.12. Manage woodland edges for birds

- We captured three studies of two experiments, of which one, a before-and-after study from the UK (1), found an increase in the local population of European nightjars following several management interventions, including the management of woodland edges for birds.
- Two studies of a replicated, controlled paired sites experiment in the USA (2,3) found that bird abundances were higher in woodland edges with border-edge cuts and that predation on artificial nests was lower than in uncut edges. Scrub- and edge-nesting species were more abundant. Overall species richness and nest success did not differ different between treatments.

A before-and-after study at Minsmere reserve (151 ha), Suffolk, UK, in 1978–1988 (1), found that the local population of European nightjars *Caprimulgus europaeus* increased following a series of management interventions, including creating crenulated woodland edges to maximise the length of edges. This study is discussed in detail in 'Clear or open patches in forests'.

A replicated, controlled and paired study in May–June 1996 on a mixed woodland-farmland site in Pennsylvania, USA (2), found that bird abundance was higher in 12 woodland edges subject to border-edge cuts than in 12 control (uncut) edges (8 species/100 m of cut edge vs. 6 species/100 m of uncut edge). Cut edges also contained more shrub and edge-nesting species (25 vs. 17 species), but contained fewer woodland species (nine vs. 23 species). Whilst 13 of 60 species recorded were only found in cut edges, 23 of 60 were only found in uncut edges. Overall species richness and nesting success estimates were no different between edge types (14 species/site and 54% success for 35 nests in cut edges vs. 15 species/site and 52% for 25 nests in controls). Cut edges consisted on felling trees 15–40 m into each woodlot and leaving the debris in place. This occurred two or three years before bird surveys were conducted.

A replicated, controlled and paired study from Pennsylvania, USA (3), on the same site as in (2), found that predation rates on artificial nests were over twice as high in five unmanaged woodlot edges, as in five border-cut edges (36% predation of 50 nests in five cut edges vs. 88% of 50 nests in five controls). The authors suggest this difference may be due to increased cover in cut edges. Nests were placed either on the ground, in low shrubs or in taller shrubs, up to 2 m above ground and contained two northern bobwhite *Colinus virginianus* eggs.

(1) Burgess, N. D., Evans, C. E., and Sorensen, J. (1990) The management of lowland heath for nightjars at Minsmere, Suffolk, Great-Britain. *Journal of Environmental Management*, 31, 351–359.
(2) Fleming, K. K., and Giuliano, W. M. (1998) Effect of border-edge cuts on birds at woodlot edges in southwestern Pennsylvania. *The Journal of Wildlife Management*, 62, 1430–1437.
(3) Fleming, K. K., and Giuliano, W. M. (2001) Reduced predation of artificial nests in border-edge cuts on woodlots. *The Journal of Wildlife Management*, 65, 351–355.

11.13. Manually control or remove mid-storey and ground-level vegetation (including mowing, chaining, cutting etc)

10.13.1. Forests

- Of fifteen studies captured, one, a replicated controlled study from the USA (4), found higher bird species richness in areas with mid-storey thinning, compared to control areas. One study from the USA (3) found similar bird species richness in areas with mid- and under-storey control, compared to other management types. A study from Canada (7) found fewer species in treated sites than controls.
- Seven studies from Europe (1,10) and the USA (2,4,6,8,13) found that total bird densities or those of some species or guilds were higher in areas with mid- or under-storey management, compared to before management or to areas without management. Four of these studies (1,2,6,10) used under-storey removal as part of a wider management regime.
- Five studies from the USA and Canada (3,7–9,11) found that densities of some species were lower in areas with midi or under-storey control, or that overall bird densities did not different between managed and unmanaged areas. Two of these studies (8,9) investigated several interventions at once.
- A replicated controlled study from the USA (13) found similar survival for black-chinned hummingbirds in areas with under-storey management, compared to areas with other interventions. Two replicated, controlled studies from Canada (7,11) found higher nest survival in forests with removal of deciduous trees, compared to controls. A controlled study (12) found that northern bobwhite chicks had greater foraging success in areas with cleared under-storey vegetation compared to burned areas, but lower than under other managements.
- A replicated, controlled study from the USA (5) found that mid-storey control did not appear to affect competition between species for nesting sites.

Background

In wooded habitats, control or removal of under-storey and/or mid-storey vegetation may be undertaken as a deliberate conservation management practice, for example to restore previously naturally occurring open woodland conditions that have been lost due to active fire suppression and/or declining populations of large mammal grazers and browsers. In terms of bird conservation, such practice may aim to improve habitat conditions to maintain and enhance populations of individual endangered species. However, such changes may make habitat unsuitable for other species.

Many studies (especially in pine forests and savannas) simultaneously control vegetation and introduce a prescribed burning regime. These studies are described in detail in 'Use prescribed burning'. Studies that describe the impact of mid- and under-storey vegetation removal in open pine woodlands, which form a continuum with pine savannas are discussed below.

A before-and-after study at Minsmere reserve (151 ha), Suffolk, UK, in 1978–1988 (1), investigated how European nightjars *Caprimulgus europaeus* responded to a series of management interventions, including the clearing of understory vegetation (1 m^2 of heather *Calluna vulgaris* at the base of 1–3 m tall birch *Betula* spp. trees). This study is discussed in detail in 'Clear or open patches in forests'.

A series of before-and-after trials in four open pine forests in Texas, USA (2), found that red-cockaded woodpecker *Picoides borealis* populations increased at all four sites in the early 1990s after management, including mid- and under-storey thinning, was intensified in 1989. Vegetation was removed from 1988–1993 from 310–1450 ha of cluster areas each year and 30 of 39 (77%) new breeding clusters (i.e. breeding groups of woodpeckers) established over the study period were in areas with extensive mid- and under-storey thinning. This study is discussed in detail in 'Provide artificial nesting sites', 'Translocate individuals' and 'Use prescribed burning'.

A replicated study in 1991 in Ocala National Forest, an area of sand pine *Pinus clausa* scrub in Florida, USA (3), found similar densities and species richness of birds in areas that were clearcut and had the under-storey mown, compared to areas that were clearcut and 'brake-seeded'. This study is discussed in detail in 'Clearcut and re-seed forests'.

A replicated controlled study in 1992–3 in 33 pine-grassland stands in Ouachita National Forest, Arkansas, USA (4), found that overall bird species richness and abundances were similar in stands with mid-storey thinning, compared to control stands. This study is discussed in more detail in 'Use prescribed burning'.

A controlled, replicated study in 1990–1 in mixed loblolly pine *Pinus taeda* and shortleaf pine *P. echinata* forests in eastern Texas, USA (5), found no differences in occupancy rates of nest cavities by red-cockaded woodpeckers and southern flying squirrels *Glaucomys volans* between stands with thinned mid-storey vegetation and control stands. This study is discussed in detail in 'Reduce inter-specific competition for nest sites by modifying habitats to exclude competitor species'.

A before-and-after study in mixed pine *Pinus* spp. forests in 1985–1996 in South Carolina, USA (6) found that a population of red-cockaded woodpeckers increased following the clearance of mid-storey vegetation amongst other interventions. The authors emphasise that hardwood mid-storey control using cutting and herbicides and prescribed burning mimicked the natural fire regime and was essential to the success of the project. The results of this study are discussed in more detail in 'Use prescribed burning'.

A replicated, controlled study from May–July in 1992–1995 in three replicate plots of mixed forest in British Columbia, Canada (7), found that bird abundance did not vary between sites with manual thinning of the mid- and under-storey vegetation and controls, although thinned areas held fewer species than controls. Abundance of birds increased annually (no significant differences between the control and treated areas) due primarily to the significant increase in numbers of common species. Nesting success was higher in manually thinned areas (46%) than in controls (28%). Manual thinning reduced the volume of deciduous trees by 90–96% by removing deciduous trees within 1 m of conifer seedlings, or that were 1 m taller than nearby conifers.

A replicated, controlled study in 1993–1994 in oak *Quercus*-hickory *Carya* forests in the Ozark Mountains, Arkansas, USA (8), found that one of the 14 species analysed was more abundant on plots from which under-storey vegetation was removed, compared to those with both under-storey and over-storey control. Three species were more abundant in plots with both over- and under-storey control, whilst three tree-nesting species and ground- and shrub-nesting species were more abundant in control stands.

A replicated study in 1995–1996 in pine savanna in South Carolina, USA (9), found that there were fewer scrub-successional species in stands managed for red-cockaded woodpeckers (including mid-storey thinning) than in stands which were clearcut to remove non-native pines and replanted with longleaf pines *Pinus palustris*. This study is discussed in 'Clearcut and re-seed forests'.

A replicated study in four oak- hazel *Corylus avellana* woodlands (average size 5.3 ha) in 1996–1999 in Uppland and Åland, Sweden (10), found that sites that were subject to brush cutting and tree thinning (see 'Thin trees within forests') had similar numbers of migrant and breeding birds as grazed sites, and more than some abandoned sites. Sites under traditional management (cleared in spring, mown in mid-late summer and grazed in autumn) had higher abundances of migrant birds. This study is discussed in detail in 'Employ grazing in natural and semi-natural habitats'.

A replicated, randomised, controlled study from 1992–1995 in nine 11–22 year old regenerating coniferous plantations (22–47 ha) in British Columbia, Canada (11), found that bird nesting density was lower, but success higher, in areas where deciduous trees and saplings were cut, compared to controls (40 nests and 28% success in treatment areas vs. 79 nests and 18% success in controls). Overall, density and success increased with increasing area of deciduous vegetation remnants. Three years after treatment removed 90–96% of deciduous vegetation, experimental areas had similar numbers of deciduous trees to controls. The effect also applying herbicide to the deciduous stumps is discussed in 'Apply herbicide to mid- and under-storey vegetation'.

A controlled study within a loblolly pine plantation in Louisiana, USA, in 2003–2005 (12) found that northern bobwhite *Colinus virginianus* chicks were significantly more likely to successfully capture arthropods in areas of forest that were mown, compared to areas that were burned. However, success was significantly lower than in areas that were both burned and treated with imazapyr herbicide. There was only a very small difference between mown and control areas.

A replicated, controlled study in riparian forest along the Middle Rio Grande, New Mexico, USA, in 2002–2004 (13), found an 18% increase in the number of black-chinned hummingbird *Archilochus alexandri* nests (from 114 to 134) across four sites where exotic shrubs and woody debris were removed and chipped before herbicide was applied to the root crowns of exotic species. However, an increase was only seen at one site, with the other three showing a 27% decline from 73 to 53 nests. This compared with an 8% increase at three sites with planting of native shrubs (see 'Plant native shrubs following fuel reduction') after fuel reduction and a 42% decrease at two sites where debris was burned ('Use prescribed burning'). Across all fuel reduction treatments, nest survival was around 67% before fuel reduction and 43% after; in three control plots it remained similar (54 vs. 57%).

(1) Burgess, N. D., Evans, C. E., and Sorensen, J. (1990) The management of lowland heath for nightjars at Minsmere, Suffolk, Great-Britain. *Journal of Environmental Management*, 31, 351–359.
(2) Conner, R. N., Rudolph, D. C., and Bonner, L. H. (1995) Red-cockaded woodpecker population trends and management on Texas National Forests. *Journal of Field Ornithology*, 66, 140–151.
(3) Greenberg, C. H., Harris, L. D., and Neary, D. G. (1995) A comparison of bird communities in burned and salvage-logged, clearcut, and forested Florida sand pine scrub. *The Wilson Bulletin*, 107, 40–54.
(4) Wilson, C. W., Masters, R. E., and Bukenhofer, G. A. (1995) Breeding bird response to pine-grassland community restoration for red-cockaded woodpeckers. *The Journal of Wildlife Management*, 59, 56–67.
(5) Conner, R. N., Rudolph, D. C., Saenz, D., and Schaefer, R. R. (1996) Red-cockaded woodpecker nesting success, forest structure, and southern flying squirrels in Texas. *The Wilson Bulletin*, 108, 697–711.
(6) Franzreb, K. E. (1997) Success of intensive management of a critically imperiled population of red-cockaded woodpeckers in South Carolina. *Journal of Field Ornithology*, 68, 458–470.
(7) Easton, W. E., and Martin, K. (1998) The effect of vegetation management on breeding bird communities in British Columbia. *Ecological Applications*, 8, 1092–1103.
(8) Rodewald, P. G., and Smith, K. G. (1998) Short-term effects of understory and overstory management on breeding birds in Arkansas oak-hickory forests. *The Journal of Wildlife Management*, 62, 1411–1417.
(9) Krementz, D. G., and Christie, J. S. (1999) Scrub-successional bird community dynamics in young and mature longleaf pine-wiregrass savannahs. *The Journal of Wildlife Management*, 63, 803–814.
(10) Hansson, L. (2001) Traditional management of forests: plant and bird community responses to alternative restoration of oak–hazel woodland in Sweden. *Biodiversity and Conservation*, 10, 1865–1873.
(11) Easton, W. E., and Martin, K. (2002) Effects of thinning and herbicide treatments on nestsite selection by songbirds in young managed forests. *The Auk*, 119, 685–694.
(12) Burke, J. D., Chamberlain, M. J., and Geaghan, J. P. (2008) Effects of understory vegetation management on brood habitat for northern bobwhites. *Journal of Wildlife Management*, 72, 1361–1368.
(13) Smith, D. M., Finch, D. M., and Hawksworth, D. L. (2009) Black-chinned hummingbird nest-site selection and nest survival in response to fuel reduction in a southwestern riparian forest. *The Condor*, 111, 641–652.

10.13.2. Remove mid-storey from savannas

- A controlled study in Argentina (1) found that in summer, but not overall, a control area had higher bird abundance and species richness than an area where shrubs were removed. There were also differences in community composition between treatments.

Background

Conservationists may wish to remove shrubs from savanna if they are dominating the habitat and excluding some open-habitat species. However, shrubs can also provide habitat complexity and additional niches for species.

A controlled study in 1998–1999 in one treatment (where shrubs were manually removed) and one control (shrubs left unmanipulated) area (both 200 ha) within the same Chacoan forest in Santiago del Estero, Argentina (1) found that, overall, there was no difference in bird abundance or species richness between the two areas. However, the treatment area contained significantly lower species richness and abundance than the control area in summer. At the guild level, bark-feeding insectivores were more abundant in the treatment area; whereas foliage-gleaning insectivores and arboreal seed-eating species were less abundant in the treatment area. In December 1998, terrestrial insectivores were less abundant in the treatment. Birds were surveyed four times at 30 points within each area. In the treatment area, saplings of species that form the tree layer were not removed.

(1) Codesido, M., Drozd, A. A., Gado, P. A., and Bilenca, D. (2009) Responses of a bird assemblage to manual shrub removal in a Chacoan subtropical semiarid forest, Argentina. *Ornitología Neotropical*, 20, 47–60.

10.13.3. Shrubland

- Of seven studies, one controlled study from the USA (3), found that overall bird diversity was similar between chained areas, burned areas and controls. A replicated and controlled study from the USA (5) found that overall diversity was lower on mown sites than controls, but that grassland-specialist species were present on managed sites.
- Five studies from the USA (1–3) and Europe (4,6) found than some study species were found at greater densities or abundances on sites with mechanical vegetation control than on sites with prescribed burning or no management, or that abundances increased after management. One study (4) investigated several interventions at once.
- One study from the USA (3) found that total bird densities were similar between chained, burned and control sites. A replicated controlled study from the USA (5) found that mown sites had lower bird abundances than control sites. Three studies from the USA (2,3,7) found that some species were less abundant on sites with mechanical vegetation removal, compared with burned or control sites, or showed smaller increases after management.
- One replicated, controlled study from the USA (5) found no differences between areas cut in winter and those cut in summer.

Background

Conservationists may wish to remove woody vegetation from shrublands if they are dominating the habitat and excluding some open-habitat species. Chaining is sometimes used to do this: a large heavy chain is dragged over the ground to clear vegetation. However, shrubs can also provide habitat complexity and additional niches for species.

A replicated before-and-after study in 1962–1963 in shrubland in Wisconsin, USA (1), found that male prairie chickens *Tympanuchus cupido* showed a preference for mown areas over controls when the original height of vegetation was over 15 cm (between four and 31 birds using each of five areas before mowing vs. 8–85 after), but not if it was shorter (7–45 birds using five areas before mowing vs. 12–69 birds after).

A replicated, controlled before-and-after study in 1968–1970 in sagebrush *Artemisia tridentate* shrubland in central Montana, USA (2), found that the number of strutting male greater sage grouse *Centrocercus urophasianus* increased by 28% at three lekking sites within 0.5 km of areas treated with herbicide and mechanical clearing, whilst numbers fell by 63% at a fourth site. Numbers increased by 323% at two leks more than 4 km from treated areas.

A controlled study in 1980 in Utah, USA (3), found that response of breeding songbirds in sagebrush habitat chained or burned 3–4 years earlier varied between species. Total bird densities and diversity were similar between a chained site, a burned site and a site without any intervention for 17 years. However, the chained site had significantly more Brewer's sparrow *Spizella breweri* (a sagebrush specialist) territories than the burned site, and horned lark *Eremophila alpestris* densities were 175–200% higher on the chained site than untreated sites. Vesper sparrow *Pooecetes gramineus* and western meadowlark *Sturnella neglecta* densities appeared unaffected by sagebrush control.

A before-and-after study at Minsmere reserve (151 ha), Suffolk, UK, in 1978–1988 (4), found that the local population of European nightjars *Caprimulgus europaeus* increased following a series of management interventions, including creating open patches in heath and removing dominant species such as bracken *Pteridium aquilinum*, birch *Betula* spp. and pines *Pinus* spp. This study is discussed in detail in 'Clear or open patches in forests'.

A replicated and controlled study in shrub dominated by saw-palmetto *Serenoa repens* (a type of palm) in 1988–1989 in Myakka River State Park, Florida, USA (5), found that total number of birds and the number of species found were significantly lower in two sites cut in winter (January 1988) or summer (June 1988), compared to control (uncut) sites (0.2–1.0 species and 0.2–1.2 individuals/site for cut sites vs. 2.0 and 2.7 for control sites). There were no differences between winter and summer-cut sites. Whilst total bird abundance (27 species recorded) was lowest in cut plots, species were mostly (management target) grassland specialists (e.g. Bachman's sparrow *Aimophila aestivalis* and loggerhead shrike *Lanius ludovicianus*).

A replicated, paired study in a mosaic of Mediterranean maquis, pasture and cropland in Ciudad Real province, Spain, in 2002–2003 (6), found that 21 maquis stands where most shrubs and saplings were removed (but taller trees retained) supported greater densities of bird species of high European conservation concern than paired stands without vegetation removal. Such species included red-legged partridge *Alectoris rufa*, woodchat shrike *Lanius senator* and woodlark *Lullula arborea*. The authors note that all three species were fairly common in the study area. Stands were at least 12 ha in size and were cleared between two and ten years before birds were surveyed.

A replicated, controlled study in 1999–2001 in 14 areas of shrublands on Nantucket Island, Massachusetts, USA (7), found that eastern towhee *Pipilo erythrophthalmus* and common yellowthroat *Geothlypis trichas* were significantly less common in areas that had been mown, compared with controls and burned areas (towhees: 0.7 birds/ha vs. 1.1 for control areas and 1.4 for burned areas; yellowthroats: <0.1 birds/ha vs. 0.40 for control and burned areas). Areas mown twice in a season had even fewer towhees (0.5 birds/ha) and no yellowthroats. Song sparrows *Melospiza melodia* were equally abundant on mown areas and other treatments (0.3 birds/ha for mown areas vs. 0.40 for controls and burned areas).

(1) Anderson, R. K. (1969) Prairie chicken responses to changing booming-ground cover type and height. *The Journal of Wildlife Management*, 33, 636–643.

(2) Wallestad, R. (1975) Male sage grouse responses to sagebrush treatment. *The Journal of Wildlife Management*, 39, 482–484.

(3) Castrale, J. S. (1982) Effects of two sagebrush control methods on nongamebirds. *The Journal of Wildlife Management*, 46, 945–952.

(4) Burgess, N. D., Evans, C. E., and Sorensen, J. (1990) The management of lowland heath for nightjars at Minsmere, Suffolk, Great-Britain. *Journal of Environmental Management*, 31, 351–359.

(5) Fitzgerald, S. M., and Tanner, G. W. (1992) Avian community response to fire and mechanical shrub control in south Florida. *Journal of Range Management*, 45, 396–400.

(6) De La Montaña, E., Rey-Benayas, J. M., and Carrascal, L. M. (2006) Response of bird communities to silvicultural thinning of Mediterranean maquis. *Journal of Applied Ecology*, 43, 651–659.
(7) Zuckerberg, B., and Vickery, P. D. (2006) Effects of mowing and burning on shrubland and grassland birds on Nantucket Island, Massachusetts. *Wilson Journal of Ornithology*, 118, 353–363.

10.13.4. Mow or cut natural grasslands

- Of six studies, two replicated and controlled studies from the USA (2,3) found higher densities of birds or nests on mown grasslands, compared to unmanaged or burned areas. Two controlled studies from the USA (2,6), one replicated, found lower nesting or population densities of some species, on mown grasslands compared to unmown areas. Two replicated and controlled studies (1,4) found no significant differences in nesting densities or community composition between mown and unmown areas.
- One study from the USA (5) found that grasshopper sparrow nesting success was higher on mown areas than grazed areas of grassland. A replicated controlled study from the USA (1) found that ducks had similar nesting success on cut and uncut areas.

Background

Cutting and mowing of grasslands can help maintain grass cover, as grasses can survive cutting, whilst herbs and woody plants may not. Cutting can also encourage grass re-growth and increase productivity. Alternatively, in improved soils, cutting and removing the cut vegetation can reduce the nutrient content of the grassland and allow species that rely on nutrient poor soils to return.

A replicated and controlled study in between five and eight meadows in the Lower Souris National Wildlife Refuge, North Dakota, USA, in 1961–1962 (1), found that duck pair density (mainly blue-winged teal *Anas discors* and gadwall *A. strepera*) was 6% lower in three areas mown in August than in five control areas. This difference was not significant and gadwall mostly nested in unmanaged areas beside mowed meadows and blue-winged teal frequently nested in mowed meadows. Nest success did not differ significantly between mowed and unmown meadows. A total of 398 nests were surveyed.

A replicated and controlled study in 1990–1994 in two intensively managed grassland sanctuaries in southeast Illinois, USA (2), found that short-eared owls *Asio flammeus* were more likely to nest on fields burned and mowed in the last 12 months than on controls (undisturbed for 12 months). Northern harriers *Circus cyaneus hudsonius* were less likely to. Mowing was conducted between 20th June and 15th July each year. This study is discussed in detail in 'Use prescribed burning'.

A replicated, controlled trial in May–June 1995–1996 in grasslands in Prairie Ridge State Natural Area, Illinois, USA (3), found that native grasslands mown between late July and October held higher average densities of five songbird species than unmanaged native and non-native grasslands and mowed, hayed and burned non-native grasslands. Mowed and hayed non-native grasslands held lower average densities than unmanaged or grazed grasslands but higher densities than burned non-native grasslands. However, species showed individual responses to different managements. The species surveyed were eastern meadowlark *Sturnella magna*, dickcissel *Spiza americana*, Henslow's sparrow *Ammodramus henslowii*, field sparrow *Spizella pusilla* and grasshopper sparrow *A. savannarum*. This study is discussed further in 'Use prescribed burning' and 'Graze grasslands'.

A replicated, randomised and controlled study in DeSoto National Wildlife Refuge, Iowa, USA (4), found that bird communities were not fundamentally different between areas of tallgrass prairies mown on a 3–4 year rotation and unmanaged or burned prairies (12 species/site for four mowed areas vs. 10 species/site for four burned and 11 species/site for four controls). This study is discussed in detail in 'Use prescribed burning'.

At Blue Grass Army Depot, Kentucky, USA (5), a site comparison study in April–August 2002–2003 found that grasshopper sparrow nesting success was significantly higher in a 3,950 ha area mown in July–August compared to a 2,100 ha cattle-grazed area (70% of 34 nests in mown areas fledging at least one young vs. 25% of 12 in grazed; overall survival estimated at 46% vs. 9%). Average clutch size in the mown area (five eggs) was significantly larger than in grazed area (four eggs).

A controlled study in 1999–2001 on Nantucket Island, Massachusetts, USA (6), found that song sparrows *Melospiza melodia* were significantly less abundant on mown grasslands (between one and three cuts annually), compared to controls (1 bird/10 ha on mown grasslands vs. 6 birds/10 ha on controls). There was no significant difference between mown and burned grasslands. Savanna sparrows *Passerculus sandwichensis* were equally abundant (7–9 birds/10 ha) on all treatments.

(1) Martz, G. F. (1967) Effects of nesting cover removal on breeding puddle ducks. *The Journal of Wildlife Management*, 31, 236–247.
(2) Herkert, J. R., Simpson, S. A., Westemeier, R. L., Esker, T. L., and Walk, J. W. (1999) Response of northern harriers and short-eared owls to grassland management in Illinois. *The Journal of Wildlife Management*, 63, 517–523.
(3) Walk, J. W., and Warner, R. E. (2000) Grassland management for the conservation of songbirds in the midwestern USA. *Biological Conservation*, 94, 165–172.
(4) Van Dyke, F., Van Kley, S. E., Page, C. E., and Van Beek, J. G. (2004) Restoration efforts for plant and bird communities in tallgrass prairies using prescribed burning and mowing. *Restoration Ecology*, 12, 575–585.
(5) Sutter, B., and Ritchison, G. (2005) Effects of grazing on vegetation structure, prey availability, and reproductive success of Grasshopper Sparrows. *Journal of Field Ornithology*, 76, 345–351.
(6) Zuckerberg, B., and Vickery, P. D. (2006) Effects of mowing and burning on shrubland and grassland birds on Nantucket Island, Massachusetts. *Wilson Journal of Ornithology*, 118, 353–363.

10.13.5. Mow or cut semi-natural grasslands/pastures

- Of four studies captured, one, a before-and-after study from the UK (3), found that local wader populations increased following the annual cutting semi-natural grasslands.
- A replicated, controlled study from the UK (1) found that ducks grazed at higher densities on cut areas, a second replicated study from the UK (2) found that goose grazing densities were unaffected by cutting frequency.
- A replicated study from the USA (4) found that Henslow's sparrows were more likely to be recaptured on unmown, compared with mown grasslands.

A replicated controlled study in 1971–1973 in an area of grazed saltmarsh Bridgewater Bay, Somerset, England (1), found that wigeon *Anas penelope* grazed at significantly higher densities on areas of red fescue *Festuca rubra* that were both grazed and cut, compared to areas that were only grazed (20–1,135 droppings/30 m^2 for eight cut areas vs. 0–15 for eight uncut areas). The cut and grazed areas were used at the same rate as areas of eight areas of saltmarsh grass *Puccinellia maritima* (32–695 droppings/30 m^2). The grazed areas contained large amounts of unpalatable rank fescue. Sheep were used to graze the marsh in May–September, but were removed before the arrival of wigeon in winter. Areas were cut short in September so that they resembled the areas of saltmarsh grass.

A series of replicated trials on grassland sites at two reserves in Essex, England, between 1990 and 1992 (2) found that brent geese *Branta bernicla* grazing densities on 24 grassland plots were not affected by the frequency of grass cutting (between two and five times a year). There were no differences between areas that were only cut, cut and grazed or only grazed. This study is discussed further in 'Fertilise grasslands' and 'Employ grazing in natural and semi-natural habitats'.

A before-and-after study on three islands (14.5 to 28 ha) in Lower Lough Erne RSPB reserve, Northern Ireland (3), found that numbers of northern lapwing *Vanellus vanellus* (one pair in 2000 vs. approximately 20 in 2005) and common redshank *Tringa totanus* (approximately 17 vs. 45) increased in response to the cutting of patches of rushes *Juncus* spp. in winter (January–February). Lapwings nested almost exclusively in cut patches, whilst redshank nested in uncut areas, but their chicks used the adjacent open areas for feeding.

A replicated study at a mine site in Ohio, USA (4), in 1999–2007 found that ten of 99 (10%) Henslow's sparrows *Ammodramus henslowii* ringed on four unmown non-native grassland were recaptured, whereas none of the 15 birds ringed on four mown grasslands were recaptured. In total, 87% of ringed birds were caught on unmown grasslands. Experimental plots were established in 1999 and the mown plots cut in mid-April every year. birds were ringed in 2000–2007 and recaptured in 2001–2007.

(1) Cadwalladr, D. A., and Morley, J. V. (1974) Further experiments on the management of saltings pasture for wigeon (*Anas penelope* L.) conservation at Bridgwater Bay National Nature Reserve, Somerset. *Journal of Applied Ecology*, 11, 461–466.
(2) Vickery, J. A., Sutherland, W. J., and Lane, S. J. (1994) The management of grass pastures for brent geese. *Journal of Applied Ecology*, 31, 282–290.
(3) Robson, B., and Allcorn, R. I. (2006) Rush cutting to create nesting patches for lapwings *Vanellus vanellus* and other waders, Lower Lough Erne RSPB reserve, County Fermanagh, Northern Ireland. *Conservation Evidence*, 3, 81–83.
(4) Ingold, D. J., Dooley, J. L., and Cavender, N. (2009) Return rates of breeding Henslow's sparrows on mowed versus unmowed areas on a reclaimed surface mine. *Wilson Journal of Ornithology*, 121, 194–197.

10.13.6. Reedbeds

- Of three studies captured, one controlled study from the Netherlands (2) found that warblers nested at lower densities in cut areas of reeds. Productivity and success did not vary between treatments.
- An unreplicated study from Denmark (1) found that geese grazed at the highest densities on reedbeds cut 5–12 years previously.
- One replicated study (3) investigated changing water levels in addition to cutting reeds in the UK and found that management did not affect great bittern breeding productivity but did appear to delay territory establishment.

Background

Reedbeds were traditionally cut in much of Europe to provide thatch for housing and this practice continues in some countries. Cutting reeds like this changes the composition of the reedbeds, resulting in higher spring water levels and higher reed biomass (due to increased regrowth) than in uncut reedbeds (Poulin & Lefebvre 2002). A comparison of cut and uncut reedbeds in southern France found that there was also a higher abundance of food arthropods in cut reeds, although lower abundances of birds in summer (Poulin & Lefebvre 2002).

A site comparison study in two wetland sites in Vejlerne, a wetland in North Jutland, Denmark (1), found that the highest densities of greylag geese *Anser anser* nests were found in reedbeds that were cut between five and eleven or five and 12 years before (3.1–3.4 nests/ha). No nests were found in beds cut that year and very few (and only in one site) in beds cut less than three years (fewer than 0.7 nests/ha) or more than eleven years before. The authors speculate that geese need an intermediate density of reed stems to nest effectively.

A controlled study in 1993–1995 in an area of peat marsh in Overijssel, the Netherlands (2), found that reed warblers *Acrocephalus scirpaceus* and sedge warblers *A. schoenobaenus* nested at significantly lower densities in areas of recently cut reedbed, compared to uncut areas (reed warblers: 0.8 nests/100 m of shore for cut areas vs. 2.0 nests/100 m for uncut; sedge warblers: 0.03 nests/100 m vs. 0.7 nests/100 m). There were no significant difference in the fledging success of unpredated nests in cut and uncut reed, but nest predation of reed warblers was higher in cut reed (33% predated in cut areas vs. 17% in uncut areas). There was no difference for sedge warblers (73% vs. 43%).

A replicated study in 1997–2001 in ten reedbed sites across England (3) investigated the impact of raising water levels in reedbeds on great bittern *Botaurus stellaris* breeding (see 'Manage water levels in wetlands'). Reeds at sites with low water levels were cut in spring (March–April), compared with winter (completed by December) for high water level sites, but the effect of cutting was not specifically investigated. Male bitterns at low-water sites established territories later than on high-water sites, but sites did not differ in productivity.

(1) Kristiansen, J. N. (1998) Nest site preference by greylag geese *Anser anser* in reedbeds of different harvest age. *Bird Study*, 45, 337–343.
(2) Graveland, J. (1999) Effects of reed cutting on density and breeding success of reed warbler *Acrocephalus scirpacaeus* and sedge warbler *A. schoenobaenus*. *Journal of Avian Biology*, 30, 469–482.
(3) Gilbert, G., Tyler, G. A., Dunn, C. J., Ratcliffe, N., and Smith, K. E. N. (2007) The influence of habitat management on the breeding success of the great bittern *Botaurus stellaris* in Britain. *Ibis*, 149, 53–66.

10.14. Replace non-native species of tree/shrub

- A replicated, controlled study from the USA (1) found that the number of black-chinned hummingbird nests increased at sites with fuel reduction and planting of native species, but that the increase was smaller than at sites without planting.

Background

A combination of fuel reduction (e.g. by burning or under-storey removal) and planting of native species could help replace an exotic mid- or under-storey with a native one.

A replicated, controlled study in riparian forest along the Middle Rio Grande, New Mexico, USA, in 2002–2004 (1), found an 8% increase in the number of black-chinned hummingbird *Archilochus alexandri* nests (from 75 to 81) on three sites where native shrubs were planted after fuel reduction measures. Exotic shrubs and woody debris were removed and chipped before herbicide was applied to the root crowns of exotic species. This compared with an 18% increase at four sites with fuel reduction but no planting and a 42% decrease at two sites where debris was burned and no shrubs planted. These results are discussed in more detail in 'Manually control/remove under-storey and mid-storey vegetation' and 'Use prescribed burning'.

(1) Smith, D. M., Finch, D. M., and Hawksworth, D. L. (2009) Black-chinned hummingbird nest-site selection and nest survival in response to fuel reduction in a southwestern riparian forest. *The Condor*, 111, 641–652.

10.15. Provide deadwood/snags in forests

Background

Snags (standing dead trees) and other dead wood can be important for nesting, roosting and feeding for many bird species. Providing this has therefore been suggested as a way to increase habitat carrying capacity and population sizes. Studies describing the effects of providing artificial snags are described in 'Provide artificial nesting sites'.

10.15.1. Use ring-barking (girdling), cutting or silvicides to produce snags

- Of five studies found, one replicated and controlled study from the USA (1) found that forest plots provided with snags had higher bird diversity and abundance than plots without snags added.
- Three studies from the USA (3,5) and UK (4) found that woodpeckers and other species used artificially-created snags for nesting and foraging. One study from the USA (5) found that use increased with how long a snag had been dead.

- A UK study (2) found that no crested tits used snags created for them, possibly because they were not rotted enough, or because they were too close to the ground.

Background

Woody debris can be created in forests by 'ring-barking' or 'girdling', a process which removes the living tissue from a tree in a ring around the trunk. This prevents water and nutrients from reaching the leaves and upper portions of the tree, normally killing the plant, which then decays to produce a snag.

A replicated, controlled study from May–June in 1977–1981 in four plots in pine-hardwood timber clearcuts in Texas, USA (1), found that plots with deadwood snags had higher bird species richness and abundance than plots without snags (5 and 4 species/plot; 166 and 135 individuals/40 ha respectively). Similarly, indices of community diversity and evenness were also significantly higher in plots with snags. Cavity-nesting birds occurred on plots with snags but were virtually absent from plots without snags (13 compared to 1 individuals/40 ha). Other species used snags for foraging and perching. Seventy-five snags (9.4/ha) were made from killing trees of nine species in each plot. Plots were 80 × 250 m, four with snags and four without and were cleared in 1975 and planted with loblolly pines *Pinus taeda* in 1976.

A before-and-after study in Highland, Scotland (2), found that, by 1994, no crested tits *Parus cristatus* used any of 30 tree stumps created between 1981 and 1989 to provide nesting habitats. Stumps were made by cutting trees 1–1.2 m from the ground and were supposed to rot to allow tits to excavate nesting sites. Investigating natural nest sites, the authors found that natural nests were both much higher (average of 7.3 m above ground) and in much larger trees (average diameter at breast height of 41 cm vs. 20–26 cm). They also argue that the trees probably needed more time to rot.

A replicated study over four years in Oregon, USA (3), found no differences in the rate of use by woodpeckers of Douglas-fir *Pseudotsuga menziesii* snags created by different methods. Instead, only the length of time that a tree had been dead for affected foraging rates. Girdling (ring-barking), injecting with two different silvicides and cutting the tree just below the crown all had similar effects, killing most trees within two years (silvicide and cut trees died after a year or so, girdled trees died after two years but a higher proportion died overall). Chopping at the middle of the crown was the least effective in killing trees, taking over three years and killing fewer trees. Eighteen trees were treated with each method and some snags were inoculated with saprophytic (decaying) fungi.

A small study on heathland at Great Ovens, Dorset, England (4), found that one of two mature Scots pine *Pinus sylvestris* trees which were ring-barked around November 2000 had a great spotted woodpecker *Dendrocopus major* nesting hole excavated in it by July 2005. Both trees died, the excavated one leaving 10 m of standing deadwood whilst the second tree fell, leaving a stump 1.25 m high.

A replicated study in 30 Douglas-fir stands in the Coast Range of Oregon, USA (5), found that 11 cavity-nesting species of bird used artificially-created snags for nesting or foraging, with 20% of 839 snags being used for nesting and 88% containing cavities. Significantly more snags were used in clearcut stands (with some trees retained) compared to stands managed with group selection cuts. Clearcut stands had significantly higher species richness and abundance of cavity-nesting species. Nest numbers were similar between snags clustered close together and those more widely spaced.

(1) Dickson, J. G., Conner, R. N., and Williamson, J. H. (1983) Snag retention increases bird use of a clearcut. *The Journal of Wildlife Management*, 47, 799–804.

(2) Denny, R. E., and Summers, R. W. (1996) Nest site selection, management and breeding success of crested tits *Parus cristatus* at Abernethy Forest, Strathspey. *Bird Study*, 43, 371–379.

(3) Brandeis, T. J., Newton, M., Filip, G. M., and Cole, E. C. (2002) Cavity-nester habitat development in artificially made Douglas-fir snags. *The Journal of Wildlife Management*, 66, 625–633.

(4) Liley, D. (2005) Ring-barking of Scots pine *Pinus sylvestris* trees to create standing deadwood on heathland at Great Ovens, Dorset, England. *Conservation Evidence*, 2, 123–124.
(5) Walter, S. T., and Maguire, C. C. (2005) Snags, cavity-nesting birds, and silvicultural treatments in western Oregon. *The Journal of Wildlife Management*, 69, 1578–1591.

10.15.2. Add woody debris to forests

- A randomised, replicated, controlled study from Australia (1) found that brown treecreeper numbers were higher in plots with large amounts of dead wood added, compared to control plots or those with less debris added.

Background
If large amounts of wood are available, then it may be possible to deliver the benefits of having woody debris in forests without having to damage the trees themselves.

In Gunbower State Forest, Victoria, Australia, a randomised, replicated, controlled study (1) found that brown treecreeper *Climacteris picumnus* numbers were consistently higher in plots of red river gum *Eucalyptus camaldulensis* forest with coarse woody debris (aged fallen wood >10 cm in diameter) added to them, compared to control plots (1.5–2.2 birds/ha in plots with 40–80 Mg/ha or more of debris added vs. 0.6 and 0.4 birds/ha for 20 Mg/ha and no debris added treatments). Birds appeared not to discriminate between logs and tree crowns.

(1) MacNally, R., Horrocks, G., and Pettifer, L. (2002) Experimental evidence for potential beneficial effects of fallen timber in forests. *Ecological Applications*, 12, 1588–1594.

10.16. Remove coarse woody debris from forests

- One of two replicated and controlled studies from the USA (1) found that overall breeding bird abundance and diversity were lower in plots where woody debris was removed, compared to control plots. Several individually-analysed species showed lower abundances. A replicated, controlled before-and-after study from the USA (2) found lower nest survival for black-chinned hummingbirds following debris removal.
- Some species in both studies increased after debris removal (1,2).

Background
Removal of coarse woody debris (i.e. dead woody plant material >5 mm diameter, including bark > 5 mm thickness) from forests is a practice most commonly undertaken in North American and Australian forests to reduce risk of intense wildfires by reducing the fuel-load. It may be undertaken as a deliberate conservation intervention to benefit certain bird species, but with the potential to adversely affect under-storey species that utilise accumulated dead wood for foraging or nesting.

A randomised, replicated controlled study in 1996–1999 in loblolly pine *Pinus taeda* stands at the Savannah River Site, South Carolina, USA (1), found that breeding bird abundance, species richness and diversity and resident bird abundance were lower in plots where coarse woody debris was removed, compared to control plots (17–21 territories and 11–13 species/9.3 ha plots with debris removal vs. 31 territories and 20 species for control plots). Mid-storey-, canopy- and cavity-nesting species such as red-headed woodpecker *Melanerpes erythrocephalus*, great crested flycatcher *Myiarchus crinitus*, eastern towhee *Piplio erythrophthalmus* and eastern wood-pewee *Contopus virens* were found at lower densities in removal plots. Pine warbler *Dendroica pinus* and indigo bunting *Passerina cyanea* were found at similar densities and summer tanagers *Piranga rubra* were found at higher densities. Debris removal did not appear affect winter bird community.

A replicated, controlled before-and-after study in riparian forest along the Middle Rio Grande, New Mexico, USA (2), found that black-chinned hummingbird *Archilochus alexandri* nest survival was lower after fuel reduction treatments, including the removal of coarse woody debris from forests, but was no lower in control plots. There were, however, population increases at sites with debris removal, compared to burned or control plots. This study is discussed in detail in 'Manually control or remove under-storey and mid-storey vegetation', 'Plant native shrubs following fuel reduction' and 'Use prescribed burning'.

(1) Lohr, S. M., Gauthreaux, S. A., and Kilgo, J. C. (2002) Importance of coarse woody debris to avian communities in loblolly pine forests. *Conservation Biology*, 16, 767–777.
(2) Smith, D. M., Finch, D. M., and Hawksworth, D. L. (2009) Black-chinned hummingbird nest-site selection and nest survival in response to fuel reduction in a southwestern riparian forest. *The Condor*, 111, 641–652.

10.17. Apply herbicide to mid- and under-storey vegetation

- Of seven studies, one replicated, controlled study in forests in Canada (4) found that bird species richness declined after the treatment of deciduous trees with herbicide.
- Two of the four studies monitoring bird populations (two replicated, controlled before-and-after studies) these found that numbers of red-cockaded woodpeckers (3) or male greater sage grouse (2) increased in all or some herbicide-treated areas. Increases of sage grouse were larger at two areas without vegetation control. One study (1) considered two species: one decreased while the other showed no response. Another (4) found that bird densities increased equally in both control and treatment areas.
- Three replicated, controlled before-and-after studies in forests (4,5,7) found that nest survival was lower where herbicide was applied to exotic shrubs or deciduous vegetation. One study also found lower nesting densities (5). One controlled study (6) found northern bobwhite chicks higher had foraging success in herbicide-treated forest areas.

Background

If it is not possible to manually remove mid- or under-storey vegetation from forests and shrublands then applying herbicide may be a useful alternative. There may, however, be other issues with widespread herbicide application.

A replicated, controlled before-and-after study in 1966–1969 in common sagebrush *Artemisia tridentata* shrublands in central Montana, USA (1), found that the number of breeding pairs of Brewer's sparrow *Spizella breweri* declined by 54% on plots completely sprayed with herbicide in 1968 (from 0.8 pairs/ha to 0.4). Vesper sparrows *Pooecetes gramineus* showed no response to herbicide, and neither species showed any changes on plots subject to strip spraying, partial spraying or on control plots. Plant foods (mostly grass seed) represented a greater portion, and invertebrate food a smaller portion, of both sparrow species diets on the sprayed than the unsprayed area.

A replicated, controlled before-and-after study in 1968–1970 in common sagebrush shrubland in central Montana, USA (2), found that the number of strutting male greater sage grouse *Centrocercus urophasianus* increased by 28% at three lekking sites within 0.5 km of areas treated with herbicide and mechanical clearing, whilst numbers fell by 63% at a fourth site. Numbers increased by 323% at two leks more than 4 km from treated areas. In June 1968, three areas (totalling 705 ha) (each with a lek within 0.5 km) were treated with herbicide (2,4-Dichlorophenoxyacetic acid) or mechanically cleared in alternate strips. A fourth area (441 ha) within 0.5 km of a lek was treated in June 1970.

A study in mixed pine *Pinus* spp. forests in 1985–1996 in South Carolina, USA (3) found that a population of red-cockaded woodpeckers *Picoides borealis* increased following the application of herbicides to midstory vegetation amongst other interventions. The authors emphasise that hardwood midstory control using cutting and herbicides and prescribed burning mimicked the natural fire regime and was essential to the success of the project. The results of this study are discussed in more detail in 'Use prescribed burning'.

A replicated, controlled study from May–July in 1992–1995 in three replicate plots of mixed forest in British Columbia, Canada (4), found that bird species richness and abundance became more homogeneous after herbicide treatment of deciduous vegetation (with cut stems sprayed with glyphosate). Bird species richness declined by 25% and 11% in herbicide-treated and control sites respectively. Abundance of birds increased annually (no significant differences between the control and treated areas) due primarily to significant increases in numbers of common species. Herbicide-treated areas showed a greater turnover of bird species. Nesting success was lower in herbicide-treated areas (8%) than in control areas (28%). Treatments reduced the volume of deciduous trees by 90–96% by removing deciduous trees within 1 m of conifer seedlings, or that were 1 m taller than nearby conifers.

A replicated, randomised, controlled study from 1992–1995 in nine 11–22 year old regenerating coniferous plantations (22–47 ha) in British Columbia, Canada (5), found that bird nesting density and success was lower in areas with manual control of vegetation and herbicide application, compared to control areas (45 nests and 12% success in treatment areas vs. 79 nests and 18% success in controls). Overall, density and success increased with increasing area of deciduous vegetation remnants. Three years after treatment removed 90–96% of deciduous vegetation, experimental areas still had few deciduous trees, compared to controls.

A controlled study within a loblolly pine *Pinus taeda* plantation in Louisiana, USA, in 2003–2005 (6) found that northern bobwhite *Colinus virginianus* chicks were significantly more likely to successfully capture arthropods in areas of forest that were both burned and treated with imazapyr herbicide, compared to areas that were burned, mown or control areas. Arthropod abundance (a measure of brood habitat quality) was also highest in burned and herbicide-treated areas.

A replicated, controlled before-and-after study in riparian forest along the Middle Rio Grande, New Mexico, USA (7), found that black-chinned hummingbird *Archilochus alexandri* nest survival was lower after fuel reduction treatments, including the application of herbicide to exotic shrubs, but did not fall in control plots.

(1) Best, L. B. (1972) First-year effects of sagebrush control on two sparrows. *The Journal of Wildlife Management*, 36, 534–544.
(2) Wallestad, R. (1975) Male sage grouse responses to sagebrush treatment. *The Journal of Wildlife Management*, 39, 482–484.
(3) Franzreb, K. E. (1997) Success of intensive management of a critically imperiled population of red-cockaded woodpeckers in South Carolina. *Journal of Field Ornithology*, 68, 458–470.
(4) Easton, W. E., and Martin, K. (1998) The effect of vegetation management on breeding bird communities in British Columbia. *Ecological Applications*, 8, 1092–1103.
(5) Easton, W. E., and Martin, K. (2002) Effects of thinning and herbicide treatments on nestsite selection by songbirds in young managed forests. *The Auk*, 119, 685–694.
(6) Burke, J. D., Chamberlain, M. J., and Geaghan, J. P. (2008) Effects of understory vegetation management on brood habitat for northern bobwhites. *Journal of Wildlife Management*, 72, 1361–1368.
(7) Smith, D. M., Finch, D. M., and Hawksworth, D. L. (2009) Black-chinned hummingbird nest-site selection and nest survival in response to fuel reduction in a southwestern riparian forest. *The Condor*, 111, 641–652.

10.18. Treat wetlands with herbicide

- Three of four studies (2–4), one replicated and controlled (3), found that numbers of terns, American coot and waders were found at higher densities on wetland areas sprayed with herbicide, compared to unsprayed areas. However, one study (4) found that wader numbers were not as high as on ploughed areas. One replicated and controlled study (1) found that songbird densities were lower on sprayed than unsprayed areas.

Background

Wetlands can become overgrown with vegetation, particularly if they have unusually high nutrient levels, or if invasive plants (with no natural predators) are present. Very dense vegetation can exclude open-water species from a wetland, meaning that treating the wetland with a herbicide may be beneficial. However, conservationists should be aware that some herbicides, such as glyphosate, can be extremely toxic to amphibians and fish (Relyea 2005).

Relyea, R. A. (2005) The lethal impact of Roundup on aquatic and terrestrial amphibians. *Ecological Applications*, 15, 1118–1124.

A controlled, replicated before-and-after study in 1990–1993 in 23 inland wetlands in North Dakota, USA (1), found that the densities of three songbird species were all significantly lower on sites sprayed with glyphosate, than on unsprayed sites (red-winged blackbirds *Agelaius phoeniceus*: 0.6 birds/ha on sites where 90% of the site was sprayed vs. 1.6 on controls; yellow-headed blackbirds *Xanthocephalus xanthocephalus*: average of 2.1 birds/ha on all treated sites vs. 3.1 on controls; marsh wrens *Cistothorus palustris*: 0.7 birds/ha on treated sites vs. 2.2). Experimental wetlands had significantly lower percentage covers of emergent vegetation. Sites were sprayed from the air with either 90%, 70% or 50% of the site treated. There were no differences between sites before herbicide application.

A randomised replicated study in 24 inland wetlands in North Dakota, USA (2), found that the number of black terns *Chlidonias niger* using sites in June 1991–1993 was positively correlated with the areas of open water and dead cattail *Typha* spp. present, following the aerial application of glyphosate during July 1990 and 1991. The numbers of mallard *Anas platyrhynchos*, blue-winged teal *A. discors*, northern shoveler *A. clypeata*, gadwall *A. strepera*, northern pintail *A. acuta*, redhead *Aythya americana* and ruddy duck *Oxyura jamaicensis* were all correlated with the amount of open water and the amount of cover present. Glyphosate was sprayed over 90%, 70% or 50% of the sites.

A controlled, replicated before and after study in 1990–1993 in 20 inland wetlands in North Dakota, USA (3), found that, two years after treatment, densities of American coot *Fulica americana* were significantly higher at wetlands sprayed with herbicide (0.8–1.0 birds/ha) than at untreated sites (0.2 birds/ha), but that sora *Porzana carolina* densities were significantly lower (0.1–0.3 birds/ha vs. 0.5). Coot densities were positively correlated with extent of open water and negatively with live emergent vegetation; sora densities were positively correlated with live emergent vegetation. Four sites had 90% coverage with glyphosate (aerially applied), eight had 70% coverage, four 50% and four were controls, with no coverage. No coots were found on any wetlands before treatment.

A controlled study in 2003–2004 on mudflats and areas of *Spartina alterniflora* meadows in Willapa National Wildlife Refuge, Washington, USA (4), found that average densities of unidentified *Calidris* spp. sandpipers (62 birds/ha vs. 7), western sandpiper *Calidris mauri* (50 vs. 5) and waterfowl (16 vs. 0.8) were significantly higher on areas of *Spartina* meadow spayed with glyphosate (9 kg/ha) in 2002–2003, and imazapyr (1.7 kg/ha) in 2004, compared to control areas. However, densities of dunlin *Calidris alpina*, grey plovers *Pluvialis squatarola* and dowitchers *Limnodromus* spp. did not differ, all birds were less common than on ploughed areas of *Spartina* and all but dowitchers and waterfowl were less common than on adjacent mudflats.

(1) Linz, G. M., Blixt, D. C., Bergman, D. L., and Bleier, W. J. (1996) Responses of red-winged blackbirds, yellow-headed blackbirds and marsh wrens to glyphosate-induced alterations in cattail density. *Journal of Field Ornithology*, 67, 167–176.

(2) Linz, G. M., and Blixt, D. C. (1997) Black terns benefit from cattail management in the northern Great Plains. *Colonial Waterbirds*, 20, 617–621.

(3) Linz, G. M., Bergman, D. L., Blixt, D. C., and McMurl, C. (1997) Response of American coots and soras to herbicide-induced vegetation changes in wetlands. *Journal of Field Ornithology*, 68, 450–457.

(4) Patten, K., and O'Casey, C. (2007) Use of Willapa Bay, Washington, by shorebirds and waterfowl after *Spartina* control efforts. *Journal of Field Ornithology*, 78, 395–400.

10.19. Employ grazing in natural and semi-natural habitats

10.19.1. Natural grasslands

- Five of 12 studies from the USA and Canada (4,5,8,9,12), four replicated, found that some species studied were found at higher densities on grazed than ungrazed sites. Eight studies from the USA, Canada and France (1–4,6,8,9,12), six replicated, found that some or all species studied were found at lower densities on grazed sites compared to ungrazed sites or those under other management, or that there were no differences.
- Two controlled studies from the USA and Canada (1,10), one replicated, found that duck nesting success was higher on grazed than ungrazed sites. Two studies from the USA (7,11) found that songbird nesting success was lower on grazed than ungrazed sites. Three replicated and controlled (one randomised) studies from the USA and Canada (2,3,10) found that grazing had little or no effect on nesting success in a variety of species.

Background

This section contains studies describing the effects of grazing on natural grasslands, such as prairies in the USA and Canada. The effects of grazing on semi-natural or artificial grasslands and pastures are discussed in 'Graze artificial grasslands/pastures'.

At Union Slough National Wildlife Refuge, Iowa, USA (1), a replicated, controlled trial in 1959–1961 found that blue-winged teal *Anas discors* nests in grazed grasslands and hayfields had greater success rates than those in ungrazed areas (47% success for grazed areas vs. 46% in hayfields and 14% in ungrazed areas; a total of 111 nests were monitored). However, nesting density was higher in ungrazed areas (7 nests/ha) than in grazed areas (4 nests/ha) or hayfields (4/ha). Areas were moderately grazed (15 June–1 October), ungrazed, or hay-cut (July–August, after the main nesting period). Most nests were in Kentucky bluegrass *Poa pratensis* and alfalfa *Medicago sativa*.

A replicated controlled study at three grassland sites in North Dakota, USA (2), found that upland sandpiper *Bartramia longicauda* nesting density was lower in areas grazed by cattle during the nesting season, but there was little evidence that grazing treatments outside the breeding season influenced nest success. From 1981 to 1987, cattle grazing treatments were each applied to a field in each of the three study areas: spring grazing, autumn grazing, autumn-and-spring grazing, season-long grazing, and ungrazed.

A randomised, controlled, replicated before-and-after study in 1980–1988 in mixed-grass prairie at Lostwood National Wildlife Refuge, North Dakota, USA (3), found that nest densities of gadwall *Anas strepera* and blue-winged teal were lower in areas and years with spring cattle grazing, or a combination of grazing and summer burning compared to control areas. Five other species did not show a response to grazing. Nest success was generally high (31–45%) and unaffected by treatment. The authors argue that grazing reduced brush cover that provided nesting habitat for ducks.

A paired site comparison in 1997–1999 at the Audubon Research Ranch and Davis Pasture in Arizona, USA (4), found no consistent effects of grazing grasslands on *Ammodramus* sparrow species. Baird's sparrows *A. bairdii* were more abundant on the grazed than ungrazed area in 1997 (1.3 vs. 0.5 captures/plot/day) but did not differ significantly thereafter (0.4–1.9 vs. 0.8–1.3). Grasshopper sparrows *A. savannarum* were more abundant on the grazed area in 1997 (4.7 vs. 0.44), but more abundant on the ungrazed area in 1998 and 1999 (1.3–4.8 vs. 3.2–15.0). The Audubon Research Ranch (3,160 ha) was ungrazed by livestock since 1968, the Davis Pasture (1,501 ha) had moderate grazing pressure (645–1,387 animal unit months/year).

A replicated, controlled trial in May–June 1995–1996 in grasslands in Prairie Ridge State Natural Area, Illinois, USA (5), found that both native and non-native grasslands held higher average densities of five songbird species when under light, late-season grazing than when mown, 'hayed', unmanaged or burned (non-native grasslands only). However, species showed individual responses to different managements. The species surveyed were eastern meadowlark *Sturnella magna* and dickcissel *Spiza americana*, Henslow's sparrow *Ammodramus henslowii*, field sparrow *Spizella pusilla* and grasshopper sparrow.

A site comparison study in the Pyrénées-Orientales, France, in 1998–1999 (6), found that four bird species with an unfavourable conservation status in Europe preferentially used burned hillsides, compared with unmanaged or grazed areas. This study is discussed in detail in 'Use prescribed burning'.

A study in 2002–2003 in a grassland in Kentucky, USA (7) found that grasshopper sparrows had significantly lower nesting success on a grazed grassland, compared to a mown one (estimated 25% success on cattle-grazed area vs. 70% on mown area). This study is discussed in detail in 'Mow or cut natural grasslands'.

A replicated study in 2000–2002 on 34 fields of dry, native, prairie in southern Alberta, Canada (8), found that grazing only affected six of 31 bird species investigated. Only soras *Porzana carolina* were more abundant in late-grazed fields than ungrazed fields, only marsh wrens *Cistothorus palustris* were more abundant in early-grazed fields compared to late-grazed and only lesser scaup *Aythya affinis* were more abundant in late-grazed fields than in those grazed early in the season.

A replicated study using 23 years of data (up to 2003) from a tallgrass prairie in Kansas, USA (9), found that three of seven species showed a significant response to grazing by American bison *Bos bison*: upland sandpipers and grasshopper sparrows were consistently more abundant on grazed sites, whilst Henslow's sparrows were almost absent. Dickcissel, eastern meadowlark, Bell's vireo *Vireo bellii* (a shrub-dependent species) and brown-headed cowbird *Molothrus ater* showed no significant response.

A replicated study in 2000–2002 in 32 mixed-grass prairie fields in southern Alberta, Canada (10), found that duck nesting success was influenced by grazing and vegetation structure (with higher nesting success in taller vegetation). Duck success was 43% lower in ungrazed fields compared with those grazed from July and northern shoveller *Anas clypeata* had 64% lower success in early-grazed fields, compared with those grazed from July. However, nesting success of all but one songbird species was not influenced by these factors. The authors conclude that managing for ducks using grazing and other interventions is unlikely to provide habitat for songbirds.

A study in Oklahoma, USA, in 2003–2004 (11), found that dickcissel reproductive success was lower in grazed and burned pastures compared to on tallgrass prairie managed by patch-burns. This study is discussed in detail in 'Use prescribed burning'.

A replicated study in 2002–2003 (12) at the same tallgrass prairie site in Kansas, USA, as in (9), found that three of seven species surveyed showed significant responses to low-intensity cattle grazing: upland sandpipers, grasshopper sparrows, and eastern meadowlarks were more abundant in grazed areas, whilst Henslow's sparrow, dickcissel, brown-headed cowbird (all grassland species) and Bell's vireo showed no response.

(1) Burgess, H. H., Prince, H. H., and Trauger, D. L. (1965) Blue-winged teal nesting success as related to land use. *The Journal of Wildlife Management*, 29, 89–95.

(2) Bowen, B. S., and Kruse, A. D. (1993) Effects of grazing on nesting by upland sandpipers in south-central North Dakota. *The Journal of Wildlife Management*, 57, 291–301.

(3) Kruse, A. D., and Bowen, B. S. (1996) Effects of grazing and burning on densities and habitats of breeding ducks in North Dakota. *The Journal of Wildlife Management*, 60, 233–246.

(4) Gordon, C. E. (2000) Fire and cattle grazing on wintering sparrows in Arizona grasslands. *Journal of Range Management*, 53, 384–389.

(5) Walk, J. W., and Warner, R. E. (2000) Grassland management for the conservation of songbirds in the midwestern USA. *Biological Conservation*, 94, 165–172.

(6) Pons, P., Lambert, B., Rigolot, E., and Prodon, R. (2003) The effects of grassland management using fire on habitat occupancy and conservation of birds in a mosaic landscape. *Biodiversity and Conservation*, 12, 1843–1860.

(7) Sutter, B., and Ritchison, G. (2005) Effects of grazing on vegetation structure, prey availability, and reproductive success of grasshopper sparrows. *Journal of Field Ornithology*, 76, 345–351.

(8) Koper, N., and Schmiegelow, F. K. A. (2006) Effects of habitat management for ducks on target and nontarget species. *The Journal of Wildlife Management*, 70, 823–834.
(9) Powell, A. F. L. A. (2006) Effects of prescribed burns and bison (*Bos bison*) grazing on breeding bird abundances in tallgrass prairie. *The Auk*, 123, 183–197.
(10) Koper, N., and Schmiegelow, F. K. A. (2007) Does management for duck productivity affect songbird nesting success? *The Journal of Wildlife Management*, 71, 2249–2257.
(11) Churchwell, R. T., Davis, C. A., Fuhlendorf, S. D., and Engle, D. M. (2008) Effects of patch-burn management on dickcissel nest success in a tallgrass prairie. *Journal of Wildlife Management*, 72, 1596–1604.
(12) Powell, A. F. L. A. (2008) Responses of breeding birds in tallgrass prairie to fire and cattle grazing. *Journal of Field Ornithology*, 79, 41–52.

10.19.2. Artificial grasslands/pastures

- Of ten studies captured, one replicated, controlled study from the USA (4) found lower species richness in grazed areas than ungrazed. Another replicated, controlled study from the USA (5) found no consistent differences in community composition between grazed and ungrazed areas.
- A small study from Canada (3) found an increase in duck populations following the start of grazing amongst other interventions.
- Five studies from the UK and USA (5–9), four replicated, found higher use of, or higher nesting densities in, grazed areas compared to ungrazed. Seven studies from the UK, Canada and the USA (1–5,8,9), five replicated, found no differences in use or nesting densities, or lower abundances of birds on grazed, compared with ungrazed areas. One (1) found that several species appeared to be excluded by grazing.
- Three studies from the UK, USA and Canada (3,4,10), two replicated, found that nesting success or productivity was similar, or lower, on grazed sites compared with ungrazed.

Background

Studies described below include those on water meadows and pastures and non-native grasslands but not those on coastal marshes. Studies that graze water meadows amongst other interventions designed to restore traditional water meadows are described in 'Restore or create traditional water meadows'.

A controlled study in 1980–1982 on the river Roding in Essex, England (1), found that flood pasture alongside a 1.2 km stretch of the river with grazed 'flood beams' (see 'Use environmentally sensitive flood management') held a similar density of territories to an adjacent 1.8 km stretch which was not grazed (8–13 territories/km for grazed stretch vs. 6–21 territories/km for ungrazed). Riparian species (sedge warbler *Acrocephalus schoenobaenus*, Eurasian reed warbler *A. scirpaceus* and reed bunting *Emberiza schoeniclus*) were largely confined to the ungrazed section, whilst channel-nesting species (little grebe *Tachybaptus ruficollis* and common moorhen *Gallinula chloropus*) were at similar densities in both stretches.

A series of replicated controlled trials on grassland sites at two reserves in Essex, England, between 1990 and 1992 (2) found that brent geese *Branta bernicla* did not graze at higher densities on plots that were areas grazed, compared to cut and grazed areas, or areas that were just cut. Goose grazing intensity was not affected by sheep grazing compared to cattle grazing. Six replicates of each treatment were used.

A small before-and-after study from May–July in 1992–1994 in river islands in Quebec, Canada (3), found that the number of dabbling ducks *Anas* spp. nesting in the area had increased from 143 to 263 nests following the establishment of rotational grazing and dense nesting cover (see 'Plant wild bird seed or cover mixture'). However, fewer nests than expected by an even distribution across habitats were found in unimproved or improved pasture in 1993. More nests than expected were found in unimproved and fewer than expected in improved pasture in 1994. Nests on improved pasture had significantly lower success than those in other habitats (15% success of 39 nests vs. 47–82% elsewhere), with 33% being trampled. Nests on unimproved pasture had similar success rates (68% of 71 nests) to other habitats. Nesting densities were no higher on grazed pastures areas than other habitat types, and were lower than on areas seeded with dense nesting cover see ('Plant wild bird seed or cover mixture').

A replicated controlled study in Washington County, Pennsylvania, USA, found that cattle-grazed stream-side riparian pasture had lower bird species richness and abundance than ungrazed areas (4). Birds, nests and vegetation were surveyed along 12 pairs (grazed and control) of streams in 1996 and ten pairs in 1997. Several wetland-riparian species (e.g. common snipe *Gallinago gallinago*, green-backed heron *Butorides striatus* and solitary sandpiper *Tringa solitaria*) were more frequent or only occurred in controls. The authors suggest differences are largely due to simplified vegetation structure and reduced cover. Nest densities were higher in controls and nest destruction by cattle occurred in grazed areas, although nest success (all species combined) was not affected by grazing.

A controlled before-and-after trial in 1981–1986 in Arapaho National Wildlife Refuge, Colorado, USA (5), found few differences in overall bird density changes between two lightly grazed and two ungrazed pastures. However, for three of the nine species studied in detail, and for all three guilds examined, there were differences between treatments. Red-winged blackbirds *Agelaius phoeniceus* and American robins *Turdus migratorius* increased more on grazed pastures, as did species able to tolerate a moderate range of environmental conditions. However, willow flycatchers *Empidonax traillii* and species able to tolerate either a wide range of conditions or a very narrow range may have increased more on ungrazed pastures. The authors note that evidence for changes in most species was weak. The grazed pastures were grazed in August and September (with 2.4–3.5 animal-unit-months/ha) and then rested for 34 month.

A before-and-after study at South Stack RSPB Reserve, Anglesey, Wales (6), found that red-billed chough *Pyrrhocorax pyrrhocorax* use of 26 ha of semi-improved grassland increased following the introduction of year-round cattle grazing in spring 2002. In winter and spring, cattle density was typically less than 1/ha, rising to 2.5/ha during summer. Monitoring undertaken from November 2001 to August 2003, revealed that grazing also greatly reduced sward height.

A replicated, controlled, paired sites study of wet pasture in Leicestershire, UK (7), found that bird visit rates were significantly higher in areas with livestock than in those where livestock had been excluded. Sampling involved 45 minute bird observations between April 2005–March 2007 (twice/month April–October; once/month November–March).

A randomised, replicated controlled study in upland fields sown with a grass-legume mix at Los Banos Wildlife Area, California, USA (8), found that dabbling ducks *Anas* spp. nested at higher densities in four grazed plots than four ungrazed plots in 1996 (2.2 nests/ha vs. 0.6/ha) but not 1997 (0.7/ha vs. ungrazed 0.4/ha). Nest success estimates did not significantly differ between grazed (5%) and ungrazed (3%) fields. In 1994, four 10–14 ha upland fields were seeded with a grass-legume mix. In 1995, each was divided in half by electric fencing and randomly assigned to rotational grazing (1 July–1 November) or ungrazed. Grazed fields had shorter vegetation than ungrazed fields through the winter, but by the start of the nesting season (late March) vegetation height did not differ. By the end of the nesting season (late May) grazed fields had taller vegetation.

A replicated trial in the UK (9) found that songbirds and invertebrate-feeding birds were recorded more often on semi-natural rough grazing than on upland improved pasture, but the opposite was true for corvids. This study is discussed in detail in 'Graze non-grassland habitats'.

A replicated site comparison study on 1,031 agricultural sites across England in 2004–2008 (10) found that investigated the impact of rough grazing on grey partridge *Perdix perdix*. However, the study did not distinguish between the impacts of grazing, scrub control and the restoration of various semi-natural habitats. There was a negative relationship between the combined intervention and the ratio of young to old partridges in 2008. This study describes the effects of several other interventions, discussed in the relevant sections.

(1) Raven, P. (1986) Changes in the breeding bird population of a small clay river following flood alleviation works. *Bird Study*, 33, 24–35.
(2) Vickery, J. A., Sutherland, W. J., and Lane, S. J. (1994) The management of grass pastures for brent geese. *Journal of Applied Ecology*, 31, 282–290.
(3) Lapointe, S., Giroux, J. F., Belanger, L., and Filion, B. (2000) Benefits of rotational grazing and dense nesting cover for island-nesting waterfowl in southern Quebec. *Agriculture, Ecosystems & Environment*, 78, 261–272.
(4) Popotnik, G. J., and Giuliano, W. M. (2000) Response of birds to grazing of riparian zones. *The Journal of Wildlife Management*, 64, 976–982.
(5) Stanley, T. R., and Knopf, F. L. (2002) Avian responses to late-season grazing in a shrub-willow floodplain. *Conservation Biology*, 16, 225–231.
(6) Ausden, M., and Bateson, D. (2005) Winter cattle grazing to create foraging habitat for choughs *Pyrrhocorax pyrrhocorax* at South Stack RSPB Reserve, Anglesey, Wales. *Conservation Evidence*, 2, 26–27.
(7) Anon (2007) Wetting up farmland for birds and other biodiversity, Defra Report BD1323.
(8) Carroll, L. C., Arnold, T. W., and Beam, J. A. (2007) Effects of rotational grazing on nesting ducks in California. *The Journal of Wildlife Management*, 71, 902–905.
(9) Vale, J. E., and Fraser, M. D. (2007) Effect of sward type and management on diversity of upland birds. 333–336 in: J. J. Hopkins, A. J. Duncan, D. I. McCracken, S. Peel, J. R. B. Tallowin (eds) *British Grassland Society Occasional Symposium No.38* British Grassland Society (BGS), Reading.
(10) Ewald, J. A., Aebischer, N. J., Richardson, S. M., Grice, P. V., and Cooke, A. I. (2010) The effect of agri-environment schemes on grey partridges at the farm level in England. *Agriculture, Ecosystems & Environment*, 138, 55–63.

10.19.3. Non-grassland habitats

- One of eight studies, a replicated, controlled study on savannas in Kenya (8) found more bird species on grazed site, compared with unmanaged sites. These differences were not present during drought years. A before-and-after study from the Netherlands (6) found the number of species in a mixed habitat wetland site declined after the number of grazing animals increased.
- Three studies (two replicated) from a variety of habitats in Sweden, the Netherlands and Kenya (2,6,8) found that the overall number of birds, or the densities of some species were higher in grazed than ungrazed sites, or increased after the introduction of grazing. The Kenyan study found differences were not present in drought years. Four studies from several habitats in Europe and Kenya (2,4,6,8) found that some species were found at lower densities, or not found at all, on grazed sites compared to ungrazed sites or those under different management. Five studies from several habitats from across the world (1,2,4,6,8) found no differences in the abundances or densities of some or all species between grazed sites and those that were ungrazed or under different management.
- Two replicated studies from the UK (3,7) found that productivity of northern lapwing and grey partridge was lower in grazed sites compared to ungrazed. One study (7) examined several interventions at the same time.
- A replicated study from the UK (5) found that songbirds and invertebrate-eating species were more common on rough-grazed habitats than intensive pasture, but that crows were less so.

Background

In many parts of the world, the native grazers that would have helped maintain a balance between woody and herbaceous plants in a habitat have been lost or greatly reduced in numbers. Introducing domestic grazers is a potential way to re-establish the ecosystem dynamics.

A replicated controlled trial in a cottonwood *Populus sargentii* bottomland in north-east Colorado, USA (1), found that moderate late-autumn cattle grazing had no impact on breeding densities of six selected migratory songbirds over three study years. Five 16 ha cottonwood floodplain plots were fenced and cattle-grazed in October–November 1982–1984, and five were unmanaged. Analysis focussed on six species dependent on the grass-herb-shrub layer for foraging, nesting, or both: house wren *Troglodytes aedon*, brown thrasher *Toxostoma rufum*, American robin *Turdus migratorius*, common yellowthroat *Geothlypis trichas*, yellow-breasted chat *Icteria virens* and rufous-sided towhee *Pipilo erythropthalmus*.

In a replicated study in four oak- *Quercus* spp. hazel *Corylus avellana* woodlands (average size 5.3 ha) in 1996–1999 in Uppland and Åland, Sweden (2), breeding and migrant birds were found to be more numerous in sites grazed from spring to autumn than in some abandoned sites (average of 12 breeding and 5 migrant birds in grazed sites vs. 2–4 in abandoned sites). However, abundances did not differ between grazed sites and those subject to brush cutting and tree thinning (average of eight breeding bird species and seven migrants), and migrants were less abundant than in sites under simulated traditional management (16 breeding birds and 12 migrants). Traditional management involved sites being cleared in spring, mown in mid-late summer and grazed in autumn. A total of 65 bird species were observed. Birds present in spring did not differ in abundance between management types.

A replicated, controlled trial in spring and summer 1997 in Kent, England (3), found that northern lapwings *Vanellus vanellus* had smaller clutches, lower nest survival and higher nest loss to predation on four coastal marshes with low-intensity (0.2–0.5 livestock units/ha) grazing, compared to four areas without grazing (higher proportion of four egg clutches on ungrazed marshes; 34% survival and 58% predation for 36 nests on grazed marshes vs. 64% survival and 36% predation for 50 nests on ungrazed sites). Three of the 15 unpredated nests on the grazed marshes were also trampled by livestock. Livestock presence was also found to have a weak impact on the density of lapwing nesters in 1995 and 1997, but not 1996.

A controlled before-and-after study on a reserve in Lincolnshire, England (4), found no significant changes in redshank *Tringa tetanus* breeding densities on two saltmarsh plots following the introduction of light (approximately 0.2 cows/ha) or medium (0.4–0.6 cows/ha) grazing in 1996–1997. In addition, redshank densities in 1998–2004 were no different on the medium-grazed plot (0.7 pairs/ha), compared to an ungrazed plot (0.8 pairs/ha) or a heavily-grazed plot (0.6 pairs/ha). The light-grazed plot, however, had significantly lower densities (0.4 pairs/ha) than the ungrazed plot.

A replicated trial in the UK (5) found that songbirds and invertebrate-feeding birds were recorded more often on semi-natural rough grazing than on upland improved pasture, but the opposite was true for crows. Bird numbers and species were recorded in plots of improved upland pasture grazed by cattle and sheep (ten with and ten without the seasonal removal of grazing in summer) and in plots of semi-natural rough grazing grazed by cattle from June to September (six replicates). The proportion of surveys where songbirds and invertebrate feeders were recorded was greater on semi-natural rough grazing than on improved pasture. However, the effect on the number of individuals varied over the year. The number of birds of invertebrate-feeding species was greater on semi-natural grassland between May and July (338 birds, compared to 52 and 41 on improved treatments, with and without seasonal grazing removal), but greater on improved treatments between October and January (5,833 and 1,458 birds on improved treatments compared to 606 birds on semi-natural grassland). There were fewer crows on semi-natural rough grazing plots at all times of year, but the difference was greatest during July to September (16 birds on rough grazing compared to 496 and 77 on improved plots).

A before-and-after study in Oostvaardersplassen reserve in Flevoland, the Netherlands (6), found significant changes in the bird community in a 1,900 ha area of wet and dry grasslands, reedbeds, scrub and small woodlands, following increased numbers of grazing animals. The number of breeding species declined from 92 to 70 and of the 41 species with more than ten breeding pairs, eight increased and 33 decreased. Shrub-dependent species declined, as did those requiring tall reeds to nest. The authors suggest that declines in some grassland species were due to increased trampling. The number of Heck cattle *Bos taurus*, Konik horses *Equus ferus* and red deer *Cervus elaphus* increased from 390, 284 and 246 respectively (in 1997) to 497, 982 and 1,898 in 2007 (with cattle peaking at 580 in 2002). This reduced the areas of reedbeds from 844 to 377 ha and of shrub and woodland from 97 to 50 ha. The area of dry grassland increased from 527 to 1,019 ha.

A replicated site comparison study on 1,031 agricultural sites across England in 2004–2008 (7) investigated the impact of rough grazing on grey partridge *Perdix perdix*. However, the study does not distinguish between the impacts of grazing, scrub control and the restoration of various semi-natural habitats. There was a negative relationship between the combined intervention and the ratio of young to old partridges in 2008. This study describes the effects of several other interventions, discussed in the relevant sections.

In Laikipia District, Kenya, a replicated controlled study in 2005–2007 (8) found that five plots of savanna which were recently abandoned after grazing had, on average, but not consistently, higher densities of birds and held more species than four unmanaged control areas, but fewer than five burned areas (5–8 birds and 4–6 species/100 m^2 for grazed areas vs. 3–17 birds and 3–8 species for burned areas; 4–6 birds and 3–4 species for controls). The authors note that drought removed differences between treatments, and that the yearly variations in burned plots was greater than in grazed plots, suggesting that grazing may have longer term benefits. In addition, some species were only recorded in unmanaged areas. Burning is further discussed in 'Use prescribed burning'.

(1) Sedgwick, J. A., and Knopf, F. L. (1987) Breeding bird response to cattle grazing of a cottonwood bottomland. *The Journal of Wildlife Management*, 51, 230–237.
(2) Hansson, L. (2001) Traditional management of forests: plant and bird community responses to alternative restoration of oak–hazel woodland in Sweden. *Biodiversity and Conservation*, 10, 1865–1873.
(3) Hart, J. D., Milsom, T. P., Baxter, A., Kelly, P. F., and Parkin, W. K. (2002) The impact of livestock on lapwing *Vanellus vanellus* breeding densities and performance on coastal grazing marsh. *Bird Study*, 49, 67–78.
(4) Ausden, M., Badley, J., and James, L. (2005) The effect of introducing cattle grazing to saltmarsh on densities of breeding redshank *Tringa totanus* at Frampton Marsh RSPB Reserve, Lincolnshire, England. *Conservation Evidence*, 2, 57–59.
(5) Vale, J. E., and Fraser, M. D. (2007) Effect of sward type and management on diversity of upland birds. 333–336 in: J. J. Hopkins, A. J. Duncan, D. I. McCracken, S. Peel, J. R. B. Tallowin (eds) *British Grassland Society Occasional Symposium No.38* British Grassland Society (BGS), Reading.
(6) Bijlsma, R. G. (2008) Broedvogels van de buitenkaadse Oostvaardersplassen in 1997, 2002 en 2007. A & W-rapport 1051. Altenburg & Wymenga, Veenwouden.
(7) Ewald, J. A., Aebischer, N. J., Richardson, S. M., Grice, P. V., and Cooke, A. I. (2010) The effect of agri-environment schemes on grey partridges at the farm level in England. *Agriculture, Ecosystems & Environment*, 138, 55–63.
(8) Gregory, N. C., Sensenig, R. L., and Wilcove, D. S. (2010) Effects of controlled fire and livestock grazing on bird communities in east African savannas. *Conservation Biology*, 24, 1606–1616.

10.20. Plant trees to act as windbreaks

- One of two before-and-after studies, from the UK (2), found that the local population of European nightjars increased following several interventions including the planting of windbreaks.
- A before-and-after study, from the USA (1), found that erecting a windbreak appeared to disrupt lekking behaviour in greater prairie chicken territories nearby.

Background

Excessive wind may reduce the suitability of open habitat patches. Planting windbreaks may overcome this risk.

A before-and-after study in shrubland in 1962–1964 in Wisconsin, USA (1), found that the erection of a windbreak of 4 m high pines *Pinus* spp. appeared to disrupt lekking behaviour in male greater prairie chickens *Tympanuchus cupido*, with several males vacating their territories after trees were erected nearby.

A before-and-after study at Minsmere reserve (151 ha), Suffolk, UK, in 1978–1988 (2), found that the local population of European nightjars *Caprimulgus europaeus* increased following a series of management interventions, including the planting of 'shelter belts' to reduce wind in woodland glades. This study is discussed in detail in 'Clear or open patches in forests'.

(1) Anderson, R. K. (1969) Prairie chicken responses to changing booming-ground cover type and height. *The Journal of Wildlife Management*, 33, 636–643.
(2) Burgess, N. D., Evans, C. E., and Sorensen, J. (1990) The management of lowland heath for nightjars at Minsmere, Suffolk, Great-Britain. *Journal of Environmental Management*, 31, 351–359.

10.21. Re-seed grasslands

- One of two studies, both from the UK and investigating grazing by geese *Branta* spp., found that geese grazed at higher densities on grasslands that were re-seeded, compared to control or fertilised areas (1).
- One study (2) found that areas sown with clover were grazed at higher densities than those sown with grass seed.

Background

Re-seeding grasslands may improve productivity and the growth of young grass, which in turn may increase the number of birds it can support. This intervention is sometimes used to attract geese to specific areas to reduce conflict with farmers, when geese graze their crops. See 'Provide sacrificial grasslands to reduce the impact of wild geese on crops' for details.

Re-seeding can be with grass species, or with legumes such as clover Trifolium spp. which may improve the nitrogen content of the forage.

A replicated, controlled trial in 1984–7 on a reserve on the island of Islay, west Scotland (1), found that more barnacle geese *Branta leucopsis* used wet pasture fields if they were reseeded, compared to if they were fertilised or untreated (reseeding increased dropping density by 60–135%; fertilisation by 17–42%, but not in all fields or years). The effect of reseeding declined over time, and as the overall area of rotational grassland on the reserve increased. Reseeding consisted of ploughing fields and sowing with a perennial rye-grass *Lolium perenne* dominated seed mix in May.

A replicated study on an arable field on Thorny Island, in Suffolk, England, in the winters of 1992–1993 and 1993–1994 (2) found that dark-bellied brent geese *Branta bernicla bernicla* preferentially foraged on plots sown with white clover *Trifolium repens*, compared to three grass species (10–13 droppings/m^2 for 12 clover plots vs. 0–5 droppings/m^2 for 36 grass plots). There were no differences between grass species (perennial ryegrass, red fescue *Festuca rubra* or timothy *Phleum pratense*). Plots were established in spring 1991 and preferences were found in both years, although more geese used grass plots in 1993–1994.

(1) Percival, S. M. (1993) The effects of reseeding, fertiliser application and disturbance on the use of grasslands by barnacle geese, and the implications for refuge management. *Journal of Applied Ecology*, 30, 437–443.
(2) McKay, H. V., Milsom, T. P., Feare, C. J., Ennis, D. C., O'Connell, D. P., and Haskell, D. J. (2001) Selection of forage species and the creation of alternative feeding areas for dark-bellied brent geese *Branta bernicla bernicla* in southern UK coastal areas. *Agriculture, Ecosystems & Environment*, 84, 99–113.

10.22. Fertilise artificial grasslands

- We captured four studies (1–4) examining the impacts of fertilising grasslands, all from the UK and investigating grazing by geese *Anser* and *Branta* spp. Two studies (2,4) found that more geese grazed on areas that were fertilised compared with control areas. Two studies (1,3) found that cut and fertilised areas were used more than control areas. One study (2) found that fertilised areas were used less than re-seeded grasslands.

- One study (3) found that fertilisation affected grazing at applications of 50 kg N/ha, but not 18 kg N/ha. One study (4) found that grazing rates only increased with applications of up to 80 kg N/ha.

Background

Fertilised grasslands are likely to have higher productivity and therefore support larger numbers of birds, particularly grazing species, such as geese. This intervention is commonly used to attract geese to specific areas to reduce conflict with farmers, when geese graze their crops. See 'Provide sacrificial grasslands to reduce the impact of wild geese on crops'.

A replicated, controlled trial in the winter of 1972–1973 at a 6 ha pasture (periodically flooded by saltwater) in Gloucestershire, UK (1), found that significantly more greater white-fronted geese *Anser albifrons* fed on fertilised and cut areas, compared to control areas (overall average of 30–35% of geese on cut, fertilised areas vs. 17–20% on control areas; maximum of 65% use of cut, fertilised areas vs. 20% for controls). Preferences decreased over time as preferred areas lost vegetation and became more crowded. Vegetation from experimental areas had a higher nitrogen content than that from control areas. Fertilisation consisted of 125 kg/ha of 'nitro-chalk' – 25% nitrogen – applied in mid-October. In mid-October, the grass was also cut to approximately 8 cm.

A replicated, controlled study in 1984–1987 on a reserve on the island of Islay, west Scotland (2), found that more barnacle geese *Branta leucopsis* used wet pasture fields if they were fertilised, compared to control fields (17–42% higher dropping densities in fertilised fields, but not in all fields or years). However, fewer geese used fertilised fields than re-seeded ones. Fertilisers were either 34.5% nitrogen in pellet form (at 125 kg/ha), or 'Nitrochalk' – 25% nitrogen in granular form – (at 175 kg/ha) and spread in October (wet and dry fields) and March (dry fields only).

A series of replicated controlled trials on grassland sites at two reserves in Essex, England, between 1990 and 1992 (3) found that brent geese *Branta bernicla* grazed at significantly higher densities on fertilised and cut areas, compared to unfertilised areas, but only at high levels of fertiliser application (50 kg N/ha used: 28–30 droppings/m^2 for fertilised areas vs. 23–28 droppings/m^2 for controls; 18 kg N/ha used: 30–35 droppings/m^2 for fertilised areas vs. 25–35 droppings/m^2 for control areas). There were no differences between trials using organic and inorganic fertiliser.

A replicated, controlled study in 1990–1993 at a reserve in Aberdeenshire, Scotland (4), found that spring fertiliser application in 1990–1 significantly increased the use of grassland fields by pink-footed geese *Anser brachyrynchus*, until applications of approximately 80 kg N/ha (1990: average of 13–14 goose droppings/m^2 with no application vs. 18–22 droppings/m^2 with 40 kg N/ha, 28 droppings/m^2 with 80 kg/m^2 and 27–31 droppings/m^2 with 120–160 kg N/ha; patterns in 1991 were similar but with fewer droppings). However, two slow-release fertilisers did not affect foraging densities in winter 1990–1992 (average of 24.5–26.7 droppings/m^2 for fertilised vs. 24 droppings/m^2 for control grasslands). Split fertiliser application did not increase field use, compared to a single application (average of 11 droppings/m^2 for fields with split applications vs. 10 droppings/m^2 for single applications), although the authors note it may reduce nitrogen leeching.

(1) Owen, M. (1975) Cutting and fertilizing grassland for winter goose management. *The Journal of Wildlife Management*, 39, 163–167.
(2) Percival, S. M. (1993) The effects of reseeding, fertiliser application and disturbance on the use of grasslands by barnacle geese, and the implications for refuge management. *Journal of Applied Ecology*, 30, 437–443.
(3) Vickery, J. A., Sutherland, W. J., and Lane, S. J. (1994) The management of grass pastures for brent geese. *Journal of Applied Ecology*, 31, 282–290.
(4) Patterson, I. J., and Fuchs, R. M. E. (2001) The use of nitrogen fertiliser on alternative grassland feeding refuges for pink-footed geese in spring. *Journal of Applied Ecology*, 38, 637–646.

10.23. Raise water levels in ditches or grassland

- Of seven studies captured, one, a before-and-after study from the UK (2) found that two wader species recolonised a site after water levels were raised. A third was found at very high levels. A review from the UK (7) found that high-level agri-environment schemes designed to provide wet habitats were effective at providing habitats for waders and two replicated studies from the UK (6) and Denmark (4) found that northern lapwings were more likely to nest or nested at higher numbers on grasslands with high water levels.
- A replicated and controlled study from Denmark (4) found that Eurasian oystercatchers did not nest at higher densities on fields with raised water levels and that raising water levels had no effect on nesting on restored grassland fields.
- A replicated study from the USA (5) found that predation rate on Cape Sable seaside sparrow nests increased as water levels increased.
- A replicated, controlled and paired sites study from the UK (3) found that birds visited grassland sites with raised water levels at higher rates than other fields. A replicated study from the UK (1) found no differences in feeding rates on sites with raised water levels, compared with control sites.

A replicated study from January–March in 2002 that observed 15 northern lapwing *Vanellus vanellus* chicks of one grassland site in the Isle of Islay, UK (1), found that raising water levels in the grassland did not affect lapwing foraging rate. Foraging rate increased with decreasing sward height and was greater in ditches than on rigs (strips of cultivated land). Soil moisture, however, did not significantly affect foraging rate after sward height and rig effects were accounted for. The timing of fertiliser application (to promote grass growth) and water level in ditches was manipulated at the field scale, which resulted in a range of soil moisture levels and sward heights. Water level was controlled through sluiced canals along that ran along field boundaries and in-field ditches. The authors point out that spring 2002 was particularly wet and may have confounded any effect of added soil moisture.

A before-and-after study at Campfield Marsh RSPB Reserve, Cumbria, England (2), found that five years after water levels were raised in August 1995, breeding common snipe *Gallinago gallinago* and northern lapwing recolonised the site and that, over the reserve as a whole, breeding curlew *Numenius arquata* densities were 6 pairs/km^2 (one of the highest UK breeding densities). Five fields comprising 23 ha of former cattle-grazed, species-poor perennial rye-grass *Lolium perenne* dominated grassland and arable cropland were restored. Over the five years vegetation also shifted towards target plant communities characteristic of wet grassland.

A replicated, controlled (paired) study of wet pasture and bunded and non-bunded drainage ditches in arable and pastoral areas in Leicestershire, UK (3), found that bird visit rates were significantly higher in wet pasture (0.2–0.3 visits) than in control dry plots (0.1), particularly in the summer months and in 2006. The authors suggested benefits due to management may increase over time. Visit rates were also higher to ditch-fed paired ponds (1.0 visit/month) than dry controls (0.5 visit/month). Sampling involved bird observations (45 minutes, 1–2/month between April 2005 and March 2007.

A replicated, controlled study in 615 grassland fields in Jutland, Denmark (4), found that permanent grasslands fields under an agri-environment scheme designed to increase water levels had significantly higher numbers of three species of waders (northern lapwing, black-tailed godwit *Limosa limosa*, common redshank) in 2004–2005 after the scheme was implemented, compared to in 1999–2001, before the scheme. However, this was only the case for fields that successfully retained water (40 pairs before and 90 after for wet fields vs. approximately two pairs before and five after for dry fields). In addition, fields that were dry before the scheme and wet after showed a greater increase (280–290% increase in lapwing numbers) than fields that were wet beforehand (130–170% increases). There were no increases on restored grasslands (formerly cropland), whether or not they were under the scheme, or on control fields (i.e. not under the scheme) that failed to retain water. Numbers did increase on control fields that retained water, but the numbers found on them were no different from those expected if increases were uniformly distributed across the landscape (i.e. birds did not appear to be selecting the fields preferentially). Eurasian oystercatchers *Haematopus ostrolagus* did not increase on any field types and the authors note that regional wader numbers were still far lower than in 1978–1988. The scheme involved blocking drainage pipes and 'rills' (drainage channels) as well as reducing the fertiliser inputs, grazing intensity and restricting when mowing could take place.

A replicated study in 1996–2001 in marl prairies in Everglades National Park, Florida, USA (5), found that as water level was increased, predation rates on Cape Sable seaside sparrow *Ammodramus maritimus mirabilis* nests increased. Of 429 nests found, 210 failed whilst 219 produced at least one fledgling. Nest predation accounted for 97% of failures. Early nests had higher success (47% chance of at least one fledgling/clutch) than those initiated later (1%).

A replicated study in 2005–2006 on 70 fields with wet features at nine lowland pastoral sites in east England (6) found that the probability of a field being used by nesting lapwing was significantly higher with an increase in footdrain floods. Fields with footdrain floods held the highest densities of nesting pairs. Nests were more likely to be located within 50 m of footdrain floods and chicks more likely to forage near footdrain floods (in wet mud patches created by receding water). Fields with footdrains, footdrain floods and isolated pools were visited at least once a week (March–July 2005–2006) and the number of lapwing pairs displaying parental behaviour within a 10-min sampling period used as a measure of brood density.

A 2009 literature review of agri-environment schemes in England (7) found studies that suggested more expensive agri-environment scheme options for wetland habitats (such as controlling water levels) were more effective at providing good habitat for waders than easier-to-implement options. This review also examines several other interventions, discussed in the relevant sections.

(1) Devereux, C. L., Mckeever, C. U., Benton, T. G., and Whittingham, M. J. (2004) The effect of sward height and drainage on common starlings *Sturnus vulgaris* and northern lapwings *Vanellus vanellus* foraging in grassland habitats. *Ibis*, 146, 115–122.
(2) Lyons, G. (2005) Botanical monitoring of restored lowland wet grassland at Campfield Marsh RSPB Reserve, Cumbria, England. *Conservation Evidence*, 2, 43–46.
(3) Defra (2007) Wetting up farmland for birds and other biodiversity, Defra Report BD1323.
(4) Kahlert, J., Clausen, P., Hounisen, J. P., and Petersen, I. K. (2007) Response of breeding waders to agri-environmental schemes may be obscured by effects of existing hydrology and farming history. *Journal of Ornithology*, 148, 287–293.
(5) Baiser, B., Boulton, R. L., and Lockwood, J. L. (2008) Influence of water depth on nest success of the endangered Cape Sable seaside sparrow in the Florida Everglades. *Animal Conservation*, 11, 190–197.
(6) Eglington, S. M., Gill, J. A., Bolton, M., Smart, M. A., Sutherland, W. J., and Watkinson, A. R. (2008) Restoration of wet features for breeding waders on lowland grassland. *Journal of Applied Ecology*, 45, 305–314.
(7) Natural England (2009) Agri-environment schemes in England 2009 A review of results and effectiveness.

10.24. Manage water level in wetlands

- Of six studies, one replicated, controlled study from the USA (5) found that bird diversity was affected by maintaining water levels at different levels.
- A study from the USA (1) found that ducks were more abundant when high water levels were maintained on a wetland site. Geese were more abundant when lower levels were maintained. Three studies from the USA and Canada (2,4,5), two replicated, found that different species showed preferences for different water levels in wetlands.
- A replicated study from the UK (6) found that great bitterns established territories earlier when deep water levels were maintained, but this had no effect on productivity.
- A review from Spain (3) found that management successfully maintained water near a greater flamingo nesting area, but the effects of this were not measured.

A study in 1958–1967 on Squaw Creek National Wildlife Refuge, Missouri, USA (1), found that, in general, ducks increased when the largest expanses of marshes were flooded, and geese were most abundant when the largest areas of marshes and forage plants were available in response to lowered water levels. The use of the wetlands is discussed in 'Habitat restoration and creation'.

A trial in Janurary–May 1991–1992 at coastal impoundments and intertidal mudflats on South Island, South Carolina, USA (2), found that significantly higher numbers of American avocets *Recurvirostra americana* used the impoundments as water was drawn down over the spring, compared to mudflats, despite mudflats being significantly larger. Avocet distribution within impoundments was highest where water was 10–17 cm deep and 1–30% of the area was exposed; but lowest where there were high daily fluctuations in water depth. Water was slowly drawn down from impoundments from November to April, creating a wide range of water depths, before reflooding in June.

A review of management at a coastal wetland in 1978–1982 in Andalusia, Spain (3), found that water management was successful in ensuring that there was always an area of water close to a greater flamingo *Phoenicopterus roseus* nesting area. However, the impact of this could not be quantified.

A replicated study at Delta Marsh, Manitoba, Canada (4), found that different species preferentially used areas of prairie wetlands with varying amounts of open water. Responses to water level (and associated vegetation) changes in ten adjacent prairie wetlands (150 × 300 m) created in 1980 were assessed. Censuses were conducted 1 May to 31 October (1980–1989). Yellow-headed blackbird *Xanthocephalus xanthocephalus* used shallow-flooded areas with a mix of open water and emergent vegetation, red-winged blackbird *Agelaius phoeniceus* preferred denser vegetation. American coot *Fulica americana* preferred deep water with interspersed vegetation. Dabbling ducks generally occupied marsh with equal amounts of vegetation cover and open water. Diving ducks used deeper water but there was variation between species and season, as to whether open or more densely vegetated areas were preferred.

A replicated partially-randomised, controlled study compared waterbird use of four experimentally drawndown wetlands with flooded wetlands at the Grasslands Ecological Area in California's Central Valley, USA (5), found that maximum bird diversity and abundance occurred at average depths of 10–20 cm on wetlands with a 30–40 cm difference between deepest and shallowest zones. There was limited availability of shallow-water habitat in winter but not spring, allowing waders, cinnamon teal *Anas cyanoptera* and American green-winged teal *A. carolinensis* to use the site. Use by deeper-water dabbling ducks and diving waterbirds declined during later stages of drawdown. Birds were monitored over winter and spring 1994–1995.

A replicated study in 1997–2001 in ten reedbed sites across England (6) found that male great bitterns *Botaurus stellaris* established territories significantly earlier in four sites with water levels maintained at 19–27 cm, compared to six with lower water levels (4–9 cm, four managed and four unmanaged). However, there was no effect on chick survival or overall productivity (1.3 chicks/female on high water sites vs. 1.2 on low water sites). Reeds at sites with low water levels were also cut in spring (March–April), compared with winter (completed by December) for high water level sites, but the effect of cutting was not specifically investigated.

(1) Burgess, H. H. (1969) Habitat management on a mid-continent waterfowl refuge. *The Journal of Wildlife Management*, 33, 843–847.
(2) Boettcher, R., Haig, S. M., and Bridges Jr, W. C. (1995) Habitat-related factors affecting the distribution of nonbreeding American avocets in coastal South Carolina. *The Condor*, 97, 68–81.
(3) Martos, M. R., and Johnson, A. R. (1996) Management of nesting sites for greater flamingos. *Colonial Waterbirds*, 19 S1, 167–183.
(4) Murkin, H. R., Murkin, E. J., and Ball, J. P. (1997) Avian habitat selection and prairie wetland dynamics: a 10-year experiment. *Ecological Applications*, 7, 1144–1159.
(5) Taft, O. W., Colwell, M. A., Isola, C. R., and Safran, R. J. (2002) Waterbird responses to experimental drawdown: implications for the multispecies management of wetland mosaics. *Journal of Applied Ecology*, 39, 987–1001.

(6) Gilbert, G., Tyler, G. A., Dunn, C. J., Ratcliffe, N., and Smith, K. E. N. (2007) The influence of habitat management on the breeding success of the great bittern *Botaurus stellaris* in Britain. *Ibis*, 149, 53–66.

10.25. Use environmentally sensitive flood management

- One of two studies, a before-and-after study from the UK (1), found that there were significantly more bird territories in a stretch of river with 'flood beams' installed, compared to a channelised river.
- A replicated site comparison study in the USA (2) found that 13 of 20 bird species increased at sites with the restoration of river dynamics and vegetation.

Background

Many rivers across the world have been channelised and managed, meaning that then natural dynamics (changes in water levels, flooding etc) are lost. This has destroyed habitat and can also lead to less frequent, but far more damaging floods. Using environmentally sensitive flood management may both provide habitats and reduce the damage caused by flooding.

A controlled before-and-after study on the river Roding in Essex, England (1), found that in 1982 there were more territories and more species of bird on a 3 km stretch of the river that was modified in 1979 to reduce flooding in the area, compared to an adjacent 500 m stretch of river that was channelised in 1974 (52 territories of nine species vs. three territories of two species). The experimental stretch had one bank excavated to create a 0.3 m high shelf (a 'flood beam') just above the level of the main channel. This meant that the main channel continued to carry water during dry periods (at a rate of 2 m^3/s) but during heavy rains, the beam would carry water as well (at up to 40 m^3/s) increasing the width and the flow capacity of the river.

A replicated site comparison trial in 1993–2003 on ten sites along the Sacramento River, California, USA (2), found that 13 of 20 bird species were increasing on plots where both riparian vegetation and the river's hydrological dynamics were restored. This study is discussed in detail in 'Habitat restoration and creation'.

(1) Raven, P. (1986) Changes in the breeding bird population of a small clay river following flood alleviation works. *Bird Study*, 33, 24–35.
(2) Gardali, T., Holmes, A. L., Small, S. L., Nur, N., Geupel, G. R., and Golet, G. H. (2006) Abundance patterns of landbirds in restored and remnant riparian forests on the Sacramento River, California, USA. *Restoration Ecology*, 14, 391–403.

10.26. Use greentree reservoir management

- A site comparison study from the USA (1) found significantly lower numbers of breeding mid- and under-storey birds at a greentree reservoir site than at a control site. Canopy nesting species were not affected. The species investigated were not gamebirds or wildfowl.

Background

A greentree reservoir is a temporary impoundment of a forested river-bottom aimed at providing wildfowl habitat.

In eastern Arkansas, USA, a site comparison study found significant differences in relative abundances of 12 of 28 non-gamebird species between a greentree site and an adjacent control area (1). Singing birds were surveyed in the 1980–1981 breeding seasons. Greentree reservoir management reduced understory vegetation and bird species that are primarily understory foragers were absent or at lower frequencies, and nesting opportunities for ground- or under-storey nesting species were reduced, compared to the control. Canopy-foraging species were not affected. There were overall fewer breeding species and bird abundance was lower at the greentree reservoir site.

(1) Christman, S. P. (1984) Breeding bird response to greentree reservoir management. *The Journal of Wildlife Management*, 48, 1164–1172.

10.27. Plough habitats

- One of four studies (of two experiments), from the USA (4), found that bird densities were higher on ploughed wetland areas, compared to unploughed areas.
- Three studies of a site comparison study from the UK (1–3) found that few whimbrels nested on ploughed and re-seeded areas of heathland (1), but these areas were used for foraging in early spring (2). There were no differences in chick survival between birds that used ploughed and re-seeded heathland and those that did not (3).

Background

Ploughing can be used to reduce the dominance of some species, such as *Spartina* spp. on wetlands, or heathers, on heathland.

A site comparison study in the Shetland Islands, Scotland (1) found that areas of heath seeded with grass to improve them for livestock grazing were mostly avoided by nesting whimbrels *Numenius phaeopus* in favour of unimproved heathland. In 1986 and 1987, this study monitored whimbrels in five areas of heathland that had been partly seeded, four on the island of Fetlar, one on Unst. Of 111 nests, 89% were found in unseeded heathland. Most nests were on hummocks and amongst heather *Calluna vulgaris*. Seeding with grass after ploughing or harrowing resulted in the loss of hummocks and most heather, and created a predominantly grassy habitat. Surface-seeding, without ploughing or harrowing, created less marked changes, with hummocks and heather retained, although hummock height was lowered, and in some areas only dead or dying heather was present.

In a study using the same Shetland Island heaths as (1), Grant *et al.* (2) found no significant difference in chick survival between chicks that used areas of heathland re-seeded with grass and those that did not. Individually marked chicks were monitored after hatching in 20, 23, and 26 broods in 1986, 1987 and 1988 respectively. In each year 35–65% of all chicks remained on heathland, while others (usually broods over 12 days old, from nests within 200 m of the alternative habitat) moved into other habitats.

At the same study sites as (1), Grant *et al.* (3) found that areas of heath seeded with grass after ploughing or harrowing, and older pastures, were the main early spring feeding areas for at least 90% of whimbrel pairs in the study. Habitat use by individually marked whimbrels was monitored during the pre-laying period in spring 1987 and 1988, on five Shetland Island heathlands. The birds made little use of unimproved heathland (where most nest) or heathland areas seeded without ploughing/harrowing. The greatest quantities of prey species (earthworms, oligochaetes, and crane-fly larvae, tipulids) were found in the soil of ploughed or harrowed seeded areas of heath and older pastures, with more recently seeded areas holding the highest masses of crane-fly larvae.

A controlled study in 2003–2004 on mudflats and areas of *Spartina alterniflora* meadows in Willapa National Wildlife Refuge, Washington, USA (4), found that average densities of waders and wildfowl were significantly higher on areas of *Spartina* meadow that were ploughed, compared to untreated areas or areas completely sprayed with herbicide (see 'Treat wetlands with herbicide'). Densities of some groups were 100 times those on the control areas, whilst some species found on the tilled meadows were never found on untreated *Spartina*. In addition, densities of unidentified *Calidris* spp. sandpipers, dowitchers *Limnodromus* spp. and waterfowl were significantly higher on tilled areas than on bare mud. The area was ploughed in winter-spring 2001, disked in winter 2002, and spot-treated with glyphosate during the summers of 2003 and 2004.

(1) Grant, M. C. (1992) The effects of re-seeding heathland on breeding whimbrel *Numenius phaeopus* in Shetland. I. Nest distributions. *Journal of Applied Ecology*, 29, 501–508.
(2) Grant, M. C., Chambers, R. E., and Evans, P. R. (1992) The effects of re-seeding heathland on breeding whimbrel *Numenius phaeopus* in Shetland. III. Habitat use by broods. *Journal of applied ecology*, 29, 516–523.
(3) Grant, M. C., Chambers, R. E., and Evans, P. R. (1992) The effects of re-seeding heathland on breeding whimbrel *Numenius phaeopus* in Shetland. II. Habitat use by adults during the pre-laying period. *Journal of Applied Ecology*, 509–515.
(4) Patten, K., and O'Casey, C. (2007) Use of Willapa Bay, Washington, by shorebirds and waterfowl after *Spartina* control efforts. *Journal of Field Ornithology*, 78, 395–400.

10.28. Create scrapes and pools in wetlands and wet grasslands

- Of six studies captured, four before- and after studies from the UK and North America (2–5) found that the use of sites, or the breeding population of birds on sites, increased following the creation of ponds and scrapes or was higher in areas with ditch-fed ponds.
- A study from the USA (1) found that dabbling ducks used newly-created ponds in large numbers, although other species preferred older ponds. Songbirds did not appear to be affected by pond-creation.
- A replicated site from the UK (6) found that northern lapwing chicks foraged in newly created wet features and that chick condition was higher in sites with a large number of footdrains.

Background

Creating scrapes and pools in wetlands and wet grasslands can help create a heterogenous habitat, with varying vegetation types and water levels.

A study in 1940–1942 at a marsh site in Iowa, USA (1), found that large numbers of dabbling ducks *Anas* spp. used pools and clearings created by blasting. Diving ducks *Aythya* spp. and ruddy ducks *Oxyura jamaicensis*, however, were rare in the newly-created pools as were American coot *Fulica americana*, rails *Rallus* spp. and sora *Porzana carolina*, instead being found in older ponds and clearings. Songbirds seemed largely unaffected by the blasting.

A before-and-after study on a marshland site near Lake Manitoba, Manitoba, Canada (2), found that between the summers of 1965 and 1966 there was a 288% increase in the number of wildfowl using 25 ponds created by blasting in August 1964 and 1965. There was a smaller increase at natural marshlands nearby. A total of 11 species were seen, mostly of dabbling ducks *Anas* spp. Ponds averaged 132 m^2 and 1.5 m deep.

A before-and-after study on a wetland reserve in Cumbria, England (3), found that the number of common snipe *Gallinago gallinago* nesting in an area of improved peat grassland increased from one pair in 2003 to 11 pairs in both 2004 and 2005 following the creation of 18 small scrapes in the 60 ha grassland, and other management interventions, discussed in 'Restore or create traditional water meadows'.

A before-and-after study on 160 ha of improved grassland at Ynys-Hir RSPB reserve, Powys, Wales (4), found that populations of northern lapwings *Vanellus vanellus* and common redshank *Tringa totanus* increased following a series of interventions including chisel ploughing, used on a two year rotation (approximately 8 ha in February 2002 and 10 ha areas thereafter) to break up the surface to create small hummocks and divots. This study is discussed in detail in 'Restore or create traditional water meadows'.

A replicated, controlled paired sites study of wet pasture and bunded and non-bunded drainage ditches in arable and pastoral areas in Leicestershire, UK (5), found that bird visit rates were significantly higher in wet pasture (0.2–0.3 visits) than in control dry plots (0.1), particularly in the summer months and in 2006. The authors suggested benefits due to management may increase over time. Visit rates were also higher to ditch-fed paired ponds (1.0 visit/month) than dry controls (0.5 visit/month). Sampling involved bird observations (45 minutes, 1–2/month between April 2005 and March 2007.

Within nine grazed wet grasslands sites in Broadland, Norfolk, England (6), a replicated site comparison study (March–July 2005 to 2007) found northern lapwing *Vanellus vanellus* chick foraging rates and estimated biomass intake (monitored May–July 2006) were 2–3 times higher in installed shallow wet features than in the grazing marsh. Late in the breeding season when water levels were low, chick body condition was significantly higher in fields with footdrain densities of more than 150 m/ha. The wet features supported more than twice the biomass of surface-active invertebrates and a greater abundance of aerial invertebrates than the grazing marsh. Each year, chicks (<100 g) were weighed and bill length measured to determine growth rates.

(1) Provost, M. W. (1948) Marsh-blasting as a wildlife management technique. *The Journal of Wildlife Management*, 12, 350–387.
(2) Hoffman, R. H. (1970) Waterfowl utilization of ponds blasted at Delta, Manitoba. *The Journal of Wildlife Management*, 34, 586–593.
(3) Holton, N., and Allcorn, R. I. (2006) The effectiveness of opening up rush patches on encouraging breeding common snipe *Galliango gallinago* at Rogersceugh Farm, Campfield Marsh RSPB reserve, Cumbria, England. *Conservation Evidence*, 3, 79–80.
(4) Squires, R., and Allcorn, R. I. (2006) The effect of chisel ploughing to create nesting habitat for breeding lapwings *Vanellus vanellus* at Ynys-Hir RSPB reserve, Powys, Wales. *Conservation Evidence*, 3, 77–78.
(5) Anon (2007) Wetting up farmland for birds and other biodiversity, Defra Report BD1323.
(6) Eglington, S. M., Bolton, M., Smart, M. A., Sutherland, W. J., Watkinson, A. R., and Gill, J. A. (2010) Managing water levels on wet grasslands to improve foraging conditions for breeding northern lapwing *Vanellus vanellus*. *Journal of Applied Ecology*, 47, 451–458.

11. Habitat restoration and creation

Habitat destruction is the largest threat to bird species and populations (BirdLife International 2004), and habitat protection remains one of the most important and frequently-used conservation interventions. However, in many parts of the world, restoring damaged habitats or creating new habitat patches may also be possible.

Restoration is often required by law as a response to mining or other activities that destroy natural habitats. Subsequent to cessation of mining, surface-mined areas may be revegetated using an array of techniques and with varying objectives. Tree planting may be undertaken aiming to restore natural vegetation communities and reinstate fauna present prior to mining (Nichols & Watkins 1984, Comer & Wooller 2002). Alternatively, achieving a rapid cover of herbaceous vegetation (native or non-native) may be the goal in order to counter erosion. For example in North America large areas of surface-mine sites have been grass-sown producing extensive grasslands, often in regions where most native prairie has been lost, thereby providing relatively undisturbed grassland refuges (Galligan *et al.* 2006). Several studies show that these can provide habitat for grassland birds (Ingold *et al.* 2009), some of which are declining species of conservation concern (Bajema *et al.* 2001).

This chapter describes the overall impact of habitat creation or restoration on bird species and communities. However, restoration and creation involve many individual interventions, for example the restoration of fire dynamics in pine forests; the grazing of prairies or the restoration of natural flow regimes in rivers. Studies describing the effects of these individual interventions are discussed in the chapter 'Threat: Natural system modifications'.

Bajema R. A., DeVault T. L., Scott P. E. & Lima S. L. (2001) Reclaimed coal mine grasslands and their significance for Henslow's sparrows in the American Midwest. *The Auk*, 118, 422–431.

BirdLife International. (2004) State of the World's Birds 2004: Indicators for Our Changing World. BirdLife International.

Comer S. J. & Wooller R. D. (2002) A comparison of the passerine avifaunas of a rehabilitated minesite and a nearby reserve in south-western Australia. *Emu*, 102, 305–311.

Galligan E. W., Devault T. L. & Lima S. L. (2006) Nesting success of grassland and savanna birds on reclaimed surface coal mines of the Midwestern United States. *Wilson Journal of Ornithology*, 118, 537–546

Ingold D. J., Dooley J. L. & Cavender N. (2009) Return rates of breeding Henslow's sparrows on mowed versus unmowed areas on a reclaimed surface mine. *Wilson Journal of Ornithology*, 121, 194–197.

Nichols O. G. & Watkins D. (1984) Bird utilization of rehabilitated bauxite minesites in Western Australia. *Biological Conservation*, 30, 109–131.

Key messages

Restore or create forests

Thirteen of 15 studies from across the world found that restored forests were similar to in-tact forests, that species returned to restored sites, that species recovered significantly better at restored than unrestored sites or that bird species richness, diversity or abundances in restored forest sites increased over time. One study also found that restoration techniques themselves improved over time. Nine studies found that some species did not return to restored forests or were less common and a study found that territory densities decreased over time. A study from the USA found that no more birds were found in restored sites, compared with unrestored. One study investigated productivity and found it was similar between restored and intact forests. A study from the USA found that planting fast-growing species appeared to provide better habitat than slower-growing trees.

Restore or create grasslands

Three of 23 studies found that species richness on restored grasslands was higher than unrestored habitats, or similar to remnant grassland, and three found that target species used restored grassland. Two studies from the USA found that diversity or species richness fell after restoration or was lower than unrestored sites. Seven studies from the USA and UK found high use of restored sites, or that such sites held a disproportionate proportion of the local population of birds. Two studies found that densities or abundances were lower on restored than unrestored sites, potentially due to drought conditions in one case. Five studies found that at least some bird species had higher productivities in restored sites compared to unrestored; had similar or higher productivies than natural habitats; or had high enough productivities to sustain populations. Three studies found that productivities were lower in restored than unrestored areas, or that productivities on restored sites were too low to sustain populations. A study from the USA found that older restored fields held more nests, but fewer species than young fields. Three studies found no differences between restoration techniques; two found that sowing certain species increased the use of sites by birds.

Restore or create traditional water meadows

Four out of five studies found that the number of waders or wildfowl on UK sites increased after the restoration of traditional water meadows. One study from Sweden found an increase in northern lapwing population after an increase in meadow management. One study found that lapwing productivity was higher on meadows than some habitats, but not others.

Restore or create shrubland

Three studies from the UK, USA and the Azores found local bird population increases after shrubland restoration. Two studies investigated multiple interventions and one found an increase from no birds to one or two pairs. One study from the UK found that several interventions, including shrubland restoration, were negatively related to the number of young grey partridges per adult bird on sites.

Restore or create savannas

We captured no evidence for the effects of savanna restoration on bird populations.

Restore or create inland wetlands

All 11 studies from North America and the Pacific found that birds used restored or created wetlands. Two found that rates of use and species richness were similar or higher than on natural wetlands, three found they were lower. One study found that use rates were higher than unrestored sites. One study found that bird productivity was similar in restored and natural wetlands. Two studies found that larger or more permanent restored wetlands were used more than smaller sites.

Restore or create coastal and intertidal wetlands

All six studies from the USA and UK found that target bird species used restored or created wetlands and three found that bird numbers or diversity were comparable or higher than natural wetlands, whilst three found that numbers increased over time. Three studies found that some species declined, or were less common, on restored sites.

Restore or create kelp forests

The single study captured found that five of nine species increased following kelp forest restoration in the USA.

Restore or create lagoons

The single study captured found that large number of birds used and bred in a newly created lagoon in the UK.

Revegetate gravel pits

We captured no evidence for the effects of gravel pit revegetation on bird populations.

11.1. Restore or create forests

- Thirteen of 15 studies from across the world (1–5,7–9,11,13–16) found that bird communities in restored forests were similar to original forests or that species returned to restored sites

(1,2,4,5,7,12,13,16), that species recovered significantly better than at unrestored sites (1,8), that species richness, diversity or abundances increased over time (1,3,8,9,11,14–16) or that restoration techniques themselves improved over time (1).

- Nine of the studies (1,2–5,8,9,11,16) found that some species did not return to restored sites (1,8,9), or were less common than in original forests (2–5,11,16). One study also found that overall territory density decreased over time (15) and another (10) found that territory densities were similar between sites planted with oak *Quercus* spp. saplings and unplanted sites.
- One study from the USA (2) found that productivity of birds was similar in restored and natural forests. Another found that productivity was lower (4).
- A study from the USA (6) found that fast-growing cottonwood forests less than ten years old held more territories and had higher diversity than similarly-aged oak forests.

Background

Some forests are the most complex terrestrial habitats, with countless species interacting. Restoring such complexity is difficult, but there is an ever-increasing amount of research and investment into the area. For example, insurance firms and shipping companies are financing a 25-year project to restore forest ecosystems along the Panama Canal (TEEB 2008); whilst mining companies in Australia are increasingly able to reconstruct the forests they destroyed after they have finished mining at a site (Nichols & Grant 2007).

Trees grow slowly and therefore the effects of forest restoration may not be evident for decades or even longer after restoration begins. Care must therefore be taken when interpreting the results of these studies.

Some studies below describe the effects of riparian forest and buffer strip creation as a habitat type. The effects of riparian buffer strips as a pollution-reducing intervention are discussed in 'Threat: Pollution – Provide buffer strips along rivers and streams'.

Nichols, O. G. & Grant, C. D. (2007) Vertebrate fauna recolonization of restored bauxite mines-key findings from almost 30 years of monitoring and research. *Restoration Ecology*, 15, S116–S126.
The Economics of Ecosystems and Biodiversity (2008) *An interim report*, Banson, Cambridge.

A replicated study in 1979–1981 in jarrah *Eucalyptus marginata* forests in Western Australia, Australia (1), found that approximately 84% of jarrah forest bird species (56 of 67) used 19 areas of forest restored after open-cut mining for bauxite for feeding, resting or breeding. Sixteen species (none jarrah specialists) were only present in low numbers, compared to on three forest plots. Some revegetated areas as young as 4–5 years age, supported similar bird species numbers, densities and diversities as undisturbed forest. Techniques were improved over time, from planting eastern Australian eucalypt species (chosen for timber quality and resistance to jarrah dieback disease and resulting in plantation-like vegetation with a sparse mid- and under-storey, and few ground species) to using a eucalypt mix comprising at least 50% native species, and seeding with native under-storey plants. Application of fresh topsoil promoted a higher diversity of under-storey plants which in turn benefited birds. Plots planted with the same eucalypt species but without under-storey planting or fresh topsoil addition had fewer bird species and lower densities.

A replicated, controlled study from 1989–1993 in four restoration sites (all established in 1989–1990; all 3–5 ha) and one natural site (16 ha) of riparian forest habitat in California, USA (2) found that least Bell's vireos *Vireo bellii pusillus* were slow to colonise restored sites and did so at lower abundances (9 pairs/site for restored sites vs. 41 for natural site) but exhibited similar reproductive output to natural areas when they did (56% nest success and 1.6 fledglings/nest for restored sites vs. 46% and 1.3 for natural sites). Vireos foraged in restored sites from the first growing season but they did not establish territories until small patches of vegetation became characteristic of natural nesting areas (colonisation rate was also correlated with the presence of adjacent mature riparian habitat).

On Bangka Island, Indonesia, a replicated, controlled before-and-after trial in 1992–1995 (3) found that bird species richness and diversity increased over time in eight 4 ha restored former strip mine sites, whilst it remained low in four 4 ha unrestored sites (13 species recorded in restored plots vs. nine in unrestored). After three years, species richness remained lower than in secondary forest (16 species), but many of the species present were forest specialists (absent from unrestored plots). Bird abundance also appeared to increase over time but this result was less certain. Black wattle *Acacia mangium* was planted for restoration in 1992–1994 (400 seedlings/ha) with some trees reaching 3–4 m tall by 1995.

A replicated, controlled study from May–June in 1989–1994 in riparian forest in California, USA (4) found that song sparrow *Melospiza melodia* nesting success, clutch size and density were lower in three restored sites, compared to four natural, mature sites (controls) and one naturally regenerating site (average of 1.5 pairs/ha for regenerating forest vs. 4 pairs/ha for mature forest and 12 pairs/ha for the naturally regenerating site). No differences in nestling mass or fledgling rate were found among stands.

A controlled study from July 1996 until January 1997 in Western Australia, Australia (5), found that bird assemblages were similar, but not identical in a former 300 ha mineral sands mine site, planted with native vegetation in 1977–83 and in a nearby *Banksia* woodland-mixed heath reserve. A total of 603 birds (36 species) were recorded in two areas (9 and 10 ha) of the reserve and 533 (33 species) in two areas (8 ha and 5 ha) of the mine site (28 species common to both). The same common insectivores were present in both areas and at similar abundance. The same species of honeyeaters occurred in both areas but at the mine site larger nectarivores were far more numerous (e.g. two commonest species: 102 vs. 29 in the reserve) whilst small spinebills much less common (e.g. western spinebill *Acanthorhynchus superciliosus*, 7 vs. 61). This was in part due to vegetation differences, but also as nectar-providing shrubs and trees in the restoration area had been planted in clumps, thus allowing the larger honeyeaters to dominate these nectar sources.

A replicated study in the summers of 1996–1997 at 20 restored bottomland forest sites in the Mississippi floodplain in Mississippi and Louisiana, USA (6), found that birds used young (less than ten year-old) restored cottonwood *Populus deltoides* forests more than similarly aged restored oak *Quercus* spp. forests (average of 380–449 territories/100 ha, 14–20 species/plot and Shannon index of 2.0–2.5 for 13 cottonwood stands vs. 257 territories/100 ha, 8 species/plot and Shannon index of 1.5 for seven oak stands). Conservation value (calculated as density multiplied by a conservation priority score) was highest for 5–9 year-old cottonwood stands but did not differ between oak stands and younger (2–4 year-old) cottonwood. Nest survival and predation rates did not differ between forest types, but brood parasitism was higher on 5–9 year-old cottonwood (23% of 580 nests) than young cottonwood (3% of 93 nests) or oak (1% of 152 nests) stands. The slower-growing oak stands were used more by open-country species.

A controlled, replicated study at the same sites as studied in (1), from Feburary–March and July–August in 1992, 1995 and 1998 in jarrah forests in south-western Australia (7) found that bird species richness and diversity was comparable between four mined and restored, and four intact sites, eight years after rehabilitation. Of 70 bird species inhabiting intact jarrah forest, 95% were recorded in the rehabilitated sites at some point in the succession. Community dissimilarity (between mined and intact sites) decreased over time. Bird recolonisation was significantly correlated with vegetation growth. The four rehabilitated sites were established in June–July 1990 by re-contouring the mining pit to natural conditions, ripping the pit floor to reduce compaction, and replacing the topsoil. Local trees and understory species were directly seeded and covered with fertiliser.

A replicated and controlled study between April 2000 and June 2001 in north-east New South Wales, Australia (8), found that eight 1 ha plots of restored eucalyptus forest contained more bird species (average of 19–31 species/plot for eight plots) than cleared plots (8 species/plot for two plots), but not as many as remnant forest patches (43 species/plot for two plots). The number of species found increased with the age of the restored forest, from 19 species/plot in two plots planted in 1998 to 31 species/plot in two restored in the 1950s. Five locally declining species were recorded restoration plots; five others were only recorded in remnant woodland.

A study in October–November 1996–1998 and May 1997 in the Atherton Tablelands, Queensland, Australia (9), found that whilst rainforest specialists were absent, ten species of 'mainly-rainforest' birds were recorded in a corridor of restored rainforest, only two or three years after planting. Counts of these species increased from 1 bird/count in 1996 to 4 birds/count in 1998. Community structure in the restored forest became more similar to rainforest and remnant vegetation sites over time, and the number of fruit-eaters (potentially important for increasing seed dispersal) recorded increased from 2 birds/count in 1996 to 4 birds/count in 1998.

A replicated site comparison trial in 1993–2003 on ten sites along the Sacramento River, California, USA (11), found that 13 of 20 bird species were increasing on plots revegetated as part of riparian reforestation, although abundances did not reach that of plots of remnant forest. Nine of these were also increasing on the remnant plots, with a further three only increasing in remnants. Three species were stable on both plot types and one, lazuli bunting *Passerina amoena*, declined on both (mirroring a regional trend). Restoration focused on revegetating with native trees, shrubs and understory plants, and restoring natural river processes.

A replicated, controlled study from March–May 2001–2 in two riparian oak forest sites in Missouri, USA (10), found that red-winged blackbird *Agelaius phoeniceus* territory area and density were similar between four blocks planted with oak seedlings and two control (unplanted) blocks (1,657–1,852 m^2/territory and 0.6 territories/ha for planted blocks vs. 1,540 m^2 and 0.2). Differences between blocks seeded with redtop grass *Agrostis gigantean* and those not seeded are discussed in 'Grassland restoration and creation'.

A study at Rawcliffe Bridge farm, East Yorkshire, UK (12), recorded 14 bird species in a 3 ha patch of woodland, 10–12 years after planting, including linnet *Carduelis cannabina* and willow warbler *Phylloscopus trochilus*. Three hectares of native broad-leaved woodland, berry-bearing shrubs and Corsican pine were planted on the farm in 1993–1994. Birds on the farm were monitored five times each year from 2003 to 2005, by walking the field boundaries. The number of breeding pairs/ha was estimated from clusters of sightings.

A 2007 study (13) reports on longer-term studies of the jarrah areas described in (1). In some restored plots, avian communities were becoming very similar to that of native (undisturbed) forest sites (with 95% of species recorded) within 10 years of restoration.

A replicated, controlled study from 1999–2005 in eight restored corridor sites (average width 60 m) and five natural sites (3 sites 2–5 ha; 2 sites 260–490 ha) of riparian forest in Queensland, Australia (14) found that restored and natural sites contained comparable numbers of species and community similarity increased over time. Overall, fewer species were found in natural than restored sites (60 vs. 71): 19 species were found only in natural sites; 30 species were found only in restored sites; 41 species were recorded in both sites. Over the study, 55% of the rainforest specialist species were recorded in restored sites. After 4–7 years, communities in restored sites were more similar to natural sites than younger restored sites (0–3 years old). Restoring habitat connectivity between remnant forest patches began in 1998 (50 000 trees planted by 2006) with 1–2 ha re-vegetated each year.

A controlled study in the summers of 1969–2007 in New York State, USA (15), found that species richness increased on a 9.3 ha forest site restored from an agricultural field (17 species increased and nine declined) and on a 10.7 ha site maintained at an early successional stage (an actively managed Christmas tree farm), but stayed constant at a 16.6 ha forest site. Total territory density declined on the restored site (from 96 territories in 1969–1973 to 57 in 1999–2003), although Neotropical migrant territories increased from zero to 30 and woodland species also increased. Territory density increased significantly in the managed plot (breeding pair density increased for 11 species and declined for three). The forest plot exhibited no significant difference in territory density or species richness over time.

A study in a restored koa *Acacia koa* forest in northern Hawaii, USA (16), found that three native birds showed long term population increases, with populations of the common 'amakihi *Hemignathus virens*, the 'i'iwi *Vestiaria coccinea* and the apapane *Mimatione sanguinea* all at least doubling between 1987 and 2007. Densities of 'amakihi were similar to those in closed forest (lower than in open forest), densities of i'iwi and apapane were much lower than in the forests. Three native, endangered species ('akiapola'au *H. munroi*, Hawaii creeper *Oreomystis bairdi* and Hawaii akepa *Loxopus coccineus*) were not seen in enough numbers to be analysed. This study also investigates the impact of grazer exclusion and removal from native vegetation, discussed in 'Threat: Invasive and other problematic species – Reduce adverse habitat alterations by excluding problematic species'.

(1) Nichols, O. G., and Watkins, D. (1984) Bird utilisation of rehabilitated bauxite minesites in Western Australia. *Biological Conservation*, 30, 109–131.
(2) Kus, B. E. (1998) Use of restored riparian habitat by the endangered least Bell's vireo (*Vireo bellii pusillus*). *Restoration Ecology*, 6, 75–82.
(3) Passell, H. D. (2000) Recovery of bird species in minimally restored Indonesian tin strip mines. *Restoration Ecology*, 8, 112–118.
(4) Larison, B., Laymon, S. A., Williams, P. L., and Smith, T. B. (2001) Avian responses to restoration: nest-site selection and reproductive success in song sparrows. *The Auk*, 118, 432–442.
(5) Comer, S. J., and Wooller, R. D. (2002) A comparison of the passerine avifaunas of a rehabilitated minesite and a nearby reserve in south-western Australia. *Emu*, 102, 305–311.
(6) Twedt, D. J., Wilson, R. R., Henne-Kerr, J. L., and Grosshuesch, D. A. (2002) Avian response to bottomland hardwood reforestation: the first 10 years. *Restoration Ecology*, 10, 645–655.
(7) Nichols, O. G., and Nichols, F. M. (2003) Long-term trends in faunal recolonization after bauxite mining in the jarrah forest of southwestern Australia. *Restoration Ecology*, 11, 261–272.
(8) Martin, W. K., Eyears-Chaddock, M., Wilson, B. R., and Lemon, J. (2004) The value of habitat reconstruction to birds at Gunnedah, New South Wales. *Emu*, 104, 177–189.
(9) Jansen, A. (2005) Avian use of restoration plantings along a creek linking rainforest patches on the Atherton Tablelands, north Queensland. *Restoration Ecology*, 13, 275–283.
(10) Furey, M. A., and Burhans, D. E. (2006) Territory selection by upland red-winged blackbirds in experimental restoration plots. *The Wilson Journal of Ornithology*, 118, 391–398.
(11) Gardali, T., Holmes, A. L., Small, S. L., Nur, N., Geupel, G. R., and Golet, G. H. (2006) Abundance patterns of landbirds in restored and remnant riparian forests on the Sacramento River, California, USA. *Restoration Ecology*, 14, 391–403.
(12) Bryson, R. J., Hartwell, G., and Gladwin, R. (2007) Rawcliffe Bridge, arable production and biodiversity, hand in hand. *Aspects of Applied Biology*, 81, 155–160.
(13) Nichols, O. G., and Grant, C. D. (2007) Vertebrate fauna recolonization of restored bauxite mines-key findings from almost 30 years of monitoring and research. *Restoration Ecology*, 15, S116–S126.
(14) Freeman, A. N., Freeman, A. B., and Burchill, S. (2009) Bird use of revegetated sites along a creek connecting rainforest remnants. *Emu*, 109, 331–338.
(15) Brooks, E. W., and Bonter, D. N. (2010) Long-term changes in avian community structure in a successional, forested, and managed plot in a reforesting landscape. *The Wilson Journal of Ornithology*, 122, 288–295.
(16) Camp, R. J., Pratt, T. K., Gorresen, P. M., Jeffrey, J. J., and Woodworth, B. L. (2010) Population trends of forest birds at Hakalau Forest National Wildlife Refuge, Hawai'i. *The Condor*, 112, 196–212.

11.2. Restore or create grasslands

- Of 23 studies found, three from the USA, Canada and Iceland (13,16,22) found that species richness on restored grassland sites was similar to remnant habitats or higher than unrestored sites. One replicated, randomised study from the USA (23) found that bird diversity was lower on restored grassland sites compared to hayfields or pastures, whilst a small American study (2) found that species richness declined at one of two fields restored to grassland from croplands. Three studies from the USA (12,15,21) found that target species used restored grasslands.

- Two studies from the USA (4,8) found that Conservation Reserve Program (CRP) fields held disproportionate proportions of total bird populations, or that local population trends were correlated with the amount of CRP land in the area. Six studies from the USA and UK (1,4,9,10,13,16) found that the abundances or densities of some, or all, species were higher on restored sites compared to unrestored sites, or were comparable to natural habitats. Two studies found lower abundances of species on restored sites compared to unrestored sites, although the authors of one (2) suggest that drought conditions may have confounded the results.

- Five studies from the USA (5,11,14,19,21) found that at least some bird species in restored areas of grassland had higher productivities than birds in unrestored areas; similar or higher productivies than natural habitats; or had high enough productivities to sustain populations. Three studies (3,5,11) found that productivities were lower in restored areas than unrestored, or that productivities on restored sites were too low to sustain populations.
- A replicated study from the USA (6) found that older CRP fields held more nests, but fewer species than young fields. Two replicated American studies (7,18) found no differences in species richness or abundances between CRP fields and riparian filter strips whether they were sown with warm- or cool-season grasses, whilst another (17) found that more grassland specialist species were found on sites sown with non-native species. A replicated study from the USA (20) found no difference in bird densities between sites seeded with redtop grass and those not seeded. A study from Iceland (22) found that very few birds were found on restored sites, unless they were sown with Nootka lupin.

Background

Several agri-environment schemes including the Environmentally Sensitive Area (ESA) arable reversion scheme (England and Wales), The Conservation Reserve Program (CRP) in USA, and Permanent Cover Program (PCP) in Canada, provided financial incentives to landowners for planting grass/legume cover as an alternative to annual crops. Primary or secondary objectives of the schemes were to provide habitat for nature conservation purposes.

This section describes the results from studies examining the restoration or creation of semi-natural or natural grasslands, but not of permanent artificial pastureland. For example, North American prairies and species-rich chalk grasslands in Europe are both discussed below, but the creation of wet grazing pasture in northern Europe is instead discussed in 'Restore or create traditional water meadows'.

A small 1967 study in Maryland, USA (1), investigated the impact of grassland restoration, as well as other interventions, on northern bobwhites *Colinus virginianus* and found that the population on the farm increased from five to 38 coveys in eight years. This study is described in detail in 'Threat: Agriculture – Plant new hedges'.

A small study over the summers of 1973–1975 in two former corn and soybean fields (16 and 12 ha respectively) in South Dakota, USA (2), found that species richness and abundance declined over the study period, after the fields were planted with six species of native grasses in 1971. In the old corn field, species richness declined from 11 to 5 species and total abundance declined from 80 to 22 individuals/ha. Similarly, total abundance declined from 71 to 28 in the old soybean field but species richness remained at 6–7 species per year. Grasshopper sparrows *Ammodramus savvanarum* were the most abundant birds in the old corn fields whereas dickcissels *Spiza americana* were most abundant in the old soybean field. The author pointed out that the results may have been confounded by drought conditions. Species richness was comparable between restored and mature grasslands.

A replicated, controlled study in 1978–1980 on a 41.5 ha reclaimed coal mine site in West Virginia, USA (3), found that clutch sizes of grasshopper, savanna *Passerculus sandwichensis*, vesper *Pooecetes gramineus* and field sparrows *Spizella pusilla* were similar to those reported for natural grasslands but nest predation rates were high and the main cause of nest failure in all years. Of 185 nests located, 80 (43%) were thought to be predated. Thus, although providing new habitat, low reproductive success suggests that the grassland may not benefit sparrow populations as immigration will be necessary to maintain breeding numbers.

A replicated, controlled study in the summers of 1992–1993 in North Dakota, USA (4), found that the 11 of 18 bird species recorded occurred at higher densities in CRP fields, compared to non-CRP fields. CRP fields only covered 7% of land in North Dakota but supported a disproportionate amount of the total state populations of sedge wren *Cistothorus platensis*

(27%), savanna sparrow (23%), grasshopper sparrow (22%), bobolink *Dolichonyx oryzivorus* (19%) and lark bunting *Calamospiza melanocorys* (18%).

A replicated, controlled study from April–June in 1990–1991 in Kansas, USA (5), found that eastern meadowlark *Sturnella magna* nest survival and productivity did not differ between four CRP fields (18–24 ha) planted with six native grass species in 1988–1989 and eight rangeland sites (14–259 ha) with a history of spring burning (93 and 95% daily survival rate; 1.9 and 0.7 fledglings/female for CRP and rangeland fields respectively). Overall, nest success averaged 14 and 24% for CRP and rangeland fields respectively. Predation was the primary source of nest failure in CRP fields (37 compared to 25% for CRP and rangeland fields respectively). Mowing CRP fields was a source of nest failure and abandonment. Mowing was conducted (without removing cut material) to control weeds in 1990 and one field was mowed only in selected strips.

A replicated study in summer 1992 in 19 CRP fields in Michigan, USA (6), found that more bird nests were found in older fields and that they had higher survival rates than those in younger fields (average of 22–23 nests/field and 28–29% survival for nine 4–5 year-old fields vs. 10 nests/field and 14% survival for three one year-old fields). However, bird species diversity was higher in one-year old fields, compared with four or five year-old fields. Bird abundance varied over time, with no clear pattern. A total of 32 bird species were recorded, with red-winged blackbirds *Agelaius phoeniceus*, song sparrows *Melospiza melodia*, bobolinks and sedge wrens being the most common. The majority of nests (83% of 166) found were red-winged blackbirds'.

A replicated study from May 1991 to March 1995 in an agricultural landscape in Nebraska, USA (7), found that species richness and abundance did not differ between five fields planted with cool-season, non-native, grasses and legumes and five planted with warm-season native grasses (all planted between 1987–1988; all between 20–40 ha). Dickcissels and grasshopper sparrows were the most abundant species during the breeding season (49–78% of total bird abundance). Common yellowthroats *Geothlypis trichas* and sedge wrens were more abundant on warm-season fields. American tree sparrows *Spizella arborea* were the most abundant native species during winter and were more abundant on warm-season fields. Bobolinks were significantly more abundant on cool-season fields, as were meadowlarks *Sturnella* spp. during winter.

A study in Illinois, USA (8), found that the population trends of Henslow's sparrow *Ammodramus henslowii* in 1987–1995 were positively correlated with the proportion of the counties' land in the CRP. Between 1975 and 1995 the Henslow's sparrow population in Illinois declined by over 7% a year, but the three counties with the highest proportion of CRP land (7–9%) showed large (>50% a year) population increases. In total, Henslow's sparrows were recorded in 27 of 102 counties.

A replicated, controlled study in the winters of 1994–1997 on farmland in southern England (9) found that Eurasian skylarks *Alauda arvensis*, corn buntings *Miliaria calandra* and meadow pipits *Anthus pratensis* were consistently more abundant on arable fields reverted to species-rich chalk grassland (36–37 fields surveyed annually) than on land reverted to permanent grassland (71–80 fields sown with agricultural grasses), intensively managed permanent grassland (12–17 fields) or winter wheat (23–33 fields) fields (25–230 birds/km^2 for skylarks on reverted chalk grassland vs. 0–11 birds/km^2 for other field types; 0.9–4.7 birds/km^2 vs. 0–1 birds/km^2 for buntings and 4–6 birds/km^2 vs. 0–4 birds/km^2 for pipits). Densities of rooks *Corvus frugilegus* did not differ across field types. Reverted chalk grassland fields were sown with species such as *Festuca* spp. and *Bromus* spp. grasses.

A replicated, controlled study in spring and summer 1994–1996 on 40 farms in southern England (10) found that arable fields reverted to species-rich chalk grassland consistently held higher densities of Eurasian skylarks than land reverted to permanent grassland (sown with agricultural grasses), intensively managed permanent grassland or winter wheat fields (12.0–22.8 skylarks/km^2 for 16–35 reverted chalk grassland fields vs. 2.6–11.9 skylarks/km^2 for 16–82 fields of other types). Densities of carrion crows *Corvus corone* and rooks *C. frugilegus* were not consistently higher on any field type (1–2 crows/km^2 and 0–14 rooks/km^2 for chalk grassland vs. 0–2 crows/km^2 and 0–89 rooks/km^2 for other fields). Reverted chalk grassland fields were sown with *Festuca* spp. and *Bromus* spp. grasses.

A replicated study from 1993–1995 in a mixed prairie-cropland landscape in Missouri, USA (11), found that some bird species appeared to be able to maintain stable populations on 16 restored grassland fields (eight sown with cool-season and eight with warm-season grasses), while others might not. Productivity exceeded levels necessary for population growth for four grassland species (average of 4. fledglings/nest and 3 female nestlings/nest), but not for two others (average of 3. fledglings/nest and 1 female nestlings/nest). Results were uncertain for one species (average of 4 fledglings/nest and 1 female nestlings/nest). Although large numbers of dickcissels and red-winged blackbirds nested in restored fields, there was little evidence that grass restoration contributed to their population expansion.

A replicated study in the breeding seasons of 1997–1998 at 19 reclaimed coal mine sites, totalling 11,500 ha of grassland, in south-west Indiana, USA (12), found that 200–300 singing male Henslow's sparrows were recorded on unmanaged grassland (density estimates averaging 0.2/ha) but they avoided areas of sparse or short vegetation prevalent in grazed pastures and hayed fields.

A replicated, controlled study from May–July in 1998 in 629 restored grassland sites and 564 cropland sites (distributed amongst 81 eco-region clusters) in prairie habitat in Alberta, Saskatchewan and Manitoba, Canada (13), found that the average species richness was significantly higher in restored grassland (PCP) sites than at cropland sites (2.5 compared to 1.3 species/site). Of the ten most common species, nine were found significantly more frequently at restored grassland sites, whereas just one occurred more frequently in cropland. Amongst restored grasslands, average species richness did not differ between hay and pasture sites but four species occurred more frequently in hay sites and two different species more often on pastures. Restored grasslands (planted in the early 1990s) comprised a combination of wheatgrass *Agropryon* spp., brome grass *Bromus* spp. and alfalfa *Medicago* spp. Cropland sites consisted of annually cultivated fields (mainly wheat).

A replicated, randomised and controlled before-and-after trial study from May–July in 1992–1997 in North Dakota, South Dakota and Montana, USA (14), found that the nest success and reproductive rate of mallard *Anas platyrhynchos*, gadwall *A. strepera*, blue-winged teals *A. discors*, northern shovelers *A. clypeata* and northern pintails *A. acuta* on 335 10.4 km^2 plots increased significantly following the conversion of plots to perennial grassland by 1992. Estimated nest success and recruitment rates of the five species were 46% and 30% higher than in croplands. Mallard and blue-winged teal showed the largest (38 and 32% respectively) and gadwall showed the smallest (21%) gains in recruitment rate after restoration. Nest success was positively correlated with total planted grass cover in plots for all species except northern shoveler. Daily survival rate of duck nests in croplands from a pre-existing dataset from 1980–1984 (pre-restoration) and 1990–1994 (restoration) were used to compare treatment effects.

A replicated study in May–July 1997–1998 in south-west Indiana, USA (15), found that a total of 28 breeding bird species were recorded on 19 reclaimed surface coal mine grasslands. The 20 'common species' (i.e. present at 68–100% of sites), included five grassland specialists. Red-winged blackbird, eastern meadowlark *Sturnella magna* and grasshopper sparrow were most abundant (the latter two being grassland specialists). Seven other grassland species were present (11–42% of sites) but were uncommon. Sites were 110–3,180 ha and seeded with non-native Eurasian grasses.

A replicated, controlled study from May–July in 1999–2000 in ten grasslands restored in 1987 and ten native tallgrass prairie fields in an agricultural landscape in Iowa, USA (16), found that bird species richness was similar between restored and native habitats (average of 7 species/site). Densities of eight common bird species were similar over the study period except for grasshopper sparrows and savanna sparrows, which were higher in restored grasslands (both 0.1 males/ha in native grassland vs. 0.7 and 0.3 males/ha in restored grasslands). Most species had lower densities in landscapes with high edge habitat density.

Grasslands contained both warm-season (switchgrass *Panicum virgatum*, big bluestem *Andropogong erardii* or both) and cool-season grass plantings (smooth brome *Bromus inermis* or grass-alfalfa *Medicago sativa* mixtures). Restored fields averaged 57 ha and prairie fields 54 ha.

A replicated study at 19 reclaimed mine sites in south-west Indiana, USA (17), found that grassland specialist species (e.g. grasshopper sparrow and Henslow's sparrow) were found in greater breeding abundance at sites dominated by non-native grasses and were less common on those rich in forbs. Non-specialist bird species showed no significant preference.

A replicated study from May–July in 2001–2002 in 33 corn and soybean fields containing riparian filter-strips (all ≥200 m long and >1 km apart) in Iowa, USA (18), found that species richness and abundance of grassland birds was similar between strips planted with warm-season grasses, compared to cool-season grasses, but that strips next to woody streamside vegetation held fewer species (6/site) than those with adjacent non-woody vegetation (8 species/site). Nest success was low in all treatments (only 27% of nests fledged at least one nestling) due chiefly to predation (85% of all nests). Cool-season species included brome grass, orchard grass *Dactylis glomerata* and timothy *Phleum pratense*; warm-season strips were planted mainly with switchgrass *Panicum virgatum*.

A replicated controlled study in the breeding seasons of 2000–1 in two restored-mine grasslands and two control (unmined) grasslands (8–18 ha) in Kentucky, USA (19), found that Henslow's sparrow territory size was generally smaller on restored grasslands (on 1 May, 0.29 ha/territory for 25 nests on restored sites vs. 0.34 ha/territory for 18 nests on control sites; 15 July: 0.33 ha/territory for 25 nests vs. 0.45 for 17). Average clutch size (3.8) and average number of fledglings per nest (of 48 nests, 28 fledged at least one young) were similar between sites. More insect prey was present on the reclaimed sites.

A replicated, controlled study from March–May 2001–2 in two riparian oak forest sites in Missouri, USA (20), found that red-winged blackbird territory area and density were similar between four blocks planted with oak seedlings and seeded with redtop grass *Agrostis gigantean* and those not seeded (1,657 m^2/territory and 0.6 territories/ha for two seeded blocks vs. 1,852 and 0.6 for unseeded blocks). This study is discussed in more detail in 'Restore or create forests'.

A study in 1999–2000 on the reclaimed Chinook mine (39–67 ha) and Universal mine sites in south-west Indiana, USA (21), found 465 bird nests of 31 species at Chinook and 446 at Universal. Red-winged blackbird, eastern meadowlark, field sparrow, dickcissel, grasshopper sparrow and Henslow's sparrow were the commonest nesting birds. Reproductive success (i.e. nests that fledged young) of key species, e.g. Henslow's sparrow (9 of 21 nests fledged young) and grasshopper sparrow (26 of 41), and of several other species was comparable with that in non-mined grassland habitats. Both sites were seeded with (mostly) non-native grasses and were situated in open grassland; shrub/savanna; and grassland with patches of shrubs.

A before-and-after study in southern Iceland in 2009 (22) found that eight species of birds were found in the study site following the revegetating of 'sandplains', but no birds were found in barren sandplains or strips of lyme grass *Leymus arenarius*. Meadow pipits *Anthus pratensis*, common snipe *Gallinago gallinago* and redshank *Tringa totanus* were the most common species and found at highest abundances in dense areas of Nootka lupin *Lupinus nootkathensis* (210 meadow pipits/km^2, 46 snipe/km^2 and 19 redshank/km^2 vs. 83 pipits/km^2 and 13 snipe/km^2 in areas of scattered lupins). Only meadow pipits were found in any habitat without lupins. Revegetation began in earnest in 1988, when areas were sown with lyme grass (65 kg seeds/ha), followed by lupin strips from 1992 (4 kg seeds/ha) and non-native grasses. All treatments except lupins were also fertilised.

A replicated, randomised, controlled study from May–July in 2004–2005 in Kansas and Oklahoma, USA (23), found that overall bird diversity and evenness was significantly higher in ten native prairie hayfields (both burned and unburned) and 18 grazed pastures than eight grass-restored fields. Seven species were recorded and three analysed: dickcissel density was highest in restored fields but nest success was highest and nest parasitism lowest in unburned hayfields (48% compared to 16% on average in other sites). Conversely, grasshopper sparrow density was highest in grazed pastures but nest success was lowest in these pastures and highest in burned hayfields (57% compared to 12% on average in other sites). Management did not influence density and nest survival of eastern meadowlarks, which were uniformly low across the region.

(1) Burger, G. V., and Linduska, J. P. (1967) Habitat management related to bobwhite populations at Remington farms. *The Journal of Wildlife Management*, 31, 1–12.

(2) Blankespoor, G. W. (1980) Prairie restoration: Effects on nongamebirds. *The Journal of Wildlife Management*, 44, 667–672.

(3) Wray, T., Strait, K. A., and Whitmore, R. C. (1982) Reproductive success of grassland sparrows on a reclaimed surface mine in West Virginia. *The Auk*, 99, 157–164.

(4) Johnson, D. H., and Igl, L. D. (1995) Contributions of the Conservation Reserve Program to populations of breeding birds in North Dakota. *The Wilson Bulletin*, 107, 709–718.

(5) Granfors, D. A., Church, K. E., and Smith, L. M. (1996) Eastern meadowlarks nesting in rangelands and Conservation Reserve Program fields in Kansas. *Journal of Field Ornithology*, 67, 222–235.

(6) Millenbah, K. F., Winterstein, S. R., Campa III, H., Furrow, L. T., and Minnis, R. B. (1996) Effects of Conservation Reserve Program field age on avian relative abundance, diversity, and productivity. *The Wilson Bulletin*, 108, 760–770.

(7) Delisle, J. M., and Savidge, J. A. (1997) Avian use and vegetation characteristics of Conservation Reserve Program fields. *The Journal of Wildlife Management*, 61, 318–325.

(8) Herkert, J. R. (1997) Population trends of the Henslow's sparrow in relation to the Conservation Reserve Program in Illinois, 1975–1995. *Journal of Field Ornithology*, 68, 235–244.

(9) Wakeham-Dawson, A., and Aebischer, N. J. (1998) Factors determining winter densities of birds on environmentally sensitive area arable reversion grassland in southern England, with special reference to skylarks (*Alauda arvensis*). *Agriculture, Ecosystems & Environment*, 70, 189–201.

(10) Wakeham-Dawson, A., Szoszkiewicz, K., Stern, K., and Aebischer, N. J. (1998) Breeding skylarks *Alauda arvensis* on Environmentally Sensitive Area arable reversion grass in southern England: survey-based and experimental determination of density. *Journal of Applied Ecology*, 35, 635–648.

(11) McCoy, T. D., Ryan, M. R., Kurzejeski, E. W., and Burger, L. W. (1999) Conservation reserve program: source or sink habitat for grassland birds in Missouri? *The Journal of Wildlife Management*, 63, 530–538.

(12) Bajema, R. A., DeVault, T. L., Scott, P. E., and Lima, S. L. (2001) Reclaimed coal mine grasslands and their significance for Henslow's sparrows in the American midwest. *The Auk*, 118, 422–431.

(13) McMaster, D. G., and Davis, S. K. (2001) An evaluation of Canada's Permanent Cover Program: Habitat for grassland birds? *Journal of Field Ornithology*, 72, 195–210.

(14) Reynolds, R. E., Shaffer, T. L., Renner, R. W., Newton, W. E., and Batt, B. D. J. (2001) Impact of the Conservation Reserve Program on duck recruitment in the U.S. Prairie Pothole region. *The Journal of Wildlife Management*, 65, 765–780.

(15) DeVault, T. L., Scott, P. E., Bajema, R. A., and Lima, S. L. (2002) Breeding bird communities of reclaimed coal-mine grasslands in the American midwest. *Journal of Field Ornithology*, 73, 268–275.

(16) Fletcher, R. J., and Koford, R. R. (2002) Habitat and landscape associations of breeding birds in native and restored grasslands. *The Journal of Wildlife Management*, 66, 1011–1022.

(17) Scott, P. E., and Lima, S. L. (2004) Exotic grasslands on reclaimed midwestern coal mines: An ornithological perspective. *Weed Technology*, 18, 1518–1521.

(18) Henningsen, J. C., and Best, L. B. (2005) Grassland bird use of riparian filter strips in southeast Iowa. *The Journal of Wildlife Management*, 69, 198–210.

(19) Monroe, M. S., and Ritchison, G. (2005) Breeding biology of Henslow's sparrows on reclaimed coal mine grasslands in Kentucky. *Journal of Field Ornithology*, 76, 143–149.

(20) Furey, M. A., and Burhans, D. E. (2006) Territory selection by upland red-winged blackbirds in experimental restoration plots. *The Wilson Journal of Ornithology*, 118, 391–398.

(21) Galligan, E. W., Devault, T. L., and Lima, S. L. (2006) Nesting success of grassland and savanna birds on reclaimed surface coal mines of the midwestern United States. *The Wilson Journal of Ornithology*, 118, 537–546.

(22) Gunnarsson, T. G., and Indridadottir, G. H. (2009) Effects of sandplain revegetation on avian abundance and diversity at Skogasandur and Myrdalssandur, South Iceland. *Conservation Evidence*, 6, 98–104.

(23) Rahmig, C. J., Jensen, W. E., and With, K. A. (2009) Grassland bird responses to land management in the largest remaining tallgrass prairie. *Conservation Biology*, 23, 420–432.

11.3. Restore or create traditional water meadows

- Four out of five before-and-after studies, all from the UK (2–5), found that the number of waders and wildfowl on sites increased following the restoration of water meadows. One before-and-after study from Sweden (1) found no increase in northern lapwing population following an increase in the area of managed meadows in the study area. This study also found that restored meadows were used less than expected by breeding lapwings.
- A before-and-after study from Sweden (1) found that hatching success of northern lapwings were higher on meadows than on spring-sown crops. There were no differences between meadows and autumn-sown crops or grasslands.

Background

Water meadows are areas of grazing land or hay meadow that have carefully controlled water levels to keep the soil damp. In Europe they provide valuable breeding habitats for waders and other biodiversity. The studies below describe instances when multiple interventions have been used to create water meadows. When the effects of multiple interventions, such as raising water levels and adding foot drains, can be separated, they are discussed under the relevant interventions in 'Threat: Natural system modifications'.

A before-and-after study in 1984–1994 in Västmanland, Sweden (1), found that there was no increase in northern lapwing *Vanellus vanellus* population in the study area despite an increase in the area of managed flood meadows from 163 ha to 530 ha over the study period (approximately 220 pairs in 1985 vs. 200 in 1994; range of 152–297 pairs). Both managed and unmanaged meadows were used less for nesting than expected based on their availability. However, average hatching success was significantly higher in meadows (78–90% for 54 nests in meadows), compared to spring-sown crops (29–50% of 1,236 nests). There were no differences between meadows and autumn sown crops or cultivated grassland (approximately 85% and 75% success respectively). Before 1984, the majority of meadows in the area were overgrown and abandoned.

A before-and-after study of grazing marshes in east England (2) found an increase in breeding wader numbers following a number of interventions. Northern lapwing numbers increased from 19 pairs in 1993 to 85 pairs in 2003 and common redshank *Tringa totanus* rose from four to 63 pairs. Numbers of winter wildfowl also increased over the period and changes in vegetation communities to those more tolerant of inundation occurred. In 1993, water levels were raised by 45 cm. Management included opening up existing footdrains; creating new ones; reconnecting drains to ditches; reducing grazing intensity (from 1.5–2 cattle/ha to 0.7) and stopping fertiliser inputs. From 1995, approximately 600 m of footdrains were opened/year; from 2000 onwards, approximately 2,000 m of footdrains were opened or added.

A study on 84 ha of former arable land adjoining Berney Marshes RSPB Reserve, Norfolk, England (3), found that breeding wader numbers increased after the land was restored to grazing marsh: 15–20 pairs of northern lapwing and 5–10 pairs of common redshank were found on the marsh, depending on year. The fields were regularly used for foraging by a large proportion of the estimated 100,000 wintering waterfowl (e.g. Eurasian wigeon *Anas penelope*) using the reserve. The fields were acquired in 1998, water levels were raised, foot drains were added, and grazing by sheep (and then cattle) was introduced. By 2003, plant communities had shifted towards those characteristic of lowland wet grassland.

A before-and-after study on a wetland reserve in Cumbria, England (4), found that the number of common snipe *Gallinago gallinago* nesting in an area of improved peat grassland increased from one pair in 2003 to 11 pairs in both 2004 and 2005 following several interventions including maintaining higher water levels, the initiation of a more intensive grazing regime, the cutting of rush *Juncus* spp. the creation of scrapes for feeding birds.

A before-and-after study on 160 ha of improved grassland at Ynys-hir RSPB reserve, Powys, Wales (5), found that, after a series of management interventions, the population of northern lapwings increased from 10 to 81 pairs and redshank increased from 11 to 29 pairs between 2000 and 2005. Management included chisel ploughing, used on a 2-year rotation (approximately 8 ha in February 2002 and 10 ha areas thereafter) to break up the surface to create small hummocks and divots (see 'Create scrapes and pools in wetlands and wet grasslands'). The water level was also increased and a seasonal sheep and cattle grazing regime introduced.

(1) Berg, A., Jonsson, M., Lindberg, T., and Källebrink, K. G. (2002) Population dynamics and reproduction of northern lapwings *Vanellus vanellus* in a meadow restoration area in central Sweden. *Ibis*, 144, E131–E140.
(2) Smart, M., and Coutts, K. (2004) Footdrain management to enhance habitat for breeding waders on lowland wet grassland at Buckenham and Cantley Marshes, Mid-Yare RSPB Reserve, Norfolk, England. *Conservation Evidence*, 1, 16–19.
(3) Lyons, G., and Ausden, M. (2005) Raising water levels to revert arable land to grazing marsh at Berney Marshes RSPB Reserve, Norfolk, England. *Conservation Evidence*, 2, 47–49.
(4) Holton, N., and Allcorn, R. I. (2006) The effectiveness of opening up rush patches on encouraging breeding common snipe *Galliango gallinago* at Rogersceugh Farm, Campfield Marsh RSPB reserve, Cumbria, England. *Conservation Evidence*, 3, 79–80.
(5) Squires, R., and Allcorn, R. I. (2006) The effect of chisel ploughing to create nesting habitat for breeding lapwings *Vanellus vanellus* at Ynys-Hir RSPB reserve, Powys, Wales. *Conservation Evidence*, 3, 77–78.

11.4. Restore or create shrubland

- Only one of the four studies captured investigated the effects of shrubland restoration in isolation. This small before-and-after study from the UK (2) found that one or two pairs of northern lapwing bred on an area of restored moorland, whereas none had previously bred in the area.
- A study from the USA (1) and one from the Azores (3) found that populations of target species (gamebirds and seabirds) increased following shrubland restoration, amongst other interventions.
- A replicated study from the UK (4) which did not distinguish between several interventions performed found a negative relationship between the combined intervention and the ratio of young-to-old grey partridges.

Background

Shrublands are extremely diverse habitats, being found from subantarctic regions, through temperate climates to tropical dry and moist shrublands such as karoo (in South Africa) and restingas (in Brazil). They are also found from sea level up to beyond the tree line in mountains. However, we found relatively few studies describing the effects of shrubland restoration on bird populations, so we have not subdivided the studies further.

A small 1967 study in Maryland, USA (1), investigated the impact of planting areas of shrub, as well as other interventions, on northern bobwhites *Colinus virginianus* and found that the population on the farm increased from five to 38 coveys in eight years. This study is described in detail in 'Threat: Agriculture – Plant new hedges'.

A small before-and-after study on an area of purple moor grass *Molina caerulea* dominated moorland in northern England (2) in 2004–5 found that one or two pairs of northern lapwing *Vanellus vanellus* bred on an area of restored moorland, whereas none had previously bred in the area. The moorland was mowed and flailed in 2004, which encouraged grass re-growth and subsequent heavy grazing by both livestock and wild red deer *Cervus elaphus*.

A before-and-after study on Praia Islet (12 ha), off Graciosa, Azores, Portugal (3), found that the breeding populations of common terns *Sterna hiundo*, roseate terns *S. dougallii* and

Madeiran storm petrel *Oceanodroma castro* increased dramatically after European rabbit *Oryctolagus cuniculus* eradication and subsequent habitat restoration. Restoration included the planting of native shrubs, the removal of non-native species and the control of soil erosion. This study is also discussed in 'Provide artificial nesting sites' for ground-nesting and burrow-nesting seabirds.

A replicated site comparison study on 1,031 agricultural sites across England in 2004–8 (4) investigated the impact of scrub restoration on grey partridge *Perdix perdix*. However, the study does not distinguish between the impacts of scrub restoration, scrub control, rough grazing and the restoration of various other semi-natural habitats. There was a negative relationship between the combined intervention and the ratio of young to old partridges in 2008.

(1) Burger, G. V., and Linduska, J. P. (1967) Habitat management related to bobwhite populations at Remington farms. *The Journal of Wildlife Management*, 31, 1–12.
(2) Smith, D., and Bird, J. (2005) Restoration of degraded *Molinia caerulea* dominated moorland in the Peak District National Park Eastern moorlands, Derbyshire, England. *Conservation Evidence*, 2, 101–102.
(3) Bried, J., Magalhaes, M. C., Bolton, M., Neves, V. C., Bell, E., Pereira, J. C., Aguiar, L., Monteiro, L. R., and Santos, R. S. (2009) Seabird habitat restoration on Praia Islet, Azores Archipelago. *Ecological Restoration*, 27, 27–36.
(4) Ewald, J. A., Aebischer, N. J., Richardson, S. M., Grice, P. V., and Cooke, A. I. (2010) The effect of agri-environment schemes on grey partridges at the farm level in England. *Agriculture, Ecosystems & Environment*, 138, 55–63.

11.5. Restore or create savannas

- We captured no evidence for the effects of savanna restoration or creation on bird populations.

11.6. Restore or create wetlands and marine habitats

11.6.1. Restore or create inland wetlands

- Of eleven studies captured, 11, from the mainland USA (1,2,5,6), Guam (3), Canada (8) and Hawaii (10), found that birds used artificially restored or created wetlands. Two found that rates of use and species richness were similar or higher than on natural wetlands (6,10). One found that use rates were higher than on unrestored wetlands (8).
- Three studies from the USA (5,7) and Puerto Rico (9) found that restored wetlands held lower densities and fewer species of birds than natural wetlands.
- A replicated study from the USA (5) found that least bittern productivity was similar in restored and natural wetlands.
- Two replicated studies examined wetland characteristics: one from the USA (11) found that semi-permanent restored wetlands were used more than temporary or seasonal ones. A study from Hawaii (10) found that larger restored wetlands were used more than smaller sites.

Background

This section includes studies describing the effects of wetland restoration or creation for all wetlands which are not coastal, or do not receive regular influxes of salt water.

A study in 1958–1967 on Squaw Creek National Wildlife Refuge, Missouri, USA (1), found that annual use of the 2,772 ha wetlands (created in 1935) varied from 6–27 million duck-days and from 7–19 million goose-days each year. Management included winter water removal to aerate the soil and eradicate carp, and spring flooding.

A study in 1986 at the Savannah River Site, South Carolina, USA (2), found that up to 94 wood storks *Mycteria americana* and over 210 other wading birds were seen on specially constructed and managed ponds at once. Ponds were created in a 14 ha depression and stocked with fish between 1985 and 1986.

A before-and-after study in 1992 on Guam, South Pacific (3), found that Mariana common moorhens *Gallinula chloropus guami* colonised a newly-created wetland within five months of its creation, with two adults and at least four chicks being seen. The wetland was 20–60 cm deep, 45 m long and up to 27 m wide and created using an excavator in January 1992. Spikerush *Eleocharis dulcis*, water lettuce *Pistia stratiotesm*, taro *Colocasia esculenta* and rusty flatsedge *Cyperus odoratus* were planted, although the taro died, probably because of excessive flooding.

A replicated, controlled study in 1992–1994 in wetlands in the Lake Ontario and St. Lawrence River plains, New York State, USA (4), found lower species richness and densities of wetland birds on restored wetlands compared with natural wetlands (6 species/ha and 15 birds/ha for 18 restored sites vs. 8 and 20 for eight natural sites). This pattern was stronger for wetland dependent species. Restored sites also had community compositions more similar to other restored sites than to natural wetlands. Birds were surveyed with an unlimited-radius point count within each wetland each year during the breeding season.

A replicated study from April–June in 1985–1991 in a 13 ha wetland site in South Carolina, USA (5), found that least bitterns *Ixobrychus exilis* nested at high densities (12 pairs/ha), had a 50% hatching rate and 55% of nests produced fledglings. The author points out that this rate is only slightly lower than that reported for natural wetlands. An average of 2.7 fledglings/nest were produced from an average 3.8 eggs/clutch. Most egg mortality was caused by nest instability.

A replicated, paired site study from May–July in 1997–1998 in 39 pairs of restored and natural wetlands in North and South Dakota, USA (6) found that restored wetlands exhibited equal, and often greater, avian abundance, species richness and diversity. There were no significant differences in overall bird abundance, species richness or diversity; waterfowl breeding pair density or upland species richness between restored and natural wetlands. However, Canada goose *Branta canadensis*, mallard *Anas platyrhynchos*, redhead *Aythya americana*, and ruddy duck *Oxyura jamaicensis* exhibited significantly higher densities on restored wetlands. Total area bird counts were performed four times on each wetland.

A replicated, controlled study from May–July in 2000 in Virginia, USA (7), found that bird species richness and diversity in artificially created wetlands were significantly lower than in natural wetlands (average of 11 species/site for six artificial wetlands vs. 17 for five natural wetlands). Although total bird abundance, and the abundance of wading birds, waterfowl, raptors, aerial feeders or woodpeckers were similar, natural wetlands had significantly higher songbird abundance. In addition, created wetlands exhibited bird communities with significantly lower conservation value (based trophic level and migratory status) but similar average habitat specificity and wetland dependency. All wetlands had similar surrounding habitats and were of similar ages (time since planting for created and since logging for natural wetlands), and sizes (5–15 ha).

A replicated, controlled study in April 1998–1999 on Prince Edward Island, Canada (8), found that six out of eight wildfowl species were found in significantly higher numbers in 22 restored wetlands than in 24 control (unrestored) wetlands. Four species also had significantly more broods at restored sites. Large wetlands, close to rivers and with a large proportion of cattails *Typha* spp. held more species than other sites. All sites were freshwater wetlands, 0.3–6.0 ha in size and restored sites were dredged, starting in 1990, to remove excess organic material.

A small controlled study from 2004–2005 in Toa Baja, Puerto Rico (9) found that fewer bird species were recorded in an 18 ha restored forested wetland than in a natural forested wetland (nine records of five species in restored site vs. 65 records of 16 species in the natural site). In addition, only one species (yellow-faced grassquit *Tiaris olivaceus*) was observed foraging in the restored wetland; 40% of records in the natural site were of foraging birds. Only two records, both of northern waterthrush *Seiurus novaboracensis* were made at a 14 ha control (unrestored) grassland site. The restored wetland was planted with 7,000 *Pterocarpus officinalis* and *Annona glabra* trees during 1997–2000.

A replicated, randomised study in spring from 2004–5 in 28 small restored wetlands in Illinois, USA (10), found that semi-permanent wetlands were used more frequently by waterbirds than temporary or seasonal wetlands and held more waterfowl broods (semi-permanent wetlands were used on 56% of days and held 1.1 broods/ha vs. 37% and 0.2 for seasonal and 7% and zero broods for temporary wetlands). Hydrologic management (passive restoration and management; active restoration through hydraulic engineering but passively managed or actively restored and managed through regulation of hydrologic regime) was the most important variable in explaining bird abundance and distributions. Of the 28 wetlands, 25 were <5 ha in size and 17 were <1 ha. Water birds included waterfowl, wading birds and shorebirds. Wetlands were classified as semi-permanent if there was surface-water throughout growing season; seasonal if there was surface water at the start, and for long periods of the growing season, but not by the end of it; and temporary if surface-water was only found for brief periods throughout the growing season.

A replicated, site comparison study from March 2002 to July 2003 in wetlands in Kohala-Mauna Kea, Hawai'i (11) found that Hawaiian ducks *Anas wyvilliana* used 16 restored wetlands more often than 32 agricultural wetlands, despite the greater availability of the latter. Restored wetlands had a significantly higher occupancy rate than agricultural wetlands (81 vs. 41% of sampled sites) and higher consistency of occupancy (13 vs. 7% of all surveys). Hawaiian ducks preferred wetlands that were larger (>0.23 ha), further from houses and surrounded by more wetland habitat. No wetland within 600 m of a house was occupied. Wetland occupancy was not affected by presence of invasive species or grazing intensity. Wetlands ranged from 0.01–1.30 ha and were surveyed every two months.

(1) Burgess, H. H. (1969) Habitat management on a mid-continent waterfowl refuge. *The Journal of Wildlife Management*, 33, 843–847.
(2) Coulter, M. C., McCort, W. D., and Bryan Jr, A. L. (1987) Creation of artificial foraging habitat for wood storks. *Colonial Waterbirds*, 10, 203–210.
(3) Ritter, M. W., and Sweet, T. M. (1993) Rapid colonization of a human-made wetland by Mariana common moorhen on Guam. *The Wilson Bulletin*, 105, 685–687.
(4) Brown, S. C., and Smith, C. R. (1998) Breeding season bird use of recently restored versus natural wetlands in new york. *The Journal of Wildlife Management*, 62, 1480–1491.
(5) Post, W. (1998) Reproduction of least bitterns in a managed wetland. *Colonial Waterbirds*, 21, 268–273.
(6) Ratti, J. T., Rocklage, A. M., Giudice, J. H., Garton, E. O., and Golner, D. P. (2001) Comparison of avian communities on restored and natural wetlands in North and South Dakota. *The Journal of Wildlife Management*, 65, 676–684.
(7) Snell-Rood, E. C., and Cristol, D. A. (2003) Avian communities of created and natural wetlands: bottomland forests in Virginia. *The Condor*, 105, 303–315.
(8) Stevens, C. E., Gabor, T. S., and Diamond, A. W. (2003) Use of restored small wetlands by breeding waterfowl in Prince Edward Island, Canada. *Restoration Ecology*, 11, 3–12.
(9) Acevedo, M. A. (2007) Bird feeding behavior as a measure of restoration success in a Caribbean forested wetland. *Ornitología Neotropical*, 18, 305–310.
(10) O'Neal, B. J., Heske, E. J., and Stafford, J. D. (2008) Waterbird response to wetlands restored through the Conservation Reserve enhancement program. *The Journal of Wildlife Management*, 72, 654–664.
(11) Uyehara, K. J., Engilis Jr, A., and Dugger, B. D. (2008) Wetland features that influence occupancy by the endangered Hawaiian duck. *Wilson Journal of Ornithology*, 120, 311–319.

11.6.2. Restore or create coastal and intertidal wetlands

- All six studies found, from the USA (1,2,5,6) and UK (3,4), found that target bird species used restored or created wetlands. Two found that numbers and/or diversity were at least as high as in natural wetlands (2,5), one that numbers were higher than in unrestored sites (1). Three found that bird numbers on wetlands increased over time (3,4,6).
- Two studies from the UK found that songbirds (3) and waders (4) decreased following wetland restoration, whilst a study from the USA (1) found that songbirds were more common on unrestored sites than restored wetlands.

Background

Coastal and intertidal wetlands are those that can be inundated with salt water. This can have profound consequences for the vegetation and animal life, as salt-intolerant species are excluded.

This form of habitat creation may form part of coastal protection measures, with habitats such as saltmarshes and mangroves being important in reducing the power of waves and so reducing coastal erosion.

A controlled study in 1972–1976 on two former and one current 'salt hay' farms in Delaware Bay, New Jersey, USA (1), found that there were significantly more species of waders, waterfowl and gulls on the former farms that were inundated by seawater following breaching of the dykes surrounding the farm. Significantly more songbird species were found on the current farm. After the dykes were breached, the plant community changed dramatically, with a 98% increase in saltmarsh plants, an 88% decrease in salt hay species and a 97% increase in surface water.

A controlled study in summer 1993–1994 in *Spartina*-dominated marsh at Barn Island Wildlife Management Area in Connecticut, USA (2), found that wetland birds quickly recolonised tidally-restored marsh, with a 21 ha marsh reopened to tidal exchange in 1982 holding a greater species richness and abundance of birds than three 'reference' marshes (1.2, 8 and 19 ha) and a 12 ha marsh under restoration since around 1984.

A before-and-after study at two sites in Essex, UK (3), found that the number of waders using the sites increased in the first two winters after the surrounding seawall was breached in August 1995. At one site (Tollesbury, 20 ha) the number of waders stabilised after increases in the first two winters, particularly in common redshank *Tringa totanus* and dunlin *Calidris alpine*; whilst the number of songbirds decreased after the first winter. Some species (e.g. knot *Calidris canutus*) did not start using the site until the third winter. At the second site (Orplands, 42 ha), the number of common redshank, dunlin and grey plover *Pluvialis squatarola* increased during the first winter and then the community composition changed across the site, with higher areas holding similar species to adjacent saltmarsh and lower areas being similar to mudflats.

A before-and-after study at Freiston Shore, Lincolnshire, England (4), found that the number of wildfowl and little egrets *Egretta garzetta* using the site increased from 426 and one, respectively to 2,659 and 14 between 2002–2003 and 2005–2006. This followed the breaching of the sea wall at the site, allowing the flooding of 66 ha of land in 2002. By September 2005, 70% of the area was covered in saltmarsh plants. However, the number of waders at the site decreased from 11,012 to 7,799 over the same period, and the authors note that the regular inundation with salt water prevents waders from breeding. Songbirds showed mixed responses: Eurasian skylarks *Alauda arvensis* increased from an average of 16 birds to 121; four species increased by smaller amounts; three species showed uncertain trends; meadow pipits *Anthus pratensis* declined.

A replicated, paired study in January–March 2002 in southern California, USA (5), found that wader diversity was higher in three out of five restored wetlands compared to on paired control (non-degraded) sites, lower on one and similar on a fifth. In addition, total density was as high or higher in four restored sites (although responses varied between species) and behaviour was similar across restored and controlled sites, with over 85% of three of the four species groups observed feeding. Densities of willets *Catoptrophorus semipalmatus* were often higher in restored sites whereas densities of marbled godwits *Limosa fedoa* were often denser in control sites. The authors conclude that wetland restoration should be considered at a landscape scale because each site is beneficial for a different assemblage of species.

A before-and-after study of a large-scale wetland restoration project in Chesapeake Bay, Maryland, USA (6), found that five out six 'priority species' colonised the site before 2005 after restoration began in 2002 (although the project was started in 1998). Snowy egret *Egretta thula*, cattle egret *Bubulcus ibis*, osprey *Pandion haliaetus*, common tern *Sterna hirundo*, and least tern *S. antillarum* all bred, with over 800 pairs of common terns. American black duck *Anas rubripes*, however, had not colonised the site by 2005. The authors note that tern reproductive success was very low, largely because of predation.

(1) Slavin, P., and Shisler, J. K. (1983) Avian utilisation of a tidally restored salt hay farm. *Biological Conservation*, 26, 271–285.
(2) Brawley, A. H., Warren, R. S., and Askins, R. A. (1998) Bird use of restoration and reference marshes within the Barn Island Wildlife Management Area, Atonington, Connecticut, USA. *Environmental Management*, 22, 625–633.
(3) Atkinson, P. W., Crooks, S., Drewitt, A., Grant, A., Rehfisch, M. M., Sharpe, J., and Tyas, C. J. (2004) Managed realignment in the UK – the first 5 years of colonization by birds. *Ibis*, 146, 101–110.
(4) Badley, J., and Allcorn, R. I. (2006) Changes in bird use following the managed realignment at Freiston Shore RSPB Reserve, Lincolnshire, England. *Conservation Evidence*, 3, 102–105.
(5) Armitage, A. R., Jensen, S. M., Yoon, J. E., and Ambrose, R. F. (2007) Wintering shorebird assemblages and behavior in restored tidal wetlands in southern California. *Restoration Ecology*, 15, 139–148.
(6) Erwin, R. M., Miller, J., and Reese, J. G. (2007) Poplar Island environmental restoration project: Challenges in waterbird restoration on an island in Chesapeake Bay. *Ecological Restoration*, 25, 256–262.

11.6.3. Restore or create kelp forests

- A before-and-study in the USA (1) found that the densities of five of the nine bird species analysed increased following kelp forest restoration.

Background

Kelp forests are unique habitats found in mostly cold, nutrient-rich waters. Large kelps such as *Macrocystis* spp. (the 'giant kelps') can grow to 45 m or more, creating 'underwater forests' that provide complex habitats and allow extremely productive ecosystems to flourish. However, they are vulnerable to pollution and ecosystem perturbations caused by, for example, the loss of keystone species such as sea otters *Enhydra lutris* (Jackson *et al.* 2001).

Jackson, J. B. C., Kirby, M. X., Berger, W. H., Bjorndal, K. A., Botsford, L. W., Bourque, B. J., Bradbury, R. H., Cooke, R., Erlandson, J., Estes, J. A. & others. (2001) Historical overfishing and the recent collapse of coastal ecosystems. *Science*, 293, 629–637.

A before-and-after study between 1969–1973 and 1984–1986 on a rocky shoreline in southern California, USA (1), found that shorebirds were significantly more numerous after kelp *Macrocystis pyrifera* forest restoration. Among nine species of shorebird analysed, the density of five (spotted sandpiper *Actitus macularia*, wandering tattler *Heteroscelus incanus*, whimbrel *Numenius phaeopus*, black turnstone *Arenaria melanocephala* and ruddy turnstone *Arenaria interpres*) increased. Territorial species (spotted sandpiper, wandering tattler and whimbrel) were twice as abundant in the second census. Species that do not forage in algal windthrow, such as the black-bellied plover *Pluvialis squatarola*, remained stable over the two census periods. Complete counts of all shorebirds encountered along a 4 km census route were recorded year-round over the years of the two censuses.

(1) Bradley, R. A., and Bradley, D. W. (1993) Wintering shorebirds increase after kelp (*Macrocystis*) recovery. *The Condor*, 95, 372–376.

11.6.4. Restore or create lagoons

- A before-and-after study in the UK (1) found that large numbers of bird species used, and bred, in a newly-created lagoon.

Background

Lagoons can form important feeding and overwintering grounds for wildfowl and waders, providing a rich supply of invertebrate food. However, they are often drained to provide farmland, or lost due to coastal management and development.

A before-and-after study in Lincolnshire, England (1), found that a 15 ha saline lagoon created as part of a flood defence scheme in 2002 was subsequently used by 38 species of overwintering waterbird, at least ten of which bred.

(1) Badley, J., and Allcorn, R. I. (2006) The creation of a new saline lagoon as part of a flood defence scheme at Freiston Shore RSPB Reserve, Lincolnshire, England. *Conservation Evidence*, 3, 99–101.

11.6.5. Revegetate gravel pits

- We captured no evidence for the effects of gravel pit revegetation on bird populations.

Background

Whilst mining can be a very damaging activity, it can also create habitats. After quarries and gravel pits become disused, they may offer valuable aquatic, terrestrial and marshland habitats, particularly if they are close to urban areas (Catchpole and Tydeman 1975). These habitats can often be bare immediately after abandonment, so restoring vegetation to them may greatly improve their value as habitats.

Catchpole, C. K. & Tydeman, C. F. (1975) Gravel pits as new wetland habitats for the conservation of breeding bird communities. *Biological Conservation*, **8**, 47–59.

12. Threat: Invasive alien and other problematic species

Invasive species, including animals, plants and diseases, are causing declines in many bird species worldwide. This chapter describes the evidence from interventions designed to reduce the threat from wild or feral animals, or from domestic predators such as cats Felis catus. Interventions to reduce damage from domestic livestock are described in 'Threat: Agriculture'.

Biosecurity: as awareness of the risks posed by invasive species has increased, biosecurity measures have improved in many parts of the world. Trade in, or transport of, many species is now banned or restricted in parts of the world and legislation has been implemented to try and prevent the accidental transport of problematic species. Such measures may reduce the transport of problematic species across the world, but we captured no studies describing the effects of biosecurity on bird populations.

Key messages – reduce predation by other species

Remove or control predators to enhance bird populations and communities

Both a meta-analysis and a systematic review (both global) found that bird reproductive success increased with predator control and that either post-breeding or breeding-season populations increased. The systematic review found that post-breeding success increased with predator control on mainlands, but not islands.

Control avian predators on islands

Five out of seven studies from across the world found that bird populations increased, had higher reproductive success, or lower mortality following avian predator control. However, although two discussed multiple interventions, so changes couldn't be linked to predator control, and one found that no birds returned to the site the following year. All five that reported positive effects were studies of seabird populations. The two that found little evidence studied the effects of crow removal on gamebirds or birds of prey.

Control mammalian predators on islands

Of the 33 studies from across the world, 16 described population increases or recolonisations in at least some of the sites studied and 18 found higher reproductive success or lower mortality (on artificial nests in one case). Two studies that investigated population changes found only partial increases, in black oystercatchers *Haematopus bachmani* and two gamebird species, respectively. Eighteen of the studies investigated rodent control; 12 cat *Felis catus* control and 6 various other predators including pigs *Sus scrofa* and red foxes *Vulpes vulpes*. The two that found only partial increases examined cat, fox and other larger mammal removal.

Control invasive ants on islands

A single study in the USA found that controlling the invasive tropical fire ant *Solenopsis geminata*, but not the big-headed ant *Pheidole megacephala*, led to lower rates of injuries and temporarily higher fledging success than on islands without ant control. The authors note that very few chicks were injured by *P. megacephala* on either experimental or control islands.

Control predators not on islands

A study from the UK found higher bird community breeding densities and fledging success rates in plots with red fox *Vulpes vulpes* and carrion crow *Corvus corone* control. Of the 25 taxa-specific studies, only five found evidence for population increases with predator control, whilst one found a population decrease (with other interventions also used); one found lower or similar survival, probably because birds took bait. Nineteen studies found some evidence for increased reproductive success or decreased predation with predator control, with three studies (including a meta-analysis) finding no evidence for higher reproductive success or predation with predator control or translocation from the study site. One other study found evidence for increases in only three of six species studied. Most studies studied the removal of a number of different mammals, although several also removed bird predators, mostly carrion crows and gulls *Larus* spp.

Reduce predation by translocating predators

Two studies from France and the USA found local population increases or reduced predation following the translocation of predators away from an area.

Key messages – reduce incidental mortality during predator eradication or control

Do birds take bait designed for pest control?

Two studies from New Zealand and Australia, one *ex situ*, found no evidence that birds took bait meant for pest control.

Distribute poison bait using dispensers

A study from New Zealand found that South Island robin survival was higher when bait for rats and mice was dispensed from feeders, compared to being scattered.

Use repellents on baits

A study in New Zealand found that repellents reduced the rate of pecking at baits by North Island robins. A study from the USA found that treating bait with repellents did not reduce consumption by American kestrels.

Use coloured baits to reduce accidental mortality during predator control

Two out of three studies found that dyed baits were consumed at lower rates by songbirds and kestrels. An *ex situ* study from Australia found that dyeing food did not reduce its consumption by bush thick-knees.

Key messages – reduce nest predation by excluding predators from nests or nesting areas

Physically protect nests from predators using non-electric fencing

Two of four studies from the UK and the USA found that fewer nests failed or were predated when predator exclusion fences were erected. Two studies found that nesting and fledging success was no higher when fences were used, one found that hatching success was higher.

Protect bird nests using electric fencing

Two of six studies found increased numbers of terns or tern nests following the erection of an electric fence around colonies. Five studies found higher survival or productivity of waders or seabirds when electric fences were used and one found lower predation by mammals inside electric fences. One study found that predation by birds was higher inside electric fences.

Physically protect nests with individual exclosures/barriers or provide shelters for chicks

Nine of 23 studies found that fledging rates or productivity were higher for nests protected by individual barriers than for unprotected nests. Two found no higher productivity. Fourteen studies found that hatching rates or survival were higher, or that predation was lower for protected nests. Two found no differences between protected and unprotected nests and one found that adults were harassed by predators at protected nests. One study found that chick shelters were not used much and a review found that some exclosure designs were more effective than others.

Can nest protection increase nest abandonment?

One of four studies (from the USA) found an increase in abandonment after nest exclosures were used. Two studies from the USA and Sweden found no increases in abandonment when exclosures were used and a review from the USA found that some designs were more likely to cause abandonment than others.

Can nest protection increase predation of adults and chicks?

Four of five studies from the USA and Sweden found that predation on chicks and adults was higher when exclosures were used. One of these found that adults were harassed when exclosures were installed and the chicks rapidly predated when they were removed. One study from Sweden found that predation was no higher when exclosures were used.

Use artificial nests that discourage predation

Three out of five studies from North America found lower predation rates or higher nesting success for wildfowl in artificial nests, compared with natural nests. An *ex situ* study found that some nest box designs prevented racoons from entering. A study found that wood ducks avoided anti-predator nest boxes but only if given the choice of unaltered nest boxes.

Use multiple barriers to protect nests

One of two studies found that plover fledging success in the USA was no higher when an electric fence was erected around individual nest exclosures, compared to when just the exclosures were present. A study from the USA found that predation on chicks was lower when one of two barriers around nests was removed early, compared to when it was left for three more days.

Use snakeskin to deter mammalian nest predators

A study from the US found that flycatcher nests were predated less frequently if they had a snakeskin wrapped around them.

Use mirrors to deter nest predators

We found no published evidence for the effects of mirrors on nest predation rates.

Use naphthalene to deter mammalian predators

A study from the USA found that predation rates on artificial nests did not differ when naphthalene moth balls were scattered around them.

Use ultrasonic devices to deter cats

We found no evidence for the effects of ultrasonic cat deterrents on bird populations.

Protect nests from ants

A study from the USA found that vireo nests protected from ants with a physical barrier and a chemical repellent had higher fledging success than unprotected nests.

Use 'cat curfews' to reduce predation

We captured no evidence for the effects of 'cat curfews' on bird populations.

Use lion dung to deter domestic cats

We found no evidence for the effects of lion dung application on the use of gardens by cats or on cat predation.

Play spoken-word radio programmes to deter predators

We found no published evidence for the effects of playing the radio on predation rates.

Key messages – reduce mortality by reducing hunting ability or changing predator behaviour

Use collar-mounted devices to reduce predation

Two replicated randomised and controlled studies in the UK and Australia found that fewer birds were returned by cats wearing collars with anti-hunting devices, compared to cats with control collars. No differences were found between different devices.

Use supplementary feeding to reduce predation

One of three studies found that fewer grouse chicks were taken to harrier nests when supplementary food was provided to the harriers, but no effect on grouse adult survival or productivity was found. One study from the USA found reduced predation on artificial nests when supplementary food was provided. Another study from the USA found no such effect.

Use aversive conditioning to reduce nest predation

Nine out of 12 studies found no evidence for aversive conditioning or reduced nest predation after aversive conditioning treatment stopped. Ten studies found reduced consumption of food when it was treated with repellent chemicals, i.e. during the treatment. Three, all studying avian predators, found some evidence for reduced consumption after treatment but these were short-lived trials or the effect disappeared within a year.

Reduce predation by translocating nest boxes

Two European studies found that predation rates were lower for translocated nest boxes than for controls.

Key messages – reduce competition with other species for food and nest sites

Reduce inter-specific competition for nest sites by removing competitor species

Eight of fourteen studies found that target populations increased after nest site competitors were removed, although three used multiple interventions at once. One study found that reintroductions were successful after competitors were removed. Two studies found no impact on target populations after competitors were removed. One study found higher nest box occupancy by the target species after competitor removal, another found no increase. Three studies found lower occupancy or occurrence of the competitor species after control efforts.

Reduce inter-specific competition for nest sites by modifying habitats to exclude competitor species

A study from the USA found that clearing mid-storey vegetation did not reduce the occupancy of red-cockaded woodpecker nesting holes by southern flying squirrels.

Protect nest sites from competitors

Two studies from the USA found that red-cockaded woodpecker populations increased after the installation of 'restrictor plates' around nest holes to prevent larger woodpeckers for enlarging them. Several other interventions were used at the same time. A study from Puerto Rico found lower competition between species after nest boxes were altered. A study from the USA found weak evidence that exclusion devices prevented house sparrows from using nest boxes and another study from the USA found that fitting restrictor plates to red-cockaded woodpecker holes reduced the number that were enlarged by other woodpeckers.

Reduce competition between species by providing nest boxes

A study from the USA found that providing extra nest boxes did not reduce the rate at which common starlings usurped northern flickers from nests.

Reduce inter-specific competition for food by removing or controlling competitor species

Three out of four studies found that at least some of the target species increased following the removal or control of competitor species. Two studies found that some or all target species did not increase, or that there was no change in kleptoparasitic behaviour of competitor species after control efforts.

Key messages – reduce adverse habitat alteration by other species

Reduce adverse habitat alterations by excluding problematic species

Three of four studies from the USA and UK on terrestrial habitats found higher species richness and greater abundances of some songbirds when deer were excluded from forests. A study from Hawaii found mixed effects of deer exclusion. A study from the USA found that more waterbirds used areas of wetlands from which grass carp were excluded, compared to control areas. Preferences decreased over time as exclusion plots became depleted.

Control or remove habitat-altering mammals

Four out of five studies from islands in the Azores and Australia found that seabird populations increased after rabbits or other species were removed, although three studied several interventions at the same time. Two studies from Australia and Madeira found that seabird productivity increased after rabbit and house mouse eradication.

Remove problematic vegetation

One of four studies (from Japan) found an increase in a bird population following the removal of an invasive plant. One study from the USA found lower bird densities in areas where a problematic native species was removed. One study from Australia found the Gould's petrel productivity was higher following the removal of native bird-lime trees, and a study from New Zealand found that Chatham Island oystercatchers could nest in preferable areas of beaches after invasive marram grass was removed.

Use buffer zones to reduce the impact of invasive plant control

A study from the USA found that no snail kite nests (built above water in cattail and bulrush) were lost during herbicide spraying when buffer zones were established around nests.

Key messages – reduce parasitism and disease

Remove/treat endoparasites and diseases

Two out of five studies found that removing endoparasites increased survival in birds and one study found higher productivity in treated birds. Two studies found no evidence, or uncertain evidence, for increases in survival with treatment and one study found lower parasite burdens, but also lower survival in birds treated with antihelmintic drugs.

Exclude or control 'reservoir species' to reduce parasite burdens

One of two studies found increased chick production in grouse when hares (carries of louping ill virus) were culled in the area, although a comment on the paper disputes this finding. A literature review found no compelling evidence for the effects of hare culling on grouse populations.

Remove/treat ectoparasites to increase survival or reproductive success

Three of fourteen studies found that removing ectoparasites from feathers (one study) or nests (two) increased bird survival or lowered rates of nest abandonment. Eight studies found that treating nests for parasites (or providing beneficial nesting material) did not increase survival, productivity or fledging rates. Two studies found nestlings from treated nests were in better condition than those from untreated nests, five found no such differences. Seven studies found lower infestation rates of nests or feathers when they were treated, or beneficial nesting material was provided; one found no difference. One study found that CO_2 was the only gas that effectively removed lice from feathers.

Remove/control adult brood parasites

One of 12 studies, all from the Americas, found that a host species population increased after control of the parasitic cowbird, two studies found no effect. Five studies found higher productivities or success rates when cowbirds were removed, five found that some or all measures of productivity were no different. Eleven studies found that brood parasitism rates were lower after cowbird control.

Remove brood parasite eggs from target species' nests

One of two studies found lower rates of parasitism when cowbird eggs were removed from host nests. One study found that nests from which cowbird eggs were removed had lower success than parasitised nests.

Use false brood parasite eggs to discourage brood parasitism

A study from the USA found that parasitism rates were lower for red-winged blackbird nests with false or real cowbird eggs placed in them, than for control nests.

Alter artificial nest sites to discourage brood parasitism

A replicated trial from Puerto Rico found that brood parasitism levels were extremely high across all nest box designs tested.

Key messages – reduce detrimental impacts of other problematic species

Use copper strips to exclude snails from nests

A study from Mauritius found no mortality from snails invading echo parakeet nests after the installation of copper strips around nest trees. Before installation, four chicks were killed by snails.

Reduce predation by other species

In many parts of the world, predation on bird populations by introduced predators has led to rapid and catastrophic declines. In addition, in some countries, it appears that native predators are negatively affecting birds, potentially through 'meso-predator release' – where the removal of top predators by humans has allowed populations of smaller species to expand, in turn threatening their prey. Reducing predation rates, therefore, is a potentially important method for enhancing declining bird populations.

Perhaps the most obvious way to reduce predation is to eradicate or control predator populations. This has been used in parts of Europe and the USA for a long time and more recently, concerted efforts have been made to remove invasive predators from oceanic islands around the world. However, this technique remains controversial as it is expensive, the benefits are not guaranteed and there are associated animal welfare issues.

In addition, if the species threatening the bird population is threatened or unique, then it may be particularly undesirable to control it, in which case translocating the predator population may be possible.

In some cases, both control and translocation of predators may not be viable options. For example, bird populations may be threatened by large populations of native predators or by domestic pets (with domestic cats accounting for the deaths of an estimated 25–29 million birds every year, Woods *et al.* 2003). In these cases, predators can be physically excluded from areas or deterred from areas using various repellents. Various non-lethal methods of reducing predation can be used, such as attaching collars to domestic cats to reduce the ability of domestic cats to hunt, providing supplementary food for predators, or moving nest boxes around so that predators cannot learn where they are.

Several studies in this section perform experiments using artificial nests: either man-made or genuine nests that have been filled with false or real eggs (normally from quail *Corturnix* spp.) and placed in experimental areas to monitor predation rates. It is important to note that there are questions over the validity of using artificial nests in this way: care must be taken over their placement and to remove all traces of human interference. It is possible that predators may react differently to artificial nests, compared with natural ones. Factors like these need to be taken into consideration when interpreting results from artificial nest studies (Major & Kendal 1996).

Woods, M., Mcdonald, R. A. & Harris, S. (2003) Predation of wildlife by domestic cats *Felis catus* in Great Britain. *Mammal review*, 33, 174–188.

Major, R. E. & Kendal, C. E. (1996) The contribution of artificial nest experiments to understanding avian reproductive success: a review of methods and conclusions, *Ibis*, 138, 298–307.

12.1. Remove or control predators to enhance bird populations and communities

- A meta-analysis (1) and a systematic review (2) both found that reproductive success increased with predator removal, but their exact findings differed.
- The meta-analysis found that post-breeding population size increased, whilst the systematic review found that this was true on mainlands, but not islands and that breeding populations also increased.

A meta-analysis of 20 published studies (1) showed that predator removal had a large, positive effect on hatching success – with removal areas showing higher hatching success on average than 75% of control areas – and led to a significant increase in post-breeding population size (i.e. autumn density), although no significant impact was detected on breeding population size.

A 2010 systematic review (2) found that removing predators tended to lead to increased reproductive success (hatching and fledging success) and breeding populations in birds. On mainlands, but not islands, predator removal also tended to increase post-breeding population size. Whether predators were native or not, the population trend of the bird population and whether the species was migratory or a game species did not affect responses to predator removal.

(1) Côté, I. M. & Sutherland, W. J. (1997) The effectiveness of removing predators to protect bird populations. *Conservation Biology*, 11, 395–405.
(2) Smith, R. K., Pullin, A. S., Stewart, G. B. & Sutherland, W. J. (2010) Effectiveness of predator removal for enhancing bird populations. *Conservation Biology*, 24, 820–829.

Predator control on islands

Due to ecological and evolutionary processes, islands tend to have fewer species than continents meaning that organisms are under different evolutionary pressures. In birds this can lead to traits such as island gigantism, flightlessness and ecological naivety, all of which make island species more vulnerable to introduced predators which they have been isolated from. This has led to the extinction of an estimated 2,000 species from Pacific islands alone (Steadman 1995) and currently 75% of threatened birds on oceanic islands are affected by invasive species (BirdLife 2008).

Conversely, the comparatively small size of islands makes the successful control of problematic species more likely. A 2007 review (Howald *et al.* 2007) found that 284 islands had been successfully cleared of invasive rodents, whilst at least 48 have been cleared of feral cats *Felis cattus* (Nogales *et al.* 2004).

Birdlife International (2008) *State of the world's birds: Indicators for our changing world*. Birdlife International.
Howald, G., Donlan, C., Galván, J. P., Russell, J. C., Parkes, J., Samaniego, A., Wang, Y., Veitch, D., Genovesi, P., Pascal, M., Saunders, A. & Tershy, B. (2007) Invasive rodent eradication on islands. *Conservation Biology*, 21, 1258–1268.
Nogales, M., Martín, A., Tershy, B. R., Donlan, C. J., Veitch, D., Puerta, N., Wood, B. & Alonso, J. (2004) A review of feral cat eradication on islands. *Conservation Biology*, 18, 310–319.
Steadman, D. W. (1995) Prehistoric extinctions of Pacific island birds: biodiversity meets zooarchaeology. *Science*, 267, 1123–1131.

12.2. Control avian predators on islands

- Out of 10 studies, six before-and-after studies from North America (1,4,7), Australia (5) and Europe (8,10) found that controlling avian predators led to increased population sizes (8,10), reduced mortality (5) or increased reproductive success (1,7) in seabirds on islands. The North American studies had several interventions, so increases could not be linked directly to predator control, and one found that increases were only at one of two sites studied (4).
- Two controlled studies in Europe found little evidence that crow control led to increased reproductive success in gamebirds (2) or raptors (6) on islands. A North American study (1) found that, despite higher reproductive success, very few birds returned to the study site after predator removal. A study from North America (3) found that an Atlantic puffin *Fratercula arctica* translocation programme, combined with the culling of predatory gulls, appeared to be successful.
- A study from the UK (9) found that the number of common terns *Sterna hirundo* and black-headed gulls *Larus ridibundus* declined on gravel islands despite the attempted control of large gulls.

Background

Invasive and introduced avian predators can be problematic: islands with more introduced birds have lost more native bird species (Case 1996), but we have captured no evidence of the effects of their control or eradication. However, eight studies describe the effects of controlling or removing native predators on target bird populations.

The control of invasive and introduced competitor species is discussed in 'Reduce competition with other species for nests sites' and 'Reduce competition with other species for food'.

Case, T. J. (1996) Global patterns in the establishment and distribution of exotic birds. *Biological Conservation*, 78, 69–96.

A before-and-after study on an island in Lake Onatario, Canada (1) found that the fledging success of common terns *Sterna hirundo* was significantly higher in May and June 1976 (0.44 chicks/egg laid for 66 eggs) when ring-billed gull *Larus delawarensis* nests were destroyed and vegetation manually removed from the site, than in May and June 1975, when no gull removal was used (0.18 chicks/egg laid for 217 eggs). Fledging success was still higher if late-laid eggs (laid after 4th June 1976) were included in the analysis, although all of these died. Despite the increases, only three pairs of terns returned to the site in 1977.

A controlled study on Karlsøy Island (7.7 km^2), Norway, in 1978–81 (2), found that removing hooded crows *Corvus cornix* and ravens *Corvus corax* from an experimental area did not decrease predation of black grouse *Tetrao tetrix* nests, compared to control areas (49 nests studied). Predation of willow ptarmigan *Lagopus lagopus* nests was lower in the first year of the experiment (21 nests studied) but not in the next three (total of 214 nests). The author suggests that compensatory predation by ermine *Mustela erminea* may have prevented corvid removal from having an effect. Corvids were removed by poisoning with alpha-chloralose-treated eggs and shooting.

An Atlantic puffin *Fratercula arctica* translocation programme in 1973–81 (3) found that almost all of the 774 puffin nestlings moved from Newfoundland, Canada, to Maine, USA, survived. In 1974–5 herring gulls *Larus argentatus* and great black-backed gulls *L. marinus* were culled and their nests destroyed during 1974–5 in an attempt to reduce predation. This study is discussed in 'Translocate individuals'.

A before-and-after studies during 1977–89 at a common tern *Sterna hirundo* colony in Lake Ontario, Canada (4), found the nesting population increased at one colony but decreased at another following several interventions, including the control of particular ring-billed gulls *L. delawarensis* that were heavily predating tern eggs. This study is discussed in 'Replace nesting substrate following severe weather'.

A before-and-after study on Cabbage Tree Island, Australia, between 1992 and 1994 (5) found that fewer Gould's petrels *Pterodroma leucoptera leucoptera* were killed by pied currawongs *Strepera graculina* following intensive control at the start of the 1993–4 breeding season. Twenty petrels were killed by currawongs in October–November 1992 and 43 more were found (83% of all adult mortalities recorded) over the 1992–3 breeding season, despite the destruction of seven currawong nests. However, only four petrel mortalities (25% of the total) could be attributed to currawongs in 1993–4 following the killing of 22 currawongs, leaving a population of four to six individuals of which none bred. This study is also discussed in 'Remove problematic vegetation'.

A randomised, replicated and controlled study in 1999–2000 on Orkney Mainland, Scotland (6) found that the breeding success of hen harriers *Circus cyaneus* was no different in nine territories where hooded crows *Corvus cornix* (previously *Corone cornix*) were removed, compared to territories without crow removal. The number of clutches/male, clutch size, hatching success and laying date were not affected, although experiments with artificial nests containing chicken eggs showed that predation had been reduced by crow removal (12 of 18 clutches surviving vs. two of 18). A total of 113 crows were removed from the nine territories. This study is also discussed in 'Provide supplementary food to increase reproductive success'.

A before-and-after study on two small islands in the Columbia River Estuary, Oregon, USA (7), found that an entire Caspian tern *Sterna caspia* colony (approximately 8,900 pairs) relocated from Rice Island to East Sand Island between 1999 and 2001. Several interventions were used to encourage movement including the culling of a 'limited number' of glaucous-winged-western gull hybrids (*Larus glaucensis x L. occidentalis*). Reproductive success was also higher on East Island. This study is discussed in detail in 'Move fish-eating birds to reduce conflict with fishermen'.

A before-and-after study on the Isle of May, south-east Scotland over a 23 year period (1975–1998) (8) found that the breeding population of Atlantic puffins *Fratercula arctica* increased from 3,000 to approximately 19,000 breeding pairs during a period of gull control (1972–89). Adult herring gulls *Larus argentatus* and lesser black-backed gulls *L. fuscus* were culled and gull nests destroyed, reducing the population from 17,000 to 2,500 pairs and reducing nesting density by 30%–100%. Following the end of the control in 1989 (and an increase in gull population to 4,100 pairs), the puffin population continued to increase, reaching 42,000 pairs in 1998. Puffin recruitment between 1989 and 1998 was significantly higher in areas with low gull density, or that were maintained as gull-free through the destruction of nests.

A before-and-after study at a former gravel pit in Kent, England (9), found that the number of common terns *Sterna hirundo* and black-headed gulls *Larus ridibundus* declined on gravel islands, despite attempts to remove the nests and eggs of large gulls (e.g. herring gulls *L. argentatus*) in the 1990s and early 2000s. This study is described in detail in 'Reduce interspecific competition for nest sites by removing or excluding competitor species' and 'Provide artificial nesting sites'.

A before-and-after study on Alborán Island (7 ha), southern Spain (10), found that the population of Audouin's gulls *Larus audouinii* increased from an average of 181 pairs in 1997–2000 to 626 pairs in 2009, following the control of yellow-legged gulls *Larus michahellis* from 2000 to 2009. Reproductive success also increased, from 0.3 chicks/pair in 1997–2000 to 0.25–0.9 chicks/pair in 2000–9. On average 106 adult yellow-legged gulls were culled each year, approximately 25% of the breeding population. Combined with the destruction of all eggs found, this reduced the population of gulls from 320–440 pairs in 1997–2000 to approximately 100 pairs in 2009.

(1) Morris, R. D., Kirkham, I. R. & Chardine, J. W. (1980) Management of a declining common tern colony. *The Journal of Wildlife Management*, 44, 241–245.

(2) Parker, H. (1984) Effect of corvid removal on reproduction of willow ptarmigan and black grouse. *The Journal of Wildlife Management*, 48, 1197–1205.

(3) Kress, S. W. & Nettleship, D. N. (1988) Re-establishment of Atlantic puffins (*Fratercula arctica*) at a former breeding site in the Gulf of Maine. *Journal of Field Ornithology*, 59, 161–170.

(4) Morris, H., Blokpoel, H. & Tessier, G. D. (1992) Management efforts for the conservation of common tern *Sterna hirundo* colonies in the Great Lakes: two case histories. *Biological Conservation*, 60, 7–14.

(5) Priddel, D. & Carlile, N. (1995) Mortality of adult Gould's petrels *Pterodroma leucoptera leucoptera* at the nesting site on Cabbage Tree Island, New South Wales. *Emu*, 95, 259–264.

(6) Amar, A. & Redpath, S. M. (2002) Determining the cause of the hen harrier decline on the Orkney Islands: an experimental test of two hypotheses. *Animal Conservation*, 5, 21–28.

(7) Roby, D. D., Collis, K., Lyons, D. E., Craig, D. P., Adkins, J. Y., Myers, A. M. & Suryan, R. M. (2002) Effects of colony relocation on diet and productivity of Caspian terns. *The Journal of Wildlife Management*, 66, 662–673.

(8) Finney, S. K., Harris, M. P., Keller, L. F., Elston, D. A., Monaghan, P. & Wanless, S. (2003) Reducing the density of breeding gulls influences the pattern of recruitment of immature Atlantic puffins *Fratercula arctica* to a breeding colony. *Journal of Applied Ecology*, 40, 545–552.

(9) Akers, P. & Allcorn, R. I. (2006) Re-profiling of islands in a gravel pit to improve nesting conditions for terns *Sterna* and small gulls *Larus* at Dungeness RSPB reserve, Kent, England. *Conservation Evidence*, 3, 96–98.

(10) Paracuellos, M. & Nevado, J. (2010) Culling yellow-legged gulls *Larus michahellis* benefits Audouin's gulls *Larus audouinii* at a small and remote colony. *Bird Study*, 57, 26–30.

12.3. Control mammalian predators on islands

- A paired sites study from Finland (1) and a literature review from the UK (2) found increased bird species richness and abundance (1) or population recoveries and recolonisations (2), following the control or eradication of mammalian predators.

- Predators removed included American mink *Mustela vison*, rats *Rattus* spp. pigs *Sus scrofa*, cats *Felis catus*, dogs *Canis familiaris* and grey fox *Dusicyon griseus*.

Background

Humans have introduced mammalian predators to hundreds of islands across the world, most frequently rats Rattus spp., mice Mus spp. and cats *Felis cattus*, but many others as well. These have had a devastating impact on bird populations on islands, with the historic probability of extinction on islands being well correlated with the number of introduced mammal species (Blackburn *et al.* 2004).

Most control efforts for rodents are through poisoning (with a 2007 review suggesting that bait stations are more effective than broadcast baiting, Howald *et al.* 2007), whilst trapping and hunting appear to be more effective for cats (Nogales *et al.* 2004) and larger mammals. However, this synopsis only discusses the impacts of eradication or control on native bird populations, not the details of eradication programmes themselves.

Below are two studies describing the impacts of island predator removal on bird communities. Studies describing the effects on individual taxa are in subsequent sections.

Blackburn, T. M., Cassey, P., Duncan, R. P., Evans, K. L. & Gaston, K. J. (2004) Avian extinction and mammalian introductions on oceanic islands. *Science*, 305, 1955–1958.

Howald, G., Donlan, C., Galván, J. P., Russell, J. C., Parkes, J., Samaniego, A., Wang, Y., Veitch, D., Genovesi, P., Pascal, M., Saunders, A. & Tershy, B. (2007) Invasive rodent eradication on islands. *Conservation Biology*, 21, 1258–1268.

Nogales, M., Martín, A., Tershy, B. R., Donlan, C. J., Veitch, D., Puerta, N., Wood, B. & Alonso, J. (2004) A review of feral cat eradication on islands. *Conservation Biology*, 18, 310–319.

A paired sites before-and-after study on four paired study areas (72–139 km^2) of >60 small islands in Archipelago National Park, south-west Finland (1) found that, following the removal of up to 63 introduced and predatory American mink *Mustela vison* each year in 1992–3 and 1998, experimental areas had significantly higher species richness and abundance, compared to control areas. There was a significant positive relationship between the degree of isolation of the islands and species richness and abundance in control, but not experimental areas. In all areas, larger islands had more pairs and more species.

A 2010 literature review (2) found that all five successful invasive mammal eradication and control programmes on United Kingdom Overseas Territories found native bird population recoveries and/or recolonisations following the programmes. Recovering species included seabirds and songbirds. The impacts of ten more eradication programmes have not been recorded or published, whilst a final eradication attempt (in the British Indian Ocean Territories) failed to remove black rats *Rattus rattus*. Eradicated/controlled species included pigs *Sus scrofa*, cats *Felis catus*, rats *Rattus* spp., dogs *Canis familiaris* and grey fox *Dusicyon griseus*.

(1) Nordstrom, M. & Korpimaki, E. (2004) Effects of island isolation and feral mink removal on bird communities on small islands in the Baltic Sea. *Journal of Animal Ecology*, 73, 424–433.

(2) Hilton, G. M. & Cuthbert, R. J. (2010) The catastrophic impact of invasive mammalian predators on birds of the UK Overseas Territories: a review and synthesis. *Ibis*, 152, 443–458.

12.3.1. Wildfowl

- A before-and-after study from Alaska (1) found that cackling geese *Branta hutchinsii* returned to a total of eight nesting islands between the 1970s and 1991, following the removal of Arctic foxes *Alopex lagopus* from the islands.

A before-and-after study from the Aleutian Islands, Alaska, USA on the cackling goose *Branta hutchinsii* recovery programme (1) found that geese bred on four islands from 1984 onwards after Arctic foxes *Alopex lagopus* were eradicated from them. Together with releases of captive-bred individuals, the programme resulted in the population increasing to 6,000 birds by 1991, compared with fewer than 1,000 in the 1970s, and birds were breeding on eight fox-free islands. The effects of different release techniques are discussed in 'Release captive-bred individuals into the wild to restore or augment wild populations', 'Release birds as adults or subadults, not juveniles' and 'Clip birds' wings on release'.

(1) USFWS (2001) *Aleutian Canada goose road to recovery*. U.S. Fish & Wildlife Service report.

12.3.2. Seabirds

- We found 16 before-and-after studies, one paired sites study and one literature review from around the world, all describing positive seabird responses to the removal or control of mammalian predators (mainly rats *Rattus* spp. and feral cats *Felis catus*) from islands.
- Of these 18 studies, seven (4,5,9,12,16–18) found either large population increases or recolonisations following predator eradication or control. Two of these found only partial population increases or recolonisations: a study from Alaska (4,16).
- Twelve studies found increases in reproductive success and survival (1–3,5,7,8,11,13,15) or decreases in predation and mortality (1,2,5,6,8,10,13,14) following predator control. In one case (13) there was also a small population increase.
- Rats and mice *Mus musculus* were controlled in twelve studies (2,3,5,7–12,15–17), mostly examining burrow-nesting seabirds; cats in eight (1,2,5,6,10,13,14,18), mostly on ground or cliff-nesting seabirds; and other species in two (2,4).

Background
Seabirds frequently nest on the ground, where they are vulnerable to predation by cats *Felis catus* or pigs *Sus scrofa*, or in burrows where rats *Rattus* spp. or even house mice *Mus musculus* can predate chicks, eggs and adults. In addition, seabirds are often poor walkers (as they are specialised for flying and swimming) and cannot evade predators. Because of this, breeding sites are often confined to offshore islands and their populations can be devastated if predators are introduced.

A before-and-after study on the sub-Antarctic Marion Island (290 km^2), South Africa (1), found that breeding success of great-winged petrels *Pterodroma macroptera*, increased from 0–21% to 56–60% following 14 years of cat *Felis catus* control. In addition, no signs of cat predation were found in 1990, but at least 28% of chicks were predated in 1983. Nests were monitored in 1979–80, 1982 and 1984 (between 17 and 53 nests studied) and in 1990 (50 nests). Control consisted of the release of the disease panleucopaenia, shooting and trapping.

Two before-and-after studies on two Galápagos Islands, Ecuador (2), found that dark-rumped petrel *Pterodroma phaeopygia* fledging success increased (on Floreana Island, 173 km^2) and predation of adults decreased (on Santiago Island, 585 km^2) following black rat *Rattus rattus* and feral animal control. On Floreana, fledging success increased (and nestling predation decreased) from 31% in 1981–2 to 46% (1983) and 72% (1984) with control in 1983–4. It declined in 1985 (to 23%) with increased feral cat *Felis catus* predation of young but increased again following cat control to 70–80% in 1986–8. Between 83 and 104 nests were studied each year. On Santiago, 55% of 510 monitored adult petrels were predated in 1985, but only six were in 1986, following an 80% reduction in the feral pig *Sus scrofa* population. Rats were intensively controlled by poisoning with coumatetralyl, whilst pigs, cats, goats *Capra hircus* and donkeys *Equus asnus* were shot. This study is also discussed in 'Provide artificial nesting sites' and 'Use vocalisations to attract birds to safe areas'.

A replicated and controlled paired sites study in 1993–4 in six areas of shrubland on Sand Island, Midway Atoll, Hawaii, USA (3), found that Bonin petrel *Pterodroma hypoleuca* reproductive success was higher in two out of three areas with black rat *Rattus rattus* removal (treatment sites), compared to paired control sites (65–92% success for 144 nests in treatment sites vs. 0–96% for 144 in controls). In addition, the number of nests that failed because of rat predation was significantly lower in treatment sites (overall, 41% of 17 failed nests in treatment areas were due to rats vs. 95% of 41 nests in control areas, 48 nests monitored at each site). Rats were controlled with the rodenticide Bromothalin dispensed from bait stations. The authors note that early in the 1993 breeding season, there was limited poisoning in the control sites as well, which may have resulted in higher than expected breeding success in these sites.

A controlled before-and-after study on Simeonof (4,000 ha) and Chernabura (3,000 ha) Islands in the Shumagin Islands, Alaska (4) found that the average number of pigeon guillemots *Cepphus columba* recorded increased from 28 to 46 individuals on Chernabura, between 1994 and 1995, following the eradication of introduced arctic foxes *Alopex lagopus*. There was no increase on Simeonof Island, with an average of four recorded in both years. Guillemot densities were significantly lower than on islands without foxes.

A before-and-after study on Madeira (741 km^2), Portugal between 1985 and 2000 (5) found that the population of Zino's petrel *Pterodroma madeira* on the five known breeding ledges increased from 12 pairs in 1987 to 29 pairs in 2000, following black rats *Rattus rattus* and cats *Felis catus* control. In 1987 (prior to cat control), ten adult petrels were killed by cats but only four were found dead between 1992 and 2000, of which two had been partially eaten by cats. Rats were controlled using 65 brodifacoum bait boxes, first set in 1986; cats were caught in traps, with eight placed in 1991, increasing to 17 by 2000.

A replicated before-and-after study on Natividad Island (7.4 ha), Baja California Sur, Mexico (6), found that mortality rates of black-vented shearwaters *Puffinus opisthomelas* in four survey plots fell by 90%, from 7.4 carcasses/month (April–July 1997) to 0.7 carcasses/month (March–June 2002) following the eradication of feral cats *Felis catus* from the island. Extrapolated to the entire colony, this suggests a decrease in mortality from over 1,000 birds/month to fewer than 90 birds/month, with each of 25 cats killing approximately 37 birds/month.

A before-and-after study on The Chicken Islands, North Island, New Zealand (7) found that nesting success of Pycroft's petrel *Pterodroma pycrofti* and little shearwater *Puffinus assimilis* increased from 20% in 1992–3 to 75% in 1994–5, following the eradication of Pacific rat *Rattus exulans* from Lady Alice Island (1.4 km^2) in 1994. Rats were eradicated using brodifacoum-impregnated cattle feed.

A before-and-after study on Congreso (25.6 ha), Chafarinas Islands, Spain, between 1997 and 2004 (8) found breeding success of Cory's shearwaters *Calonectris diomedea* at two subcolonies increased when black rats *Rattus rattus* were intensively controlled and fell when control was relaxed. Success was 27–44% and 51% during two periods of little or no control (1997–8, 2000), compared with 70% and 71% during intensive control (2000 and 2002–4). Increases were due to decreased chick mortality from 52% and 23% in 1999 and 2001 to 11% in 2000. The increase in breeding success after control was greater at the more easily accessible colony, although that colony had lower breeding success overall. Between 208 and 280 nests were studied each year.

A before-and-after study on Lundy Island (445 ha), south-west UK (9) found that Manx shearwaters *Puffinus puffinus* and Atlantic puffins *Fratercula arctica* both returned to breed on the island after an absence of 45 and 20 years respectively, following the successful eradication (by poisoning) of brown rats *Rattus norvegicus* and black rats *Rattus rattus* in 2004.

A before-and-after study on Isla Isabelle (194 ha), western Mexico (10) found that mortality rates of adult sooty terns *Onychoprion fuscata* fell from an estimated 23–33% of the nesting population over the breeding season in 1991–4 to 5% in 1996 and then to 2% in 2002–4, following the eradication of cats *Felis catus* through trapping, poisoning (with 1080 sodium monofluoroacetate) and shooting. An attempted eradication of black rats *Rattus rattus* at the same time, using brodifacoum poisoning failed.

A before-and-after study on Mokoli'i Island (1.6 ha), Hawaii, USA (11), found that the reproductive output of wedge-tailed shearwaters *Puffinus pacificus* increased from a single chick between 1999 and 2001 to 126 chicks in 2002 and 185 in 2003, following the eradication of black rats *Rattus rattus* between March and October 2002. Rats were eradicated using diphacinone bait and snap and cage traps.

A controlled before-and-after study on Langara Island (3,270 ha), British Columbia, Canada (12), found that the estimated population of ancient murrelets *Synthliboramphus antiquus* increased from 13,000 (1999) to 24,000 (2004), following the eradication of brown rats *Rattus norvegicus* in 1995 through brodifacoum poisoning. This followed a period of population decline between 1981 and 1999. On islands without eradication programmes (and on Langara Island between 1981–99), ancient murrelet populations declined if rats were present (22% over ten years on Lyell Island, 50% over seven years followed by extirpation on Kunghit Island) and were stable or increasing on islands without rats. A population of Cassin's auklets *Ptychoramphus aleuticus* (previously extirpated on Langara) also re-established following rat eradication.

A before-and-after study over 17 years (1990–2007) on Ascension Island (88 km^2), South Atlantic (13), found that the sooty tern *Onychoprion fuscata* population increased (though not significantly) from 302,000–417,000 birds (1990 and 2001) to 420,000 (2007), following feral cat *Felis catus* eradication in 2002–4. The authors note that no increase would be expected until 2008 due to the breeding cycle of the terns. Predation by cats fell from 33 birds/night (early 1990s) to zero birds/night (2003), and overall nesting success rose from 54% of 233 nests to 68% of 656 nests. Following cat eradication, there was a significant increase in the number of tern chicks being predated by rats, from zero prior to cat eradication (473 days of monitoring) to 46% of 200 chicks ringed in 2005 (40 days of monitoring). Cats were removed through poisoning (with 1080 sodium monofluoroacetate), trapping and other methods (started in 2002, completed 2004).

A before-and-after study on Juan de Nova Island (4.4 km^2), Mozambique Channel in July–August 2006 (14) found that predation rates on sooty terns *Sterna fuscata* fell from an estimated 2,205 terns/week to 416 terns/week following the control of feral cats *Felis catus*. Forty three cats were removed through trapping and shooting, leaving an estimated ten remaining. Predation rates were based on the recovery of 122 tern carcasses, of which 89 (73%) showed signs of predation and consumption and 27 (22%) showed signs of predation but not consumption ('surplus killing').

A before-and-after study on Selvagem Grande (245 ha), Madeira Archipelago, Portugal (15), found that breeding success and productivity of Cory's shearwaters *Calonectris diomedia borealis* was significantly higher, following the eradication of rabbits *Oryctolagus cuniculus* and house mice *Mus musculus* from the island. This paper is discussed in detail in 'Control or remove habitat-altering mammals'.

A review of rat *Rattus* spp. eradication programmes on five islands around the UK between 1959 and 2006 (16) found that bird populations increased on four of the islands: Atlantic puffins *Fratercula arctica* have recolonised Ailsa Craig, south-west Scotland and their range and population increased on Handa, north Scotland. Manx shearwaters *Puffinus puffinus* increased from 700 to 2,000 pairs on Ramsey, west Wales and 308 to 1,120 pairs on Lundy, south-west England (see (9)). However, they did not recolonise Cardigan Island, south-west Wales.

A before-and-after study on Feno Islet (1.6 ha), Azores, Portugal between 1997 and 2009 (17) found that both roseate tern *Sterna dougallii* and common tern *S. hirundo* populations returned to the islet following the eradication of black rats *Rattus rattus* in March–April 2007. Pre-rat populations of roseate and common terns were estimated at 340 and 280 pairs respectively, but none bred in 2005. Common terns returned in 2007 and increased to approximately 120 pairs by 2009. Few roseate terns returned in 2007–8, but 260 pairs bred in 2009. A total of nine rats were trapped on the island.

A before-and-after study on Ascension Island (88 km^2), South Atlantic (18) found that five species of ground-nesting seabird recolonised the island in small but increasing numbers, following the eradication of cats *Felis catus* in 2004: white-tailed tropicbird *Phaethon lepturus* (25 pairs in 2007), red-billed tropicbird *Phaethon aethereus* eight pairs), brown noddy *Anous stolidus* (79 pairs), masked booby *Sula dactylatra* (152 pairs) and brown booby *Sula leucogastor*

(29 pairs). All recolonising populations were small (less than 18% of the population breeding in cat-inaccessible sites in 2002) and breeding success for boobies and brown noddy was lower than other populations. The study also found that sooty tern *Onychoprion fuscata* numbers increased (see reference (13)). Cats were poisoned with 1080 sodium monofluoroacetate and trapped (488 cats poisoned, 70 captured in live traps, nine caught in various methods).

(1) Cooper, J. & Fourie, A. (1991) Improved breeding success of great-winged petrels *Pterodroma macroptera* following control of feral cats *Felis catus* at subantarctic Marion Island. *Bird Conservation International*, 1, 171–175.

(2) Cruz, J. & Cruz, F. (1996) Conservation of the dark-rumped petrel *Pterodroma phaeopygia* of the Galapagos Islands, 1982–1991. *Bird Conservation International*, 6, 23–32.

(3) Seto, N. W. & Conant, S. (1996) The effects of rat (*Rattus rattus*) predation on the reproductive success of the Bonin petrel (*Pterodroma hypoleuca*) on Midway Atoll. *Colonial Waterbirds*, 19, 171–185.

(4) Byrd, G. V., Bailey, E. P. & Stahl, W. (1997) Restoration of island populations of black oystercatchers and pigeon guillemots by removing introduced foxes. *Colonial Waterbirds*, 253–260.

(5) Zino, F., Oliveira, P., King, S., Buckle, A., Biscoito, M., Neves, H. C. & Vasconcelos, A. (2001) Conservation of Zino's petrel *Pterodroma madeira* in the archipelago of Madeira. *Oryx*, 35, 128–136.

(6) Keitt, B. S. & Tershy, B. R. (2003) Cat eradication significantly decreases shearwater mortality. *Animal Conservation*, 6, 307–308.

(7) Parrish, R. (2005) Pacific rat *Rattus exulans* eradication by poison-baiting from the Chickens Islands, New Zealand. *Conservation Evidence*, 2, 74–75.

(8) Igual, J. M., Forero, M. G., Gomez, T., Orueta, J. F. & Oro, D. (2006) Rat control and breeding performance in Cory's shearwater (*Calonectris diomedea*): effects of poisoning effort and habitat features. *Animal Conservation*, 9, 59–65.

(9) Lock, J. (2006) Eradication of brown rats *Rattus norvegicus* and black rats *Rattus rattus* to restore breeding seabird populations on Lundy Island, Devon, England. *Conservation Evidence*, 3, 111–113.

(10) Rodríguez, C., Torres, R. & Drummond, H. (2006) Eradicating introduced mammals from a forested tropical island. *Biological Conservation*, 130, 98–105.

(11) Smith, D. G., Shiinoki, E. K. & VanderWerf, E. A. (2006) Recovery of native species following rat eradication on Mokoli'i Island, O'ahu, Hawai'i. *Pacific Science*, 60, 299–303.

(12) Regher, H. M., Rodway, M. S., Lemon, M. J. F. & Hipfner, J. M. (2007) Recovery of the Ancient Murrelet *Synthliboramphus antiquus* colony on Langara Island, British Columbia, following eradication of invasive rats. *Marine Ornithology*, 35, 137–144.

(13) Hughes, B. J., Martin, G. R. & Reynolds, S. J. (2008) Cats and seabirds: effects of feral domestic cat *Felis silvestris catus* eradication on the population of sooty terns *Onychoprion fuscata* on Ascension Island, South Atlantic. *Ibis*, 150, 122–131.

(14) Peck, D. R., Faulquier, L., Pinet, P., Jaquemet, S. & Le Corre, M. (2008) Feral cat diet and impact on sooty terns at Juan de Nova Island, Mozambique Channel. *Animal Conservation*, 11, 65–74.

(15) Zino, F., Hounsome, M. V., Buckle, A. P. & Biscoito, M. (2008) Was the removal of rabbits and house mice from Selvagem Grande beneficial to the breeding of Cory's shearwaters *Calonectris diomedea borealis*? *Oryx*, 42, 151–154.

(16) Ratcliffe, N., Mitch, I., Varnham, K., Verboven, N. & Higson, P. (2009) How to prioritize rat management for the benefit of petrels: a case study of the UK, Channel Islands and Isle of Man. *Ibis*, 151, 699–708.

(17) Amaral, J., Almeida, S., Sequeira, M. & Neves, V. (2010) Black rat *Rattus rattus* eradication by trapping allows recovery of breeding roseate tern *Sterna dougallii* and common tern *S. hirundo* populations on Feno Islet, the Azores, Portugal. *Conservation Evidence*, 7, 16–20.

(18) Ratcliffe, N., Bell, M., Pelembe, T., Boyle, D., Benjamin, R., White, R., Godley, B., Stevenson, J. & Sanders, S. (2010) The eradication of feral cats from Ascension Island and its subsequent recolonization by seabirds. *Oryx*, 44, 20–29.

12.3.3. Gamebirds

- A single replicated and controlled study on two Swedish islands (1) found that four species of gamebirds had larger broods, and more females had chicks, when predators were controlled. Two of the species also showed population-level responses.

A replicated, controlled study on two islands (18 km^2 and 23.5 km^2) in the Gulf of Bothnia, Sweden between 1976 and 1984 (1) found that gamebird brood sizes were significantly larger and a higher proportion of females had chicks over a four year period when predators were controlled, compared to when predators were not removed (with predator control: 5.5 chicks/

brood, 77% of 378 hens had chicks; without predator control: 3.3 chicks/brood, 59% of 314 hens had chicks). Species studied were capercaillie *Tetrao urogallus*, black grouse *Tetrao tetrix*, hazel grouse *Bonasa bonasia* and willow ptarmigan *Lagopus lagopus*, with adult capercaillie and black grouse counts increasing by 56–80% after predators had been controlled for two years, and counts at leks increasing by 166–174%. Predators (European pine martens *Martes martes* and red foxes *Vulpes vulpes*) were trapped and shot.

(1) Marcstrom, V., Kenward, R. E. & Engren, E. (1988) The impact of predation on boreal tetraonids during vole cycles: an experimental study. *Journal of Animal Ecology*, 57, 859–872.

12.3.4. Rails

- Two before-and-after studies from Australia (1) and the Galapagos Islands (2) found increases in survival or population density of rails on islands following the removal of feral pigs *Sus scrofa*.

Background

Rails have been especially badly affected by introduced predators and other human impacts, with an estimated 800 species becoming extinct in the Pacific alone (Steadman 1995). They have diversified on islands across the world, with many species losing the ability to fly, thus rendering them extremely vulnerable to invasive predators.

Steadman, D. W. (1995) Prehistoric extinctions of Pacific island birds: biodiversity meets zooarchaeology. *Science*, 267, 1123–1131.

A before-and-after study in 1979–84 on Lord Howe Island (56 km^2), Australia (1), found that the Lord Howe Island woodhen *Tricholimnas sylvestris* population increased after feral pigs *Sus scrofa* were controlled and a captive breeding programme was launched. Before pig control, the woodhen population was a maximum of ten breeding pairs, with adult mortality higher than juvenile recruitment. Following pig control, 56 released, captive-bred birds were found to survive for up to two years and 19 young were successfully raised between 1982 and 1984. Additionally, woodhens have started to expand their range from the unfavourable territories they were previously confined to, to more favourable, previously pig-infested regions. A total of 186 were destroyed between 1979 and 1981. This study is also discussed in 'Release captive-bred individuals' and 'Use captive breeding to increase or maintain populations'.

A controlled before-and-after study on Santiago Island (585 km^2), Galapagos, Ecuador (2) found that densities of Galapagos rails *Laterallus spilonotus* increased following the eradication of feral mammals between 1998 and 2006 (279 rails found at 8.5–17.9 rails/ha in 2004–5 vs. 18 rails at 0–1.4 rails/ha in 1986–7). Over the same period, there was a smaller increase in rails detected on Ferdandina Island, which has remained free from invasive mammals (no rails detected during surveys in 1986–7 although some were heard outside survey times, 11 rails detected in 2004–5) and a decrease on Isabella Island, which retains feral goats, pigs and donkeys (13 rails in 247 survey plots at eight sites in 2004–5 vs. 24 rails in 60 survey plots in 1986–7). Donkeys *Equus asinus*, 17,000 pigs *Sus scrofa* and 70,000 goats *Capra hircus* were removed.

(1) Miller, B. & Mullette, K. J. (1985) Rehabilitation of an endangered Australian bird – the Lord Howe Island woodhen *Tricholimnas sylvestris* (Sclater). *Biological Conservation*, 34, 55–95.
(2) Donlan, C., Campbell, K., Cabrera, W., Lavoie, C., Carrion, V. & Cruz, F. (2007) Recovery of the Galápagos rail (*Laterallus spilonotus*) following the removal of invasive mammals. *Biological Conservation*, 138, 520–524.

12.3.5. Waders

- A controlled before-and-after study in New Zealand (2) found that the Chatham Island oystercatcher *Haematopus chathamensis* population increased following the removal of feral cats *Felis catus* and other species.

- A second controlled before-and-after study in Alaska, USA (1), found small increases in black oystercatcher *Haematopus bachmani* breeding populations on two islands, but the overall population only increased on one, declining on the other.

A controlled before-and-after on Simeonof (4,000 ha) and Chernabura (3,000 ha) Islands in the Shumagin Islands, Alaska (1), found that the probable breeding populations of black oystercatchers *Haematopus bachmani* increased following the eradication of introduced arctic foxes *Alopex lagopus* (Simeonof: four pairs in 1994 vs. five in 1995; Chernabura: three and five pairs). Total estimated population increased on Simeonof (34 to 41 birds) but decreased on Chernabura (25 to 19 birds). Oystercatcher densities were significantly lower than on islands without foxes.

A controlled before-and-after study on Chatham Island (899 km^2), New Zealand between 1999 and 2005 (2) found that the number of Chatham Island oystercatchers *Haematopus chathamensis* breeding in a 14 km stretch of beach increased from 16 to 35 pairs over the study period, following the initiation of control (trapping and shooting) of predators, principally feral cats *Felis catus*, but also other introduced mammals and weka *Gallirallus australis*. Birds in the management area fledged 18–35 chicks/year, compared with very low fledging success in unmanaged areas.

(1) Byrd, G. V., Bailey, E. P. & Stahl, W. (1997) Restoration of island populations of black oystercatchers and pigeon guillemots by removing introduced foxes. *Colonial Waterbirds*, 20, 253–260.
(2) Moore, P. (2005) Predator control to increase breeding success of Chatham Island oystercatcher *Haematopus chathamensis*, Chatham Island, New Zealand. *Conservation Evidence*, 2, 80–82.

12.3.6. Raptors

- A study in Mauritius (1) found that numbers of Mauritius kestrel *Falco punctatus* may have increased following the trapping of predators near nests. However, the authors do not provide any data to support this observation.

A study of an integrated conservation programme for the endangered Mauritius kestrel *Falco punctatus* from 1973–1994 in montane forest habitat and a captive breeding centre in Black River, Mauritius (1) found that trapping and removing nest predators may have significantly increased breeding pair productivity. Two introduced predators (small Indian mongoose *Herpestes auropunctatus* and feral cat *Felis catus*) were trapped near some wild nests and at all artificial nestboxes. The authors provide no data on the numbers of predators trapped or experimental data on the effects of their removal.

(1) Jones, C. G., Heck, W., Lewis, R. E., Mungroo, Y., Slade, G. & Cade, T. (1995) The restoration of the Mauritius kestrel *Falco punctatus* population. *Ibis*, 137, S173–S180.

12.3.7. Pigeons

- Two before-and-after trials on Mauritius found that fewer pink pigeon *Columba mayeri* nests were predated (1) and more chicks were fledged (2) following systematic and intensive rat control.

Background

The pink pigeon *Nesoenas mayeri* is an endangered species from Mauritius that declined to just ten wild individuals in 1990 before intensive captive breeding and *in situ* management. The species is still under pressure from several threats, including predation by introduced mammals: crab-eating macaque *Macaca fascicularis*, Indian mongoose *Herpestes auropunctatus*, rats *Rattus* spp. and feral cats *Felis catus* (Reese Lind 1994; Swinnerton 2001).

Reese Lind, C. (1994) Management of the EEP Pink Pigeon *Columba (Nesoenas) mayeri* population. *Dodo: Journal of the Jersey Wildlife Preservation Trust*, 30, 106–113.
Swinnerton, K. (2001) Ecology and conservation of the Pink Pigeon *Columba mayeri* on Mauritius. *Journal of the Jersey Wildlife Preservation Trust*, 37, 99.

A before-and-after trial in 'Pigeon Wood' (mixed forests), Black River Gorges National Park, Mauritius in 1989–91 (1) found that fewer pink pigeon *Nesoenas mayeri* (formerly *Columba mayeri*) nest were predated by rats *Rattus* spp. in 1992 (12% of eight nests predated), compared to in 1989–90 (32% of 22 nests predated, following the initiation of systematic rat control (using brodifacoum bait stations). This study is also discussed in 'Provide supplementary food to increase adult survival', 'Use captive breeding to increase or maintain populations', 'Release captive bred individuals' and 'Provide supplementary food after release' and 'Predator control on islands'.

A before-and-after trial in Brise Fer, Mauritius (2) found that four pink pigeon *Nesoenas mayeri* (formerly *Columba mayeri*) chicks fledged in 2006, during the trialling of a new 'hockey stick' bait station to control rats *Rattus* spp., compared to no successful fledgings in 2005 and 'few to no chicks' in the previous few years.

(1) Jones, C., Swinnerton, K., Taylor, C. & Mungroo, Y. (1992) *The release of captive-bred pink pigeons* Columba mayeri *in native forest on Mauritius*. A progress report July 1987–June 1992. *Dodo*, 28, 92–125.
(2) Tatayah, R., Haverson, P., Wills, D. & Robin, S. (2007) Trial of a new bait station design to improve the efficiency of rat *Rattus* control in forest at Black River Gorges National Park, Mauritius. *Conservation Evidence*, 4, 20–24.

12.3.8. Parrots

- Two before-and-after studies in New Zealand (1,2) found reduced nest predation and successful recolonisation of an island following invasive mammal eradication or control.

Background

The kakapo *Strigops habroptila* is a large, flightless parrot, once common across New Zealand but almost wiped out by the late 20th century due to predation of adults by feral cats *Felis catus*, dogs *Canis familiaris* and mustelids and the loss of eggs and chicks to rats Rattus spp. Starting in the 1970s, the remaining birds were translocated to islands free from cats and populations were very intensively managed (see 'General responses to small and declining populations').

The Kermadec red-crowned parakeet *Cyanoramphus movazelandiae*, also from New Zealand was the first parrot to recolonise an island after predator removal.

A small before-and-after study on Codfish Island (1,500 ha), New Zealand (1) found that none of six kakapo *Strigops habroptilus* nests were lost to rats *Rattus* spp. in 1997, when there was intensive trapping and poisoning of rats close to nests (in conjunction with remotely operated detonators to scare rats, see 'Guard nests to prevent predation'). In contrast, there were potentially unsustainable predation rates before 1997. The study does not fully describe the impact of predator control on the species – the translocations and subsequent population stabilisation would only have been possible with prior control.

A before-and-after study on Raoul Island (2,938 ha), Kermadec Islands, New Zealand (2) found that the island was recolonised by Kermadec red-crowned parakeets *Cyanoramphus movazelandiae* in 2008, following the eradication of goats *Capra hircus* by hunting (in 1986) and cats *Felis cattus*, brown rats *Rattus norvegicus* and black rats *R. rattus* by poisoning and hunting (between 2002 and 2004). In 2008 the parakeet population was at least 100 individuals, of which 44 were born in 2008. Before this, parakeets had been absent for 172 years.

(1) Jansen, W. (2005) Rat *Rattus* control at nests of the endangered kakapo *Strigops habroptilus* on Codfish Island, New Zealand. *Conservation Evidence*, 2, 1–2.

(2) Ortiz-Catedral, L., Ismar, S., Baird, K., Brunton, D. & Hauber, M. (2009) Recolonization of Raoul Island by Kermadec red-crowned parakeets *Cyanoramphus novaezelandiae cyanurus* after eradication of invasive predators, Kermadec Islands archipelago, New Zealand. *Conservation Evidence*, 6, 26–30.

12.3.9. Songbirds

- Two before-and-after trials in the Seychelles (1) and Cook Islands (3) describe population increases in magpie robins and monarch flycatchers following cat and rat control. A before-and-after study from New Zealand (2) found that the population of South Island robins *Petroica australis australis* was almost identical before and after rat control.
- Two studies (3,5) found higher reproductive success in monarch flycatchers and shrikes in areas with rodent control, compared to areas without control. However, this was climate dependent in shrikes.
- A before-and-after study from Hawaii (4) found lower predation on artificial nests after intensive rodent control.

A before-and-after study on Frégate (210 ha), Seychelles, between 1981 and 1984 (1), found that the population of Seychelles magpie robins *Copsychus sechellarum* (a species confined solely to Frégate at the time) increased from 18 to 25 birds following the removal of at least 56 feral cats *Felis cattus* in 1981–2 and the probable eradication of cats from the island. The authors argue that a reduction in foraging habitat due to agricultural abandonment hindered further population growth.

A before-and-after study in 1987–9 on Breaksea Island (170 ha), South Island, New Zealand (2) found an almost identical number of South Island robins *Petroica australis australis* after a rat eradication programme as before. This study is discussed in detail in 'Do birds take bait designed for pest control?'

A paired before-and-after study in Totokoitu Valley, Rarotonga (6,700 ha), Cook Islands (3), found that the nesting success of kakerori (Rarotonga flycatchers) *Pomarea dimidiate* was significantly higher in areas with an intensive black rat *Rattus rattus* poisoning programme (62% success for 71 nests in 1988–93), compared to the same areas in the 1987–8 breeding season (no nesting attempted) and to other areas without a poisoning programme (26% success for 47 nests in 1987–93). In areas with rat control adult mortality also fell significantly, from 24% in 1989–90 to 6% in 1989–93. The population of kakerori increased from 29 birds in 1989 to 60 in 1993. This study also used nest guards, discussed in 'Physically protect nests with individual exclosures/barriers'.

A before-and-after study on O'ahu, Hawaii, USA, in 1998 (4), found that predation rates on artificial O'ahu 'elepaio *Chasiempis sandwichensis ibidis* nests were significantly lower following intensive rodent control compared to before control began. There was a 55% reduction in predation rate for 40 nests placed on the ground and a 45% reduction for 40 nests placed in trees. Egg survival rates over 15 days were approximately 80% following rodent control and 20–40% before control. Survival of artificial nests in trees was not significantly different from genuine 'elapaio nests. Control was through trapping and poisoning (with diphacinone).

A controlled, replicated study on San Clemente Island, California, USA, between 2000 and 2006 (5) found that 172 pairs of San Clemente loggerhead shrikes *Lanius ludovicianus mearnsi* produced 1.1 more fledglings when rodents were controlled in their territories during April–July, compared to control pairs. In drier-than-average years, experimental pairs also raised 1.1 extra fledglings to independence (40 days old). However, there was no effect of rodent control on fledgling success in wetter-than-average years. Management in December–March did not increase either measure of productivity. Rodents were controlled with cholecalciferol rodenticide. This study also investigated the impact of supplementary feeding in addition to rodent control, which is discussed in 'Provide supplementary food to increase reproductive success' and the success of captive-bred individuals, in 'Release captive-bred individuals'.

(1) Watson, J., Warman, C., Todd, D. & Laboudallon, V. (1992) The Seychelles magpie robin *Copsychus sechellarum* – ecology and conservation of an endangered species. *Biological Conservation*, 61, 93–106.
(2) Taylor, R. H. & Thomas, B. W. (1993) Rats eradicated from rugged Breaksea island (170 ha), Fiordland, New Zealand. *Biological Conservation*, 65, 191–198.

(3) Robertson, H. A., Hay, J. R., Saul, E. K. & McCormack, G. V. (1994) Recovery of the kakerori: an endangered forest bird of the Cook Islands. *Conservation Biology*, 8, 1078–1086.
(4) VanderWerf, E. A. (2001) Rodent control decreases predation on artificial nests in O'ahu 'elepaio habitat. *Journal of Field Ornithology*, 72, 448–457.
(5) Heath, S. R., Kershner, E. L., Cooper, D. M., Lynn, S., Turner, J. M., Warnock, N., Farabaugh, S., Brock, K. & Garcelon, D. K. (2008) Rodent control and food supplementation increase productivity of endangered San Clemente Loggerhead Shrikes (*Lanius ludovicianus mearnsi*). *Biological Conservation*, 141, 2506–2515.

12.4. Control invasive ants on islands

- A replicated, randomised and controlled, before-and-after paired sites study (1) in the USA found temporarily increased fledging success, but no decrease in injuries inflicted by *Solenopsis geminata* on wedge-tailed shearwaters *Puffinus pacificus* following ant control. However, there was no change in fledging success or injury rate on an island dominated by the big-headed ant *Pheidole megacephala* following its eradication, either on the experimental or control island.

Background

Several species of ants, including fire ants Solenopsis spp., yellow crazy ants *Anoplolepis gracilipes* and Argentine ants *Linepithema humile* have been spread over the world by human activities. Ants are key species in many ecosystems and can have enormous direct and indirect impacts on native species.

A replicated, randomised and controlled, before-and-after paired sites study over three breeding seasons in 2002–2004 on two pairs of offshore islets (<5 ha) in Hawaii, USA (1), found that wedge-tailed shearwaters *Puffinus pacificus* from an island previously dominated by the invasive tropical fire ant *Solenopsis geminata* showed temporarily increased fledging success, but no decrease in injuries inflicted by *S. geminata* following ant control (27–38% of chicks injured in all seasons), whilst fledging rates remained constant and injuries increased on an untreated island (8% injured in 2002, 80–100% in 2003–4). There was no change in fledging success or injury rate on an island dominated by the big-headed ant *Pheidole megacephala* following its eradication, either on the experimental or control island, but very few chicks were injured by ants. Severely injured chicks (20% of tissue on their feet lost) weighed significantly less than uninjured chicks and did not fledge. Between 15 and 43 chicks were monitored on each islet each year. Following a year of baseline data collection, ant populations were controlled with granular protein-based ant bait in February 2003 on one randomly selected islet of each pair.

(1) Plentovich, S., Hebshi, A. & Conant, S. (2009) Detrimental effects of two widespread invasive ant species on weight and survival of colonial nesting seabirds in the Hawaiian Islands. *Biological Invasions*, 11, 289–298.

12.5. Control predators not on islands

- A single replicated and randomised, paired sites study from the UK (1) found that plots with predator control had increased density and fledgling success of breeding birds.

Background

When predators are controlled on continental sites, rather than islands, the chances of eradicating predators are far smaller and may not be desirable – many of the studies we have captured describe the control of native predators.

We have classed landmasses above approximately 5,000 km^2 as 'non-islands' because the largest island where eradication has so far been attempted (as of July 2011) is South Georgia, South Atlantic, at 3,718 km^2.

One study discussing the impacts of predator control on bird communities is described below; studies describing the effects on individual taxa are in subsequent sections.

A replicated, randomised, paired site study from March–July in 2000–2008 in 2 pairs of plots (9.3–14.4 km^2) in Northumberland, UK (1) found that plots where predators were experimentally controlled displayed increased density and fledgling success of breeding birds. Reductions in foxes *Vulpes vulpes* and carrion crows *Corvus corone* led to an average threefold increase in the percentage of pairs fledging young of lapwing *Vanellus vanellus*, golden plover *Pluvialis apricaria*, curlew *Numenius arquata*, red grouse *Lagopus lagopus scoticus* and meadow pipit *Anthus pratensis*; and subsequently led to increases in breeding numbers (≥14%/year) of lapwing, curlew, golden plover and red grouse, all of which declined in the absence of predator control (≥17%/year). There was no significant effect of predator culling for any wader species. Predator culling reduced the abundance of fox by 43% and crow 78%.

(1) Fletcher, K., Aebischer, N. J., Baines, D., Foster, R. & Hoodless, A. N. (2010) Changes in breeding success and abundance of ground-nesting moorland birds in relation to the experimental deployment of legal predator control. *Journal of Applied Ecology*, 47, 263–272.

12.5.1. Seabirds

- A before-and-after study from New Zealand (2) found an increase in a tern population following intensive trapping of invasive mammals.
- A before-and-after study from Canada (1) found increases in tern fledging success following gull control.

Background

Seabirds frequently nest on the ground, where they are vulnerable to predation by both mammals and other birds. In addition, they are often poor walkers, as they are specialised for flying and swimming, and are poor at evading predators.

A before-and-after study at a common tern *Sterna hirundo* colony in eastern Canada (1) found that fledging success was higher in 1994 when chick-predating gulls (four herring gulls *Larus argentatus* and one great black-backed gull *Larus marinus*) were selectively shot, compared with 1993 and 1995, when no gulls were culled (16% of 115 eggs fledged vs. no chicks fledging from 165 eggs).

A before-and-after study at three sites in northern North Island, New Zealand (2), found that the population of New Zealand fairy terns *Sterna nereis davisae* increased from a low of five breeding pairs in 1987 and an annual decline of 1.5% to between 35 and 40 individuals in 2005 and an annual increase of 1.4%, following the continual trapping of introduced mammalian predators (feral cats *Felis catus*, hedgehogs *Erinaceus europaeus*, stoats *Mustela erminia*, ferrets *M. putorius*, weasels *M. nivalis*, Australian brush-tailed possums *Trichosurus vulpecula* and rats *Rattus* spp.) from 1992 onwards. On average 100 hedgehogs and 12 cats were trapped each year.

(1) Guillemette, M. & Brousseau, P. (2001) Does culling predatory gulls enhance the productivity of breeding common terns? *Journal of Applied Ecology*, 38, 1–8.
(2) Wilson, T. & Hansen, K. (2005) Predator control to enhance breeding success of the New Zealand fairy tern *Sterna nereis davisae*, North Island, New Zealand. *Conservation Evidence*, 2, 89.

12.5.2. Wildfowl

- Six out of seven studies (1–3,5–7), mostly from North America found higher reproductive success of ducks when mammalian predators were removed. A before-and-after study found higher survival of captive-bred brown teal *Anas chlorotis* following feral cat *Felis catus* control (8).
- One meta-analysis (4) from the USA and Canada found that ducks on sites with mammalian predator removal did not have higher reproductive success and trends in reproductive success were no more positive than on sites without predator control.

A controlled study at two prairie and forest sites in Minnesota, USA, between 1959 and 1964 (1) found that using predator control significantly increased the survival of duck nests (59% of 247 nests vs. 29% of 112) and the number of ducklings produced (from 4,858 to 7,571) compared to the same sites when control was not used. The 'survival' of artificial wildfowl nests also increased (81% of 654 nests vs. 34% of 699). The duck species were blue-winged teal *Anas discors*, mallard *A. platyrhynchos*, and gadwall *A. strepera*. Raccoons *Procyon lotor*, striped skunks *Mephitis mephitis* and red fox *Vulpes vulpes* were removed using strychnine-treated eggs and trapping. The authors note that the two sites may not have been far enough apart to prevent predator immigration from control areas to treatment areas, so results may be conservative.

A controlled study on two wetland sites in south Manitoba, Canada, in June 1966 (2) found that a higher proportion of artificial nests survived in an area where nests contained strychnine-treated eggs, than in an area with non-poisoned eggs over a 16 day period (84% survival of 215 nests, with predated nests replaced every four days vs. 66% of 225 nests, predated nests replaced every four days). A total of 33 striped skunks *Mephitis mephitis* and 15 Franklin's ground squirrels *Poliocitellus franklinii* were killed.

A controlled, replicated study in South Dakota, USA, between April and August 1971 (3), found that duck egg hatching success was significantly higher, and more ducklings were produced, on both idle fields and active agricultural land when predators were removed, compared to control sites (with predator removal: 85–92% hatching success of 324 nests, producing 22 ducklings/ha on idle fields and 0.7 on active farmland vs. without predator removal: 51–68% of 245 nests and 4.7 ducklings/ha on idle fields and 0.5 on active farmland). Dabbling duck *Anas* spp., diving duck *Aythya* spp. and ruddy duck *Oxyura jamaicensis* nests were studied. Predators removed through poisoning, trapping and shooting were red fox *Vulpes vulpes*, raccoon *Procyon lotor*, striped skunk *Mephitis mephitis* and American badger *Taxidea taxus*.

A replicated, randomised, paired site study from May–June in 1987–1990 in 15 pairs of waterfowl production areas (61–201 ha) consisting of equal wetland and grassland habitats in Minnesota and North Dakota, USA (4), found that four duck species exhibited higher nest success and daily survival rate in sites where predators were removed. Mean daily survival rate of nests was significantly higher in predator-removal sites than control sites (0.94 compared to 0.91). Mean hatching rate was 13.5% for predator-removal sites and 5.6% for control sites but there was considerable variation in both treatments (1–58% and 1–62% respectively). Nest predation rate was significantly lower in predator-removal sites than control sites (91% compared to 96%). However, hatch rate was not correlated to the number of predators removed. Ducks species analysed were mallard *Anas platyrhynchos*, blue-winged teal *Anas discors*, gadwall *Anas strepera* and northern pintail *Anas acuta*.

A 1996 meta-analysis of 58 studies from the Prairie Pothole Region of the USA and Canada between 1935 and 1992 (5) found that the nesting success of dabbling ducks *Anas* spp. did not differ significantly between sites where predator removal was practiced and those without removal. In addition, there was a significant decline in nesting success over the study period, but the rate of this decline did not differ between sites with predator removal and those without. Two of the studies analysed are described above (1,3).

A randomised, replicated and controlled study in northern North Dakota, USA, in 1995–6 (6) found that survival rates of dabbling duck *Anas* spp. nests was higher on eight mixed agriculture and wetland sites (each 41.5 km^2) with predator removal than on eight sites without predator removal (42% survival for 1,584 nests vs. 23% for 1,122). Trapping and shooting removed 2,404 predators between 1994 and 1996, most of which were raccoons *Procyon lotor*, striped skunks *Mephitis mephitis* and red fox *Vulpes vulpes*.

A replicated study in southern Saskatchewan, Canada, in 2000–1 (7), found that survival rates of mallard *Anas platyrhynchos* ducklings was 41–50% higher in four 41 km^2 grassland-wetland sites where predators were removed (average survival rate of 59% for 686 ducklings from 78 broods), compared with four sites without predator removal (40% survival). Survival was measured until 30 days old, with a total of 686 ducklings from 78 broods studied. A total of 509 predators were removed: red foxes *Vulpes vulpes*, striped skunks *Mephitis mephitis*, raccoons *Procyon lotor*, coyotes *Canis latrans*, American badgers *Taxidea taxus* and American mink *Neovison vison*.

A before-and-after study of a reintroduction programme in the north of North Island, New Zealand (8) found far higher survival rates of captive-bred brown teal *Anas chlorotis* in 2004 than in 2003, following a more extensive feral cat *Felis catus* control programme in between releases. This study is discussed in more detail in 'Release captive-bred individuals'.

(1) Balser, D. S., Dill, H. H. & Nelson, H. K. (1968) Effect of predator reduction on waterfowl nesting success. *The Journal of Wildlife Management*, 32, 669–682.
(2) Lynch, G. M. (1972) Effect of strychnine control on nest predators of dabbling ducks. *The Journal of Wildlife Management*, 36, 436–440.
(3) Duebbert, H. F. & Kantrud, H. A. (1974) Upland duck nesting related to land use and predator reduction. *The Journal of Wildlife Management*, 38, 257–265.
(4) Sargeant, A. B., Sovada, M. A. & Shaffer, T. L. (1995) Seasonal predator removal relative to hatch rate of duck nests in waterfowl production areas. *Wildlife Society Bulletin*, 23, 507–513.
(5) Beauchamp, W. D., Nudds, T. D. & Clark, R. G. (1996) Duck nest success declines with and without predator management. *The Journal of Wildlife Management*, 60, 258–264.
(6) Garrettson, P. R. & Rohwer, F. C. (2001) Effects of mammalian predator removal on production of upland-nesting ducks in North Dakota. *The Journal of Wildlife Management*, 65, 398–405.
(7) Pearse, A. T. & Ratti, J. T. (2004) Effects of predator removal on mallard duckling survival. *The Journal of Wildlife Management*, 68, 342–350.
(8) O'Connor, S. (2005) Captive breeding and release of brown teal *Anas chlorotis* into the Moehau Kiwi Sanctuary, Coromandel, New Zealand. *Conservation Evidence*, 2, 72–73.

12.5.3. Gamebirds

- Four controlled studies in Europe (1,3–5) found increased populations (1,4,5) or productivity (3) of grouse and partridges on sites with predator removal. One study (1) tested multiple interventions simultaneously.
- A fifth replicated UK study (2) found no increase in grouse densities or reproductive success on sites with gamekeepers, compared to those without.

A replicated, controlled study in the spring of 1974 on a cereal farm in Villers-Cotterêts, France (1), found that grey partridges *Perdix perdix* were significantly more abundant in areas provided with 'partridge cafeterias' than in control areas. These 'cafeterias' included stoat *Mustela ermine* box-traps. This study is discussed in 'Provide supplementary food to increase adult survival'.

A replicated study (2) at five sites in northern England and Scotland, found that densities and breeding success of black grouse *Tetrao tetrix* was no higher on moors with a gamekeeper (and associated predator control) than on moors without a game keeper (average of 1.6 males/km^2, 2.3 females/km^2 and 1.8 chicks/female on 11 keepered moors vs. 1.9 males/km^2, 2.7 females/km^2 and 1.9 chicks/female on nine unkeepered moors, n = 9). Moors were an average of 50 km^2.

A replicated, controlled study at two farmland and woodland sites in southern England between 1985 and 1990 (3) found that grey partridge *Perdix perdix* breeding success and brood sizes were significantly higher when predators were controlled, compared to years without removal. This led to August partridge numbers being 75% higher and breeding numbers the next year being 36% higher. Over three years this led to breeding densities that were 2.6 times greater when predators were removed. Predators removed through trapping and shooting were predominantly red foxes *Vulpes vulpes*, carrion crows *Corvus corone* and black-billed magpies *Pica pica*.

A controlled before-and-after study years in northern Scotland between 1989 and 1999 (4) found that the breeding productivity of western capercaillie *Tetrao urogallus* and 'survival' rates of 48 artificial nests were higher during the last three years (1994–6) of predator removal, compared to nine sites without predator removal. However, in the previous two years of predator removal (1992–3) and years without removal (1989–91, 1997–9), productivity was lower on the experimental site. In non-removal years, productivity averaged 0.1 chicks/female, compared with 1.4 chicks/female in removal years. Predator removal involved trapping carrion crows *Corvus corone* (a total of 368) and shooting red foxes *Vulpes vulpes* (a total of 22 adults and 52 cubs).

A controlled study in 2002–9 on mixed farmland in Hertfordshire, England (5), found that the number of grey partridges *Perdix perdix* increased significantly on an experimental site, where predators were controlled (along with several other interventions), but only slightly on a control site without predator control. This increase was apparent in spring (from fewer than 3 pairs/km^2 in 2002 to 12 pairs/km^2 in 2009, with a high of 18 pairs/km^2 vs. approximately 1 pair/km^2 on the control site in 2002, increasing to approximately 4 pairs/km^2 in 2009) and autumn (from fewer than 10 birds/km^2 in 2002 to approximately 65 birds/km^2 in 2009, with a high of 85 birds/km^2 vs. approximately 4 birds/km^2 on the control site in 2002, increasing to approximately 15 birds/km^2 in 2009). Predators controlled were red fox *Vulpes vulpes*, stoats *Mustela erminea*, brown rats *Rattus norvegicus*, carrion crows *Corvus corone* and black-billed magpies *Pica pica*. The experimental site also had supplementary food provided and habitat creation (see 'Provide supplementary food to increase adult survival' and 'Plant wild bird seed or cover mixture'). The effects of agri-environment schemes and the provision of set-aside are also discussed in 'Pay farmers to cover the costs of conservation measures' and 'Provide or retain set-aside'.

(1) Westerskov, K. E. (1977) Covey-oriented partridge management in France. *Biological Conservation*, 11, 185–191.
(2) Baines, D. (1996) The implications of grazing and predator management on the habitats and breeding success of black grouse *Tetrao tetrix*. *Journal of Applied Ecology*, 33, 54–62.
(3) Tapper, S. C., Potts, G. R. & Brockless, M. H. (1996) The effect of an experimental reduction in predation pressure on the breeding success and population density of grey partridges *Perdix perdix*. *Journal of Applied Ecology*, 33, 965–978.
(4) Summers, R. W., Green, R. E., Proctor, R., Dugan, D., Lambie, D., Moncrieff, R., Moss, R. & Baines, D. (2004) An experimental study of the effects of predation on the breeding productivity of capercaillie and black grouse. *Journal of Applied Ecology*, 41, 513–525.
(5) Aebischer, N. J. & Ewald, J. A. (2010) Grey Partridge *Perdix perdix* in the UK: recovery status, set-aside and shooting. *Ibis*, 152, 530–542.

12.5.4. Rails

- A single study from the USA (1) found more California clapper rails *Rallus longirostris obsoletus* on sites with higher numbers of foxes removed.

An analysis of data from 24 sites in south San Francisco Bay, USA, between 1991 and 1996 (1) found that the number of California clapper rails *Rallus longirostris obsoletus* surveyed each winter was positively correlated with the capture rate of red foxes *Vulpes vulpes* the previous year. At one site, the rail population increased from 40 in 1989 to 104 in 1994. Over the study period, the number of foxes trapped remained relatively constant (66–94/year) despite increased trapping effort, suggesting population decline. However, the authors suggest that fox immigration into the area meant that predator control would only be effective in the short term.

(1) Harding, E. K., Doak, D. F. & Albertson, J. D. (2001) Evaluating the effectiveness of predator control: the non-native red fox as a case study. *Conservation Biology*, 15, 1114–1122.

12.5.5. Cranes

- A single trial from the USA (1) found that greater sandhill cranes *Grus canadensis tabida* had higher hatching and fledging success in years with predator control, compared to years without control.

A trial in south-east Oregon, USA, between 1966 and 1989 (1), found that greater sandhill crane *Grus canadensis tabida* hatching success and fledging success were both higher in ten years when predators were controlled than in nine years without predator control (average hatching/fledging success of 55% and 9.1% for 662 clutches in predator control years vs. 42% and 5.1% for 434 in non-control years). Coyotes *Canis latrans* and ravens *Corvus corax* were controlled by poisoning and shooting. The main nest predators were coyotes (predating 17% of clutches in predator control years and 27% in non-control years), ravens (14% and 20%) and raccoons *Procyon lotor* (11% and 8%, the increase possibly related to a reduction in coyote numbers).

(1) Littlefield, C. D. (2003) Sandhill crane nesting success and productivity in relation to predator removal in southeastern Oregon. *The Wilson Bulletin*, 115, 263–269.

12.5.6. Waders

- Three out of four controlled studies in the UK (1,4) and the USA (3) found some evidence for higher reproductive success or lower predation rates for waders in areas or years with predator removal, although one UK study (1) found that only three of six species investigated had increased reproductive success in years with predator removal.
- Predators removed were carrion crows *Corvus corone*, gulls *Larus* spp., red foxes *Vulpes vulpes* and cats *Felis catus*.

A controlled before-and-after study at a site in east Scotland in 1981–9 (1) found that nesting success for three out of six wader species was significantly higher in years when avian predators were controlled, than in years with no control. Success was higher for curlew *Numenius arquata* (26% of 79 nests vs. 82% of 50), redshank *Tringa tetanus* (0% of 14 nests vs. 75% of 20) and lapwing *Vanellus vanellus* (29% of 88 nests vs. 75% of 49), but not for golden plover *Plucialis apricaria* (0% of eight nests vs. 0–54% of 21), snipe *Gallinago gallinago* (32% of 11 nests vs. 57% of 32) or oystercatcher *Haematopus otralegus* (0% of 22 nests vs. 29% of 16). Nesting success for golden plover and oystercatcher was higher in a control site with no predator control. The proportion of golden plover nests predated by crows and gulls fell between 1981–5 and 1986–9, but red foxes *Vulpes vulpes* predated all nests from 1987–9. Carrion crows *Corvus corone* and common gulls *Larus canus* were controlled with alpha-chloralose treated eggs.

A review of management at a coastal wetland in Bouches-du-Rhône, France (2) described the culling of yellow-legged gulls *Larus cachinnans* from 1960 until 1980. However, the impact of culling on greater flamingo *Phoenicopterus roseus* populations could not be separated from that of other management interventions, discussed in 'Provide artificial nest sites', 'Use decoys to attract birds to safe areas' and 'Manage water levels in wetlands'.

A replicated before-and-after study on beaches in Monterey Bay, California, USA (3), found that the proportion of snowy plover *Charadrius alexandrinus* nests lost to predation each year fell from an average of 28% of 833 during 1984–1992 to 9% of 577 during 1993–1999 following the initiation in 1993 of predator removal targeting red foxes *Vulpes vulpes* and feral cats *Felis catus*. This study also used predator exclosures and is discussed in more detail in 'Physically protect nests with individual exclosures/barriers', 'Can nest protection increase nest abandonment?' and 'Can nest protection increase predation of adult and chick waders?'.

A replicated, controlled trial at 13 lowland wet grassland sites in England and Wales between 1996 and 2003 (4) found no overall increase in the success of 3,139 northern lapwing *Vanellus vanellus* nests during four years with predator control, compared to four years without. However, when differences in initial predator densities were accounted for, control did improve survival, having a greater impact at sites with higher predator densities. At two sites where predators were controlled for all eight years, nesting success was not significantly different from the 11 other sites. Predators were red fox *Vulpes vulpes* and carrion crow *Corvus corone*, with average declines of 40% for foxes and 56% for crows.

(1) Parr, R. (1993) Nest predation and numbers of golden plovers *Pluvialis apricaria* and other moorland waders. *Bird Study*, 40, 223–231.
(2) Martos, M. R. & Johnson, A. R. (1996) Management of nesting sites for greater flamingos. *Colonial Waterbirds*, 20, 167–183.
(3) Neuman, K. K., Page, G. W., Stenzel, L. E., Warriner, J. C. & Warriner, J. S. (2004) Effect of mammalian predator management on snowy plover breeding success. *Waterbirds*, 27, 257–263.
(4) Bolton, M., Tyler, G., Smith, K. & Bamford, R. (2007) The impact of predator control on lapwing *Vanellus vanellus* breeding success on wet grassland nature reserves. *Journal of Applied Ecology*, 44, 534–544.

12.5.7. Parrots

- A replicated, controlled trial in New Zealand (1) found increased kaka *Nestor meridionalis* nesting success and lower predation at sites with mammal predator removal than at unmanaged sites.

A replicated, controlled trial in New Zealand (1) found that the nesting success of kakas *Nestor meridionalis* was significantly higher at three sites with control (80–87% of 70 nests successful) than at three unmanaged sites (10–38% of 43 nests). Predation rate of nesting females was also significantly lower at sites with predator control (5% of 38 tracked females vs. 65% of 17). Stoats *Mustela erminea*, common brushtail possums *Trichosurus vulpecula* and rats *Rattus* spp. were controlled with trapping and poisoning with 1080 sodium monofluoroacetate and anti-coagulents. Data comes from 1996–2000 for five sites and 1984–1996 for one (unmanaged) site.

(1) Moorhouse, R., Greene, T., Dilks, P., Powlesland, R., Moran, L., Taylor, G., Jones, A., Knegtmans, J., Wills, D., Pryde, M., Fraser, I., August, A. & August, C. (2003) Control of introduced mammalian predators improves kaka *Nestor meridionalis* breeding success: reversing the decline of a threatened New Zealand parrot. *Biological Conservation*, 110, 33–44.

12.5.8. Songbirds

- A before-and-after study from New Zealand (8) found that a reintroduced population of New Zealand robins *Petroica australis* declined without predator control and increased with rat poisoning. Two UK studies (7,10), one non-experimental, found increased populations of some species following control of bird and mammal predators.
- One replicated, controlled study from New Zealand (2) found lower New Zealand robin survival in areas where rodent bait was broadcast, but no difference with controls when dispensers were used.
- Six studies from New Zealand (1), Australia (4), UK (5,6,9) found increased nest success or survival (in one case, (4), with artificial nests) following bird and mammal predator control.
- One randomised, replicated and controlled study from the USA (3) found no difference in nest survival in a site with mammalian predator removal.

A replicated, controlled, paired sites study in Fiordland, New Zealand, in 1990–3 (1) found that mohua (yellowhead) *Mohoua ochrocephala* nests produced significantly more chicks in a site where stoats *Mustela erminea* were trapped than in a site without trapping for the first breeding season (1990–1) (80% fledging success, 2.1 fledglings/breeding group for ten groups vs. 36%, 1.1 fledglings/breeding group for 14 groups). In subsequent years, as the stoat population fell, success increased in both areas and the difference between sites became non-significant (87–90% fledging success, 2.6–2.7 fledglings/breeding group for 19 groups at the trapped sites vs. 66–75%, 1.9–2.5 fledglings/breeding group for 15 groups at the untrapped site). A total of 62 stoats were removed from the trapped site in 1990–1.

A replicated, controlled study in three southern beech *Nothofagus* stands on South Island, New Zealand between August and November 1996 (2) found that survival of South Island robins *Petroica australis australis* was not significantly higher when brodifacoum bait was dispensed from bait feeders (29/30 birds surviving, 97%) than in a control site (18/21, 86%), but was significantly lower when the bait was broadcast (12/23, 52%). This study is described in 'Distribute poison bait using dispensers'.

A randomised, replicated and controlled trial conducted over two breeding seasons in North Dakota, USA (3), found no significant difference in the daily survival rates of songbird nests at eight sites where medium-sized mammalian nest predators had been removed compared with eight control sites. Species removed were red fox *Vulpes vulpes*, striped skunk *Mephitis mephitis*, raccoon *Procyon lotor* and American badger *Taxidea taxus*. Observations from artificial nests suggest that compensatory predation by smaller mammal species (such as ground squirrels *Spermophilus* spp.) may have counteracted any effects of target predator removal.

A controlled before-and-after study in November–December 1999 in a 240 ha eucalypt forest in New South Wales, Australia (4), found that 104 artificial nests survived for significantly longer following the removal of pied currawongs *Strepera graculina* from an experimental grid (average survival of 1.9 days before cull vs. 3.0 days afterwards). There was no change in a nearby control grid, without currawong removal (3.1 days before vs. 3.6 afterwards). Before the cull, survival was significantly higher on the control grid, but there was no difference following the cull. Three pairs of currawongs were culled, with three more nests having either fledged or failed by the time of the cull in early December 1999.

A study at three farmland sites in central England in 1992–1998 (5) found that nest survival rates of four songbirds were negatively related to the breeding density of carrion crows following the control of nest predators. These species were Eurasian blackbird *Turdus merula*, song thrush *T. philomelos*, dunnock *Prunella modularis* and yellowhammer *Emberiza citrinella*. Non-significant negative relationships were also found for whitethroat *Sylvia communis* and chaffinch *Fringilla coelebs* nesting success and predator densities. Brown rats *Rattus norvegicus*, red foxes *Vulpes vulpes*, stoats *Mustela erminea*, weasels *M. nivalis*, carrion crows *Corvus corone* and magpies *Pica pica* were controlled through trapping and shooting. Between 151 and 951 nests of each species were studied.

A small replicated, controlled study from May–June in 1992–1998 in 1 experimental (3 km^2) and four unmanaged arable farms in Leicestershire, England (7) found that the abundance of nationally declining songbird species and species of conservation concern significantly increased through time in the sites at which predators were controlled. Although there was no overall difference in bird abundance, species richness or diversity between the experimental and control sites, numbers of nationally declining species rose by 102% (except for skylark *Alauda arvensis* and yellowhammer *Emberiza citrinella*). Nationally stable species rose (insignificantly) by 47% (with 8 species exhibiting net increases, especially greenfinch *Carduelis chloris* 68%, and 4 species exhibiting net decreases). The author concluded that controlling nest predators (from April–July each year), as part of an integrated management package, provided the greatest benefits to species of conservation concern, but did not affect species diversity at the farm scale.

A before-and-after study between 1996 and 1998 at a farmland site in eastern England (6) found that daily survival rates of Eurasian skylark *Alauda arvensis* nests in non-rotational set-aside areas were significantly higher (96% daily survival for 168 nests) following the introduction of intensive control of mammalian predators than when predator control was either 'light' (95.6% survival for 51 nests) or absent (91% survival for 192 nests). There was no significant difference between light control and no control. These differences resulted in average overall survival rates of 40.7%, 23.3% and 12.3% for heavy, light and no control, respectively. The main species targeted were mustelids, hedgehogs *Erinaceus europaeus* and red foxes *Vulpes vulpes*. This study also discusses the impact of set-aside plots, described in 'Provide or retain set-aside areas in farmland'.

A before-and-after study in a forest remnant in the south of New Zealand's North Island between September 1999 and September 2004 (8) found that the population of 40 New Zealand robins *Petroica australis* (referred to as North Island robins *P. longipes*) reintroduced to the study site in March 1999 declined to 18 individuals by September 1999 and then to 11 birds by September 2000. Following the control of introduced predators (black rat *Rattus rattus* and brush-tailed possum *Trichosurus vulpecula*) with brodifacoum baits between mid-2000 and March 2002, rat populations (measured with tracking tunnels) fell and the robin population increased to 19 birds. Poisoning was stopped in March 2002, the rat population increased and the robin population decreased to eight birds by September 2004.

A before-and-after study on a mixed farmland-woodland site in central England (9) found that the fledging success of spotted flycatcher *Muscicapa striata* nests was significantly higher when predators (grey squirrels *Sciurus carolinensis*, brown rats *Rattus norvegicus*, red foxes *Vulpes vulpes*, black-billed magpies *Pica pica* and carrion crows *Corvus corone*) were controlled (77% for 11 nests in 1997–2001) than when there was no control (16% for 28 in 2002–4).

A before-and-after study on a mixed farm in central England (10) between 1992 and 2007 (a continuation of the data series used in (5)), found that controlling predator (carrion crow *Corvus corone*, magpie *Pica pica*, red fox *Vulpes vulpes* and other mammals) populations appeared to increase blackbird *Turdus merula* breeding population. However, the authors caution that the study is not experimental and that other explanations for the trends seen cannot be eliminated.

(1) O'Donnell, C., Dilks, P. & Elliott, G. (1996) Control of a stoat (*Mustela erminea*) population irruption to enhance mohua (yellowhead) (*Mohoua ochrocephala*) breeding success in New Zealand. *New Zealand Journal of Zoology*, 23, 279–286.
(2) Brown, K. P. (1997) Impact of brodifacoum poisoning operations on South Island robins *Petroica australis australis* in a New Zealand *Nothofagus* forest. *Bird Conservation International*, 7, 399–408.
(3) Dion, N., Hobson, K. A. & Larivière, S. (1999) Effects of removing duck-nest predators on nesting success of grassland songbirds. *Canadian Journal of Zoology*, 77, 1801–1806.
(4) Fulton, G. R. & Ford, H. A. (2001) The pied currawong's role in avian nest predation: A predator removal experiment. *Pacific Conservation Biology*, 7, 154–160.
(5) Stoate, C. & Szczur, J. (2001) Could game management have a role in the conservation of farmland passerines? A case study from a Leicestershire farm. *Bird study*, 48, 279–292.
(6) Donald, P. F., Evans, A. D., Muirhead, L. B., Buckingham, D. L., Kirby, W. B. & Schmitt, S. I. A. (2002) Survival rates, causes of failure and productivity of Skylark *Alauda arvensis* nests on lowland farmland. *Ibis*, 144, 652–664.
(7) Stoate, C. (2002) Multifunctional use of a natural resource on farmland: wild pheasant (*Phasianus colchicus*) management and the conservation of farmland passerines. *Biodiversity and Conservation*, 11, 561–573.
(8) Armstrong, D. P., Raeburn, E. H., Lewis, R. M. & Don Ravine (2006) Estimating the viability of a reintroduced New Zealand robin population as a function of predator control. *The Journal of Wildlife Management*, 70, 1020–1027.
(9) Stoate, C. & Szczur, J. (2006) Potential influence of habitat and predation on local breeding success and population in spotted flycatchers *Muscicapa striata*. *Bird Study*, 53, 328–330.
(10) White, P. J. C., Stoate, C., Szczur, J. & Norris, K. (2008) Investigating the effects of predator removal and habitat management on nest success and breeding population size of a farmland passerine: a case study. *Ibis*, 150, 178–190.

12.6. Reduce predation by translocating predators

- Two studies from France (2) and the USA (3) found local population increases (2) or reduced predation (3) following the translocation of predators away from an area. A study in Saudi Arabia (1) found that predation was no lower when predators were translocated from the bird release site.

Background

If suitable habitats exist for predators away from sensitive bird populations then translocating predators may represent a solution to high levels of predation.

A before-and-after study of a houbara bustard *Chalmydotis undulata macqueenii* release programme in desert steppe in south-west Saudi Arabia (1) found that the rate of predation of bustards did not change from 1993, with no predator control, to 1994, when feral cats *Felis catus* and red foxes *Vulpes vulpes* were caught and translocated away from the release site (47% of 25 subadult bustards predated in 1993 vs. 45% of 34 in 1994). However, the pattern of predation did differ, with fewer birds predated near the release site (predation occurring an average of 3 km from the release site in 1993 vs. 8 km in 1994) and birds surviving for longer on average (average of four days before predation in 1993 vs. 14 days in 1994). The release programme itself is discussed in 'Release captive bred individuals', 'Release birds as sub-adults or adults, not juveniles', 'Use holding pens at release sites' and 'Use holding pens at release sites and clip birds' wings'.

A paired sites before-and-after trial at two urban locations in the Paris area, France, between 2003 and 2005 (2) found that the number of juvenile blue tits *Parus caeruleus* increased by 40% and the number of adult long-tailed tits *Aegithalos caudatus* increased 50 fold following the removal of 91 black-billed magpies *Pica pica* from experimental sites (a 58% reduction in density), with no corresponding increase in control sites where magpies were not removed. However, removal appeared to cause a 70% reduction in the number of adult blackcaps *Sylvia atricapilla*. There was no change in the number of juveniles or adults in seven other species.

A before-and-after trial on Santa Barbara Island, California, USA (3), found that the proportion of Xantus's murrelet *Synthliboramphus hypoleucus* eggs predated by mice was significantly lower in 2004, when approximately 1,650 deer mice *Peromyscus maniculatus elusus* (endemic to the island) were translocated away from the colony, compared to the 1993–2005 average, excluding 2004 (21% of 73 eggs predated vs. an average 37% of 64). In addition, the productivity/nest was higher (1.11 vs. 0.93), but hatching success was not significantly different (56% vs. 54%).

(1) Combreau, O. & Smith, T. R. (1998) Release techniques and predation in the introduction of houbara bustards in Saudi Arabia. Biological Conservation, 84, 147–155.
(2) Chiron, F. & Julliard, R. (2007) Responses of songbirds to magpie reduction in an urban habitat. *The Journal of Wildlife Management*, 71, 2624–2631.
(3) Millus, S. A., Stapp, P. & Martin, P. (2007) Experimental control of a native predator may improve breeding success of a threatened seabird in the California Channel Islands. *Biological Conservation*, 138, 484–492.

Reduce incidental mortality during predator eradication or control

- Eradication and control programmes run the risk of damaging the bird populations they are meant to protect as birds may take poison bait or get caught in traps. To maximise the beneficial effects of eradication and control programmes they must therefore be as effective as possible whilst minimising their impact on non-target species.

12.7. Do birds take bait designed for pest control?

- Two studies, one randomised, replicated and controlled, from New Zealand (1) and Australia (2) found no evidence that birds took bait meant for pest control.

Background

If non-target species take poison bait designed for problematic species control then extra care will be needed during control programmes, for example, the use of bait feeders, repellents or dyed bait (see interventions below).

A before-and-after study on Breaksea Island (170 ha), South Island, New Zealand (1) found that there was no significant difference in the number of South Island robins *Petroica australis australis* counted in 1987, prior to a rat eradication campaign, compared to after the eradication of rats in 1988 and 1989 (130 robins in 1987, 127 and 129 in 1988 and 1989 respectively; 192 birds counted at 133 bait stations in 1988, 194 at 140 stations in 1989). Rats were eradicated using brodifacoum baits in both briquettes and plastic bags. The lack of change in the robin population implies that birds were not adversely affected by the poisoning and did not take the bait.

A randomised, replicated and controlled study over eight days in Adelaide Zoo, Australia (2), found that eight bush stone-curlews did not consume untreated bait (consisting of 50–100 g pieces of sun-dried horsemeat and dried oats) when also provided with their normal food (consisting of beef mince, fruit and 'Woombaroo insectivore mix' – a commercially available feed mix), which they continued to eat.

(1) Taylor, R. H. & Thomas, B. W. (1993) Rats eradicated from rugged Breaksea island (170 ha), Fiordland, New Zealand. *Biological Conservation*, 65, 191–198.
(2) Johnston, G. & McCarthy, P. (2007) Susceptibility of bush stone-curlews (*Burhinus grallarius*) to sodium fluoroacetate (1080) poisoning. *Emu*, 107, 69–73.

12.8. Distribute poison bait using dispensers

- A controlled study in New Zealand (1) found that survival of South Island robins *Petroica australis australis* was higher when brodifacoum was dispensed from bait feeders compared to where bait was scattered.

Background
If bait is easily visible, birds may be more likely to feed on it opportunistically. Therefore providing it in dispensers, which partially hide bait or make it difficult for non-target species to access it, may reduce incidental mortality of non-target species.

A controlled study in three southern beech *Nothofagus* stands on South Island, New Zealand between August and November 1996 (1) found that survival of South Island robins *Petroica australis australis* during predator removal operations was higher when brodifacoum was dispensed from bait feeders than in a site where 3 kg/ha poison was scattered and left exposed on the forest floor (29/30 birds surviving, 97% vs. 12/23 birds surviving, 52%). Survival in a control site, with no poisoning, was higher than in the broadcast site but not significantly different from the bait feeder site (18/21 birds surviving, 86%). Feeders were designed to allow black rats *Rattus rattus* and house mice *Mus musculus* to enter and retrieve baits, but not brush-tailed possums *Trichosurus vulpecula*.

(1) Brown, K. P. (1997) Impact of brodifacoum poisoning operations on South Island robins *Petroica australis australis* in a New Zealand *Nothofagus* forest. *Bird Conservation International*, 7, 399–408.

12.9. Use repellents on baits

- A replicated, randomised and controlled experiment in the USA (1) found that methyl anthanilate and aminoacetophenone did not reduce consumption of baits by American kestrels *Falco sparverius*.
- A replicated, randomised and controlled experiment in New Zealand (2) found that treating baits with pulegone or Avex™ reduced pecking rates in North Island robins *Petroica australis longipes*.

Background
If repellents that discourage birds from taking bait but do not affect bait consumption by target species can be identified then the risk of incidental mortality during eradication programmes may be greatly reduced.

A replicated, randomised and controlled, *ex situ* experiment (1) found that on three of four test days, 33 captive American kestrels *Falco sparverius* were no less likely to choose to consume chicks with feeding repellent (dead day-old chicks treated with either methyl anthanilate or aminoacetophenone) compared to untreated chicks. Kestrels fed on treated chicks consumed less over the study (with fewer methyl anthanilate-treated chicks consumed), but not to the point of losing body condition (body weights were similar across treatments).

Treating chicks with repellents did not affect consumption in comparison to dyeing them blue or green (see 'Use coloured baits to reduce accidental mortality during predator control'). This study is also discussed in 'Use aversive conditioning to reduce nest predation'.

A replicated, randomised and controlled experiment on Tiritiri Matangi Island, North Island, New Zealand in June 2000 (2) found that wild North Island robins *Petroica australis longipes* pecked significantly less at dough baits treated with (either sprayed with or dipped in) a combination of green dye, pulegone and Avex^{TM} (two avian repellents) than at control baits containing green dye and cinnamon. In addition, rate of pecking at treated baits declined over time (sprayed baits: 5 pecks on first day, 4 on second day, 2 on third day vs. 9.5, 10 and 9.5 pecks for control baits, n = 17 birds; dipped baits: 1.5 pecks on first day, 0.5 pecks on second, 0.25 on third, zero pecks on fourth vs. 6, 5, 3 and 3.25 for control baits, n = 21 birds).

(1) Nicholls, M. K., Love, O. P. & Bird, D. M. (2000) An evaluation of methyl anthranilate, aminoacetophenone, and unfamiliar coloration as feeding repellents to American kestrels. *Journal of Raptor Research*, 34, 311–318.
(2) Day, T. D., Matthews, L. R. & Waas, J. R. (2003) Repellents to deter New Zealand's North Island robin *Petroica australis longipes* from pest control baits. *Biological Conservation*, 114, 309–316.

12.10 Use coloured baits to reduce accidental mortality during predator control

- Two replicated and controlled trials in the USA (1,2) found that dyed baits were consumed at significantly lower rates than control baits.
- A replicated, randomised and controlled trial in Australia (3) found no differences in consumption rates of dyed and control baits.

Background

Most birds (unlike most mammals) are visual hunters and foragers, meaning that strangely coloured baits may not be taken as readily, and so using them during eradication programmes may reduce incidental mortality.

A review of three replicated, controlled trials in Day County, South Dakota, USA, and another in Routt County, Colorado (1), in 1945, found that bird assemblages (mainly songbirds and doves) took a higher proportion of uncoloured grain (88% taken) than of yellow (73%) or green (23%) grain, and when uncoloured was not available, birds took more yellow (87% and 15% in two sites) than green (39% and 9%). In addition, more dead birds were found with uncoloured poison grain in their stomachs than with yellow. Very few birds were found with green grain.

A replicated, randomised and controlled, *ex situ* experiment (2) found that consumption of day-old chicks by 33 American kestrels *Falco sparverius* was greatly reduced by dying chicks green or blue, with no birds consuming blue-dyed chicks and two birds also avoiding green-dyed chicks. All birds reduced food intake significantly. Treating chicks with two repellents (discussed in 'Use repellents on baits') did not affect consumption in addition to dyeing. This study is also discussed in 'Use aversive conditioning to reduce nest predation'.

A replicated, randomised and controlled trial in Adelaide Zoo, South Australia (3), found that dyeing food (minced beef, fruit and 'Wombaroo insectivore mix' – a commercially available food mix) blue had no effect on its consumption by six captive bush stone-curlews *Burhinus grallarius* over a ten day period.

(1) Kalmbach, E. R. & Welch, J. F. (1946) Colored rodent baits and their value in safeguarding birds. *The Journal of Wildlife Management*, 10, 353–360.
(2) Nicholls, M. K., Love, O. P. & Bird, D. M. (2000) An evaluation of methyl anthranilate, aminoacetophenone, and unfamiliar coloration as feeding repellents to American kestrels. *Journal of Raptor Research*, 34, 311–318.
(3) Johnston, G. & McCarthy, P. (2007) Susceptibility of bush stone-curlews (*Burhinus grallarius*) to sodium fluoroacetate (1080) poisoning. *Emu*, 107, 69–73.

Reduce nest predation by excluding predators from nests or nesting areas

- A 2011 systematic review (1) found that excluding predators from nests significantly increased hatching success, although individual barriers around nests sometimes had adverse impacts.

Background

As well as direct predation on adults (see previous section), predators can have a devastating impact on bird populations through predating eggs and chicks too young to defend themselves or run away. Species ranging from hedgehogs to pigs to other birds can all affect bird populations in this way and in many cases it is not desirable or practical to remove these species. Therefore the use of barriers and cages to prevent predators from attacking nests is widespread.

These can take the form of barriers or cages around individual nests, fences around groups of nests or suitable nesting areas, or repellents to either disguise the presence of eggs or discourage predators from approaching. We found one systematic review (Smith *et al.* 2010) which compares these approaches and also investigates potential adverse effects of nest protection.

As in the section on predator removal, several studies perform experiments using artificial nests: either man-made or genuine nests that have been filled with false or real eggs (normally from quail Corturnix spp.) and placed in experimental areas to estimate predation rates.

A 2011 systematic review (1) found that excluding predators using fences (see 'Physically protect nests from predators using non-electric fencing') or barriers around individual nests ('Physically protect nests with individual exclosures/barriers') significantly increased hatching success. Individual barriers appeared to be slightly (non-significantly) more effective than fences, but some studies found that they increased predation on adults (see 'Can nest protection increase predation of adult and chick waders?').

(1) Smith, R. K., Pullin, A. S., Stewart, G. B. & Sutherland, W. J. (2011) Is nest predator exclusion an effective strategy for enhancing bird populations? *Biological Conservation*, 144, 1–10.

12.11. Physically protect nests from predators using non-electric fencing

- Two studies (1,3) from the USA and UK found that fewer nests were predated or failed when predator exclusion fences were erected.
- Two studies from the USA found that nesting success (2) or fledging success (4) did not differ between areas with fences erected and those without fences; although one (4) found that hatching rates were higher with fences.

Background

The simplest way to exclude predators from a nesting area is to erect a fence around it. This is likely to be relatively cheap and easy, but may also prevent non-predators from entering the area, with potentially unforeseen results.

A controlled, replicated before-and-after study in Massachusetts, USA (1) found that the proportion of least tern *Sterna antillarum* nests lost to predation was significantly lower in two colonies protected in 1990–1 by 1.2 m high wire-mesh fencing (<1% nests predated, 87% hatched successfully, 191 nests monitored), compared to either three unprotected colonies over the same time period (46% of 69 nests predated, 41% hatched successfully) or the study colonies and nine additional colonies without fencing between 1987 and 1991 (52% of 833 nests predated, 16% hatched successfully).

A 1996 meta-analysis of 58 studies from the Prairie Pothole Region of the USA and Canada between 1935 and 1992 (2) found that the nesting success of dabbling ducks *Anas* spp. declined over the study period, and that the rate of this decline did not differ between fenced sites and sites with no predator control. However, the intercept of the regression slope did differ significantly; with nesting success being higher in fenced sites than in sites without management or where predators were removed. There was no difference between fenced and island sites.

A small paired site study in 1998 on South Uist, north-west Scotland (3) found that fewer wader nests failed at two sites where fences were erected and hedgehogs *Erinaceus europaeus* removed, compared to two control, unfenced areas (38% of 52 nests lost, three to hedgehogs vs. 55% of 53 nests failing and 15 to hedgehogs respectively). Therefore, a smaller proportion of failures were attributable to European hedgehogs. Species in this study included the lapwing *Vanellus vanellus*, dunlin *Calidris alpina*, redshank *Tringa totanus* and snipe *Gallinago gallinago*. There was no evidence of compensatory predation by other species following hedgehog removal. Fences successfully excluded hedgehogs from one experimental site, but rabbit *Oryctolagus cuniculus* burrows allowed 33 hedgehogs to re-enter the second site.

A randomised, controlled study in 1996 at one site on a sandbar in Washington State, USA (4) found that egg survival and hatching success of gull pairs in the western gull *Larus occidentalis* × glaucous-winged gull *Larus glaucescens* hybrid complex were significantly higher for nests with makeshift, 30 cm tall wooden exclusion fences (54% egg survival, 38% hatching success for ten pairs) than for control nests with no screening or 'natural screening' e.g. driftwood etc. (14% egg survival, 13% hatching success for 54 pairs). The fledging rate, however, was not significantly higher for protected nests (29% vs. 8% respectively) and the distribution of nests that failed to produce any fledglings did not differ from a uniform distribution across protected and control nests.

(1) Rimmer, D. W. & Deblinger, R. D. (1992) Use of fencing to limit terrestrial predator movements into least tern colonies. *Colonial Waterbirds*, 15, 226–229.
(2) Beauchamp, W. D., Nudds, T. D. & Clark, R. G. (1996) Duck nest success declines with and without predator management. *The Journal of Wildlife Management*, 60, 258–264.
(3) Jackson, D. B. (2001) Experimental removal of introduced hedgehogs improves wader nest success in the Western Isles, Scotland. *Journal of Applied Ecology*, 38, 802–812.
(4) Good, T. P. (2002) Breeding success in the western gull x glaucous-winged gull complex: The influence of habitat and nest-site characteristics. *The Condor*, 104, 353–365.

12.12. Protect bird nests using electric fencing

- One before-and-after study from the UK (1) found an increase in tern numbers after the erection of an electric fence, whilst a study from the USA (2) found an increase in the number of nests.
- Five studies from the USA (2–6) found higher survival or productivity at wader or seabird colonies with electric fencing, compared to areas without fencing, although one study (6) found that hatching rates were no different, whilst nesting success was only higher in one of two years.
- One study from the USA (6) found lower predation by mammalian predators inside electric fence exclosures, whilst predation by birds was higher.

Background

For fencing to be effective it has to physically prevent predators from entering an area, which is likely to be very difficult for species such as racoons *Procyon lotor*. Electric fences may offer a solution as animals will be unlikely to attempt to climb through them after receiving electric shocks.

A before-and-after study in 1973 and 1984 on a sand spit in eastern Scotland (1) found that the number of sandwich terns *Sterna sandvicensis* nesting in a colony increased from approximately 80 pairs in 1973 to approximately 450 pairs in 1974, following the erection of a 45 cm high electric fence to separate the colony from the mainland. Previous low numbers were attributed to red fox *Vulpes vulpes* predation, but after the fence was erected only a single fox was recorded breaching the fence and this animal did not approach the terns.

A before-and-after study in 1978 on a beach in Massachusetts, USA (2) found that the number of least tern *Sterna antillarum* nests in a colony decreased from 138 to 45 between the 20th and 23rd June (red fox *Vulpes vulpes* tracks were found in the colony), before the erection of an electric fence around the colony on the 24th June. The number of nests increased to 85 following the erection of the fence and no new fox tracks were found within the colony. No nests outside the fence survived. In total, 27 chicks fledged from the colony; the authors estimate that all, or nearly all, came from eggs laid after the erection of the fence.

A replicated, controlled trial from 1986–1988 in wetlands in North Dakota, USA (3) found that nest survival of 54 piping plover *Charadrius melodus* nests on four beaches protected by a combined wire mesh and electric fence (1.2 m high, designed to stop mammalian predators) was 71% higher than for 234 nests on 21 unfenced beaches. Chick survival and the fledging rate were 55% and 82% greater on fenced than unfenced beaches, but these increases were not significant.

A replicated, controlled trial from 1991–1994 on alkaline flats in Oklahoma, USA (4) found that the nesting success (i.e. at least one egg hatching in a nest) of least terns *Sterna antillarum* was significantly higher inside two electric fence exclosures than outside (81% of 60 nests vs. 56% of 129 nests respectively). The same pattern was seen for snowy plovers *Charadrius alexandrinus*, but the difference was not significant (79% of 22 nests vs. 62% of 26 nests). The proportion of both tern and plover eggs predated (mainly by coyotes *Canis latrans*) was lower inside the fence (10% vs. 20% predation for terns; 6% vs. 11% for plovers. The fence was 86 cm high and designed to prevent coyotes from entering. This study is also discussed in 'Provide nesting habitat for birds that is safe from extreme weather'.

A replicated, controlled trial from 1987–1991 in three wetland-grassland sites in North Dakota and Minnesota, USA (5) found that using fencing (1.8 m tall with an electrified top wire and with ground-level openings to allow broods to leave) to exclude mammalian predators from 25 ha of nesting habitat significantly increased the nesting success of dabbling ducks *Anas* spp. (75% of 452 nests inside exclosures), compared to those nesting outside exclosures (no data provided for control). The proportion of nests inside exclosures compared with control areas increased significantly for mallard *A. platyrhynchos*, gadwall *A. strepera*, blue-winged teal *A. discors* and northern pintail *A. acuta*, but not for northern shoveler *A. clypeata* and dabbling ducks. The authors note that there was a local and regional decline in dabbling duck numbers over the study period, probably due to an ongoing drought.

A replicated, controlled trial on the same study site as (4) in 1995–6 (6) found that the hatching success of snowy plover *Charadrius alexandrinus* nests was not significantly different (for either year of monitoring) between nests inside three electric fence exclosures (4.5, 24 and 20 ha) and outside exclosures (1995: 44% of nests inside vs. 34% nests outside; 1996: 61% vs. 57%). However, apparent nesting success did differ in 1996 (71% of 17 monitored nests were successful vs. 49% of 160 nests) but not in 1995 (37% of 70 nests inside vs. 38% of 168). The proportion of eggs lost to mammalian predators (mainly coyotes *Canis latrans*) was lower inside the exclosures (1% vs. 6%), but more eggs were predated by birds, mainly ring-billed gulls *Larus delawarensis* (11% vs. 3%).

(1) Forster, J. A. (1975) Electric fencing to protect sandwich terns against foxes. *Biological Conservation*, 7, 85.
(2) Minsky, D. (1980) Preventing fox predation at a least tern colony with an electric fence. *Journal of Field Ornithology*, 51, 180–181.
(3) Mayer, P. M. & Ryan, M. R. (1991) Electric fences reduce mammalian predation on piping plover nests and chicks. *Wildlife Society Bulletin*, 19, 59–63.
(4) Koenen, M. T., Utych, R. B. & Leslie Jr, D. M. (1996) Methods used to improve least tern and snowy plover nesting success on alkaline flats. *Journal of Field Ornithology*, 67, 281–291.

(5) Cowardin, L. M., Pietz, P. J., Lokemoen, J. T., Sklebar, H. T. & Sargeant, G. A. (1998) Response of nesting ducks to predator exclosures and water conditions during drought. *The Journal of Wildlife Management*, 62, 152–163.
(6) Winton, B. R., Leslie Jr., D. M. & Rupert, J. R. (2000) Breeding ecology and management of snowy plovers in north-central Oklahoma. *Journal of Field Ornithology*, 71, 573–584.

12.13. Physically protect nests with individual exclosures/barriers or provide shelters for chicks

Background

If fencing does not work to exclude predators (for example, predatory birds), or is not a viable option, it may be possible to protect individual nests using a variety of cages and exclosures. These must be able to allow chicks and adults to get in and out, but not predators and should be quick to install to minimise the chances of parents abandoning nests (see 'Can nest protection increase nest abandonment?').

Unfortunately, because each cage is over a nest, it is possible that predators will learn the association and that providing the exclosures will actually increase predation on either adults or chicks (see 'Can nest protection increase predation of adult and chick waders?')

12.13.1. Ground nesting seabirds

- A before-and-after study from Japan (3) found an increase in fledging rates of little terns *Sterna albifrons* following the provision of chick shelters and other interventions.
- Two studies from the USA (1) and Canada (2) found reduced predation of tern chicks following the provision of chick shelters.
- A small study from the USA (4) found low levels of use of chick shelters, except when predators were present.

A before-and-after study from 1978–1980 at seven least tern *Sterna antillarum* colonies on Nantucket Island, Massachusetts, USA (1) found that predation rates on chicks were greatly reduced following the provision of chick shelters (43 cm high cones made from 11 slats, with a 66 cm basal diameter), compared to previous years. In 1978, a pair of American kestrels *Falco sparverius* 'greatly reduced' tern productivity at a colony by removing a tern chick approximately every 15 minutes for two hours, whereas a pair of northern harriers *Circus cyaneus* reduced productivity at another colony by 80% in 1979 (four chicks from 20 nests escaping). In 1980, with shelters present, no kestrels or harriers were seen hunting within the tern colonies, although they were present in the vicinity.

A small before-and-after study in 1990 study on a breakwater in Lake Erie, Canada (2) found that no common tern *Sterna hirundo* chicks were predated by herring gulls *Larus argentatus* or ring-billed gulls *L. delawarensis* over 12 days following the provision of small plywood shelters (two 12.5 × 25 cm rectangles attached to form a 10 cm high triangular shelter), compared with ten chicks being predated in the eight days between first hatching and shelter provision. A total of 29 chicks were studied, with 11 disappearing (six before shelter provision and five after) in addition to those predated.

A before-and-after study in 2001–2002 in Tokyo, Japan (3) found that the provision of 200 chick shelters on the roof of a sewage plant in 2002, combined with the provision of nesting substrate, appeared to increase the fledging rate of a little tern *Sterna albifrons* colony, compared with 2001 when birds were first observed and before habitat alterations (23% of 2,665 eggs fledged in 2002 vs. 1.5–2.1% of 335 eggs in 2001). The nesting substrate consisted of fine-grained (2–3 mm) 'dried sludge' spread over 2 ha, with 30 tonnes of shell fragments, while 38% of the rooftop was painted white. Chick shelters consisted of a sheet of wire mesh spread across two bricks.

A small study in 2003 on two warehouse roofs in Texas, USA (4) found that least tern *Sterna antillarum* chicks did not use wooden shelters more than would be expected by chance. However, on 18% of occasions when adults were observed mobbing predators (n = 39), chicks were seen using either artificial plants or skylights as cover (on two occasions chicks ran towards structures, on five occasions chicks were already in cover).

(1) Jenks-Jay, N. (1982) Chick shelters decrease avian predation in least tern colonies on Nantucket Island, Massachusetts. *Journal of Field Ornithology*, 53, 58–60.
(2) Burness, G. P. & Morris, R. D. (1992) Shelters decrease gull predation on chicks at a common tern colony. *Journal of Field Ornithology*, 63, 186–189.
(3) Hayashi, E., Hayakawa, M., Satou, T. & Masuda, N. (2002) Attraction of little terns to artificial roof-top breeding sites and their breeding success. *Strix*, 23, 143–148.
(4) Butcher, J. A., Neill, R. L. & Boylan, J. T. (2007) Survival of least tern chicks hatched on gravel-covered roofs in north Texas. *Waterbirds*, 30, 595–601.

12.13.2. Waders

- Three of 13 studies from the USA (3,4,7) found higher productivity from nests protected by individual barriers than unprotected nests. Two studies from the USA (8) and Sweden (13) found no higher productivity from protected nests.
- Eight studies from the USA (2–5,8) and Europe (10–13) found higher hatching rates, or survival, or low predation of nests protected by individual barriers, although two of these (8,13) found that higher hatching rates did not result in higher productivity. Two small studies from North America (1,6) found no differences in predation or survival rates between protected and unprotected nests. One replicated and controlled study from the USA (9) found that initial survival was higher on protected nests but that adults were harassed by predators. Exclosures were then removed and the formerly protected nests suffered high predation rates (see 'Can nest protection increase predation of adults and chicks?').
- A meta-analysis from the USA (2) found that there were differences in the effectiveness of different exclosure designs.

A small randomised, replicated and controlled trial in 1978 on a beach on Lake Erie, Canada (1) found that predation rates of killdeer *Charadrius vociferous* nests were not significantly different between 12 nests protected with a novel predator exclosure ('H' shaped frame with the nest in the centre and eight 7 × 12 cm openings, covered in 1.4 cm mesh hardware cloth) and 17 control nests (75% of protected nests and 71% of unprotected nests predated). However, no protected nests were lost to avian predators (gulls *Larus* spp. or American crow *Corvus brachyrhynchos*). Instead they were predated by raccoons *Procyon lotor* or mustelids which could enter exclosures.

A 1992 meta-analysis (2) analysed data from 211 nest exclosures across eight US states and three Canadian provinces to determine exclosure features that led to lowest predation rates of piping plover *Charadrius melodus* nests. Overall, exclosures were effective (10% (21) of nests being predated, mainly by red foxes *Vulpes vulpes* but also American crows *Corvus brachyrhynchos* and other predators). Estimated predation probabilities revealed that: mid-sized exclosures (3–6 m^2) suffered higher predation (26% of 48 exclosures) than small (<3 m^2, 5% of 23) or large (>6 m^2, 8% of 140) exclosures. Square enclosures were predated at a higher rate (72% of 19) than circular (8% of 166) or triangular (0% of 26) ones. Exclosures supported by 'tomato stakes' (thin gardening stakes) were predated more (80% predation of 18 exclosures) than unsupported (3% of 35) or metal/wood supported (8% of 158). Exclosures with mid-height posts (122 cm) were predated more (29% of 40) than short (<122 cm, 3% of 41) or tall (>122, 9% of 130) posts and exclosure with low fences (<122 cm, 42% of 27) were more likely to be predated than those with high (>122 cm, 8% of 184) fences. Fences buried to less than 10 cm were more likely (27% of 62) to be predated than those buried to more than 10 cm (6% of 149). There were no significant differences between different mesh sizes (5 × 10 cm vs. 5 × 5 cm) or whether exclosures were covered or not.

A replicated, controlled before-and-after study from 1988–1989 at six beaches on Cape Cod, Massachusetts, USA (3), found that the daily survival rates, overall hatching success and number of chicks fledged/pair of 29 piping plover *Charadrius melodus* nests protected by wire predator exclosures (circular wire fences, 1 m tall, around nests with 5 × 10 cm mesh) were significantly higher than for 24 unprotected nests (daily survival rates: 99% for protected vs. 93% and unprotected nests; overall hatching success: 74% of 104 eggs vs. 19% of 59 eggs; 1.96 chicks/pair vs. 0.12 chicks/pair). Before exclosures were used, between 1985 and 1987, 79% of 126 plover nests were destroyed by predators (red foxes *Vulpes vulpes*, striped skunks *Mephitis mephitis*, American crows *Corvus brachyrhynchos* and gulls *Larus* spp.).

A controlled, replicated study from 1986–1989 on a beach in Massachusetts, USA (4) found that hatching rates of 26 piping plover *Charadrius melodus* nests protected by triangular wire mesh fences (5 cm wire mesh, 30.5 m perimeter, placed around individual nests) were higher than for unprotected nests (92% and 25% of nests hatching at least one egg, respectively). On average, protected nests produced significantly more nestlings than unprotected nests (3.5 and 1 chicks/nest respectively). All but one of the losses of unprotected nests was due to predation; no protected nests were predated.

A replicated, controlled study in 1992 in the North Slope of Alaska, USA (5) found that the average daily survival rate of 13 pectoral sandpiper *Calidris melanotos* nests protected by wire mesh cages was significantly higher than that for 39 unprotected nests (survival rates of 0.98 for caged and 0.72 for uncaged nests). The mesh of the 31 cm tall and 66–69 cm diameter cages was sufficiently large (5 × 10 cm) to allow female plovers to enter and exit, but prevented arctic foxes *Alopex lagopus* from digging under or entering nests. No protected nests were lost to predation.

A small replicated, controlled study from 1994–95 on beaches and alkaline flats in Colorado, USA (6) found that daily survival rates of 27 snowy plover *Charadrius alexandrinus*, 16 killdeer *C. vociferous* and 9 piping plover *C. melodus* nests (in 1994) and 28 snowy plover nests (in 1995) were no higher for nests protected by predator exclosures (61 cm tall, 122 cm in diameter cylinders of 5 × 5 cm or 5 × 10 cm wire mesh, for snowy and piping plover or killdeer respectively) than for unprotected nests (daily survival rates in 1994: 0.98 vs. 0.98 for snowy plovers; 0.97 vs. 0.99 for killdeer; 0.98 vs. 0.98 for piping plovers; in 1995: 0.98 vs. 0.97 for snowy plovers). Five protected nests (20%) were predated in 1994 and three (21%) in 1995, by snakes, rodents or skinks. The authors cite small sample sizes, unbalanced experimental design and 'inappropriate statistical analyses' as possible reasons for the lack of a significant result.

A replicated, controlled study from 1996–1997 at three alkali lakes in North Dakota and Montana, USA (7) found that the average number of fledglings produced by piping plover *Charadrius melodus* pairs provided with fences (0.9 m tall and made from 5 cm poultry wire) around individual nests was significantly higher than for unprotected pairs (1.7 chicks/pair for 46 pairs with nest fences vs. 0.7 chicks/pair for 43 unprotected nests). This study is described further in 'Use multiple barriers to protect nests'.

A replicated before-and-after study from 1984–90 and 1991–99 on beaches in California, USA (8) found that the mean hatching rate and hatching rate/male of snowy plover *Charadrius alexandrinus* nests increased following the protection of nests with 1.5 m high wire exclosures (hatching rate in 1984–90: 43% of 728 nests vs. 68% of 682 nests in 1991–99; hatching rate/male: 2 chicks/male vs. 2.7 chicks/male). However, the mean number of chicks fledged per male did not change (0.86 and 0.81 chicks fledged/male for 1984–90 and 1991–99 respectively). Between 1993 and 1999, there was also the targeted removal of red foxes *Vulpes vulpes* and feral cats *Felis catus*, described in 'Control predators not on islands'. The study also discusses nest abandonment and adult mortality, see 'Can nest protection increase nest abandonment?' and 'Can nest protection increase predation of adults and chicks?'.

A small, replicated, before-and-after study at a site in eastern England (10) found that the average productivity of little ringed plover *Charadrius dubius* increased from 0.6 chicks/pair during 1984–95 to 1.6 chicks/pair during 1996–2005 following the protection of nests with wire cages (61 × 61 × 30.5 cm cages with 5 × 5 cm mesh). The average number of nesting pairs at the site increased from 1.3 pairs/year prior to nest protection to 7.6 pairs/year after protection.

A small trial in eastern England (11) found that neither of two little ringed plover *Charadrius dubius* nests that were protected by wire cages (61 × 61 × 30.5 cm cages with 5 × 5 cm mesh) in 2005 and 2006 were lost to predation.

A replicated, randomised and controlled trial in 2002 and 2004 at three grazed pasture sites in south-west Sweden (*12*) found that nests provided cages (truncated cone steel cages with 6.5–8.5 cm spacings between vertical bars and 4 × 4 cm steel netting on top) had significantly higher average daily survival rates than unprotected nests in both common redshank *Tringa totanus* (99.7% for 34 protected nests vs. 96% for 32 unprotected nests in 2002) and northern lapwing *Vanellus vanellus* (99% for 37 protected nests vs. 97 for 153 unprotected nests in 2002 and 2004). However, there was higher predation of adult redshank on protected nests, and possibly higher abandonment by lapwings, see 'Can nest protection increase predation of adults and chicks?'.

A replicated, controlled before-and-after study from 1999–2004 on pastures in south-west Sweden (*13*) found that the average hatching rate of southern dunlin *Calidris alpina schinzii* nests was significantly higher for nests protected by steel cages (20 cm high truncated cones with 7.5 cm gaps between vertical bars and 4 × 4 cm steel mesh covering the top) than for unprotected nests (67% of 25 protected nests survived to hatching vs. 41% of 61 unprotected nests). Moreover, protected nests were more likely to hatch more than one chick (80% of 25 protected nests vs. 57% of 60 unprotected nests). However, comparing 1993–98 (when no nests were protected) with 1999–2004 (when some nests were protected) revealed that there was no significant change in either the number of fledglings/breeding adult or the number of new recruits/breeding adult produced by the study sites. This study is also discussed in 'Can nest protection increase predation of adults and chicks'?

(1) Nol, E. & Brooks, R. J. (1982) Effects of predator exclosures on nesting success of killdeer. *Journal of Field Ornithology*, 53, 263–268.
(2) Deblinger, R. D., Vaske, J. J. & Rimmer, D. W. (1992) An evaluation of different predator exclosures used to protect Atlantic coast piping plover nests. *Wildlife Society Bulletin*, 20, 274–279.
(3) Melvin, S. M., MacIvor, L. H. & Griffin, C. R. (1992) Predator exclosures: a technique to reduce predation at piping plover nests. *Wildlife Society Bulletin*, 20, 143–148.
(4) Rimmer, D. W. & Deblinger, R. D. (1992) Use of fencing to limit terrestrial predator movements into least tern colonies. *Colonial Waterbirds*, 15, 226–229.
(5) Estelle, V. B., Mabee, T. J. & Farmer, A. H. (1996) Effectiveness of predator exclosures for pectoral sandpiper nests in Alaska. *Journal of Field Ornithology*, 67, 447–452.
(6) Mabee, T. J. & Estelle, V. B. (2000) Assessing the effectiveness of predator exclosures for plovers. *The Wilson Bulletin*, 112, 14–20.
(7) Murphy, R. K., Greenwood, R. J., Ivan, J. S. & Smith, K. A. (2003) Predator exclusion methods for managing endangered shorebirds: are two barriers better than one? *Waterbirds*, 26, 156–159.
(8) Neuman, K. K., Page, G. W., Stenzel, L. E., Warriner, J. C. & Warriner, J. S. (2004) Effect of mammalian predator management on snowy plover breeding success. *Waterbirds*, 27, 257–263.
(9) Niehaus, A. C., Ruthrauff, D. R. & McCaffery, B. J. (2004) Response of predators to western sandpiper nest exclosures. *Waterbirds*, 27, 79–82.
(10) Gulickx, M. M. C. & Kemp, J. B. (2007) Provision of nest cages to reduce little ringed plover *Charadrius dubius* nest predation at Welney, Norfolk, England. *Conservation Evidence*, 4, 30–32.
(11) Gulickx, M. M. C., Kemp, J. B., Beecroft, R. C. & Green, A. C. (2007) Provision of nest cages to reduce predation of little ringed plovers *Charadrius dubius* at Kingfishers Bridge, Cambridgeshire, England. *Conservation Evidence*, 4, 49–50.
(12) Isaksson, D., Wallander, J. & Larsson, M. (2007) Managing predation on ground-nesting birds: the effectiveness of nest exclosures. *Biological Conservation*, 136, 136–142.
(13) Pauliny, A., Larsson, M. & Bloqvist, D. (2008) Nest predation management: effects on reproductive success in endangered shorebirds. *Journal of Wildlife Management*, 72, 1579–1583.

12.13.3. Storks and ibises

- A randomised, replicated and controlled study from Cambodia (1) found that giant ibis *Thaumatibis gigantean* fledgling rates were higher for nests in protected trees than controls.

A randomised, replicated and controlled study from 2004–2006 in northern Cambodia (1) found daily survival rates of nests during the nestling period and average fledging rates were significantly higher for 24 giant ibis *Thaumatibis gigantea* nests in trees fitted with predator exclusion devices (an 80 cm wide strip of hard, smooth plastic, fitted at least 1.5 m from the ground) than for 28 nests in unprotected trees (daily survival rates of 99.9% for protected vs. 99.3% for unprotected nests, leading to overall survival rates of 90% vs. 61% respectively; and average fledging rates of 1.9 chicks/nest vs. 1.25 chicks/nest respectively). Protected trees were also more likely to be re-used in the next year (73% vs. 9%, 22 trees monitored).

(1) Keo, O., Collar, N. J. & Sutherland, W. J. (2009) Nest protectors provide a cost-effective means of increasing breeding success in giant ibis *Thaumatibis gigantea*. *Bird Conservation International*, 19, 77–82.

12.13.4. Songbirds

- Three studies from across the world (1,3,4) found increased fledging success for nests in trees protected by individual barriers. A replicated controlled study from the USA (2) also found higher success for ground-nests protected by individual barriers.
- Two studies from the UK (5) and Japan (4) found lower predation rates on nests protected by individual barriers.

A small, replicated and controlled study between 1984 and 1988 in northern South Island and Little Barrier Island, New Zealand (1) found that moving chicks from natural nests into caged artificial nests appeared to increase fledging success in rifleman *Acanthisitta chloris* (5% of 44 caged chicks died; 94% natural nests destroyed by predators), whitehead *Mohoua albicilla* (none of 10 caged chicks died; an estimated 0.68 chicks/successful natural nest died, n = 41 nests), grey gerygone (warbler) *Gerygone igata* (13% mortality for 15 caged chicks; 33% for nine uncaged chicks) and shining bronze-cuckoo *Chrysococcyx lucidus* (*Chalcites lucidus*) (both caged chicks survived, one uncaged chick died, number of uncaged chicks was not given). In addition, a single chaffinch *Fringilla coelebs* (an introduced species) was caged and survived. In 1987–88, all chicks from one of six caged grey fantail *Rhipidura fuliginosa* nests died, whereas no uncaged nests were lost. Cages were made of wire-mesh, stretched to ensure that food could be passed through by the parents but that birds' heads could not be caught.

A replicated, controlled study in 1979 on saltmarshes in Florida, USA (2) found that protecting seaside sparrow *Ammodramus maritimus* nests (located near the ground) with cylindrical metal barriers (1.5 m tall with a flexible wire canopy) increased the proportion of nests fledging young from 6% (34 nests) to 48% (42 nests). The main predators excluded were rice rat *Oryzomys palustris*.

A before-and-after study on Rarotonga (6,700 ha), Cook Islands (3) found that the nesting success of kakerori (Rarotonga flycatchers) *Pomarea dimidiate* was higher in 1990–92 when nests were protected with predator guards (30 cm aluminium bands around nesting trees to prevent introduced black rats *Rattus rattus* from climbing to the nests) than in 1987–89 when guards were not used. However this increase was not significant (nesting success of 36% for 48 nests in 1990–92 vs. 16% for 42 in 1987–89). This study also investigated the use of predator removal through poisoning, discussed in 'Control mammalian predators on islands'. Using predator guards in predator removal areas did not appear to affect nest survival.

A replicated before-and-after study from 1999–2000 in a mixed forest in Kyushu, Japan (4) found that the proportion of nest box broods (unprotected nest boxes were predated in 2006 either, but that these were all further away (>300 of varied tits *Parus varius* and great tits *P. major*) predated by Japanese martens *Martes melampus* fell from 12% (92 broods) to 2% (118 broods) and the overall success rate of nests increased from 29% to 44% following the installation of small wooden blocks directly below the entrance hole of the nest boxes in 2000. Overall predation rates fell from 46% to 28% and the number of broods lost to birds fell from three to zero (although this was not significant). The number of broods lost to snakes did not change between years.

A replicated before-and-after study from 2005–2006 in north-west England (5) found that the proportion of blue tit *Cyanistes caeruleus* nests in nestboxes that were predated by great spotted woodpeckers *Dendrocopos major* fell from 25% in 2005 (57 nests) to 2% in 2006 (48 nests), after the sides of 31 nest boxes were covered in 13 × 13 mm square wire mesh. The authors note that none of the unprotected nests were predated in 2006 either, but that all these boxes were at least 300 m from active woodpecker nests.

(1) Hamel, R. de & McLean, I. G. (1989) Caging as a technique for rearing wild passerine birds. *The Journal of Wildlife Management*, 53, 852–856.
(2) Post, W. & Greenlaw, J. S. (1989) Metal barriers protect near-ground nests from predators. *Journal of Field Ornithology*, 60, 102–103.
(3) Robertson, H. A., Hay, J. R., Saul, E. K. & McCormack, G. V. (1994) Recovery of the kakerori: an endangered forest bird of the Cook Islands. *Conservation Biology*, 8, 1078–1086.
(4) Yamaguchi, N., Kawano, K. M., Yamaguchi, Y. & Saito, T. (2005) Small protection plates against marten predation on nest boxes. *Applied Entomology and Zoology*, 40, 575–577.
(5) Mainwaring, M. C. & Hartley, I. R. (2008) Covering nest boxes with wire mesh reduces great spotted woodpecker *Dendrocopos major* predation of blue tit *Cyanistes caeruleus* nestlings, Lancashire, England. *Conservation Evidence*, 5, 45–46.

12.14. Can nest protection increase nest abandonment?

- A replicated before-and-after study from the USA (2) found that nest abandonment increased after nest exclosures were installed. Two replicated studies in Sweden (3) and the USA (4) found small levels of abandonment, or non-significant increases in abandonment following nest exclosure installation.
- A meta-analysis from the USA (1) found that some designs of nest exclosures were more likely to lead to abandonment than others.

Background

Placing individual barriers over nests is likely to disturb incubating parents, and in extreme cases, this may lead to them abandoning the nest. Unless nests are abandoned, studies are unlikely to mention it and so will be discussed in 'Physically protect nests with individual exclosures/barriers'.

A 1992 meta-analysis (1) analysed data from 211 nest exclosures across eight US states and three Canadian provinces to determine which factors increased or decreased the likelihood of nesting piping plovers *Charadrius melodus* abandoning their nests. Twenty two (10%) of the nests were abandoned, and the estimated probability of abandonment was significantly higher for exclosures with covers (12%, of 178 nests) than for uncovered exclosures (0% of 33). Exclosures without supporting posts were also more likely to be abandoned (40% of 35) compared to those with metal or wood posts (7% of 176). Exclosures with short posts (<122 cm) were more likely to be abandoned (32% of 41) than those with taller posts (122 cm: 5% of 40; >122 cm: 8% of 130). Finally, nests in Canada were more likely to be abandoned (40% of 35) than those in the north (5% of 121) or mid-Atlantic (12% of 55) USA. No factors related to exclosure construction, size, shape, fence height or depth (buried beneath the ground to prevent predators digging under), nor mesh size significantly altered the probability of nest abandonment.

A replicated before-and-after study from 1984–90 and 1991–99 on beaches in California, USA (2) found that nest abandonment by adult snowy plovers *Charadrius alexandrinus* increased in 1991–99 following the protection of 49% of nests (n = 682) with predator exclosures (1.5 m high triangular wire fences), compared to 1984–90 when none of the 728 monitored nests were protected (4% for 1984–90 vs. 8% for 1991–99). This study is also discussed in 'Predator control not on islands', 'Physically protect nests with individual exclosures/barriers' and 'Can nest protection increase predation of adult and chick waders?'.

A replicated, randomised and controlled trial in 2002 and 2004 at three grazed pasture sites in south-west Sweden (3) found that there was a slight trend towards higher nest abandonment in northern lapwing *Vanellus vanellus* with protected nests (truncated cone steel cages with 6.5–8.5 cm spacing between vertical bars and 4 × 4 cm steel netting on top), but that the trend was not significant (3 of 37 caged nests abandoned vs. 2 of 153 non-caged nests). This study is also discussed in 'Physically protect nests with individual exclosures/barriers' and 'Can nest protection increase nest abandonment?'

A replicated trial at three black tern *Chlidonias niger* colonies in wetlands in Maine, USA (4) found that surrounding nests with both a chick retention fence (a 30 cm high, 1 m diameter circular fence with an overhead 'concealment flap' of wire covered in landscaping cloth, 15 cm off the ground) and a predator exclosure fence (1.4 m high, 4.6 m diameter) appeared to cause the abandonment of three nests (of 17) immediately after fences were erected in 2001, however, no nests were abandoned in 2002 (14 nests). This study is also discussed in 'Use multiple barriers to protect nests'.

(1) Vaske, J. J., Rimmer, D. W. & Deblinger, R. D. (1994) The impact of different predator exclosures on piping plover nest abandonment. *Journal of Field Ornithology*, 65, 201–209.
(2) Neuman, K. K., Page, G. W., Stenzel, L. E., Warriner, J. C. & Warriner, J. S. (2004) Effect of mammalian predator management on snowy plover breeding success. *Waterbirds*, 27, 257–263.
(3) Isaksson, D., Wallander, J. & Larsson, M. (2007) Managing predation on ground-nesting birds: the effectiveness of nest exclosures. *Biological Conservation*, 136, 136–142.
(4) Heath, S. R. & Servello, F. A. (2008) Effects of predation and food provisioning on black tern chick survival. *The Wilson Journal of Ornithology*, 120, 167–175.

12.15. Can nest protection increase predation of adults and chicks?

- Three replicated and controlled studies from North America (1,2) and Sweden (4) found higher levels of predation on adult birds with nest exclosures, one study from Sweden (5) found that predation was no higher.
- A replicated and controlled study from Alaska (3) found that long-tailed jaegers *Stercorarius longicaudus* learned to associate exclosures with birds, targeting adult western sandpipers *Calidris mauri* and quickly predating chicks when exclosures were removed.

Background

Nest cages and individual barriers are distinctive and may attract predators if they learn to associate the cages with potential prey. As with the previous section, unless studies specifically mention increased predation, they are discussed in 'Physically protect nests with individual exclosures/barriers'.

A replicated, controlled study from 1993–2002 at five alkali lakes in Alberta and Saskatchewan (Canada), Montana and North Dakota, USA (1) found that adult piping plovers *Charadrius melodus* were more likely to be predated if their nests were protected by exclosures (5% of 1,355 nests suffering mortality) than if their nests were unprotected (no adults predated at 420 nests). Predation rates were highest (up to 48%) at sites with 4–15% tree cover within 2 km of the nests and zero in areas with few trees (across the study period, 393 nests monitored). At one site, when small (1–1.7 m diameter) exclosures were replaced with large (3–4 m diameter) ones with netting tops, predation rate fell from 34% in 1999 (55 nests) to 11% in 2000 (39 nests) In areas where large cages only were used, predation rates were 0.7% (303 nests). Most (78%) losses were to raptors.

A replicated before-and-after study on beaches in California, USA (2) found that nest abandonment rates of snowy plovers *Charadrius alexandrinus* combined with adult mortality increased between 1984–90 and 1991–99 (1% of 728 nests in 1984–90 vs. 4% of 682 in 1991–99) following the protection of 49% of nests with predator exclosures (1.5 m high triangular wire fences) after 1991. In addition, although only 49% of nests were protected, 75% of adult disappearances (assumed to be due to predation) were from protected nests (significantly more than expected by chance). This study is also discussed in 'Predator control not on islands', 'Physically protect nests with individual exclosures/barriers' and 'Can nest protection increase nest abandonment?'.

A replicated and controlled study in 2001 in the Yukon Delta, Alaska, USA, (3) found that survival of western sandpiper *Calidris mauri* nests was higher when they were protected by exclosures (see 'Physically protect nests with individual exclosures/barriers'). However, after 17 days, long-tailed jaegers (skuas, which predate on sandpiper adults, chicks and eggs) *Stercorarius longicaudus* began associating exclosures with nests and targeting them (whilst ignoring control nests), causing sandpipers to flush, sometimes colliding with the exclosures. One chick died from cold exposure whilst adults were being harassed by jaegers and exclosures were removed after 19 days. Following exclosure removal, chicks from exclosure nests were less likely to survive than those from control nests, with some chicks being predated minutes after the removal of exclosures.

A replicated, randomised and controlled trial in 2002 and 2004 at three grazed pasture sites in south-west Sweden (4) found that there were significantly higher predation rates on adult common redshank *Tringa tetanus* with protected nests (protected by truncated cone steel cages with 6.5–8.5 cm spacing between vertical bars and 4 × 4 cm steel netting on top) than for birds brooding at unprotected nests (nine adults from eight protected nests predated, from a total of 37 nests vs. a single bird from 31 unprotected nests). This study is also discussed in 'Physically protect nests with individual exclosures/barriers'.

A replicated, controlled trial between 1999 and 2004 on pastures in south-west Sweden (5) found that protecting southern dunlin *Calidris alpina schinzii* nests with cages (20 cm high truncated cones with 7.5 cm gaps between vertical bars and 4 × 4 cm steel mesh covering the top) did not significantly affect the predation rates on brooding adults (7% of 57 adults at protected nests predated vs. 13% of 16) adults at unprotected nests). This study is also discussed in 'Physically protect nests with individual exclosures/barriers'.

(1) Murphy, R. K., Michaud, I. M., Prescott, D. R., Ivan, J. S., Anderson, B. J. & French-Pombier, M. L. (2003) Predation on adult piping plovers at predator exclosure cages. *Waterbirds*, 26, 150–155.
(2) Neuman, K. K., Page, G. W., Stenzel, L. E., Warriner, J. C. & Warriner, J. S. (2004) Effect of mammalian predator management on snowy plover breeding success. *Waterbirds*, 27, 257–263.
(3) Niehaus, A. C., Ruthrauff, D. R. & McCaffery, B. J. (2004) Response of predators to western sandpiper nest exclosures. *Waterbirds*, 27, 79–82.
(4) Isaksson, D., Wallander, J. & Larsson, M. (2007) Managing predation on ground-nesting birds: the effectiveness of nest exclosures. *Biological Conservation*, 136, 136–142.
(5) Pauliny, A., Larsson, M. & Bloqvist, D. (2008) Nest predation management: effects on reproductive success in endangered shorebirds. *Journal of Wildlife Management*, 72, 1579–1583.

12.16. Use artificial nests that discourage predation

- Three trials in North America (2–4) found lower predation or higher nesting success of wildfowl in nest boxes or nesting 'tubs' than natural nests in tree cavities or on the ground.
- A trial in captivity (5) found that raccoons could be prevented from entering nest boxes if they were topped with a metal cone with a 7.6 cm overhang and the distance between entrance hole and the roof was increased from 30 to 60 cm.
- A replicated study in the USA (1) found that fewer woods duck *Aix sponsa* used nest boxes with predator guards on when given the choice of unaltered boxes, but that both designs were used with equal frequency when only one design was available.

Background

Nest boxes are distinctive and often easier to see than natural nests. Therefore there is the possibility that predators will learn to associate boxes with nests and target them specifically. This could even lead to nest boxes acting as 'ecological traps' – with birds using them preferentially but having very low reproductive success in them. Therefore, 'anti-predator' devices and boxes may be desirable.

A replicated trial in woodland on Rhode Island, USA, in 1955–56 (1), found that installing predator guards on wood duck *Aix sponsa* nest boxes reduced the usage of nest boxes, compared to unguarded nest boxes, when birds were given a choice of boxes (55% of 40 boxes with guards used vs. 93% of 40 unguarded boxes). However, in areas where all boxes were either guarded or unguarded, there was no significant difference in usage (51% of 55 guarded nests used vs. 56% of 52 unguarded). This study is also discussed in 'Provide artificial nesting sites'.

A before-and-after study on a marshland site in Montana, USA (2), found that Canada goose *Branta canadensis* nests on raised artificial nesting platforms were significantly less likely to be predated than the population average (7% of 14 failed platform nests predated vs. population average of 61% of 404 failed nests). This study is discussed in detail in 'Provide artificial nest sites'.

A replicated, controlled study from 1958–1961 in multiple woodlots in Illinois, USA (3) found that wood ducks *Aix sponsa* had higher nesting success in nest boxes, compared to natural cavities (71% success for 574 metal nest boxes vs. 37% for 116 natural cavities). As racoon *Procyon lotor* predation accounted for most nest loss, the authors conclude that metal nest boxes provided some protection from predation. This study is also discussed in 'Provide artificial nest sites'.

A replicated study on a wetland site in Missouri, USA (4), found that 2% of 268 Canada goose *Branta canadensis* clutches in artificial nest 'tubs' were predated in 1962–5, compared 32% of 106 clutches laid on the ground. Artificial nests consisted of 'No. 2 round washtubs' placed in trees, on fence posts or specially constructed stands and partially filled with sawdust and straw to allow geese to bury unattended eggs. Tubs were placed 1–20 m off the ground (height did not affect predation rate). Some nests were still considered useable after nine years.

A trial in captivity in Illinois, USA, in the 1960s (5) found that raccoons *Procyon lotor* could enter metal wood duck *Aix sponsa* houses (cylindrical metal houses with a 33 cm cone on top and a 10 cm circular entrance hole) more easily if the cone on top of the house was increased beyond the standard 33 cm. However, if the cone was extended to provide a 7.6 cm overhanging roof and the distance between entrance hole and the roof was increased from 30 to 60 cm, then none of the five raccoons tested could access the nest hole. Reducing the size of the entrance hole to 8.9 × 7 cm also prevented all raccoons from entering, but the authors warn that it is not certain whether wood ducks would use such entrances.

(1) Cronan, J. M. (1957) Effects of predator guards on wood duck box usage. *The Journal of Wildlife Management*, 21, 468.
(2) Craighead, J. J. & Stockstad, D. S. (1961) Evaluating the use of aerial nesting platforms by Canada geese. *The Journal of Wildlife Management*, 25, 363–372.
(3) Bellrose, F. C., Johnson, K. L. & Meyers, T. U. (1964) Relative value of natural cavities and nesting houses for wood ducks. *The Journal of Wildlife Management*, 28, 661–676.
(4) Brakhage, G. K. (1966) Tub nests for Canada geese. *The Journal of Wildlife Management*, 30, 851–853.
(5) Eaton, R. L. (1966) Protecting metal wood duck houses from raccoons. *The Journal of Wildlife Management*, 30, 428–430.

12.17. Use multiple barriers to protect nests

- A replicated, controlled study from the USA (1) found no evidence that erecting an electric fence around nests protected by individual barriers increased fledging success in piping plovers *Charadrius melodus*.
- A replicated study from the USA (2) found that removing the outer of two nest protection fences after 15 days appeared to reduce predation compared to when both fences were left for 18 days.

A replicated, controlled study in 1996 and 1997 at three alkali lakes in North Dakota and Montana, USA (1) found that piping plover *Charadrius melodus* fledging rates were higher with mesh fences erected around individual nests (see 'Physically protect nests with individual exclosures/barriers'). When an electric fence (1.1 m tall) was erected around study sites there was a non-significant increase in fledging success, compared with sites where only individual nest fences were used (2.1 chicks/pair with electric fence and nest fences, n = 50 vs. 1.7 chicks/pair with only nest fences, n = 46).

A replicated study from 2001–2002 at three black tern *Chlidonias niger* colonies in wetlands in Maine, USA (2) found that surrounding nests with both a chick retention fence and a predator exclosure fence but removing the chick retention fence 15 days after hatching in 2002 appeared to reduce predation (three chicks from one nest predated, n = 33 chicks from 14 nests), compared to when the retention fence was left until chicks were 18 days old in 2001 (17 chicks from seven nests predated, n = 36 chicks from 14 nests). The chick retention fence was 30 cm high and 15 cm off the ground, and consisted of a 1 m diameter circular fence with an overhead 'concealment flap' of wire covered in landscaping cloth; the predator exclosure fence was 1.4 m high and 4.6 m in diameter. The study did not include control (unprotected) nests, so the overall effectiveness of the fences cannot be judged. This study also discussed nest abandonment; see 'Can nest protection increase nest abandonment?'

(1) Murphy, R. K., Greenwood, R. J., Ivan, J. S. & Smith, K. A. (2003) Predator exclusion methods for managing endangered shorebirds: are two barriers better than one? *Waterbirds*, 26, 156–159.
(2) Heath, S. R. & Servello, F. A. (2008) Effects of predation and food provisioning on black tern chick survival. *The Wilson Journal of Ornithology*, 120, 167–175.

12.18. Plant nesting cover to reduce nest predation

Background
Studies relevant to this intervention are in 'Threat: Agriculture'.

12.19. Use snakeskin to deter mammalian nest predators

- A randomised, replicated and controlled trial in the USA (1) found that artificial nests were less likely to be predated if they had snake skin wrapped around them than control nests.

Background
Some bird species such as great crested flycatchers *Myiarchus crinitus* and tufted titmice *Baeolophus bicolor* use snake skins in their nests, possibly to reduce predation.

A randomised, replicated and controlled trial in May–June 2004 in Arkansas, USA (1) found that artificial great crested flycatcher *Myiarchus crinitus* nests placed inside 60 nest boxes were less likely to be predated if there was black rat snake *Elaphe obsolete* skin inside the nest box (0/20 nests predated) or both inside and outside the nest box (0/20 predated) than if there was no snake skin present (5/20 predated). Predation was mainly by southern flying squirrels *Glaucomys volans*. Snake skins were treated by being placed in proximity to a live rat snake for five hours prior to installation.

(1) Medlin, E. C. & Risch, T. S. (2006) An experimental test of snake skin use to deter nest predation. *The Condor*, 108, 963–965.

12.20. Use mirrors to deter nest predators

- We found no published evidence for the effects of mirrors on nest predation rates.

Background

Anecdotal evidence suggests that fixing mirrors to nest boxes may reduce predation rates by species such as great spotted woodpeckers *Dendrocopos major*.

12.21. Use naphthalene to deter mammalian predators

- A replicated, controlled study from the USA (1) found that scattering naphthalene moth balls near artificial nests did not affect predation rates.

Background

Naphthalene is a strong-smelling and potentially harmful organic compound, frequently used to deter many animals, such as moths and rodents.

A replicated, controlled study in July 1986 in a cord grass *Spartina alterniflora* marsh in South Carolina, USA (1) found that eggs placed in 40 abandoned red-winged blackbird *Agelaius phoneiceus* nests (mostly in southern red cedar *Juniperus silicicola* or marsh elder *Iva frutescens*) were as likely to be predated if six moth balls (treated with 100% naphthalene) were scattered in the vegetation within a 2 m radius the nest (50% of 20 nests predated), as if no moth balls were used (35% of 20 nests predated).

(1) Gawlik, D. E., Hostetler, M. E. & Bildstein, K. L. (1988) Naphthalene moth balls do not deter mammalian predators at red-winged blackbird nest. *Journal of Field Ornithology*, 59, 189–191.

12.22. Use ultrasonic devices to deter cats

- We found no evidence for the effects of ultrasonic cat deterrents on bird populations.

Background

Ultrasonic devices emit high-pitched (normally above 20 kHz) noise, generally above the hearing of humans, but audible to cats. The devices aim to discourage cats from approaching them and therefore keep cats away from sensitive areas. Two randomised, replicated and controlled trials in the UK (Nelson *et al.* 2006) found that an ultrasonic cat deterrent in gardens reduced the number of visits by cats by 32% in 63 gardens across an 18 week trial but had no effect over 96 gardens in a 33 week trial.

Nelson, S. H., Evans, A. D. & Bradbury, R. B. (2006) The efficacy of an ultrasonic cat deterrent. *Applied Animal Behaviour Science*, 96, 83–91.

12.23. Protect nests from ants

- A randomised, replicated and controlled study from the USA (1) found higher fledging success from white-eyed vireo *Vireo griseus* nests protected from ants with a physical barrier and a chemical repellent, compared to control nests.

Background

Several ant species including the red imported fire ant *Solenopsis invicta*, the big-headed ant *Pheidole megacephala* and yellow crazy ants *Anoplolepis gracilipes* have been introduced across the world and have often become invasive. Whilst too small to damage most adult birds, they can have severe impacts on chicks. Due to their huge numbers and adaptability eradication is rarely an option (although see 'Control invasive ants on islands'). Therefore preventing them from accessing nests may be the best intervention.

A randomised, replicated and controlled study in March–July 2006–7 in a grassland/oak-juniper woodland mosaic in Texas, USA (1) found that 18 white-eyed vireo *Vireo griseus* nests, protected from red imported fire ants *Solenopsis invicta* with a physical barrier and a chemical repellent, had significantly higher fledging success than 26 unprotected nests (31% vs. 10%). The same effect was seen in 13 experimental and 14 control black-capped vireo *V. atricapilla* nests, but this difference (13% vs. 7%) was non-significant. The physical barrier was Tanglefoot – a gum resin that traps crawling insects, applied to the branch >25 cm from each nest; the repellent was Arinix™ spiral wrap – a permethrin releasing plastic wrapped around the branch on top of the Tanglefoot.

(1) Campomizzi, A. J., Morrison, M. L., Farrell, S. L., Wilkins, R. N., Drees, B. M. & Packard, J. M. (2009) Red imported fire ants can decrease songbird nest survival. *The Condor*, 111, 534–537.

12.24. Guard nests to prevent predation

Background

If populations are reduced to extremely low levels and continue to be threatened by nest predation, then extremely intensive monitoring can be used to 'guard nests' and protect nests from predators through direct interference. Due to the intensive nature of this work it is only likely to be viable if there are volunteers to do it, and the population being monitored is very small. Nest guarding can be used as a response to range of threats and is therefore discussed in 'General responses to small/declining populations'.

12.25. Use 'cat curfews' to reduce predation

- We captured no evidence for the effects of 'cat curfews' on bird populations.

Background

'Cat curfews' have been started in some cities in Australia, which are aimed at keeping domestic cats *Felis catus* inside for either the entire day, or during the night, when they do the majority of their hunting.

12.26. Use lion dung to deter domestic cats

- We found no evidence for the effects of lion dung application on the use of gardens by cats or on cat predation.

Background

Lion *Panthera leo* dung is sold by some companies as a way of deterring domestic cats from entering gardens.

12.27. Play spoken-word radio programmes to deter predators

- We found no published evidence for the effects of playing the radio on predation rates.

Background

An anecdotal account from Abbotsbury Swannery, Dorset, UK, claimed that playing spoken-word radio programmes deterred red foxes *Vulpes vulpes* from attacking mute swans *Cygnus olor*.

Reduce mortality by reducing hunting ability or changing predator behaviour

Background

In many circumstances it may be unfeasible or undesirable to control predator numbers or exclude predators from areas. However, it may still be possible to reduce the impact of predators in a variety of ways.

12.28. Use collar-mounted devices to reduce predation

- Two replicated randomised and controlled studies in the UK (1) and Australia (2) found that significantly fewer birds were returned by cats wearing collars with various anti-hunting devices, compared to controls.
- A replicated, randomised and controlled study from the UK (1) found no significant differences between different devices.
- Both UK studies (1) found that collars were easily lost.

Background

Cat collars can be modified to carry either simple bells or more complex sonic alerts which scare birds before cats can catch them. Another option is a 'bib' which hangs in front of the cat's legs, preventing them from hunting effectively.

A replicated, randomised and controlled study across the UK between April and August 2002 (1) found that for a total of 89 cats, fewer birds were returned by those fitted with a collar and bell (41% reduction and 74 birds) or a collar with a 'CatAlert™' sonic device (51% reduction and 59 birds) than by cats with a plain collar (117 birds). The difference between bell and ultrasound was not significant. A second replicated and randomised study between May and September 2003 found that, for a total of 67 cats, the number of birds returned by cats was not significantly affected by whether cats were wearing collars with one bell, two bells or 'CatAlert™' sonic devices. In both trials, the authors note that collars were easily lost, with a total of 55 sonic device, 39 one-bell, 16 two-bell and 21 plain collars lost (and replaced) over the study. The authors also note that that this study does not support the assertion that cats can learn to adapt hunting to render bells less effective.

A replicated, randomised and controlled study in Perth, Australia in November–December 2005 (2) found that wearing a 'CatBib™' 'pounce protector' (a neoprene flap that hangs from a collar in front of a cat's front legs, acting either as a visual warning or as a barrier to pouncing) for three weeks, reduced the number of cats catching birds by 81% compared to when the same cats were not wearing the 'CatBib™' (5 vs. 26; n = 56 cats). The average number of birds captured per cat was also significantly lower (0.29 versus 0.88). Adding a bell to the 'CatBib™' did not further reduce hunting.

(1) Nelson, S. H., Evans, A. D. & Bradbury, R. B. (2005) The efficacy of collar-mounted devices in reducing the rate of predation of wildlife by domestic cats. *Applied Animal Behaviour Science*, 94, 273–285.
(2) Calver, M., Thomas, S., Bradley, S. & McCutcheon, H. (2007) Reducing the rate of predation on wildlife by pet cats: the efficacy and practicability of collar-mounted pounce protectors. *Biological Conservation*, 137, 341–348.

12.29. Use supplementary feeding to reduce predation

- A controlled cross-over experiment from the UK (2) found that there was no difference in grouse adult survival or productivity when supplementary food was provided to hen harrier *Circus cyaneus* compared to in control areas.

- This study (2) and another from the USA that used artificial nests (1) found that nest predation rates were reduced in areas when supplementary food was provided to predators. A second study from the USA (3) found no such effect.

Background

If predators are dependent on threatened birds then providing supplementary food for the predators may remove some predation pressure whilst also supporting the predator population. However, it is also possible that providing supplementary food will allow an increase in the predator population and therefore increase predation pressure.

Studies describing the effects of supplementary food on fed populations are described in a separate chapter.

A randomised, replicated and controlled experiment on eight Conservation Reserve Program sites in 1993–94 in Texas, USA (1) found that the predation rates on artificial nests (containing three chicken *Gallus gallus domesticus* eggs with 1 nest/4.3 ha), were 45% lower in plots where supplementary predator food was provided (details of food provided are not given), compared to nests in control plots. A total of 1,735 artificial nests were used.

A controlled cross-over experiment, on moorland in south-west Scotland, UK, in 1998 and 1999 (2) found that adult red grouse *Lagopus lagopus scoticus* survival was no higher in 13 hen harrier *Circus cyaneus* territories that were provided with a total of 256 kg of food in spring (over two years), than in control (unfed) territories (78% survival for 94 birds in fed areas vs. 74% of 97 in control areas). Supplementary feeding in the summer (when harriers are provisioning young) reduced the number of grouse chicks being brought to 14 fed broods, compared to ten unfed broods (an average of 0.5 chicks/100 hr, seven in total vs. 3.7 chicks/100 hr, 32 in total). However, there was no corresponding improvement in grouse breeding success in fed areas.

A replicated, randomised and controlled study in May–July 2000 in 28 longleaf pine *Pinus palustris* forest plots in Georgia, USA (3) found no differences in predation rates on artificial nests in areas provided with supplementary food (commercial dry dog food supplied *ad libitum* from feeders) and control areas (nest predation over one week: 62% for prey-supplemented areas vs. 55% for control plots; 770 nests tested). Birds and small mammals were responsible for more predation events in food-supplemented plots, whilst unknown predators were responsible for more in non-supplemented plots. Nests were placed on the ground and contained two Japanese quail *Corturnix japonica* eggs and one wax covered wooden egg. This study also evaluated the impact of prescribed burning on nest survival, discussed in 'Use prescribed burning – pine forests'. There was no interaction between feeding and burning.

(1) Vander Lee, B. A., Lutz, R. S., Hansen, L. A. & Mathews, N. E. (1999) Effects of supplemental prey, vegetation, and time on success of artificial nests. *The Journal of Wildlife Management*, 63, 1299–1305.

(2) Redpath, S. M., Thirgood, S. J. & Leckie, F. M. (2001) Does supplementary feeding reduce predation of red grouse by hen harriers? *Journal of Applied Ecology*, 38, 1157–1168.

(3) Jones, D. D., Conner, L. M., Warren, R. J. & Ware, G. O. (2002) The effect of supplemental prey and prescribed fire on success of artificial nests. *The Journal of Wildlife Management*, 66, 1112–1117.

12.30. Use aversive conditioning to reduce nest predation

Background

If predators can be taught to associate unpleasant tastes with eggs and/or chicks, then they may stop predating them. This can be achieved either through treating eggs in nests (spraying eggs or the nest) or placing artificial nests in the environment with eggs injected with unpleasant chemicals (not possible with natural nests as it will kill eggs).

12.30.1. Avian predators

- Five studies from the USA (1–3) and Europe (4,7) found reductions in consumption of eggs treated with various chemicals. A further *ex situ* study from the USA (5) found that American kestrels *Falco sparverius* consumed fewer chicks when they were treated, but not to the point of losing body condition.
- Three studies from the USA (1–3) found some evidence that treating eggs with some chemicals may have reduced predation of eggs after treatment stopped, or of untreated eggs, although two of these (2,3) were only short term experiments and the third (1) found that the effect was lost after a year.
- Four studies from the Europe (4,6,7) and the USA (5) found no evidence for conditioning, or a reduction in predation of wild (untreated) eggs.

A randomised, replicated and controlled before-and-after experiment at 21 sites in Illinois and Iowa, USA, in summer 1986 (1), found that predation of dyed-green chicken eggs by American crows *Corvus brachyrhynchos* over a 23 day period, was significantly reduced when 50% or 100% of green eggs (eight provided each day in total) were treated with Landrin (a tasteless but illness-inducing chemical). There was no corresponding reduction in consumption of green eggs at sites where they were not treated with Landrin. Sites where 12.5% of green eggs were treated had intermediate levels of predation (100% sites: 7.8 attacks/day before treatment vs. 1.2 attacks/day after provision of Landrin-treated eggs; 50% sites: 5.6 vs. 1.4; 12.5% sites: 6.0 vs. 3.4; control sites: 7.6 vs. 7.2). At 50% sites, crows also stopped predating un-dyed eggs and consumption was reduced at 12.5% and 100% sites but remained unchanged at control sites. Post-test trials (when green eggs were again distributed but did not contain Landrin) in 1986 found that crows resumed predation at 100% sites but not at 12.5% or 50% sites. Further tests in 1987 found that crows at all sites except 50% ones resumed predating green eggs.

A replicated *ex situ* trial in the USA (2) found that consumption of Japanese quail *Corturnix japonica* eggs by 30 fish crows *Corvus ossifragus* (in five treatment groups) was affected by the injection of different chemicals into the eggs. Topically applying methyl anthranilate to the outside of eggs (alone or in conjunction with injecting 18 mg of methiocarb) reduced consumption compared to other treatments (injection of 18 mg of carbachol or methiocarb or 18 mg methiocarb plus 100 mg methyl anthranilate). Post-treatment tests with untreated eggs found that only crows from the topical methyl anthranilate groups did not consume eggs on the first day following treatment stopping (only two of 12 birds from these two groups resumed consumption in the post-treatment phase, although seven others moved but did not eat eggs). In a separate experiment, 16 crows in two groups were given eggs injected with either 30 mg methiocarb or 40 mg carbachol for five days. These crows consumed more eggs than those exposed to topical methyl anthranilate treated eggs, but fewer than other previous treatments. Five of eight crows exposed to 30 mg methiocarb and three of eight exposed to 40 mg carbachol ate eggs in the post-treatment test period.

A replicated, randomised and controlled before-and-after experiment in California, USA, in 1991 (3) found that significantly fewer Japanese quails' *Coturnix japonica* eggs were taken by ravens *Corvus corax* from artificial nest scrapes in a 'test period' following a 'training period' (approximately two weeks), when eggs were treated with methiocarb than during the training period itself (0–33% of eggs taken during the test vs. 9–67% during training). Fewer eggs were taken from four sites that contained treated eggs over the test period, compared to sites where eggs were untreated in the test period, but had been treated during training (3% of eggs taken at one of four treated sites only vs. 0–33% taken at untreated sites). A follow-up experiment found that one of the eight raven pairs previously conditioned to avoid quails' eggs resumed predation of eggs (both treated and control eggs) when they were placed in a simulated Californian least tern *Sterna antillarum browni* colony. A further experiment found that when methiocarb-treated eggs were presented at ten sites within three least tern colonies in 1992, a total of 20 eggs were removed or broken over 1,450 'egg days' and no tern eggs were predated by ravens at any of the colonies in 1992.

A controlled, replicated before-and-after study at a heronry in northern Italy in 1994 (4) found that the percentage of greenish-brown hens' eggs predated by hooded crows *Corvus cornix* was significantly lower than blue hens' eggs when they were treated with carbachol. There was no difference in predation rates either before or after the 12 day treatment period (before treatment: 100% of both blue and brown eggs consumed within one day, n = 40; treatment: 61% of untreated blue eggs consumed vs. 38% of treated brown eggs, n = 480; after treatment: 90% of both blue and brown eggs consumed, n = 40). During the conditioning period, consumption rates were similar until the tenth day and then only a single brown egg was consumed over three days.

A replicated, randomised and controlled *ex situ* experiment with 33 American kestrels *Falco sparverius* (5) found that control (untreated) day-old chicks were preferentially chosen, compared with chicks treated with methyl anthranilate on two out of four experimental days (ten birds choosing controls first vs. one choosing methyl anthranilate treated chicks and nine choosing controls vs. two choosing treated). Birds showed a preference for controls over aminoacetophenone-treated chicks on one day (nine choosing controls vs. two choosing treated chicks). On all other days there was no difference in preference for treatment or control chicks. The total amount of food consumed was highest for control kestrels, intermediate for those fed on aminoacetophenone-treated chicks and lowest for those fed methyl anthranilate-treated chicks, however, kestrels did not appear to lower consumption to the point of threatening body condition: there were no significant differences between kestrel weights at the end of the trial. A further replicated, randomised and controlled experiment found that treating cockerels with the two chemicals *and* dyeing them either green or blue greatly reduced food intake. However, there was no difference between consumption of dyed and treated chicks and controls that were only dyed, suggesting that aversive conditioning was not occurring. This study is also discussed in 'Use coloured baits to reduce accidental mortality during predator control' and 'Use repellents on baits'.

A controlled before-and-after study in a least tern *Sterna albifrons* colony in west Portugal (6) found that 15 artificial nests containing methiocarb (an illness-inducing chemical) treated quails' eggs were predated by carrion crows *Corvus corone* at the same rate as 15 artificial nests containing untreated quails' eggs. During pre-treatment (no eggs treated), first treatment phase (six days with treated eggs in experimental nests and untreated eggs in control nests) and second treatment phase (a further eight days of treatment) all eggs were destroyed within 24 hours of placement, although many treated eggs were not consumed following removal.

A replicated before-and-after study on Vila Islet, Azores, Portugal, in 2003 (7), found that the number of methiocarb-treated domestic quails' *Coturnix coturnix* eggs (11.25 mg methiocarb/egg) predated by yellow-legged gulls *Larus michahellis* from artificial tern nests in a mixed common tern *Sterna hirundo* and roseate tern *S. dougalli* colony, over six days was significantly lower than the number of untreated eggs taken in the previous three days (2.5 treated eggs predated/day vs. 10.6 untreated eggs predated/day). Once terns started laying, treated eggs were placed in 18 artificial nests at the colony and replaced if predated. No gulls were observed removing eggs over 13 days, but European starlings *Sturnus vulgaris* took both treated eggs and tern eggs. Predation of treated eggs declined over time, but there was no corresponding decline in predation on genuine tern eggs (days 1–6: 13% of tern eggs and 9.3% of treated eggs predated; days 7–13: methiocarb concentration increased to 22.5 mg/egg, 12.3% of tern eggs predated vs. 5.6% of treated eggs).

(1) Dimmick, C. R. & Nicolaus, L. K. (1990) Efficiency of conditioned aversion in reducing depredation by crows. *Journal of Applied Ecology*, 27, 200–209.

(2) Avery, M. L. & Decker, D. G. (1994) Responses of captive fish crows to eggs treated with chemical repellents. *The Journal of Wildlife Management*, 58, 261–266.

(3) Avery, M. L., Pavelka, M. A., Bergman, D. L., Decker, D. G., Knittle, C. E. & Linz, G. M. (1995) Aversive conditioning to reduce raven predation on California least tern eggs. *Colonial Waterbirds*, 18, 131–138.

(4) Bogliani, G. & Bellinato, F. (1998) Conditioned aversion as a tool to protect eggs from avian predators in heron colonies. *Colonial Waterbirds*, 21, 69–72.

(5) Nicholls, M. K., Love, O. P. & Bird, D. M. (2000) An evaluation of methyl anthranilate, aminoacetophenone, and unfamiliar coloration as feeding repellents to American kestrels. *Journal of Raptor Research*, 34, 311–318.
(6) Catry, T. & Granadeiro, J. P. (2006) Failure of methiocarb to produce conditioned taste aversion in carrion crows consuming little tern eggs. *Waterbirds*, 29, 211–214.
(7) Neves, V. C., Panagiotakopoulos, S. & Furness, R. W. (2006) A control taste aversion experiment on predators of roseate tern (*Sterna dougallii*) eggs. *European Journal of Wildlife Research*, 52, 259–264.

12.30.2. Mammalian predators

- One study from the USA (1) and three *ex situ* experiments (2–4) found evidence for lower consumption of eggs treated with repellent chemicals.
- However, when untreated eggs were provided simultaneously with (1) or after (2–4) treated eggs, no studies found evidence for continued lower predation. I.e. aversive conditioning did not occur. In addition, a study from the USA (5) found no effect of repellent chemicals on predation rates of genuine nests.

A replicated and controlled before-and-after experiment in southern Connecticut, USA, in June–September 1986 (1), found that distributing 40 eggs treated with 20–25 mg of emetine dihydrochloride along 0.7–1.0 km transects at three second growth deciduous forest sites each week for three weeks reduced consumption of eggs by mammalian predators (raccoons, opossums, skunks and rodents) by >80% during treatment and for the following three week period (from >75% of eggs predated daily to <15%). There was no corresponding decrease at five control sites, where only untreated eggs were presented (daily predation rates rose from 3% to 90%). However, a randomised, replicated and controlled paired sites study in July–September 1987 found that egg predation was not significantly different at four experimental sites, where 10 eggs treated with 20–25 mg of emetine dihydrochloride and 10 untreated eggs were placed in set locations twice a week, compared to control sites, where only untreated eggs were provided.

A randomised, replicated and controlled experiment on ten captive coyotes *Canis latrans* (each tested with ten treatments over ten consecutive three day trials) in Utah, USA (2), found that no differences in food consumption, time delay before eating or amount of time spent eating when one of ten volatile chemicals was placed adjacent to food (so that they could be smelt but not ingested) at either the first or second exposure or in post-exposure trials. However, it also found that injecting eggs with 1 ml of one of ten volatile chemicals reduced the amount of egg consumed during both first and second exposures, compared to control eggs, for all chemicals except ammonia. However, egg consumption during post-treatment trials was unchanged following treatment and all eggs in post-treatment trials were opened. The ten chemicals tested were: allyl sulphide (garlic), ammonia, capsaicin (chilli pepper), chloroacetophenone (chemical mace), cinnamaldehyde (hot cinnamon), ethyl acetate, isoamyl nitrite (smelling salts), napthaldehyde (mothballs), pulegone (mint extract) and undecanone (commercially available dog repellent). When injected, pulegone, allyl sulphide and cinnamaldehyde reduced the amount of egg consumed significantly more than the other chemicals.

A replicated, controlled experiment with 12 captive coyotes *Canis latrans* (3) found that they preferentially consumed eight untreated eggs from untreated nests, compared to four untreated eggs from nests sprayed with pulegone (mint extract) or four eggs sprayed with pulegone, over a three day period. A second trial with 29 coyotes found that, during a five-day conditioning period when coyotes were presented with eggs injected with 1 ml pulegone, they opened and consumed fewer eggs each day (from 100% to <40% opened, <8% consumed). However, after the conditioning period, coyotes continued to eat 100% of untreated eggs when presented with them, either singly or alongside pulegone injected and sprayed eggs.

A randomised, replicated and controlled *ex situ* experiment in the UK (4) found that administering thiabendazole orally to 33 rats after they ate either a chicken *Gallus gallus domesticus* or quail *Cortunix coturnix* egg reduced the rate that they subsequently fed on either chicken or quail eggs, compared to control rats. Experimental rats ate 83% fewer eggs over eight post-conditioning tests and spent 80% less time eating eggs. No rats offered the same type (chicken or quail) of egg as in the experiment ate it in the first post-conditioning trial and only 20% of those offered the alternative egg ate it. All effects grew weaker over the eight post-conditioning tests, with most experimental combinations being indistinguishable from controls after eight tests.

Two randomised, replicated and controlled experiments in April–July 1996 at a mixed ring-billed gull *Larus delawarensis* and California gull *Larus californicus* colony in Idaho and California, USA (5), found that none of three aversive conditioning strategies reduced the number of eggs from experimental nests predated, compared to control nests. At a colony in Idaho, a trial with 110 repeats of each treatment found that, following a two week conditioning phase, where 45 pulegone (mint extract) injected chicken eggs were distributed and replaced around the colony, neither placing two drops (1 ml each) of pulegone on the edge of gull nests, nor spraying 2 ml of pulegone around the periphery of a nest reduced the number of eggs predated (26–38% of nests with pulegone drops were predated, 25–34% of sprayed nests and 22–37% of controls). At a colony in Idaho and two in California, a trial with 275 replicates found that, after a similar conditioning phase, burying a cup containing 2 ml of pulegone (mint extract) so that its lip was flush with the ground of the nest did not reduce the number of eggs predated, compared to control nests (30–31% predation for treated nests vs. 33–35% for controls).

(1) Conover, M. R. (1990) Reducing mammalian predation on eggs by using a conditioned taste aversion to deceive predators. *The Journal of Wildlife Management*, 54, 360–365.
(2) Hoover, S. E. & Conover, M. R. (1998) Effectiveness of volatile irritants at reducing consumption of eggs by captive coyotes. *The Journal of Wildlife Management*, 62, 399–405.
(3) Hoover, S. E. & Conover, M. R. (2000) Using eggs containing an irritating odor to teach mammalian predators to stop depredating eggs. *Wildlife Society Bulletin*, 28, 84–89.
(4) Massei, G., Lyon, A. J. & Cowan, D. P. (2002) Conditioned taste aversion can reduce egg predation by rats. *The Journal of Wildlife Management*, 66, 1134–1140.
(5) Conover, M. R. & Lyons, K. S. (2005) Will free-ranging predators stop depredating untreated eggs in pulegone-scented gull nests after exposure to pulegone-injected eggs? *Applied Animal Behaviour Science*, 93, 135–145.

12.31. Reduce predation by translocating nest boxes

- Two studies from Europe (1,2) found that predation rates were lower for relocated nest boxes, compared to controls.

Background

Many predators optimise their hunting by searching areas where they have previously been successful. This potentially makes birds nesting in nest boxes vulnerable – they cannot move nesting sites and can lose their clutches year after year if a predator learns their location. Moving nest boxes between years may, therefore, reduce predation and increase reproductive success. This is not the same as translocating birds from an area of high predation to a safer location. Studies describing this intervention are discussed in a separate section within 'General responses to small/declining populations'.

A small, randomised and controlled cross-over study from 1975–1990 in boreal forest in south-east Norway (1) found that relocated Tengmalm's owl *Aegolius funereus* nest boxes were predated less by European pine martens *Martes martes* than unmoved boxes. Nest boxes moved in 1983 by 50–200 m suffered significantly lower predation in 1984–5, compared to before relocation (1975–83), or to control (unmoved) boxes (40% of five nests predated after relocation vs. 100% of 13 nests before and 83% of six controls). Treatments were reversed in 1988–90: the 14 control boxes moved by 110–370 m and previously moved boxes were kept in the same place. Predation rates on newly moved boxes fell (0% predation for four nests after relocation vs. 77% for 22 nests in 1975–85). No statistical comparison was possible with boxes moved in 1983, as only two nesting attempts were made in 1988–90 (of which one was predated).

A replicated, controlled study from 1995–1998 in oak *Quercus* spp. forests in west central Italy (2) found that predation on nest boxes by European pine martens *Martes martes* increased significantly with age, with 76% of clutches being predated when boxes were six years old. Relocating nest boxes to 800–2000 m away significantly reduced predation rates, compared to nest boxes moved by approximately 100 m (10 of 188 clutches and 37 of 147 clutches predated respectively).

(1) Sonerud, G. A. (1993) Reduced predation by nest box relocation: differential effect on Tengmalm's owl nests and artificial nests. *Ornis Scandinavica*, 24, 249–253.
(2) Sorace, A., Petrassi, F. & Consiglio, C. (2004) Long-distance relocation of nestboxes reduces nest predation by pine marten *Martes martes*. *Bird Study*, 51, 119–124.

Reduce competition with other species for food and nest sites

Intense competition can have detrimental impacts on populations by reducing access to food and nesting sites. This can be especially damaging if competitor species are introduced from elsewhere, as has happened in many parts of the world. Native species may not be able to adapt to new competitors and may suffer as a result.

Control of competitor species is a possible intervention (discussed below) but is very controversial with both cost-effectiveness and animal welfare issues to consider.

12.32. Reduce inter-specific competition for nest sites by removing competitor species

Background

As humans modify ever-increasing amounts of habitat across the world, the number of nesting sites for many species is becoming limited, potentially increasing competition and reducing reproductive output. Two potential solutions are to increase the number of nesting sites (see separate section in 'General responses to small/declining populations') or to reduce competition by removing or controlling competitor species (see below).

12.32.1. Ground nesting seabirds

- Four studies from Canada (1,2,4) and the UK (7) found increased tern *Sterna* spp. populations following the control or exclusion of gulls *Larus* spp. In two studies (1,2) many interventions were used, making it impossible to tell which was responsible. One study from the UK (5) and one from Canada (2) found that controlling large gulls had no impact on smaller species.
- Two studies from the USA (3) and UK (6) found that exclusion devices successfully reduced the numbers of gulls at sites, although one (3) found that they were only effective at small colonies and the other (6) found that methods varied in their effectiveness and practicality.

A before-and-after study over 13 years at an offshore common tern *Sterna hirundo* colony in Lake Erie, Canada (1) found that the number of breeding pairs steadily increased between 1977 and 1989 following various management interventions but this increase could not be linked clearly to any of them. Interventions were: the erection of signs informing people to avoid disturbing nesting birds (1981), the replacement of nesting substrate following flooding (1988), preventing gulls from nesting and destroying gull nests (1977 onwards) and shooting particular ring-billed gulls *Larus delawarensis* that were heavily predating tern eggs (three gulls shot in 1987). This study is also discussed in 'Control avian predators on islands' and 'Manually remove vegetation from wetlands'.

Two before-and-after studies in 1977–89 at two common tern *Sterna hirundo* colonies in Lake Ontario, Canada (2), found that the nesting population increased at one colony but decreased at the second following the use of several interventions, including the exclusion of ring-billed gulls *Larus delawarensis*. This study is discussed in 'Replace nesting substrate following severe weather'.

A replicated study in the summers of 1992–5 at four lake sites in Minnesota, USA (3), found that coloured nylon string or monofilament line strung between electric fence posts (2 m apart, 0.8 m above ground) were effective at preventing ring-billed gulls *Larus delawarensis* from breeding (and so out-competing common terns *Sterna hirundo*) at small or new colonies (with 60–70 gulls abandoning two sites after erection of strings and destruction of several nests). However, strings were not effective at larger, denser colonies (even when the distance between wires was reduced to 1.2 m), unless placed in a grid structure (and some gulls still nested at the site). A combination of chicken wire at ground level and wires 1 m apart visible wires above ground was not effective at the large colony it was tested at. Adding monofilament line to this structure did not deter gulls but several Caspian terns *S. caspia* (a non-target species) became entangled in some of the monofilament wires.

A before-and-after study from 1989–1996 in the upper St. Lawrence River, Canada (4) found that excluding ring-billed gulls *Larus delawarensis* from a 0.17 ha island increased the number of common terns *Sterna hirundo* from zero in 1989 to 135 pairs in 1993 (compared with 121 pairs in 1976 before gulls colonised the island). Monofilament lines (70 cm apart over approximately 60% of the island), combined with destroying gull nests and eggs during 1990–3, reduced the number of gulls breeding from 181 pairs in 1989 to zero in 1990 and 1991. The exclosure was not erected in 1994–6 and fewer gull nests were destroyed, leading to a recovery in gull numbers from two (1994) to 250 nests (1996). Tern numbers remained high in 1994–5 (141 and 149 pairs) but fell to three pairs in 1996.

A before-and-after study at a former gravel pit in Kent, England (5), reports that removing the eggs and nests of large gulls *Larus* spp. in the 1990s did not have any effect on halting the decline of smaller *Larus* spp. and terns *Sterna* spp. It also reports that the management was disturbing great cormorants *Phalacrocorax carbo* nesting nearby and was stopped in 2003. A grid of 'exposure lines' deployed over the islands was also unsuccessful in either excluding large gulls or attracting terns. Other management on the island is described in 'Provide artificial nesting sites' and 'Manually remove vegetation from wetlands'.

A study on Coquet Island (5.4 ha) in 2000–5 in north-east England (6) found that a gas gun, scarecrows, bird-scaring rockets, taped distress calls and human disturbance were all effective in deterring herring gulls *Larus argentatus* and lesser black-backed gulls *L. Fuscus* from disrupting common tern *Sterna hirundo* breeding. However, a 'humming line' (two strips of plastic that vibrate in the wind), a gird of plastic string designed to stop birds settling and a 'scarer rope' (a slow burning rope with a series of explosives attached) all had practical issues: the humming line broke, gulls became entangled in the string grid and the rope was difficult to light in wet weather and scared common eiders *Somateria mollissima* from their nests. The consequences of these measures are discussed below.

A before-and-after study from 1998–2001 and 2004–2008 (7) on Coquet Island, between 2000 and 2009 found that the disturbance regimes employed in (6) successfully reduced the number herring gulls *Larus argentatus* and lesser black-backed gulls *L. fuscus* nesting (approximately 250 pairs in 2002 vs. <20 in 2009) and allowed the recovery of four tern *Sterna* spp. populations (roseate terns *S. dougallii*: 36 pairs in 1998–2001 vs. 80 pairs in 2004–8; arctic tern *S. paradisaea*: 770 pairs vs. 1,070 pairs; common tern *S. hirundo*: 970 pairs vs. 1,100 pairs). The number of gulls remained at 10–20 pairs/year from 1980–96 but, following disturbance at the Isle of May and Farne Islands, increased by 445–920% from 1997–2000, whilst tern numbers declined. Sandwich terns *S. sandvicensis* declined between 1998–2001 (1,500 pairs) and 2004–8 (1,000 pairs) but the authors note that this is thought to be in the normal range of variation in the species.

(1) Morris, R. D., Kirkham, I. R. & Chardine, J. W. (1980) Management of a declining common tern colony. *The Journal of Wildlife Management*, 44, 241–245.
(2) Morris, H., Blokpoel, H. & Tessier, G. D. (1992) Management efforts for the conservation of common tern *Sterna hirundo* colonies in the Great Lakes: two case histories. *Biological Conservation*, 60, 7–14.
(3) Maxson, S. J., Mortensen, S. A., Goodermote, D. L. & Lapp, C. S. (1996) Success and failure of ring-billed gull deterrents at common tern and piping plover colonies in Minnesota. *Colonial Waterbirds*, 242–247.
(4) Blokpoel, H. & Tessier, G. D. (1997) Successful restoration of the Ice Island common tern colony requires on-going control of ring-billed gulls. *Colonial Waterbirds*, 20, 98–101.
(5) Akers, P. & Allcorn, R. I. (2006) Re-profiling of islands in a gravel pit to improve nesting conditions for terns *Sterna* and small gulls *Larus* at Dungeness RSPB reserve, Kent, England. *Conservation Evidence*, 3, 96–98.
(6) Morrison, P. & Allcorn, R. I. (2006) The effectiveness of different methods to deter large gulls *Larus* spp. from competing with nesting terns *Sterna* spp. on Coquet Island RSPB reserve, Northumberland, England. *Conservation Evidence*, 3, 84–87.
(7) Booth, V. & Morrison, P. (2010) Effectiveness of disturbance methods and egg removal to deter large gulls *Larus* spp. from competing with nesting terns *Sterna* spp. on Coquet Island RSPB reserve, Northumberland, England. *Conservation Evidence*, 7, 39–43.

12.32.2. Woodpeckers

- All four studies we captured describe the management of red-cockaded woodpeckers *Picoides borealis* in open pine forests in the USA.
- One small study (1) found an increase in woodpecker population following the removal of southern flying squirrels *Glaucomys volans*, whilst a second (2) found a population increase following squirrel removal, along with other interventions and a third (3) found that reintroductions were successful when squirrels were controlled.
- A randomised, replicated and controlled before-and-after study found fewer holes were occupied by squirrels (4) following control efforts, but that occupancy by red-cockaded woodpeckers was no higher.

A small before-and-after study in a loblolly *Pinus taeda* and longleaf *P. palustris* pine forest in South Carolina, USA (1) found that by the end of 1991, the local population of red-cockaded woodpeckers *Picoides borealis* consisted of six breeding pairs and 15 other birds after the removal of southern flying squirrels *Glaucomys volans* (compared with one breeding pair and two other birds in 1986). As the population grew, 26–108 southern flying squirrels were removed from potential woodpecker nest cavities each month. This study is discussed in detail in 'Translocate individuals').

A study in mixed pine *Pinus* spp. forests in South Carolina, USA (2) found that a population of red-cockaded woodpeckers *Picoides borealis* increased from four individuals in 1985 to 99 in 1996, whilst the number of breeding pairs increased from one to 19 following intensive management. Management included the removal of 2,304 southern flying squirrels *Glaucomys volans* (a competitor for, and kleptoparasite of, woodpecker cavities) from the site. Of the squirrels removed, 1,511 (66%) were from artificial cavities, 652 (28%) from natural cavities and 141 (6%) from nest boxes. Other interventions included the provision of artificial cavities and nest boxes (see 'General responses to small/declining populations – Provide artificial nesting sites'), fitting artificial cavities with restrictor plates to prevent them being enlarged by other woodpeckers (see 'Protect nest sties from competitors'). Other interventions included translocations of adults and fledglings and habitat management and are discussed in 'General responses to small/declining populations – Translocate individuals' and 'Threat: Natural system modifications – Forest modifications'.

A study between 1994 and 1996 in a mixed pine *Pinus* spp. forest in eastern Texas, USA (3) found that reintroducing red-cockaded woodpeckers *Picoides borealis* into parts of their former range was successful when southern flying squirrels *Glaucomys volans* (a competitor for, and kleptoparasite of, nesting cavities) were removed before woodpecker release. This study is discussed in detail in 'Translocate individuals'.

A randomised, replicated and controlled before-and-after study in 2001–2 in 16 sites in longleaf pine *Pinus palustris* forests in northern Florida, USA (4) found that culling 168 southern flying squirrels *Glaucomys volans* from potential red-cockaded woodpecker *Picoides borealis* breeding cavities over a 12 month period significantly reduced the number of cavities occupied by squirrels (squirrels occupied 0.46 cavities/territory, red-cockaded woodpeckers occupied approximately 2 cavities/territory), compared to control territories but that there was no corresponding increase in cavity occupancy by red-cockaded woodpeckers (squirrels occupied 0.96 cavities/territory, red-cockaded woodpeckers occupied approximately 2 cavities/territory). Instead, there was an increase in occupancy by red-bellied woodpeckers *Melanerpes carolinus*, another cavity kleptoparasite (1 cavity/territory occupied vs. 0.69 cavities/territory for experimental and control sites respectively). This increase was most noticeable between July–December (a 103% increase compared to controls), when most fledgling red-cockaded woodpeckers acquire cavities.

(1) Allen, D. H., Franzreb, K. E. & Escano, R. E. F. (1993) Efficacy of translocation strategies for red-cockaded woodpeckers. *Wildlife Society Bulletin*, 21, 155–159.

(2) Franzreb, K. E. (1997) Success of intensive management of a critically imperiled population of red-cockaded woodpeckers in South Carolina. *Journal of Field Ornithology*, 68, 458–470.

(3) Carrie, N. R., Conner, R. N., Rudolph, D. C. & Carrie, D. K. (1999) Reintroduction and postrelease movements of red-cockaded woodpecker groups in eastern Texas. *The Journal of Wildlife Management*, 63, 824–832.

(4) Kappes Jr, J. J. & Davis, J. M. (2008) Evidence of positive indirect effects within a community of cavity-nesting vertebrates. *The Condor*, 110, 441–449.

12.32.3. Songbirds

- Two studies from Australia (1,3) found increases in bird populations and species richness after the control of noisy miners *Manorina melanocephala* – a native but hyper-competitive species.
- A controlled study from Italy (2) found that blue tits *Parus caeruleus* nested in more nest boxes when hazel dormice *Muscardinus avellanarius* were excluded from nest boxes over winter.

A paired site study of patches of remnant eucalypt woodland in Victoria, Australia (1), found a significant increase in bird abundance and species richness after reduction in the numbers of noisy miners *Manorina melanocephala* in two of three sites. The differences were attributable to an influx of honeyeaters and other small insectivorous birds. In a third site, possibly as the result of the presence of under-storey vegetation, there was only a small starting population of noisy miners. The reduction in their numbers influenced the species composition but not bird abundance.

A controlled trial in 2001–2 in beech, holly and oak forests on Sicily, Italy (2), found that blue tits *Parus caeruleus* (also *Cyanistes caeruleus*) occupied a higher proportion of nest boxes in an experimental area where hazel dormice *Muscardinus avellanarius* were excluded from nest boxes over winter, compared to a control area where dormice were not excluded, but this difference was not significant. The authors argue that the lack of significance may be due to the small sample size (25 nest boxes in each treatment). Dormice were excluded by blocking nest box entrances between November 2001 and March 2002. This study is also discussed in 'Provide artificial nesting sites'.

A before-and-after study of bird species in privately owned remnant eucalypt woodland in New South Wales, Australia (3), found a decline in small and medium songbirds after a dense colony of noisy miners *Manorina melanocephala* became established. The number of bird species increased after a cull of the noisy miners, and improved further as new planting of native trees and shrubs became established. The results are consistent with noisy miners causing a decline in small woodland bird diversity by competitive exclusion, released by culling. The restoration of a shrub layer is likely to have played a part in the maintained increase in the diversity of bird species, but the relative contributions of the cull and planting cannot be quantified. The study was not replicated or controlled, and the cull was unofficial and unsanctioned.

(1) Grey, M. J., Clarke, M. F. & Loyn, R. H. (1997) Initial changes in the avian communities of remnant eucalypt woodlands following a reduction in the abundance of noisy miners, *Manorina melanocephala*. *Wildlife Research*, 24, 631–648.

(2) Sarà, M., Milazzo, A., Falletta, W. & Bellia, E. (2005) Exploitation competition between hole-nesters (*Muscardinus avellanarius*, Mammalia and *Parus caeruleus*, Aves) in Mediterranean woodlands. *Journal of Zoology*, 265, 347–357.

(3) Debus, S. J. S. (2008) The effect of noisy miners on small bush birds: an unofficial cull and its outcome. *Pacific Conservation Biology*, 14, 185–190.

12.33. Reduce inter-specific competition for nest sites by modifying habitats to exclude competitor species

- A replicated controlled study from the USA (1) found no impact of mid-storey clearance on the occupation of red-cockaded woodpecker *Picoides borealis* nesting cavities by southern flying squirrels *Glaucomys volans*.

Background

Because of the potential controversies over control and eradication efforts, it may be more appropriate to attempt to modify habitats so that they suit one species more than the others. Studies specifically designed to favour one competitor are discussed below, whilst more general habitat modifications are discussed in 'Natural system modifications'.

A controlled, replicated study in 1990–1 in mixed loblolly pine *Pinus taeda* and shortleaf pine *Pinus echinata* forests in eastern Texas, USA (1) found that red-cockaded woodpecker *Picoides borealis* nest cavities were occupied by southern flying squirrels *Glaucomys volans* and woodpeckers at approximately the same rates. Nest cavity occupation by both species was unaffected by the clearance of midstory hardwood vegetation in woodpecker territories (17 sites cleared of hardwood: 51% of cavities occupied by woodpeckers, 22% by squirrels; seven sites not cleared: 52% occupied by woodpeckers, 27% occupied by squirrels). Midstory vegetation is often assumed to encourage flying squirrels.

(1) Conner, R. N., Rudolph, D. C., Saenz, D. & Schaefer, R. R. (1996) Red-cockaded woodpecker nesting success, forest structure, and southern flying squirrels in Texas. *The Wilson Bulletin*, 108, 697–711.

12.34. Protect nest sties from competitors

- Two replicated studies from the USA (3,4) found that red-cockaded woodpecker *Picoides borealis* populations increased in five forests after several interventions, including the installation of restrictor plates around nesting holes, were implemented.
- A study from Puerto Rico (1) found evidence for lower competition between Puerto Rican parrots *Amazona vittata* and pearly-eyed thrashers *Margarops fuscatus* after modifications were made to nest boxes.
- A replicated, controlled study from the USA (2) found weak evidence for the effects of exclusion devices on house sparrows *Passer domesticus* nesting in nest boxes and a study from the USA (5) found that fitting restrictor plates to red-cockaded woodpecker holes reduced the number that were enlarged.

In Luquillo municipality, Puerto Rico, in the mid-1970s, provision of nest boxes and nest-hole modifications reduced competition from non-native, pearly-eyed thrashers *Margarops fuscatus* for Puerto Rican parrot *Amazona vittata* nest cavities (1). Nest boxes for thrashers were erected near parrot nest cavities, and baffles and other modifications were installed such that the bottom of cavities used by parrots could not be viewed from the entrance, a change that parrots tolerated, but that made the cavities unacceptable as nesting sites for thrashers. Two benefits were apparent, the local pair of thrashers ceased to disturb the parrot cavity and their territorial presence prevented other thrashers from approaching.

A replicated, controlled study in 1990–1991 in 73 experimental and 73 control nestboxes in Nebraska, USA (2) found that, in 1990, house sparrows *Passer domesticus* (an introduced species) delayed nesting in nestboxes that had three monofilament wires (37 cm apart) held in front of nest boxes on wire 'prongs', resulting in a shorter breeding period (broods started an average of 25 days earlier in control boxes). The modified nest boxes also resulted in fewer successful clutches compared to control boxes (29% of experimental boxes with successful clutches vs. 53% of controls). However, in 1991, there were no such differences, (65% of 60 control clutches successful vs. 55–70% for 120 experimental boxes). All box types were used for roosting over winter.

A replicated before-and-after study in four National Forests in Texas, USA (3), found that red-cockaded woodpecker *Picoides borealis* populations increased at all four sites in the early 1990s after management, including the installation of restrictor plates around nesting holes, was intensified in 1989. The plates were installed in order to prevent enlargement of cavities by pileated woodpeckers *Dryocopus pileatus*. This study is discussed in 'Provide artificial nesting sites', 'Translocate individuals' and 'Use prescribed burning'.

A study in mixed pine *Pinus* spp. forests in 1985–96 in South Carolina, USA (4) found that a population of red-cockaded woodpeckers *Picoides borealis* increased following intensive management including fixing artificial cavities with restrictor plates to prevent them being enlarged by other woodpeckers. Other interventions were the provision of artificial cavities and nest boxes (see 'General responses to small/declining populations – Provide artificial nesting sites'), translocations of adults and fledglings and habitat management (control of mid-storey vegetation and prescribed burning) and are discussed in 'General responses to small/declining populations – Translocate individuals' and 'Threat: Natural system modifications – Forest modifications'.

A replicated study in 1996 in a longleaf pine *Pinus palustris* forest in eastern Texas, USA (5) found that fitting restrictor plates (steel plates that stop entrance holes being enlarged) to red-cockaded woodpecker *Picoides borealis* nesting cavities significantly reduced the proportion of holes that were enlarged (and rendered unsuitable) by larger pileated woodpeckers *Dryocopus pileatus*, compared to control cavities (2% of restrictor plate-fitted cavities enlarged, n = 54; 41% of control cavities enlarged, n = 276). The authors note that preventing hole enlargement may prevent other species, such as American kestrels *Falco sparverius* and eastern screech-owls *Megascops asio*, from nesting in woodpecker cavities too.

(1) Snyder, N. F. R., Wiley, J. W. & Kepler, C. B. (1987) The parrots of Luquillo: natural history and conservation of the Puerto Rican parrot.
(2) Pochop, P. A., Johnson, R. J. & Eskridge, K. M. (1993) House sparrow response to monofilament lines at nest boxes and adjacent feeding sites. *The Wilson Bulletin*, 504–513.
(3) Conner, R. N., Rudolph, D. C. & Bonner, L. H. (1995) Red-cockaded woodpecker population trends and management on Texas National Forests. *Journal of Field Ornithology*, 66, 140–151.
(4) Franzreb, K. E. (1997) Success of intensive management of a critically imperiled population of red-cockaded woodpeckers in South Carolina. *Journal of Field Ornithology*, 68, 458–470.
(5) Saenz, D., Conner, R. N., Shackelford, C. E. & Rudolph, D. C. (1998) Pileated woodpecker damage to red-cockaded woodpecker cavity trees in eastern Texas. *The Wilson Bulletin*, 110, 362–367.

12.35. Reduce competition between species by providing nest boxes

- A replicated, controlled study from the USA (1) found that providing extra nest boxes did not reduce the rate at which common starlings *Sturnus vulgaris* usurped northern flickers *Colaptes auratus* from nests.

Background

If several species are competing for nest sites then it might be possible to reduce this competition by providing a surplus of nesting sites. The more general effects of nest box provision are discussed in 'General responses to small/declining populations'.

A replicated, controlled study from March–July in 1994–1996 in 40 experimental and 14 control sites of northern flicker *Colaptes auratus* nest cavity and nest box pairs in Ohio, USA (1) found that the provision of nest boxes do not deter common starlings *Sturnus vulgaris* from usurping flicker nest cavities. Overall, 68% of experimental flicker sites lost a total of 42 cavity-nests to starlings in spite of the presence of a nearby flicker nest box, and nine of these pairs lost two or more cavities to starlings. Flicker pairs with starlings fledged significantly less young than pairs without starlings (20% compared to 36% respectively). Flicker pairs without starlings produced significantly larger clutches than pairs with starlings (7.4 compared to 5.4 eggs/nest). Only one flicker pair nested in a nest box rather than a nest cavity. Nest boxes were installed within 0.5–2.0 m of all flicker cavities. Starlings were present in experimental but not control sites.

(1) Ingold, D. J. (1998) The influence of starlings on flicker reproduction when both naturally excavated cavities and artificial nest boxes are available. *The Wilson Bulletin*, 110, 218–225.

12.36. Reduce inter-specific competition for food by removing or controlling competitor species

- Two controlled before-and-after studies from the UK (2,3) found that six species of wildfowl showed significant increases following the removal of fish from lakes. Three other species did not show increases (2).
- A study from France (1) found that grey partridges *Perdix perdix* increased at a site with several interventions, including the control of competitor species.
- A before-and-after study from Spain (4) found no change in the rate of kleptoparasitic attacks on herons after the culling of gulls at a colony.

Background

If food resources are limiting then competition with other species can limit reproduction or survival. Conservationists can either provide supplementary food (see 'General responses to small/declining populations') or control competitor species (discussed below).

A replicated, controlled study in the spring of 1974 on a cereal farm in Villers-Cotterêts, France (1), found that grey partridges *Perdix perdix* were significantly more abundant in areas provided with 'partridge cafeterias' than in control areas. These 'cafeterias' included mouse poison dispensers. This study is discussed in 'Provide supplementary food to increase adult survival'.

A controlled before-and-after study in southern England between the winters of 1984/5 and 1990/1 (2) found that following the removal of 6.5 tonnes of coarse fish from a gravel pit lake (17 ha) in November 1987, there was a significant increase in the average populations of mute swan *Cygnus olor* (3.6 individuals in 1984–1987 vs. 69.3 in 1987–91), gadwall *Anas strepera* (1.1 vs. 19.6), shoveler *A. clypeata* (4.3 vs. 36.3), pochard *Aythya ferina* (40.0 vs. 82.2) and coot *Fulica atra* (2.1 vs. 203.1). There was a non-significant increase in tufted duck *Aythya fuligula* population, concurrent with a general increase in the area, and no change in populations of mallard *Anas platyrhynchos*, teal *A. crecca* or wigeon *A. penelope*. There were no corresponding changes at a control (13 ha) lake that did not have coarse fish removed. Increases were thought to be due to increases in benthic invertebrate and macrophyte abundance.

A controlled, before-and-after study from 1986–1990 on one gravel pit lake in Great Linford, UK (3) found that tufted duck *Aythya fuligula* feeding success and abundance increased significantly following fish removal from a selected area of the lake in 1987–8 (from 0 to 149 brood observations), while numbers declined in areas where fish were not removed or were reintroduced (from 92 to 4 brood observations). Average brood size increased from 3 to 4 post fish-removal. Additionally, both invertebrate and plant-eating wintering waterfowl increased their use of the lake and species new to the lake began to nest and produce young. In total, 396 kg/ha of fish biomass was removed from the lake during 1987–1988 through seine netting and electro-fishing.

A before-and-after study from 1993–1995 at a heronry on an island off north-east Spain (4) found that there was no difference in kleptoparastic attacks by yellow-legged gulls *Larus cachinnans* on either little egrets *Egretta garzetta* or night herons *Nycticorax nycticorax* following the culling of gulls herons (0.30 attacks/hour before the cull, 54 hours, 1081 heron flights and 16 attacks recorded; 0.32 attacks/hour after the cull, 98 hours and 3581 heron flights and 13 attacks recorded). The gull population declined from approximately 13,500 pairs in 1993 to approximately 7,500 pairs in early April 1995, following the poisoning of breeding adults every year starting in 1992. All attacks were on herons in flight by subadult gulls (which would not be affected by the culling) and unsuccessful attacks were recorded.

(1) Westerskov, K. E. (1977) Covey-oriented partridge management in France. *Biological Conservation*, 11, 185–191.
(2) Phillips, V. E. (1992) Variation in winter wildfowl numbers on gravel pit lakes at Great Linford, Buckinghamshire, 1974–79 and 1984–91, with particular reference to the effects of fish removal. *Bird Study*, 39, 177–185.
(3) Giles, N. (1994) Tufted duck (*Aythya fuligula*) habitat use and brood survival increases after fish removal from gravel pit lakes. *Hydrobiologia*, 279–280, 387–392.
(4) Bosch, M. (1996) The effects of culling on attacks by yellow-legged gulls (*Larus cachinnans*) upon three species of herons. *Colonial Waterbirds*, 19, 248–252.

Reduce adverse habitat alteration by other species

In many parts of the world, humans have altered the species present in a habitat, or altered their abundances. These changes can lead to detrimental changes in ecosystems. For example, the increased abundances of native and introduced deer in the UK, caused by the removal of top predators and the introduction of new species, has led to changes in the composition of forests; whilst introduced goats have caused enormous changes in habitat on many oceanic islands.

12.37. Reduce adverse habitat alterations by excluding problematic species

Background

One possible response to problematic, habitat-altering species is to create exclusion areas, allowing natural habitats to regenerate away from the problematic species.

12.37.1. Terrestrial species

- Three studies from the USA (1,2) and the UK (4) found higher numbers of certain songbird species and a higher species richness in these groups when deer were excluded from forests. Intermediate canopy-nesting species in the USA and common nightingales *Luscinia macrorhynchos* in the UK were the species to benefit.
- A study from Hawaii found mixed effects of grazer exclusion (3), with some species showing population increases, some declines and other different long- and short-term trends.

A replicated study in four hardwood forest sites in Pennsylvania, USA, between 1980 and 1991 (1) found higher species richness and abundances of intermediate canopy-nesting songbirds in plots with lower densities of white-tailed deer *Odocoileus virginianus* across sixteen experimental plots (deer density of 3.7 deer/km^2: averages of 17.5 individuals and 11 intermediate canopy-nesting species in each plot; 7.9 deer/km^2: 16 individuals and 11.5 intermediate canopy-nesting species; 14.9 deer/km^2: averages of 13 individuals and seven ICN species; 24.9 deer/km^2: averages of 10.5 individuals and 7.5 intermediate canopy-nesting species). Plots were 13 or 26 ha and contained between one and four deer. There were no changes in either species richness or abundance of ground-nesting or upper canopy-nesting species. Threshold densities for songbirds not found with high deer densities (eastern wood pewee *Contopus virens*, indigo bunting *Passerina cyanea*, least flycatcher *Empidonax minimus*, yellow-billed cuckoo *Coccyzus americanus*, cerulean warbler *Dendroica cerulea*, eastern phoebe *Sayornis phoebe* and American robin *Turdus migratorius*) appeared to be between 7.9 and 14.9 deer/km^2.

A replicated, controlled study in northern Virginia, USA (2) found significantly higher numbers of 16 species of ground and intermediate canopy-dwelling songbirds in four 4 ha plots of deciduous forest with deer excluded between 1990 and 1998, compared to five control plots. However, there was no corresponding increase in bird diversity as species were gained and lost as understory vegetation developed. There was also no significant difference in the number of resident birds (eight songbird species and one woodpecker) caught.

A study in two koa *Acacia koa* forests in northern Hawaii, USA (3), found that all seven native birds recorded in an area of open forest from which feral grazers (cows and pigs) had been excluded showed long-term population stability or growth. However, all but two showed short-term declines. In a closed forest from which grazers were excluded, only two species showed an increase, with the rest either stable or declining. Birds were monitored between 1987 and 2007 in the open forest and 1999 and 2007 in the closed forest. This study is also discussed in 'Habitat restoration and creation – Forest restoration'.

A replicated and controlled paired study in south-east England (4) found a significantly higher density of common nightingale *Luscinia megarhynchos* territories (monitored in 2000–8) in a coppiced woodland from which deer were excluded between 1999 and 2003, compared to control plots protected by an easily-breached brushwood fence (0.60 territories/ha in eight exclusion plots vs. 0.04 territories/ha in eight controls). The proportion of territories in exclusion plots also increased, from 0% in 2000 to 70–80% in 2005–7. A total of 48 territories were mapped. Native roe deer *Capreolus capreolus* and introduced fallow deer *Dama dama* and Reeves' muntjac *Muntiacus reevesi* were excluded by erecting 1.8 m steel fences. The authors argue that differences are due to the area of optimal-age coppice (3–8 year old) within plots.

(1) deCalesta, D. S. (1994) Effect of white-tailed deer on songbirds within managed forests in Pennsylvania. *The Journal of Wildlife Management*, 58, 711–718.
(2) McShea, W. J. & Rappole, J. H. (2000) Managing the abundance and diversity of breeding bird populations through manipulation of deer populations. *Conservation Biology*, 14, 1161–1170.
(3) Camp, R. J., Pratt, T. K., Gorresen, P. M., Jeffrey, J. J. & Woodworth, B. L. (2010) Population trends of forest birds at Hakalau Forest National Wildlife Refuge, Hawai'i. *The Condor*, 112, 196–212.
(4) Holt, C. A., Fuller, R. J. & Dolman, P. M. (2010) Experimental evidence that deer browsing reduces habitat suitability for breeding common nightingales *Luscinia megarhynchos*. *Ibis*, 152, 335–346.

12.37.2. Aquatic species

- A replicated paired study in the USA (1) found that waterbirds preferentially used wetland plots from which grass carp *Ctenopharyngodon idella* were excluded but moved as these became depleted over the winter.

Background

Many of the most destructive invasive species are aquatic, for example zebra mussels *Dreissena polymorpha* and grass carp *Ctenopharyngodon idella*. Excluding such species has unique difficulties, but it may be possible to maintain small areas without them.

A replicated, paired study in the winters of 1993–1995 in six sites in an open-water refuge in Alabama, USA (*1*) found that waterbird density in October was higher in plots with grass carp *Ctenopharyngodon idella* excluded, but that birds moved to plots with carp present over the winter. Exclusion plots had higher levels of native vegetation (dominated by muskgrass *Chara* spp. and sago pondweed *Potamogeton pectinatus*) but birds left as this became depleted. Control plots were dominated by (non-native) milfoil *Myriophyllum spicatum*. Exclusion plots were established to re-establish native species using netting (1.27 cm^2 mesh, 1.2 m tall) held in place by buoyant ropes and weighted with steel rods to create 0.1 ha plots (25 × 50 m). Identical plots (without netting) were created (≤25 m away from the native plot) in areas of 100% milfoil coverage.

(1) Benedict, R. J. & Hepp, G. R. (2000) Wintering waterbird use of two aquatic plant habitats in a southern reservoir. *The Journal of Wildlife Management*, 64, 269–278.

12.38. Control or remove habitat-altering mammals

- Four studies from the Azores (1,2,5) and Australia (3) found that seabird populations increased following the eradication of European rabbits *Oryctolagus cuniculus* or other species, although in three studies (2,3,5) there were several other interventions used as well.
- Two studies from Australia (3) and the Madeira archipelago, Portugal (4), found that seabird populations' productivities increased following rabbit and house mouse *Mus musculus* (4) eradications, with several other interventions used in the Australian study.

Background

On continents and large islands, the chances of eradicating problematic species is very low, but we have captured five studies that describe the effects of successfully eradicating habitat-altering invasive species, most commonly European rabbits *Oryctolagus cuniculus* from small islands across the world. Such control efforts are often accompanied by attempts to restore native vegetation and habitats, described in 'Habitat restoration and creation'.

A before-and-after study on Ilhéu da Praia, Azores, Portugal (1), found that the breeding populations of common terns *Sterna hirundo* and roseate terns *S. dougallii* increased significantly from 1997 (approximately 100 pairs of common and fewer than ten pairs of roseate terns) to 2002 (641 pairs of common and 133 pairs of roseate terns), following the eradication of European rabbits *Oryctolagus cuniculus* in 1997 by brodifacoum poisoning. Common tern populations initially increased to approximately 950 pairs in 2000, before falling again.

A controlled before-and-after study on Santiago Island (585 km^2), Galapagos, Ecuador (2) found that densities of Galapagos rails *Laterallus spilonotus* increased following the eradication of feral mammals (including habitat-altering species) between 1998 and 2006. This study is discussed in 'Control mammalian predators on islands'.

A before-and-after study on Cabbage Tree Island, south-east Australia (3) found that the population of Gould's petrel *Pterodroma leucoptera leucoptera* has increased since the eradication of European rabbits *Oryctolagus cuniculus* in the austral winter of 1997, with more breeding pairs (186–599 pairs in 1989–97, before rabbit eradication vs. 818–1025 pairs in 1997–2006, after eradication), higher breeding success (17–59% success, average of 33% vs. 46–57%, average of 51%) and more fledglings produced (31–331 fledglings/year vs. 374–488 fledglings/year). Rabbit removal (by sequential epidemics of myxomatosis, rabbit haemorrhagic disease and brodifacoum application) was, however, only one of several conservation interventions and the conservation of Gould's petrels is also discussed in 'Provide artificial nesting sites', 'Provide supplementary food', 'Translocate individuals', 'Artificially incubate and hand-rear birds in captivity', 'Remove problematic vegetation' and 'Control avian predators on islands'.

A before-and-after study on Selvagem Grande (2.45 km^2), Madeira Archipelago, Portugal (4), found that the breeding success and productivity of Cory's shearwaters *Calonectris diomedia borealis* was significantly higher in 2002–6 (54–56% of 2,075 nests producing fledglings), following the eradication of European rabbits *Oryctolagus cuniculus* and house mice *Mus musculus* (a potential predator) than in 13 years during 1982–2001, before eradication (36–45% of 4,952 nests producing fledglings). Eradications were simultaneous, through the application of brodifacoum in July–September 2002 and it was not, therefore, possible to determine which species was constraining the shearwater population. The authors argue that because of the timing of the eradications and the instantaneous impact on productivity, the two species must have been impacting on chick, rather than egg survival.

A before-and-after study on Praia Islet (12 ha), off Graciosa, Azores, Portugal (5), found that the breeding populations of common terns *Sterna hiundo*, roseate terns *S. dougallii* and Madeiran storm petrel *Oceanodroma castro* increased dramatically following the eradication of European rabbits *Oryctolagus cuniculus* in 1997, subsequent habitat restoration (see 'Restore shrubland' for details) and the provision of artificial nest sites on the island from 1996 (tern boxes) and 2000 (petrel boxes) (see 'Provide artificial nesting sites' for details). Common terns increased from no pairs in 1996 to over 1,000 pairs in 2006; roseate terns from zero before 2000 to over 400 pairs; storm petrels from no pairs before 2000 to almost 800 breeding pairs in 2006. Rabbits were eradicated using brodifacoum bait stations and broadcasting pellets.

(1) Pitta Groz, M. & Pereira, J. C. (2005) Invasive alien species as a threat to seabird populations: an account of habitat restoration on 'Ilhéu da Praia' (Graciosa, Azores) Special Protection Area. *Airo*, 15, 3–9.
(2) Donlan, C., Campbell, K., Cabrera, W., Lavoie, C., Carrion, V. & Cruz, F. (2007) Recovery of the Galápagos rail (*Laterallus spilonotus*) following the removal of invasive mammals. *Biological Conservation*, 138, 520–524.
(3) Priddel, D. & Carlile, N. (2007) Population size and breeding success of Gould's Petrel *Pterodroma leucoptera leucoptera* on Cabbage Tree Island, New South Wales: 1996–97 to 2005–06. *Corella*, 31, 79–82.
(4) Zino, F., Hounsome, M. V., Buckle, A. P. & Biscoito, M. (2008) Was the removal of rabbits and house mice from Selvagem Grande beneficial to the breeding of Cory's shearwaters *Calonectris diomedea borealis*? *Oryx*, 42, 151–154.
(5) Bried, J., Magalhaes, M. C., Bolton, M., Neves, V. C., Bell, E., Pereira, J. C., Aguiar, L., Monteiro, L. R. & Santos, R. S. (2009) Seabird habitat restoration on Praia Islet, Azores Archipelago. *Ecological Restoration*, 27, 27–36.

12.39. Remove problematic vegetation

- A before-and-after study from Japan (4) found higher numbers of long-billed plovers *Charadrius placidus* after the removal of invasive black locust *Robinia pseudoacacia*.
- A study from Australia (2) found lower mortality of Gould's petrels *Pterodroma leucoptera leucoptera* following the removal of most of an island's (native) bird-lime tree *Pisonia umbellifera* population, whilst a study from New Zealand (3) found that Chatham Island oystercatchers *Haematopus chathamensis* could nest in preferable areas following invasive marram grass *Ammophila arenaria* control.
- A site comparison from the USA (1) found lower densities of several birds in areas with (native) velvet mesquite *Prosopis juliflora* control.

Background

We captured four studies describing the effects of species-specific control programmes for plant species. Studies describing more general interventions, such as thinning forests or cutting grasslands are described in 'Natural system modifications'.

A site comparison study in desert grassland in Arizona, USA (1), found that more Gambel's quail *Lophortyx gambelii* were seen and more Gambel's quail, mourning dove *Zenaida macroura*, white-winged dove *Zenaida asiatica*, and scaled quail *Callipepla squamata* were heard in undisturbed velvet mesquite *Prosopis juliflora* var. *velutina* and mesquite with clearings than on mesquite-free pasture. There were no significant differences between undisturbed mesquite and mesquite with clearings.

A before-and-after study on Cabbage Tree Island, south-east Australia (2) found that of 122 adult Gould's petrels *Pterodroma leucoptera leucoptera* specimens in the Australian National

Wildlife Collection, Canberra, 48% were thought to have been killed by becoming entangled in the fruits of the bird-lime tree *Pisonia umbellifera*. In addition, between 1989 and 1991, several petrels were seen entangled and dying in *P. umbellifera* fruits every year. However, following the eradication of most of the population of *P. umbellifera* trees in 1992 and their failure to flower in 1993, no petrels were found dead due to entanglement in the 1992–3 or 1993–4 breeding seasons. This study is also discussed in 'Control avian predators on islands'.

A small before-and-after on Chatham Island, New Zealand between 2001 and 2004 (3) found that three pairs of Chatham Island oystercatchers *Haematopus chathamensis* at two beach sites (each approximately 100 m long and 40 m wide) nested higher up the beach and further from the storm tide zone (thus reducing the chances of nests being washed away) following the removal of invasive marram grass *Ammophila arenaria* through the application of broad-scale (Roundup™ – glyphosate) and grass-specific (Gallant™) herbicides and physical removal. The impact on nesting success or productivity was not reported.

A before-and-after study on the middle reach of the Tama River, Honshu, Japan (4), found that the number of long-billed plovers *Charadrius placidus* observed in the study area increased significantly following the removal of invasive black locust *Robinia pseudoacacia* from river islands in 2001–2 to expose bare ground, which was then covered in gravel to create gravel and sand bars (0.4 plovers/observation trip, n = 15 trips in 2001 vs. 4 plovers/trip, n = 25 in 2002; 19 plovers/trip, n = 19 in 2004; 11 plovers/trip, n = 7 in 2006. The increase from 2001 to 2002 was significant). Hatching rates were extremely variable in the study area (19% in 2006 vs. 100% in 2003) and could not be compared with earlier studies or studies from other sites due to incomplete data and different survey methods.

(1) Germano, D. J., Hungerford, R. & Martin, S. C. (1983) Responses of selected wildlife species to the removal of mesquite from desert grassland. *Journal of Range Management*, 36, 309–311.
(2) Priddel, D. & Carlile, N. (1995) Mortality of adult Gould's petrels *Pterodroma leucoptera leucoptera* at the nesting site on Cabbage Tree Island, New South Wales. *Emu*, 95, 259–264.
(3) Moore, P. & Davis, A. (2004) Marram grass *Ammophila arenaria* removal and dune restoration to enhance nesting habitat of Chatham Island oystercatcher *Haematopus chathamensis*, Chatham Islands, New Zealand. *Conservation Evidence*, 1, 8–9.
(4) Katayama, N., Amano, T. & Ohori, S. (2010) The effects of gravel bar construction on breeding long-billed plovers. *Waterbirds*, 33, 162–168.

12.40. Use buffer zones to reduce the impact of invasive plant control

- A study from the USA (1) found that having buffer zones around snail kite *Rostrhamus sociabilis* nests, where no herbicides were sprayed, resulted in no nests being lost during a vegetation control programme.

Background

Control efforts for invasive or problematic species can be very destructive and if not performed carefully may damage native species. We captured one study attempting to reduce the impact of invasive water plant control on snail kite *Rostrhamus sociabilis* nests, which are built above water in cattail Typha spp. and bulrush *Scripus validus* and are therefore vulnerable to herbicide applications, which can kill supporting vegetation, collapsing the nests.

A study from March–July in 1988 in one wetland area in Florida, USA (1), found that creating 'no-spray' buffer zones extending 68 m around the perimeter of kite colonies and 23–46 m around individual nests resulted in none of the 19 snail kite nests monitored being adversely affected by an aquatic plant control programme: nests averaged 0.73 fledglings/nest and no nests collapsed. This fledging rate was similar to the overall rate of 0.74 fledglings/nest recorded in other regions of Lake Okeechobee during 1987–1993. Buffer zones were used in the breeding season, whilst water hyacinth *Eichhornia crassipes* and water lettuce *Pistia stratiotes* control measures were ongoing. The initiative was expanded to other lakes supporting kite nests in March 1989.

(1) Rodgers, J. A. (1998) Fate of artificially supported snail kite *Rostrhamus sociabilis* nests in central Florida, U.S.A. *Bird Conservation International*, 8, 53–57.

Reduce parasitism and disease

12.41. Remove/treat endoparasites and diseases

- Three studies from across the world investigating a range of taxa and parasites found that birds had higher productivity (2) or survival (4,5) if either chicks or adults were treated for endoparasites.
- One small study from Spain (1) found no effect of *Staphylococcus aureus* treatment on eagle survival, while a study from Mauritius (4) found uncertain evidence as to whether trichomoniasis treatment increased survival of pink pigeons *Nesoenas mayeri* after fledging.
- A randomised, replicated and controlled trial from the Netherlands (3) found lower parasite burdens but also lower survival in Eurasian oystercatcher *Haematopus ostralegus* chicks treated with anthelmintic drugs.

Background

Diseases such as such as trichomoniasis or bird pox can have severe impacts on bird populations in areas where they are not native and where birds do not have immunity, but native diseases can also be important limiting factors for populations.

A small study in 1976–88 in the wetlands of the Doñana National Park, Spain (1) found that survival rates of Spanish imperial eagle *Aquila adalberti* chicks did not appear to be affected by treating them for *Staphylococcus aureus* infections (two of 19 untreated chicks died, probably from infections vs. none of the nine treated chicks died before fledging). The authors note that this may be due to the small sample size in the study. This study also discusses other interventions, in 'Add perches to electricity pylons to reduce electrocution', 'Bury or isolate power lines', 'Use signs and access restrictions to reduce disturbance at nest sites' and 'Foster eggs or chicks with wild conspecifics'.

A controlled cross-over experiment during 1996–2000 on two moors in northern England (2) found that red grouse *Lagopus lagopus scoticus* in an area provided with quartz grit treated with anthelmintic drugs raised between 38% and 100% more chicks than grouse in a control area (treatment areas: 4.9–7.1 chicks/hen estimated from 36 radio-tagged birds and 4.9–6.7 chicks/hen estimated from 125 birds seen on counts using pointing dogs vs. control areas: 1.9–4.8 chick/hen from 36 tagged birds and 2.8–4.5 chicks/hen from 117 on dog counts) and had significantly lower levels of infection of the parasitic nematode *Trichostrongylus tenuis* (34% fewer worms over five years). This was despite the fact that the medicated areas did not have larger broods or higher hatching success (medicated areas: 9.6 eggs/clutch, 90% hatching success for 161 clutches; control areas: 9.4 eggs/clutches, 94% hatching rate for 153 clutches). Survival rates of adults did not vary between medicated and control areas.

A randomised, replicated and controlled trial in 1996 on an island in the northern Netherlands (3) found that fledging success of Eurasian oystercatcher *Haematopus ostralegus* broods was significantly lower when chicks were treated with anthelminthic drugs, compared to controls (18–20% success for treated broods vs. 29–38% for controls). This was despite there being no significant differences in clutch size or hatching success between groups and treated hatchlings having significantly lower incidence of gut parasites (41% of 17 treated chicks infected vs. 60% of 20 untreated chicks). The authors suggest that interference with chicks' immune systems could have driven this pattern.

A controlled, replicated study from 1992–1999 on Ile aux Aigrettes, Mauritius (4) found that pink pigeon *Nesoenas mayeri* (formerly *Columba mayeri*) chicks with suspected trichomoniasis

Trichomonas gallinae had significantly higher survival if treated with carnidazole compared with untreated chicks showing symptoms (54% survival for 89 treated chicks vs. 0% survival for 19 untreated). Across both symptomatic and asymptomatic chicks, treated chicks had higher survival rates than untreated chicks (62% survival of 129 treated vs. 27% of 123 untreated). However, treatment did not affect subsequent juvenile survival up to 150 days old. Across sites on both the island and mainland Mauritius, survival of treated birds (both juveniles and adults) was significantly higher than untreated birds (74% survival of 19 treated birds vs. 25% survival for 24 untreated). Providing medicated water to all individuals on Ile aux Aigrettes did not reduce the incidence of trichomoniasis in the subpopulation, with birds becoming re-infected after treatment stopped.

A randomised, replicated and controlled experiment in wetlands in Manitoba, Canada (5) found that survival of 322 American coot *Fulica americana* chicks was higher in 2004, when they were treated with fenbendazole (an anthelmintic drug), compared to untreated chicks (51% survival to 40 days for treated chicks vs. 39% for untreated chicks). In 2005, survival of 340 chicks was again increased by treatment, but chicks with parents that were treated whilst incubating also had higher survival rates, despite there being no detectable change in parasite burden in adult birds (58% if both parents and chicks treated; 46% if only chicks treated; 45% if only parents treated vs. 33% if neither treated).

(1) Ferrer, M. & Hiraldo, F. (1991) Evaluation of management techniques for the Spanish imperial eagle. *Wildlife Society Bulletin*, 19, 436–442.
(2) Newborn, D. & Foster, R. (2002) Control of parasite burdens in wild red grouse *Lagopus lagopus scoticus* through the indirect application of anthelmintics. *Journal of Applied Ecology*, 39, 909–914.
(3) Van Oers, K., Heg, D. & Le Drean Quenec'hdu, S. (2002) Anthelminthic treatment negatively affects chick survival in the Eurasian oystercatcher *Haematopus ostralegus*. *Ibis*, 144, 509–517.
(4) Swinnerton, K. J., Greenwood, A. G., Chapman, R. E. & Jones, C. G. (2005) The incidence of the parasitic disease trichomoniasis and its treatment in reintroduced and wild pink pigeons *Columba mayeri*. *Ibis*, 147, 772–782.
(5) Amundson, C. L. & Arnold, T. W. (2010) Anthelmintics increase survival of American coot (*Fulica americana*) chicks. *The Auk*, 127, 653–659.

12.42. Exclude or control 'reservoir species' to reduce parasite burdens

- A literature review (3) found no compelling evidence that culling mountain hares *Lepus timidus* (a carrier of the ticks that carry louping ill virus) increased red grouse *Lagopus lagopus scoticus* populations.
- A controlled before-and-after study from the UK (1) did find that there was a significant increase in chick production on grouse moors with hare culling, compared to control sites but no change in population density. A comment on this paper (2) argued that the controls used in it were not adequate.

Background

If a bird population is threatened by a parasite or disease that can infect multiple species, it may not be enough to treat the population, because of the risk of constant re-infection from the other host, which can act as a 'reservoir species'. Instead it may be necessary to treat the other hosts or to exclude them from particular habitats or control their numbers.

While we found studies describing the impact of controlling reservoir species through culling, we found no intervention-based studies describing the impact of excluding reservoir species from an area. However, a correlative study in Scotland in 2007 found that areas of forest that were fenced to exclude deer had fewer ticks (carriers of louping ill virus which attacks red grouse *Lagopus lagopus scoticus*) than unfenced areas, but that areas of adjacent moorland (where the grouse were actually found) were not affected (Ruiz-Fons & Gilbert, 2010).

Ruiz-Fons, F. & Gilbert, L. (2010) The role of deer as vehicles to move ticks, *Ixodes ricinus*, between contrasting habitats, *International Journal for Parasitology*, 40, 1013–1020.

A controlled before-and-after study in the Scottish Highlands between 1993 and 2001 (1) found that there was no significant increase in the population density of red grouse *Lagopus lagopus scoticus* at a site with mountain hare *Lepus timidus* (a carrier of the ticks that carry louping ill virus) culling, compared to a control site without hare culling (approximately 25 grouse/km^2 in 1993 and 100/km^2 in 2001 at the experimental site vs. 140/km^2 and 275/km^2 at the control). However, there was a significant increase in the number of chicks produced/female at the treatment site, compared to the control (approximately 1.2 chicks/female in 1991 and 5 in 2001 at the experimental site vs. 3.5 and 3.0 at the control) and a significant reduction of louping ill virus at the treatment site, compared to a second control site. Hare densities were reduced from 8/km^2 in 1993 to 0 in 1998. A comment on this paper in 2004 (2) argues that the control sites were not adequate, as they differed in either the pre-existing incidence of louping ill virus or in various environmental conditions.

A 2010 literature review (3) found 'no compelling evidence' that culling mountain hares *Lepus timidus* (a carrier of the ticks that carry louping ill virus) increased red grouse *Lagopus lagopus scoticus* populations. The authors note that there is some evidence for an effect of culling on the prevalence of louping ill virus (e.g. in (1)) but that evidence for population-level effects is uncertain, partly due to a lack of understanding of the population dynamics of both hares and grouse.

(1) Laurenson, M. K., Norman, R. A., Gilbert, L., Reid, H. W. & Hudson, P. J. (2003) Identifying disease reservoirs in complex systems: mountain hares as reservoirs of ticks and louping-ill virus, pathogens of red grouse. *Journal of Animal Ecology*, 72, 177–185.
(2) Cope, D. R., Iason, G. R. & Gordon, I. J. (2004) Disease reservoirs in complex systems: a comment on recent work by Laurenson *et al.* *Journal of Animal Ecology*, 73, 807–810.
(3) Harrison, A., Newey, S., Gilbert, L., Haydon, D. T. & Thirgood, S. (2010) Culling wildlife hosts to control disease: mountain hares, red grouse and louping ill virus. *Journal of Applied Ecology*, 47, 926–930.

12.43. Remove/treat ectoparasites to increase survival or reproductive success

Background

Blood-sucking insects such as lice and Protocalliphora flies can have severe impacts on birds, especially nestlings, and greatly reduce reproductive success.

12.43.1. Remove ectoparasites from feathers

- A replicated and controlled study in the UK (1) found that red grouse *Lagopus lagopus scoticus* treated with spot applications had lower tick and disease burdens and higher survival than controls, whilst birds with impregnated tags had lower tick burdens only.
- A replicated *ex situ* study in Hawaii (2) found that CO_2 was the most effective way to remove lice from feathers, although this treatment did not kill the lice.

Background

Ectoparasites have the greatest impacts on young birds, but can also reduce the fitness or condition of adults. In addition, they can be important carriers of diseases, such as ticks carrying louping ill virus LIV.

A replicated and controlled study in 1995–6 on a grouse moor in Morayshire, Scotland (1), found that red grouse *Lagopus lagopus scoticus* treated with spot applications of deltamethrin had significantly lower tick burdens, compared to control birds, significantly lower louping ill virus (LIV) infections and higher survival rates at ten weeks old compared with controls (82% survival for seven treated broods vs. 64% for seven controls). Chicks treated with lambdacyhalothrin-impregnated (tick-removal) tags also had lower tick burdens than controls, as did chicks from broods where mother only was treated (0–0.1 immature ticks/chick for broods with treated chicks vs. 0.5–0.7 for broods with treated hens and 1.4–2.3 for controls), but they did not have lower LIV infection rates or increased survival to ten weeks old.

A replicated *ex situ* study in Hawaii (2) tested four different fumigants (chloroform, diethyl ether, ethyl acetate and carbon dioxide, CO_2) to assess the efficiency and speed with which they immobilised and detached slender pigeon lice *Columbicola columbae* from the feathers of feral pigeons *Columba livia*. CO_2 was fastest in immobilising lice (average of 61 s for three trials of ten lice), followed by chloroform (122 s), ethyl acetate (198 s) and ether (201 s). CO_2 was also fastest at detaching lice from feathers (average of 181 s for 30 lice) followed by chloroform (192 s), ethyl acetate (293 s) and ether (307 s). However, CO_2 only detached 22% of lice, compared to 33% for ethyl acetate, 56% for ether and 76% for chloroform. In addition, CO_2 (unlike the other chemicals) did not kill the lice, which quickly recovered when given fresh air.

(1) Laurenson, M. K., Hudson, P. J., McGuire, K., Thirgood, S. J. & Reid, H. W. (1997) Efficacy of acaricidal tags and pour-on as prophylaxis against ticks and louping-ill in red grouse. *Medical and Veterinary Entomology*, 11, 389–393.

(2) Visnak, R. M. & Dumbacher, J. P. (1999) Comparison of four fumigants for removing avian lice. *Journal of Field Ornithology*, 70, 42–48.

12.43.2. Remove ectoparasites from nests

- Six of the seven studies that investigated infestation rates (3,5,7–9,11) found lower rates in nests treated for ectoparasites, one (that used microwaves to treat nests) did not find fewer parasites (6).
- Two studies from the USA (4,7) found higher survival or lower abandonment in nests treated for ectoparasites, whilst seven studies from across the world (1–3,5–7,6,9) found no differences in survival, fledging rates or productivity between nests treated for ectoparasites and controls.
- Two studies from the USA (1) and the UK (3) found that chicks from nests treated for ectoparasites were in better condition than those from control nests. Four studies found no such effect (2,5,9,10).

Background

Nestlings are more vulnerable than adults to parasites, so reducing parasite burdens in nests may have a larger impact on reproductive success than treating adults.

A replicated, controlled experiment during 1987–8 in New York State, USA (1) found that survival rates and fledging age of both eastern bluebird *Sialis sialis* and tree swallow *Tachycineta bicolor* chicks did not differ significantly between nests where blowflies *Protocalliphora* spp. were experimentally removed, control nests or nests with blowfly larvae added (bluebirds: 78% survival for removal nests, n = 25 vs. 87% for controls, n = 26; fledging age of 18 days for both treatments; tree swallows: 97% for removal nests, n = 29 vs. 90% for addition nests, n = 30 and 94% for controls, n = 32; fledging ages of 19.4, 19.7 and 19.8 days for removal, addition and control nests respectively). Bluebird nestlings were significantly heavier at 14 days old in removal nests, but the authors note that this was a small change (averages of 28.2 g in removal and 27.0 g in control nests).

A replicated, controlled study in a mixed deciduous woodland in Uppsala, Sweden, in late spring 1988–91 and 1994 (2), found that great tit *Parus major* nest boxes from which hen fleas *Ceratophyllus gallinae* were removed did not have higher hatching or fledging success compared to control nests (83–96% hatching and 69–100% fledging success for 46–47 experimental nests vs. 83–95% and 85–99% for 82–85 controls). In two out of five years, nests treated with pyrethrin had heavier and larger chicks in better condition, but in the remaining three years there were no differences. In another year, treated nests had shorter nestling periods. The authors argue that the impact of fleas is greater in cooler years with higher precipitation.

A replicated, paired site study over two summers in 1991–92 in central Scotland (3) found that sand martin *Riparia riparia* chicks from nests treated with Alugan pesticide (4.25% Bromocyclen) had fewer ectoparasites than chicks from untreated nests and fledged at higher weights (average of 18.8 g for 44 treated nests vs. 18.0 g for 44 control nests; within pairs, chicks from treated nests were heavier in 61% of cases). However, there was no difference in nestling survival between treated and control nests.

A replicated, controlled study from 1982–1989 in Nebraska, USA (4) found that survival of cliff swallow *Petrochelidon pyrrhonota* chicks from nests treated with Naled (an organophosphate insecticide) did not decline with later laying date, unlike control nests. The number of chicks surviving to ten days old was also higher for fumigated nests than control nests (1.75–2.80 chicks/nest survived to ten days old in 15 fumigated nests vs. a maximum of 2.50 chicks/nest, declining to <1.25 chicks/nest for clutches laid after 25th June for non-fumigated nests).

A small controlled study in 1996 in Corsica, France (5) found that there were significantly fewer blowfly *Protocalliphora* spp. larvae on blue tit *Parus caeruleus* broods that were repeatedly moved to microwave-treated nests (at two, five, seven, nine, twelve and fifteen days old), than in control broods that were not treated (0.2 parasites/chick for six treated broods vs. 15.0 parasites/chick for nine control broods). Fledging weights, survival and other indicators of reproductive success were not reported.

A small randomised, controlled study over three breeding seasons in north-eastern Algeria (6) found that blue tit *Parus caeruleus* nests that were heat treated with a microwave (three minutes heating at 830 W) had significantly fewer blowflies *Protocalliphora* spp. than control (unheated) nests (2.0 flies/nest for ten experimental nests vs. 54.9 flies/nest for eleven controls). Experimental nests also had lower numbers of ticks *Ixodes ricinus* and hen fleas *Ceratophyllus gallinae* (8.8 ticks/nest and 3.1 fleas/nest for experimental nests vs. 20.5 ticks/nest and 15.1 fleas/nest for controls), but these differences were not significant after controlling for confounding factors. Growth rates of chicks did not differ between treatments but more chicks fledged from treated nests compared to controls (6.0 chicks/nest vs. 4.6 chicks/nest).

A replicated, controlled study at two brown pelican *Pelecanus occidentalis* colonies in North Carolina, USA (7), found that pelican chicks in nests sprayed with insecticide (1% dilution of Rabon® 50 WP applied three times during the 1998 and 1999 breeding seasons) had significantly fewer immature ticks on them, compared to those in control nests (0.4–4.2 ticks/chick in 64 treated nests vs. 10.4–20.1 ticks/chick in 60 control nests). In the first year of the study, significantly more control nests were abandoned than treated nests, but there was no difference in the second study year (1998: none of 30 treated nests abandoned vs. 27% and 40% of 30 control nests on the two islands, 1999: two treated nests, three control nests sprayed with water and three undisturbed nests were abandoned). There were no significant differences in chick survival between treatment and control nests in either colony in either year.

A randomised, replicated and controlled trial over two breeding seasons in 2001–2 in British Columbia, Canada (8) found that tree swallow *Tachycineta bicolor* nests treated with insecticide (in 2001: 38.1% diatomaceous earth, 0.2% pyrethrins and 1.0% piperonyl butoxide; 2002: 80% silicon dioxide) contained significantly fewer bird fleas *Ceratophyllus idius* and blow flies *Protocalliphora* spp. than untreated nests (approximately 25 fleas and 28 blowflies in 23 treated nests vs. 350 fleas and 50 blowflies in 32 control nests).

A randomised, replicated and controlled trial at two brown pelican *Pelecanus occidentalis* colonies in South Carolina, USA (9) found that treating pelican nests with insecticide (175 ml of a 0.5% dilution of Rabon® 50 WP sprayed on to nests) significantly reduced the number of ticks *Carios capensis* on pelican chicks compared to control nests at one colony in 2004 but not on the other (Marsh Island: 1.1 ticks/chick for 45 treated chicks vs. 10.4 ticks/chick for 50 untreated chicks; Crab Bank: 0.01 ticks/chick for 48 treated chicks vs. 0.1 ticks/chick for 50 untreated nests). There were no differences in tick burden between treatment and control nests in 2005. There were also no differences between treatment and control nests in terms of survival of chicks to 21 days or chick growth rates.

A replicated, controlled study from June–July in 2007 in 36 experimental and 19 control nestboxes in an agricultural habitat in Alberta, Canada (10), found that tree swallow *Tachycineta bicolor* nestlings subject to ectoparasite removal through feather removal and insecticide did not grow faster or fledge earlier than control nestlings. Nestlings in control nests were larger than those in nests from which feathers were removed and insecticide applied (17.5 compared to 16.2 and 16.1 g/unit time respectively). Growth rate was positively related to number of feathers in the nest. Time between hatching and fledging and number of chicks fledged did not differ (20.4, 20.8 and 20.6 days between hatching and fledgling; 5.8, 5.4 and 5.5 nestlings fledged for control, feathers removed and insecticide nests respectively). The abundance and composition of parasitic arthropods in the nest did not differ between treatments. The authors conclude that feathers did not serve as an ectoparasite barrier, though they affected nestlings' growth rates positively.

A randomised, replicated and controlled study in 2006–7 in a mixed grassland-wetland-forest ecosystem in British Columbia, Canada (11) found that northern flicker *Colaptes auratus* nests that were fumigated (with diatomaceous earth and 0.5% pyrethrin) had fewer ectoparasitic flies *Carnus hemapterus* than control nests (fewer than five parasites/nestling for 33 fumigated nests vs. 10–17 parasites/nestling for 44 control nests). Chicks from control nests also fledged at lower weights than those from fumigated nests (129–132 g for females and 133–136 g for males in fumigated nests vs. 124–126 g for females and approximately 129 g for males in control nests). These relationships held for both new and reused nests.

(1) Roby, D. D., Brink, K. L. & Wittmann, K. (1992) Effects of bird blowfly parasitism on eastern bluebird and tree swallow nestlings. *The Wilson Bulletin*, 104, 630–643.
(2) Dufva, R. & Allander, K. (1996) Variable effects of the hen flea *Ceratophyllus gallinae* on the breeding success of the great tit in relation to weather conditions. *Ibis*, 138, 772–777.
(3) Alves, M. A. (1997) Effects of ectoparasites on the sand martin *Riparia riparia* nestlings. *Ibis*, 139, 494–496.
(4) Brown, C. R. & Brown, M. B. (1999) Fitness components associated with laying date in the cliff swallow. *The Condor*, 101, 230–245.
(5) Hurtrez-Boussès, S., Renaud, F., Blondel, J., Perret, P. & Galan, M.-J. (2000) Effects of ectoparasites of young on parents' behaviour in a Mediterranean population of blue tits. *Journal of Avian Biology*, 31, 266–269.
(6) Bouslama, Z., Lambrechts, M. M., Ziane, N., Djenidi, R. & Chabi, Y. (2002) The effect of nest ectoparasites on parental provisioning in a north-African population of the blue tit *Parus caeruleus*. *Ibis*, 144, E73–E78.
(7) Norcross, N. L. & Bolen, E. G. (2002) Effectiveness of nest treatments on tick infestations in the eastern brown pelican. *The Wilson Bulletin*, 114, 73–78.
(8) Dawson, R. D. (2004) Efficacy of diatomaceous earth at reducing populations of nest-dwelling ectoparasites in tree swallows. *Journal of Field Ornithology*, 75, 232–238.
(9) Eggert, L. M. & Jodice, P. G. (2008) Growth of brown pelican nestlings exposed to sublethal levels of soft tick infestation. *The Condor*, 110, 134–142.
(10) Stephenson, S., Hannon, S. & Proctor, H. (2009) The function of feathers in tree swallow nests: insulation or ectoparasite barrier? *The Condor*, 111, 479–487.
(11) Wiebe, K. L. (2009) Nest excavation does not reduce harmful effects of ectoparasitism: an experiment with a woodpecker, the northern flicker *Colaptes auratus*. *Journal of Avian Biology*, 40, 166–172.

12.43.3. Reduce nest ectoparasites by providing beneficial nesting material

- A randomised, replicated and controlled experiment in Canada (1) found lower numbers of some, but not all, parasites in nests provided with beneficial nesting material, but that there was no effect on fledging rates or chick condition.

Background

Some birds incorporate sprigs of volatile-rich plants into their nests, possibly as a method of reducing parasite burdens. Adding similar material may therefore represent a low-cost method of reducing the impact of parasites on nestlings.

A randomised, replicated and controlled experiment in 2000 at four sites in New Brunswick and Nova Scotia, Canada (1) found that the number of purple martin fleas *Ceratophyllus idius* was significantly lower in tree swallow *Tachycineta bicolor* nests with fresh yarrow *Achillea millefolium* foliage added every two days whilst clutches were being laid, compared to control nests with no added foliage (419 fleas/nest for 23 experimental nests vs. 773 fleas/nest for 44 controls). There were no corresponding differences in the number of blowfly *Protocalliphora* spp. pupae (3.3 and 2.5 pupae/nest for experimental and control nests respectively), nestling mass (23.5 g in experimental nests vs. 23.8 g in controls), nestling leukocyte profiles, or proportion of young fledging (5.3 fledglings/nest for experimental nests vs. 5.1 fledglings/nest for controls). The authors speculate that adult tree swallows may increase provisioning rate to compensate for flea parasitism.

(1) Shutler, D. & Campbell, A. A. (2007) Experimental addition of greenery reduces flea loads in nests of a non-greenery using species, the tree swallow *Tachycineta bicolor*. *Journal of Avian Biology*, 38, 7–12.

12.44. Guard nests to reduce risk of ectoparasites

Background

If populations are reduced to extremely low levels and are threatened by parasitism, then extremely intensive monitoring can be used to 'guard nests' and remove parasites if they increase to damaging levels. Due to the intensive nature of this work it is only likely to be viable if there are volunteers to do it, and the population being monitored is very small. Nest guarding is also used as a response to several other threats, such as predation and poaching and is therefore discussed in 'General responses to small/declining populations'.

12.45. Remove/control brood parasites

Background

Brood parasites are species that lay their eggs in other species' nests, with the chicks cared for by the hosts. They can have extremely detrimental impacts on bird species, with parasite chicks frequently evicting all host eggs or chicks. However, some brood parasite species, such as common cuckoos *Cuculus canorus* are themselves declining or threatened, making control or the reduction of parasitism undesirable.

However, in North America, brown-headed cowbirds Molothrus ater are increasingly abundant and expanding in their range. Brown-headed cowbirds have also been implicated in the near extinction of at least one species, the pale-headed brush finch *Atlapetes pallidiceps* in South America (Oppel *et al.* 2004), meaning that their control is seen as a conservation priority. All the studies we captured concern cowbird species.

Oppel, S., Schaefer, H. M., Schmidt, V. & Schroeder, B. (2004) Cowbird parasitism of pale-headed brush-finch *Atlapetes pallidiceps*: implications for conservation and management. *Bird Conservation International*, 14, 63–75.

12.45.1. Remove/control adult brood parasites

- All 11 studies from across the world that investigated parasitism rates (1–6,8–12) found that they were lower following cowbird *Molothrus* spp. control.

- One study from Ecuador (12) found an increase in host species population after cowbird control, but two American studies found no such effect (1,3).
- Five studies from the Americas (1,3,7,9,11) found higher productivities or success rates of host nests when cowbirds were removed, five (4,5,7,10,11) found that at least some measures of productivity did not change with cowbird control.

A before-and-after study in pine forests in Michigan, USA (1) found that the population of Kirtland's warbler *Dendroica kirtlandii* did not significantly increase during 1972–81 when a total of 33,536 brown-headed cowbirds *Molothrus ater* (a brood parasite) were removed, compared to 1931–71, when no control was in place (average population of 207 individuals during 1975–81 vs. 201 for 1931–71). Population size remained similar despite parasitism rates being significantly lower and warbler productivity being significantly higher during the period with cowbird control (3.4% of nests parasitised and 2.8 chicks/nest fledged with cowbird control, number of nests not provided vs. 59% and less than 1 chick/nest fledged before control, 91 nests examined).

A replicated trial in 1980 in Puerto Rico (2) as part of the same study as in (6,8) found that parasitism of yellow-shouldered blackbird *Agelaius xanthomus* nests was significantly lower in two mangrove forest sites where shiny cowbirds *Molothrus bonariensis* were removed, compared to sites where cowbirds were not removed (45% of 11 nests parasitised where all cowbirds were removed vs. 30% of ten where female cowbirds were removed; 67% of nine where males were removed and 92% of 12 in control sites). This study also investigated the impact of different nest boxes on parasitism, discussed in 'General responses to small/declining populations – Provide artificial nesting sites'.

A before-and-after study in young jack pine *Pinus banksiana* forests in Michigan, USA (3) found that, following the trapping and removal of a total of 84,937 brown-headed cowbirds *Molothrus ater* between 1972 and 1992, parasitism rates of Kirtland's warbler *Dendroica kirtlandii* nests by cowbirds fell significantly and productivity increased, compared to 1944–57 or 1966–71, before cowbird control was started (1944–57: 55% of 137 nests parasitised vs. 1966–71: 69% of nests parasitised, 0.8 fledglings/nest, sample size not provided; 1972–1977: 6% of 230 nests parasitised, 2.7 fledglings/nest; 1989–91: 2% of 48 nests parasitised). However, these changes did not result in increased numbers of singing males, with seasonal numbers fluctuating from 167–243 during 1971–89.

A before-and-after study at four coastal scrub sites in California, USA (4), found that levels of parasitism of California gnatcatcher *Polioptila californica* nests (initiated after 5th May) by brown-headed cowbirds *Molothrus ater* was significantly lower in 1994–5, following the removal of a total of 507 cowbirds over the two years, compared to in 1992–3 (10% of 132 nests parasitised in 1994–5 vs. 46% of 107 nests in 1992–3). However, nest success over the whole breeding season was no higher in years with cowbird control (21.7% of nests successful in 1994–5 vs. 11.2% in 1992–3).

A replicated, controlled before-and-after study between 1984 and 1995 at three 50–100 ha (2 experimental and 1 control) hardwood forest sites in Pennsylvania, USA (5) found that the proportion of hooded warbler *Wilsonia citrina* nests parasitised by brown-headed cowbirds *Molothrus ater* was significantly lower during years when cowbirds were removed (11% of 241 nests parasitised in 1991–5 vs. 64% of 28 nests parasitised in 1984–90). When comparing sites, parasitism rates were lower on sites with cowbird removal (0–11% of 280 nests parasitised at removal sites vs. 38–58% of 32 nests at controls). However, there were no changes in warbler nesting success between sites with low (≤5%) and high (>30%) levels of parasitism (average of 1.7 fledglings/nest in sites with low parasitism vs. 1.6 fledglings/nest in sites with high parasitism). The authors suggest that nesting success may be more determined by high rates (22–52%) of predation, than by parasitism. An average of 11–20 female and 7.5–17 male cowbirds were removed each year.

A study in 1996–9 in coastal forests on Puerto Rico (6) found that only a single yellow-shouldered blackbird *Agelaius xanthomus* nest was parasitised by shiny cowbirds *Molothrus bonairensis* in the study period. The authors argue that this was due to a widespread cowbird eradication programme initiated in 1984. This study is discussed in more detail in 'Provide artificial nesting sites'.

A replicated, controlled study in 1995–9 at three riparian sites in British Columbia, Canada (7), found that the success rates of song sparrow *Melospiza melodia* nests were higher at sites where female brown-headed cowbirds *Molothrus ater* were removed (51 in 1996, 163 in 1997 at one site; 24 in 1998 at another) than in a control site with no cowbirds removed, both across all years (44% success for 296 nests in removal sites vs. 32% success for 615 in control sites) and just in years when cowbirds were removed (44% success for 296 nests in removal sites vs. 34% success for 399 in control sites). Nest survival rates were higher for song sparrow eggs in removal sites (96.5% daily survival vs. 94.7%), but there was no significant difference in nestling survival (97.4% daily survival vs. 96.6%). Significantly fewer nests were abandoned after being parasitised by cowbirds in removal sites than control sites (9% abandoned vs. 16.5%).

A controlled before-and-after study in 2000, 2001 and 2003 in mangrove forests in Puerto Rico (8) found that a significantly lower proportion of yellow-shouldered blackbird *Agelaius xanthomus* and yellow warbler *Dendroica petechia* nests were parasitised by shiny cowbirds *Molothrus bonariensis* following the removal of adult cowbirds (from 1982 onwards), cowbird eggs and chicks from artificial nests used by blackbirds (from 1991 onwards) (blackbirds: 91–95% of 202 nests studied in 1975–83 vs. 3% of 927 nests in 2000, 2001 and 2003; warblers: 63% of nests in 1975–83 vs. 37% of 165 nests in 2000, 2001 and 2003). Decreases in areas without cowbird control were either smaller or non-existent (44% of 32 blackbird nests and 85% of 13 warbler nests in reference areas parasitised). The effect of egg and chick removal is discussed in 'Remove brood parasite eggs from target species' nests'.

A controlled before-and-after cross-over study between 2003 and 2005 at five tall-grass prairie sites (24–83 ha) in Kansas, USA (9) found that brood parasitism by brown-headed cowbirds *Molothrus ater* on Bell's vireos *Vireo bellii* was significantly lower at sites where a total of 980 cowbirds (171 adult females, 724 adult males and 85 juveniles) were removed in 2004–5, compared to before removals (47 and 58% parasitism in 130 nests in removal plots in 2004 and 2005 respectively vs. 64–81% parasitism in 130 nests in 2003). There was no corresponding decrease in areas when cowbirds were not removed (77–85% parasitism in 278 nests in non-removal plots in 2004–5). Vireo productivity was higher at removal plots (2.6 compared to 1.2 vireo fledglings/pair) and nest success was higher for non-parasitised nests (64% of nests producing at least one chick vs. 51% of parasitised nests). However, cowbird productivity was also higher for removal plots than control plots (0.3 compared to 0.1 cowbird chicks/vireo pair).

A controlled cross-over experiment at four 24–36 ha tall-grass prairie sites in Kansas, USA (10) found that, in 2004, parasitism of dickcissel *Spiza americana* nests by brown-headed cowbirds *Molothrus bonariensis* was significantly lower in two plots where 346 cowbirds were removed (76 adult females, 231 adult males and 39 juveniles) than in two control plots (51% of 53 treatment nests and 85% of 27 control nests respectively). However, in 2005 when treatments were reversed and 634 cowbirds (95 adult females, 493 adult males and 46 juveniles) removed from the remaining two plots, there was no such difference (78% of 45 nests and 82% of 44 nests respectively). In neither year were there differences in dickcissel productivity between experimental and control plots (2004: 0.32 and 0.29 chicks/nest respectively; 2005: 0.06 and 0.04 chicks/nest respectively). The authors suggest that nest survival was very low (34% in 2004, 7% in 2005) due to predation and other causes, not because of parasitism.

A before-and-after study in Puerto Rico (11) using some of the data from (8) found that the proportion of yellow warbler *Dendroica petechia* nests parasitised by shiny cowbirds *Molothrus bonariensis* fell following cowbird control (see (8)). Nesting productivity of warblers also increased following cowbird control (0.7 warbler chicks/nest in 62 nests in 2001–2 vs. 0.35 warbler chicks/nest in 1977–80). However, the proportion of nests fledging chicks appeared to be lower following control (31% of nests in 2001–2 vs. 45% of nests in 1977–80). The majority of nests lost were predated (80% of 41 failed nests).

A before-and-after trial in arid scrubland in the Yunguilla Valley, Ecuador (12), found a large increase in the number of territorial male pale-headed brush finches *Atlapetes pallidiceps* following the control of brown-headed cowbirds *Molothrus ater*, starting in 2003 (12–29 birds recorded in 1999–2002 vs. 27–50 in 2003–7). In addition, parasitism rates appeared lower, with only a single nest being parasitised in 2003 and none in 2005. Between 18 and 69 cowbirds were shot each year at the site.

(1) Kelly, S. T. & DeCapita, M. E. (1982) Cowbird control and its effect on Kirtland's Warbler reproductive success. *The Wilson Bulletin*, 94, 363–365.
(2) Wiley, J. W., Post, W. & Cruz, A. (1991) Conservation of the yellow-shouldered blackbird *Agelaius xanthomus*, an endangered West Indian species. *Biological Conservation*, 55, 119–138.
(3) Kepler, C. B., Irvine, G. W., DeCapita, M. E. & Weinrich, J. (1996) The conservation management of Kirtland's Warbler *Dendroica kirtlandii*. *Bird Conservation International*, 6, 11–22.
(4) Braden, G. T., McKernan, R. L. & Powell, S. M. (1997) Effects of nest parasitism by the brown-headed Cowbird on nesting success of the California gnatcatcher. *The Condor*, 99, 858–865.
(5) Stutchbury, B. J. (1997) Effects of female cowbird removal on reproductive success of hooded warblers. *The Wilson Bulletin*, 109, 74–81.
(6) López-Ortiz, R., Ventosa-Febles, E. A., Reitsma, L. R., Hengstenberg, D. & Deluca, W. (2002) Increasing nest success in the yellow-shouldered blackbird *Agelaius xanthomus* in southwest Puerto Rico. *Biological Conservation*, 108, 259–263.
(7) Smith, J. N., Taitt, M. J., Zanette, L. & Myers-Smith, I. H. (2003) How do brown-headed cowbirds (*Molothrus ater*) cause nest failures in song sparrows (*Melospiza melodia*)? A removal experiment. *The Auk*, 120, 772–783.
(8) López-Ortiz, R., Ventosa-Febles, E. A., Ramos-Álvarez, K. R., Medina-Miranda, R. & Cruz, A. (2006) Reduction in host use suggests host specificity in individual shiny cowbirds (*Molothrus bonariensis*). *Ornitología Neotropical*, 17, 259–269.
(9) Kosciuch, K. L. & Sandercock, B. K. (2008) Cowbird removals unexpectedly increase productivity of a brood parasite and the songbird host. *Ecological Applications*, 18, 537–548.
(10) Sandercock, B. K., Hewett, E. L. & Kosciuch, K. L. (2008) Effects of experimental cowbird removals on brood parasitism and nest predation in a grassland songbird. *The Auk*, 125, 820–830.
(11) Vincenty, M., Tossas, A. G., Bird-Pico, F. J., Lopez-Ortiz, R. & Ventosa, E. A. (2009) Yellow warbler (*Dendroica petechia*) breeding success in relation to shiny cowbird (*Molothrus bonariensis*) brood parasitism in Boquerón, Puerto Rico. *Ornitología Neotropical*, 20, 523–533.
(12) Krabbe, N., Juiña, M. & Sornoza, A. F. (2011) Marked population increase in pale-headed brush-finch *Atlapetes pallidiceps* in response to cowbird control. *Journal of Ornithology*, 152, 219–222.

12.45.2. Remove brood parasite eggs from target species' nests

- A controlled before-and-after study on Puerto Rico (1) found lower rates of parasitism of yellow-shouldered blackbird *Agelaius xanthomus* nests when shiny cowbird *Molothrus bonariensis* eggs were removed from nests.
- A replicated, controlled study from 1997–9 in grassy fields in New York State, USA (2) found that song sparrow *Melospiza melodia* nests that had cowbird eggs removed from them had lower success than nests which were parasitised and that did not have eggs removed.

Background

Brood parasites will only have detrimental impacts on host species if their chicks hatch in host nests. Therefore, removing parasite eggs from host nests before they hatch may protect host populations.

A controlled before-and-after study in mangrove forests on Puerto Rico in 2000, 2001 and 2003 (1) found that a significantly lower proportion of yellow-shouldered blackbird *Agelaius xanthomus* nests were parasitised by shiny cowbirds *Molothrus bonariensis*, compared to yellow warbler *Dendroica petechia* nests, following the removal of cowbird eggs and chicks from artificial nests used by blackbirds from 1991, and the control of adult cowbirds from 1983 (3% of 927 blackbird nests vs. 37% of 165 warbler nests). Prior to cowbird control, parasitism rates had been higher for blackbirds (91–95% of 202 blackbird nests in 1975–83 vs. 63% of warbler nests). Parasitism rates in areas without cowbird control were lower for blackbirds (44% of 32 nests) and higher for warblers (85% of 13 nests). The authors suggest that removing eggs and nestlings reduces the proportion of cowbirds that imprint on specific hosts, reducing future parasitism. The effect of adult cowbird removal is discussed in 'Remove adult brood parasites'.

A replicated, controlled study from 1997–9 in grassy fields in New York State, USA (2) found that song sparrow *Melospiza melodia* nests parasitised by brown-headed cowbirds *Molothrus ater* had lower productivity when cowbird eggs were removed, compared to parasitised nests when cowbird eggs were not removed (median of 0% of eggs from nests with eggs removed produced nestlings vs. 75% of eggs from nests where cowbird eggs were not removed). There were no differences in the number of song sparrow nestlings surviving to five days old between parasitised nests, non-parasitised nests and parasitised nests with cowbird eggs removed.

(1) López-Ortiz, R., Ventosa-Febles, E. A., Ramos-Álvarez, K. R., Medina-Miranda, R. & Cruz, A. (2006) Reduction in host use suggests host specificity in individual shiny cowbirds (*Molothrus bonariensis*). *Ornitología Neotropical*, 17, 259–269.

(2) Hauber, M. E. (2009) Does the removal of avian brood parasite eggs increase host productivity? A case study with brown-headed cowbirds *Molothrus ater* and song sparrows *Melospiza melodia* near Ithaca, New York, USA. *Conservation Evidence*, 6, 83–88.

12.46. Use false brood parasite eggs to discourage brood parasitism

- A replicated, controlled experiment in the USA (1) found lower parasitism rates for red-winged blackbird *Agelaius phoeniceus* nests with false or real brown-headed cowbird *Molothrus ater* eggs added to them.

Background

Brood parasites choose to lay eggs in nests that have not previously been parasitised, to minimise intra-specific competition. Therefore, placing false parasite eggs in nests may protect nests by discouraging parasites from laying.

A replicated, controlled experiment in 1985 and 1991 on seven wetland sites in Colorado, USA (1) found that the proportion of red-winged blackbird *Agelaius phoeniceus* nests parasitised by brown-headed cowbirds *Molothrus ater* in 1985 was significantly lower for nests that had artificial or real cowbird eggs placed in them, than for control nests (5% of 57 and 32% of 54 nests parasitised respectively). In 1991, the rate of parasitism was again lower for nests with artificial eggs added but was not significantly different (6% of 40 and 16% of 25 of nests parasitised respectively). The authors suggest this may be due to the small sample size in 1991. Artificial egg size (26.1 × 17.2 mm, resembling blackbird eggs vs. 20.1 × 16.1 mm, resembling cowbird eggs) or the use of artificial or real cowbird eggs did not affect parasitism rates. Adding eggs did not alter clutch size or hatching success of blackbirds (average clutch size of 3.8 eggs/clutch for 97 experimental nests vs. 3.7 eggs/clutch for 79 controls; average hatching success of nests that hatched at least one egg: 3.2 eggs/nest for 48 experimental nests vs. 3.1 eggs/nest for 42 controls).

(1) Ortega, C. P., Ortega, J. C. & Cruz, A. (1994) Use of artificial brown-headed cowbird eggs as a potential management tool in deterring parasitism. *The Journal of Wildlife Management*, 58, 488–492.

12.47. Provide supplementary food to increase parental presence and so reduce brood parasitism

Background

Brood parasites only tend to lay in host nests when the parents are absent, to reduce the risk of eggs being rejected. Therefore, maximising the time parent birds spend at their nest may reduce the incidence of parasitism. One possible way of doing this would be to provide supplementary food, meaning that birds have to spend less time foraging. Studies discussing this and other uses of supplementary feeding are discussed in 'General responses to small/declining populations'.

12.48. Alter artificial nest sites to discourage brood parasitism

- A replicated trial from Puerto Rico (*1*) found that brood parasitism levels were extremely high across all nest box designs tested.

A replicated trial in 1980 in Puerto Rico (*1*) found that, shiny cowbirds *Molothrus bonariensis* parasitised yellow-shouldered blackbird *Agelaius xanthomus* nests in all 16 nest box types tested with 96% of 103 nests parasitised. The effect of cowbird control is discussed in 'Threat: Invasive and other problematic species – Remove/control brood parasites'.

(1) Wiley, J. W., Post, W. & Cruz, A. (1991) Conservation of the yellow-shouldered blackbird *Agelaius xanthomus*, an endangered West Indian species, *Biological Conservation*, 55, 119–138.

Reduce detrimental impacts of other problematic species

12.49. Use copper strips to exclude snails from nests

- A single small, before-and-after study in Mauritius (1) found no snail-caused chick mortality in 2004–7 after the installation of copper strips at seven echo parakeet *Psittacula eques* nest holes, compared to four fatalities in 2003–4.

Background

In Mauritius, the critically endangered echo parakeet *Psittcaula eques* has a very small breeding population and low reproductive output. In 2002–4, four chicks were found dead having been covered in slime and suffocating after African giant land snails Achatina spp. (introduced to Mauritius and many other islands for food) moved other them.

Snails are deterred from moving over copper due to a reaction between their slime and the metal, and trials in Mauritius showed that Achatina individuals did not move over copper strips. A snail exclusion barrier was therefore tested.

A small before-and-after study in a forest in Black River Gorges National Park, Mauritius (1) found that African giant land snails *Achatina* spp. did not enter seven nest holes used by echo parakeets *Psittacula eques* between the 2004/5 breeding season and 2007 following the installation of a 50 mm wide strip of copper around the trunk of each occupied tree. Before copper strip installation, four chicks were found dead in two nest holes between 2003 and 2004 due to asphyxiation by snail slime on the snail's foot.

(1) Tatayah, R. V. V., Malham, J. & Haverson, P. (2007) The use of copper strips to exclude invasive African giant land-snails *Achatina* spp. from echo parakeet *Psittacula eques* nest cavities, Black River Gorges National Park, Mauritius. *Conservation Evidence*, 4, 6–8.

13. Threat: Pollution

Pollution, of many diverse types, has direct and indirect impacts on birds – an indication of the wider problems it creates for humans and biodiversity alike. Water-borne pollutants can devastate otherwise productive wetland and coastal habitats. Many pesticides linked to bird deaths are still in widespread use, especially in developing countries. Oil spills remain a threat to some seabirds, while solid waste is an increasing problem. Little is known of the long-term effects of many pollutants, including those that persist and accumulate in the environment.

Key messages – Industrial pollution

Clean birds following oil spills

Three studies from South Africa and Australia found high survival of oiled-and-cleaned penguins and plovers, but a large study from the USA found low survival of cleaned common guillemots. Two studies found that cleaned birds bred and had similar success to un-oiled birds. After a second spill, one study found that cleaned birds were less likely to breed. Two studies found that cleaned birds had lower breeding success than un-oiled birds.

Relocate birds following oil spills

A study from South Africa found that a high percentage of penguins relocated following an oil spill returned to and bred at their old colony. More relocated birds bred than oiled-and-cleaned birds.

Use visual and acoustic 'scarers' to deter birds from landing on pools polluted by mining or sewage

Two studies from Australia and the USA found that deterrent systems reduced bird mortality on toxic pools. Four of five studies from the USA and Canada found that fewer birds landed on pools when deterrents were used, one found no effect. Two studies found that radar-activated systems were more effective than randomly-activated systems. One study found that loud noises were more effective than raptor models.

Use repellents to deter birds from landing on pools polluted by mining

An *ex situ* study from the USA found that fewer common starlings consumed contaminated water laced with chemicals, compared to untreated water.

Key messages – Agricultural pollution

Reduce pesticide, herbicide and fertiliser use generally

One of nine studies found that the populations of some species increased when pesticide use was reduced and other interventions used. Three studies found that some or all species were found at higher densities on reduced-input sites. Five found that some of all species were not at higher densities. A study from the UK found that grey partridge chicks had higher survival on sites with reduced pesticide input. Another found that partridge broods were smaller on such sites and there was no relationship between reduced inputs and survival or the ratio of young to old birds.

Restrict certain pesticides or other agricultural chemicals

A before-and-study from Spain found an increase in the regional griffon vulture population following the banning of strychnine, amongst several other interventions.

Provide food for vultures to reduce mortality from diclofenac

A before-and-after trial in Pakistan found that oriental white-backed vulture mortality rates were significantly lower when supplementary food was provided, compared to when it was not.

Make selective use of spring herbicides

We captured no evidence for the effects of selective use of spring herbicides on bird populations.

Use organic rather than mineral fertilisers

We captured no evidence for the effects of using organic, not mineral, fertilisers on bird populations.

Reduce chemical inputs in permanent grassland management

A study from the UK found that no more foraging birds were attracted to pasture plots with no fertiliser, compared to control plots.

Leave headlands in fields unsprayed (conservation headlands)

Three studies from Europe found that several species were strongly associated with conservation headlands; two of these found that other species were not associated with them. A review from the UK found larger grey partridge populations on sites with conservation headlands. Three studies found higher grey partridge adult or chick survival on sites with conservation headlands, one found survival did not differ. Four studies found higher grey partridge productivity on sites with conservation headlands, two found similar productivities and one found a negative relationship between conservation headlands and the number of chicks per adult partridge.

Provide unfertilised cereal headlands in arable fields

We did not find any studies describing the effects of unfertilised cereal headlands on bird populations.

Plant riparian buffer strips

We did not find any studies describing the effects of riparian strips on bird populations through their impact on water pollution levels. Their use as a habitat type is discussed in 'Habitat restoration and creation'.

Provide buffer strips around in-field ponds

We did not find any studies describing the effects of buffer strips around in-field ponds on bird populations through their impact on water pollution levels. Their use as a habitat type is discussed in 'Habitat restoration and creation'.

Key message – Air-borne pollutants

Use lime to reduce acidification in lakes

A study from Sweden found no difference in osprey productivity during a period of extensive liming of acidified lakes compared to two periods without liming.

Key messages – Excess energy

Turn off lights to reduce mortality from artificial lights

A study from the UK found that fewer seabirds were downed when artificial (indoor and outdoor) lighting was reduced at night, compared to nights with normal lighting.

Reduce the intensity of lighthouse beams

We captured no evidence for the effects of reducing the intensity of lighthouse beams on bird mortality.

Shield lights to reduce mortality from artificial lights

A study from the USA found that fewer shearwaters were downed when security lights were shielded, compared to nights with unshielded lights.

Use flashing lights to reduce mortality from artificial lights

A study from the USA found that fewer dead birds were found beneath aviation control towers with only flashing lights, compared to those with both flashing and continuous lights.

Use lights low in spectral red to reduce mortality from artificial lights

Two studies from Europe found that fewer birds were attracted to low-red lights (including green and blue lights), compared with the number expected, or the number attracted to white or red lights.

Use volunteers to collect downed birds and rehabilitate them

We found no studies that report on the effectiveness of using volunteers to collect and rehabilitate downed birds.

Industrial pollution

13.1. Clean birds following oil spills

- Three studies from South Africa (2,4) and Australia (8) found high survival of rehabilitated penguins and plovers or similar survival to un-oiled birds. However a large study from the USA and Canada (1) found that rehabilitated common guillemots *Uria aalge* had significantly lower survival than untreated birds.
- Three studies from South Africa (5,6) and Australia (8) found that rehabilitated birds bred, with one (5) finding that rehabilitated birds had similar breeding success to un-oiled birds. However, this study found that birds rehabilitated after a second spill were less likely to breed, whilst two other studies (3,6) found that rehabilitated birds had lower success than un-oiled birds.

Background

Oil spills at sea can kill large numbers of seabirds and have the potential to wipe out entire populations where these are small or localised. Oil can stick to birds' feathers, making them lose their water-proofing and potentially leading to hypothermia. When birds try to clean their feathers they ingest oil and are likely to become poisoned.

Birds can be taken in and cleaned, but this is an expensive operation, with the cleaning operations after the Treasure oil spill in South Africa in 2000 costing an estimated $100/bird (Whittington 2003). In addition, whereas cleaning may prevent adults from dying, there is evidence that the offspring of rehabilitated birds have lower survival than normal (Barham *et al.* 2008). This means that the hand-rearing of offspring may be an important intervention after oil spills (Barham *et al.* 2008, see 'Artificially incubate and hand-rear birds in captivity').

In addition to the direct effects described in the studies below, one study (Ryan 2003) estimated that the population of African penguins *Spheniscus demersus* was 19% larger than it would have been without rehabilitation efforts (approximately 163,000 adults, compared to 137,000).

Barham, P. J., Underhill, L. G., Crawford, R. J. M. & Leshoro, T. M. (2007) Differences in breeding success between African penguins (*Spheniscus demersus*) that were and were not oiled in the MV Treasure oil-spill in 2000. *Emu*, 107, 1–7.

Barham, P. J., Underhill, L. G., Crawford, R. J. M., Altwegg, R., Mario Leshoro, T., Bolton, D. A., Dyer, B. M. & Upfold, L. (2008) The efficacy of hand-rearing penguin chicks: evidence from African Penguins (*Spheniscus demersus*) orphaned in the Treasure oil spill in 2000. *Bird Conservation International*, 18, 144–152.

Ryan, P. G. (2003) Estimating the demographic benefits of rehabilitating oiled African penguins, pp. 25–29 in eds. Nel, D. C. & Whittington, P. A. *Rehabilitation of oiled African Penguins: a conservation success story*. BirdLife South Africa and Avian Demography Unit, Cape Town.

Whittington, P. A. (2003) Post-release survival of rehabilitated African penguins, pp. 8–17 in eds. Nel, D. C. & Whittington, P. A. *Rehabilitation of oiled African Penguins: a conservation success story*. BirdLife South Africa and Avian Demography Unit, Cape Town.

A replicated controlled study in Canada and the USA (1) found that ringed seabirds that were oiled, cleaned and released ('treated') were found dead (recovered) much sooner than birds that were not oiled (between six and 111 days before 98 treated birds were recovered vs. 216–1,019 days for 700 non-oiled birds). In addition, estimated survival rates of oiled common guillemots *Uria aalge* were just 13% over 20 days (resulting in negligible annual survival), much lower than the 90–95% annual survival for adults (20–40% for juveniles) commonly seen.

A replicated study of African penguin *Spheniscus demersus* survival between 1994 and 1996 following a 1994 oil spill near Cape Town, South Africa (2) found that 65% of 4,076 penguins collected, cleaned, banded and released were re-sighted within two years of release. The majority of these (73%) were seen in the first year but new sightings continued until the end of the study period. The number of dead birds reported (24 from monitoring teams, 25 by the public) was very close to the number expected from previous studies and the authors argue that large-scale mortality of penguins was unlikely to have occurred.

A replicated study in Tasmania, Australia (3), in the 1995–6 and 1996–7 breeding seasons found that pre-fledging masses of chicks from rehabilitated oiled little penguins *Eudyptula minor* were significantly lower than those from non-oiled birds (approximately 700–800 g for chicks from 65 pairs with rehabilitated birds vs. 850–900 g for 167 un-oiled pairs). Hatching success did not differ between groups but the number of chicks produced/egg and fledging success were significantly lower among rehabilitated birds in 1995–6, especially for nests that had a rehabilitated female (with 22% lower fledging success), and laying date was also delayed (eggs in nests from early October for un-oiled birds, but first appeared on the 4 November for rehabilitated birds). These differences were not apparent in 1996–7.

A replicated, controlled study in the Western Cape, South Africa, in 1994–9 (4), found that average annual survival of African penguins *Spheniscus demersus* that were oiled, cleaned and released following four oil spills birds was estimated at 79%, compared with 81% for non-oiled birds, a non-significant difference. Between 40 and 87% of rehabilitated birds were recorded back at their breeding colonies after being released (with between 101 and 2,962 birds rehabilitated each time). The low number of birds recorded for one spill (40% after four years) may have been due to penguins being found a long way from their colonies and therefore released in an inappropriate place (72% of birds that were seen were recorded at a different colony). This study also discusses the survival of hand-reared penguins, orphaned by oil spills, described in 'Artificially incubate and hand-rear birds in captivity'.

A replicated, controlled study on Dassen Island, Western Cape, South Africa (5), found that at least 60% of African penguins *Spheniscus demersus* that were rehabilitated following the 1994 *Apollo Sea* oil spill had bred within six years of the spill. Productivity of these birds was no different from un-oiled birds (0.32 chicks/egg for 599 oiled birds vs. 0.30 for 558 un-oiled) and their chicks showed identical growth patterns. However, the authors note that during some periods of stress, the rehabilitated birds had significantly lower productivity than un-oiled birds. Of 2,744 birds rehabilitated after the *Treasure* spill in 2000, 75% were seen two years later, but only 17% had bred. Rehabilitated birds were more likely than controls to change breeding partners (67% keeping mates vs. 80–94%), but this difference appeared to be temporary. This study is also discussed in 'Relocate birds away from oil spills'.

A controlled, replicated study on Robben Island, South Africa, between 2001 and 2005 (6), found that African penguin *Spheniscus demersus* pairs with at least one parent that had been oiled and rehabilitated (i.e. cleaned and returned to the wild) following an oil spill in 2000 had significantly lower fledging success, compared either to pairs without rehabilitated birds (control pairs), or those with birds banded either for research or following rehabilitation from earlier oil spills (43% of 321 chicks fledging from pairs with rehabilitated birds vs. 61% of 170 from controls and 61% of 114 from previously-banded pairs). Hatching success and clutch size were not significantly different between groups and the differences in fledging success were due to high levels of mortality in older chicks from rehabilitated pairs.

A small study in South Africa (7) examined the survival and reproduction of hand-reared African penguins *Spheniscus demersus* orphaned after the *Treasure* oil spill in 2000. This is discussed in 'Artificially incubate and hand-rear birds in captivity'.

A small study in Victoria, Australia, in 2003–6 (8) found that two hooded plovers *Thinornis rubricollis* that were oiled following an oil spill in 2003 and captured, cleaned and released, survived for at least two years, bred and raised at least one chick, which also bred. This study is also discussed in 'Use signs and access restrictions to reduce disturbance at nest sites'.

(1) Sharp, B. E. (1996) Post-release survival of oiled, cleaned seabirds in North America. *Ibis*, 138, 222–228.
(2) Underhill, L. G., Bartlett, P. A., Baumann, L., Crawford, R. J. M., Dyer, B. M., Gildenhuys, A., Nel, D. C., Oatley, T. B., Thornton, M., Upfold, L., Williams, A. J., Whittington, P. A. & Wolfaardt, A. C. (1999) Mortality and survival of African Penguins *Spheniscus demersus* involved in the Apollo Sea oil spill: an evaluation of rehabilitation efforts. *Ibis*, 141, 29–37.
(3) Giese, M., Goldsworthy, S. D., Gales, R., Hamill, J. & Brothers, N. (2000) Effects of the Iron Baron oil spill on little penguins (*Eudyptula minor*). III. Breeding success of rehabilitated oiled birds. *Wildlife Research*, 27, 583–591.
(4) Whittington, P. A. (2003) Post-release survival of rehabilitated African penguins. 8–17 in: D. C. Nel, P. A. Whittington (eds) *Rehabilitation of oiled African penguins: a conservation success story* BirdLife South Africa and Avian Demography Unit, Cape Town.
(5) Wolfaardt, A. C. & Nel, D. C. (2003) Breeding productivity and annual cycle of rehabilitated African penguins following oiling. 18–24 in: D. C. Nel, P. A. Whittington (eds) *Rehabilitation of oiled African Penguins: a conservation success story*. BirdLife South Africa and Avian Demography Unit, Cape Town.
(6) Barham, P. J., Underhill, L. G., Crawford, R. J. M. & Leshoro, T. M. (2007) Differences in breeding success between African penguins (*Spheniscus demersus*) that were and were not oiled in the MV Treasure oil-spill in 2000. *Emu*, 107, 1–7.
(7) Barham, P. J., Underhill, L. G., Crawford, R. J. M., Altwegg, R., Mario Leshoro, T., Bolton, D. A., Dyer, B. M. & Upfold, L. (2008) The efficacy of hand-rearing penguin chicks: evidence from African Penguins (*Spheniscus demersus*) orphaned in the Treasure oil spill in 2000. *Bird Conservation International*, 18, 144–152.
(8) Weston, M. A., Dann, P., Jessop, R., Fallaw, J., Dakin, R. & Ball, D. (2008) Can oiled shorebirds and their nests and eggs be successfully rehabilitated? A case study involving the threatened hooded plover *Thinornis rubricollis* in south-eastern Australia. *Waterbirds*, 31, 127–132.

13.2. Relocate birds following oil spills

- A replicated study in South Africa (1) found that a higher percentage of African penguins *Spheniscus demersus* that were relocated following an oil spill bred at their old colonies, compared to birds which were rehabilitated after being oiled, despite fewer relocated birds being seen at their home colony.

Background

If an oil spill is definitely going to affect a large number of birds and there is sufficient warning, then it may be possible to temporarily relocate birds away from the danger. This avoids having to clean birds, which can be expensive, stressful and may not work (see 'Clean birds following oil spills'). However, it is still likely to be expensive and carries the risk that birds will not be able to return to the original site. Alternatively, there is the possibility, with fast-flying species, that they will return before the clean-up operations are complete, thus becoming oiled anyway.

A replicated trial following the *Treasure* oil spill in 2000 between Robben and Dassen Islands, Western Cape, South Africa (1), found that 62% of 1,130 African penguins *Spheniscus demersus* that were moved 800 km east and released were recorded on Dassen Island, two years after the event, with 41% breeding. This compared with a higher number of rehabilitated (oiled, cleaned and released) birds being seen (75% of 2,744), but fewer breeding (17%). Relocated birds began arriving at Dassen Island 11 days after being released, with most arriving 18 days or so after relocation. In total, 19,500 birds were relocated. This study is also discussed in 'Clean birds following oil spills'.

(1) Wolfaardt, A. C. & Nel, D. C. (2003) Breeding productivity and annual cycle of rehabilitated African penguins following oiling. 18–24 in: D. C. Nel, P. A. Whittington (eds) *Rehabilitation of oiled African Penguins: a conservation success story*. BirdLife South Africa and Avian Demography Unit, Cape Town.

13.3. Deter or prevent birds from landing on toxic pools

Background

Various techniques have been employed to deter birds from landing in hazardous water bodies, especially contaminated mine-tailings ponds. Tailing ponds store water-borne, often toxic, waste material derived from mining activities, power plant evaporation ponds can contain sodium decahydrate (which can crystallise on feathers) and oil spills can also have catastrophic effects on seabirds (see 'Clean birds following oil spills').

Numerous 'conventional' deterrent techniques are used e.g. scarecrows, and propane-powered gas guns (that produce periodic loud explosions). However, their effectiveness often declines over time as birds habituate to deterrent stimuli (Bomford & O'Brien 1990). Therefore, radar-activated on-demand systems (which activate deterrents located in/around ponds only upon the approach of flying birds) have been developed in an attempt to reduce the problem of habituation (Ronconi *et al.* 2004).

Repellents are also used to deter birds, although the literature on the effectiveness of these is smaller.

Similar interventions are discussed in the chapter on aquaculture, with the aim of reducing conflict with fish-eating birds.

Bomford, M. & O'Brien, P. H. (1990) Sonic deterrents in animal damage control: a review of device tests and effectiveness. *Wildlife Society Bulletin*, 18, 411–422.

Ronconi, R. A., St. Clair, C. C., O'Hara, P. D. & Burger, A. E. (2004) Waterbird deterrence at oil spills and other hazardous sites: potential applications of a radar-activated on-demand deterrence system. *Marine Ornithology*, 32, 25–33.

13.3.1. Use visual and acoustic 'scarers' to deter birds from landing on pools polluted by mining or sewage

- Two studies (3,4) found lower bird mortality or fewer birds rescued from toxic ponds when deterrent systems were used. Four of five studies (2–4,7) found that fewer birds landed on pools with deterrents than controls, although one of these (3) found that the effect was weaker for grebes compared to wildfowl and absent for waders. One study that used regular broadcasts of different sounds (5) found that it had no impact on bird behaviour.
- Two studies (6,7) investigated different systems and found that radar-operated systems were more effective than systems that worked at random intervals. One of these studies (7) also found that loud noises were more effective than moving peregrine falcons *Falco peregrinus* models.

A trial of an early design of radar-activated system (that triggered a stereo, fire alarm, propane cannons and firecrackers) aimed at scaring birds from mine tailing facilities in Nevada, USA (1), showed that the system had promise. Subsequent trials of the system are described in (2,4,6).

Two controlled experiments at a 18.2 ha desulphurisation pond in north-central USA (2) found that significantly fewer birds landed on ponds where a scaring system was in place (autumn and spring 1993–1994: 17 birds used a pond when the system was active for 48 hour periods vs. 125 when it was inactive for 48 hour periods; autumn 1994: 16 of 43,964 bird landings, 2% of the expected number, occurred on the pond with the system). The radar-activated system (a BirdAvert© system) used various deterrents (broadcasts of recordings of e.g. dogs barking, guns firing and falcons screaming; strobe light; plastic falcons with flapping wings, and automated scarecrows).

A randomised, replicated and controlled trial in South Australia in 1996–7 (3), found that the number of wildfowl on sewerage ponds was 90% lower when a slowly rotating beacon with an intermittent, low-angled beam was floated in the centre of the ponds, compared to control ponds with no beacon (average of approximately 2 ducks/night on experimental ponds vs. 36 ducks/night on control ponds). There was no change in the behaviour of waders, and grebes dived, rather than dispersing. A follow-up, before-and-after experiment found that the number of wildfowl and hoary-headed grebes *Poliocephalus poliocephalus* on toxic and acidic tailing ponds were over 66% lower in the 12 months following the installation of a slowly rotating beacon with an intermittent, low-angled beam that floated in the centre of the ponds, compared to the 12 months before installation. Over half of the casualties were when the device was not fully operational and mortality rates were reduced to one-sixth of post-installation levels in the 12 months after the device became fully operational. Of the 15 mortalities following installation, four were hoary-headed grebes.

A controlled trial at the Jim Bridger Power Plant, Wyoming, USA, in 1996–7 (4) found that waterfowl were 12.5 times less likely to fly over and 4.2 times less likely to land on two ponds (36.5 and 80.8 ha) when a radar-activated deterrent system was used, compared to an adjacent freshwater pond (93.2 ha) with no deterrent; non-waterfowl were seven times less likely to land. Bird rescues per year decreased by more than 400 (>70% fewer rescues) in the first year of full operation. Between 685 and 714 rescues occurred in preceding years, 859 in the transition year, and 210 in the first year of full operation (mortality reduced by more than 77% relative to previous years). When flying birds were detected, the system broadcast alarm and distress calls of a variety of animals, let off 'screamer' cartridges, and finally a bird aerosol tear gas was triggered (only used if birds were still detected after initial deterrents were activated).

A replicated, controlled trial at two sites in San Francisco Bay, California (USA) found that the Breco Bird Scarer (an orange buoy designed to drift with an oil slick) did not alter waterbird behaviour when it was broadcasting sounds as opposed to non-broadcasting (5). The buoy broadcasts up to 30 different sounds at up to 130 dB at 1 m, at varying intervals (30 sec to 5 min, dependent on how programmed). Alternating 2-day treatment (device 'on') and control ('off') periods were conducted. No significant deterrent effect was noted on numbers of three common wintering duck species (greater and lesser scaup *Aythya affinis* and *A. marila*, surf scoter *Melanitta perspicillata*) and all other waterbirds.

A literature review (6) suggests that the use of radar-activated deterrent systems both increases effectiveness at deterring waterbirds at contaminated sites and reduces the cost necessary (compared to conventional methods) to achieve success.

A randomised, replicated and controlled trial at a tar sands mine in Alberta, Canada, in 2003 (7), found that a lower proportion of 372 groups of birds landed on three tailing ponds when an on-demand bird deterrent system was used, compared to control periods when the system was not used or to periods when industry standard deterrents were used (1–5% of bird groups landing with on-demand system vs. 5–16% for industry standard and 8–23% for controls). The on-demand system used radar-activated propane cannons, high-intensity strobe lights, moving models of peregrine falcons *Falco peregrinus* and broadcasts of peregrine calls; the industry standard system used human effigies and cannons that fired at random intervals. A further trial found that birds were significantly more likely to change direction when propane cannons were fired on demand, compared to when a peregrine model was moved and peregrine calls played (11% of 28 bird groups responded when peregrine models were activated vs. 40% of 30 bird groups responding when cannons were used).

(1) Weber, R. A. & Filas, B. A. (1993) Experimental radar-activated hazing system. *The Journal of the Acoustical Society of America*, 93, 2377–2378.

(2) Johansson, C. A., Hardi, P. J. & White, C. M. (1994) *An inexpensive fully automated hazing system reduces avian landings on a 45 acre 'defended' pond by 97%*. Unpublished report to Region 6. Washington, DC: US Fish and Wildlife Service, USA.

(3) Read, J. L. (1999) A strategy for minimising waterfowl deaths on toxic waterbodies. *Journal of Applied Ecology*, 36, 345–350.

(4) Stevens, G. R., Rogue, J., Weber, R. & Clark, L. (2000) Evaluation of a radar-activated, demand-performance bird hazing system. *International Biodeterioration & Biodegradation*, 45, 129–137.

(5) Whisson, D. A. & Takekawa, J. Y. (2000) Testing the effectiveness of an aquatic hazing device on waterbirds in the San Francisco Bay estuary of California. *Waterbirds: The International Journal of Waterbird Biology*, 23, 56–63.

(6) Ronconi, R. A., St Clair, C. C., O'Hara, P. D., Burger, A. E., Day, R. H., Cooper, B. A. & Burger, A. E. (2004) Waterbird deterrence at oil spills and other hazardous sites: potential applications of a radar-activated on-demand deterrence system. *Marine Ornithology*, 32, 25–33.
(7) Ronconi, R. A. & St. Clair, C. C. (2006) Efficacy of a radar-activated on-demand system for deterring waterfowl from oil sands tailings ponds. *Journal of Applied Ecology*, 43, 111–119.

13.4. Use repellents to deter birds from landing on pools polluted by mining

- A randomised, replicated and controlled *ex situ* trial from the USA (1) found that fewer common starlings *Sturnus vulgaris* consumed contaminated water when it was treated with repellents, compared to untreated water.

Background
If the threat from pollution is not from birds landing on ponds but from consuming the water, then repellents may be a more effective or inexpensive method of reducing harm.

A randomised, replicated and controlled trial in captivity in Philadelphia, USA (1), found that 36 common starlings *Sturnus vulgaris* consumed less mine-pond water if it was treated with bird repellents, compared to untreated mine-pond water. The repellents tested were: o-aminoacetophenone (OAP), 2-amino-4,5-dimethoxyacetophenone, methyl anthranilate, 4-ketobenztriazine (4KBT) and veratryl amine, all at concentrations of 0.5% by volume. The most effective repellent was OAP and the least effective 4KBT; there were no significant differences between effectiveness of any of the other chemicals. All repellents were effective for five weeks on water which was pH 10.6 and contained 150 ppm sodium cyanide.

(1) Clark, L. & Shah, P. S. (1993) Chemical bird repellents: possible use in cyanide ponds. *The Journal of Wildlife Management*, 57, 657–664.

Agricultural pollution

13.5. Reduce pesticide or herbicide use generally

- A single small study from the UK (2) investigated population level effects of reduced chemical inputs, and found that the populations of some species increased when pesticide use was restricted alongside other interventions.
- Three studies (3,6,7), two replicated, one controlled, from the UK found that some or all species were found at higher densities on sites with reduced pesticide inputs, in one case with other interventions as well. Five studies from the UK (3–5,7,9), four replicated, four controlled, found that some or all species were not found at higher densities on fields or sites with reduced chemical inputs, or were not associated with reduced inputs.
- A controlled before-and-after study from the UK (1) found that grey partridge *Perdix perdix* chicks had higher survival on sites with reduced pesticide applications. A replicated study from the UK (8) found that reduced chemical inputs had a negative relationship with partridge brood size and no relationship with survival or the ratio of young to old birds.

Background
Pesticides and herbicides are likely to reduce the plant and insect diversity on farmland and this, in turn, can reduce food supplies for birds. Reducing these inputs may therefore help bird populations. Reductions can be accomplished by using smaller quantities or fewer applications. See also 'Pay farmers to cover the costs of conservation measures' and 'Reduce management intensity on permanent grassland'.

A controlled before-and-after study in 1989–94 in a 28 km^2 area of arable farmland in Sussex, England (1), found significantly higher survival rates of grey partridge *Perdix perdix* chicks on 21 km^2 with irregular insecticide applications, compared to a 7 km^2 farm with insecticide applications four times a year (average of 34% survival on low application farms vs. 22% on high application farm). Before the start of intensive insecticide application (1970–88), there was no difference between areas (27% survival on low application farms vs. 30% on intensive application farm).

A small replicated controlled study from May–June in 1992–8 in Leicestershire, England (2), found that the abundance of nationally declining songbirds and species of conservation concern significantly increased on a 3 km^2 site where pesticide use was restricted (alongside several other interventions), although there was no overall difference in bird abundance, species richness or diversity between the experimental and three control sites. Numbers of nationally declining species rose by 102% (except for Eurasian skylark *Alauda arvensis* and yellowhammer *Emberiza citrinella*). Nationally stable species rose (insignificantly) by 47% (eight species increased, four decreased). The other interventions employed were: 'Manage hedges to benefit wildlife', 'Plant wild bird seed cover strips', 'Provide supplementary food', 'Control predators' and 'Create beetle banks'.

A replicated study in 1999 and 2003 on farms in East Anglia and the West Midlands, England (3), found that five of twelve farmland bird species analysed were positively associated with a general reduction in herbicide use and conservation headlands. This study is discussed in more detail in 'Leave headlands in fields unsprayed (conservation headlands)'.

A randomised, replicated, controlled trial on four farms in south-west England, in 2003–6 (4), found that no more foraging birds were attracted to 12, 50 × 10 m plots of permanent pasture with no fertiliser impact, compared to 12 control (conventionally managed) plots. Experimental plots were managed in the same way as control plots except for the lack of fertiliser, and all plots were cut twice in May and July, and grazed in autumn/winter. This study is also discussed in 'Reduce management intensity on permanent grassland', 'Undersow spring cereals', 'Raise mowing height on grasslands', 'Reduce grazing intensity on permanent grasslands' and 'Plant wild bird seed or cover mixture'.

A replicated trial in 2004–2006 in north-west England (5) found no differences in bird numbers between conventional and minimum input barley fields. Sixteen trial fields were sown with spring barley each on a separate dairy or mixed farm in Cheshire, Staffordshire and north Shropshire. One half of each barley field was managed conventionally, the other half managed with minimum pesticide inputs (no insecticide after 15 March, no broad-leaved herbicide after 31 March, limited graminicides). Birds were monitored on each field, in summer 2005 and winter 2005/06.

A replicated site comparison on 186 overwinter stubble fields in Devon, England (6), found that cirl buntings *Emberiza cirlus*, foraged at significantly higher densities on stubble fields under a 'Special Project' agri-environment option, compared to stubbles under standard agri-environment schemes (approximately 0.45 birds/ha for 102 special project stubble fields vs. 0.05 birds/ha for 52 conventional wheat stubbles and 0.15 birds/ha for 32 conventional barley stubbles). The special project stubbles were also preferentially selected to some extent by four other species of songbird. The special project was designed to encourage cirl buntings and allowed the use of fungicides, growth regulators and specified herbicides to control grass weeds, but prohibited the use of insecticides and herbicides to control broad-leaved weeds.

A controlled study in 2000–5 on 61 ha of farmland in Bedfordshire, England (7), found that both winter and summer densities of most farmland bird species studied were higher on areas with no pesticide input, compared to areas with conventional levels of pesticides (higher summer densities with no pesticides for 10 of 14 species, although only Eurasian skylark *Alauda arvensis*, yellow wagtail *Motacilla flava* and linnet *Carduelis cannabina* showed a significant increase; all songbirds and 16 of 19 species recorded in winter were at higher densities on zero-fertiliser fields). Skylarks were also found in significantly higher numbers on areas with zero fertiliser inputs, but no other species were affected by fertiliser reduction.

This study also investigated the impact of set-aside provision (see 'Provide or maintain set-aside') and spring sowing wheat (see 'Sow crops in spring, not autumn').

A replicated site comparison study on 1,031 agricultural sites across England in 2004–8 (8) found that grey partridge *Perdix perdix* brood size was negatively related to reduced chemical inputs, when combined with overwinter stubbles. However, this combination of interventions was positively associated with year-on-year changes in partridge density. There were no relationships with overwinter survival or the ratio of young to old birds. This study describes the effects of several other interventions, discussed in the relevant sections.

A replicated, controlled study from April–July and November–February in 2004–6 on 16 livestock farms in the West Midlands, England (9), found that there were no differences in the usage of barley fields between fields sprayed with only a narrow-spectrum herbicide (amidosulfuron, at 25–40 g/ha) and those sprayed with both a narrow- and a broad-spectrum herbicide. Insect-eating songbirds and crows showed reduced use of broad-spectrum-sprayed fields in summer and late summer respectively, but all other groups used fields at equal rates. Barley fields on the farms were split, with half being used for each treatment. Narrow-spectrum herbicide was applied in April–May and broad-spectrum in July.

(1) Aebischer, N. J. & Potts, G. R. (1998) Spatial changes in grey partridge (*Perdix perdix*) distribution in relation to 25 years of changing agriculture in Sussex, UK. *Gibier faune sauvage, Game Wildlife*, 15, 293–308.
(2) Stoate, C. (2002) Multifunctional use of a natural resource on farmland: wild pheasant (*Phasianus colchicus*) management and the conservation of farmland passerines. *Biodiversity and Conservation*, 11, 561–573.
(3) Stevens, D. K. & Bradbury, R. B. (2006) Effects of the Arable Stewardship Pilot Scheme on breeding birds at field and farm-scales. *Agriculture, Ecosystems & Environment*, 112, 283–290.
(4) Defra (2007) *Potential for enhancing biodiversity on intensive livestock farms (PEBIL)*. Defra Report BD1444.
(5) Mortimer, S., Westbury, D., Dodd, S., Brook, A., Harris, S., Kessock-Philip, R., Chaney, K., Lewis, P., Buckingham, D. & Peach, W. (2007) Cereal-based whole crop silages: potential biodiversity benefits of cereal production in pastoral landscapes. *Aspects of Applied Biology*, 81, 77–86.
(6) Bradbury, R., Bailey, C., Wright, D. & Evans, A. (2008) Wintering cirl buntings *Emberiza cirlus* in southwest England select cereal stubbles that follow a low-input herbicide regime. *Bird Study*, 55, 23–31.
(7) Henderson, I. G., Ravenscroft, N., Smith, G. & Holloway, S. (2009) Effects of crop diversification and low pesticide inputs on bird populations on arable land. *Agriculture, Ecosystems & Environment*, 129, 149–156.
(8) Ewald, J. A., Aebischer, N. J., Richardson, S. M., Grice, P. V. & Cooke, A. I. (2010) The effect of agri-environment schemes on grey partridges at the farm level in England. *Agriculture, Ecosystems & Environment*, 138, 55–63.
(9) Peach, W. J., Dodd, S., Westbury, D. B., Mortimer, S. R., Lewis, P., Brook, A. J., Harris, S. J., Kessock-Philip, R., Buckingham, D. L. & Chaney, K. (2011) Cereal-based wholecrop silages: A potential conservation measure for farmland birds in pastoral landscapes. *Biological Conservation*, 144, 836–850.

13.6. Restrict certain pesticides or other agricultural chemicals

- A small scale study (1) found that Pyrazophos reduced chick food abundance more than other foliar fungicides. A before-and-after study from Spain (2) found that the population of griffon vultures *Gyps fulvus* increased in the study area following multiple interventions including the banning of strychnine.

Background

Certain agricultural chemicals such as DDT, diclofenac and strychnine are now know to be extremely harmful to birds, either by direct poisoning, or interfering with reproduction. Many of these chemicals are now restricted or prohibited, but many others still remain in use. For example, some organophosphates such as monocrotophos are used to control insect pests and have caused mass poisoning incidents in raptors and cranes Grus spp. (Goldstein *et al.* 1999; Pain *et al.* 2004).

Goldstein, M. I., Lacher, T. E., Woodbridge, B., Bechard, M. J., Canavelli, S. B., Zaccagnini, M. E., Cobb, G. P., Scollon, E. J., Tribolet, R. & Hooper, M. J. (1999) Monocrotophos – Induced mass mortality of Swainson's Hawks in Argentina, 1995–96. *Ecotoxicology*, 8, 201–214.

Pain, D. J., Gargi, R., Cunningham, A. A., Jones, A. & Prakash, V. (2004) Mortality of globally threatened Sarus cranes *Grus antigon* from monocrotophos poisoning in India. *Science of the Total Environment*, 326, 55–61.

A small scale study of cereal fields treated with foliar fungicides in the UK (1) found that chick food insect abundance was reduced to a greater extent following applications of Pyrazophos compared to other fungicides. Compared to untreated crops, chick food insects were reduced by 31–70% in crops treated with Pyrazophos, 10% with Propiconazole and 3% with Triadimefon applications. The effect of Pyrazophos was greater when applied at an earlier growth stage (GS37: 70%; GS50: 45%; GS60: 31% reduction). Following Pyrazophos applications, total predatory arthropods were reduced by 25–48%; aphid specific predators 35–84% (17% with Triadimefon) and parasitoids 34–55%. Fungicides were sprayed at GS 50 in winter wheat in 1984. Pyrazophos was also sprayed at GS60 in spring barley (1984) and GS37 in winter barley (1985). Chick food insects were sampled by sweep nets or suction sampling.

A before-and-after between 1969 and 1989 in the western Pyrenees, Spain (2), found that the population of griffon vultures *Gyps fulvus* increased from 282 pairs (in 23 colonies) in 1969–75 to 1,097 pairs (46 colonies) in 1989 following the initiation of multiple conservation interventions including the banning of strychnine, a major cause of vulture mortality, in 1984. Additional surveys in 1979 and 1984 found 364 pairs (in 26 colonies), 589 pairs (32 colonies) respectively. This study is also discussed in 'Habitat protection', 'Use legislative regulation to protect wild populations' and 'Provide supplementary food to increase adult survival'.

(1) Sotherton, N. W. & Rands, M. R. W. (1987) Predicting, measuring and minimising the effects of pesticides on farmland wildlife on intensively managed arable land in Britain. *Pesticide science and biotechnology. Proceedings of the 6th international congress of pesticide chemistry*, 433–436.

(2) Donazar, J. A. (1990) Population trends of the griffon vulture *Gyps fulvus* in northern Spain between 1969 and 1989 in relation to conservation measures. *Biological Conservation*, 53, 83–91.

13.7. Provide food for vultures to reduce mortality from diclofenac

- A before-and-after trial in Pakistan (1) found that oriental white-backed vulture *Gyps bengalensis* mortality rates were significantly lower when supplementary food was provided, compared to when it was not.

Background

Diclofenac, a non-steroidal anti-inflammatory drug has caused widespread and dramatic declines in vulture populations in South Asia by inducing renal failure and visceral gout in birds feeding on the carcasses of livestock previously treated with the drug (Oaks *et al.* 2004). Whereas, there are measures to reduce its use and replace it with non-toxic alternatives (Pain *et al.* 2008), it has so far remained on sale. Vulture 'restaurants' have therefore been proposed as a method of reducing mortality until the drug is phased out, by providing a safe, uncontaminated source of food for the remaining vulture populations.

Other studies on the provision of food (for reasons other than diclofenac avoidance) are discussed in 'Supplementary feeding'.

Pain, D. J., Bowden, C. G. R., Cunningham, A. A., Cuthbert, R., Das, D., Gilbert, M., Jakati, R. D., Jhala, Y., Khan, A. A., Naidoo, V., Lindsay Oaks, J., Parry-Jones, J., Prakash, V., Rahmani, A., Ranade, S. P., Sagar Baral, H., Ram Senacha, K., Saravanan, S., Shah, N., Swan, G., Swarup, D., Taggart, M. A., Watson, R. T., Virani, M. Z., Wolter, K. & Green, R. E. (2008) The race to prevent the extinction of South Asian vultures. *Bird Conservation International*, 18, S30–S48.

Oaks, J. L., Gilbert, M., Virani, M. Z., Watson, R. T., Meteyer, C. U., Rideout, B. A., Shivaprasad, H. L., Ahmed, S., Chaudhry, M. J. I., Arshad, M., Mahmood, S., Ali, A. & Khan, A. A. (2004) Diclofenac residues as the cause of vulture population decline in Pakistan. *Nature*, 427, 630–633.

A before-and-after trial in 2003–4 (1) found that daily mortality of oriental white-backed vultures *Gyps bengalensis* at a colony in Punjab province, Pakistan, was significantly lower during two periods when supplementary food (diclofenac-free donkey carcasses) was provided at a nearby 'vulture restaurant', compared to two control periods (0.072 birds/day dying over 111 days when food was provided vs. 0.387 birds/day over 116 days without food). Of the 30 dead vultures examined (eight from supplemented periods), 29 showed signs of diclofenac poisoning. Home range size of three radio-tagged vultures appeared to contract when they discovered the 'restaurant' (thus reducing the possibility of contact with diclofenac) but two further tagged vultures did not use the restaurant at all.

(1) Gilbert, M., Watson, R. T., Ahmed, S., Asim, M. & Johnson, J. A. (2007) Vulture restaurants and their role in reducing diclofenac exposure in Asian vultures. *Bird Conservation International*, 17, 63.

13.8. Make selective use of spring herbicides

- We captured no evidence for the effects of selective use of spring herbicides on bird populations.

13.9. Use organic rather than mineral fertilisers

- We captured no evidence for the effects of using organic, not mineral, fertilisers on bird populations.

Background

Organic fertilisers include farmyard manure and soil treatments derived from sewage sludge, whilst mineral fertilisers are manufactured preparations of nitrate, phosphate or 'NPK'. Organic fertilisers have the potential benefit of contributing to better soil structure and therefore reducing soil erosion, whilst also, potentially reducing the risk of downstream eutrophication.

13.10. Reduce chemical inputs in permanent grassland management

- A randomised, replicated, controlled study from the UK (1) found that no more foraging birds were attracted to pasture plots with no fertiliser, compared to control plots.

Background

Reducing chemical inputs into permanent grasslands is often used in conjunction with reducing the mowing height or delaying the first mowing or grazing date on grasslands. Studies that examine several of these interventions at once are discussed in 'Threat: Agriculture – Reduce management intensity on permanent grasslands'.

A randomised, replicated, controlled trial on four farms in south-west England, in 2003–6 (1), found that no more foraging birds were attracted to 12, 50 × 10 m plots of permanent pasture with no fertiliser impact, compared to 12 control (conventionally managed) plots. Experimental plots were managed in the same way as control plots except for the lack of fertiliser, and all plots were cut twice in May and July, and grazed in autumn/winter. This study also discusses several other interventions including 'Reduce management intensity on permanent grasslands'.

(1) Defra (2007) *Potential for enhancing biodiversity on intensive livestock farms (PEBIL)*. Defra Report BD1444.

13.11. Leave headlands in fields unsprayed (conservation headlands)

- Three studies from Europe (6,9,12), two replicated, found that conservation headlands were frequently used by some of all of the bird species studied, or were strongly associated with species. A review from the UK (8) found that grey partridge *Perdix perdix* populations were far larger on farms with conservation headlands and other interventions in place than other farms. Two replicated studies from Europe (6,9) found that species were not associated with, or were no more abundant on, conservation headlands, compared with control fields.
- All four studies, three replicated, that investigated survival (3,5,7,11) found higher grey partridge *Perdix perdix* chick or adult survival on sites with conservation headlands than control sites. One found that this difference was not significant.
- Five studies from Europe (1–3,5,10), four replicated, found larger grey partridge broods on farms with conservation headlands, one study (5) found that differences were not significant. One replicated study from the UK (2) found that fewer broods were found in fields with conservation headlands. Another replicated study from the UK (11) found no relationship between conservation headlands and partridge brood size or young to adult ratio.

Background

Conservation headland management (under European agri-environment schemes) involves restricted fertiliser, herbicide and insecticide spraying in a 6 m margin of sown arable crop. The prescription allows selected herbicide applications to control certain weeds or invasive species.

A replicated, controlled study of cereal headlands on an arable farm in north-east Hampshire, UK (1), found that grey partridge brood size was significantly larger on unsprayed compared to sprayed headlands (6.4 chicks/brood on unsprayed vs. 2.2 on sprayed headlands). Abundance of chick food species ('true bugs', caterpillars and sawfly larvae, leaf beetles and weevils) was significantly greater in unsprayed headlands compared to sprayed headlands (180 individuals/50 sweeps for unsprayed vs. 62 for sprayed). Three areas were split into two treatment plots: sprayed with conventional pesticides or 6 m headlands left unsprayed. Grey partridge brood size was recorded from August–September 1983. Insects were sampled using a sweep net (50 sweeps in June).

A replicated, controlled study in 1980–3 in arable fields on a farm in Hampshire, England (2) (the same site as in (1)), found that grey partridge *Perdix perdix* broods were significantly larger in 1983 on plots with conservation headlands, compared to controls, with headlands sprayed with fungicides and herbicides (averages of 5.1–10.3 chicks/brood for 29 broods in unsprayed areas vs. 1.8–2.4 chicks/brood for 39 broods on controls). No differences were found in 1980–1, before conservation headlands were implemented. However, more broods were found on conventional fields, reflecting more pairs (49 vs. 37) in the spring. The author argues that larger broods were the result of higher chick survival, with conservation headland plots contain significantly more food insects than controls.

A replicated, controlled study in 1984 on the same farm as in (1,2) and on eight sites in East Anglia, England (3), found that grey partridge *Perdix perdix* broods had significantly higher survival, and were significantly larger on plots with conservation headlands, compared to control plots with conventionally-sprayed headlands (average of 75% survival and 7.8–10.0 chicks/brood for five broods on conservation headland plots vs. 60% and 4.7–7.5 chicks/brood for four broods on conventional plots; 196 broods surveyed). This paper also describes similar, although less conclusive effects on two non-native gamebirds.

A paired, replicated, controlled study in the 1980s in cereal fields in southern and eastern England, UK (4), found that in each year 1983–6, the brood size of grey partridge *Perdix perdix* and/or pheasant *Phasianus colchicus* was higher on blocks of cereal fields with conservation headlands (6–10 and 4–7 chicks respectively) compared with normally sprayed headlands (3–8 and 2–3 chicks) (Sotherton & Robertson 1990). Breeding density of grey partridges on the Hampshire farm increased from 4 to 12 pairs/km^2 between 1979 and 1986. No such increases were recorded on adjacent farms where pesticide regimes remained unchanged.

A replicated, controlled study of cereal fields on ten pairs of farms in central and southern Sweden (5) found that grey partridge brood size, chick survival and abundance of invertebrates tended to be higher on farms with unsprayed headlands (6 m wide) compared to those sprayed conventionally. Mean brood size tended to be higher on experimental farms (half headlands unsprayed; 7–9) than on control farms (sprayed; 3–8). Numbers of broods (10–19 vs. 4–16), chick survival rate (26–54% vs. 11–47%) and numbers of partridge pairs in the spring (20–30 vs. 15–24) also tended to be higher on experimental farms. However, none of these differences was statistically significant. Mean density of chick food insect groups (Heteroptera, Homoptera, Curculionidae, Chrysomelidae, larvae of Lepidoptera and Tenthredinidae) was significantly higher on unsprayed (25–74) compared to sprayed headlands of wheat (5–32). Farm pairs (control and experimental) were within 5 km of each other and of similar size, cropping and agricultural practice. On the experimental farm, the headlands left unsprayed (50%) were swapped each year (1991–1993). Partridge counts were undertaken in spring and after harvest using dogs to flush birds. Ten invertebrate samples (0.5 m^2) were taken from each headland during the first week in July using vacuum-suction.

A replicated, controlled study of arable fields on eight farms in the Netherlands (6) found that unsprayed field margins had a higher abundance of blue-headed wagtail *Motacilla flava flava* than sprayed edges. Blue-headed wagtails made 1.5–2.4 visits/km to unsprayed margins compared to just 0.5 visits/km for sprayed margins. Numbers of Eurasian skylarks *Alauda arvensis* and meadow pipits *Anthus pratensis* did not differ significantly in sprayed and unsprayed margins (skylark: 0.4 vs. 0.2–0.4; pipits: 0.1 vs. 0.1). Blue-headed wagtails and skylarks visited field margins more than field centres and sprayed edges bordering ditches more than sprayed edges adjacent to a second plot. Strips 6 m wide along field edges were left unsprayed by herbicides and insecticides (total length 2,560–3,790 m/year) and were compared to sprayed edges in the same field and to the sprayed field in 1992–1993. Farmland birds were sampled using a linear transect census, with all birds visiting field margins recorded and a similar size strip in the centre of each field recorded. Birds were sampled 10–12 times between April and mid-July.

A 1998 literature review (7) found three studies, two in the UK ((2,3) described above) and one in Sweden, showing that gamebird (grey partridge *Perdix perdix*) chick survival rates were significantly higher in conservation headlands with reduced pesticide inputs compared to controls receiving the usual pesticide application.

A literature review of studies in the UK (8) found that the populations of grey partridge *Perdix perdix* was 600% higher on farms with conservation measures aimed at partridges in place, compared to farms without these measures. Measures included the provision of conservation headlands, planting cover crops, using set-aside and creating beetle banks.

A replicated study in 1999 and 2003 on farms in East Anglia and the West Midlands, England (9), found that five of twelve farmland bird species analysed were positively associated with conservation headlands and a general reduction in herbicide use (see separate intervention). These were corn bunting *Miliaria calandra* (a field-nesting species) and chaffinch *Fringilla coelebs*, greenfinch *Carduelis chloris*, whitethroat *Sylvia communis*, and yellowhammer *Emberiza citrinella* (all boundary-nesting species). The study did not distinguish between conservation headlands and a general reduction in herbicide use, classing both as interventions reducing pesticide use. A total of 256 arable and pastoral fields across 84 farms were surveyed. Several other interventions are also analysed and discussed in the relevant sections.

A 2009 literature review of agri-environment schemes in England (10) found evidence (including in studies reviewed in this section) that grey partridge *Perdix perdix* broods were significantly larger in cereal fields with a 6 m unsprayed margin around them, compared to conventional fields. This review also examines several other interventions, discussed in the relevant sections.

A 2009 literature review of European farmland conservation practices (11) found that gamebirds made frequent use of conservation headlands, for shelter and foraging. The authors note that the effects on non-gamebirds are less certain.

A replicated site comparison study on 1,031 agricultural sites across England in 2004–8 (12) found that grey partridge *Perdix perdix* overwinter survival was positively correlated with the proportion of a site under conservation headlands in 2007–8, and with year-on-year density changes in 2006–7. There were no relationships with brood size or the ratio of young to old birds. This study describes the effects of several other interventions, discussed in the relevant sections.

(1) Rands, M. W. R., Sotherton, N. W. & Moreby, S. J. (1984) Some effects of cereal pesticides on gamebirds and other farmland fauna. 98–113 *Proceedings of the recent developments in cereal production* University of Nottingham.
(2) Rands, M. R. W. (1985) Pesticide use on cereals and the survival of grey partridge chicks: a field experiment. *Journal of Applied Ecology*, 22, 49–54.
(3) Rands, M. R. W. (1986) The survival of gamebird (Galliformes) chicks in relation to pesticide use on cereals. *Ibis*, 128, 57–64.
(4) Sotherton, N. W. (1991) Conservation Headlands: a practical combination of intensive cereal farming and conservation. 373–397 in: L. G. Firbank, N. Carter, J. F. Derbyshire, G. R. Potts (eds) *The Ecology of Temperate Cereal Fields* Blackwell Scientific Publications.
(5) Chiverton, P. A. (1994) Large-scale field trials with conservation headlands in Sweden. *British Crop Protection Council Monographs*, 58, 185–190.
(6) De Snoo, G. R., Dobbelstein, R. & Koelewijn, S. (1994) Effects of unsprayed crop edges on farmland birds. *British Crop Protection Council Monographs*, 58, 221–226.
(7) Sotherton, N. (1998) Land use changes and the decline of farmland wildlife: an appraisal of the set-aside approach. *Biological Conservation*, 83, 259–268.
(8) Aebischer, N. J., Green, R. E. & Evans, A. D. (2000) From science to recovery: four case studies of how research has been translated into conservation action in the UK. 43–54 in: N. J. Aebischer, A. D. Evans, P. V. Grice, J. A. Vickery (eds) *Ecology and Conservation of Lowland Farmland Birds* British Ornithologists Union, Tring.
(9) Stevens, D. K. & Bradbury, R. B. (2006) Effects of the Arable Stewardship Pilot Scheme on breeding birds at field and farm-scales. *Agriculture, Ecosystems & Environment*, 112, 283–290.
(10) Natural England (2009) *Agri-environment schemes in England 2009. A review of results and effectiveness.* Natural England, Peterborough.
(11) Vickery, J. A., Feber, R. E. & Fuller, R. J. (2009) Arable field margins managed for biodiversity conservation: a review of food resource provision for farmland birds. *Agriculture, Ecosystems & Environment*, 133, 1–13.
(12) Ewald, J. A., Aebischer, N. J., Richardson, S. M., Grice, P. V. & Cooke, A. I. (2010) The effect of agri-environment schemes on grey partridges at the farm level in England. *Agriculture, Ecosystems & Environment*, 138, 55–63.

13.12. Provide unfertilised cereal headlands in arable fields

- We did not find any studies describing the effects of unfertilised cereal headlands on bird populations.

Background

Unfertilised cereal headlands are designed to provide food for birds during the cropping year and consist of a 6–24 m headland which (under UK Entry-Level Stewardship requirements) cannot be sprayed with insecticides between 15th March and harvest, and which can only have limited herbicide application.

13.13. Plant riparian buffer strips

- We did not find any studies describing the impact of riparian strips on reducing water pollution and how this affected bird populations. However, riparian strips also provide valuable habitats in their own right. Studies describing the use of riparian strips by birds are described in 'Habitat restoration and creation'.

Background

Buffer strips along rivers and streams are designed to reduce pollution from agricultural or other activities from reaching waterways. They should reduce soil erosion and ensure that water remains clearer and less polluted.

Buffer strips can consist of one or more of: a zone of native, river-side trees; a second zone of native shrubs and a third zone of native grasses and forbs.

13.14. Provide buffer strips around in-field ponds

- We found no studies describing the effect of buffer strips around in-field ponds on pollution levels and bird populations.

Background

Buffer strips around in-field ponds serve the same purpose as those along rivers – reducing pollution and providing habitat. Studies describing the use of buffer strips by birds are described in 'Habitat restoration and creation'.

Air-borne pollutants

13.15. Use lime to reduce acidification in lakes

- A before-and-after study from Sweden (1) found no difference in osprey *Pandion haliaetus* productivity during a period of extensive liming of acidified lakes compared to two periods without liming.

Background

Acid rain, caused by airborne pollutants such as sulphur dioxide and nitrogen oxides, can acidify lakes and reduce the number of fish they hold. Adding lime (calcium and magnesium-rich minerals) to lakes can help reduce acidity.

A before-and-after study at three acidified lakes in south-west Sweden (1) found that the average number of large (over four week-old) osprey *Pandion haliaetus* nestlings produced in successful breeding attempts was not significantly different during a period of extensive liming of acidified lakes (1980–7), compared to either of two periods without liming (1961–71: 2.0 nestlings/successful breeding attempt; 1972–9: 2.0 nestlings/successful attempt; 1980–7: 1.9 nestlings/successful attempt).

(1) Eriksson, M. O. G. & Wallin, K. (1994) Survival and breeding success of the osprey *Pandion haliaetus* in Sweden. *Bird Conservation International*, 4, 263–277.

Excess energy

13.16. Reduce incidental mortality from birds being attracted to artificial lights

Background

Many species of birds fly at night and can be attracted to artificial lights. This can be detrimental both through direct collision mortality, but also because seabirds especially can be vulnerable to predators or dehydration if they are downed by collisions, even if they are not hurt.

Dangerous lights can be designed specifically to be seen, for example on communications towers or lighthouses, or can simply be lights used at night, for example, domestic lighting.

13.17. Turn off lights to reduce mortality from artificial lights

- A before-and-after study from the UK (1) found that fewer seabirds (Manx shearwaters *Puffinus puffinus*, European storm petrels *Hydrobates pelagicus* and Leach's storm petrels *Oceanodroma leucorhoa*) were attracted to artificial lighting and downed when lighting was reduced at night, compared to when normal lighting was in place.

Background

The simplest way to prevent birds from being attracted to and colliding with lights is to turn them off. However this might not be practical for economic, safety or other considerations.

A before-and-after study on St Kilda, Scotland, between 2005 and 2008 (1) found that fewer seabirds were attracted to artificial lighting and downed when lighting was reduced at night, compared to when normal lighting was in place (27 birds found when lighting was reduced for the whole of autumn 2007 and most of 2008 vs. 54 birds downed and two dead when lighting was not reduced in 2005–6 and 24 downed in 20 days when lighting was not reduced in 2008). Downed birds included 59 Manx shearwater *Puffinus puffinus* (including all birds downed by reduced lighting), one European storm petrel *Hydrobates pelagicus* and 45 Leach's storm petrel *Oceanodroma leucorhoa*. Lighting consisted of outdoor lights and indoor lights left on overnight with no window coverings.

(1) Miles, W., Money, S., Luxmoore, R., and Furness, R. W. (2010) Effects of artificial lights and moonlight on petrels at St Kilda. *Bird Study*, 57, 244–251.

13.18. Reduce the intensity of lighthouse beams

- We captured no evidence for the effects of reducing the intensity of lighthouse beams on bird mortality.

Background

Less intense lighthouse beams may be seen by birds from less far away, or prove less attractive. However, as lighthouse beams are designed to be extremely bright and visible, dimming them may not be feasible.

13.19. Shield lights to reduce mortality from artificial lights

- A replicated, controlled study in Hawaii (1) found that fewer Newell's shearwaters *Puffinus newelli* were found grounded when security lights were shielded, compared to nights when they were not.

Background

Lights that are not designed to be seen as warnings can be shielded, so that the light from them is directed down (or in the required direction). This can reduce the light projected into the sky and therefore the visibility and attractiveness of the lights to birds.

A replicated, controlled study at a resort on Hawaii, USA, in 1980 and 1981 (1), found that significantly fewer Newell's shearwaters *Puffinus newelli* were found grounded under security lights on nights when twelve of the brightest lights had 'hoods' placed on them, compared to alternate nights when lights were not shielded (272 birds on 32 'shielded' nights vs. 444 birds on 32 control nights). The reduction was greater in 1981 (52%) than 1980 (29%), possibly because peak shearwater fledging coincided with a full moon (and so higher ambient light) in 1980, but with a new moon in 1981. The shields reduced upwards radiation of light and during the experiment most other lights on the resort were permanently shielded.

(1) Reed, J. R., Sincock, J. L. & Hailman, J. P. (1985) Light attraction in endangered procellariiform birds: reduction by shielding upward radiation. *The Auk*, 102, 377–383.

13.20. Use flashing lights to reduce mortality from artificial lights

- A randomised, replicated and controlled trial from the USA (1) found that fewer dead birds were found beneath control towers that used only flashing lights, as opposed to those using both flashing and continuous lights.

Background

Flashing lights may prove less attractive to birds but remain equally visible to humans, making them suitable for warning lights, on communications towers or similar.

A randomised, replicated and controlled trial on 24 control towers in Michigan, USA, during May and September 2005 (1) found that there were significantly fewer bird carcasses found beneath towers lit with red or white flashing lights, compared with control towers using the Federal Aviation Administration standard of red flashing lights combined with non-flashing red lights (average mortality of 3.7 birds/tower for experimental towers vs. 13 birds/tower for controls). There were no differences between three different flashing-light treatments. Three tall (>305 m) towers with non-flashing lights caused significantly more fatalities than any of the smaller towers. The majority of birds killed were night-migrating songbirds but also included gamebirds and woodpeckers.

(1) Gehring, J., Kerlinger, P. & Manville II, A. M. (2009) Communication towers, lights, and birds: successful methods of reducing the frequency of avian collisions. *Ecological Applications*, 19, 505–514.

13.21. Use lights low in spectral red to reduce mortality from artificial lights

- Two studies from the North Sea (1) and the Netherlands found that fewer birds were attracted to low-red lights (including green and blue lights), compared with the number expected (1), or the number attracted to white or red lights (2).

Background

Birds are most sensitive to red light, meaning that if bulbs used at night emit less red light then they may prove less attractive to birds, whilst remaining equally visible to humans.

A study on an oil rig in the southern North Sea on three nights in October 2007 (1) found that the number of migrating birds circling the rig was 10–50% of the number expected when the majority of external lights were replaced with 'low red' bulbs (150–2,500 birds observed circling vs. 750–5,000 birds expected). Low red bulbs emit lower levels of red light than standard bulbs. Attracted birds were mainly songbirds, waders and wildfowl and expected numbers were estimated based on previous observations and calculated from the number of migrating birds recorded on nearby islands, the number observed from the rig and the weather conditions.

A replicated, controlled study from Friesland, the Netherlands (2), in September–November 2003, found that on clear nights, significantly fewer migrating birds were attracted to two 1,000 W lamps when they were covered with opaque white or red filters (61% of 38 birds and 54% of 13 birds reacting to each), compared with green (13% of eight) or blue (3% of 37) filters. The same pattern, but with higher overall levels of attraction were detected on overcast nights (white: 81% of 156 birds reacting; red: 54% of 24; green: 27% of 77; blue: 5% of 38).

(1) van de Laar, F. J. T. (2007) Green light to birds: Investigation into the effect of bird friendly lighting. Report NAM locatie L15-FA-1. NAM, Assen, The Netherlands.
(2) Poot, H., Ens, B. J., de Vries, H., Donners, M. A., Wernand, M. R. & Marquenie, J. M. (2008) Green light for nocturnally migrating birds. *Ecology and Society*, 13, 47.

13.22. Use volunteers to collect downed birds and rehabilitate them

- We found no studies that report on the effectiveness of using volunteers to collect and rehabilitate downed birds.

Background

A lot of the mortality caused by birds crashing into lights is not direct collision mortality, but caused by stunned and injured birds being predated (e.g. by gulls after crashing onto oil rigs, van de Laar 2007) or dying of dehydration. If volunteers or conservationists can collect downed birds soon after they crash and rehabilitate them, they may be able to greatly reduce collision mortality.

van de Laar, F. J. T. (2007) *Green light to birds: Investigation into the effect of bird friendly lighting*. Report NAM locatie L15-FA-1. NAM, Assen, The Netherlands.

14. Threat: Climate change, extreme weather and geological events

Climate change, extreme weather and geological events are all very large-scale threats. Most interventions used in response to them, therefore, are general conservation interventions such as artificial nest sites, translocations and captive breeding, and as such are discussed in their own sections ('General responses to small/declining populations' and 'Captive breeding, rearing and releases (*ex situ* conservation)').

Key messages

Water nesting mounds to increase incubation success in malleefowl

A single small trial in Australia found that watering malleefowl nests increased their internal temperature but that a single application of water did not prevent the nests drying out and being abandoned during a drought.

Replace nesting substrate following severe weather

A before-and-after study found that a common tern colony increased following the replacement of nesting habitats, whilst a second found that a colony decreased. In both cases, several other interventions were used at the same time, making it hard to examine the effect of habitat provision.

14.1. Water nesting mounds to increase incubation success in malleefowl

- A small controlled in Australia (1) found that two malleefowl *Leipoa ocellata* nests were abandoned after they dried out, despite being watered, although unwatered nests were abandoned much earlier.

Background

Malleefowl *Leipoa ocellata* and other megapodes build large nesting mounds, filled with vegetation. This vegetation rots and produces heat which then incubates the eggs. This process requires the vegetation to be damp, meaning that eggs may die in dry conditions. Artificially adding water to the mounds may help prevent this.

A small controlled trial in mallee scrub in South Australia, Australia, in October–December 1981 (a drought year) (1) found that two malleefowl *Leipoa ocellata* nest mounds which were watered to promote microbial decomposition were abandoned around the 6th December, after birds constructed egg chambers but did not lay eggs. However, the internal temperature rose to approximately 35°C following the addition of 400 litres of water (equivalent to approximately 57 mm of rain on the mounds), compared to a maximum of approximately 25°C for two control (unwatered) mounds. However, watered mounds dried out in late November, the temperature fell and birds abandoned them. Control nests were abandoned around 12th November.

(1) Booth, D. T. & Seymour, R. S. (1984) Effect of adding water to malleefowl mounds during a drought. *Emu*, 84, 116–118.

14.2. Replace nesting substrate following severe weather

- Two before-and-after studies from Canada (1) found that common tern *Sterna hirundo* populations increased at one colony where nesting substrates were replaced, but decreased at a second. Several other interventions were used at both sites, making it difficult to evaluate the effects of substrate replacement.

Background

Many birds require specific substrates to nest on, and if flooding or heavy rains remove these then they may not nest. Replacing the substrate following removal may therefore help a colony to survive. Most studies describing the reconstruction of nesting habitats or the protection of nests from flooding are described in 'Provide artificial nesting sites'.

Two before-and-after studies at two common tern *Sterna hirundo* colonies between 1977–89 in Ontario, Canada (1), found the nesting population increased at one colony but decreased at another following a combination of interventions, including the replacement of nesting substrate following flooding. Other interventions included: erecting signs highlighting the presence of nesting birds, vegetation control, preventing gulls *Larus* spp. from nesting, destroying gull nests and shooting particular ring-billed gulls *L. delawarensis* that were heavily predating tern eggs. Gulls were only culled at the site with population increase, whilst the authors attribute declines to continued high levels of disturbance, vegetation growth and mammalian predators.

(1) Morris, H., Blokpoel, H. & Tessier, G. D. (1992) Management efforts for the conservation of common tern *Sterna hirundo* colonies in the Great Lakes: two case histories. *Biological Conservation*, 60, 7–14.

15. General responses to small/ declining populations

Population size is largely determined by mortality and reproductive rates. Most of the chapters in this book are aimed at minimising threats, but there are also many general interventions that can be implemented, aimed largely at increasing reproductive rates.

This chapter describes the general interventions that can be put in place in the wild, whilst those that are performed in captivity (*ex situ* conservation) are described in the next.

Key messages – inducing breeding, rehabituation and egg removal

Use artificial visual and auditory stimuli to induce breeding in wild populations

A small study from the British Virgin Islands found an increase in breeding behaviour after the introduction of visual and auditory stimulants.

Rehabilitate injured birds

Two studies of four studies from the UK and USA found that 25–40% of injured birds taken in by centres were rehabilitated and released. Three studies from the USA found that rehabilitated birds appeared to have high survival. One found that mortality rates were higher for owls than raptors.

Remove eggs from wild nests to increase reproductive output

A study from Canada found that whooping crane reproductive success was higher for nests with one or two eggs removed than for controls. A study from the USA found that removing bald eagle eggs did not appear to affect the wild population and a replicated study from Mauritius found that removing entire Mauritius kestrel clutches appeared to increase productivity more than removing individual eggs as they were laid.

Key messages – provide artificial nesting sites

Provide artificial nesting sites

Thirty-seven of the 209 studies reviewed that investigated artificial nesting sites found that populations increased (or remained constant as other areas decreased) after artificial nests were used, or that populations relocated to artificial nesting sites. One of these found an increase in the UK population of loons. Three studies found no increases in populations with nest provision. One study found higher songbird species richness in areas with artificial nests. Only 18 studies reported low or no use of artificial nest sites, three found that artificial sites were used preferentially to natural sites and eight found that use of nests increased over time. Sixty-one studies found that productivity, success or survival were similar or higher in artificial nests, compared to natural nests. Twenty found lower productivity or survival than natural nests, two because of high poaching rates from artificial nests. Forty-six studies found that use or productivities differed between particular positions, designs or orientations of artificial nests, nine studies found no such differences. One study found that nesting success was as high in old nest boxes as new. One study found that modifying artificial nesting cavities to allow easy access did not disrupt woodpecker behaviour.

Clean artificial nests to increase occupancy or reproductive success

Five out of ten studies from North America and Europe found that songbirds preferentially nested in cleaned nest boxes or those sterilised using microwaves, compared to used nest boxes. One study found that the preference was not strong enough for birds to switch nest boxes after they were settled. One study found that birds avoided heavily-soiled nest boxes. Two studies birds had a preference for used nest boxes and one found no preference for cleaned or uncleaned boxes. None of the five studies that examined it found any effect of nest box cleanliness on nesting success or parasitism levels.

Use differently-coloured artificial nests

A study from the USA found that two bird species (a thrush and a pigeon) both showed colour preferences for artificial nests, but that these preferences differed between species. In each case, clutches in the preferred colour nest were less successful than those in the other colour.

Provide nesting material for wild birds

One of two studies found that wild birds took nesting material provided; the other found only very low rates of use.

Repair/support nests to support breeding

A study from Puerto Rico found that no chicks died from chilling after nine nests were repaired to prevent water getting in.

Artificially incubate eggs or warm nests

One study found that no kakapo chicks or eggs died of cold when they were artificially warmed when females left the nest. A study from the UK found that great tits were less likely to interrupt their laying sequence if their nest boxes were warmed, but there was no effect on egg or clutch size.

Provide nesting habitat for birds that is safe from extreme weather

Two of three studies found that nesting success of waders and terns was no higher on raised areas of nesting substrate, with one finding that similar numbers were lost to flooding. The third study found that Chatham Island oystercatchers used raised nest platforms, but did not report on nesting success.

Remove vegetation to create nesting areas

Two out of six studies found increases in population sizes at seabird and wader colonies after vegetation was cleared and a third found that an entire colony moved to a new site that was cleared of vegetation. Two of these studies found that several interventions were used at once. Two studies found that gulls and terns used plots cleared of vegetation, one of these found that nesting densities were higher on partially-cleared plots than totally cleared, or uncleared, plots. One study found that tern nesting success was higher on plots after they were cleared of vegetation and other interventions were used.

Guard nests to increase nest success

We captured four studies describing the effects of guarding nests. One, from Costa Rica, found an increase in scarlet macaw population after nest monitoring and several other interventions. Two studies from Puerto Rico and New Zealand found that nest success was higher, or mortality lower, when nests were monitored. A study from New Zealand found that nest success was high overall when nests were monitored.

Key messages – foster chicks in the wild

Foster eggs or chicks with wild conspecifics

Twenty-three studies out of 26 from around the world reported high fledging rates or other measures of success for chicks and eggs fostered to wild individuals of the same species. One study found that fostered gannet chicks were lighter and had lower hatching and fledging rates than control chicks, one found low fertility of fostered bald eagle eggs (and therefore low success rates) and one found that a plover chick died soon after fostering.

Foster eggs or chicks with wild non-conspecifics (cross-fostering)

Two of eight studies found high fledging success for eggs or chicks fostered to non-conspecific nests, four studies found low levels of success for cross-fostering. One study on tit species in Europe found different levels of success depending on which species were used as foster parents. A study from New Zealand found that cross-fostering stilts increased productivity, but fledging and recruitment rates of cross-fostered chicks was lower than for chicks fostered to conspecifics' nests.

Key messages – provide supplementary food

Provide supplementary food to increase reproductive success

Ten of 85 studies (of 79 experiments and a literature review) found that local populations or population densities of birds increased following the provision of supplementary food, or were higher than unfed areas (one investigated multiple interventions). Three studies found that populations declined or densities remained similar. Twenty studies of 19 experiments found that breeding success or chick survival increased with food supplementation, 19 found no such effects. Seventeen studies found that chicks and eggs were heavier or in better condition, or grew faster with supplementary food, ten found no evidence for this. Ten studies found that clutches were larger when parents were fed, 15 found they were not. Twenty-nine studies found that fed birds laid earlier, were more likely to lay or re-nest, had shorter intervals between clutches or fledged their chicks earlier. Seven studies found that fed birds fledged chicks later, were no more likely to nest or re-nest, or did not lay any earlier than controls. Eight studies found that fed birds showed positive behavioural responses to feeding (more likely to nest in fed nest boxes, spending more time incubating etc), three found no response or negative responses. Two studies found evidence that feeding had a larger impact in years when natural food was scarcer.

Provide supplementary food to allow the rescue of a second chick

A study from Spain found that second chicks from lammergeier nests survived longer if nests were provided with food, in one case allowing a chick to be rescued.

Provide supplementary food to increase adult survival

Seven studies of 64 (from 62 experiments) found that populations or densities of some species were higher in areas supplied with food for adult birds. Six found similar population trends in fed and unfed areas for some species. Six studies found higher adult survival with feeding, five found similar or lower survival. Eleven studies found that birds supplied with food were heavier or in better condition, five found no such effects in some species. One study found that fed birds spent more time displaying than unfed, one study found no behavioural changes. Overall, ten studies found high use of supplementary food, two found low use and 19 found that use of feeders varied depending on position, timing or what was in them. One study found no evidence for feeding leading to dependence on supplementary food.

Can supplementary feeding increase predation or parasitism?

Two of three replicated and controlled studies, both from Spain, found potentially deleterious effects of supplementary feeding. One found higher predation on artificial nests near carcasses provided for vultures, the other found higher levels of potentially dangerous gut microflora when fed on livestock carrion, compared to those fed on wild rabbits. A replicated, controlled study from the USA found that supplying seeds in a predictable manner did not increase predation on songbirds.

Provide supplementary food through the establishment of food populations

One of four studies that established prey populations found that wildfowl fed on specially-planted rye grass. Two studies found that cranes in the USA and owls in Canada did not respond to established prey populations. A study from Sweden found that attempts to increase macroinvertebrate numbers for wildfowl did not succeed.

Provide perches to improve foraging success

One of four studies, from Sweden, found that raptors used clearcuts provided with perches more than clearcuts without perches. Two studies found that birds used perches provided, but a controlled study from the USA found that shrikes did not alter foraging behaviour when perches were present.

Place feeders close to windows to reduce collisions

A randomised, replicated and controlled study in the USA found that fewer birds hit windows, and fewer were killed, when feeders were placed close to windows, compared to when they were placed further away.

Provide supplementary water to increase survival or reproductive success

A controlled study from Morocco found that northern bald ibises provided with supplementary water had higher reproductive success than those a long way from water sources.

Provide calcium supplements to increase survival or reproductive success

Eight of 13 studies (including a literature review) from across the world found some positive effects of calcium provisioning on birds' productivities (six studies) or health (two studies). Six studies (including the review) found no evidence for positive effects on some of the species studied. One study from Europe found that birds at polluted sites took more calcium supplement than those at cleaner sites.

Key messages – translocations

Translocate birds to re-establish populations or increase genetic variation

We captured 59 studies of 80 translocation programmes and a review of 239 programmes. Thirty-six studies of 59 translocation programmes found that translocations established populations or resulted in population growth. In addition, the review found that between 63% of 134 programmes in 1987 and 67% of 105 in 1993 were deemed successful. Six studies of nine programmes found that translocated birds had high productivity and seventeen programmes found that birds had high survival or remained in the release area. One study found that translocated birds had higher survival than released captive-bred birds. Nine studies of nine programmes found that translocated birds had high mortality, low breeding success or failed to establish a population. One found that breeding success was lower than for the source population of the birds. One study found that birds were more likely to breed after translocation if there was a smaller latitudinal distance between source population and release site. One study found that wing-clipping female ducks when moving them prevented them from abandoning their ducklings.

Use techniques to increase the survival of species after capture

A study from the US found that providing dark, quiet environments with readily-available food and water increased the survival of small songbirds after capture and the probability that they would adapt to captivity. A study from the USA found that keeping birds warm during transit increased survival.

Ensure translocated birds are familiar with each other before release

Two studies from New Zealand found no evidence that ensuring birds were familiar with each other increased translocation success.

Ensure genetic variation to increase translocation success

We did not capture any studies on the effects of ensuring genetic variation in translocated birds.

Translocate nests to avoid disturbance

All five studies captured found some success in relocating nests while they were in use, but one found that fewer than half of the burrowing owls studied were moved successfully; a study found that repeated disturbance caused American kestrels to abandon their nest and a study found that one barn swallow abandoned its nest after it was moved.

Use vocalisations to attract birds to new sites

Seven out of ten studies from around the world found that seabirds were more likely to nest or land to areas where vocalisations were played, or moved to new nesting areas after vocalisations were played. Four of these studied multiple interventions at once. Three studies found that birds were no more likely to nest or land in areas where vocalisations were played.

Use decoys to attract birds to new sites

Ten studies found that birds nested in areas where decoys were placed or that more birds landed in areas with decoys than control areas. Six studies used multiple interventions at once. One study found that three-dimensional models appeared more effective than two-dimensional ones, and that plastic models were more effective than rag decoys.

Alter habitats to encourage birds to leave on area

A study from Canada found that an entire Caspian tern population moved after habitat was altered at the old colony site, alongside several other interventions.

15.1. Use artificial visual and auditory stimuli to induce breeding in wild populations

- A single small study from the British Virgin Islands (1) found that there was an increase in breeding behaviour in a small population of Caribbean flamingos *Phoenicopterus ruber* following the introduction of visual and auditory stimulants.

Background

Some species that nest in colonies may require a 'critical mass' of individuals to be present before breeding behaviour is stimulated. If this is the case, it might be possible to help very small populations (beneath this threshold) to breed by using decoys and vocalisations to stimulate courtship behaviours.

A small before-and-after study in the British Virgin Islands, Caribbean in 1992 (1) found there was an increase in group display and nest-building behaviour in a population of six (two females, four males) Caribbean flamingos *Phoenicopterus ruber*, following the introduction of ten life-sized flamingo decoys, eight artificially constructed mud nests (some with artificial eggs) and the playback of recordings of display vocalisations (3.6% of behavioural records in the two weeks after stimuli introduction were related to group display vs. no records in 12 hours before stimuli introduction).

(1) O'Connell-Rodwell, C. E., Rojek, N., Rodwell, T. C. & Shannon, P. W. (2004) Artificially induced group display and nesting behaviour in a reintroduced population of Caribbean flamingo *Phoenicopterus ruber ruber*. *Bird Conservation International*, 14, 55–62.

15.2. Rehabilitate injured birds

- Two replicated studies from the USA (1) and UK (4) found that 40% and 25% of raptors were released following rehabilitation. The USA study also found that 32% of owls were released.
- Three replicated studies from the USA (1–3) all found relatively high survival of released raptors, with only 2.4% of birds being recovered (i.e. found dead, 1) and 66–68% survival over two weeks (2) and six weeks (3). One study found that mortality rates were higher for owls than raptors (1).

Background

Relatively large and slow-reproducing species, such as raptors, can be badly affected by increases in adult mortality. Rehabilitating birds that have been injured may therefore be an important conservation action, particularly if it occurs on a large scale. In 1978, there were approximately 225 active wildlife or raptor rehabilitation programmes in the USA (Hamilton *et al.* 1988), showing that if they are effective, such programmes could make a real difference to bird populations.

Hamilton, L. L., Zwank, P. J. & Olsen, G. H. (1988) Movements and survival of released, rehabilitated hawks. *Journal of Raptor Research*, 22, 22–26.

A replicated study of raptors, owls and vultures brought into a rehabilitation centre in Minnesota, USA, between 1974 and 1980 (1), found that 452 of 1133 raptors (40%) brought to the centre were released back into the wild. Of these, 2.4% were recovered (i.e. were injured or killed), with 55% of these recoveries being within six weeks of release. Release rates for owls were lower (175 of 551 birds, 32%) and a higher proportion of owls (8%) were recovered after release. However, only 21% of these were within six weeks of release. Two of nine turkey vultures *Cathartes aura* released and neither was recovered. Size of bird did not seem to affect possibility of release and the severity of the original injury did not appear to affect post-release survival.

A small study in mixed croplands, forests and pastures in Louisiana, USA (2), found that, of eight red-tailed hawks *Buteo jamaicensis* and one red-shouldered hawk *B. lineatus* rehabilitated and released over six occasions in 1985–6, one red-tailed hawk died 17 days after release, the red-shouldered hawk was shot and had to be rehabilitated again and four other red-tailed hawks survived for more than two weeks after release. This implies that these four releases were successful as starvation normally occurs within two to three weeks if hawks do not feed. The remaining three red-tailed hawks could not be successfully tracked. The birds had been admitted to a rehabilitation centre for a range of reasons, from confiscation by officials to gunshot wounds and had been in the centre from a few weeks to over a year.

A replicated study in Minnesota, USA (3), found that, of 19 bald eagles *Haliaeetus leucocephalus* that were rehabilitated and released with radiotrackers in the winters (November–February) of 1987–90 from a rehabilitation centre, 13 (68%) definitely survived for more than six weeks, three (16%) definitely died and contact with three was lost within ten days of release. One female bred and fledged a chick in both 1989 and 1990. Eagles ranged from 2–610 km from their release sites, which were located along the Mississippi River. Eagles were admitted to the centre for reasons ranging from starvation to bone fractures.

A retrospective study of admissions and releases at a rehabilitation centre in Cheshire, England, between January 2000 and December 2004 (4), found that 50 of the 205 (25%) Eurasian sparrowhawks *Accipiter nisus* admitted to the centre were released following treatment. No data were available on post-release survival.

(1) Duke, G. E., Redig, P. T. & Jones, W. (1981) Recoveries and resightings of released rehabilitated raptors. *Raptor Research*, 15, 97–107.
(2) Hamilton, L. L., Zwank, P. J. & Olsen, G. H. (1988) Movements and survival of released, rehabilitated hawks. *Journal of Raptor Research*, 22, 22–26.
(3) Martell, M., Redig, P., Nibe, J., Buhl, G. & Frenzel, D. (1991) Survival and movements of released rehabilitated bald eagles. *Journal of Raptor Research*, 25, 72–76.
(4) Kelly, A. & Bland, M. (2006) Admissions, diagnoses, and outcomes for Eurasian sparrowhawks (*Accipiter nisus*) brought to a wildlife rehabilitation centre in England. *Journal of Raptor Research*, 40, 231–235.

15.3. Remove eggs from wild nests to increase reproductive output

- A replicated study from Mauritius (2) found that harvesting entire clutches appeared to increase Mauritius kestrels *Falco punctatus* productivity more effectively than removing individual eggs as they were laid.
- A replicated study over 30 years in Canada (3) found that wild whooping cranes *Grus americana* reproductive success was higher for nests with one or two eggs removed than for control nests.
- A single study from the USA (1) found that removing bald eagle *Haliaeetus leucocephalus* eggs from wild nests for hand-rearing did not appear to greatly affect the wild population.

Background

If a bird population is at a very low level then removing eggs from nests can be used to encourage females to continue laying, with some species replacing eggs as they are removed. The removed eggs can then be hand-reared (see 'Artificially incubate and hand-rear birds in captivity') or fostered ('Foster eggs or chicks with wild conspecifics' and 'Foster eggs or chicks with non-conspecifics'), increasing the productivity of the species.

Alternatively, removing eggs from wild nests may actually increase the success of these nests, possibly due to competition between siblings. Studies discussing this aspect of the intervention are described below.

A replicated, controlled study in two marshland sites in Florida, USA, between 1985 and 1990 (1) found that 78% of 58 bald eagle *Haliaeetus leucocephalus* pairs that had their first clutch removed for hand-rearing ('donor nests') between 1985 and 1988 laid replacement clutches within two months. Replacement clutches were slightly smaller than first clutches (58 first clutches averaged 2.1 eggs/clutch vs. 1.8 eggs/clutch for 45 second clutches). In one study area, donor nests produced fewer fledglings than control pairs (1.0 fledgling/nest for 16 donor nests vs. 1.5 fledglings/nest for 39 controls), but this was not true in a second area (1.2 fledgling/nest for 26 donor nests vs. 1.1 fledglings/nest for 41 controls). Donor nests were more productive in the year before eggs were removed than the year after donation (approximately 1.3 fledglings/clutch for 32 pairs the year before donation vs. 0.85 fledglings/clutch for 34 pairs the year after). A demographic model suggested that a donor population would be very slightly smaller than a control population after 25 years. Timing of clutch removal did not affect the speed or probability of replacement clutches being laid.

A replicated 1995 study on the Mauritius kestrels *Falco punctatus* conservation programme (2) found that harvesting whole clutches rather than single eggs was more successful in increasing wild pair productivity: 95% of females re-laid within 14 days of clutch removal, but fertility fell rapidly in clutches where eggs were removed as they were laid. Females laid up to 4 clutches/season as a result of harvesting, but clutch fertility decreased to zero by the fourth clutch. Clutch size was an average of 3.4 eggs/clutch for 96 first clutches, compared with 3.3 for 63 second clutches.

A replicated controlled study in north-west Territories and Alberta, Canada, between 1967 and 1996 (4) found that the reproductive success of wild whooping cranes *Grus americana* was higher for nests that had one of two eggs removed, compared to control nests. Both recruitment of juveniles to the population and survival until August (eggs were removed in May) were higher (50% chance of recruitment from nests with eggs removed vs. 39% for unmanipulated nests). A total of 496 eggs were removed from wild nests in the study period, representing 62% of all crane nests during this time period. The success of artificially incubating and rearing the removed eggs is discussed in (3) in 'Captive breeding, rearing and releases (*ex situ* conservation)'.

(1) Wood, P. B. & Collopy, M. W. (1993) Effects of egg removal on bald eagle productivity in northern Florida. *The Journal of Wildlife Management*, 57, 1–9.

(2) Jones, C. G., Heck, W., Lewis, R. E., Mungroo, Y., Slade, G. & Cade, T. (1995) The restoration of the Mauritius kestrel *Falco punctatus* population. *Ibis*, 137, S173–S180.

(3) Kuyt, E. (1996) Reproductive manipulation in the whooping crane *Grus americana*. *Bird Conservation International*, 6, 3–10.

(4) Boyce, M. S., Lele, S. R. & Johns, B. W. (2005) Whooping crane recruitment enhanced by egg removal. *Biological Conservation*, 126, 395–401.

15.4. Provide artificial nesting sites

Background

Providing additional nest sites for birds is a frequently used intervention, both as a conservation method and to aid the study of birds.

However, it is unclear how effective it is as a method of increasing populations. Nest boxes are most likely to be effective in increasing a population if the population is limited by nesting sites and, as with all interventions in this synopsis, unless population-level effects are studied then any use of nest boxes or apparent increase in population may not represent an increase in numbers but instead a redistribution of the same birds, and could even hide a population decline.

It is also worth noting that occupancy rates for nest boxes may not be a good measure of their effectiveness as it is highly dependent on the number of boxes available and the population in the study site. For example, if 100 boxes are erected in a site with 30 occupied by Species A and 50 by Species B, then it may appear that Species B relies on the boxes more. However, if the local populations of the two species are 30 and 100, respectively, then, in fact, a far higher proportion of Species A used the boxes (100% vs. 50%). Care, therefore, must be taken in interpreting results from nest box studies.

15.4.1. Divers/loons

- A replicated before-and-after study from the UK (3) found that there was a very large increase in loon productivity on lakes provided with nesting rafts, with a corresponding increase in productivity across the whole country.
- Two studies from the USA (1,4) found higher nesting success on lakes with floating nesting rafts, compared to sites without rafts, but no new territories were established on lakes without loons but with rafts (1).
- A replicated study from the UK (2) found that loons used nesting rafts and artificial islands in some areas of the UK, but not others.

Background

Divers or loons (Gaviidae) are specialised for swimming and diving and cannot move quickly on land. This means they are vulnerable to predators and frequently nest on small islands and patches of vegetation. Providing these habitats where they are absent may increase the number of birds using a site, increase their reproductive success, or allow the colonisation of new lakes.

A small before-and-after study at two small lakes in Minnesota, USA (1), found that common loons (great northern divers) *Gavia immer* used floating nesting platforms in 83% of breeding attempts in 1970–3, following their installation in 1970. Nesting success was 67%, compared with 60% for loons on four lakes with natural islands and older platforms (the number of nesting attempts is not given). However, no new territories were established at three additional lakes without loons after platforms were installed in 1970. Platforms were made from sedge mats and logs and anchored with concrete blocks between 3 m and several hundred metres from shore.

A 1992 replicated study of the use of artificial islands and floating platforms at 17 wetland nature reserves across the UK (2) found that red-throated loons (divers) *Gavia stellata* and arctic loons (black-throated divers) *G. arctica* used both well-vegetated and bare shingle-covered islands and platforms at inland sites (i.e. on lakes and lochs) in Scotland. However, neither species used islands or rafts at Scottish coastal sites, or any sites in England or Wales. The review also examines island and platform use by grebes, rails, ground-nesting seabirds, waders and wildfowl.

A replicated before-and-after trial on lochs in Scotland between 1980 and 1997 (3) found that installing 63 nesting rafts in arctic loon (black-throated diver) *Gavia arctica* territories in 1992–5 increased chick productivity of the British population by an estimated 44% (from approximately 0.24 large chicks/territory to 0.35 large chicks/territory, 60–100 territories monitored each year), with an estimated 170% increase in the 44 territories where rafts were used (representing approximately 25% of the British population). Rafts consisted of a 3.6 × 2.4 m polystyrene rectangle with sides sloped to allow access from the water. They were covered in hessian, netting and turf and anchored to concrete blocks in shallow (1–3 m) water close to natural nest sites known to be susceptible to flooding and hidden from public roads to avoid illegal egg collecting. Rafts lasted at least ten years with annual maintenance.

A replicated, controlled trial at 52 lakes in Wisconsin, USA, in 1996–8 (4), found that common loons (great northern divers) *Gavia immer* had significantly higher hatching and fledging success in 26 lakes provided with nesting platforms, compared to 26 control lakes, without platforms (83% of clutches incubated until hatching date and 0.74 chicks fledged/clutch for 23 clutches on platforms vs. 49% of 41 clutches incubated and 0.56 chicks fledged/clutch for 59 clutches on natural sites). Increases were found across all lakes, irrespective of previous productivities. Rate of platform use increased each year, from 15% in 1996 to 50% in 1998 and was high across all lake qualities. Platforms were 1 m^2 and made from Styrofoam blocks surrounded by logs and covered in soil and moist vegetation. They were anchored in water 0.5–5 m deep and 6–15 m from shore, inside existing territories.

(1) McIntyre, J. W. & Mathisen, J. E. (1977) Artificial islands as nest sites for common loons. *The Journal of Wildlife Management*, 41, 317–319.
(2) Burgess, N. D. & Hirons, G. J. M. (1992) Creation and management of artificial nesting sites for wetland birds. *The Journal of Environmental Management*, 34, 285–295.
(3) Hancock, M. (2000) Artificial floating islands for nesting black-throated divers *Gavia arctica* in Scotland: construction, use and effect on breeding success. *Bird Study*, 47, 165–175.
(4) Piper, W. H., Meyer, M. W., Klich, M., Tischler, K. B. & Dolsen, A. (2002) Floating platforms increase reproductive success of common loons. *Biological Conservation*, 104, 199–203.

15.4.2. Grebes

- A single study from the UK (1) found that grebes used nesting rafts in some areas of the UK but not others, and that the characteristics of used rafts differed geographically.

Background

Like divers/loons, grebes are highly specialised for swimming and diving and are therefore vulnerable on land. Providing artificial islands or floating rafts may increase both the number of birds nesting at a site, and their reproductive success.

A 1992 review of the use of artificial islands and floating platforms in 17 wetland nature reserves across the UK (1) found that great-crested grebes *Podiceps cristatus* and little grebes *Tachybaptus ruficollis* used well-vegetated islands and platforms at inland sites in the south of the UK. In addition, great-crested grebes used sparsely covered islands and platforms at inland sites, and well-vegetated ones at coastal sites. Neither species used either type of nesting site in Scotland. The review also examines island and platform use by divers, rails, ground-nesting seabirds, waders and wildfowl.

(1) Burgess, N. D. & Hirons, G. J. M. (1992) Creation and management of artificial nesting sites for wetland birds. *Journal of Environmental Management*, 34, 285–295.

15.4.3. Ground and tree-nesting seabirds

- Three studies from the UK (3,10) and the Azores (11) found increases in gull and tern populations following the provision of rafts/islands (3) or providing nest boxes alongside other interventions (10,11).
- A controlled, replicated study from the USA (9) found that terns had higher nesting success on nesting rafts in one of two years monitored and a before-and-after study from Japan (7) found that nesting success increased after the provision of nesting substrate.
- Five studies from Canada (1,2,4) and Europe (5,8) found that terns used re-profiled or artificial islands or nesting rafts, but pelicans did not (5).
- A small study from Hawaii (6) found that red-footed boobies *Sula sula* preferentially nested in an artificial 'tree-style' nesting structure, compared to other designs.

Background

Many seabirds require open areas of sparse vegetation to nest. These can be created either through the provision on artificial rafts and islands, described below, or through the clearing of vegetation on existing sites, which is discussed in 'Natural system modifications'.

A small replicated, controlled study from May–August in 1982 on a concrete breakwater in Port Colborne, Canada (1), found that common terns *Sterna hirundo* nested at higher densities on two plots enhanced with clumps of mossy stonecrop and driftwood added (62% of 166 clutches in these plots), compared to plots layered with gravel (29% of clutches) or control plots of bare concrete (9% of clutches). Enhanced plots were also colonised earlier. Average clutch size and hatching rates were similar between plots (2.4–2.5 eggs/clutch and 76–86% hatching success), but the average number of chicks fledged per pair was significantly higher in enhanced (1.6) and control (1.3) plots than in gravel-layered plots (0.6). The breakwater was divided into six 5 × 7 m plots, with two plots for each treatment.

A replicated trial in 1990 at Lake Ontario, Canada (2), found that common terns *Sterna hirundo* successfully nested on four floating wooden rafts the same season that they were installed, with at least 170 fledglings being produced (average of 1.3 fledglings/nest). Terns successfully defended the rafts from Canada geese *Branta canadensis* and ring-billed gulls *Larus delawarensis* and used all four rafts. Rafts were 5 × 5 m, covered with sand and gravel and each had six decoy terns on (see 'Attract birds to safe areas using decoys' for more studies on decoys).

A 1992 review of the use of artificial islands and floating platforms in 17 wetland nature reserves across the UK (3) found that all seven species of gull and tern investigated used sparsely-vegetated islands and platforms at southern, coastal sites, but that nesting sites elsewhere were not used by four of the species. Sandwich terns *S. sandvicensis* used vegetated nesting sites at southern coastal sites, whilst black-headed gulls *L. ridibundus* and common terns *S. hirundo* nested at all sites. At one site in Kent, the provision of 20 shingle islands has attracted 350 pairs of Sandwich and common terns and 1,000 pairs of black-headed gulls. The review also examines island and platform use by grebes, divers, rails, waders and wildfowl.

A small trial in 1993–5 in western Lake Ontario, Canada (4), found that the number of Caspian terns *Sterna caspia* nesting on an artificial raft increased from one pair in 1993 (raising two chicks) to 50 pairs (raising 97 chicks) in 1995. In 1995 the raft produced the majority of young in the area, due to heavy predation on mainland nests by red foxes *Vulpes vulpes*. The raft was 3.6 × 9.8 m, covered in sand and gravel, was anchored adjacent to a mainland subcolony and covered with a tarpaulin between April and May to discourage ring-billed gulls *Larus delawarensis* from nesting. Eight tern decoys were also placed on the raft (discussed in 'Use decoys to attract birds to safe areas'), a sound system played vocalisations from a Caspian tern colony in 1993 (see 'Use vocalisations to attract birds to safe areas') and in 1995, eight chick shelters were added to the raft.

A replicated study in 1987–1990 of a managed wetland in Macedonia, Greece (5) found that the target species, Dalmation pelicans *Pelecanus crispus*, did not benefit consistently from artificial habitats although other waterbirds did. Two constructed rafts and one artificial island were used extensively by a variety of waterbirds as resting and foraging sites. Common terns *Sterna hirundo* colonised the rafts in both years (average 12 nests and 14 fledglings/raft). Dalmatian pelicans did not colonise the rafts. Many waterbirds, including pelicans, were observed roosting on the island but no successful breeding took place in 1988–1989. In April 1990, 26 pelicans colonised the islands. Thirteen nests contained 1–2 eggs each. By June, however, the pelicans had deserted the island, no eggs remained and some nests had been destroyed. The authors speculate that fisherman landed on the island and removed the eggs. Pelicans did not return to the island.

A small study at a shrubland site on O'ahu, Hawaii, USA (6), found that red-footed boobies *Sula sula* preferentially nested in a 'tree-style' artificial nest platform, compared to a transported tree, a 'tripod-style' platform or a linear platform. A total of approximately 15 young were produced between 1992 and 1998 from nests on the tree-style platform, which consisted of a 6 m tall beam with cross beams, providing nine potential nest sites. A transported kiawe *Prosopis pallida* tree and kiawe branches were used for perching and several booby nests, five tripod platforms (each providing seven nesting sites) were erected in 1992, but only three nests were built over three breeding seasons and a single linear platform providing 50 nest sites was used only once.

A before-and-after study in 2001–2 in Tokyo, Japan (7) found that the fledging rates in a little tern *Sterna albifrons* colony was higher following the provision of nesting substrate and chick shelters. This study is discussed in 'Physically protect nests with individual exclosures/barriers or provide shelters for chicks'.

A before-and-after study at a former gravel pit in Kent, England (8), found that one pair of common terns *Sterna hirundo*, five pairs of black-headed gulls *Larus ridibundus* and approximately 100 pairs of herring gulls *L. argentatus* nested on a series of gravel islands after they were re-profiled and lowered in February 2005 to encourage winter flooding. Vegetation was also removed from the islands (see 'Manually remove vegetation from wetlands'). The number of birds nesting on the islands was originally high but declined until none nested there in 2002.

A controlled, replicated trial in 2003–4 at a wetland site in Wisconsin, USA (9), found that black terns *Chlidonias niger* occupied 63–66% of 41 floating nest platforms provided each year (34–35% of the local population used them). Platform nests had significantly higher hatching success and nest survival rates in 2004, but not 2003. Eggs laid on platforms were significantly larger than those on natural substrates, suggesting that platforms were occupied by high-quality birds (and were therefore preferred). Platforms were 46 × 46 cm polystyrene and plywood squares, covered in hardware cloth and anchored to the lake bottom with a metal pipe (allowing vertical movement). Platforms were spaced 10–15 m apart in clusters of 5–10 and were positioned in the same location during both years of the study.

A before-and-after trial in north-east England (10) found that the number of roseate terns *Sterna dougallii* at an island site increased following the creation in 2000 of an artificial nesting terrace and the provision of additional nest boxes (94 pairs in 2006 vs. an average of approximately 28 pairs in 1975–1999). Since 2003, all breeding pairs have used nest boxes. Before 2000 there were up to 12 nest boxes on the island, but 25 were installed in 2000 and more added each year until 200 boxes were present in 2006. Boxes were 15 × 30 × 45 cm with a 15 cm doorway; the terrace was 25 m long originally (it was extended in 2001), with three tiers, each protected by flagstones to prevent burrowing birds undermining its structure.

A before-and-after study on Praia Islet (12 ha), off Graciosa, Azores, Portugal (11), found that the breeding populations of common terns *Sterna hiundo* and roseate terns *S. dougallii* increased dramatically (from no pairs to over 1,000 and 400 pairs respectively) following the installation of nest boxes in 1996, combined with the eradication of rabbits (see 'Control or remove habitat-altering mammals') and habitat restoration ('Shrubland restoration'). Fifty wooden boxes were installed in 1996 in the area with the least vegetation and the proportion of the common terns nesting in the boxes increased between 1996 and 2006. The effect of nest boxes for burrow-nesting seabirds is also discussed.

(1) Richards, M. H. & Morris, R. D. (1984) An experimental study of nest site selection in common terns. *Journal of Field Ornithology*, 55, 457–466.
(2) Dunlop, C. L., Blokpoel, H. & Jarvie, S. (1991) Nesting rafts as a management tool for a declining common tern (*Sterna hirundo*) colony. *Colonial Waterbirds*, 14, 116–120.
(3) Burgess, N. D. & Hirons, G. J. M. (1992) Creation and management of artificial nesting sites for wetland birds. *Journal of Environmental Management*, 34, 285–295.
(4) Lampman, K. P., Taylor, M. E. & Blokpoel, H. (1996) Caspian terns (*Sterna caspia*) breed successfully on a nesting raft. *Colonial Waterbirds*, 135–138.
(5) Pyrovetsi, M. (1997) Integrated management to create new breeding habitat for Dalmatian pelicans (*Pelecanus crispus*) in Greece. *Environmental Management*, 21, 657–667.
(6) Rauzon, M. J. & Drigot, D. (1999) Red-footed booby use of artificial nesting platforms. *Waterbirds*, 22, 474–477.
(7) Hayashi, E., Hayakawa, M., Satou, T. & Masuda, N. (2002) Attraction of little terns to artificial rooftop breeding sites and their breeding success. *Strix*, 23, 143–148.
(8) Akers, P. & Allcorn, R. I. (2006) Re-profiling of islands in a gravel pit to improve nesting conditions for terns *Sterna* and small gulls *Larus* at Dungeness RSPB reserve, Kent, England. *Conservation Evidence*, 3, 96–98.
(9) Shealer, D. A., Buzzell, J. M. & Heiar, J. P. (2006) Effect of floating nest platforms on the breeding performance of black terns. *Journal of Field Ornithology*, 77, 184–194.
(10) Morrison, P. & Gurney, M. (2007) Nest boxes for roseate terns *Sterna dougallii* on Coquet Island RSPB reserve, Northumberland, England. *Conservation Evidence*, 4, 1–3.
(11) Bried, J., Magalhaes, M. C., Bolton, M., Neves, V. C., Bell, E., Pereira, J. C., Aguiar, L., Monteiro, L. R. & Santos, R. S. (2009) Seabird habitat restoration on Praia Islet, Azores Archipelago. *Ecological Restoration*, 27, 27–36.

15.4.4. Burrow-nesting seabirds

- Four studies from across the world found evidence for population increases (10,11,14) or new populations being established (13) in petrel species following the provision of nest boxes. In two cases nest boxes were combined with the translocation of chicks (13) or other interventions (14).
- Six studies from across the world (1,3,4,7,8,10) found high occupancy rates for artificial burrows by seabirds, with three (4,7,10) finding that occupancy increased over time, taking years to build up. Three studies from across the world found very low occupancy rates for artificial burrows used by petrel species.
- Eight studies from across the world (1,3–5,9–12) found that the productivity of birds using artificial burrows was high, in many cases as high or higher than in natural burrows. One replicated study from the USA (2) and a small study from the Galapagos (6) found low productivity of petrels using artificial burrows.

Background

Many petrels and shearwaters and auks nest underground in burrows dug themselves or burrowing mammals such as rabbits. This means that they require very different artificial nests to the ground- and tree-nesting species discussed in the previous section.

A replicated and controlled study in 1980–84 on an island in Washington State, USA (1), found that rhinoceros auklet *Cerorhinca monocerata* nesting success was at least as high for individuals nesting in artificial burrows as it was for those in natural burrows (50–89% of 10–20 artificial burrows producing chicks vs. an estimated 53–56% success for natural burrows). No differences in growth rates of chicks in artificial and natural burrows were detected. Burrows were plywood chambers (23 × 25 × 71 cm) with a 91 cm long, 13–15 cm diameter entrance tunnel. Chambers were buried 15 cm underground after being filled with fresh soil and arranged so entrances faced the sea.

A replicated trial in 1980–3 on four islands in Maine, USA (2), found that only two Leach's storm petrel *Oceanodroma leucorhoa* chicks fledged successfully from 264 artificial burrows over three years (both in the third year). Between 14 and 46 burrows (5–17%) were colonised, with 6–10 being used for breeding. A total of 22 eggs were produced over three years, of which ten hatched, producing two chicks. Burrows had a 40–50 cm entrance tunnel (10 cm diameter) and a 25 × 25 × 25 cm nest chamber set to one side and covered with a large flat rock (for inspection). This study also describes the impact of playing vocalisations to attract petrels, described in ('Use vocalisations to attract birds to safe areas').

A replicated trial on an island in Washington State, USA, in 1989–91 (3), found that rhinoceros auklets *Cerorhinca monocerata* used an average of 91% of 40 artificial burrows provided, with 58% of burrows (64% of occupied burrows) producing chicks. There were no differences in use or productivity between burrows in areas occupied by glaucous-winged gulls *Larus glaucescens* and those without gulls. Artificial burrows were of the same design as in (1), with 20 spread over approximately 300 m of cliff.

A replicated controlled trial in 1992–4 on Cabbage Tree Island, New South Wales, Australia (4), found that Gould's petrel *Pterofroma leucoptera leucoptera* breeding success was comparable or higher in artificial burrows, compared to natural cavities (36–63% breeding success for 14–19 pairs in artificial burrows vs. 23–31% success for natural cavities in previous years). The number of birds using artificial burrows increased over time, from 53% of 90 boxes visited but no eggs laid in 1992 to 19 eggs laid, and 13 chicks fledging in 1994. Burrows consisted of a 20 × 25 × 52 cm polyethylene box, accessed through a 40 cm long tunnel of 10 cm diameter PVC piping. In 1993, many eggs rolled into the entrance tunnel, so a 4 cm barrier was added at the nest chamber entrance in 1994.

A controlled, replicated study 1993–4 on Mousa, Shetland, northern Scotland (5), found that hatching and fledging rates of European storm petrels *Hydrobates pelagicus* in artificial nests was not significantly different from those in natural nests (93% of 29 eggs in artificial nests hatched and 81% of 27 nestlings fledged vs. 81% of 42 eggs and 62% of 34 chicks in natural nests). Overall, 36% of 81 boxes were used each year, with 26 nests on a boulder beach used more often than 55 nests in dry stone walls (46% vs. 31–33% respectively). Nests had a nesting chamber of 10 cm long, 15.2 cm diameter PVC piping, an observation chamber and a 6 cm diameter entrance tunnel.

A replicated study on Santa Cruz in the Galápagos Islands, Ecuador (6), found that the use of artificial burrows by dark-rumped petrels *Pterodroma phaeopygia phaeopygia* increased each year from 1988–90, although only one chick was fledged from the burrows. In 1988, 68% of the eighty burrows were prospected by petrels, although none nested. In 1989, 39 petrels staying in burrows overnight and in 1990, four pairs laid eggs in burrows. However, three of the four chicks produced were predated, probably by rats. Petrel vocalisations were played at the nest site each night, with the results discussed in 'Use vocalisations to attract birds to safe areas'. Predator control on the Galápagos Islands is discussed in 'Control mammalian predators on islands'.

A replicated study in 1986–95 on an island in British Columbia, Canada (7), found that ancient murrelets *Synthliboramphus antiquus* used artificial nest boxes installed in May 1986, but that it took several years for them to regularly use them (7–14% of 26 nest boxes occupied in 1987–9, with eggs laid in only a single burrow each year vs. 67% of 21 boxes occupied in 1995, at least seven rearing young). Nest boxes were 40 × 40 × 13 cm and buried 100–150 m from the sea, within an area used by approximately 5,000 ancient murrelets.

A replicated study from February–April in 1999 in 21 Chatham petrel *Pterodroma axillaris* nest sites in the Chatham archipelago, New Zealand (8), found that artificial nest sites were used by petrels. No details are provided on productivity.

A replicated, controlled study on Cabbage Tree Island, New South Wales, Australia, in 1995 (9), found that the fledging rate of 30 Gould's petrel *Pterodroma leucoptera* chicks translocated from their burrows to artificial nests nearby and hand-fed was not significantly different from control (unmoved, parent-fed) birds (100% of translocated chicks fledging vs. 29/30 controls). Nest burrows were of the type described in (4). This study is also described in 'Provide supplementary food to increase reproductive success', 'Translocate individuals' and 'Artificially incubate and hand-rear birds in captivity'.

A controlled before-and-after study in 1997–2001 in two sea caves near Benidorm, Spain (10), found that European storm petrels *Hydrobates pelagicus* nesting in artificial nest boxes had significantly higher nesting success than petrels in natural nests, except in the year boxes were provided (36–49% for 803 natural nests vs. 40–75% for 72 occupied nest boxes). Occupancy rates increased over time (6% of 86 boxes in 1997 to 29% of 83 in 2001) and were higher in a cave not illuminated at night by city lights and for boxes placed over old nest sites. There was no decrease in the number of petrels at natural nesting sites over the study, so the birds in nest boxes probably represented new breeders. Nest boxes were 25 × 12 cm PVC boxes with drainage holes, lined with sand and fitted with a small tunnel preventing access by gulls.

A controlled study in 2000–1 on Praia Islet in the Azores, Portugal (11), found that Maderian storm petrels *Oceanodroma castro* nesting in artificial nest chambers had higher overall productivity than those in natural burrows in two out of three breeding seasons (0.42–0.64 chicks/pair for birds in artificial nests vs. 0.15–0.29 chicks/pair for natural burrows). Between 40 and 49 of 115–147 chambers were used and the authors argue that most birds using them were new breeders, meaning that the early-breeding population and late-breeding populations would have increased by 28% over two years and 11% in one year respectively. Chambers consisted of drainable plastic plant pots lined with stones, soil and dry grass, buried, covered with flexible lids and led to by 6 cm entrance. The pot had holes to ensure water drained through it. This study also used recorded petrel vocalisations to attract petrels to the site, discussed in 'Use vocalisations to attract birds to safe areas'.

A replicated study on Mana Island, North Island, New Zealand (12), found that 49% of 239 common diving petrels *Pelecanoides urinatrix* fledged successfully after being translocated to the island in 1997–9 from two other islands and hand-reared in artificial nests. However, 94%

of 53 breeding attempts on the island until 2003 were in natural burrows, rather than artificial nests. The nests consisted of two buried chambers (30 × 15 cm and 20 × 20 cm) reached by a 60 cm section of 10 cm diameter PVC pipe. This study is discussed in more detail in 'Artificially incubate and hand-rear birds in captivity', 'Use vocalisations to attract birds to safe areas' and 'Translocate individuals'.

A replicated study on Boondelbah Island, New South Wales, Australia (13), found that 98% of Gould's petrel *Pterodroma leucoptera leucoptera* chicks fledged successfully from artificial nests, after being translocated from Cabbage Tree Island, 1.4 km away, leading to the establishment of a new breeding colony. One hundred plastic boxes of the type described in (4) were installed over a 150 m^2 and chicks placed directly in them and supplied with food. This study is discussed in more detail in 'Translocate individuals'.

A before-and-after study on Praia Islet (12 ha), off Graciosa, Azores, Portugal (14), found that the breeding population of storm petrel *Oceanodroma castro* increased from no breeding pairs to almost 800 (before 2000 and in 2006, respectively), following the installation of artificial nesting burrows in 2000. Burrows consisted of a plastic plant pot (with drainage holes), buried and with a 6 cm entrance burrow leading to a hole in the side. Rabbits were also eradicated from the island (see 'Control or remove habitat-altering mammals' and habitat restored 'Shrubland restoration'). The effect of nest boxes for ground-nesting seabirds is also discussed.

(1) Wilson, U. W. (1986) Artificial rhinoceros auklet burrows: a useful tool for management and research. *Journal of Field Ornithology*, 57, 295–299.
(2) Podolsky, R. H. & Kress, S. W. (1989) Factors affecting colony formation in Leach's storm petrel. *The Auk*, 106, 332–336.
(3) Wilson, U. W. (1993) Rhinoceros auklet burrow use, breeding success, and chick growth: gull-free vs. gull-occupied habitat. *Journal of Field Ornithology*, 64, 256–261.
(4) Priddel, D. & Carlile, N. (1995) An artificial nest box for burrow-nesting seabirds. *Emu*, 95, 290–294.
(5) Bolton, M. (1996) Energy expenditure, body-weight and foraging performance of storm petrels *Hydrobates pelagicus* breeding in artificial nesting chambers. *Ibis*, 138, 405–409.
(6) Cruz, J. & Cruz, F. (1996) Conservation of the dark-rumped petrel *Pterodroma phaeopygia* of the Galapagos Islands, 1982–1991. *Bird Conservation International*, 6, 23–32.
(7) Gaston, A. J. (1996) A nest box for ancient murrelets. *Colonial Waterbirds*, 19, 116–120.
(8) Sullivan, W. J., Wilson, K.-J. & Paterson, A. (2000) Influence of artificial burrows and microhabitat on burrow competition between Chatham petrels *Pterodroma axillaris* and broad-billled prions *Pachyptila vittata*. *Emu*, 100, 329–333.
(9) Priddel, D. & Carlile, N. (2001) A trial translocation of Gould's petrel (*Pterodroma leucoptera leucoptera*). *Emu*, 101, 79–88.
(10) de León, A. & Mínguez, E. (2003) Occupancy rates and nesting success of European storm-petrels breeding inside artificial nest-boxes. *Scientia Marina*, 67, 109–112.
(11) Bolton, M., Medeiros, R., Hothersall, B. & Campos, A. (2004) The use of artificial breeding chambers as a conservation measure for cavity-nesting procellariiform seabirds: a case study of the Madeiran storm petrel (*Oceanodroma castro*). *Biological Conservation*, 116, 73–80.
(12) Miskelly, C. M. & Taylor, G. A. (2004) Establishment of a colony of common diving petrels (*Pelecanoides urinatrix*) by chick transfers and acoustic attraction. *Emu*, 104, 205–211.
(13) Priddel, D., Carlile, N. & Wheeler, R. (2006) Establishment of a new breeding colony of Gould's petrel (*Pterodroma leucoptera leucoptera*) through the creation of artificial nesting habitat and the translocation of nestlings. *Biological Conservation*, 128, 553–563.
(14) Bried, J., Magalhaes, M. C., Bolton, M., Neves, V. C., Bell, E., Pereira, J. C., Aguiar, L., Monteiro, L. R. & Santos, R. S. (2009) Seabird habitat restoration on Praia Islet, Azores Archipelago. *Ecological Restoration*, 27, 27–36.

15.4.5. Wildfowl

- Six studies from North America (4,5,8,27) and Europe (14,25) found that wildfowl populations increased with the provision of artificial nests, although one study from Finland (25) found that there was no increase in the number of broods or chicks in areas with nest boxes.
- Twelve studies from North America (4–6,8,10,12,14,17–19,24,27) investigated the success of nests in artificial nests with nine (4–6,8,10,12,14,18,24) finding that success and productivity was

high, sometimes higher than or similar to natural nests. Two studies (19,24) found that success for some species in nest boxes was lower than for natural nests. Two studies investigated the impact of nest box location, finding that hidden nests had higher success (17) and that nests over water were more successful than those in trees over land (27).

- Nineteen studies from across the world (1–4,6,9–16,18–21,23,27) investigated occupancy rates of artificial nests, finding that rates varied from no use of 25 nest boxes in a single site in Indonesia (23) to 100% occupancy across 20 sites in the USA (3) with one study (13) finding that nest boxes were used more than natural cavities. Two studies found that occupancy rates increased over time (4,6), whilst four studies found that occupancy rates appeared to be affected by design (9,11,16) or positioning (9,15).
- Three studies from North America (7,17,26) found that nest boxes could have other impacts on reproduction and behaviour, with common starlings *Sturna vulgaris* (a nest site competitor) avoiding some nest box designs (7); hidden nest boxes having lower intra-specific nest parasitism than easily visible boxes (17) and female common eiders *Somateria mollissima* losing less weight over incubation if they were nesting in shelters, compared to birds nesting in the open (26), although they lost weight quicker after nesting.

Background

Wildfowl nest both in trees (e.g. wood ducks *Aix sponsa* and goldeneyes *Bucephala* spp.) and on the ground (e.g. dabbling ducks *Anas* spp.). These different strategies require very different interventions if conservationists are to provide nesting sites. We have therefore split studies describing the provision of artificial islands and floating rafts for wildfowl (described in the next section) from those describing the provision of nest boxes and other artificial nests.

One study, Divoky & Suydam (1995), describes the use of a nesting shelter for ground-nesting common eiders *Somateria mollissima*, and is included in this section as it is providing a shelter, rather than nesting substrate.

Divoky, G. J. & Suydam, R. (1995) An artificial nest site for arctic nesting common eiders. *Journal of Field Ornithology*, 66, 270–276.

A replicated study in 1941–6 at a wetland site in Connecticut, USA (1), found that wood ducks *Aix sponsa* nested in only 12 of 274 nest boxes (4%) erected between 3.6 m and 7.3 m off the ground. In contrast, up to 67% of boxes were occupied by grey squirrels *Sciurus carolinensis* at once, with other rodents and insects also occupying boxes.

A replicated study in wetland habitats in Washington State, USA, in 1944–53 (2), found that 12 of 18 artificial nests made from willow were used by Canada geese *Branta canadensis* in 1944. Over the next four years, a further 13 nests of various types were placed in trees and by 1951 there were 53 available nests in the area. Twelve of these were inspected in 1951 and found to contain eggs. The author also states that placing logs on river islands increases their attractiveness to nesting geese, but no data was provided to support this.

A replicated trial in 20 woodland sites on Rhode Island, USA, in 1955–56 (3) found that wood ducks *Aix sponsa* used 36–100% of nest boxes installed in study areas. A total of 102 nest boxes were observed in 1955 and 85 in 1956. This study is discussed in more detail in 'Use artificial nests that discourage predation'.

A before-and-after study on a marshland site in Montana, USA (4), found that the breeding population of Canada geese *Branta canadensis* did not increase consistently following the installation of raised nesting platforms in 1954 (200 pairs in 1953, increasing to 285 in 1955, falling to 154 in 1956–8). The authors note that the population decrease after 1956 appeared to be due to overhunting. Platform use increased from 5% of the breeding population in 1954 to 18% in 1958. Clutch size and nest success were similar on and off platforms, but significantly more goslings hatched in platform nests (average of 3.6 goslings/clutch for 49 nests on platforms vs. population average of 3.1 goslings/clutch for 1,113 nests). Platforms consisted of a wooden tray 76 cm × 66 cm, 15 cm deep and filled with soil and decaying vegetation. They were placed at heights of 1.2–13.7 m in trees on islands in a lake and on the lake's shoreline. This study is also discussed in 'Use artificial nests that discourage predation'.

A replicated, controlled study from 1958–1961 in multiple woodlots containing nestboxes (114 ha in total) and 1 woodlot containing all natural cavities (9.3 ha) in Illinois, USA (5), found that wood duck *Aix sponsa* breeding pair density increased from 10–15 to over 90 pairs during the study period. Ducks exhibited higher nest success in nestboxes (71% success for 574 metal nest boxes vs. 37% for 116 natural cavities), although a smaller proportion were occupied (48% compared 23% occupation), probably due to lower racoon *Procyon lotor* predation (see 'Use artificial nests that discourage predation'). Female wood ducks usually returned to the nesting areas where they last bred successfully, so the authors suggest that nest boxes should be grouped into units (2–3 per ha in high-quality habitat were recommended). Most nestboxes were metal cylinders with elliptical entrances.

A replicated study on a total of 11 marshland sites in Iowa, USA, in 1964–9 (6) found that mallards *Anas platyrhynchos* used 33% of 705 artificial nests over the study period. The percentage of the mallard population using artificial nests increased from 31% in 1966 to 46% in 1969. Four other species (blue-winged teal *A. discors*, gadwall *A. strepera*, redheads *Arytha americana* and Canada geese *Branta canadensis*) used a total of 12 nests over the study period. Nesting success in artificial nests was 87%, far higher than previous records of mallard nest success (normally 27–52%). Nests were cone-shaped, hardware-cloth baskets, 18 cm deep and erected on poles sunk into marshland, dry land and in vegetation. Not all 11 sites were used every year.

A small replicated, controlled study from 1963–1970 in 15 sites of marshy or wooded duck habitat in Maryland, USA (7) found that wood ducks *Aix sponsa* had no significant preference for next box design but that common starlings *Sturna vulgaris* avoided horizontal nest boxes with large entrances. Starlings showed greater preference for vertical boxes with small entrances (3 × 4 inches) and avoided horizontal boxes with large entrances (semicircular, 11 inches in diameter). In 1965, when all vertical boxes were removed and all horizontal boxes had larger openings, there was an abrupt decrease in starling nest box use, despite no change in the starling population size. Starlings preferred boxes in open sites than those in wooded sites, whereas wood ducks showed no preference. The basic experimental nest box was a horizontal cylinder (24 inches long and 12 inches in diameter) made of metal or wire netting covered with roofing paper and had wooden ends.

A before-and-after trial in a mixed forest and wetland site in Mississippi, USA (8), found that the population of wood ducks *Aix sponsa* in the study site increased dramatically following the installation of 253 nest boxes between 1966 and 1969 (average of 30–35 pairs in 1960–5 vs. 231 nest boxes used in 1969). Between 1966 and 1969, 15,273 eggs were laid in nest boxes, with 6,036 ducklings leaving nest boxes. In contrast, ten belt transects (20 m × 1,128–5,486 m) detected 27 natural nesting cavities, none of which were used by ducks. Hatching success was 67% in nest boxes, with 10% of eggs being predated in 1969. A further study at a wetland site in Louisiana, USA, found that, in 1969, 53% of 30 nest boxes erected were used by wood ducks, with 243 eggs laid, of which 47% hatched (34% were destroyed by predators).

A replicated study in Australia (9), found that wildfowl in Western Australia used only 1% of 1,999 artificial nests in 1974, whereas 36% of 2,440 artificial nests in Victoria were used in 1975–6. The majority of records from Victoria were of chestnut teal *Anas castanea*, which are uncommon in Western Australia. All Western Australian nests were made from plastic drums and erected at 23 wetland sites in 1969–74; Victorian nests were made from various wooden and metal boxes and plastic drums and were erected at 26 wetland sites between 1975–6. All nests were attached to poles and trees at heights of up to 3.6 m. Nests below 50 cm off the ground were mostly avoided by birds.

A replicated study between 1964 and 1975 in six wetland sites in Texas, USA (10), found that black-bellied whistling ducks *Dendrocygna autumnalis* used an average of 81% of nest boxes erected in trees. On average, 52 nest boxes were available each year and were monitored an average of 14 times a year. A total of 778 clutches were laid over the study period, with 40% incubated and 75% of these hatching at least one egg successfully (210 nests, 28% of all nests). Sixty three percent of eggs in successful nests hatched, compared with a population average of 20%.

A series of replicated studies at river and lake sites in northern Ontario, Canada in 1974–9 (11), found that cavity-nesting ducks (mainly common goldeneyes *Bucephala clangula*) preferentially used nest boxes with large (13 × 10 cm) entrance holes high (33 cm) above the floor of nest boxes with dark interiors. Nest boxes with large entrance holes were used more than those with medium (10.5 × 8 cm) holes; boxes with small (7.5 × 6 cm) holes were not used by goldeneyes or hooded mergansers *Mergus cucullatus* (318 sets of boxes tested). Boxes with entrance holes 18 or 25.5 cm above the base of the box were not used by goldeneyes (201 sets) and boxes with dark-stained interiors were used more than those with unstained interiors (39 breeding attempts vs. 13 attempts, 73–5 sets for each of six years). Differences between tree species were minimal and orientation had no impact.

A replicated study in 1978 in a forested marshland site in South Carolina, USA (12), found that wood ducks *Aix sponsa* used 89% of 55 nest boxes erected between 1974 and 1978. Five-gallon plastic buckets were used slightly more often than wooden nest boxes and 'fiber cylinders' (95% of 20 buckets used, compared with 86% of 35 boxes and cylinders). Hatching rates did not vary between nest types, with 28% of the 847 eggs laid being predated or deserted and 79% of the remaining 608 eggs hatching. Bucket nests were five-gallon buckets with a 7.6 cm diameter entrance hole and a secured lid. All nests were placed at a variety of heights and in a variety of vegetation types.

A replicated controlled study 1977–9 in riverine forests in Louisiana, USA (13), found that wood ducks *Aix sponsa* used nest boxes more frequently than natural cavities (0.4% of 5,374 nest boxes inspected contained nests vs. 0.03% of 3,993 natural cavities). The most frequently used nest boxes were large (60 × 60 × 30 cm), with a circular or oval entrance of less than 140 cm^2. This study also examined nest box use by other birds (owls, woodpeckers and songbirds).

A before-and-after study in northern Scotland (14) found that a breeding population of common goldeneyes *Bucephala clangula* established itself in a forested landscape following the installation of a total of 83 nest boxes between 1961 and 1982. Goldeneye numbers were monitored from 1960, with a single female nesting in a natural cavity in 1970. Nest boxes were first used in 1974 (two breeding attempts), with 41 breeding attempts in 1982 and the percentage of occupied boxes increasing from 6% to 49% over the same period. Occupancy rates and nesting success were highest for boxes close to rivers (72% occupation for 13 boxes, 78% success for 36 attempts), compared with those by small lakes (30% occupancy, 57% success for 37 boxes and 58 attempts), large lakes (26% occupancy, 50% success for 25 boxes and 44 attempts) or marshes (4% occupancy and 50% success for nine boxes and two attempts).

A small single-site study from March–May in 1981–1982 in an island (1.2 ha) within Kentucky Lake in Tennessee, USA (15) found that wood ducks *Aix sponsa* nested in 44–63% of the nest boxes provided although in 1982, 11 of 44 wood duck nests were destroyed, probably by common grackles *Quiscalus quiscula* and nestlings in one nest were preyed upon by grackles. According to the authors, the presence of vacant nest boxes in both years suggests that grackles and wood ducks were not competing for nest sites. This study also discusses the use of nest boxes by grackles.

A series of replicated studies in 1977–84 at two lakes in eastern Ontario, Canada (16), found that the lining nest boxes with wood shavings significantly increased their use by four duck species, whilst entrance hole size and height above the ground had uncertain effects. All 18 ducks nesting in 100 pairs of nest boxes over two years chose boxes lined with wood shavings over those without. Entrance hole size did not significantly influence box choice by goldeneyes *Bucephala clagnula* but hooded mergansers *Lophodytes cucullatus* and wood ducks *Aix sponsa* used small entrances more (ten and four boxes used, compared with three and zero boxes with large holes), whilst common mergansers *Mergus merganser* only nested in four boxes with large entrances. Boxes 6 m above the ground were used more often by goldeneyes than those at 4.5 m or 3 m (14 breeding attempts vs. nine and three attempts, 20 sets of boxes in each of eight years) but the authors argue that occupancy rates would not change in the absence of choice.

A replicated study over 12 years between 1976 and 1987 in deciduous woodlands and wetlands in northeast Illinois, USA (17) found that successful wood duck *Aix sponsa* clutches in well-hidden nest boxes had significantly higher hatching success than those in conspicuous boxes (82% hatching success for 28 successful well-hidden nests vs. 74% for 150 successful conspicuous nests), probably due to lower levels of intraspecific brood parasitism (30% of 47 hidden clutches parasitised vs. 50% for 198 conspicuous ones). However, visible nests were more likely to raise at least one duckling (60% of 47 hidden nests successful vs. 76% of 198 conspicuous nests), and hatched more ducklings (7.1 ducklings/successful nest for hidden nests vs. 9.3–9.9 ducklings/nest for conspicuous nests) possibly due to larger clutch sizes caused by brood parasitism (12.4 eggs/clutch for hidden nests vs. 15.7–16.3 eggs/clutch for conspicuous nests).

A replicated study in 1987 at four ranches in Tamaulipas, Mexico (18), found that Muscovy ducks *Cairina moschata* only used 13 of 407 nest boxes (3%), with 77% of these successfully hatching eggs and fledging 96 ducklings. Overall hatching success was 54% of 177 eggs. Three black-bellied whistling ducks also fledged from the ten successful nests. Nest boxes were 42 × 42 × 62 cm with a 21 cm diameter entrance hole and were erected on metal or wooden poles or trees either on islands in a lake (168 boxes) or close to ponds and waterways.

A replicated study over four breeding seasons in 1985–8 at two lakes in Veracruz, Mexico (19), found that black-bellied whistling ducks *Dendrocygna autumnalis* using nest boxes had very low reproductive success (11.1% success for nine attempts in 1986, 6.6% for 30 attempts in 1987–8), mainly because of predation by opossums *Didelphis* spp., raccoons *Procyon lotor* and humans. Occupancy rates varied, with one of 16 pairs using boxes in 1985; 17–30% of 30 pairs in 1986 and 40–75% of 20 pairs in both 1987 and 1988. Nest boxes were made from either liana baskets or hollowed palm trunks, with the latter being the only nests used (except for a single basket in 1985). Thirteen baskets were provided in 1985 and ten in 1986; ten trunks were provided in 1986, 16 in 1987 and 17 in 1988. Boxes were placed in positions similar to naturally occurring nests and checked every two weeks during the breeding seasons. Opossums also frequently occupied nest boxes.

A replicated study at 16 areas in an open pine forest in South Carolina, USA (20), found that over nine breeding seasons (1982–90), between 19 and 44 female wood ducks *Aix sponsa* used nest boxes, with 120 (1982–3) or 150 (1984–90) boxes provided each year.

A small trial on dunes on an island in northern Alaska, USA, in 1992–3 (21), found that 16 out of 20 nesting structures provided for common eiders *Somateria mollissima* were destroyed the year after installation, but that the remaining four provided seven potential nest sites, of which three (in two structures) were used, hatching at least one egg successfully. The structures consisted of a wooden cross providing four uncovered, semi-sheltered nesting quadrants protected on two sides by 20 × 61 cm boards. All three nests were in the south-facing quadrants.

A replicated study at three sites in southern Finland over four breeding seasons in 1993–7 (22) found that breeding common goldeneyes *Bucephala clangula* showed a significant preference for nest boxes erected on the shoreline of lakes compared to those 14–140 m into the surrounding forest (shore boxes occupied before forest boxes for 73–95% of 50 pairs of boxes, with 8% of pairs occupied in the same season). Female goldeneye inspected shore and forest boxes equally and therefore appeared to actively choose shore boxes.

A replicated one-year study in 2001 in Sumatra, Indonesia (23), found that no white-winged ducks *Cairina scutulata* were observed entering any of 25 nest boxes erected in trees in a swamp forest site. Boxes were 53 × 43 × 41–48 cm, with a 20 × 17 cm entrance hole and four drainage holes, and were erected 1.5–4.0 m above the ground (mainly in durian trees *Durio zibethinus*). The author argues that nest boxes may take several years to be accepted and so may be used in the future.

A replicated, controlled study in a deciduous forest in British Columbia, Canada, in 1997–9 (24), found that Barrow's goldeneyes *Bucephala islandica* laid larger clutches but had lower nesting success in nest boxes, compared to natural nest cavities (10.5 eggs/clutch and 45–50% success for 174 clutches in nest boxes vs. 7.5 eggs/clutch and 54–86% for 41 clutches in natural cavities). There were no differences for buffleheads *B. albeola* (8.4 eggs/clutch and 75–90% success for 46 clutches in boxes vs. 8.5 eggs/clutch and 55–90% success for 100 clutches in natural cavities). Hatching dates did not differ for either species between nest types. Goldeneye nests in boxes were predated mainly by black bears *Ursus americanus* compared with small mammals and common starlings *Sturnus vulgaris* in natural nests. Predation of all bufflehead nests was low and mainly by American red squirrel *Tamiasciurus hudsonicus* and American pine marten *Martes americana*. The authors suggest that differences in goldeneye nests were due to nest boxes being concentrated in highly visible locations, whilst natural nests were widely dispersed. Natural bufflehead nests, however, were positioned similarly to nest boxes.

A replicated and controlled before-and-after study in southern Finland in 1988–99 (25) found that the number of common goldeneye *Bucephala clangula* breeding pairs increased on 35 lakes following the provision of 50 nest boxes (average of 0.8 pairs/lake in 1988–91, before nest box provision vs. 1.1 pairs/lake in 1995–99, afterwards). There were no increases on 17 lakes without nest boxes provided (average of 0.8 pairs/lake in 1988–91 vs. 0.9 pairs/lake in 1995–9). However, there was no increase in the number of broods at experimental lakes (0.17 broods/lake vs. 0.19 broods/lake) and the increase in the number of chicks fledging was not significant (0.59 young fledged/lake in 1988–91 vs. 0.86 young/lake in 1995–9). Nest boxes were 25 × 26 × 70 cm with a 9 cm diameter entrance hole. The authors suggest nest site availability may limit the number of breeding goldeneyes, but that other factors appear to limit reproductive output.

A replicated, randomised and controlled paired study in tundra on Mitivik Island, Hudson Bay, Nunavut, Canada, in 2001 and 2003 (26), found that female common eider *Somateria mollissima* provided with shelters whilst nesting lost less weight over the incubation period, compared to control females without shelters (average weight at end of incubation of 1,312 g for 34 sheltered birds vs. 1,266 g for 31 controls). Sheltered birds, however, appeared to lose weight more rapidly at the end of incubation. Sheltered nests had more stable temperatures than controls (average temperature range was 6.1°C lower for sheltered nest). Paired nests were less than 10 m apart (to ensure similar microclimates) and were at similar stages of incubation when a shelter consisting of a 46 × 46 cm roof and two 25 × 46 cm walls (with 12, 2.5 cm holes in) was placed over one of the nests. The walls were positioned facing east-west.

A replicated before-and after study in 1999–2004 in boreal forests surrounding 60 lakes in Québec, Canada (27), found that the number of breeding pairs of common goldeneyes *Bucephala clangula* and Barrow's goldeneyes *B. islandica* increased from ten and 28 pairs in 1999 to 46 and 43 pairs in 2003 following the provision of 105–133 nest boxes each year from 1998–9. The number of broods increased in 2000, but not subsequently. Goldeneyes used 23–43% of nest boxes, with 37–67% hatching success for 261 nests. Three nests were erected at each lake: those above water or on trees on the shore had higher success rates than those in clearcuts 25–160 m away from shore (50% success for 56 nests above water, 56% for 63 nests on the shore and 40% for 86 nests in clearcuts). Boxes were 24 × 22 × 60 cm with a 10 × 13 cm entrance hole. American kestrels *Flaco sparverius* also used nest boxes, although their use declined over time.

(1) Frank, W. J. (1948) Wood duck nesting box usage in Connecticut. *The Journal of Wildlife Management*, 12, 128–136.

(2) Yocom, C. F. (1952) Techniques used to increase nesting of Canada geese. *The Journal of Wildlife Management*, 16, 425–428.

(3) Cronan, J. M. (1957) Effects of predator guards on wood duck box usage. *The Journal of Wildlife Management*, 21, 468.

(4) Craighead, J. J. & Stockstad, D. S. (1961) Evaluating the use of aerial nesting platforms by Canada geese. *The Journal of Wildlife Management*, 25, 363–372.

(5) Bellrose, F. C., Johnson, K. L. & Meyers, T. U. (1964) Relative value of natural cavities and nesting houses for wood ducks. *The Journal of Wildlife Management*, 28, 661–676.

(6) Bishop, R. A. & Barratt, R. (1970) Use of artificial nest baskets by mallards. *The Journal of Wildlife Management*, 34, 734–738.

(7) McGilvrey, F. B. & Uhler, F. M. (1971) A starling-deterrent wood duck nest box. *The Journal of Wildlife Management*, 35, 793–797.
(8) Strange, T. H., Cunningham, E. R. & Goertz, J. W. (1971) Use of nest boxes by wood ducks in Mississippi. *The Journal of Wildlife Management*, 35, 786–793.
(9) Norman, F. I. & Riggert, T. L. (1977) Nest boxes as nest sites for Australian waterfowl. *The Journal of Wildlife Management*, 41, 643–649.
(10) McCamant, R. E. & Bolen, E. G. (1979) A 12-year study of nest box utilization by black-bellied whistling ducks. *The Journal of Wildlife Management*, 43, 936–943.
(11) Lumsden, H. G., Page, R. E. & Gauthier, M. (1980) Choice of nest boxes by common goldeneyes in Ontario. *The Wilson Bulletin*, 92, 497–505.
(12) Griffith, M. A. & Fendley, T. T. (1981) Five-gallon plastic bucket: an inexpensive wood duck nesting structure. *The Journal of Wildlife Management*, 45, 281–284.
(13) McComb, W. C. & Noble, R. E. (1981) Nest-box and natural-cavity use in three mid-south forest habitats. *The Journal of Wildlife Management*, 45, 93–101.
(14) Dennis, R. & Dow, H. (1984) The establishment of a population of goldeneyes *Bucephala clangula* breeding in Scotland. *Bird Study*, 31, 217–222.
(15) Spero, V. M. & Pitts, T. D. (1984) Use of wood duck nest boxes by common grackles. *Journal of Field Ornithology*, 55, 482–483.
(16) Lumsden, H. G., Robinson, J. & Hartford, R. (1986) Choice of nest boxes by cavity-nesting ducks. *The Wilson Bulletin*, 98, 167–168.
(17) Semel, B., Sherman, P. W. & Byers, S. M. (1988) Effects of brood parasitism and nest-box placement on wood duck breeding ecology. *The Condor*, 90, 920–930.
(18) Markum, D. E. & Baldassarre, G. A. (1989) Breeding biology of Muscovy ducks using nest boxes in Mexico. *The Wilson Bulletin*, 101, 621–626.
(19) Feekes, F. (1991) The black-bellied whistling duck in Mexico–from traditional use to sustainable management? *Biological Conservation*, 56, 123–131.
(20) Hepp, G. R. & Kennamer, R. A. (1992) Characteristics and consequences of nest-site fidelity in wood ducks. *The Auk*, 109, 812–818.
(21) Divoky, G. J. & Suydam, R. (1995) An artificial nest site for arctic nesting common eiders. *Journal of Field Ornithology*, 66, 270–276.
(22) Pöysä, H., Milonoff, M., Ruusila, V. & Virtanen, J. (1999) Nest-site selection in relation to habitat edge: experiments in the common goldeneye. *Journal of Avian Biology*, 30, 79–84.
(23) Drilling, N. (2001) Pilot nest-box project for white-winged ducks in Sumatra. *TWSG News*, 13, 16–18.
(24) Evans, M. R., Lank, D. B., Boyd, W. S. & Cooke, F. (2002) A comparison of the characteristics and fate of Barrow's goldeneye and bufflehead nests in nest boxes and natural cavities. *The Condor*, 104, 610–619.
(25) Pöysä, H. & Pöysä, S. (2002) Nest-site limitation and density dependence of reproductive output in the common goldeneye *Bucephala clangula*: implications for the management of cavity-nesting birds. *Journal of Applied Ecology*, 39, 502–510.
(26) Fast, P. L., Grant Gilchrist, H. & Clark, R. G. (2007) Experimental evaluation of nest shelter effects on weight loss in incubating common eiders *Somateria mollissima*. *Journal of Avian Biology*, 38, 205–213.
(27) Savard, J. P. & Robert, M. (2007) Use of nest boxes by goldeneyes in eastern North America. *Wilson Journal of Ornithology*, 119, 28–34.

15.4.6. Wildfowl – artificial/floating islands

- Two studies from North America (1,2) found that a variety of wildfowl used artificial islands and floating rafts, and had high (70–80%) nesting success.

- A replicated study from across the UK (3) found that wildfowl preferentially nested on well vegetated islands, compared to bare ones.

Background

Some species of wildfowl nest on the ground and so as well as providing nest boxes (see separate intervention), conservationists may be able to increase the survival of wildfowl broods by providing artificial island or floating rafts in water bodies.

A replicated study on two marshland sites in Pennsylvania, USA, in 1976–8 (1), found that 56% of 20–34 artificial nesting rafts were used by wildfowl, with mallards *Anas platyrhynchos* using up to 50% of nest rafts, blue-winged teal *A. discors* up to 9% and Canada geese *Branta canadensis* up to 5%. Hatching success on rafts was 80%. Rafts had a wooden frame and Styrofoam centre, an arching roof of wire mesh with two anchors of different weights allowing the raft to float up and down with changing water levels. The authors estimate the cost at $0.85/duckling (in 1979 dollars).

A replicated study at seven prairieland impoundments in Alberta, Canada (2), found that in 1976–8, 1,349 nests from 13 species of wildfowl were found on 75 artificial islands (75 islands searched in 1976–7 and 53 in 1978). Ducks (12 species) nested at densities of 1.8–29.1 nests/ha, with 43–59% success. Canada geese *Branta canadensis* nested at densities of 0.2–7.1 nests/ha, with 70% success (144 nests). Islands were most productive when small, far from shore and with high vegetation cover. Islands were created before flooding of the impoundments by raising some areas about to be flooded or isolating peninsulas with ditches and were between 0.13 ha and 6.6 ha in size.

A replicated 1992 study of the use of artificial islands and floating platforms in 17 wetland nature reserves across the UK (3) found that 11 species of wildfowl nested with greater frequency on well vegetated islands and platforms than on sparsely vegetated ones. This pattern was strongest at inland northern reserves, where all 11 species used well-vegetated sites, but none used sparsely covered ones. At coastal sites and southern reserves the pattern was weaker, but well-vegetated sites were always used by more species. The species studied were eight species of ducks, Canada geese *Branta canadensis*, feral greylag geese *Anser anser* and mute swans *Cygnus olor*. At four sites, the provision of vegetated islands or rafts resulted in the establishment of new populations of five duck species. The review also examines island and platform use by grebes, divers, ground-nesting seabirds, waders and rails.

(1) Brenner, F. J. & Mondok, J. J. (1979) Waterfowl nesting rafts designed for fluctuating water levels. *The Journal of Wildlife Management*, 43, 979–982.
(2) Giroux, J.F. (1981) Use of artificial islands by nesting waterfowl in southeastern Alberta. *The Journal of Wildlife Management*, 45, 669–679.
(3) Burgess, N. D. & Hirons, G. J. M. (1992) Creation and management of artificial nesting sites for wetland birds. *The Journal of Environmental Management*, 34, 285–295.

15.4.7. Gamebirds

- A replicated study in China (1) found that an estimated 36–41% of the local population of Cabot's tragopans *Tragopan caboti* used nesting platforms.

A replicated trial in subtropical conifer, deciduous and mixed forests in eastern China (1) found that a large percentage of female Cabot's tragopans *Tragopan caboti* in the region used some of the 200 artificial nesting platforms erected at 11 sites in 2002–3 (12–16 platforms used, an estimated 36–41% of the local population). Platforms consisted of 30 cm diameter bowls (15 cm deep) made from bamboo strips and fixed to trees across a variety of habitats, with platforms more likely to be used if they were in areas of mixed forest and close to the edge of forest patches.

(1) Deng, W., Zheng, G., Zhang, Z., Garson, P. J. & McGowan, P. J. K. (2005) Providing artificial nest platforms for Cabot's tragopan *Tragopan caboti* (Aves: Galliformes): a useful conservation tool? *Oryx*, 39, 158–163.

15.4.8. Rails

- A replicated study from across the UK (1) found that common moorhens *Gallinula chloropus* and common coot *Fulica atra* readily used artificial islands for nesting.

A replicated study in 1992 reviewed the use of artificial islands and floating platforms in 17 wetland nature reserves across the UK (1) found that common moorhens *Gallinula chloropus* and common coot *Fulica atra* used both well-vegetated and bare shingle-covered islands and platforms at both inland and coastal sites. The only exception was moorhens not using sparsely covered platforms and islands at northern coastal sites. The review also examines island and platform use by grebes, divers, ground-nesting seabirds, waders and wildfowl.

(1) Burgess, N. D. & Hirons, G. J. M. (1992) Creation and management of artificial nesting sites for wetland birds. *Journal of Environmental Management*, 34, 285–295.

15.4.9. Waders

- Two replicated studies from the UK (1) and the USA (2) found that waders used artificial islands and nesting sites.
- The UK study found that sparsely vegetated islands at coastal sites were used more than well vegetated and inland sites.

A replicated study in 1992 reviewing the use of artificial islands and floating platforms in 17 wetland nature reserves across the UK (1) found that six species of wader were more likely to use sparsely-vegetated islands and platforms for nesting on than well-vegetated ones. Platforms at inland sites were used less than those at coastal sites. Pied avocets *Recurvirostra avosetta* only bred on sparsely-covered islands and platforms at southern, coastal sites, with the provision of shingle islands leading to a significant increase in the avocet population at one site and the establishment of a population at another. Common ringed plovers *Charadrius hiaticula* only used islands at coastal sites, whilst little ringed plovers *C. dubius* only used sparsely-vegetated islands, but at both inland and coastal sites. Common redshank *Tringa tetanus*, northern lapwing *Vanellus vanellus* and Eurasian oystercatchers *Haematopus ostralegus* nested on islands at almost all sites. The review also examines island and platform use by grebes, divers, ground-nesting seabirds, rails and wildfowl.

A replicated, controlled study from March–August in 1994–1998 in 5 newly created site, 1 older artificial site and 1 natural site in coastal habitats in California, USA (2) found that snowy plovers *Charadrius alexandrius nivosus* used sites created for common terns *Sterna antillarum browni* but fledge rates declined steadily over the study period. The number of plover nests increased from 5 in 1994 to 38 in 1997, and were found on 4 of the 5 created areas. The natural site has the highest number of nests in total compared to the newly and older created sites (39, 25 and 8 nests respectively). Fledge rate in 1995 was higher at the newly created site (1.4 fledglings/nest) than at the control sites in any year but declined to 0.27 fledglings/nest in 1998. Average fledge rates were similar amongst sites (0.57, 0.47 and 0.52 fledglings/nest for newly created sites, old site and natural site respectively). Created sites were dredge spoils of coarse material.

(1) Burgess, N. D. & Hirons, G. J. M. (1992) Creation and management of artificial nesting sites for wetland birds. *Journal of Environmental Management*, 34, 285–295.
(2) Powell, A. N. & Collier, C. L. (2000) Habitat use and reproductive success of western snowy plovers at new nesting areas created for California least terns. *The Journal of Wildlife Management*, 64, 24–33.

15.4.10. Ibises and flamingos

- A study in Turkey (1) found that northern bald ibises *Geronticus eremite* moved to a site with artificial breeding ledges.
- A before-and-after study from France and Spain (2) found that large numbers of greater flamingos *Phoenicopterus roseus* used artificial nesting islands.

A study in southeast Turkey in 1977–88 (1) found that a northern bald ibis *Geronticus eremite* population moved from a breeding site threatened by development to an artificial breeding site provided 3 km away within an artificial breeding station (discussed in 'Use captive breeding to increase or maintain populations') and consisting of wooden ledges approximately 20 m away from where captive birds were. This study is also described in 'Release captive-bred individuals'.

A before-and-after study reviewing management at two coastal wetland sites in Bouches-du-Rhône, France and in Andalusia, Spain (2), found that large numbers of greater flamingos *Phoenicopterus roseus* used artificial nesting islands that were created at the sites. At the site in France, over 12,000 pairs used the island in one year, with 94,000 chicks raised between 1974 and 1993. At least 2,300 pairs used the nesting site in Spain. Islands were created from mud and later reinforced with stones and sand to reduce erosion. Decoy nests were placed on the islands (see 'Use decoys to attract birds to safe areas' for details) and 5–10% of the 700 which were erected in France in 1973–4 were used as nests in 1974. This study is also described in 'Manage water levels in wetlands' and 'Control predators not on islands'.

(1) Akçakaya, R. (1990) Bald ibis *Geronticus eremita* population in Turkey: an evaluation of the captive breeding project for reintroduction. *Biological Conservation*, 51, 225–237.
(2) Martos, M. R. & Johnson, A. R. (1996) Management of nesting sites for greater flamingos. *Colonial Waterbirds*, 167–183.

15.4.11. Raptors

- Nine studies from North America (1–8) and Spain (9) found that raptors used artificial nesting platforms, although one (2) describes low levels of use and another describes use increasing over time (4).
- Two studies from the USA (1,4) describe increases in populations or population densities of raptors following the installation of artificial nesting platforms.
- Three studies (6,7,9) describe successful use of platforms, whilst three describe lower productivity (1) or failed nesting attempts (3,5), although these studies only describe a single nesting attempt each.

Background

Many birds of prey return every year to the same nesting site, adding material so that the nest gets larger and larger. In extreme cases, this can damage the tree or structure that the nest is on, resulting in the nest falling. Providing platforms or other robust structures for birds to build on may, therefore, help to increase reproductive success.

A before-and-after study at a marshland site in Maryland, USA (1), found that the number of osprey *Pandion haliaetus* nests at the site increased from 4–6 before 1968 to 22 in 1971, and chick production tripled, following the erection of 24 artificial nesting platforms in 1968–72. Platforms had an 82% occupancy rate (59 nesting attempts out of 72 available nest-years) and more nests were found on platforms than at natural nest sites (59 nesting attempts on platforms vs. 12 at other sites). Nests on platforms produced an average of 1.3 chicks/nest, whilst natural nests produced 1.8 chicks/nest. Platforms consisted of a 122 × 122 cm platform of planks and wire on a 6.1 m wooden pole. The platform was braced, sunk 150 cm into the ground and designed to withstand hurricane-force winds.

A small study in 1976–9 in three scrub and grassland habitats in Idaho, USA (2), found that ferruginous hawks *Buteo regalis* nested on 24 nesting platforms provided in 1976, with one attempt in 1977 and three attempts in both 1978 and 1979. An average of 1.7 chicks/nest fledged. Platforms were provided in shaded/un-shaded pairs, and hawks only used unshaded platforms, with one pair moving platforms when the shade was moved to the platform they had used. This study also discusses platform use by common ravens *Corvus corax*, discussed in 'Provide artificial nest sites for songbirds'.

A small study at a reservoir in Arizona, USA (3), found that a bald eagle *Haliaeetus leucocephalus* pair used an artificial nesting structure in the breeding season of 1978–9, but failed to fledge any chicks. The structure consisted of a tripod of aluminium pipes supporting an existing nest which had failed in 1976 (when it fell in the water) and 1977 (when it was blown down in high winds). The nest was thought to have failed due to thin egg-shells.

A controlled before-and-after study over nine years in two pastoral sites in Canada (4) found that the breeding density of ferruginous hawks *Buteo regalis* increased following the provision of 98 nesting platforms in 1975 in an experimental area (nine nests in 1975 vs. 14 in 1983, increased populations in all five subareas). There was no increase in a control area, without platforms. Swainson's hawk *B. swainsoni* populations also increased from 0.1 pairs/km^2 in 1975 to 0.15 pairs/km^2 in 1983, but there were no differences between experimental and control areas. Less than 40% of hawk populations used platforms for the first two years, but use increased with time. Platforms were either 120 × 60 × 20 cm wooden boxes, or wire baskets, 60–90 cm in diameter and 20 cm deep. Both were lined with shrubs or grasses and mounted on wooden poles, buried 60–90 cm in the ground. After 17 poles fell, the authors recommended burying them deeper. Platforms provided with shade were used more than un-shaded ones.

A small study at a lake in Saskatchewan, Canada (5), found that one out of two artificial nesting platforms were used by bald eagles *Haliaeetus leucocephalus*. The platforms were erected in 1980 and a nest was built in 1986, although no chicks fledged from it.

A replicated study in 1978–90 at a lake in Saskatchewan, Canada (6), found that osprey *Pandion haliaetus* pairs fledged significantly more chicks from nests built on artificial platforms than from those in trees (1.3 chicks/breeding attempt for 70 attempts on platforms vs. 0.9 chicks/attempt for 205 attempts in trees). This difference was due to higher success rates on platforms (63% of 70 attempts on platforms vs. 46% of 205 attempts in trees), with no significant differences between productivities of successful nests (2.1 chicks/nest for 44 successful attempts on platforms vs. 2.0 chicks/nest for 94 successful attempts in trees). Nests were erected between 1978 and 1985 (ten platforms, 33% usage) and 1986–1990 (ten nests, 95% usage) and were made of wood.

A small study in Florida, USA (7), found that a bald eagle *Haliaeetus leucocephalus* pair successfully used a nest on an artificial platform built in June 1989. The pair fledged two chicks from the nest on the platform in 1990 and attempted nesting again in 1991, although the second attempt was disrupted by heavy traffic below the platform and was not successful. The platform was made of plywood, measured 1.5 × 1.5 m and was erected on a power pylon, <1.5 m to the site of several unsuccessful nesting attempts. A nest of loblolly pine *Pinus taeda* branches was also provided on the platform.

A replicated study reviewing an osprey *Pandion haliaetus* translocation programme in an urban area of Minnesota, USA (8), found that all but three of 26 nest sites used by 143 translocated ospreys and their young were artificial nesting platforms provided for the birds. Of these, 20 nests were productive, with only one not being situated on a nesting platform. This study is discussed in more detail in 'Translocate individuals'.

A small study at a reservoir in southern Spain in 2005 (9) found that a pair of ospreys *Pandion haliaetus* successfully raised two chicks that were fostered to them in a nest on an artificial nesting platform. This study is discussed in more detail in 'Foster eggs or chicks with wild conspecifics'.

(1) Rhodes, L. I. (1972) Success of osprey nest structures at Martin National Wildlife Refuge. *The Journal of Wildlife Management*, 36, 1296–1299.
(2) Howard, R. P. & Hilliard, M. (1980) Artificial nest structures and grassland raptors. *Raptor Research*, 14, 41–45.
(3) Grubb, T. G. (1983) Bald eagle activity at an artificial nest structure in Arizona. *Raptor Research*, 17, 114–121.
(4) Schmutz, J. K., Fyfe, R. W., Moore, D. A. & Smith, A. R. (1984) Artificial nests for ferruginous and Swainson's hawks. *The Journal of Wildlife Management*, 48, 1009–1013.
(5) Bortolotti, G. R., Dzus, E. H. & Gerrard, J. M. (1988) Bald eagle nest on an artificial tree-top platform. *Journal of Raptor Research*, 22, 66–67.
(6) Houston, C. S. & Scott, F. (1992) The effect of man-made platforms on osprey reproduction at Loon Lake, Saskatchewan. *Journal of Raptor Research*, 26, 152–158.
(7) Marion, W. R., Quincy, P. A., Cutlip, C. G. & Wilcox, J. R. (1992) Bald eagles use artificial nest platform in Florida. *Journal of Raptor Research*, 26, 266.
(8) Martell, M. S., Englund, J. V. & Tordoff, H. B. (2002) An urban osprey population established by translocation. *Journal of Raptor Research*, 36, 91–96.

(9) Muriel, R., Ferrer, M., Casado, E. & Schmidt, D. (2006) First breeding success of osprey (*Pandion haliaetus*) in mainland Spain since 1981 using cross-fostering. *Journal of Raptor Research*, 40, 303–304.

15.4.12. Falcons

- Four studies from the USA (5,20) and Europe (13,16) found that local populations of falcons increased following the installation of artificial nesting sites, with one (13) reporting that there was no decline in natural nest use following the installation and use of nest boxes. A replicated study from Canada (7) found that the local population of American kestrels *Falco sparverius* did not increase following the erection of nest boxes.
- Eight studies from across the world (1,2,5,6,9,13,15,17) found that the success and productivity of falcons in nest boxes was high and equal to, or higher than those in natural nests. Four studies from across the world (18,20–22) found that productivities in nest boxes were lower than in natural nests or in previously published results, or that some falcons were evicted from their nests by barn owls *Tyto alba* (22).
- Four studies from across the world (7,11,21,22) found no differences in productivity between nest box designs or positions, whilst two, from Spain (12) and Israel (17) found that productivity in boxes varied between designs and habitats.
- Twenty-one studies from across the world found nest boxes were used (1–3,5–22) by falcons, with one in the UK finding that nest boxes were not used at all (4). One study from Canada (7) found that falcons preferentially nested in nest boxes over natural nest sites; a study from Mauritius (8) found that most breeding attempts were in nest boxes.
- Four studies (3,14,16,18) found that use increased over time. Seven studies found that position (3,5,10,11) or design (7,21,22) affected use, whilst three found no differences between design (11,21) or positioning (19).

Background

There is a large literature on artificial nest use by falcons, falcons are known to use relatively small nest boxes, compared to other birds of prey and some species (e.g. the lesser kestrel *Falco naumanni*) nest communally. Therefore we have separated out studies investigating artificial nest use by falcons from those investigating other birds of prey.

A small study in a pine forest in California, USA (1), found that a pair of prairie falcons *Falco mexicanus* successfully used a nesting ledge installed on a nesting cliff in autumn 1978. Four eggs were laid on the artificial ledge in 1979 and two chicks hatched and fledged. The platform was made of steel, with rock added to it to encourage use and was held in place by expansion bolts.

A replicated study in 1976–80 in juniper and pine forests in the Great Basin, California, USA (2), found that 31% of 208 nest boxes examined were used by American kestrels *Falco sparverius* and that 82% of these (53 nest boxes) successfully fledged at least one chick. Clutches contained an average of four eggs, with an estimated fledging rate of 3.1 chicks/active nest box. Nest boxes were 18 × 20 × 33 cm in size, with a 7.6 cm diameter entrance hole and erected at 2–6 m from the ground in trees. The use of boxes increased year on year, from 20% in 1976 to 38% in 1980.

A replicated study at reclaimed surface mine sites in West Virginia and Pennsylvania, USA (3), found that American kestrels *Falco sparverius* preferentially used nest boxes sited away from a woodland edge; 10 of 65 (15%) woodland edge boxes were used compared with 47% located 50 m or more from a woodland edge. In 1980, 60 nest boxes were erected at 18 mines, and in 1981 a further 91 at 24 mines. Kestrels used 14 (23%) boxes on 10 of 18 (56%) mines in 1980, and 33 of 91 (36%) on 19 of 24 (79%) mines in 1981. Mine sites where boxes were used had significantly less bare ground cover and a deeper litter layer.

A replicated study in two upland pine forests in Wales and England (4) found that Eurasian kestrels *Falco tinnunculus* did not use any of the 41 nest boxes provided between 1973–8. Twenty seven nest boxes were erected on 2 m posts in Wales between 1973 and 1976 and lined with peaty turf, whereas in northern England, 14 boxes of the same design were attached to mature spruce trees (five to six whorls from the top). Only one pair of kestrels bred in the 20 km^2 around the Welsh site, with two pairs breeding in the 10 km^2 around the English site.

A controlled before-and-after study in Missouri, USA (5), found a large increase in nesting and overwinter population densities of American kestrels *Falco sparverius* in a 78 km^2 area, where 125 nest boxes were erected in 1982–3 (0.05 birds/km^2 in 1977–81 vs. 0.32 birds/km^2 in 1984). There was no increase in a 90 km^2 control area, without nest boxes (0.02 birds/km^2 in 1977–81 vs. 0.03 birds/km^2 in 1984), but there was in an urban control area (0.13 birds/km^2 in 1977–81 vs. 0.23 birds/km^2 in 1984), possibly due to increased food availability. Overall, kestrels used 53% of the 125 nest boxes available. Nesting success was significantly higher in boxes mounted on man-made structures (64–78% on buildings and utility poles) than on trees (33%), but still lower than in natural sites (86–88%). However, they produced as many young through double broods and replacement clutches.

A small before-and-after study in a pine forest in northern California, USA (6), found that a pair of peregrine falcons *Falco peregrinus* reproduced successfully following the enlargement of their nesting ledge through the use of ditching dynamite in December 1983. In 1984–8, 13 chicks fledged from the site (2.7 chicks/year), whereas all previous nesting attempts had failed due to eggs and chicks falling from the small ledge. The ledge was 36 m up on a dolomitic limestone cliff and the authors caution that only two of four sticks of dynamite used detonated and, had all four exploded, the ledge may have been destroyed. Previous attempts to enlarge the ledge with hand tools had not worked.

A replicated study in 1988–93 in boreal forests in Saskatchewan, Canada (7), found that American kestrels *Falco sparverius* nested preferentially in large nest boxes over small (81–94% of 66 kestrels in nest boxes nesting in large boxes when given the choice). Nest boxes were also preferred over natural cavities (a maximum of 5–15% of natural cavities used vs. 53–88% of 17–19 nest boxes used each year). There were no differences in reproductive success or predation rates between large and small nest boxes (40–87% success for 54 clutches in large nest boxes vs. 33–86% success for 23 clutches in small nest boxes). Comparisons with natural cavities were not possible due to small sample sizes. The author argues that providing 345 nest boxes over the study period did not increase the local population. Nest boxes had a basal area of 241 cm^2 (small boxes) or 469 cm^2 (large boxes) and a 7.5 cm diameter entrance hole.

A study of an integrated conservation programme for the endangered Mauritius kestrel *Falco punctatus* from 1973–1994 in montane forest habitat and a captive breeding centre in Black River, Mauritius (8) found that nestboxes in areas where natural nest sites were limited were used by released birds, with 90% of 105 documented nesting attempts, during 1988–1989 and 1993–1994, occurring in nestboxes. In the 1993–1994 breeding season, 49% of all monitored wild pairs used nestboxes and several returned to nest in the same ones.

A replicated trial at nine mixed agricultural sites in Iowa, USA (9), found that American kestrels *Falco sparverius* occupied 66% of 56 nest boxes for at least one year between 1989 and 1992, with a maximum of 42% occupied in any one year. Clutches contained an average of 4.4 eggs (49 clutches) and 4.2 chicks fledged on average from each successful box (33 boxes, average of 2.7 chicks/box). These values are similar to previously recorded productivities for American kestrels. Four wooden nest boxes were erected in 1988–9 at each site on pylons, windmills, barns or wooden posts. An additional two PVC boxes were erected in 1990 at each site, plus a final two at one of the sites.

A replicated trial in 1987–91 in mixed agricultural habitats and woodland in Pennsylvania, USA (10), found that American kestrels *Falco sparverius* used 76% of 130 nest boxes at least once over the five-year study period, with 49% of 259 nesting attempts raising at least one offspring. Kestrels most frequently used unconcealed nest boxes in open habitats away from forested areas, and with a lot of light entering. Nest boxes with south-east orientations were used most frequently. Nesting success was also higher in nest boxes with high light intensities. Nest boxes were 26 × 24 × 33 cm with a 7.6 cm diameter entrance hole and were attached 2.0–6.5 m above the ground.

A replicated study in agricultural sites in southern Finland (11) found that Eurasian kestrels *Falco tinnunculus* occupied 18–22% of 161 nest boxes between 1985 and 1995, with no differences between small, intermediate and large boxes. Boxes sheltered from prevailing weather were more likely to be occupied than exposed boxes (25% of 80 sheltered boxes used vs. 17% of 81 exposed boxes). There were no significant differences in clutch size or number of fledglings produced between nest boxes types and orientations, with success related to laying date and vole abundance. Occupied boxes were, on average, further from forest edges, roads and inhabited houses, and closer to grassy ditches than unoccupied boxes. Boxes were 25 × 27.5 × 25 cm, with a 12.5 × 25 cm entrance (small); 34 × 35 × 20 cm, with a 12 × 34 cm entrance (intermediate); or 33.5 × 45 × 30 cm, with a 12 × 33.5 cm hole (large).

A replicated study in Badajoz, Spain, in 1989 (12), found that European kestrels *Falco tinnunculus* used 16% of 567 nest boxes placed in seven agricultural and woodland habitats. There were no significant differences in laying date or productivity between habitats. However, when only habitats with more than 15 occupied boxes were analysed, nests in pastures were found to have significantly larger clutch and higher breeding success than those in cereal fields (4.4 eggs/clutch and 4.2 fledglings/nest for 39 nests in pastures vs. 3.7 eggs/clutch and 3.5 fledglings/nest for 19 nests in cereal fields). Nest boxes were erected on power pylons across the habitat types in spring 1989.

A before-and-after study in mixed farmland and oak woodlands in Avila and Segovia, central Spain (13), found that the local population of Eurasian kestrels *Falco tinnunculus* more than doubled between 1993 (23 pairs) and 1998 (55 pairs) following the installation of 47 nest boxes over the same period. The number of kestrels in natural nests remained approximately constant (15–25 pairs), whilst the number in nest boxes increased from three (1993) to 35 (1998). Birds in nest boxes fledged more chicks and experienced less nest predation than those in natural sites (3.6–3.8 fledglings/clutch and 12% predation for 79 nest box clutches vs. 2.4–2.8 fledglings/clutch and 37% predation for 37 clutches in natural nests). Nest box chicks had more ectoparasites, but this difference was not significant. Nest boxes were installed in winter: 14 in 1993–4, 11 in 1994–5, 16 in 1996–7 and six in 1998.

A small study in mixed farmland and woodlands in Alentejo, Portugal (14), found that lesser kestrels *Falco naumanni* used 25% of 36 nesting cavities in two 'breeding towers' in 2003. The towers were constructed in 1997 and 1999 but not occupied until 2002, after modifications were made to the nest chambers to create an enlarged nest cavity. Three pairs bred successfully in 2002 in addition to common kestrels *F. tinnunculus*, rollers *Coracias garrulus*, barn owls *Tyto alba* and jackdaws *Corvus monedula*.

A replicated study in a 1500 km^2 area of mixed deciduous forests in Pennsylvania, USA (15), found that American kestrels *Falco sparverius* used an average of 86 nest boxes each year between 1993 and 2002 (32% of the approximately 270 boxes in the area). Pairs laid an average of 4.6 eggs/clutch and fledged 2.7 nestlings/box (171 boxes monitored). First breeding attempts were successful 69% of the time. These productivity levels are similar to those recorded elsewhere. Boxes were 26 × 24 × 33 cm, with a 7.6 cm diameter entrance hole. They were erected 3–6 m off the ground (usually on trees or utility poles, but sometimes on sheds and barns).

A before-and-after study in mixed farmland and woodlands in Alentejo, Portugal (16), found that the local population of lesser kestrels *Falco naumanni* increased by 36% between 2003 and 2006, following the provision of over 450 artificial nest sites over the same period. The number of pairs nesting in artificial sites increased from 29 in 2003 (150 nests available and a total local population of 268 birds) to 121 in 2006 (450 nests and 364 birds in total). Nests included nest boxes, clay pots and multi-cavity 'breeding walls' and 'breeding towers' (discussed in (14)).

A replicated, controlled study in mixed agricultural habitats in the North District of Israel, in 1999–2006 (17), found that Eurasian kestrel *Falco tinnunculus* nesting in small nest boxes produced more chicks than those in large boxes (3.0 chicks/clutch for 37 in small boxes vs. 1.9 chicks/clutch for 44 clutches in large), with no differences between boxes and natural nests (2.1 chicks/clutch for 56 attempts). Large boxes had higher failure rates (48% of 44 attempts) compare to small (20% of 37) and natural nests (20% of 56 in natural nests). When only successful nests were analysed, all boxes fledged more chicks than natural nests (3.6–3.9 chicks/clutch for 52 clutches in nest boxes vs. 2.7 chicks/clutch for 44 in natural nests). Boxes were either: 50 × 75 × 50 cm with a 25 × 15 cm entrance hole and mounted 2.5–3.0 m above ground or 50 × 30 × 30 cm with a 22 × 15 cm hole and 5–6 m above ground. Sixty large and eleven small boxes were erected.

A replicated, controlled trial in five towns in Apulia, southern Italy (18), found that lesser kestrels *Falco naumanni* nested in artificial nest boxes in the first two years after installation. Two hundred nest boxes were placed on flat roofs in 2007, 8% were occupied in 2007 and 17.5% used in 2008. Nest box breeders fledged an average of 1.8 chicks/clutch in 2007 (17 clutches), similar to pairs nesting in attics (1.7 chicks/clutch for 18 clutches) but significantly lower than pairs nesting in wall cavities (2.7 chicks/clutch for ten clutches). Productivity was lower in 2008 (1.5 chicks fledged/clutch, 35 clutches) but no comparison was possible with other nest types.

A replicated study in 2005 in sagebrush steppe and agricultural fields in Idaho, USA (19), found that 71% of 59 nest boxes were occupied by American kestrels *Falco sparverius*. Box orientation did not significantly affect occupancy rates (although no east-facing boxes were occupied) but did affect hatching success (43% of 21 south-west-facing nest boxes unsuccessful vs. 25% of 12 south-east-facing boxes and 0% for nine facing north-west). West-facing boxes were approximately 0.6°C cooler on average than east- and south-facing boxes, and also less humid. Nest boxes were 21 × 21 × 46 cm and erected at 2.5–3.0 m off the ground on utility poles.

A controlled before-and-after study in 1989–93 in Florida, USA (20), found that the population of south-eastern American kestrels *Falco sparverius paulus* in a 3,600 km^2 experimental area of dry mixed forests and agricultural land increased following the installation of 388 nest boxes in 1990–3 (5.0 birds/100 km^2 in 1989 vs. 32.3 birds/100 km^2 in 1992). There was no corresponding increase in a similar area without boxes (34.4 birds/100 km^2 in 1989 vs. 34.9 birds/km^2 in 1992). The number of boxes used increased each year, reaching 158 in 1993 and a total of 365 nesting attempts (39 of which were re-nesting). Nesting success averaged 67%, with 2.4 fledglings/nest. This is relatively low compared with previously recorded productivities. Boxes had a base of 19.7 × 23.5 cm, with a 8.9 cm diameter entrance hole.

A replicated study in eucalyptus stands in farmland in Lower Galilee, Israel (21), in 2008–9, found that Eurasian kestrels *Falco tinnunculus* nested with equal frequency and equal success in nest baskets of two different sizes (13 of 76 nests used, average of 1.8 chicks fledged/clutch for six clutches in small baskets vs. 2.5 chicks/clutch for six in large baskets). Overall productivity in this study was lower than previously recorded in nest boxes in the same region (2.2 fledglings/breeding attempt vs. 3.2 fledglings/attempt in previous studies). Nest baskets were metal bowls filled with coconut fibre and were either 30 cm in diameter and 16 cm deep (small) or 40 cm in diameter and 20 cm deep (large). The positions of large and small nests exchanged in 2008. The study also discusses nest box use by long-eared owls *Asio otus*.

A replicated study in eucalyptus stands in farmland in Lower Galilee, Israel (22), in 2008–9, found that Eurasian kestrels *Falco tinnunculus* nested more frequently in nest boxes with large entrance holes than in boxes with small holes (17% of 51 large-entrance nest boxes occupied vs. approximately 8% of 49 small-entrance boxes). Breeding success of kestrels did not vary between nest box types although 22% of kestrels in boxes with large holes abandoned them because of barn owl *Tyto alba* interference. There was no such interference in small nest boxes. Nest boxes were 50 × 75 × 50 cm with either 15 × 30 cm (large) or 7.5 cm diameter (small) entrances. In 2008, 27 large and 25 small boxes were erected, with 24 of each in 2009. The positions of large and small boxes were exchanged between years. This study also discusses nest box use by owls and songbirds.

(1) Boyce Jr, D. A., Fisher, L., Lehman, W. E., Hipp, B. & Peterson, J. (1980) Prairie falcons nest on an artificial ledge. *Raptor Research*, 14, 46–50.
(2) Bloom, P. H. & Hawks, S. J. (1983) Nest box use and reproductive biology of the American kestrel in Lassen County, California. *Raptor Research*, 17, 9–14.
(3) Wilmers, T. J. (1983) Kestrel use of nest boxes on reclaimed surface mines in West Virginia and Pennsylvania.
(4) Petty, S. J. (1985) A negative response of kestrels *Falco tinnunculus* to nestboxes in upland forests. *Bird Study*, 32, 194–195.
(5) Toland, B. R. & Elder, W. H. (1987) Influence of nest-box placement and density on abundance and productivity of American Kestrels in central Missouri. *The Wilson Bulletin*, 99, 712–717.
(6) Pagel, J. E. (1989) Use of explosives to enhance a peregrine falcon eyrie. *Journal of Raptor Research*, 23, 176–178.
(7) Bortolotti, G. R. (1994) Effect of nest-box size on nest-site preference and reproduction in American kestrels. *Journal of Raptor Research*, 28, 127–133.
(8) Jones, C. G., Heck, W., Lewis, R. E., Mungroo, Y., Slade, G. & Cade, T. (1995) The restoration of the Mauritius kestrel *Falco punctatus* population. *Ibis*, 137, S173–S180.
(9) Craft, R. A. & Craft, K. P. (1996) Use of free ranging American kestrels and nest boxes for contaminant risk assessment sampling: a field application. *Journal of Raptor Research*, 30, 207–212.
(10) Rohrbaugh Jr, R. W. & Yahner, R. H. (1997) Effects of macrohabitat and microhabitat on nest-box use and nesting success of American kestrels. *The Wilson Bulletin*, 109, 410–423.
(11) Valkama, J. & Korpimaki, E. (1999) Nestbox characteristics, habitat quality and reproductive success of Eurasian kestrels. *Bird Study*, 46, 81–88.
(12) Aviles, J. M., Sanchez, J. M. & Parejo, D. (2001) Breeding rates of Eurasian kestrels (*Falco tinnunculus*) in relation to surrounding habitat in southwest Spain. *Journal of Raptor Research*, 35, 31–34.
(13) Fargallo, J. A., Blanco, G., Potti, J. & Vinuela, J. (2001) Nestbox provisioning in a rural population of Eurasian kestrels: breeding performance, nest predation and parasitism. *Bird Study*, 48, 236–244.
(14) Franco, A. M., Marques, J. T. & Sutherland, W. J. (2005) Is nest-site availability limiting lesser kestrel populations? A multiple scale approach. *Ibis*, 147, 657–666.
(15) Katzner, T., Robertson, S., Robertson, B., Klucsarits, J., McCarty, K. & Bildstein, K. L. (2005) Results from a long-term nest-box program for American kestrels: implications for improved population monitoring and conservation. *Journal of Field Ornithology*, 76, 217–226.
(16) Catry, I., Alcazar, R. & Henriques, I. (2007) The role of nest-site provisioning in increasing lesser kestrel *Falco naumanni* numbers in Castro Verde Special Protection Area, southern Portugal. *Conservation Evidence*, 4, 54–57.
(17) Charter, M., Izhaki, I., Bouskila, A. & Leshem, Y. (2007) The effect of different nest types on the breeding success of Eurasian kestrels (*Falco tinnunculus*) in a rural ecosystem. *Journal of Raptor Research*, 41, 143–149.
(18) Bux, M., Giglio, G. & Gustin, M. (2008) Nest box provision for lesser kestrel *Falco naumanni* populations in the Apulia region of southern Italy. *Conservation Evidence*, 5, 58–61.
(19) Butler, M. W., Whitman, B. A. & Dufty, A. M. (2009) Nest box temperature and hatching success of American kestrels varies with nest box orientation. *The Wilson Journal of Ornithology*, 121, 778–782.
(20) Smallwood, J. A. & Collopy, M. W. (2009) Southeastern American kestrels respond to an increase in the availability of nest cavities in north-central Florida. *Journal of Raptor Research*, 43, 291–300.
(21) Charter, M., Izhaki, I. & Leshem, Y. (2010) Does nest basket size affect breeding performance of long-eared owls and Eurasian kestrels? *Journal of Raptor Research*, 44, 314–317.
(22) Charter, M., Izhaki, I. & Leshem, Y. (2010) Effects of the risk of competition and predation on large secondary cavity breeders. *Journal of Ornithology*, 151, 791–795.

15.4.13. Owls

- Three studies from the UK (6,7,11) appeared to show increases in local populations of owls following the installation of artificial nests, although the authors from one note that they could not rule out birds merely switching from natural nest sites (11). Another UK study found that providing nesting sites when renovating buildings maintained barn owl *Tyto alba* populations, whilst they declined at sites without nests (12).
- Four studies from the USA (2,5) and the UK (7,11) found high levels of breeding success in artificial nests, three finding equal or higher productivity than natural nests (2,5,7). A replicated, controlled study from the USA (10) found lower productivity from artificial nests, whilst a replicated, controlled study from Finland (9) found that artificial nests were only successful in the absence of larger owls and a replicated, controlled study from Hungary (16) found that fledglings from artificial nests were less likely to be found alive after one year.

- Four studies from the USA (1) and Europe (3,4,6) found that artificial nests were used at least as frequently as natural nesting sites. Five studies from across the world (2,8,9,14,15) found that owls used artificial nests, with one finding that use increased over time, although only for one of two species (14).
- Three studies found that owls differentiated between nests in different positions (1,3,15), whilst five studies found that different designs of nests differed in occupancy (3,4,13,17) or productivity (18). Three studies found occupancy did not differ between designs (5,11,18) and two found no differences in productivity for different designs (5,13).

A controlled study 1977–9 in riverine forests in Louisiana, USA (1), found that Eastern screech owls *Megascops asio* (formerly *Otus asio*) used nest boxes more frequently than natural cavities (0.7% of 5,374 nest boxes inspected contained nests vs. 0.03% of 3,993 natural cavities). Frequently used nests faced north and were situated under tree limbs, in trees with lianas. Barred owls *Strix varia* also used nest boxes, but at very low frequencies. This study also examined nest box use by other birds (wildfowl, woodpeckers and songbirds).

A short 1984 review of several nest box programmes in the USA (2) found that barred owls *Strix varia* appeared to successfully use nest boxes provided in a range of habitats. Owls nesting in nest boxes in Minnesota produced more chicks than those in natural nests (2.75 fledglings/nesting attempt for 12 attempts in nest boxes vs. 2.00 fledglings/nesting attempt for six attempts in natural nests); in Wisconsin, two nest boxes produced ten chicks over four years; in Michigan, a single box produced young in three of the four years it was monitored (1979–82). Boxes varied in design, but most were modified wood duck boxes, approximately 33 × 34 × 36 cm, with 18–22 cm entrance holes.

A replicated, controlled study in 1970–83 in boreal forests in Hedmark, Norway (3), found that three species of owl appeared to nest preferentially in nest boxes, compared to natural cavities. Pygmy owls *Glaucidium passerinum* showed the weakest preference (55% of 20 nests were in nest boxes), followed by hawk owls *Surnia ulula* (75% of 12 nests in boxes) and Tengmalm's owls *Aegolius funereus* (97% of 167 nests in boxes). The number of nesting cavities available is not recorded. Tengmalm's owls used boxes on isolated trees in clearcuts most, and those closed mature forest the least. Pygmy owl boxes were smaller (with a 45 mm entrance hole) than other boxes (with a 58 mm entrance) and only one (5%) was predated, compared to 69 (37%) Tengmalm's owl clutches and four (33%) hawk owl clutches.

A replicated, controlled study in 1979–85 in west Finland (4) found that Tengmalm's owls *Aegolius funerus* nested in nest boxes at least as frequently as natural cavities (3–12% of nest boxes used, depending on design vs. 8.5% of natural cavities). Small boxes (internal diameter <26 cm, entrance holes <15 cm) were used more frequently than larger nest boxes (10–12% of 894 small boxes used vs. 3% of 165 larger boxes). Boxes made from hollowed logs (17–20 cm internal diameter, 8–10 cm entrance hole) and natural cavities were used less than small nest boxes, but not significantly so (7% of 677 log boxes and 9% of 177 natural cavities used). The study area was increased each year of the study, until it reached 1,300 km^2, with 450 nest sites. A total of 1,736 nest boxes were used and 177 natural cavities searched.

A replicated, controlled study in 1967–75 in urban woodland in central Texas, USA (5), found that eastern screech owls *Megascops asio* (formerly *Otus asio*) had equal nesting success in nest boxes as in natural nests (average of 3.9 eggs/clutch and 51% of eggs producing fledglings in nine nest boxes vs. 3.8 eggs/clutch and 57% of eggs producing fledglings in nine natural nests). The wood used in nest boxes (plywood, pine or cedar) and nest box size did not appear to affect use or reproductive success. Nest boxes had basal areas of 225, 400 or 625 cm^2, with a 6.8 cm entrance hole 25 cm above the base. Boxes were erected 3–4 m above ground on large trees.

A before-and-after study at a 150 km^2 in Norfolk, England (6), found that barn owl *Tyto alba* population density increased from 15 pairs/100 km^2 in 1989 to 27 pairs/100 km^2 in 1993, following the provision of 60 nest boxes. Nest boxes were used at the same rate as natural nest sites, and pairs using boxes in trees produced more eggs (but not significantly more fledglings) than other nest types. Nest boxes were located in buildings (43 boxes, a maximum of 11 used in a single year) and on trees (17 boxes, a maximum of five used in a single year).

Two before-and-after studies in pine forests in the UK (7) found local population increases in tawny owls *Strix aluco* and barn owls *Tyto alba* following the provision of nest boxes, although the authors note that the tawny owl population may have responded to an increased food supply. In 1980–91, 90–160 boxes were erected in an area of north-east England. All local birds used boxes by 1983 and the population increased from 40 to 66 pairs. At a site in south-west Scotland, 33–87 nest boxes were provided for barn owls in 1984–90. Resident birds did not move nest sites, but new breeders moved into the area and used boxes (37 pairs using boxes in 1988), increasing the population from five to approximately 42 pairs by 1993.

A replicated study in California, USA, between 1988 and 1993 (8) found that 18 burrowing owls *Athene cunicularia* that were 'evicted' (using one-way doors) from their original burrows at five grassland sites, apparently occupied artificial burrows created 7–75 m away from original burrows. A pair provided with three burrows 165 m from their original burrow did not use them. The authors note that owls were not ringed, so those in artificial burrows could not be confirmed as the evicted birds.

A replicated and controlled trial in 1989–94 in boreal forests in Central Finland, Finland (9), found that Tengmalm's (boreal) owls *Aegolius funereus* successfully used nest boxes provided. However, of 15 possible breeding attempts in Ural owl *Strix uralensis* territories, only four were made (27%) and all failed. In contrast, 15 of 20 possible attempts in eagle owl *Bubo bubo* territories were made (75%) and all but two were successful. The presence of Ural owls was therefore found to significantly reduce both the probability of occupation and the chances of success for the smaller Tengmalm's owl.

A replicated, controlled trial in 1993–5 in arid shrubland in New Mexico, USA (10), found that burrowing owls *Athene cunicularia* (formerly *Speoty cunicularia*) nesting in artificial burrows produced significantly more nestlings, but significantly fewer fledglings than pairs in natural burrows (3.5 nestlings/pair and 1.5 fledglings/pair for eight pairs in artificial burrows vs. 2.2 nestlings/pair and 1.9 fledglings/pair for 59 natural burrows). Only 12 of 28 nestlings (43%) in artificial nests survived to fledging, with most being predated or cannibalised. Artificial burrows were constructed from a 19 l plastic bucket buried and connected to the surface with 5 m of 10 cm diameter PVC pipes. Both bucket and pipes had holes drilled in to ensure drainage.

A replicated study in 1981–96 in a reed-dominated wetland site in Cambridgeshire, England (11), found that long-eared owls *Asio otus* readily used two designs of wicker baskets, with 77 nesting attempts over the study period. Of the 71 nest monitored, 42 (59%) hatched eggs and 36 (51%) fledged at least one chick. Between one and nine baskets were used each year, with three to 23 baskets available. It was not possible to confirm whether the apparent population increase was genuine or caused by owls switching from natural nest sites. Baskets were either local 'fruit-picker' baskets or dog baskets (30 cm diameter and 15 cm deep) and replaced every 4–5 years. Baskets were placed in trees, mostly hawthorn, 3.5–5.0 m above the ground.

A small controlled study in 1990–3 in Devon and Cornwall, England (12), found that activity in buildings used by barn owls as nesting and/or roosting sites dropped by 68% in nine areas following the conversion or demolition of the building, but was maintained in three other areas where a cavity and access hole were incorporated into the conversion or another nearby (<50 m away) building. There were no changes in eight control areas.

A randomised, replicated study in prairie, shrubland and farmland in south-west Idaho, USA, in 1997–8 (13) found that western burrowing owls *Athene cunicularia hypugaea* preferentially used artificial burrows with large (1,750 cm^3) nest chambers, compared to small or medium (707 cm^3 or 900 cm^3) chambers (31 large chambers selected vs. six medium and seven small, a total of 81 burrows of each type available). Burrows with small (10 cm diameter) tunnels were also preferred, compared to those with large (15 cm diameter) tunnels (30 burrows with small tunnels occupied vs. 14 with large burrows, 72 of each type available). However, there were no differences in reproductive output between nest types. Burrows were arranged in clusters containing all burrow types and designed to resemble natural nests. Chambers were lined with soil and natural burrows nearby were blocked with rocks to ensure owls used artificial burrows.

A replicated study in boreal forests in Gansu, China (14), found that boreal owls *Aegolius funereus*, but not Sichuan wood owls *Strix uralensis* (formerly *S. davidi* also known as Ural owls) used nest boxes provided in 2002–4. A total of 120 nest boxes were provided, 50 for boreal owls and 70 for Sichuan wood owls. No boreal owls used nest boxes in 2002, four bred in 2003 (fledging eight chicks from three nests) and six bred in 2004 (fledging eight chicks from four clutches). Several species of songbird also use boxes. Boxes were erected at least 3 m up on tree trunks, lined with grass and mosses and of the same design as used in Europe for Ural owls. From 2003, some were encased in bark to make them appear more natural.

A replicated study on four agricultural sites in north-east Arkansas, USA (15), found that, in 2001, barn owls *Tyto alba* nested in four of 48 nest boxes erected in 2000, making up 29% of nests in the area. All occupied boxes were on a single site with a high owl density and were on artificial structures, although boxes were equally distributed on trees and artificial structures. No nest boxes were used in 2000, possibly because they were erected after birds had settled for the year (boxes erected between January and March). Boxes were made from 39 cm diameter PVC piping with the ends blocked (one only partially, leaving half the end as an entrance hole) and drainage holes drilled in. Boxes were placed 2.5–6.8 m above the ground.

A replicated, controlled study in Hungary in 1995–2003 (16) using ring-recapture data found that juvenile barn owls *Tyto alba* fledged from nest boxes were significantly less likely to be recovered alive than those reared in church towers (25% of 75 nest box-reared birds recovered alive one year after fledging vs. 40% of 116 church tower-raised birds). This difference in survival was only apparent in the first year after fledging, with similar proportions being recovered six years after fledging (28% of nest box-reared birds vs. 41% of church tower-raised birds).

A replicated study in eucalyptus stands in farmland in Lower Galilee, Israel (17), in 2008–9, found that barn owls *Tyto alba* nested in 67% of 51 nest boxes with large (15 × 30 cm) entrance holes, but none of 49 nest boxes with small (7.5 cm diameter) entrances. In contrast, common scops owls *Otus scops* only nested in boxes with small entrances, using approximately 10% of 49 small entrance boxes, but no large entrance ones, possibly due to competition with the larger barn owls. Boxes were attached to shaded parts of eucalyptus trees and the positions of large and small-entranced boxes were exchanged between years. This study also investigates nest box use by kestrels and songbirds.

A replicated study in eucalyptus stands in farmland in Lower Galilee, Israel (18), in 2008–9, found that long-eared owls *Asio otus* fledged more chicks from small nest baskets, compared to large ones (1.25 fledglings/breeding attempt for eight attempts in large baskets vs. 3.7 fledglings/attempt for seven in small baskets). Differences in clutch size and hatching success were non-significant (3.3 eggs/clutch and 60% hatching success for large nests vs. 5.0 eggs/clutch and 64% hatching success for small nests) and owls occupied nest types with equal frequency. Nest baskets were metal bowls filled with coconut fibre and were either 30 cm in diameter and 16 cm deep (small) or 40 cm in diameter and 20 cm deep (large). Thirty eight of each type of nest were erected in 2007, with the positions of large and small nests exchanged in 2008. The study also discusses nest box use by Eurasian kestrels *Falco tinnunculus*.

(1) McComb, W. C. & Noble, R. E. (1981) Nest-box and natural-cavity use in three mid-south forest habitats. *The Journal of Wildlife Management*, 45, 93–101.
(2) Johnson, D. H. & Follen, Sr, D. G. (1984) Barred owls and nest boxes. *Raptor Research*, 18, 34–35.
(3) Sonerud, G. A. (1985) Risk of nest predation in three species of hole nesting owls: influence on choice of nesting habitat and incubation behaviour. *Ornis Scandinavica*, 16, 261–269.
(4) Korpimäki, E. (1987) Selection for nest-hole shift and tactics of breeding dispersal in Tengmalm's owl *Aegolius funerus*. *Journal of Animal Ecology*, 56, 185–196.
(5) Gehlbach, F. R. (1994) Nest-box versus natural-cavity nests of the eastern screech-owl: an exploratory study. *Journal of Raptor Research*, 28, 154–157.

(6) Johnson, P. N. (1994) Selection and use of nest sites by barn owls in Norfolk, England. *Journal of Raptor Research*, 28, 149–153.
(7) Petty, S. J., Shaw, G. & Anderson, D. I. K. (1994) Value of nest boxes for population studies and conservation of owls in coniferous forests in Britain. *Journal of Raptor Research*, 28, 134–142.
(8) Trulio, L. A. (1995) Passive relocation: a method to preserve burrowing owls on disturbed sites. *Journal of Field Ornithology*, 99–106.
(9) Hakkarainen, H. & Korpimäki, E. (1996) Competitive and predatory interactions among raptors: An observational and experimental study. *Ecology*, 77, 1134–1142.
(10) Botelho, E. S. & Arrowood, P. C. (1998) The effect of burrow site use on the reproductive success of a partially migratory population of western burrowing owls. *Journal of Raptor Research*, 32, 233–240.
(11) Garner, D. J. & Milne, B. S. (1998) A study of the long-eared owl *Asio otus* using wicker nesting baskets. *Bird Study*, 45, 62–67.
(12) Ramsden, D. J. (1998) Effect of barn conversions on local populations of barn owl *Tyto alba*. *Bird Study*, 45, 68–76.
(13) Smith, B. W. & Belthoff, J. R. (2001) Effects of nest dimensions on use of artificial burrow systems by burrowing owls. *The Journal of Wildlife Management*, 65, 318–326.
(14) Fang, Y. (2005) Conservation action on the endemic owls at Lianhuashan Mountains. *Final Report of Project OWL 2002*.
(15) Radley, P. M. & Bednarz, J. C. (2005) Artificial nest structure use and reproductive success of barn owls in northeastern Arkansas. *Journal of Raptor Research*, 39, 74–79.
(16) Klein, Á., Nagy, T., Csörgő, T. & Mátics, R. (2007) Exterior nest-boxes may negatively affect barn owl *Tyto alba* survival: an ecological trap. *Bird Conservation International*, 17, 273–281.
(17) Charter, M., Izhaki, I. & Leshem, Y. (2010) Effects of the risk of competition and predation on large secondary cavity breeders. *Journal of Ornithology*, 151, 791–795.
(18) Charter, M., Izhaki, I. & Leshem, Y. (2010) Does nest basket size affect breeding performance of long-eared owls and Eurasian kestrels? *Journal of Raptor Research*, 44, 314–317.

15.4.14. Oilbirds

- A before- and after-study in Trinidad and Tobago (1) found an increase in size of an oilbird colony following the creation of artificial nesting ledges.

Background

Oilbirds *Steatornis capripensis* are a unique, nocturnal species in their own family Steatornithidae, related to nightjars, potoos and frogmouths. However, unlike their relatives oilbirds feed entirely on fruit and nest and roost in caves in northern and central South America. It is thought that the availability of nesting and roosting sites in these caves may limit populations, although increasing habitat destruction is also threatening the species (Thomas 1999).

Thomas, B. T. (1999) Family Steatornithidae (oilbird). *Handbook of the birds of the world: Volume 5, Barn owls to hummingbirds* (eds Hoyo, J., Elliott, A. & Sargatal, J.), pp. 244–252. Lynx Edicions, Barcelona.

A before- and after-study at a colony of oilbirds *Steatornis capripensis* nesting in a cave at the Asa Wright Nature Center, Trinidad (1), found that the population increased from 25–30 birds to more than 100 individuals and 43 nests by 1977 following the installation of artificial concrete ledges in 1967–1968. Before this, the colony appeared limited in size by the 15 or 16 ledge nest sites available. In 1977, 21 of the nests were on artificial ledges.

(1) Lambie, I. (1993) Good news about oilbirds. *Bellbird*, 2.

15.4.15. Pigeons

- Two replicated studies from the USA (1) and the Netherlands (2) found high use rates and high nesting success of pigeons and doves using artificial nests.

A replicated study in 1945–6 in garden habitats in Ohio, USA (1), found that 63% of 31 mourning dove *Zenaida macroura* (formerly *Zenaidura macroura*) nesting attempts in artificial nests were successful. It was not possible to compare this success with natural nests, because the author was not confident they had found all unsuccessful natural nests. Nests consisted of cones of black or green roofing paper 17.8 cm at the widest and 5.1 cm deep. This study also examines nest use by American robins *Turdus migratorius*, and the effect of different coloured nests in 'Use differently-coloured artificial nests'.

A replicated study in 1988–2000 in Noord-Brabant, The Netherlands (2), found that stock pigeons *Columba oenas* used nest boxes provided in mixed agricultural habitats, laying an average of 118 eggs laid/year with 52% hatching and 84% of chicks fledging (an average of 52 chicks/year). Boxes were 20 × 20 × 50 cm, with an 8 × 8 cm entrance hole and placed 3–5 m above the ground in trees, 20–30 m apart. Jackdaws *Corvus monedula* also used the nest boxes, but were removed from 1995 onwards.

(1) Calhoun, J. B. (1948) Utilization of artificial nesting substrate by doves and robins. *The Journal of Wildlife Management*, 12, 136–142.
(2) Potters, H. (2009) Broedbiologie van een kleine populatie nestkastbewonende Holenduiven in westelijk Noord-Brabant [Breeding biology of a small population of stock pigeon *Columba oenas* in North-Brabant]. *Limosa*, 82, 1–12.

15.4.16. Trogons

- A small study from Guatemala (1) found that at least one resplendent quetzal *Pharomachrus mocinno* nested in nest boxes provided.

A small study in a tropical cloud forest in south-west Guatemala (1) found that, in 1973, at least one, and possibly more, resplendent quetzals *Pharomachrus mocinno* nested in eight nest boxes installed between 1968 and 1973. A total of 16 nest boxes were installed.

(1) LaBastille, A. (1974) Use of artificial nest-boxes by quetzals in Guatemala. *Biological Conservation*, 6, 64–65.

15.4.17. Rollers

- A before-and-after study from Spain (2) found that the use of nest boxes by European rollers *Coracias garrulous* increased over time and that use varied between habitats.
- A replicated controlled trial from Spain (1) found no difference in success rates between new and old nest boxes, although birds in old boxes began nesting earlier.

A replicated, controlled study in pastureland in 1989–90 in Extramadura, Spain (1), found that European rollers *Coracias garrulus* nesting in new nest boxes did not have higher reproductive success than those nesting in older nest boxes (average of 4.3 eggs/clutch, 69% breeding success and 4.0 fledglings/successful nest for 16 pairs in old boxes vs. 4.1 eggs/clutch, 83% and 3.6 fledglings/nest for 49 pairs in new boxes) and began laying significantly later (average laying date of 25th May for clutches in 16 old boxes vs. 30th May for 49 new boxes). Rollers did not appear to preferentially use either old or new boxes. Boxes were 32 × 18 × 19 cm with a 6 × 18 cm entrance hole and were installed on power line pylons at an average density of 9.5 boxes/km of power line.

A before-and-after study in agricultural habitats in Extramadura, Spain (2), found that the number of European rollers *Coracias garrulus* using artificial nest boxes increased from 29 in 1988 (76% of the 38 boxes available) to 350 in 1991 (55% of 641 available boxes). Nest boxes use varied with habitat: from 68% use in unwooded pasture (256 boxes available) to only 34% in cereal fields with holm oaks (32 boxes). Nest boxes were the same design as in (1).

(1) Aviles, J. M., Sanchez, J. M. & Parejo, D. (2000) The roller *Coracias garrulus* in Extremadura (southwestern Spain) does not show a preference for breeding in clean nestboxes. *Bird Study*, 47, 252–254.
(2) Avilés, J. M. & Parejo, D. (2004) Farming practices and roller *Coracias garrulus* conservation in southwest Spain. *Bird Conservation International*, 14, 173–181.

15.4.18. Swifts

- A study from the USA (1) found that Vaux's swifts *Chaetura vauxi* successfully used nest boxes provided.

A study in 1999–2002 in pine forests in Oregon, USA (1), found that Vaux's swifts *Chaetura vauxi* successfully nested in nest boxes, with 53% of 51 nesting attempts successful and fledging an average of 3.5 chicks/nest. A total of 103 nest boxes were erected, with 12–17 used each year and a total of 30 boxes used. Boxes had a 30 × 30 × 350 cm with a 10 × 15 cm entrance hole and attached 10–15 m above the ground in trees.

(1) Bull, E. L. (2003) Use of nest boxes by Vaux's swifts. *Journal of Field Ornithology*, 74, 394–400.

15.4.19. Woodpeckers

- Four studies from the USA (5,7,10,11) found local increases in red-cockaded woodpecker *Picoides borealis* populations or the successful colonisation of new areas following the installation of 'cavity inserts' (described above). One study (10) also found that the productivity of birds using the inserts was significantly higher than the regional average.
- Two studies from the USA (5,6) found that red-cockaded woodpeckers *Picoides borealis* used cavity inserts, in one case more frequently than making their own holes or using natural cavities (5). One study from the USA (1) found that woodpeckers roosted, but did not nest, frequently in nest boxes.
- Five studies from the USA found that some woodpeckers excavated holes in artificial snags (2–4,7,8) but only ever roosted in excavated holes or in nest boxes provided.
- A small study in the USA (9) found that modifying artificial nests to allow easy access did not alter the behaviour of birds using them.

Background

Woodpeckers typically nest in hollows in trees. These are normally hollowed out from soft wood. Therefore, if this substrate is lacking in a site, providing it may be beneficial. Providing dead wood and 'snags' is described in 'Natural system modifications', whilst providing artificial snags, typically made of polystyrene cylinders is described below.

Alternatively, 'cavity inserts' can be provided. These consist of a nest box inserted into a tree. Full details of their installation can be found in Allen (1991), but basically consist of cutting a 10 × 25 × 6 cm block out of a live tree and gluing a nest box into place to give the impression of a tree with a hole and a nesting cavity. These inserts are used most frequently to provide nesting habitats for red-cockaded woodpeckers *Picoides borealis*.

Allen, D. H. (1991) An insert technique for constructing artificial red-cockaded woodpecker cavities. *Southeastern Forest Experiment Station General Technical Report SE-73*, United States Forest Service.

A controlled study 1977–9 in riverine forests in Louisiana, USA (1), found that neither northern flickers *Colaptes auratus* nor red-bellied woodpeckers *Melanerpes carolinus* (formerly *Centurus carolinus*) nested with any frequency in nest boxes provided, with only a single woodpecker nest found in 5,374 inspections of 235 boxes. Both species, as well as red-headed woodpeckers *M. erythrocephalus* and hairy woodpeckers *Picoides villosus* (formerly *Dendrocopus villosus*) used nest boxes for roosting. Boxes were of three sizes between 30 × 15 × 15 cm with a 5.0 × 7.0 cm entrance hole and 60 × 30 × 30 cm with a 13 cm diameter entrance. All boxes had 5–10 cm of pine shavings in the bottom. This study also examined nest box use by other birds (wildfowl, owls and songbirds).

A replicated trial in 1980 in woodlots in Ohio, USA (2), found that downy woodpeckers *Picoides pubescens* were more likely to excavate cavities in artificial snags of intermediate height (242 cm tall, ten of 16 snags used), compared to tall (363 cm tall, five of 16 used) or small (121 cm tall, one of 16 used) snags. There was some evidence that males preferentially excavated holes in intermediate or tall snags, whilst females preferred small or intermediate ones. Snags were polystyrene cylinders, 22.5 cm in diameter, painted brown and mounted on metal poles.

A replicated trial in 1979–80 in a deciduous forest in Ohio, USA (3), found that downy woodpeckers *Picoides pubescens* excavated 51 roosting cavities in 42 artificial snags. Raccoons *Procyon lotor* destroyed 18 cavities, with woodpeckers excavating new holes near nine of these. Two species of songbird used cavities excavated by woodpeckers (see 'Provide artificial nesting sites for songbirds' for details). Snags were polystyrene cylinders 242 cm high, 22 cm diameter and were erected 10 cm above ground on metal poles. A total of 50 cylinders were erected. Laboratory tests showed that polystyrene did not have a negative impact on woodpecker health.

A replicated trial in 1982–3 in an area of forest clearcut 12 years previously in Ohio, USA (4), found that a total of 34 cavities were excavated in 99 artificial snags erected in autumn 1982. Thirty one of these were probably excavated by downy woodpeckers *Picoides pubescens*, with the remaining three probably being excavated by hairy woodpeckers *P. villosus*, red-bellied woodpeckers *Melanerpes carolinus* or northern flickers *Colaptes auratus*. Only downy woodpeckers were found roosting in cavities. Excavation rates were highest within 35 m of the edge of the clearcut. Snags were polystyrene cuboids, 21 × 21 × 237 cm, erected vertically on a fibreglass stake on a 16 m grid.

A replicated, controlled, paired site study from April–July in 1988–9 in 20 experimental and 20 control sites of nesting cavities in a forest reserve in North Carolina, USA (5) found that red-cockaded woodpeckers *Picoides borealis* used artificial nesting cavities significantly more than creating their own or using abandoned cavities. Woodpeckers were significantly more likely to occupy vacant experimental sites (nine occupied) than vacant control sites (zero occupied). Similarly, abandoned experimental sites were occupied more (nine occupied) than control sites (none occupied). Abandoned sites lacking artificial cavities were never occupied. The 18 experimental sites occupied corresponded to a net addition of 12 social units to the population. Out of six breeding pairs, four nested successfully, raising seven young, while 2 failed. Vacant (previously unoccupied) experimental sites (provisioned with two cavities) were paired with ten control (no cavities provided) sites and were matched in habitat characteristics. Similarly, ten abandoned experimental sites were matched with abandoned control sites.

A small trial at two open pine woodland sites in Texas, USA (6), found that all four translocated red-cockaded woodpeckers *Picoides borealis* that remained at two release sites (see 'Translocate individuals' for details) used artificial nesting cavities provided at the release sites.

A series of before-and-after trials in four open pine forests in Texas, USA (7), investigated the impact of multiple interventions, including the provision of 736 artificial cavities (mostly 'inserts'), on red-cockaded woodpecker *Picoides borealis* populations. Following declines throughout much of the 1980s, numbers of active clusters stabilised, with increases apparent at all four sites in the early 1990s and a total 39 new clusters established, 22 of which were in areas with artificial nesting sites. Restrictor plates were also installed around red-cockaded woodpecker nesting holes to prevent enlargement of cavities by pileated woodpeckers *Dryocopus pileatus*. This study is discussed in more detail in 'Translocate individuals' and 'Use prescribed burning'.

A replicated trial in 1986–91 in five forest types in Texas, USA (8), found that downy woodpeckers *Picoides pubescens* excavated cavities in all ten artificial snags installed in upland hardwood forest, and 13 of 17 snags provided in pine-hardwood habitats. However, they did not use any of ten artificial snags installed in pine-only forest, whilst ten in bottomland hardwood forest were not used in 1989, after which they were washed away in floods. Downy woodpeckers did not use snags for nesting and none of the other six woodpeckers in the area used the snags at all. Snags were brown-painted polystyrene cylinders, 26 cm in diameter, 242 cm tall and mounted on iron posts.

A small study in open pine forests in South Carolina, USA (9), found that modifying seven red-cockaded woodpecker *Picoides borealis* inserts did not alter the behaviour of the male woodpeckers using the inserts. They were modified by drilling a 7.7 cm hole in the front, approximately 12 cm below the entrance and fitting a plug in it. The purpose was to allow easy inspection of the inserts.

A single site study from 1985–1996 in a pine forest in South Carolina, USA (10) found that the number of breeding pairs of red-cockaded woodpeckers *Picoides borealis* increased from one to 19, producing 43 fledglings, and the overall population size grew from four to 99 individuals following the provision of artificial nest cavities amongst other interventions. The mean fledging success for 1985–96 was 2.3 fledglings/nest, which was significantly higher than the regional average (1.7 fledglings/nest). A total of 305 artificial nest cavities, fitted with metal plates to prevent enlargement by other species (see 'Protect nest sties from competitors'), were installed over the study period. In addition, the forest mid-storey was thinned and prescribed burning used (see 'Threat: Natural system modifications – Forest modifications'), and birds. Nests were monitored monthly throughout the year except over the breeding season (April–July) when they were monitored weekly.

A before-and-after trial in a longleaf pine *Pinus palustris* forest in Louisiana, USA (11), found that the number of groups of red-cockaded woodpeckers *Picoides borealis* with breeding pairs increased from 22 (68% of all groups) to 28 (93%) between 1993 and 1995, following the installation of 44 artificial nesting cavities in 1993–5 (55 cavities were already available in 1993). In 1993–5, the number of groups in the area decreased from 33 to 30, but average group size increased, as did the number of groups with helpers and the number of cavities occupied. Most breeding males (77% of 30 birds) continued to use natural cavities, but 71% of 28 breeding females and 65% of 23 helper birds used artificial cavities. Cavities consisted of wooden boxes 10 × 15 × 25 cm with 4.5 cm diameter, 6.5 cm long entrance tunnel and a cylindrical 20 cm × 7.5 cm diameter cavity inside. Cavities were inserted into holes carved out of live pine trees, at either 4 or 7.5 m above ground.

(1) McComb, W. C. & Noble, R. E. (1981) Nest-box and natural-cavity use in three mid-south forest habitats. *The Journal of Wildlife Management*, 45, 93–101.

(2) Grubb Jr, T. C. (1982) Downy woodpecker sexes select different cavity sites: an experiment using artificial snags. *The Wilson Bulletin*, 94, 577–579.

(3) Peterson, A. W. & Grubb, T. C. (1983) Artificial trees as a cavity substrate for woodpeckers. *The Journal of Wildlife Management*, 47, 790–798.

(4) Petit, D. R., Petit, K. E., Grubb Jr, T. C. & Reichhardt, L. J. (1985) Habitat and snag selection by woodpeckers in a clear-cut: an analysis using artificial snags. *The Wilson Bulletin*, 97, 525–533.

(5) Copeyon, C. K., Walters, J. R. & III, J. H. C. (1991) Induction of red-cockaded woodpecker group formation by artificial cavity construction. *The Journal of Wildlife Management*, 55, 549–556.

(6) Rudolph, D. C., Conner, R. N., Carrie, D. K. & Schaefer, R. R. (1992) Experimental reintroduction of red-cockaded woodpeckers. *The Auk*, 109, 914–916.

(7) Conner, R. N., Rudolph, D. C. & Bonner, L. H. (1995) Red-cockaded woodpecker population trends and management on Texas National Forests. *Journal of Field Ornithology*, 66, 140–151.

(8) Conner, R. N. & Saenz, D. (1996) Woodpecker excavation and use of cavities in polystyrene snags. *The Wilson Bulletin*, 108, 449–456.

(9) Edwards, J. W., Stevens, E. E. & Dachelet, C. A. (1997) Insert modifications improve access to artificial red-cockaded woodpecker nest cavities. *Journal of Field Ornithology*, 68, 228–234.

(10) Franzreb, K. E. (1997) Success of intensive management of a critically imperiled population of red-cockaded woodpeckers in South Carolina. *Journal of Field Ornithology*, 68, 458–470.

(11) Carrie, N. R., Moore, K. R., Stephens, S. A. & Keith, E. L. (1998) Influence of cavity availability on red-cockaded woodpecker group size. *The Wilson Bulletin*, 93–99.

15.4.20. Parrots

- A before-and-after study from Costa Rica (5) found that the local population of scarlet macaws *Ara macao* increased following the installation of nest boxes along with several other interventions.
- Five studies from South (2,7,10) and Central (4) America and Mauritius (9) that nest boxes were used by several species of parrots, with one finding an increase in use over time until the majority

of the population used them (9). One replicated study from Peru found that blue-and-yellow macaws *Ara ararauna* only used modified palms, not 'boxes' (6), whilst another replicated study found that scarlet macaws *Ara macao* used both PVC and wooden boxes, but that PVC lasted much longer (1).

- Four studies from Venezuela (3,11) and Columbia (7,8) found that several species very rarely, if ever, used nest boxes.
- Six studies from Central (4) and South (2,6–8,10) America found that parrots nested successfully in nest boxes, with two species showing higher levels of recruitment into the population following nest box erection (7) and another finding that success rates for artificial nests were similar to natural nests (6).
- Three studies from South America (2,3,11) found that artificial nests had low success rates, in two cases due to poaching (3,11).

A replicated trial of two nest box designs in 1992–2000 at a tropical rainforest site in south-west Peru (1) found that both designs were used by scarlet macaws *Ara macao*, with PVC tubes proving more durable than wooden boxes. Eight boxes (45 × 45 × 160 cm with a 15 cm diameter entrance hole) made from tropical cedar *Cedrella odorata* wood were hung on the tall emergent trees in March 1992, with seven used by macaws (one was occupied by bees). No wooden boxes were useable by September 1999. Five PVC tube nest boxes (30 or 35 cm diameter, two 17 × 15 cm entrance holes, lined with caulk and with wooden or metal top and bottom) were hung in August–September 1992, of which four were occupied in 1993–4. More nest boxes were hung in later years and nine of 12 available boxes were used in the 1999–2000 breeding season.

A replicated, controlled study (2) in 1999–2003 found that scarlet macaws *Ara macao* occupied a large number of nest boxes at the same rainforest site as in (1), with 13 of 14 PVC nest boxes and one of four wooden boxes occupied. Eggs were laid in ten; eggs hatched in six and four (all PVC tubes) fledged at least one chick. Four natural nests were used and three fledged at least one chick. When data were combined with that from 1999–2001, chick survival was similar (75–6%) in natural nests and nest boxes. Hatching rates were lower in PVC nest boxes (of the same type as in (1) than in natural nests (41% vs. 65%), possibly due to higher temperature fluctuations. Hatching rates in wooden boxes were very high (80%) but the authors note the small sample size may make this result unreliable.

A replicated study in 1997–9 in a tropical forest on an island in Venezuela (3) found that yellow-headed Amazons *Amazona barbadensis* used repaired natural cavities more often than artificial nest boxes. One box of 16 was used in 1997, a second in 1998, and three in 1999 compared with all 15 repaired natural cavities being repeatedly occupied following repair. Fledging rates were also low for nest boxes; with four out of five clutches being removed by poachers (the remaining clutch fledged three chicks). Nest boxes were wooden, 160 × 30 × 30 cm with a 20 × 15 cm entrance hole, had 10 cm of woodchips inside and grooves to allow parrots to climb out. Boxes were placed 2–4 m up in trees, with all used boxes being on *Bulnesia arborea* trees.

A replicated study in 1996–2000 in four tropical forest sites in central Costa Rica (4) found that scarlet macaws *Ara macao* successfully hatched eleven clutches from six of 38 (16%) nest boxes provided between 1995 and 2000. Three of these boxes were 1 m plastic barrels (14 erected in total), two were 35 cm PVC tubes (15 erected in total) and one was in a 100 × 60 cm wooden box (nine erected in total). All boxes were erected 10–20 m above the ground in trees. Three clutches were laid in PVC tubes in 2000 but were destroyed by monkeys.

A before-and-after study in western Costa Rica (5) found an increase in a scarlet macaw *Ara macau* population from 185–225 individuals in 1990–4 to 225–265 in 1997–2003, following the provision of artificial nests and several other interventions (see 'Use education programmes and local engagement to reduce pressures on species', 'Promote sustainable alternative livelihoods based on species', and 'Guard nests to increase nest success').

In 1990–4 the population had been showing a 4%/year decline. This study is discussed in more detail in 'Increase "on-the-ground" protection to reduce unsustainable levels of exploitation'.

A replicated study in 1992–2004 in a palm swamp in south-east Peru (6) found that blue-and-yellow macaws *Ara ararauna* nested in modified Mauritia palms *Mauritia flexuosa*, but not in five PVC nest boxes (a pair of scarlet macaws *A. macao* used one box). A total of 41 palms had their crowns removed over the study period and slowly rotted to produce nesting cavities, with 12 nesting attempts by blue-and-yellow macaws in these cavities. Productivity in 1995 of two pairs of blue-and-yellow macaws and three pairs of red-bellied macaws *Orthopsittaca manilata* nesting in the palms was comparable to previous estimates for the region (50% success, 0.5 chicks/nest). As palms rotted and productivity declined over time, the authors recommend a rotation system to maintain both the structure of the palm swamp and provide adequate nesting cavities.

A replicated study in 2004–6 in four tropical montane sites in Colombia (7) found that four of five threatened parrot species bred and roosted in 240 nest boxes provided. Indigo-winged parrots *Hapalopsittaca fuertesi*, golden-plumed parakeets *Leptosittaca branickii*, flame-winged parakeets *Pyrhura calliptera* and Santa Marta parakeets *P. viridicata* all used the nest boxes, with an increase in the numbers of young individuals of indigo-winged parrots and golden-plumed parakeets entering natural populations. There was no evidence that rusty-faced parrots *H. amazonina* used boxes. Boxes were 100 × 25 × 25 cm with a 10–15 cm entrance hole, except for those designed for rusty-faced parrots, which were only 60 cm tall. All boxes had grooves to allow parrots to climb the inside.

A replicated study in 2006–7 at five montane sites in Colombia (8) found that yellow-eared parrots *Ognorhynchus icterotis* rarely used artificial nest boxes provided, with one box occupied in 2006 and one in 2007, raising two fledglings each. A total of 42 nest boxes were built and hung 10–20 m above ground on Quindío wax palms *Ceroxylum quindiuense* lacking suitable nest cavities. Boxes were wooden and hexagonal, with 140 × 20 cm sites and a 10 cm entrance hole, with grooves inside to allow parrots to climb out.

A replicated study in 1997–2007 in tropical forest in south-west Mauritius (9) found that the number of echo parakeet *Psittacula eques* pairs using nest boxes increased from none in 1997–2000 (with two nest boxes available) to 41 in 2006–7 (65 boxes available). In 2006–7, 73% of all parakeet nests (56 nests in total) were in nest boxes, with 71% of these successfully fledging chicks. The nest boxes were wooden, 65 cm high and made from untreated wood with a metal roof and a perch outside. The authors suggest that the number of captive-bred parakeets released into the population may have helped with the uptake of nest boxes, as they were more familiar with artificial structures than wild-bred individuals.

A replicated study in 2005 at a tropical forest site and nearby ranches in Colombia (10) studied a subset of the nest boxes used in (7) and found that indigo-winged parrots *Hapalopsittaca fuertesi* used 13 of 120 nest boxes provided. A total of 39 eggs were laid, 32 (82%) hatched (the remaining seven were infertile) and 25 chicks (78% of those that hatched) fledged. Overall, 10 (91%) nests in boxes successfully fledged one or more chick. The boxes were of the same design as those in (7).

A replicated study in 2007–2009 (part of a longer study from 2000–2009) in 11 monitored yellow-shouldered parrot *Amazona barbadensis* nests in tropical forest habitat on Margarita Island, Venezuela (11) found that artificial nests exhibited low occupancy rates and did not significantly hinder poachers. Of the 12 artificial nests used to supplement existing natural nest, six were used: 1 was used every year, 1 was used in 2007 and 2008, and 4 were used only once (which equated to 25% nest use rate). Moreover, only 40% of the nestlings succeeded in fledgling, the rest being subject to an armed group of poachers raiding the site designated as an assisted breeding programme. The artificial nests were made from the preferred natural nesting tree of yellow-shouldered parrots, verawood (*Bulnesia arborea*). This study is also discussed in 'Relocate nestlings to reduce poaching', 'Use education programmes and local engagement to help reduce pressures on species', 'Employ locals as biomonitors' and 'Foster eggs or chicks with wild conspecifics'.

(1) Brightsmith, D. (2000) Macaw reproduction and management in Tambopata, Peru II: nest box design and use. Unpublished report, Duke University, Durham, North Carolina.
(2) Brightsmith, D. & Figari, A. (2003) Breeding ecology and clay bank usage by macaws from Madre de Dios. *Report to INRENA.*
(3) Sanz, V., Rodriguez-Ferraro, A., Albornoz, M. & Bertsch, C. (2003) Use of artificial nests by the yellow-shouldered parrot (*Amazona barbadensis*). *Ornitología Neotropical*, 14, 345–351.
(4) Vaughan, C., Nemeth, N. & Marineros, L. (2003) Ecology and management of natural and artificial scarlet macaw (*Ara macao*) nest cavities in Costa Rica. *Ornitología Neotropical*, 14, 381–396.
(5) Vaughan, C., Nemeth, N. M., Cary, J. & Temple, S. (2005) Response of a scarlet macaw *Ara macao* population to conservation practices in Costa Rica. *Bird Conservation International*, 15, 119–130.
(6) Brightsmith, D. & Bravo, A. (2006) Ecology and management of nesting blue-and-yellow macaws (*Ara ararauna*) in Mauritia palm swamps. *Biodiversity and Conservation*, 15, 4271–4287.
(7) Quevedo, A., Salaman, P., Mayorquin, A., Osorno, N., Valle, H., Solarte, C., Reinoso, R., Sanabria, J., Carantón, D., Díaz, V. & others (2006) Loros amenazados de la Cordillera Central de los Andes de Colombia: una iniciativa de conservación basada en la investigación y la educación ambiental. *Conservación Colombiana*, 1, 21–57.
(8) Salaman, P., Quevedo, A. & Verhelst, J. C. (2006) Proyecto Loro Orejiamarillo: una iniciativa de conservación. *Conservación Colombiana*, 2, 7–11.
(9) Tatayah, R. V. V., Malham, J., Haverson, P., Reuleaux, A. & Van de Wetering, J. (2007) Design and provision of nest boxes for echo parakeets *Psittacula eques* in Black River Gorges National Park, Mauritius. *Conservation Evidence*, 4, 16–19.
(10) Tovar Martinez, A. E. (2009) Use of nest boxes by indigo-winged parrots *Hapalopsittaca fuertesi* in and around the Reserva Municipal El Mirador, Quindío, Colombia. *Ornitología Neotropical*, 20, 357–368.
(11) Briceño-Linares, J. M., Rodríguez, J. P., Rodríguez-Clark, K. M., Rojas-Suárez, F., Millán, P. A., Vittori, E. G. & Carrasco-Muñoz, M. (2011) Adapting to changing poaching intensity of yellow-shouldered parrot (*Amazona barbadensis*) nestlings in Margarita Island, Venezuela. *Biological Conservation*, 144, 1188–1193.

15.4.21. Songbirds

- Only three studies out of 66 from across the world (2,36,39) found low rates of nest box occupancy, although this may be partially the result of publishing biases. Thrushes, crows, swallows and New World warblers were the target species with low rates of use. Thrushes, crows, finches, swallows, wrens, tits, Old World and tyrant flycatchers, New World blackbirds, sparrows, waxbills, starlings and ovenbirds all used nest boxes. One study from the USA (6) found that wrens used nest boxes more frequently than natural cavities.
- Five studies from across the world (12,20,24,48,53,62) found higher population densities or population growth rates in areas with nest boxes, whilst one study from the USA (20) found higher species richness in areas with nest boxes. One study from Chile (53) found that breeding populations (but not non-breeding populations) were higher for two species when next boxes were provided.
- Twelve studies from across the world (9,10,14,16,17,27,31,37,38,40,58,61) found that productivity of birds in nest boxes was higher or similar to those in natural nests. One study (26) found there were more nesting attempts in areas with more nest boxes, although a study from Canada (51) found no differences in behaviour or productivity between areas with high or low densities of nest boxes. Two studies from Europe (8,40) found lower predation of some species using nest boxes. However, three studies from the USA (1,17,39) found low production in nest boxes, either in absolute terms or relative to natural nests.
- Thirteen studies from across the world founds that use (11,13,18,29,41,46,55,56,59,63), productivity (10,32,33,56) or usurpation (55) varied with nest box design, whilst seven found no difference in occupation rates (15,23,25) or success (4,5,13,15,23) with different designs.
- Similarly, fourteen studies found different occupation (3,12,16,18,19,25,42,43,45,47–49,57) or success (16,34,45,50,57) rates depending on the position or orientation of artificial nest sites. Two studies found no difference in success (49,57) with different positions.

A replicated study in 1945–6 in garden habitats in Ohio, USA (1), found that American robins *Turdus migratorius* nesting in artificial nests had lower success rates than those natural nests

(33% success for 24 nesting attempts in artificial nests vs. 50% of 48 in natural nests). Fourteen pairs of robins used the nests, but only seven successfully raised chicks. Nests consisted of cones of black or green roofing paper 17.8 cm at the widest and 5.1 cm deep. This study also examines nest use by mourning doves *Zenaida macroura* (formerly *Zenaidura macroura*), and the effect of different coloured nests in 'Use differently-coloured artificial nests'.

A controlled study in mixed farmland in north-east Scotland in 1971 (2) found that carrion crows *Corvus corone* did not nest in artificial trees, irrespective of whether they were provided with supplementary food or not. In one experiment, a line of 15 artificial trees (3–6 m branches tied to fence posts and provided with an old crow's nest) were set up, approximately 70 m apart. Two pairs of crows established territories, but neither attempted to breed. A second experiment provided a single artificial tree in two occupied territories, 70 m from the tree used by the resident pair. Neither artificial tree was used, as the resident pairs successfully defended their territories. This study also investigated the effects of supplementary feeding on crow reproduction, discussed in 'Provide supplementary food to increase reproductive success'.

A small study in 1976–9 in three scrub and grassland habitats in Idaho, USA (3), found that common ravens *Corvus corax* nested on nesting platforms provided, with four pairs using them in 1976, but only a single attempt in 1979. An average of 2.8 chicks/nest were produced. Twenty four platforms were provided in shaded/un-shaded pairs, with 23 out of 29 young fledged from shaded platforms. This study also discusses platform use by ferruginous hawks *Buteo regalis*, discussed in 'Provide artificial nest sites for raptors'.

A replicated study in 1972–8 in Texas, USA (4), found that purple martin *Progne subis* reproductive output did not differ between aluminium and wooden bird 'houses' (82% of chicks fledging from 275 pairs in aluminium houses vs. 86% from 116 pairs in wooden houses; average of 28 day nestling period for 82 pairs in aluminium houses and 67 pairs in wooden houses). Aluminium houses had six, 12 or 24 compartments and were commercially available, wooden houses were homemade but of similar designs, with 15 × 15 × 15 cm compartments.

A small replicated study in 1976–8 in a mixed forest in Washington State, USA (5), found that mountain bluebird *Sialia currucoides* reproductive success (clutch size, eggs number hatched and chicks number) did not differ between two nest box designs. Of 22 marked females that successfully bred, 68% returned to the same territory and nest box type and 71% of those that changed territory selected the same type of nest box. Unsuccessful breeders, however, were more likely to change territory and box type in subsequent years (60% five unsuccessful breeders changed territories, 40% also changed nest box type). The majority of boxes had a 30 × 15 cm base and a 4.4 cm diameter entrance hole 10 cm above the floor; the rest had a 12.7 × 12.7 cm based and a 3.8 cm hole 14 cm above the floor. All were erected 1.5 m above the ground on fence posts and 400 m apart.

A controlled study 1977–9 in riverine forests in Louisiana, USA (6), found that Carolina wrens *Thryothorus ludovicianus* nested more frequently in nest boxes than in natural cavities (nests found in 0.50% of 5,374 nest box inspections vs. 0.07% of 3,993 natural cavities). Two hundred and thirty five nest boxes were erected, of three sizes: large (60 × 30 × 30 cm with a 13 cm diameter entrance); medium (45 × 20 × 20 cm with a 7.5 cm diameter entrance) and small (30 × 15 × 15 cm with a 5.0 × 7.0 cm entrance). Large boxes had a 15 × 13 × 2.5 cm shelf 8 cm below the entrance. All boxes had 5–10 cm of pine shavings in the bottom. Other songbirds also used nest boxes and cavities, were too infrequent to be analysed. This study also examined nest box use by other birds (wildfowl, woodpeckers and owls).

A replicated trial in 1979–80 in a deciduous forest in Ohio, USA (7), found that two pairs of Carolina chickadees *Parus carolinensis* and 28 pairs of house wrens *Troglodytes aedon* nested in cavities excavated from polystyrene cylinders by downy woodpeckers *Picoides pubsecens*. This study is discussed in more detail in 'Provide artificial nesting sites for woodpeckers'.

A replicated study in 1972–80 (8) found that predation of great tit *Parus major* and pied flycatcher *Ficedula hypoleuca* clutches in a mixed forest in Kronoberg, Sweden, was significantly lower in nest boxes than in natural cavities (5% predation for 112 great tit clutches and 4% for 112 flycatcher clutches in boxes vs. 17% of 76 clutches and 23% for 31 clutches in natural nests). There were no differences in predation of blue tit *P. caeruleus* (alternatively *Cyanistes caeruleus*) or marsh tit *P. palustris* (alternatively *Poecile palustris*) clutches. Three wood nuthatches *Sitta europaea* nested in boxes, none of which were predated (compared to 6% predation of 113 clutches in natural nests). For the three tit species, woodpeckers were responsible for 48% of predation in boxes, compared with 17% in natural cavities.

A small single-site study from March–May in 1981–1982 in an island (1.2 ha) within Kentucky Lake in Tennessee, USA (9) found that common grackles *Quiscalus quiscula* nested in 20–21% of nest boxes provided, fledging 71% of nestlings, which was higher than in other areas (average of 27%). This study also discusses the use of nest boxes by wood ducks *Aix sponsa*.

Three replicated trial in 1982–3 in woodlands in Sweden (10), found that pied flycatchers *Ficedula hypoleuca* and collared flycatchers *F. albicollis* had significantly larger clutches in larger nest boxes, whereas the effect on fledging success was less clear. A study with four nest box designs in a mixed wood in Kronobergs found that pied flycatchers laid larger clutches in larger nest boxes (6.4 eggs/clutch for 36 clutches in boxes with 57 cm^2 basal area; 6.5 eggs/clutch for 62 in 87 cm^2; 6.6 eggs/clutch for 16 in 104 cm^2 and 7.0 for eight in 125 cm^2). Fledging success was lowest in the smallest boxes (51%), highest (79%) in the second smallest and intermediate in the other designs (57% and 62%). Two trials in deciduous woods on Gotland found that collared flycatchers had smaller clutches in small boxes in 1982 only (5.6–5.7 eggs/clutch vs. 6.1–6.2 eggs/clutch). Fledging success was higher in normal boxes in one wood, but higher in smaller boxes in the other wood.

A replicated study in May–August 1984 in flooded riparian habitats in Tennessee, USA (11), found that prothonotary warblers *Protonotaria citrea* nested in 39% of 301 nest boxes provided, with tree swallows *Tachycineta bicolor* using three boxes and tufted titmice *Parus bicolor* one. Three box types were used and prothonotary warblers showed a significant preference for milk cartons (55% of 145 used, 68% of all warbler nests) and an avoidance of wooden boxes (7% of 84 used, 5% of all nests). PVC pipes were used in approximate proportion to their availability (45% of 71 used, 27% of all nests). The authors suggest that the lower volume of milk cartons and PVC tubes might explain these differences. All boxes were erected 1.5–2.0 m above the water surface on trees.

A replicated study from 1980–1984 in 5 sites of spruce, alder or mixed deciduous woodland in Trondheim, Norway (12), found that pied flycatchers (*Ficedula hypleuca*) prefer to nest in larger, upright nest boxes. Pied flycatchers lay larger clutches in larger (base area 180 cm^2) rather than smaller (base area 70 cm^2) nestboxes (6.4 compared to 6.7 eggs/nest respectively). When nestboxes were of the same small size within a plot, females laid larger clutches in nestboxes placed in normal, upright position than in tilted ones (6.9 compared to 6.3 eggs/nest). Normal, upright nest boxes were used preferentially across sites (80% of nestboxes occupied). There was no difference in fledgling success between nestboxes. Between 15 and 60 artificial nest boxes (diameter of entrance was 32 mm and 15 cm from base of entrance to bottom) were installed in each site each year.

A controlled before-and-after study in pine-oak woodlands in Arizona, USA (13), found the population density of six cavity-nesting songbirds more than doubled on two out of three experimental plots (one thinned, one with 75% of the oak and pine foliage removed), following the installation of 60 nest boxes on each plot in 1979 (21–46 pairs/40 ha in 1973–5 and 1979 vs. 64–108 pairs/40 ha in 1980–3). A third plot with no management showed a small but non-significant increase in population density. Violet-green swallows *Tachycineta thalassina*, pygmy nuthatches *Sitta pygmaea* and western bluebirds *Sialia mexicana* all showed significant increases in population density. Nest box use by five of the species was significantly higher in open and thinned forest plots, with more than 85% of nests of violet-green swallow, pygmy nuthatch and mountain chickadee *Parus gambeli* in nest boxes, compared to 30–35% in dense forests. White-breasted nuthatches *S. carolinensis* used a lower proportion of boxes in all habitats (approximately 63% of nests in open and thinned forests in nest boxes and almost none in dense forest). Bluebirds nested almost exclusively in nest boxes in all habitats. Boxes were 1,900 cm^3, made of woodcrete with entrance holes 3.2 or 3.8 cm in diameter and placed 5–11 m above ground.

A replicated, controlled study over two years in deciduous woodland in Oxfordshire, England (14), did not find consistent effects of nest box provision on tit *Parus* spp. reproduction. Great tits *P. major* and blue tits *P. caeruleus* (alternatively *Cyanistes caeruleus*) had significantly higher fledging success in nest boxes in 1983, but not 1984 (77% and 74% of eggs in nest boxes producing fledglings in 1983 vs. 35% and 46% of those in natural nests; 73% and 45% for nest boxes in 9184 vs. 56% and 51% for natural nests). In 1984, the percentage of nests fledging at least one chick was significantly higher for great tits in boxes (82% for boxes vs. 27% for natural nests), but not blue tits (62% and 69%). Differences were smaller in 1983 (91% for both species in nest boxes vs. 75% and 81% for great tits and blue tits in natural nests).

A replicated paired study in the summers of 1977–84 in farmland in Tennessee, USA (15), found that eastern bluebirds *Sialis sialis* did not show any preference for, or have higher reproductive success in, large nest boxes, compared to small ones (50 nests built in large boxes, 4.6 eggs/clutch and 2.9 fledglings/nest vs. 44 nests built in small boxes, 4.7 eggs/clutch and 3.0 fledglings/nest). There was some evidence that more chicks from large boxes returned to the study area, compared to those from small boxes (9% vs. 3% respectively). Ten pairs of boxes were erected 75 cm apart on metal supports. Small boxes had a 71.5 cm^2 basal area, large had 143 cm^2 and both had a 3.8 cm diameter entrance hole.

A replicated study in May–June 1983–5 in three riparian woodland sites in Wyoming, USA (16), found that 37% of 65 nest boxes erected in 1982 were used by house wrens *Troglodytes aedon*, with 20% contained unused nests. Wrens successfully fledged chicks from 73% of the 73 nests, were more likely to use boxes in open habitats and more likely to fledge chicks from these boxes. Boxes were 14 × 14 × 28 cm, with a 3.8 cm entrance hole.

A replicated trial in the summers of 1983–6 at three streamside sites in Wyoming, USA (17), found that house wrens *Troglodytes aedon* used nest boxes erected at all three sites, whilst tree swallows *Tachycineta bicolor* used boxes on one site. Clutch mortality was higher in swallows (87% of 29 clutches) than wrens (37–67% for 99 clutches), with interference by wrens occurring in 45% of 29 swallow nests. Wren clutch loss was largely due to predation, although the rate of clutch loss was lower on the site where swallows did not use nest boxes.

A replicated trial in woodlands, hedges and around villages on Nakano-shima island, Kyshu, Japan (18), found that Ryukyu robins *Erithacus komadori* occupied approximately 5% of nest boxes in 1989, but 35% in 1990, possibly because more boxes in 1990 had entrance holes larger than 4 cm (in 1989 only 1% of 175 boxes with entrances <3 cm in diameter were occupied vs. 44% of 25 with entrances over 3.5 cm). Occupation rates varied across habitats, from 17% of 24 boxes in village vegetation to 42% in some bamboo stands.

A replicated trial in a tidal swamp forest in Virginia, USA (19), found that prothonotary warblers *Protonotaria citrea* nested in 27–34% of nest boxes erected in 1987–8 (140–214 boxes available). Warblers appeared to favour boxes on trees near open water that were surrounded by relatively large trees, but orientation and height above ground did not appear to have an effect. Wasps occupied 29–42% of nest boxes, presumably preventing colonisation by warblers. Boxes were 28 × 9 × 6 cm, with a 3.8 cm diameter entrance hole and erected on trees 20–280 cm above ground.

A replicated, controlled, paired sites study in 1985–6 in slash pine *Pinus elliotti* plantations in Florida, USA (20), found that bird species richness and abundance were significantly higher on three sites with snags and nest boxes installed, compared to control plots without them (6.8 species and 9.5 pairs of birds on experimental plots vs. 3.5 species and 4.5 pairs on control plots). Eastern bluebirds *Sialia sialis* and great crested flycatchers *Myiarchus crinitus* were both significantly more abundant on experimental plots, with flycatchers using more small boxes (15 × 15 × 20 cm, nine boxes used) than large (25 × 25 × 50 cm, three used). Sixteen snags (8 m tall pine trunks) were erected evenly across plots, each with a small or large nest box attached.

A replicated, controlled study from May–August in 1985–7 in two woodland ranches consisting sites in Wyoming, USA (21) found that house wren *Troglodytes aedon* males nesting in nest boxes exhibited higher frequencies of polygyny than those males in natural cavities. Polygyny was greater amongst males in nest boxes (controlling >1 nest boxes/territory) than those in natural cavities (53 compared to 10% of males respectively). Furthermore, 83% of nest box males and 47% of cavity males made consistent attempts at attracting second mates. Song output (songs sung/hr) was significantly higher for males attempting polygyny but were similar between polygynous nestbox and cavity males. The authors conclude that, although nestboxes do not cause polygyny in wrens, access to nest boxes appears to facilitate higher rates of polygyny. In 15 experimental sites two or three nest boxes (25–40 m apart) were erected, 31 control sites were also studied.

A replicated study from February–September in 146 (in 1990) and 180 (in 1991) artificial nest boxes in the USA (22) found that 74% of nest boxes were used for egg laying by house sparrows *Passer domesticus*. Sparrows laid an average of 4.5 eggs per clutch. Average hatching success (ratio of eggs laid to nestlings hatched) was 51%. Average breeding success (ratio of eggs hatched to nestlings fledged) was 32%. In 1991, eight of the nest boxes were used by house wrens *Troglodytes aedon*. By 1991, only 8 of the 120 second-year boxes (but none of the 60 first-year boxes) required replacement. Nestboxes withstood severe fluctuations in temperature, humidity, wind and rain. Entrance holes were 3.8 cm in diameter, roofs and floors had 0.8 cm diameter holes (3 and 5 respectively) for ventilation and drainage.

A replicated trial in 1988–90 in wetlands in Michigan, USA (23), found that adult and two year-old female tree swallows *Tachycineta bicolor* used large and small nest boxes with equal frequency (adults used 27% of large boxes available and 19% of small; two year-olds used 19% and 17% respectively), and that there were no significant differences in reproductive success between nest box types. Boxes were all 19.5 cm tall with a 12.4 cm diameter entrance whole and erected on metal poles (greased to deter mammalian predators). Large boxes had a 14 × 14 cm base, small were 10 × 14 cm.

A randomised, replicated and controlled before-and-after study at two pine forest sites in Colorado, USA (24), found that population densities of cavity-nesting birds on 27 experimental plots increased by 500–550% in 1992–3 following the installation of four nest boxes on each plot (7.5–10.0 birds/plot in 1990–1 vs. 41.5–52.5 birds/plot in 1992–3). There were significantly smaller increases on 27 control plots, which had had similar densities before nest box addition (8.5–16.0 birds/plot in 1990–1 vs. 15–20 birds/plot in 1992–3). Birds used 33–55% of the 108 boxes, with the most common species being pygmy nuthatch *Sitta pygmaea* (25% of box uses), house wren *Troglodytes aedon* (21%), mountain chickadee *Parus gambeli* (18%), white-breasted nuthatch *Sitta carolinensis* (14%) and western bluebird *Sialia mexicana* (12%). Open-nesting birds also increased in both experimental and control plots. All plots were circles with a 50 m radius.

Two replicated trials in 1992–4 in woodlots in Ohio, USA (25), found that Carolina chickadees *Parus carolinensis* showed no preference for artificial snags filled with sawdust over those that weren't filled, but nested more frequently in snags located high above ground (nine snags occupied) and in non-shrub habitat (18% of 38 snags occupied), compared to low snags (two occupied) and in shrub cover (8% of 38 snags occupied). Snags were placed in groups of four (two filled with sawdust and two empty; two in shrub habitat and two in other habitats) or in pairs (one with the entrance 1.2 m above ground, one 3 m above ground). Snags consisted of 1.2–3.08 m sections of 7.8 cm PVC tubing with a nest chamber attached at the top, 'planted' in the ground.

A randomised, replicated trial in 1985 in South Carolina, USA (26), found that a 59% of 94 eastern bluebird *Sialia sialis* nesting attempts were in 23 territories provided with two nest boxes, compared to 41% in 20 territories supplied with a single box. There were no significant differences in the number of eggs, nestlings or fledglings between territory types. When two boxes were used, they were erected less than 24 m apart, allowing a single male to defend both.

A replicated, controlled study on mixed farmland and suburban habitats in Punjab, India, in 1992 (27), found that Indian mynahs *Acridotheres tristis* nesting in nest boxes had larger clutches and lower nestling mortality than those nesting in natural sites (4.8 eggs/clutch for 22 clutches and 49% mortality for 68 chicks in boxes vs. 3.9 eggs/clutch for 16 clutches and 68% mortality for 41 chicks in natural sites). However, whilst nesting success and overall productivity were higher in nest boxes, these differences were not significant (64% success and 1.6 fledglings/pair for nest boxes vs. 50% and 0.8 chicks/pair for natural nests) and hatching success was similar (approximately 65%) Mynahs also appeared more likely to lay second and third broods in boxes than natural sites. Thirty boxes were erected, half wooden (22 × 22 × 34 cm) and half PVC tubes (16 cm diameter with a wooden base).

A replicated study in marshland in 1991–2 in British Columbia, Canada (28), found that tree swallows *Tachycineta bicolor* occupied 90–100% of 125 nest boxes provided. The effect of cleaning boxes is described in 'Clean nest boxes to increase occupancy or reproductive success'.

Three replicated trials in pine forests in north and east Scotland (29) found that tits *Parus* spp., with the exception of crested tits *P. cristatus* preferentially nested in deep nest boxes over shallow ones and empty boxes over those filled with wood shavings. In 1991, trials at two sites found that, of 50 pairs of nest boxes erected (one 'deep': 12 × 8 × 25 cm; one 'shallow': 11.5 × 10.5 × 15 cm), 15 of the deep boxes were occupied (eight by great tits *P. major*, five by blue tits *P. caeruleus* and two by coal tits *P. ater*) with only a single shallow box occupied by a pair of crested tits. In 1993–4, a second trial at one site found that, of 83 pairs of nest boxes erected (one empty, one with wood shavings, all of the 'deep' design), 23 empty boxes were occupied (16 by great tits, four by blue tits and three by coal tits), compared to 12 filled boxes (eleven by crested tits and one by great tits).

A replicated study in tropical forest in Kanungu District, Uganda (30), found that two or three nest boxes provided in March–April 1996 were used by stripe-breasted tits *Parus fasciiventer* for nesting, hatching one or two chicks each. Fledging success was unknown and no boxes were used in 1997. Boxes were wooden, 30 cm deep and attached 3–10 m up in trees.

A replicated, controlled study in 1989–94 in grazed and ungrazed pine-oak woodlands in California, USA (31), found that the benefits of nesting in nest boxes, compared to natural cavities, varied between songbird species. Western bluebirds *Sialia mexicana* gained the most advantage, with higher nesting success, lower predation rates and marginally more young fledged in boxes. Plain titmice *Parus inornatus* and house wrens *Troglodytes aedon* marginally benefited from nesting in boxes, with marginally lower predation rates, more eggs hatched and more young fledged (titmice); or lower predation rates, larger clutches, more eggs hatched, more young fledged and marginally higher nesting success (wrens). Ash-throated flycatchers *Myiarchus cinerascens* experienced no apparent benefits from nesting in boxes. Nest boxes had a basal area of 137 cm^2 with 3.2 or 3.8 cm entrance holes, and placed 2 m above ground on trees. Between 44 (1989–91) and 92 (1992–4) boxes were monitored annually.

A replicated study between 1968 and 1994 in orchards and fields in Wisconsin, USA (32), found that eastern bluebirds *Sialis sialis* were less likely to suffer total clutch loss in nest boxes with screen openings in the roof than in three other designs (16% of 1,506 nesting attempts lost in open-top boxes vs. 28% of 1,066 in standard boxes, 33% of 36 in one gallon tin-cans and 31% of 29 in hollowed posts). In addition, clutches in open-top boxes were larger than those in standard boxes (4.4 eggs/clutch for open-top vs. 4.3 eggs/clutch for standard boxes) and had higher hatching success (82% vs. 72%), chick survival (93% vs. 87%) and overall survival (76% vs. 62%). Open-top boxes were 10 × 13 × 29 cm, with a 9 cm screened opening in the roof; standard boxes were 10 × 13 × 18–21 cm with no roof hole.

A replicated trial in summer 1992 in a wetland in Ontario, Canada (33), found that tree swallow *Tachycineta bicolor* clutches were significantly smaller in small nest boxes than in large (average of 5.8 eggs/clutch for 11 small boxes and 6.5 eggs/clutch for 12 large) and fewer eggs hatched, although this difference was not significant (72% of eggs hatching in small boxes vs. 84% in large). Once brood sizes were standardised by transferring chicks between nests, there were no differences in the timing of reproduction, chick mass or size or fledging rates (91% for small boxes vs. 97% for large), although nestling flight feathers were shorter after 15 days in small boxes, possibly due to overcrowding. All nest boxes were the same when installed, but after swallows began nesting, small boxes had an insert installed, which reduced the basal area from 178 cm^2 to 75 cm^2.

A replicated, randomised study from March–August in 1986–93 in five sites within a suburban park containing 25 solitary (spaced 50 m apart) nest boxes and 25 colonial nest boxes (3–5 m apart) in Budapest, Hungary (34), found that the reproductive patterns of tree sparrows *Passer montanus* were affected by nest box spacing. Most fledglings were produced by females that moved from colonial to solitary nests (an average of 0.2 fledglings/brood) and fewest by females that retained colonial nests in subsequent seasons (0.07 fledglings/brood). Overall, female fledglings were significantly more likely to have been hatched from solitary nests (65% of all females) whereas males were more frequent in colonial nest broods (67% of all males). Both female and male recruits were most likely to breed for the first time in colonial nests. The distance between colonial and solitary nest boxes within each study plot was 100 m. The distance between neighbouring study plots was 500 m.

A replicated study in 1996–8 in a longleaf pine forest in South Carolina, USA (35), found that great crested flycatchers *Myiarchus crinitus* nested in 20% of 330 nest boxes provided, laying eggs in 88% of these (59 boxes in total). Differences in occupation and reproduction between areas burned in different seasons are described in 'Use prescribed burning'.

A replicated trial in June–August 1997–8 at 12 sites with differing tree densities in Utah, USA (36), found that only 2% of 120 nest boxes erected were occupied by birds: four by tree swallows *Tachycineta bicolor* and one by mountain bluebirds *Sialia currucoides*. All five occupied boxes were in sparsely-treed meadows, with none in densely-treed meadows or forests. The low uptake may have been due to large numbers of suitable natural cavities, with 15% of 271 natural cavities surveyed containing bird nests.

A replicated, controlled study in 1996–9 in coastal forests on Puerto Rico (37) found that 65% of yellow-shouldered blackbird *Agelaius xanthomus* nests were successful in artificial nests, compared to 48% of those in natural nests. In addition, natural nests suffered higher egg loss (74% vs. 43%) and chick loss (52% vs. 22%) than artificial nests. Nests were PVC 'elbows' on top of fence posts, with wire baskets inside and protected from rats by a metal cone attached to the fence post. This study occurred at the same time as a shiny cowbird *Molothrus bonairensis* eradication programme, discussed in 'Remove/control brood parasites'.

A replicated, controlled trial in 14 slash pine *Pinus elliotti* plantations in Florida, USA (38), found that, over 1997–8, estimates of nesting success of great crested flycatchers *Myiarchus crinitus* were almost identical between nest boxes (37% survival for 32 nests) and tree cavities (38% for 27 nests). Success was higher in boxes in 1997 (53%), but lower in 1998 (26%), due to increased predation during the incubation period. The height of nest boxes and habitat variables did not appear to influence occupation rates of nest boxes, with 32 of 160 boxes across eight plots occupied over the two years.

A replicated trial in 2000 in a suburban wetland park in Indiana, USA (39), found that 37% of 67 milk-carton nest boxes were occupied by house wrens *Troglodytes aedon* (24 boxes) and Carolina chickadees *Poecile carolinensis* (one box). Only 23% of wrens and no chickadees successfully fledged young, with failures due to predation, mostly by mammals that ripped open the cardboard nest boxes. Wrens preferentially nested in boxes on small trees, possibly as a defence against climbing predators. No prothonotary warblers *Protonotaria citrea*, the target species of the study, nested in any of the boxes.

A replicated, controlled study in old growth forest in Podlaskie, Poland, between 1993 and 1999 (40) found no differences in reproductive output of collared flycatchers *Ficedula albicollis* nesting in artificial nest boxes, compared to those in natural nests (average of 4.9 fledglings/nest for 122 nests in boxes and 150 nests in natural sites). There were no differences in clutch size, laying date or the percentage of nests losing eggs or chicks, but artificial nests were less likely to be completely predated (0–28% of broods completely lost to predation), compared to natural nests (20–54%). The authors suggest this may be due to predators not yet adapting to nest boxes.

A replicated paired study in 1996–7 in North Carolina, USA (41), found that eastern bluebirds *Sialia sialis* showed an overwhelming preference for nest boxes made from woodcrete over those made from wood (90% of 102 pairs picking woodcrete boxes). When given the choice of re-using a soiled woodcrete box or switching to a clean wooden box, 73% of 26 pairs remained in the woodcrete box. The effect of cleaning boxes is discussed in more detail in 'Clean nest boxes to increase occupancy or reproductive success'. Approximately 100 pairs of boxes were mounted in pairs, on poles 1 m apart, one of woodcrete and one of wood.

A replicated, paired site study from March–August in 2000–2003 in 20 paired nest box groups (10 placed along wetland edges and 10 in farmlands) in Rutland, England (42) found that tree sparrows *Passer montanus* showed a significant preference for nest boxes in wetland habitat, compared to those in farmland sites (eight wetland nest boxes colonised vs. no farmland sites). Nest box groups consisted of 5 nest boxes placed 2–20 apart; sunflower seeds were randomly provided to one nest box group within each pair, with the results discussed in 'Provide supplementary food to increase reproductive success'.

A replicated study in mixed forests in eastern Ontario, Canada, between 1999 and 2002 (43), found that black-capped chickadees *Parus atricapillus* (also called *Poecile atricapillus*) only nested in 20–27% of 176 nest boxes provided (compared with 99 nests in natural cavities in the study area). Boxes were erected in sets of four, facing 90° from each other on the same tree, and chickadees showed no preferences for particular orientations. Boxes were made from PVC tubing and filled with sawdust.

A replicated study from over 800 nest boxes across the eastern USA and Canada from 1998–2002 (44) found that eastern bluebirds *Sialia sialis* used nest boxes across the study area (between 28° and 48° N). However, breeding pairs exhibited higher net reproductive rates in nest boxes at lower latitudes, having on average 17–33% higher likelihood of repeated egg-laying, multiple brooding and successful fledging events than boxes in the northern range (an average of 1.8 broods/box/pair in the south vs. 1.3 in the north). The average number of fledglings was also significantly greater in the south (1.7 vs. to 1.3 fledglings/box/pair).

A replicated trial in 1995–2000 in both deciduous and coniferous forests in Pärnu County, Estonia (45), found that great tits *Parus major* laying in nest boxes had significantly higher breeding success in coniferous forests, compared to those in deciduous woods, with heavier fledglings and higher recruitment (8.1 fledglings/pair, 17.5 g/chick and 2% recruitment for coniferous forest vs. 7.7 fledglings/pair, 16.8 g/chick and 1.1% recruitment for deciduous). However, tits laid earlier and had larger clutches and eggs in the deciduous forest (first egg on 29th April, 10.9 eggs/clutch and 1.7 ml/egg for deciduous vs. 30th April, 10.6 eggs/clutch and 1.6 ml/egg for coniferous). They also occupied a higher proportion of nest boxes in deciduous forest, although it should be noted that only half the number of boxes were erected (approximately 20% occupancy of 500–600 boxes in deciduous vs. 9% of 1,200–1,300 in conifers).

A replicated trial in 1998–2003 in beech, holly and oak forests on Sicily, Italy (46), found that blue tits *Parus caeruleus* (also *Cyanistes caeruleus*) showed a significant preference for small nest boxes with a 3.2 cm entrance hole, compared to large boxes with a 5 cm entrance. This study also describes the impact of hazel dormouse *Muscardinus avellanarius* exclusion on blue tit nesting success (see 'Reduce inter-specific competition for nest sites by removing or excluding competitor species').

A replicated trial between 1997 and 2004 in a reedbed in Lancashire, England (47), found that bearded tits (bearded reedlings) *Panurnus biarmicus* used 13–66% (average of 42%) of the 40–73 nest boxes provided, using boxes more frequently if they were placed over 10–15 cm of standing water. Nest boxes consisted of bundles of reeds tied on to wooden poles and pulled apart to allow tits to enter the bundle and build nests. There were positioned approximately 10 m inside the reedbed and installed in early February.

A replicated study in open farmland in Massachusetts, USA (48), found that tree swallows *Tachycineta bicolor* occupied 55% of 153 nest boxes provided in 2004, with an average of five eggs/clutch. Swallows preferentially selected east- and south-facing nest boxes before June, but used nest boxes based on availability later in the summer. East- and south- facing boxes heated up faster in the morning, but only early in the breeding season.

A replicated trial in arable farming landscapes in Norfolk, England, in the summers of 1997–2001 (49) found that tits *Parus* spp. nested in a higher proportion of hanging woodcrete boxes (38% 48 boxes occupied), compared to tree-mounted woodcrete boxes (25% of 48) or thick and thin wooden boxes (20% and 16% of 48 boxes respectively). Patterns were the same for great tits *Parus major*, blue tits *P. caeruleus* (also *Cyanistes caeruleus*) and all species combined (also including coal tits *P. ater* (also *Periparus ater*) and marsh tits *P. palustris* (also *Poecile palustris*)), although a higher proportion of great tits used woodcrete boxes (91% of great tits vs. 47% of blue tits). Clutch size, brood size and number of young fledged by blue tits and great tits did not differ significantly between box types. Woodcrete boxes were either attached to a tree trunk (18 cm high, base 18 cm diameter) or free-hanging (19 cm high, base 11 cm diameter). Wooden boxes were 16.5 × 15 × 19.5 cm, and of either 1.9 cm or 2.4 cm thick wood. All designs had a 3.2 cm diameter entrance. Another trial found that a higher proportion of tit *Parus* spp. nests were in 50 green nest boxes (72% of 41 nests) than in 50 brown boxes (28%), and in 50 boxes with circular entrances (68%) compared to those with a wedge-shaped entrance (32%).

A replicated trial in oak-savanna and vineyards in California, USA (50), in 2003–4 , found that western bluebirds *Sialia mexicana* occupied 56–67% of 120 nest boxes with entrance holes large enough to admit bluebirds. Clutches were larger (5.3 eggs/clutch vs. 5.0) and started earlier in vineyards than savanna. Tree swallows *Tachycineta bicolor*, house wrens *Troglodytes aedon*, ash-throated flycatchers *Myiarchus cinerascens* and violet-green swallows *Tachycineta thalassina* also occupied small numbers of boxes (all <5%).

A study in summer 2001 in a hayfield in south-east Ontario, Canada (51), found there were no differences in tree swallow *Tachycineta bicolor* nest-building behaviour or reproductive success between areas with a high density of nest boxes (boxes 20 m apart, average of 5.6 eggs/clutch and 3.4 chicks fledged/nest) or a low density (boxes 28 m apart, average of 5.3 eggs/clutch and 3.7 chicks fledged/nest). A total of 52 nest boxes were erected in a grid, 1.5 m above ground on metal poles, although only 22 were used for analysis.

A small study, as part of a large study in mixed agricultural and woodland habitats in Arkansas, USA (52), found that, in 2003, a female blue grosbeak *Passerina caerulea* successfully reared four chicks in a nest box designed for eastern bluebirds *Sialis sialis*. The authors indicate that, to their knowledge, this represented the first record of cavity-nesting (and hence nest-box use) by blue grosbeak. A total of 200 bluebird nest boxes (with 10 × 15 cm bases) were erected 2 m above ground.

A replicated, controlled trial in 1999 at a southern beech *Nothofagus* forest in Maule Region, Chile (53), found that the breeding population densities of thorn-tailed rayaditos *Aphrastura spinicauda* and house wrens *Troglodytes aedon* were significantly on higher ten experimental plots, each with five nest boxes installed, compared to ten control plots (4.2 rayaditos/ha and 4.0 wrens/ha vs. 1.0/ha and 1.4/ha for control plots). However, there were no differences in non-breeding season populations. Overall, 22% of boxes were used by rayaditos and 14% by wrens. A further 28% had nesting material in, but no active nest.

A trial from 2003 to 2005 on a single farm, Rawcliffe Bridge, East Yorkshire, UK (54) found that nest boxes were 54%, 50% and 68% occupied in 2003, 2004 and 2005 respectively. In 2003, all five boxes designed for tree sparrow were occupied. In 2005, 20 tree sparrow *Passer montanus* boxes (70% of the 28 provided) were occupied. The number of breeding tree sparrows on the farm increased from 6 to 20 pairs between 2003 and 2005. In the years 2003, 2004 and 2005, 32, 60 and 84 bird nest boxes were put up, including some designed for tree sparrows. They were inspected in February each year. Birds on the farm were monitored five times each year from 2003 to 2005, by walking the field boundaries. The number of breeding pairs/ha was estimated from clusters of sightings.

A before-and-after trial at a wetland site in Cambridgeshire, England (55), found that between eight and 50 pairs of sand martins *Riparia riparia* (and one pair of common kingfishers *Alcedo atthis*) nested annually in 130 artificial burrows drilled in a limestone cliff in 1995. No martins nested in 1996 because the burrows were too small. They were enlarged in 1997 (using a high pressure water jet) to include a nest chamber. The cliff is 3 m high and 80 m long and the burrows 5 cm in diameter and 60–90 cm long.

A replicated study in the summers of 2005–7 in an area of swamp forest in New York State, USA (56), found that black-capped chickadees *Parus atricapillus* (also *Poecile atricapillus*) nested in a higher proportion of artificial snags than nest boxes filled with wood shavings (60–70% of 20 snags excavated by chickadees each year and 25–30% used for nests vs. 40–50% and 15% for filled boxes). Chickadees also used more nest boxes filled with wood shavings than unfilled boxes. Nests in artificial snags were less likely to be usurped by mice and no more likely to be usurped by house wrens *Troglodytes aedon* than nest boxes. Twenty sites were used, 12 with snags and filled boxes and eight with unfilled boxes as well. Snags consisted on 10.2 cm diameter PVC pipes with a 2.8 cm entrance hole and filled with wood shavings.

A replicated trial in the summers of 2002–6 in suburban habitats in Toledo, Spain (57), found higher tree sparrow *Passer montanus* occupancy rates, and higher reproductive success in woodcrete nest boxes, compared to wooden ones (average of 76.5% of 50 woodcrete boxes occupied and 81% success for 152 clutches vs. 33.5% of 50 wooden boxes occupied and 79% success for 68 clutches). Differences in success were due to earlier clutches, a shorter incubation period and more reproductive attempts per season. Clutch size and nestling condition did not differ between box types. The authors suggest that higher temperatures in woodcrete boxes could explain the differences. One hundred boxes were placed in pairs (one of each material) less than 5 m apart and hung from trees. Wooden boxes were larger than woodcrete ones (2,057 cm^3 vs. 1,869 cm^3).

A replicated study in an oak-dominated forest in Gloucestershire, England, between 1990 and 2004 (58) found that significantly fewer pied flycatcher *Ficedula hypoleuca* chicks fledged from south-west-facing nest boxes, compared to other orientations. There were no differences for great tits *Parus major* and blue tits *Parus caeruleus* in productivity, but great tits showed a preference for certain nest box orientations, with significantly fewer nests in south-west-facing nest boxes. There were no preferences in the other two species.

A replicated study in acacia stands and shrubland in New South Wales, Australia (59), found that, after the installation of 400 nest boxes in July–September 2005, nearly all breeding attempts by zebra finches *Taeniopygia guttata* were in nest boxes, with 572 clutches laid in 2005–7 and over 90% of boxes being used at least once. Clutch size (average of 4.9 eggs/clutch for 559 clutches) and fledging success (58% of 522 clutches successful) in boxes was higher than in natural nests (average of 4.0 eggs/clutch for 110 natural nests and 13% success for 84 clutches). Small numbers of southern whitefaces *Aphelocephala leucopsis* and chestnut-rumped thornbills *Acanthiza uropygialis* also used nest boxes. Boxes were 14 × 9.3 × 12 cm with a 3 cm entrance hole and erected at 1–1.85 m above ground on trees or steel posts. The only unused boxes were high, far from natural cover and on poles, not trees.

A replicated study at an ancient temple site in rainforest on Java, Indonesia (60), found that Java sparrows *Padda oryzivora* nested successfully in 20 wooden nest boxes provided (two pairs in 2007, three in 2008), but not in 25 made from bamboo, or 25 of coconut shell. Wooden boxes were 32 × 15 × 23 cm with a 5 cm diameter entrance; bamboo nests were 40 cm sections of bamboo, 10–12 cm diameter with a 5 cm entrance hole; coconut nests were coconut shells with an 8 cm entrance hole. All nests were placed 8–20 m above ground in trees, either on the trunk of amongst the canopy. Two pairs of Javan mynas *Acridotheres javanicus* also used wooden boxes in 2007.

A study on Motuihe Island, New Zealand (61), found that only two of 150 nest boxes provided were used by 20 North Island saddleback *Philesturnus rufusater* translocated from Matangi Island in August 2005. Saddlebacks also preferred natural cavities over the 110 roosting boxes put out prior to release. The success of the translocation is discussed in 'Translocate individuals'.

A small replicated, controlled study from October–January in 2003–6 in two cattle ranches with predominantly *Celtis tala* woodlots in Buenos Aires Province, Argentina (62) found the southern house wrens *Troglodytes aedon bonariae* had higher reproductive rates in four nest box sites than in five natural cavity sites (72% of nest box nests produced at least one fledgling vs. 41% of natural cavities). The overall probability that at least one chick would fledge from nests was 66% in nest boxes and 25% in cavities. Nest box breeding pairs produced more nestlings in each attempt (3 nestlings/nest for nest boxes vs. 1 for natural cavities). The main cause of nest failure was predation. There was no difference between clutch size, brood size, or fledglings produced. The social mating system was unaffected by nest boxes. Nestboxes were wood (30.5 × 16.5 × 12.7 cm) with an entrance-hole 38 mm in diameter.

A long-term controlled paired sites study (63) in the same Estonian sites as in (45) found that the breeding density of great tits *Parus major* was significantly higher in areas with nest boxes in, compared to areas with no boxes. This held for all 13 pairs of transects in both deciduous (5.2 pairs/km for six transects in nest box areas vs. 1.8 pairs/km for six in control areas) and coniferous forests (3.3 pairs/km for seven transects in nest box areas vs. 0.2 pairs/km for seven in control areas). Over 1,000 nest boxes were erected in the experimental areas in the 1970s, on trees 1.5–2.0 m above ground.

A replicated study in eucalyptus stands in farmland in Lower Galilee, Israel (64), in 2008–9, found both jackdaws *Corvus monedula* and house sparrows *Passer domesticus* nested more frequently in nest boxes with small (7.5 cm diameter) entrances, compared to nest boxes with large (15 × 30 cm) entrance holes (jackdaws: nested in approximately 25% of 49 small-entrance boxes vs. 2% of 51 large-entrance boxes; sparrows: nested in 10% of small-entrance boxes and no large-entrance boxes). This may have been due to competition with barn owls *Tyto alba* which were able to enter the large-entrance boxes only. Boxes were attached to shaded parts of eucalyptus trees and the positions of large and small boxes were exchanged between years. This study also investigated nest box use by kestrels and owls.

A replicated, randomised study from January–June in 2007 in 58 semi-arid rural gardens in Jezreel Valley, Israel (65) found that great tits *Parus major* 47% of the nest boxes and succeeded in fledging at least one young in 74% of the 20 nest boxes in which they laid eggs, with an average of five chicks/pair. Breeding density was 5.4 pairs/10 ha, with more nests built in areas of higher tree density and tree species richness.

(1) Calhoun, J. B. (1948) Utilization of artificial nesting substrate by doves and robins. *The Journal of Wildlife Management*, 12, 136–142.
(2) Yom-Tov, Y. (1974) The effect of food and predation on breeding density and success, clutch size and laying date of the crow (*Corvus corone* L.). *Journal of Animal Ecology*, 43, 479–498.
(3) Howard, R. P. & Hilliard, M. (1980) Artificial nest structures and grassland raptors. *Raptor Research*, 14, 41–45.
(4) Brown, C. R. (1981) Reproductive success of purple martins in aluminum versus wooden birdhouses. *Journal of Field Ornithology*, 52, 148–149.
(5) Herlugson, C. J. (1981) Nest site selection in mountain bluebirds. *The Condor*, 83, 252–255.
(6) McComb, W. C. & Noble, R. E. (1981) Nest-box and natural-cavity use in three mid-south forest habitats. *The Journal of Wildlife Management*, 45, 93–101.
(7) Peterson, A. W. & Grubb, T. C. (1983) Artificial trees as a cavity substrate for woodpeckers. *The Journal of Wildlife Management*, 47, 790–798.
(8) Nilsson, S. G. (1984) The evolution of nest-site selection among hole-nesting birds: the importance of nest predation and competition. *Ornis Scandinavica*, 15, 167–175.
(9) Spero, V. M. & Pitts, T. D. (1984) Use of wood duck nest boxes by common grackles. *Journal of Field Ornithology*, 55, 482–483.
(10) Gustafsson, L. & Nilsson, S. G. (1985) Clutch size and breeding success of pied and collared flycatchers *Ficedula* spp. in nest-boxes of different sizes. *Ibis*, 127, 380–383.
(11) Petit, L. J., Fleming, W. J., Petit, K. E. & Petit, D. R. (1987) Nest-box use by prothonotary warblers (*Protonotaria citrea*) in riverine habitat. *The Wilson Bulletin*, 99, 485–488.
(12) Slagsvold, T. (1987) Nest site preference and clutch size in the pied flycatcher *Ficedula hypoleuca*. *Ornis Scandinavica*, 18, 189–197.
(13) Brawn, J. D. & Balda, R. P. (1988) Population biology of cavity nesters in northern Arizona: do nest sites limit breeding densities? *The Condor*, 90, 61–71.
(14) East, M. L. & Perrins, C. M. (1988) The effect of nestboxes on breeding populations of birds in broad-leaved temperate woodlands. *Ibis*, 130, 393–401.
(15) Pitts, T. D. (1988) Effects of nest box size on eastern bluebird nests. *Journal of Field Ornithology*, 59, 309–313.
(16) Finch, D. M. (1989) Relationships of surrounding riparian habitat to nest-box use and reproductive outcome in house wrens. *The Condor*, 91, 848–859.
(17) Finch, D. M. (1990) Effects of predation and competitor interference on nesting success of house wrens and tree swallows. *The Condor*, 674–687.
(18) Higuchi, H., Kawakubo, N., Koura, T., Santanda, F., Mizoguchi, F. & Hamaya, S. (1990) Frequency of use of nest boxes and breeding ecology of the Ryukyu robin *Erithacus komadori* on Nakano-shima in the Tokara Islands. *Strix*, 9, 1–13.
(19) Blem, C. R. & Blem, L. B. (1991) Nest-box selection by prothonotary warblers. *Journal of Field Ornithology*, 62, 299–307.
(20) Caine, L. A. & Marion, W. R. (1991) Artificial addition of snags and nest boxes to slash pine plantations. *Journal of Field Ornithology*, 62, 97–106.

(21) Johnson, L. S. & Kermott, L. H. (1991) Effect of nest-site supplementation on polygynous behavior in the house wren (*Troglodytes aedon*). *The Condor*, 93, 784–787.
(22) Pochop, P. A. & Ron J. Johnson (1993) Pentagon milk-carton nest box. *Journal of Field Ornithology*, 64, 239–243.
(23) Lombardo, M. P. (1994) Nest architecture and reproductive performance in tree swallows (*Tachycineta bicolor*). *The Auk*, 111, 814–824.
(24) Bock, C. E. & Fleck, D. C. (1995) Avian response to nest box addition in two forests of the Colorado Front Range. *Journal of Field Ornithology*, 66, 352–362.
(25) Grubb Jr, T. C. & Bronson, C. L. (1995) Artificial snags as nesting sites for chickadees. *The Condor*, 97, 1067–1070.
(26) Plissner, J. H. & Gowaty, P. A. (1995) Eastern bluebirds are attracted to two-box sites. *The Wilson Bulletin*, 107, 289–295.
(27) Dhanda, S. K. & Dhindsa, M. S. (1996) Breeding performance of Indian mynah *Acridotheres tristis* in nest boxes and natural sites. *Ibis*, 138, 788–791.
(28) Rendell, W. B. & Verbeek, N. A. (1996) Old nest material in nestboxes of tree swallows: effects on reproductive success. *The Condor*, 98, 142–152.
(29) Summers, R. W. & Taylor, W. G. (1996) Use by tits of nest boxes of different designs in pinewoods. *Bird Study*, 43, 138–141.
(30) Perrins, C. M. (1997) Stripe-breasted tits use nest boxes. *Bulletin of the African Bird Club*, 4, 67–68.
(31) Purcell, K. L., Verner, J. & Oring, L. W. (1997) A comparison of the breeding ecology of birds nesting in boxes and tree cavities. *The Auk*, 646–656.
(32) Radunzel, L. A., Muschitz, D. M., Bauldry, V. M. & Arcese, P. (1997) A long-term study of the breeding success of eastern bluebirds by year and cavity type. *Journal of Field Ornithology*, 68, 7–18.
(33) Stewart, L. M. & Robertson, R. J. (1999) The role of cavity size in the evolution of clutch size in tree swallows. *The Auk*, 116, 553–556.
(34) Sasvari, L. & Hegyi, Z. (2000) Sex-related local recruitment in colonial and solitary breeding European tree sparrows *Passer montanus* L. *Ibis*, 142, 119–122.
(35) White, D. H. & Seginak, J. T. (2000) Nest box use and productivity of great crested flycatchers in prescribed-burned longleaf pine forests. *Journal of Field Ornithology*, 71, 147–152.
(36) Lawler, J. J. & Edwards Jr, T. C. (2002) Composition of cavity-nesting bird communities in montane aspen woodland fragments: the roles of landscape context and forest structure. *The Condor*, 104, 890–896.
(37) López-Ortiz, R., Ventosa-Febles, E. A., Reitsma, L. R., Hengstenberg, D. & Deluca, W. (2002) Increasing nest success in the yellow-shouldered blackbird *Agelaius xanthomus* in southwest Puerto Rico. *Biological Conservation*, 108, 259–263.
(38) Miller, K. E. (2002) Nesting success of the great crested flycatcher in nest boxes and in tree cavities: are nest boxes safer from nest predation? *The Wilson Bulletin*, 114, 179–185.
(39) Dailey, T. B. (2003) Nest box use and nesting success of house wrens (*Troglodytes aedon*) in a midwestern wetland park. *The Ohio Journal of Science*, 103, 25–28.
(40) Mitrus, C. (2003) A comparison of the breeding ecology of collared flycatchers nesting in boxes and natural cavities. *Journal of Field Ornithology*, 74, 293–299.
(41) Stanback, M. T. & Rockwell, E. K. (2003) Nest-site fidelity in eastern bluebirds (*Sialia sialis*) depends on the quality of alternate cavities. *The Auk*, 120, 1029–1032.
(42) Field, R. H. & Anderson, G. Q. (2004) Habitat use by breeding tree sparrows *Passer montanus*. *Ibis*, 146, 60–68.
(43) Mennill, D. J. & Ratcliffe, L. M. (2004) Nest cavity orientation in black-capped chickadees *Poecile atricapillus*: do the acoustic properties of cavities influence sound reception in the nest and extra-pair matings? *Journal of Avian Biology*, 35, 477–482.
(44) Cooper, C. B., Hochachka, W. M. & Dhondt, A. A. (2005) Latitudinal trends in within-year reoccupation of nest boxes and their implications. *Journal of Avian Biology*, 36, 31–39.
(45) Mänd, R., Tilgar, V., Lõhmus, A. & Leivits, & Agu (2005) Providing nest boxes for hole-nesting birds – Does habitat matter? *Biodiversity and Conservation*, 14, 1823–1840.
(46) Sarà, M., Milazzo, A., Falletta, W. & Bellia, E. (2005) Exploitation competition between hole-nesters (*Muscardinus avellanarius*, Mammalia and *Parus caeruleus*, Aves) in Mediterranean woodlands. *Journal of Zoology*, 265, 347–357.
(47) Wilson, J. (2005) Nest box provision to provide additional nesting sites for bearded tits *Panurus biarmicus* at Leighton Moss RSPB Reserve, Lancashire, England. *Conservation Evidence*, 2, 30–32.
(48) Ardia, D. R., Pérez, J. H. & Clotfelter, E. D. (2006) Nest box orientation affects internal temperature and nest site selection by tree swallows. *Journal of Field Ornithology*, 77, 339–344.
(49) Browne, S. (2006) Effect of nestbox construction and colour on the occupancy and breeding success of nesting tits *Parus* spp. *Bird Study*, 53, 187–192.

(50) Fiehler, C. M., Tietje, W. D. & Fields, W. R. (2006) Nesting success of Western Bluebirds (*Sialia mexicana*) using nest boxes in vineyard and oak-savannah habitats of California. *Wilson Journal of Ornithology*, 118, 552–557.
(51) Male, S. K., Jones, J. & Robertson, R. J. (2006) Effects of nest-box density on the behavior of tree swallows during nest building. *Journal of Field Ornithology*, 77, 61–66.
(52) Risch, T. S. & Robinson, T. J. (2006) First observation of cavity nesting by a female blue grosbeak. *Wilson Journal of Ornithology*, 118, 107–108.
(53) Tomasevic, J. A. & Estades, C. F. (2006) Stand attributes and the abundance of secondary cavity-nesting birds in southern beech (Nothofagus) forests in South-Central Chile. *Ornitología Neotropical*, 17, 1–14.
(54) Bryson, R. J., Hartwell, G. & Gladwin, R. (2007) Rawcliffe Bridge, arable production and biodiversity, hand in hand. *Aspects of Applied Biology*, 81, 155.
(55) Gulickx, M. M., Beecroft, R. & Green, A. (2007) Creation of artificial sand martin *Riparia riparia* burrows at Kingfishers Bridge, Cambridgeshire, England. *Conservation Evidence*, 4, 51–53.
(56) Cooper, C. & Bonter, D. (2008) Artificial nest site preferences of black-capped chickadees. *Journal of Field Ornithology*, 79, 193–197.
(57) Garcia-Navas, V., Arroyo, L., Sanz, J. J. & Dias, M. (2008) Effect of nestbox type on occupancy and breeding biology of tree sparrows *Passer montanus* in central Spain. *Ibis*, 150, 356–364.
(58) Goodenough, A., Maitland, D., Hart, A. & Elliot, S. (2008) Nestbox orientation: a species-specific influence on occupation and breeding success in woodland passerines. *Bird Study*, 55, 222–232.
(59) Griffith, S. C., Pryke, S. R. & Mariette, M. (2008) Use of nest-boxes by the Zebra Finch (*Taeniopygia guttata*): implications for reproductive success and research. *Emu*, 108, 311–319.
(60) Kurniandaru, S. (2008) Providing nest boxes for Java sparrows *Padda oryzivora* in response to nest site loss due to building restoration and an earthquake, Prambanan Temple, Java, Indonesia. *Conservation Evidence*, 5, 62–68.
(61) Parker, K. A. & Laurence, J. (2008) Translocation of North Island saddleback *Philesturnus rufusater* from Tiritiri Matangi Island to Motuihe Island, New Zealand. *Conservation Evidence*, 5, 47–50.
(62) Llambias, P. E. & Fernandez, G. J. (2009) Effects of nestboxes on the breeding biology of southern house wrens *Troglodytes aedon bonariae* in the southern temperate zone. *Ibis*, 151, 113–121.
(63) Mänd, R., Lei, A., Lei, M. & Rodenhouse, N. L. (2009) Provision of nestboxes raises the breeding density of great tits *Parus major* equally in coniferous and deciduous woodland. *Ibis*, 151, 487–492.
(64) Charter, M., Izhaki, I. & Leshem, Y. (2010) Does nest basket size affect breeding performance of long-eared owls and Eurasian kestrels? *Journal of Raptor Research*, 44, 314–317.
(65) Charter, M., Leshem, Y., Halevi, S. & Izhaki, I. (2010) Nest box use by great tits in semi-arid rural residential gardens. *The Wilson Journal of Ornithology*, 122, 604–608.

15.5. Clean nest boxes to increase occupancy or reproductive success

- Five studies from Spain (2) and North America (5,7,9,10) found that various songbirds preferentially nested in cleaned nest boxes, compared to used ones. One study from the USA (10) found that eastern bluebirds showed this preference, but most did not switch from a soiled to a cleaned nest box. One study from the USA (3) found that birds showed an avoidance of heavily-soiled boxes and one from Canada (5) found that tree swallows *Tachycineta bicolor* preferentially selected nests which were sterilised by microwaving.

- Two studies from the USA (1,4) found that eastern bluebirds *Sialia sialis* and house wrens *Troglodytes aedon* preferentially nested in uncleaned nest boxes, and one study (8) found that prothonotary warblers *Protonotaria citrea* showed no preference for cleaned or uncleaned boxes.

- None of the five studies that investigated it (3,4,6–8) found any difference in success or parasitism levels between cleaned and uncleaned nest boxes.

Background

Old nest boxes may contain parasites and so reduce breeding success in them, or they may provide a suitable base for building new nests. Cleaning them out may, therefore, have a positive or a negative impact on nest site choice and reproductive success.

A replicated controlled study in 1993 in wooded pasture in Kentucky, USA (1), found that eastern bluebirds *Sialia sialis* preferentially occupied nest boxes containing old nests when given the choice (93% of 41 boxes occupied contained old nests). One hundred boxes were presented in pairs, one empty and one containing an old nest at 50 sites previously used by bluebirds. Boxes had a 10 × 10 cm base and a 2.9 × 10 cm entrance and were erected 1.5 m above ground.

A replicated, controlled study in 1993–4 in two areas of montane oak *Quercus pyrenaica* forest in the Community of Madrid, Spain (2), found that pied flycatchers *Ficedula hypoleuca* preferentially nested in nest boxes which had no old nesting material in them, although this preference was only significant in 1994 (approximately 75% of clean nest boxes occupied vs. 18–50% of boxes with old material inside). The authors note that 55% of the boxes with old nests in were completely cleaned in 1993, suggesting an aversion to old nesting material. In 1993, nest material was removed from all boxes before being replaced in 20 boxes (leaving 39 clean); in 1994, nest material was removed from 17 boxes, with 13 left with old material inside.

A replicated, controlled study in 1993–4 in woodland patches in Wyoming, USA (3), found that house wrens *Troglodytes aedon* showed no preference for cleaned nest boxes over controls with old nests in (46% of 59 pairs used cleaned boxes, 54% used controls). However, only 27% of heavily soiled boxes (with thick layers of dried faeces in) were used if they contained old nests, compared to 68% of used boxes which were only lightly soiled. There were no differences in reproductive output or blowfly infestations between nest box types. Forty (in 1993) or 50 (1994) pairs of boxes, one cleaned and one with an old nest in, were erected less than two metres apart across the study area.

A replicated, controlled study in 1993 in a floodplain and forest site in Illinois, USA (4), found that house wrens *Troglodytes aedon* nested in a lower proportion of cleaned nest boxes, compared to control boxes containing old nests (49% of 111 cleaned nest boxes used vs. 62% of 107 uncleaned boxes). There were no differences in reproductive output (5.5 nestlings/clutch surviving until 12 days old in cleaned nests vs. 5.2 in uncleaned nests, total of 24 nests examined) or mite infestation rates between box types. All boxes had been successfully used in 1992, with old nesting material removed from approximately half of them.

A replicated, controlled trial in marshland in 1991 in British Columbia, Canada (5), found that tree swallows *Tachycineta bicolor* preferentially nested in empty and clean boxes, but also preferred those where the old material had been microwaved to those with old, untouched nesting material (40 of 54 cleaned boxes used vs. 25 of 50 microwaved boxes and 13 of 54 untouched boxes). Pairs of boxes from different treatments were erected 3 m apart in tree swallow territories and which box was used was recorded.

A replicated, controlled study in marshland in 1991–2 in British Columbia, Canada (6), found that tree swallow *Tachycineta bicolor* reproductive success was not affected by removing old nesting material from nest boxes, removing nesting material and adding a raised 'floor' to simulate old nesting material or microwaving old nesting material. In 1992, cleaned birds using cleaned boxes laid and hatched eggs significantly earlier than those using other nest types (first eggs laid on 15th May and hatched on the 2nd June for 37 cleaned boxes vs. 18–20th May and 4–6th June for 68 others). Bird fleas *Certaophyllys idius* were more numerous in boxes with old nesting material. Use of boxes is discussed in 'Provide artificial nesting sites'.

A replicated, controlled study in mixed farmland in South Carolina, USA (7), found that, in 1988, eastern bluebirds *Sialia sialis* were more likely to reuse nest boxes cleaned after the season's first breeding attempt, compared to control (uncleaned) boxes (72% of 12 cleaned boxes reused vs. 57% of 12 controls). However, there were no differences in nesting success or overall number of nesting attempts between cleaned and control boxes (44% nesting success, 1.7 fledglings/second clutch and 24 nesting attempts for cleaned boxes vs. 50% success, 2.1 fledglings/second clutch and 26 attempts in control boxes), or in the likelihood of nest boxes being reused in 1989 (92% of 12 cleaned boxes used vs. 75% of controls). Alternative nest boxes were erected 200 m from previously used boxes, with 50% of new and 50% of old boxes being cleaned.

A replicated, controlled study in a tidal swamp in Virginia, USA (8), found that prothonotary warblers *Protonotaria citrea* showed no preference for cleaned nest boxes compared to control boxes with old nests in (32–38% of 164 cleaned boxes vs. 26–41% of 136 controls). The presence of an old nest did not affect laying date, clutch size, nestling mortality or brood parasitism by brown-headed cowbirds *Molothrus ater* (first eggs laid on 4th May, average of 5.1 eggs/clutch and 4% parasitism for both cleaned and control boxes; 14% egg and nestling loss for cleaned boxes vs. 10% for controls). Overall, warblers built 207 nests in 300 nest boxes provided.

A replicated study in North Carolina, USA (9), found that eastern bluebirds *Sialis sialis* preferentially selected cleaned nest boxes over uncleaned boxes, with 71% of 45 pairs switching from a previously-used box to an unused one. However, if successful nest boxes were cleaned then 75% of 32 pairs remained in the same box, rather than moving to an identical, cleaned box.

A replicated paired study in 1996–7 in North Carolina, USA (10), found that eastern bluebirds *Sialia sialis* preferentially used clean woodcrete nest boxes over woodcrete boxes that had already been used once in the year, with 71% of 45 pairs switching boxes. However, 73% of 26 pairs did not switch from a soiled woodcrete box to a clean wooden box. The preference for different box types is discussed in detail in 'Provide artificial nest sites'.

(1) Davis, W. H., Kalisz, P. J. & Wells, R. J. (1994) Eastern bluebirds prefer boxes containing old nests. *Journal of Field Ornithology*, 65, 250–253.
(2) Merino, S. & Potti, J. (1995) Pied flycatchers prefer to nest in clean nest boxes in an area with detrimental nest ectoparasites. *The Condor*, 97, 828–831.
(3) Johnson, L. S. (1996) Removal of old nest material from the nesting sites of house wrens: effects on nest site attractiveness and ectoparasite loads. *Journal of Field Ornithology*, 67, 212–221.
(4) Pacejka, A. J. & Thompson, C. F. (1996) Does removal of old nests from nestboxes by researchers affect mite populations in subsequent nests of house wrens? *Journal of Field Ornithology*, 67, 558–564.
(5) Rendell, W. B. & Verbeek, N. A. (1996) Old nest material in nest boxes of tree swallows: effects on nest-site choice and nest building. *The Auk*, 113, 319–328.
(6) Rendell, W. B. & Verbeek, N. A. (1996) Old nest material in nestboxes of tree swallows: effects on reproductive success. *The Condor*, 98, 142–152.
(7) Gowaty, P. A. & Plissner, J. H. (1997) Breeding dispersal of eastern bluebirds depends on nesting success but not on removal of old nests: an experimental study. *Journal of Field Ornithology*, 68, 323–330.
(8) Blem, C. R., Blem, L. B. & Berlinghoff, L. S. (1999) Old nests in prothonotary warbler nest boxes: effects on reproductive performance. *Journal of Field Ornithology*, 70, 95–100.
(9) Stanback, M. T. & Dervan, A. A. (2001) Within-season nest-site fidelity in eastern bluebirds: Disentangling effects of nest success and parasite avoidance. *The Auk*, 118, 743–745.
(10) Stanback, M. T. & Rockwell, E. K. (2003) Nest-site fidelity in eastern bluebirds (*Sialia sialis*) depends on the quality of alternate cavities. *The Auk*, 120, 1029–1032.

15.6. Use differently-coloured artificial nests

- A replicated study from the USA (1) found that two species showed different colour preferences for nest boxes, but that in each case, the preferred colour had lower nesting success than the less preferred colour.

Background

There is a possibility that birds will preferentially select nests of different colours, or that different coloured nests will have different success rates, possibly due to variable rates of predation.

A replicated study in 1945–6 in garden habitats in Ohio, USA (1), found that American robins *Turdus migratorius* made more nesting attempts in 62 green nests (16) than 59 black nests (eight), but there were an equal number of successes (four successful attempts in each: 25% success in green nests vs. 50% success in black). Mourning doves *Zenaida macroura* (formerly *Zenaidura macroura*) made 18 nesting attempts in 59 black nests compared to nine in 62 green nests, but there were four successful attempts in each colour (i.e. 22% success for black nests and 44% for green). The use of the nests (cones of black or green roofing paper 17.8 cm at the widest and 5.1 cm deep) by both species is discussed further in 'Provide artificial nest sites'.

(1) Calhoun, J. B. (1948) Utilization of artificial nesting substrate by doves and robins, *The Journal of Wildlife Management*, 12, 136–142.

15.7. Provide nesting material for wild birds

- A replicated study in the UK (2) found that songbirds used feathers provided at a very low rate and nest construction did not appear to be resource limited.
- A replicated, controlled study from Australia (1) found that four species of egrets used supplementary nesting material provided, preferentially taking material from raised platforms over water compared to plots on dry land.

Background

In some habitats nesting sites may be abundant, but material to create or line nests may be lacking. In these situations, conservationists may wish to provide this material.

A replicated, controlled study from September–January in 1989–1990 in 6 experimental and 3 control plots in a wetland in New South Wales, Australia (1) found that four species of egret (great white egret *Ardea alba*, intermediate egret *A. intermedia*, little egret *Egretta garzetta* and cattle egret *A. ibis*) collected supplementary nest material preferentially from raised platforms over water than from plots on dry land. At all locations over nineteen weeks there was a strong preference for material presented on platforms compared with that presented in supplementation plots (80% compared <20% of supplementary sticks respectively). The author suggests that nest material supplementation may reduce tree defoliation and lead to enhanced breeding success through fewer eggs and chicks falling out and greater thermal insulation. Sticks (0.3–2 cm diameter, 15–40 cm in length) were provided weekly on 2 × 1.3 m platforms over water and 2 m^2 plots on dry land. Control plots (2 m^2, 5–15 m away from experimental plots) were left with 'naturally present' sticks.

A replicated study from March–July in 1995–1997 in a mixed woodland area containing 20 experimental plots near Glasgow, Scotland (2) found that songbird species used supplementary feathers at very low rates and that nest construction is not resource limited. The mean feather loss per week from experimental plots was 14.4% and only in one week (mid-May) of the study did it rise above 40%. The proportion of marked feathers recovered from nests was 2.8%. A total of 41 nests (from 10 different songbird species) were found. Plots contained 50 marked (unique 2 mm diameter waterproof paint spot/site) feathers (30–50 mm contour feathers from woodpigeons *Columba palumbus*) placed directly on the ground each week.

(1) Baxter, G. S. (1996) Provision of supplementary nest material to colonial egrets. *Emu*, 96, 145–150.
(2) Hansell, M. & Ruxton, G. D. (2002) An experimental study of the availability of feathers for avian nest building. *Journal of Avian Biology*, 33, 319–321.

15.8. Repair/support nests to support breeding

- A small study from Puerto Rico (1) found that nine Puerto Rican parrot *Amazona vittata* nests were repaired, resulting in no chicks dying of cold.

Background

If nesting sites are limited then birds may use nests which are not suitable. Rather than move the nests or eggs (see 'Translocations'), it may be more appropriate to repair or support the nests.

A small study on nest success of Puerto Rican parrots *Amazona vittata* in the Luquillo Mountains, Puerto Rico, between 1973 and 1989 (1) found that nest guarding and repair work by volunteers and biologists/technicians prevented the failure of nine nests in nest holes that had wet cavities, which, without repair, would have resulted in the eggs becoming chilled. A total of 71 nests were guarded. This study is also discussed in 'Guard nests to increase nest success'.

(1) Lindsey, G. D. (1992) Nest guarding from observation blinds: strategy for improving Puerto Rican parrot nest success. *Journal of Field Ornithology*, 63, 466–472.

15.9. Artificially incubate eggs or warm nests

- A replicated, controlled trial in the UK (1) found that great tits *Parus major* were less likely to interrupt their laying sequence if their nest box was heated, although there was no effect on egg or clutch size.
- A small study in New Zealand (2) found that no kakapo *Strigopus habroptilus* eggs or chicks died from chilling following the use of nest warmers. Before this a nest had been lost to chilling.

Background

If incubating parents spend a long time away from the nest then the eggs may cool and potentially develop abnormally or die before hatching. For very intensively-managed species with very low populations, it may therefore make sense to warm eggs gently whilst parents are away from the nest.

A replicated, controlled trial in 1991 in woodland in Oxfordshire, England (1), found that blue tits *Parus caeruleus* nesting in heated nest boxes did not have significantly heavier eggs or larger clutches than those in unheated boxes. However, birds were less likely to interrupt their laying sequence in heated boxes (33% of 16 heated nests had interruptions vs. 67% of 14 unheated nests). Heat was provided by a small 'night light' candle, 8 cm below the bottom of the box, which raised the temperature in the box by an average of 6°C, saving roosting blue tits approximately 0.77 kcal/night, comparable to 35% of the energetic cost of producing an egg.

A small study on Codfish Island, South Island, New Zealand (2) found that no kakapo *Strigopus habroptilus* eggs or chicks died from chilling between 1997 and 2005, following the use of specially designed nest heat pads to keep eggs and chicks warm while the female is off the nest. Before pads were used, a nest containing three eggs lost failed, apparently due to chilling of the eggs and chicks as the female spent large periods of time away from the nest.

(1) Yom-Tov, Y. & Wright, J. (1993) Effect of heating nest boxes on egg laying in the blue tit (*Parus caeruleus*). *The Auk*, 110, 95–99.
(2) Jansen, W. P. (2005) Artificial incubation of kakapo *Strigops habroptilus* eggs and brooding of chicks while in the nest, Codfish Island, New Zealand. *Conservation Evidence*, 2, 6–7.

15.10. Provide nesting habitat for birds that is safe from extreme weather

- A small from New Zealand (3) found Chatham Island oystercatchers *Haematopus chathamensis* used raised nest platforms made from car tyres (designed to raise nests above the level of storm surges). The success of these nests is not reported.

- Two replicated, controlled studies from the USA (1,2) found that the nesting success of terns and waders was no higher on specially raised areas of nesting substrate, compared to unraised areas, with one study (1) finding that a similar proportion of nests were lost to flooding in raised and unraised areas.

Background

In habitats prone to flooding, inundation with water may be a significant cause of mortality. Providing nests or nesting habitat that is protected from water (e.g. by being raised) may therefore increase reproductive success.

A replicated, controlled trial in on alkaline flats in Oklahoma, USA, between 1991 and 1994 (1), found that nesting success of least terns *Sterna antillarum* and snowy plovers *Charadrius alexandrinus* was not higher on nesting ridges designed to protect nests from flooding, compared to nests not on ridges (terns: 53% of 32 nests on ridges vs. 53% of 28 nests off ridges; plovers: 79% of 22 nests on ridges vs. 62% of 26 nests off ridges). The proportions of nests lost to flooding were similar on and off the ridges for both species. This study is also discussed in 'Protect bird nests using electric fencing'.

A randomised, replicated and controlled paired study on three saltmarsh islands in Virginia, USA, in 2002 (2) found that hatching success of four ground-nesting bird species (American oystercatchers *Haematopus palliatus*, common terns *Sterna hirundo*, gull-billed terns *S. nilotica*, and black skimmers *Rynchops niger*) was not higher on sections of oyster shell piles artificially raised by 15–20 cm than on control (unraised) areas of piles (common terns: 60% of 15 nests hatching at least one egg on raised areas vs. 42% of 26 on control areas; gull-billed terns: 62% of 13 nests on raised areas vs. 89% of nine on control areas). Too few oystercatchers or skimmers nested for comparisons to be made. No species showed a significant preference for either raised or control areas. The authors note that whilst there were no significant differences between hatching successes, raised nests at three of the five shell piles studied survived flooding whilst unraised nests did not.

A small study on Chatham Island, New Zealand between 1998 and 2004 (3) found that up to seven pairs of Chatham Island oystercatcher *Haematopus chathamensis* used raised nest platforms made from car tyres (designed to raise nests above the level of storm surges). The success of these nests is not reported. The effect of moving the nests up the beach is discussed in 'Move nests whilst birds are using them'.

(1) Koenen, M. T., Utych, R. B. & Leslie Jr, D. M. (1996) Methods used to improve least tern and snowy plover nesting success on alkaline flats. *Journal of Field Ornithology*, 67, 281–291.

(2) Rounds, R. A., Erwin, R. M. & Porter, J. H. (2004) Nest-site selection and hatching success of waterbirds in coastal Virginia: some results of habitat manipulation. *Journal of Field Ornithology*, 75, 317–329.

(3) Moore, P. (2005) Storm surge protection of Chatham Island oystercatcher *Haematopus chathamensis* nests using tyre nest-platforms, Chatham Island, New Zealand. *Conservation Evidence*, 2, 78–79.

15.11. Remove vegetation to create nesting areas

- Two out of six studies found that the number of waders and terns nesting in an area increased following the removal of vegetation (3,5), and another (4) found that a tern colony moved to an area prepared by removing vegetation. Two of these studies (3,4) used multiple interventions at once. One study (3) found a decrease in colony size after several interventions, including vegetation control.

- A study from the UK (6) found that gulls and terns nested in an area cleared of vegetation and a controlled study from Puerto Rico (2) found that although no terns nested in plots cleared completely of vegetation, more nested in partially-cleared plots than in uncleared plots.

- A before-and-after study from Canada (1) found that tern nesting success was higher after plots were cleared of vegetation and other interventions were used.

Background

Many birds nest on bare substrate or in areas of sparse vegetation and may abandon a site if it becomes overgrown. Removing vegetation may therefore help to maintain a breeding colony, or attract birds to a new area.

This intervention describes the general removal of vegetation, studies discussing the removal of specific species are described in 'Threat: Invasive alien and other problematic species – Remove problematic vegetation'.

A before-and-after study on an island in Lake Ontario, Canada (1), found that the fledging success of common terns *Sterna hirundo* was significantly higher when ring-billed gull *Larus delawarensis* nests were destroyed and vegetation manually removed from the site, compared to before management. This study is discussed in detail in 'Control avian predators on islands'.

A control trial conducted over two seasons (1986–7) at two sooty tern *Sterna fuscata* breeding colonies in grasslands in the Culebra Archipelago of eastern Puerto Rico (2) found that terns did not nest in six experimental plots that had been entirely cleared of vegetation, but did nest in three plots in which partial removal of vegetation had resulted in 25%, 50% or 75% vegetation cover. In 1987, there were more nests in partially-cleared areas than in control (un-cleared) plots (58 nests in three regrowth plots vs. 40 in four controls).

Two before-and-after studies in 1977–89 at two common tern *Sterna hirundo* colonies in Lake Ontario, Canada (3), found that the nesting population increased at one colony but decreased at the second following the use of several interventions, including the removal of vegetation from the nesting area. This study is discussed in 'Replace nesting substrate following severe weather'.

A before-and-after study on two small islands in the Columbia River Estuary, Oregon, USA (4), found that an entire Caspian tern *Sterna caspia* colony (approximately 8,900 pairs) relocated from Rice Island to East Sand Island between 1999 and 2001, following the creation and maintenance of 1.5–3.0 ha suitable nesting habitat on East Sand Island in 1999–2001. A bulldozer was used to clear debris and vegetation, tractors smoothed the bare sand and plants were removed physically and through spraying with herbicide each year. In addition, various other interventions were used to encourage birds to move to East Sand Island (see 'Use decoys to attract birds to new nesting areas', 'Use vocalisations to attract birds to new nesting areas' and 'Control avian predators on islands') and habitat on Rice Island was modified to encourage birds to leave (see 'Alter habitat to encourage birds to leave'). The impact on conflict reduction (the purpose of the translocation) is discussed in 'Move fish-eating birds to reduce conflict with fishermen'.

A before-and-after study in 2002–3 in Lancashire, England (5), found that the number of waders nesting on limestone slag banks doubled in the summer after vegetation removed from 465 m^2 of the banks, compared to the summer before vegetation removal (2002: three pairs of ringed plovers *Charadrius hiaticula*, four northern lapwing *Vanellus vanellus*, four oystercatcher *Haematopus ostralegus*; 2003: six, nine and seven pairs respectively). Previously, in 1999, 2,390 m^2 of vegetation had also been removed. Vegetation was scrapped from the banks using the front bucket of a JCB in the winter, before any nesting birds arrived.

A before-and-after study at a former gravel pit in Kent, England (6), found that one pair of common terns *Sterna hirundo*, five pairs of black-headed gulls *Larus ridibundus* and approximately 100 pairs of herring gulls *L. argentatus* nested on a series of gravel islands after vegetation was removed from them. This study is discussed in 'Provide artificial nesting sites'.

(1) Morris, R. D., Kirkham, I. R. & Chardine, J. W. (1980) Management of a declining common tern colony. *The Journal of Wildlife Management*, 44, 241–245.

(2) Saliva, J. E. & Burger, J. (1989) Effect of experimental manipulation of vegetation density on nest-site selection in Sooty Terns. *The Condor*, 91, 689–698.

(3) Morris, H., Blokpoel, H. & Tessier, G. D. (1992) Management efforts for the conservation of common tern *Sterna hirundo* colonies in the Great Lakes: two case histories. *Biological Conservation*, 60, 7–14.

(4) Roby, D. D., Collis, K., Lyons, D. E., Craig, D. P., Adkins, J. Y., Myers, A. M. & Suryan, R. M. (2002) Effects of colony relocation on diet and productivity of Caspian terns. *The Journal of Wildlife Management*, 66, 662–673.
(5) Wilson, J. (2005) Removal of grass by scraping to enhance nesting areas for breeding waders at Leighton Moss RSPB Reserve, Lancashire, England. *Conservation Evidence*, 2, 60–61.
(6) Akers, P. & Allcorn, R. I. (2006) Re-profiling of islands in a gravel pit to improve nesting conditions for terns *Sterna* and small gulls *Larus* at Dungeness RSPB reserve, Kent, England. *Conservation Evidence*, 3, 96–98.

15.12. Guard nests to increase nest success

- A before-and-after study from Costa Rica (4) found an increase in scarlet macaw *Ara macau* population following the monitoring of nests, along with several other interventions.
- Two studies from Puerto Rico (1) New Zealand (2) found that parrot nest success was higher or mortality reduced or nest success higher with intensive monitoring of nests, compared to periods without monitoring. A study from New Zealand (3) also found high overall nest success when nests were monitored.

Background

If populations are reduced to extremely low levels and have low reproductive success then extremely intensive monitoring can be used to 'guard nests' and protect them from a range of threats through direct intervention. Due to the intensive nature of this work it is only likely to be viable if there are volunteers available to do it, and the population being monitored is very small.

A time-series study on nest success of Puerto Rican parrots *Amazona vittata* in the Luquillo Mountains, Puerto Rico, between 1973 and 1989 (1) found that the nest success of 71 Puerto Rican parrot *Amazona vittata* nests was 66% following the instigation of intensive nest monitoring in 1973, compared with an estimated 11–26% success of 19 nests before nest guarding (1955–72) and a predicted 38% success in 1973–80 had guarding not occurred. Threats to nests included the natural deterioration of nest cavities, predation, exoparasites, poor parental care, unviable eggs (replaced with captive-bred eggs), poor growth or unsuccessful fledging of chicks, human intrusion (three nests) and competition from other pairs. Some nests were affected by multiple threats.

A small before-and-after study in 1997 on Codfish Island (1,500 ha), New Zealand (2) found that remotely operated detonators successfully scared rats *Rattus* spp. from kakapo *Strigops habroptilus* nests on two occasions. In conjunction with intensive trapping and poisoning of rats on a grid system around six nests (see 'Control mammalian predators on islands'), this ensured that no nests were lost to rats in 1997, compared with potentially unsustainable predation in the years preceding 1997. No adverse effects on kakapos were found.

A study on Codfish Island (1,500 ha), South Island, New Zealand, in 2002 (3) found that 24 kakapo *Strigopus habroptilus* nests that were monitored by volunteers produced 26 chicks, of which 24 fledged. Volunteers followed a strict set of protocols and reported to 'controllers' frequently to ensure eggs survived. The authors argue that reproduction would have been lower without the intensive monitoring that volunteers provided (and which would not have been possible financially with paid staff).

A before-and-after study in western Costa Rica (4) found an increase in a scarlet macaw *Ara macau* population from 185–225 individuals in 1990–4 to 225–265 in 1997–2003, following the protection of artificial and natural nesting cavities and several other interventions (see 'Use education programmes and local engagement to reduce pressures on species', 'Promote sustainable alternative livelihoods based on species', and 'Provide artificial nesting sites'). In 1990–4 the population had been showing a 4%/year decline. This study is discussed in more detail in 'Increase "on-the-ground" protection to reduce unsustainable levels of exploitation'.

(1) Lindsey, G. D. (1992) Nest guarding from observation blinds: strategy for improving Puerto Rican parrot nest success. *Journal of Field Ornithology*, 63, 466–472.
(2) Jansen, W. (2005) Rat *Rattus* control at nests of the endangered kakapo *Strigops habroptilus* on Codfish Island, New Zealand. *Conservation Evidence*, 2, 1–2.
(3) Jansen, W. P. (2005) Using conservation volunteers to assist in monitoring of nests of the critically endangered kakapo *Stigops habroptilus*, on Codfish Island, New Zealand. *Conservation Evidence*, 2, 8–10.
(4) Vaughan, C., Nemeth, N. M., Cary, J. & Temple, S. (2005) Response of a scarlet macaw *Ara macao* population to conservation practices in Costa Rica. *Bird Conservation International*, 15, 119–130.

Foster chicks in the wild

15.13. Foster eggs or chicks with wild conspecifics

Background

Natural variations in reproductive output can be detrimental when populations are very small, for example if pairs fail to produce fertile eggs or some pairs repeatedly fail to raise offspring successfully. One way to minimise this problem is to foster eggs and chicks between nests. Eggs and chicks from nests with more offspring than they are likely to be able to raise can be moved to those with infertile eggs. Alternatively, if a pair produces fertile eggs or healthy chicks but consistently fails to raise chicks then it may be possible to transfer offspring to a more successful pair.

In other circumstances it may be possible to foster chicks with other species ('cross-fostering'). Studies describing this intervention are discussed in the following section 'Foster eggs or chicks with wild non-conspecifics (cross-fostering).'

15.13.1. Gannets and boobies

- A small controlled study in Australia (1) found that Australasian gannet chicks *Morus serrator* were lighter, and hatching and fledging success lower in nests which had an additional egg or chick added. However, overall productivity was (non-significantly) higher in experimental nests.

A small controlled study at a marine reserve in Queensland, Australia, in the breeding seasons of 1997–8 and 1998–9 (1) found that Australasian gannet chicks *Morus serrator* were significantly lighter, and hatching and fledging success significantly lower in nests where a second egg or chick was added to the nest ('experimental nests'), compared to control nests (maximum weight of approximately 2,500 g for experimental nests in 1997–8, n = 4 vs. approximately 3,250 g for controls, n = 8; data not provided for 1998–9; 1997–9: hatching success of 35% for experimental nests vs. 70% for controls; fledging success of 63% for experimental nests vs. 90% for control). Over both years, the number of chicks fledged by experimental nests was higher than control nests, but this was not significant (1.3 chicks/nest for experimental nests, n = 8 vs. 0.9 chicks/nest for controls, n = 8). This study also investigated the impact of supplementary feeding on gannet chicks (see 'Provide supplementary food to increase reproductive success').

(1) Bunce, A. (2001) Effects of supplementary feeding and artificial twinning on nestling growth and survival in Australasian gannets (*Morus serrator*). *Emu*, 101, 157–162.

15.13.2. Waders

- Two small trials in North America (1,2) found that piping plovers *Charadrius melodus* accepted chicks introduced into their broods, although in one case (1) the chick died later the same day.
- A replicated study from New Zealand (3) found that survival of fostered black stilts *Himantopus novasezelandiae* was higher for birds fostered to conspecifics rather than a closely related species.

A small trial on a beach in Nova Scotia, Canada in July 1983 (1) found that a pair of piping plovers *Charadrius melodus* adopted a chick introduced to their brood after it was abandoned by its parents when one day old. The foster parents brooded the chick but, following heavy rains on the day of introduction, it disappeared and is assumed to have died. The author suggests that its weakened condition, due to being abandoned by its parents meant that it could not survive the rainstorm, whereas its 'stepsiblings' could and were later seen flying.

A small trial on a beach in Connecticut, USA, in May 1990 (2) successfully introduced an orphaned piping plover *Charadrius melodus* chick into a foster family with four similarly-aged chicks. No aggressive behaviour was observed towards the foster chicks and all five young were seen flying in July. The chick was originally released within 11 m of the foster family when it was one day old.

A replicated study in South Island, New Zealand (3), investigated the survival of black stilts *Himantopus novasezelandiae*, fostered by both conspecifics and black-winged stilts *H. himantopus*. This study found that there was higher recruitment into the local population from chicks fostered by conspecifics. The study is discussed in more detail in 'Foster eggs or chicks with wild non-conspecifics (cross-fostering)' and 'Artificially incubate and hand-rear birds in captivity'.

(1) Flemming, S. P. (1987) Natural and experimental adoption of piping plover chicks. *Journal of Field Ornithology*, 58, 270–273.
(2) Midura, A. N. & Beyer, A. M. (1991) An observation of human-induced adoption in piping plovers. *Journal of Field Ornithology*, 62, 429–552.
(3) Reed, C. E. M., Ron J. Nilsson & Murray, D. P. (1993) Cross-fostering New Zealand's black stilt. *The Journal of Wildlife Management*, 57, 608–611.

15.13.3. Vultures

- Two small studies, one a New World vulture (1) and one of an Old World species (2) found that single chicks were successfully adopted by foster conspecifics, although in one case (1) this led to the death of one of the foster parents' chicks.

A small study on a farm in North Carolina, USA, in June 1975 (1) found that transferring a 35–40 day-old (American) black vulture *Coragyps atratus* chick from a nest that was about to be destroyed to a nest containing two 30–35 day-old chicks led to the successful rearing of the fostered chick. However, the smaller of the two chicks originally in the nest was neglected by its parents and died soon after the foster chick was introduced. No data on the fledging success or subsequent survival of the surviving chicks is provided.

A small study on Sicily, Italy (2), found that a captive-bred Egyptian vulture *Neophron percopterus* chick fostered into a wild nest in July 2003 was accepted by the foster parents and their two chicks and fledged successfully when approximately 90 days old. The chick was placed in the nest when 60 days old and competed successfully for food. The parents were supplied with supplementary food to ensure that the burden of feeding three chicks was not excessive (vultures tend to raise one or two chicks).

(1) Stewart, P. A. (1983) Adoption of introduced young and neglect of own by nesting black vultures. *The Wilson Bulletin*, 310–311.
(2) Di Vittorio, M., Falcone, S., Diliberto, N., Cortone, G., Massa, B. & Sarà, M. (2006) Successful fostering of a captive-born Egyptian vulture (*Neophron percnopterus*) in Sicily. in: Fabrizio Sergio (eds) *Journal of Raptor Research*, 40, 247–248.

15.13.4. Raptors

- Ten out of 11 studies from across the world (2–11) found that fostering raptor chicks to wild conspecifics had high success rates.
- A single study from the USA (1) found that only one of six eggs fostered to wild bald eagle *Haliaeetus leucocephalus* nests were hatched and raised. The authors suggest that the other eggs may have been infertile.

- A replicated study from Spain (4) found that Spanish imperial eagle *Aquila adalberti* chicks were no more likely to survive to fledging if they were transferred to foster nests from three chick broods (at high risk from siblicide), compared to chicks left in three-chick broods.
- A replicated study from Spain (7) found that young (15–20 years old) Montagu's harrier *Circus pygargus* chicks were successfully adopted, but three older (27–29 day old) chicks were rejected.

A replicated study in the eastern USA in 1977–80 (1) found that, of 12 captive bred bald eagle *Haliaeetus leucocephalus* nestlings (ten hand-reared and two parent-reared) fostered to wild nests when they were 2.5–5 weeks old, 11 were accepted by the foster parents. The remaining nestling was killed by a foster parent shortly after being introduced to the nest. Wild nests all had either eggs (including dummy eggs designed to induce continued incubation) or nestlings, which sometimes remained in the nest and were sometimes transferred to other wild nests. All foster pairs had histories of reproductive failure. In addition, in 1978–9, six captive-produced eggs were transferred to wild nests in the Chesapeake Bay area of Virginia and Delaware. Only one of these hatched and the eaglet was successfully raised and fledged. It is not known whether the other eggs were fertile, but the authors suggest that they may have been chilled after foster parents took a long time to return to the nest after the introduction of the new eggs. A further two eaglets were removed from the wild as eggs, hand-reared (described in 'Artificially incubate and hand-rear birds in captivity') and returned to more successful wild nests. Both eaglets were seen in advanced stages of development and were presumed to have fledged. This study also describes several *ex situ* interventions, discussed in 'Use captive breeding to increase or maintain populations', 'Use artificial insemination in captive breeding' and 'Release captive-bred individuals'.

A small study in wetlands in the Doñana National Park, southern Spain, in summer 1984 (2), found that an orphaned fledgling Spanish imperial eagle *Aquila adalberti* was successfully supported by foster parents. The orphan was found after leaving the nest (aged approximately 91 days) and initially taken into captivity before being released, with some supplementary food, into the home range of a family with two young of approximately the same age (which had left the nest but which were still dependent on parental feeding). All young were fed and chased by parents at approximately the same frequency, suggesting the foster fledgling had been accepted. This study also discusses other interventions, in 'Add perches to electricity pylons to reduce electrocution', 'Bury or isolate power lines', 'Use signs and access restrictions to reduce disturbance at nest sites' and 'Remove/treat endoparasites'.

A small study at a lake in Pennsylvania, USA, in 1988 (3), found that two osprey *Pandion halieatus* chicks were successfully adopted by a breeding pair that had lost their two chicks to predation. The foster chicks were 5.5 weeks old and were placed in the nest two weeks after the original young were last seen and were accepted later that day. No information is provided on fledging or subsequent survival but the author notes that 'the nestlings were fed well and protected by their foster parents throughout the nesting period'.

A replicated study in 1977–88 in wetlands in the Doñana National Park, Andalucia, Spain (4), found that there were no differences in survival between chicks relocated from nests were siblicide was likely to occur into nests thought to be safer (68% of 19 chicks surviving) compared to unmoved chicks (82% of 77 chicks surviving from all unmanipulated nests and 73% of 18 chicks from broods of three). Twelve of the moved chicks were from within the park and seven were moved in from outside. This study also discusses other interventions, in 'Add perches to electricity pylons to reduce electrocution', 'Bury or isolate power lines', 'Use signs and access restrictions to reduce disturbance at nest sites' and 'Remove/treat endoparasites'.

A 1993 review (5) found that 69% of 71 captive-bred Mauritius kestrel *Falco punctatus* chicks fostered to wild nests on Mauritius between 1986 and 1992 survived until independence. This study is discussed in more detail in 'Use captive breeding to increase or maintain populations', 'Artificially incubate and hand-rear birds in captivity' and 'Release captive-bred individuals'.

A 1995 update (6) of the same conservation programme studied in (5), found that fostering of hand-reared nestlings had a probability of surviving to independence. A total of 331 birds were released into various sites from 1984–1985 and 1993–1994 of which 78% became independent and 61% survived their first winter. Of these, 105 young were placed into 46 different broods (5–18 day old nestlings were placed in the nests of wild pairs that had been incubating for >2 weeks containing 1–5 other nestlings). Overall, 96 (91%) were fledged and 78 (81%) of these survived to independence. A total of 44 wild breeding pairs successfully raised at least one young to independence. The remainder of the released young were hacked in nestboxes. At the end of the 1993–1994 breeding season, the natural population had recovered to 222–286 birds (containing at least 56 breeding pairs and 40–70 non-breeding birds).

A replicated study in cereal fields in Madrid province, central Spain, in the breeding seasons of 1991–6 (7) found that 15 Montagu's harrier *Circus pygargus* nestlings introduced into foster nests at the age of 15–20 days were all successfully 'adopted', whereas three fledglings introduced at the age of 27–29 days were rejected by their intended foster parents and attacked when they begged for food. All nests already had nestlings in and all but two received a single nestling. One of these one was given two 15–20 day-old nestlings, the other two 27–29 day-olds.

A replicated study in pine forests in Slovakia in summers between 1993 and 2000 (8) found that golden eagle *Aquila chrysaetos* chicks removed from nests with two chicks, hand-reared or fostered in captivity and then fostered by wild conspecifics were successfully raised 74% of the time (of 35 fostering attempts). Failures were due to siblicide, predation or unknown causes. Without fostering, second chicks in eagle nests are frequently killed by siblings. In 2000 a chick was removed from a nest with two chicks in and initially placed in the nest used by the foster parents that year. However the parents ignored the chick, so it was moved to a nest used in previous years, 690 m away. Once in the second nest, the chick was fed and cared for. The authors suggest that the second nest was more obvious and so the parent eagles could see the foster chick more easily.

A replicated 2004 study of a Mauritius kestrel *Falco punctatus* release programme (9) found no difference in survival between birds 'hacked' as fledglings (and provided with supplementary food until independence), those fostered to wild breeding pairs or wild-bred birds (80% for 42 fostered birds; 80% for 46 hacked birds and 75% for 284 wild-bred birds). This study is discussed in detail in 'Release captive-bred individuals into the wild to restore or augment wild populations'.

A small study in Maharashtra, western India, in March 2003 (10) found that a Bonelli's eagle *Aquila fasciatus* (also known as *Hieraaetus fasciatus*) nestling that was removed from its nest when 40–42 days-old was repeatedly ejected from its nest by its parents after being returned. However, when transferred to a nest 250 km away occupied by a pair with two fledglings that had already fledged, the nestling was fed by both parents and fledglings for a week until it fledged in late March.

A small study at a reservoir in southern Spain in 2005 (11) found that a pair of ospreys *Pandion haliaetus* successfully raised two chicks that were fostered in an artificial nest (see 'Provide artificial nesting sites') when 12 and 15 days-old. The suitability of the pair as parents was tested by temporarily fostering a black kite *Milvus migrans* with them. Only after the kite had been fed and looked after were osprey chicks introduced. Both chicks fledged aged 53 and 55 days and left on migration 47 and 48 days after fledging. One chick was monitored with a GPS locator and reached typical wintering grounds in Senegal.

(1) Wiemeyer, S. N. (1981) Captive propagation of bald eagles at Patuxent Wildlife Research Center and introductions into the wild, 1976–80. *Raptor Research*, 15, 68–82.
(2) Gonzalez, J. L., Heredia, B., Gonzalez, L. M. & Alonso, N. (1986) Adoption of a juvenile by breeding Spanish imperial eagles during the postfledging period. *Raptor Research*, 20, 77–78.
(3) Rymon, L. M. (1990) Osprey nestlings fostered by hacked adults two weeks after predation of their young. *Journal of Raptor Research*, 24, 71–72.
(4) Ferrer, M. & Hiraldo, F. (1991) Evaluation of management techniques for the Spanish imperial eagle. *Wildlife Society Bulletin*, 19, 436–442.
(5) Cade, T. J. & Jones, C. G. (1993) Progress in restoration of the Mauritius kestrel. *Conservation Biology*, 7, 169–175.

(6) Jones, C. G., Heck, W., Lewis, R. E., Mungroo, Y., Slade, G. & Cade, T. (1995) The restoration of the Mauritius kestrel *Falco punctatus* population. *Ibis*, 137, S173–S180.
(7) Arroyo, B. E. & García, J. T. (2002) Alloparental care and kleptoparasitism in the semicolonial Montagu's Harrier *Circus pygargus*. *Ibis*, 144, 676–679.
(8) Kornan, M., Majda, M., Macek, M. & Kornan, J. (2003) An unusual case of adoption of a golden eagle (*Aquila chrysaetos*) chick in the Mala Fatra mountains, northwestern Slovakia. *Journal of Raptor Research*, 37, 259–260.
(9) Nicoll, M. A. C., Jones, C. G. & Norris, K. (2004) Comparison of survival rates of captive-reared and wild-bred Mauritius kestrels (*Falco punctatus*) in a re-introduced population. *Biological Conservation*, 118, 539–548.
(10) Pande, S. A., Pawashe, A. P. & Pednekar, B. (2004) How long is too long? a case of fostering nestling Bonelli's eagles (*Hieraaetufsa sciatus*). *Journal of Raptor Research*, 38, 381–382.
(11) Muriel, R., Ferrer, M., Casado, E. & Schmidt, D. (2006) First breeding success of osprey (*Pandion haliaetus*) in mainland Spain since 1981 using cross-fostering. *Journal of Raptor Research*, 40, 303–304.

15.13.5. Owls

- A replicated study in the USA (1) found high fledging rates for barn owl *Tyto alba* chicks fostered to wild pairs.
- A replicated controlled study from Canada (2) found that captive-reared burrowing owl *Athene cunicularia* chicks fostered to wild nests did not have significantly lower survival or growth rates than wild chicks.

A replicated study in Utah, USA (1), found that eight of ten barn owl *Tyto alba* chicks fostered to wild owl pairs in 1978 fledged successfully, with one male being confirmed as breeding in 1979, 60 km from the fledging site. Six young (all of which fledged) were placed in existing broods, with either one or two chicks in each, and four chicks were used to replace a clutch of four infertile eggs (two of these later died after falling from the nest).

A replicated, controlled trial in mixed grasslands in Saskatchewan, Canada, in 2001–3 (2), found that captive-reared burrowing owl *Athene cunicularia* chicks fostered to wild nests appeared to have lower survival rates than their wild foster siblings, but that this difference was not significant (six of nine foster owls died before migration vs. two of nine wild chicks). There were no differences in growth rates between wild chicks and captive chicks fostered at two to four days after hatching, three weeks after hatching or six weeks after hatching. In total, 54 birds were fostered, but not all could be monitored. Foster parents were supplied with one dead mouse a day for each fostered chick in their brood. This study is also discussed in 'Release captive bred individuals' and 'Use holding pens at release sites'.

(1) Marti, C. D. & Wagner, P. W. (1980) Successful releases of captive barn owls. *Raptor Research*, 14, 61–62.
(2) Poulin, R. G., Danielle Todd, L., Wellicome, T. I. & Brigham, R. M. (2006) Assessing the feasibility of release techniques for captive-bred burrowing owls. *Journal of Raptor Research*, 40, 142–150.

15.13.6. Cranes

- A small study in Canada (1) found high rates of fledging for whooping crane *Grus americana* eggs fostered to first time breeders (which normally have very low fertility).

A small study between 1986 and 1991 (1) found that at least three 'novice' breeding pairs of whooping cranes *Grus americana* in a population in north-west Territories and Alberta, Canada, successfully raised chicks when their own eggs were substituted for other eggs which were definitely fertile. Novice pairs normally have lower reproductive success than more experienced pairs. At least one pair with low breeding success was also provided with a fertile egg several days from hatching and successfully raised the chick. This study is also discussed in 'Use captive breeding to increase or maintain populations', 'Release captive-bred individuals' and 'Foster eggs or chicks with wild non-conspecifics (cross-fostering)'.

(1) Kuyt, E. (1996) Reproductive manipulation in the whooping crane *Grus americana*. *Bird Conservation International*, 6, 3–10.

15.13.7. Bustards

- A small study in Saudi Arabia (1) found that a captive-bred egg was successfully fostered to a female in the wild.

A small trial in a desert site in desert steppe in south-west Saudi Arabia in 1995 (1) found that a released, captive-bred female houbara bustard *Chalmydotis undulata macqueenii* successfully raised a captive-laid egg fostered into her nest. The chick hatched and fledged at 41 days old. The female had originally laid a single, infertile egg. The release programme is discussed in 'Release captive-bred individuals'.

(1) Gelinaud, G., Combreau, O. & Seddon, P. J. (1997) First breeding by captive-bred houbara bustards introduced in central Saudi Arabia. *Journal of Arid Environments*, 35, 527–534.

15.13.8. Woodpeckers

- Three studies from the USA (1–3) found that red-cockaded woodpecker *Picoides borealis* chicks fostered to conspecifics had high fledging rates.
- One small study (1) found that fostered chicks survived better than chicks translocated with their parents.

A small study in loblolly *Pinus taeda* and longleaf *P. palustris* pine forests in South Carolina, USA (1), found that all three red-cockaded woodpecker *Picoides borealis* nestlings translocated with their parents died, whereas two nestlings fostered to wild pairs in the release site were successfully raised. One (a female) disappeared after months, the other (a male) successfully bred. This study is discussed in more detail in 'Translocate individuals'.

A small study in a pine forest site in Mississippi, USA, in 1996 (2), found that two orphaned red-cockaded woodpecker *Picoides borealis* nestlings introduced into two foster nests fledged successfully (along with the non-fostered nestlings) and that at least one survived to the following breeding season (when it remained at its foster cluster as a helper). The chicks were both male and fostered when approximately 11 days old into broods containing a single nestling. One was added to the nest with the nestling still present, the other was added whilst the nestling was temporarily removed, to ensure the parents fed the foster chick. Between removal from their nest holes and fostering (later the same day), the chicks were supplied with mealworms and crickets.

A replicated, paired site study from April–July in 1997–1998 in 20 experimental and 18 control (containing 22 nestlings) red-cockaded woodpecker *Picoides borealis* nests in 5 forest sites in Louisiana, USA (3), found that fostered nestlings exhibited similar fledging rates to native nestlings in the same nests (85% of 20 fostered and 86% of 22 native nestlings fledged) and nestlings in control nests (68% of 22 control nestlings fledging). On average, fostered nests produced more fledglings than control nests (1.8 compared to 1.3 fledglings/nest). Feeding rates for fostered and native nestlings were similar. Cross-fostered nestlings were matched by age. Native and control nestlings were handled and returned to their native nests.

(1) Franzreb, K. E. (1999) Factors that influence translocation success in the red-cockaded woodpecker. *The Wilson Bulletin*, 111, 38–45.
(2) Richardson, D. M., Copeland, M. & Bradford, J. W. (1999) Translocation of orphaned red-cockaded woodpecker nestlings. *Journal of Field Ornithology*, 400–403.
(3) Wallace, M. T. & Buchholz, R. (2001) Translocation of red-cockaded woodpeckers by reciprocal fostering of nestlings. *The Journal of Wildlife Management*, 65, 327–333.

15.13.9. Parrots

- A replicated study from Venezuela (1) found that yellow-shouldered Amazon *Amazona barbadensis* chicks had high fledging rates when fostered to conspecific nests in the wild.

- A second replicated study from Venezuela (2) found significantly lower poaching rates of yellow-shouldered Amazons *Amazona barbadensis* when chicks were moved to foster nests closer to a field base.

A 1998 review of a yellow-shouldered Amazon *Amazona barbadensis* release programme on Margarita Island, Venezuela (1), found that, of 53 nestlings fostered to wild nests between 1989 and 1996, 44 (83%) chicks eventually fledged. The population on the island increased from 750 to approximately 1900 individuals between 1989 and 1996 as a result of recruitment increasing from zero in 1989 to 53 birds/year following management. This study is discussed in more detail in 'Release captive-bred individuals', 'Artificially incubate or hand-rear birds in captivity' and 'Use education programmes and local engagement to help reduce pressures on species'.

A replicated study in 2005–2006 (part of a longer study from 2000–2009) in 18 monitored yellow-shouldered Amazons *Amazona barbadensis* nests in tropical forest habitat on Margarita Island, Venezuela (2) found that fostering fledglings and assisted breeding significantly decreased poaching rates. The use of foster nests and assisted breeding in 2005 decreased poaching from 56% at the end of 2004 to 18% in 2005 and 0% poaching of monitored nests in 2006. Fledglings from high-risk nests (further away from the base) in the study area were moved to foster nests (possessing similarly aged fledglings) near the field base. All fledglings from each nest were then removed and placed them in a labelled wooden box in a secure facility after sunset, and returned to the nest at sunrise. This strategy was initiated once the parents stopped spending the night inside the nests, around 30–40 days after hatching. This study is also discussed in 'Relocate nestlings to reduce poaching', 'Use education programmes and local engagement to help reduce pressures on species', 'Employ locals as biomonitors' and 'Foster eggs or chicks with wild conspecifics'.

(1) Sanz, V. & Grajal, A. (1998) Successful reintroduction of captive-raised yellow-shouldered amazon parrots on Margarita Island, Venezuela. *Conservation Biology*, 12, 430–441.
(2) Briceño-Linares, J. M., Rodríguez, J. P., Rodríguez-Clark, K. M., Rojas-Suárez, F., Millán, P. A., Vittori, E. G. & Carrasco-Muñoz, M. (2011) Adapting to changing poaching intensity of yellow-shouldered parrot (*Amazona barbadensis*) nestlings in Margarita Island, Venezuela. *Biological Conservation*, 144, 1188–1193.

15.14. Foster eggs or chicks with wild non-conspecifics (cross-fostering)

Background

If the wild populations of a species are very small then it may not be possible to foster offspring to conspecifics. However, it may be possible to foster chicks and eggs to a similar, but more abundant species, if one is present. This can increase the reproductive output of the wild population of the endangered species, or even allow for the reintroduction of a population into parts of its former range.

15.14.1. Petrels and shearwaters

- A replicated and partially controlled study from Hawaii (1) found that Newell's shearwater *Puffinus newelli* eggs fostered to wedge-tailed shearwater *P. pacificus* nests had high fledging rates.

A replicated partially controlled study at two sites on Kaua'i, Hawaii, USA, in 1978–80 (1) found that Newell's shearwater *Puffinus newelli* (formerly *P. puffinus newelli*) eggs transferred to wedge-tailed shearwater *P. pacificus* had high hatching and fledging rates, with an average of 74% of 90 fostered eggs producing a fledgling. This is similar to the highest recorded rates for Manx shearwaters *P. puffinus* and slightly higher than those of wedge-tailed shearwaters. The main cause of mortality was egg predation by introduced common mynas *Acridotheres tristis*, with only one chick being evicted by foster parents. Fostered chicks were slightly heavier and larger than Newell's shearwaters raised by their natural parents. Data was not available on the return rates or breeding success of fostered chicks.

(1) Byrd, G. V., Sincock, J. L., Telfer, T. C., Moriarty, D. I. & Brady, B. G. (1984) A cross-fostering experiment with Newell's race of Manx shearwater. *The Journal of Wildlife Management*, 48, 163–168.

15.14.2. Waders

- A replicated and controlled study from the USA (1) found that killdeer *Charadrius vociferus* eggs incubated and raised by spotted sandpipers *Actitis macularia* had similar fledging rates to parent-reared birds.
- A replicated and controlled study from New Zealand (2) found that cross-fostering black stilt *Himantopus novasezelandiae* chicks to black-winged stilt *H. himantopus* nests significantly increased nest success, but that cross-fostered chicks had lower success than chicks fostered to conspecifics' nests.

A replicated and controlled experiment on two islands in Lake Michigan, USA, in 1987–9 (1) found that killdeer *Charadrius vociferus* eggs incubated and raised by spotted sandpipers *Actitis macularia* did not have significantly different hatching or fledging rates, compared to parent-reared eggs and chicks (47% hatching success, 0.8 fledglings/pair and 48% fledging success for cross-fostered chicks, n = 16 broods vs. 54%, 0.6 fledglings/pair and 27% for parent-reared chicks, n = 24 broods). There were no significant behavioural differences between parent-reared and cross-fostered chicks and one cross-fostered chick was seen two years after fledging, when it courted and mated with wild killdeer. No parent-reared chicks were seen again but the authors note that killdeer have low site-fidelity and so may not be seen again.

A replicated and controlled study in mountain streams and rivers in South Island, New Zealand, in the austral springs of 1981–7 (2) found that fledging success of managed black stilt *Himantopus novasezelandiae* nests was at least ten times that reported from unmanaged nests (13–27 chicks fledging in the population each year, a 20–42% fledging rate vs. 2% reported in another study for unmanaged nests). Eggs were removed from black stilt nests and artificially incubated (see 'Artificially incubate and hand-rear birds in captivity'), before being returned as they were hatching. If replacement in the original nest was not possible then eggs were placed in a foster nest, either another black stilt nest or a black-winged stilt *H. himantopus* nest. Fledging rates and recruitment to the local population were higher for chicks fostered by black stilts than cross-fostered chicks (66% of 50 chicks fostered by black stilts were resighted and five recruited locally vs. 19% of 21 cross-fostered chicks, with a single recruit). The authors note that cross-fostered chicks followed their foster parents on migration, probably leading to low recruitment.

(1) Powell, A. N. & Cuthbert, F. J. (1993) Augmenting small populations of plovers: an assessment of cross-fostering and captive-rearing. *Conservation Biology*, 7, 160–168.
(2) Reed, C. E. M., Ron J. Nilsson & Murray, D. P. (1993) Cross-fostering New Zealand's black stilt. *The Journal of Wildlife Management*, 57, 608–611.

15.14.3. Ibises

- A 2007 literature review (1) describes attempting to foster northern bald ibis *Geronticus eremite* chicks with cattle egrets *Bubulcus ibis* as unsuccessful.

A 2007 review of northern bald ibis (waldrapp) *Geronticus eremite* conservation (1) found that raising ibis chicks with cattle egrets *Bubulcus ibis* was not successful. This study discusses several *ex situ* interventions, described in the relevant sections.

(1) Bowden, C. G. R., Boehm, C., Jordan, M. J. R. & Smith, K. W. (2007) Why is reintroduction of northern bald ibis *Geronticus eremita* so complicated? An overview of recent progress and potential. 27–35 in: M. M. Lamont (eds) *The Proceedings of the IV International Symposium on Breeding Birds in Captivity; 2007 Sept 12–Sept 16* Toronto, Ontario, Canada.

15.14.4. Cranes

- Two studies from the USA (1,2) found low fledging success for cranes fostered to non-conspecifics' nests.

Background

There two species of cranes Grus spp. are resident in North America: the endangered whooping crane *G. americana* which migrates from central Canada to the southern USA; and the sandhill crane *G. canadensis* which contains several subspecies, some of which are migratory and some sedentary. Sandhill cranes offer the potential to foster whooping cranes, but there is uncertainty over the ability of whooping cranes to migrate successfully following if they are raised by another species.

As part of the planning for a whooping crane *Grus americana* reintroduction programme, a replicated study in Florida, USA, in 1982–7 (1) found that 22% of 23 wild Florida sandhill crane *G. canadensis pratensis* pairs successfully fledged chicks from captive-laid greater sandhill crane *G. c. tabida* eggs fostered in their nests. A further 35% hatched at least one egg but failed to fledge any chicks, 26% began incubation but then abandoned the substituted eggs and 17% immediately abandoned the eggs. Overall, survival of 34 cross-fostered eggs was 39% (from hatching to leaving the territory), lower than the 56% survival of captive-bred and released cranes, discussed in 'Release captive-bred individuals'. The eggs came from a combination of wild birds in Idaho, USA, and captive birds from Florida. Greater sandhill cranes are migratory, whilst Florida sandhill cranes are not. Migratory movements of fostered birds were larger than a control group of Florida sandhill cranes, but not significantly so.

A study in Idaho, USA, between 1975 and 1991 (2) found that 215 wild-sourced and 73 captive-bred whooping crane *Grus americana* eggs that were cross-fostered into sandhill crane *G. canadensis* nests had high hatching success (210 eggs hatching, 73% of total) but low fledging success (85 birds fledging, 30%), low survival (13 individuals alive in 1991, 5%) and no pairs formed between fostered individuals. Causes of mortality included predation by coyote *Canis latrans* and birds, collisions with fences and powerlines and disease. This study is also discussed in 'Use captive breeding to increase or maintain populations', 'Foster eggs or chicks with wild conspecifics' and 'Release captive-bred individuals'.

(1) Nesbitt, S. A. & Carpenter, J. W. (1993) Survival and movements of greater sandhill cranes experimentally released in Florida. *The Journal of Wildlife Management*, 57, 673–679.
(2) Kuyt, E. (1996) Reproductive manipulation in the whooping crane *Grus americana*. *Bird Conservation International*, 6, 3–10.

15.14.5. Songbirds

- A replicated study from the USA (1) found that the survival of cross-fostered yellow warbler *Dendroica petechia* chicks was lower than previously-published rates for the species, although incubation and nestling periods were very similar.
- A replicated and controlled study from Norway (2) found that the success of cross-fostering small songbirds varied depending on the species of chick and foster birds. However, only great tits *P. major* raised by blue tits *P. caeruleus* had lower pairing success than control birds, whilst blue tits raised by coal tits *P. ater* had higher recruitment than controls, or those raised by great tits.

A replicated study in 1978 and 1980 in a parkland site in Michigan, USA (1), transferred yellow warbler *Dendroica petechia* eggs and nestlings (two to six days old) to chipping sparrow *Spizella passerina* nests and found that four of six clutches transferred in 1978 produced fledglings that were fed by foster parents 16–26 days after hatching. In 1980, 34 fledglings were produced from a total of 64 eggs and 13 nestlings that were transferred into 26 nests. Eleven reached an age of at least 24 days and one male was seen returning to the study area in 1981 and showing normal yellow warbler behaviour. Incubation and nestling periods in cross-fostered chicks were identical to previously published results for yellow warblers, but survival of cross-fostered chicks was lower than previously published results. In 1978, nestlings and eggs were also transferred to field sparrow *S. pusilla*, and house wren *Troglodtyes aedon*, but the results were not provided. This study was used to investigate the possibility of cross-fostering Kirtland's warblers *D. kirtlandii*, which was endangered at the time of the study.

A replicated and controlled study in woodlands in southern Norway in 1998–2000 (2) found that cross-fostering did not affect the recruitment of great tits *Parus major* (12% of birds raised by blue tits *P. caeruleus* observed the following year, n = 155 chicks vs. 13% of control birds, n = 196) or pied flycatchers *Ficedula hypoleuca* (4% recruitment for birds raised by great or blue tits, n = 573 vs. 6% for control birds, n = 935). However, blue tits raised by coal tits *P. ater* had higher recruitment than controls, or those raised by great tits (18% recruitment for birds raised by coal tits, n = 38 chicks vs. 10% for birds raised by great tits, n = 242 and 7% for control birds, n = 175). Cross-fostering reduced pairing success in great tits (27% pairing success for cross-fostered chicks, n = 11 vs. 95% for controls, n = 20) but not in blue tits (100% pairing success for both cross-fostered and control chicks, n = 17 and 11 respectively) or flycatchers (95% pairing success for cross-fostered chicks, n = 19 vs. 95% for controls, n = 39). All fostering occurred during incubation, with eggs moved between nests before they hatched.

(1) Brewer, R. & Morris, K. R. (1984) Cross-fostering as a management tool for the Kirtland's warbler. *The Journal of Wildlife Management*, 48, 1041–1045.
(2) Slagsvold, T., Hansen, B. T., Johannessen, L. E. & Lifjeld, J. T. (2002) Mate choice and imprinting in birds studied by cross-fostering in the wild. *Proceedings of the Royal Society of London. Series B: Biological Sciences*, 269, 1449–1455.

Provide supplementary food

- A replicated, controlled study from Europe (1) found that overall, gardens with supplementary food did not contain more species than those without. However, there was some evidence that gardens with supplementary food in five countries did contain more species than unfed ones, when countries were analysed separately.

Background

Food supply is one of the key factors determining mortality and reproductive rates. Providing supplementary food is therefore often used as a technique to support small populations. However, feeding is only likely to have a positive effect on a population if the food supply is limiting either reproduction or survival. Because of differences in population responses to food during the breeding season and at other times, we have divided studies into those investigating the impact of feeding on reproduction and those investigating adult survival.

As with all interventions in this synopsis, studies that investigate population-level effects are most useful for conservationists. This is especially true for supplementary feeding, as many birds have large foraging ranges and the appearance of increased numbers at a feeding station, or even in the habitat surrounding feeders may not represent an increase in numbers but a redistribution of the same birds, and could even hide a population decline.

It is also important to note that the effect of providing food can be confounded by many factors. For example, variations in natural food supplies due to population cycles or irregular fruiting, whilst droughts or other extreme weather and pollution levels can also affect how populations respond to food.

Providing supplementary food can also be used to improve the success of release programmes for captive-bred birds. Studies describing this intervention are discussed in 'Captive breeding, rearing and releases (*ex situ* conservation) – Provide supplementary food after release'.

A replicated controlled study of 440 gardens across 14 countries in Western Europe (excluding the UK) from October 1988 until May 1989 (1) found that 264 gardens frequently provided with supplementary food were visited by an average of 21 species of birds, compared with 22 species for 40 moderately-fed gardens and 21 species for 116 infrequently-fed gardens. Differences were not significant. There was considerable variation across the study area, and feeding frequency appeared to affect the number of species visiting gardens in France and Switzerland, with 17.5 species visiting four infrequently-fed gardens, 13.3 species visiting four moderately-fed gardens and 20.7 species visiting 104 frequently-fed gardens. There was a similar, weaker effect for West Germany, the Netherlands and Belgium. Frequently-fed gardens were provided with food in more than two-thirds of the weeks studied, moderately-fed ones were provided with food for between one and two thirds of the weeks and infrequently-fed ones were provided with food in fewer than one third of the weeks studied.

(1) Thompson, P. S., Greenwood, J. J. D. & Greenaway, K. (1993) Birds in European gardens in the winter and spring of 1988–89. *Bird Study*, 40, 120–134.

15.15. Provide supplementary food to increase reproductive success

15.15.1. Petrels

- A replicated controlled study in Australia (1) found that Gould's petrel *Pterodroma leucoptera* chicks provided with supplementary food had very similar fledging rates to both control and hand-reared birds, but were significantly heavier than other birds.

A replicated, controlled study on Cabbage Tree Island, New South Wales, Australia, in 1995 (1), found that the fledging rate of 30 Gould's petrel *Pterodroma leucoptera* chicks provided with supplementary food was identical to that of control (unmoved, parent-fed) birds and not significantly different from translocated and hand-reared chicks (29/30 fed chicks fledged vs. 30/30 translocated chicks and 29/30 controls). Fed chicks were also heavier than both translocated and control chicks. Approximately 25 g of supplementary food was provided every three days, in addition to parent-provided food, starting at approximately three months old and continued until fledging. This study is also discussed in 'Provide artificial nesting sites', 'Translocate individuals' and 'Artificially incubate and hand-rear birds in captivity'.

(1) Priddel, D. & Carlile, N. (2001) A trial translocation of Gould's petrel (*Pterodroma leucoptera leucoptera*). *Emu*, 101, 79–88.

15.15.2. Gannets and boobies

- A small controlled study in Australia (1) found that Australasian gannet *Morus serrator* chicks were significantly heavier if they were supplied with supplementary food, but only in one of two years. Fledging success of fed nests was also higher, but not significantly so.
- A randomised replicated and controlled study in the Galapagos Islands (2) found that fed female Nazca boobies *Sula granti* were more likely to produce two-egg clutches if they were fed, and that second eggs were significantly heavier.

A small controlled study at a marine reserve in Queensland, Australia, in the breeding seasons of 1997–8 and 1998–9 (1) found that Australasian gannet *Morus serrator* chicks were reached significantly heavier weights in 1997–8 when they were fed every 2–3 days (starting at five days old and continuing until 40 days old) with approximately 5% of their bodyweight in pilchards *Sardinops sagax*, compared to control (unfed) chicks, however there were differences in weight in 1998–9 were not significant, although trends were in the same direction (1997–8: maximum weight of approximately 3900 g for fed chicks, n = 4 vs. approximately 3250 g for controls, n = 8). Over both years fledging success was higher for fed nests, but this was not significant (100% fledging success for fed nests vs. 90% for controls). This study also investigated the impact of adding foster chicks to gannet nests (see 'Foster eggs or chicks with wild conspecifics').

A randomised, replicated and controlled trial on Isla Española, Galapagos Islands, Ecuador, in the 1997–8 breeding season (2) found that female Nazca boobies *Sula granti* were more likely to produce two-egg clutches if they were fed at least 200 g of mullet *Mugil cephalus* twice daily, compared with control (unfed) females (92% of 49 fed females produced two eggs vs. 70% of 50 control females). Second eggs were also slightly larger from fed females, compared to controls (68 mm^3 for 44 eggs from fed females vs. 66 mm^3 for 32 from control females), first-laid eggs were no different between groups. Egg laying date and laying interval were similar between treatments. Females were fed until ten days had passed without laying an egg.

(1) Bunce, A. (2001) Effects of supplementary feeding and artificial twinning on nestling growth and survival in Australasian gannets (*Morus serrator*). *Emu*, 101, 157–162.
(2) Clifford, L. D. & Anderson, D. J. (2001) Food limitation explains most clutch size variation in the Nazca booby. *Journal of Animal Ecology*, 70, 539–545.

15.15.3. Auks

- Two replicated and controlled studies from the UK (1,2) found that Atlantic puffin *Fratercula arctica* chicks provided with supplementary food were significantly heavier than control chicks. One study (1) found differences between populations, suggesting some are more food-limited than others.
- The two UK studies found that fed chicks fledged at the same time as controls, whilst a randomised, replicated and controlled study from Canada (3) found that tufted puffin *Fratercula cirrhata* chicks supplied with supplementary food fledged later than controls.
- The Canadian study (3) found that fed chicks had faster growth by some, but not all, metrics.

A controlled, replicated study on St. Kilda, western Scotland, in 1975 (1), found that Atlantic puffin *Fratercula arctica* chicks fledged at significantly higher weights if they were provided daily with 50 g of sprats *Sprattus sprattus*, compared to control chicks (average weight of 316 g for 11 fed chicks vs. 301 g for 37 controls). Five chicks removed from burrows and fed sprats *ad libitum* were even heavier (365 g) while three unfed chicks with a single parent were lighter (240 g). All removed chicks and fed chicks fledged, 37 of 39 controls fledged and three of six single-parent chicks fledged. There was no difference in fledging age between fed and control chicks (40 days), but single-parent chicks took longer to fledge (45 days). A parallel study on the Isle of May (eastern Scotland) found smaller differences between treatments (removed: 367 g for six chicks; fed: 344 g for ten; controls: 331 g for 70; single-parent: 303 g for four), suggesting that St. Kilda puffins are to some degree food limited.

A replicated, controlled trial on the Isle of May, eastern Scotland, in June and July 1995 (2), found that Atlantic puffin *Fratercula arctica* chicks attained greater peak and fledging weights if they were fed daily (starting at eight days old and continued until fledging) with 25 g of sardines *Sardina pilchardus*, compared with control (unfed) chicks (peak weight of 319 g and fledging weight of 287 g for fed chicks, n = 22 vs. 305 g and 271 g for controls, n = 22). There was no difference in growth rates between treatments until chicks were 32 days old and no differences between growth rates of wing, head or tarsi. Fed chicks reached peak weights later than controls (peak weight at 35 days for fed chicks vs. 32 days for controls) but fledged at the same age (41 days old). Fed chicks were supplied with food by their parents significantly less often than control chicks. The authors suggest that additional weight was due to significantly less feeding by parents shortly before fledging, meaning that supplementary food provided a higher proportion of food received.

A randomised, replicated and controlled experiment on Triangle Island, British Columbia, Canada, in 1999 and 2000 (3) found that growth rates of tufted puffin *Fratercula cirrhata* chicks' culmen (upper beak) and tarsi (lower leg) were significantly higher when chicks were fed daily with either 58 g of herring *Clupea* sp. (in 1999) or 50 g of sand lance *Ammodytes* sp. (in

2000), compared to control (un-fed) chicks, although analysis revealed the effect on culmen growth was only apparent late in chick development. There was no effect of feeding on the rates of either wing growth or weight gain, and, in both years, fed chicks fledged later than unfed chicks (47–48 days at fledging for 32 fed chicks vs. 43–46 days for 32 controls). Parents of fed chicks made fewer provisioning trips but did not change the number, or species, of prey delivered each time.

(1) Harris, M. P. (1978) Supplementary feeding of young puffins, *Fratercula arctica*. *Journal of Animal Ecology*, 47, 15–23.
(2) Cook, M. I. & Hamer, K. C. (1997) Effects of supplementary feeding on provisioning and growth rates of nestling puffins *Fratercula arctica*: evidence for regulation of growth. *Journal of Avian Biology*, 28, 56–62.
(3) Gjerdrum, C. (2004) Parental provisioning and nestling departure decisions: a supplementary feeding experiment in tufted puffins (*Fratercula cirrhata*) on Triangle Island, British Columbia. *The Auk*, 121, 463–472.

15.15.4. Gulls, terns and skuas

- Four studies of three experiments from Europe (1,2) and Alaska (3,4) found that providing supplementary food increased fledging success or chick survival in two gull species, although a study from the UK (1) found that this was only true for one island, with abnormally low breeding success. A second island with higher success was not affected by feeding. Two of the experiments (2–4) fed parent birds and one (1) fed the chicks directly.
- One study from the Antarctic (5) found no effect of feeding parent skuas on productivity.
- One study from Alaska (4) found increased chick growth when parents were fed; one study from the Antarctic (6) found no increase in chick growth.

A replicated, controlled experiment in summer 1989 on Flat Holm and Skomer islands, south Wales (1), found that lesser black-backed gulls *Larus fuscus* provided with supplementary food laid significantly larger clutches and eggs on Flat Holm but not on Skomer (2.7 eggs/clutch for 21 fed clutches vs. 2.4 eggs/clutch for 34 unfed clutches on Flat Holm; 2.9 for 23 fed clutches on Skomer vs. 2.7 for 42 unfed). Average clutch size of unfed gulls on Flat Holm was also lower in 1989 than in previous years, whilst unfed clutches on Skomer were similar to the national average. This suggests that the population on Flat Holm was to some extent food limited in 1989, although laying date did not differ between fed and unfed clutches. Birds were provided with 200 g of fish each day.

A randomised, replicated and controlled study at a mixed gull colony on the island of Terschelling, the Netherlands, in April–July 1992 (2) found that pairs of lesser black-backed gulls *Larus fuscus* whose chicks were provided with supplementary food until fledging had significantly higher fledging success than control pairs (1.9 fledglings/nest and 87% of nests fledging at least one chick for 12 fed nests vs. 0.9 fledglings/nest and 56% success for 14 control nests). Fledging success of pairs whose chicks were fed until seven days old was intermediate (1.3 fledglings/nest and 67% success, 12 nests). In addition, significantly fewer chicks were predated in fed nests (0.5 chicks/nest for fully fed nests; 0.8 chicks/nest for partially fed and 1.3 chicks/nest for control nests). There were no significant differences in clutch size, egg size or hatching success between groups. Food was provided at an average of 46 g/day increasing to 76 g/day after a week and 150 g after three weeks, continuing until approximately 40 days old.

A replicated, controlled study during the breeding seasons of 1996–7 in the northern Gulf of Alaska, USA (3) using the same data as (4), found that black-legged kittiwake *Rissa tridactyla* pairs that were provided with supplementary food had significantly higher fledging success in both years than pairs fed until laying or hatching, or than control (unfed) pairs (81–85% fledging success and 1.0–1.3 chicks/pair for 43 fed pairs vs. 51–53% and 0.3–0.6 chicks/pair for 128 controls). Fed pairs also had larger clutch sizes and higher hatching success in 1997 but not 1996 (1996: 1.9 eggs/clutch and 76% hatching success for 73 and 49 clutches from fed pairs vs. 1.9 eggs/clutch and 65% for 59 and 83 clutches for controls; 1997: 1.8 eggs/clutch and 74% hatching success, 76 and 50 clutches from fed pairs vs. 1.6 eggs/clutch and 50%, for 59 and 85 control clutches). There was no effect of feeding on laying success (92–97% success for 157 fed pairs vs. 91–94% for 128 controls). Supplementary food consisted of 163 g/day of small pieces of herring, provided two or three times daily beginning in May, three weeks before first laying and ending in mid-August.

A replicated, controlled study during the breeding seasons of 1996–7 in the northern Gulf of Alaska, USA (4) using the same data as (3), found that black-legged kittiwake *Rissa tridactyla* pairs that were provided with supplementary food had earlier laying and hatching dates, faster growing chicks and higher chick survival than control pairs (for 157 fed pairs, first eggs laid on 3–4th June, hatched 30th June–1st July, 79–82% chick survival to 40 days vs. first eggs laid on 7th June, hatched 4th July, 51–53% chick survival to 40 days for 128 control pairs). However, there were no significant differences in egg volume, incubation period or chick fledging weight between treatments. Supplementary food consisted of 163 g/day of small pieces of herring, provided two or three times daily beginning in May, three weeks before first laying and ending in mid-August.

A randomised replicated and controlled trial on King George Island, Antarctic Peninsula, in the boreal summer of 2000–1 (5) (as part of the same study as (6)) found that south polar skua *Catharacta maccormicki* (also known as MacCormick's Skua) pairs provided with supplementary food did not raise significantly more chicks than control (unfed) pairs (average of 1.3 chicks/pair for 27 fed pairs vs. 1.5 chicks/pair for 27 controls). Supplementary food consisted of 25 g of fish provided to adults every other day when chicks were 6–35 days old and 100 g of fish when 35–55 days old. This corresponds to approximately 20% of a chick's daily energy needs.

A randomised, replicated and controlled trial on King George Island, Antarctic Peninsula, in the boreal summer of 2000–1 (6) (as part of the same study as (5)) found that male south polar skuas *Catharacta maccormicki* (also known as *Stercorarius maccormicki*) from 27 pairs provided with supplementary food were present at nests more often than males from 27 control (unfed) pairs (males present for 83% of observations when chicks were 7–50 days old for fed pairs vs. 74% for controls, a total of 955 observations). There was no difference in female attendance (81% attendance for fed pairs vs. 80% for controls, 955 observations). Chicks from fed pairs were not significantly larger than chicks from control (unfed) pairs, although wing growth was slightly faster in fed chicks (there were no changes in mass, head size or tarsus growth rates). Supplementary food consisted of 25 g of fish provided to adults every other day when chicks were 6–35 days old and 100 g of fish when 35–55 days old. This corresponds to approximately 20% of a chick's daily energy needs. This study also investigated the impact of feeding on adult condition, discussed in 'Provide supplementary food to increase adult survival'.

(1) Hiom, L., Bolton, M., Monaghan, P. & Worrall, D. (1991) Experimental evidence for food limitation of egg production in gulls. *Ornis Scandinavica*, 22, 94–97.

(2) Bukacinski, D., Bukacinska, M. & Spaans, A. L. (1998) Experimental evidence for the relationship between food supply, parental effort and chick survival in the lesser black-backed gull *Larus fuscus*. *Ibis*, 140, 422–430.

(3) Gill, V. A. & Hatch, S. A. (2002) Components of productivity in black-legged kittiwakes *Rissa tridactyla*: response to supplemental feeding. *Journal of Avian Biology*, 33, 113–126.

(4) Gill, V. A., Hatch, S. A. & Lanctot, R. B. (2002) Sensitivity of breeding parameters to food supply in black-legged kittiwakes *Rissa tridactyla*. *Ibis*, 144, 268–283.

(5) Ritz, M. S., Hahn, S. & Peter, H. U. (2005) Factors affecting chick growth in the South Polar skua (*Catharacta maccormicki*): food supply, weather and hatching date. *Polar Biology*, 29, 53–60.

(6) Ritz, M. S. (2006) Sex-specific mass loss in chick-rearing South Polar skuas *Stercorarius maccormicki* – stress induced or adaptive? *Ibis*, 149, 156–165.

15.15.5. Wildfowl

- A small randomised and controlled *ex situ* study from Canada (1) found faster growth and higher weights for fed greater snow goose *Chen caerulescens atlantica* chicks than unfed ones, but no differences in mortality rates.

A small randomised and controlled *ex situ* study on Bylot Island, north-west Territories, Canada, in 1991 (1), found that greater snow goose *Chen caerulescens atlantica* goslings provided with commercial duck food *ab libitum* grew faster and heavier than control goslings, which, except in bad weather when they were in danger of starvation, only had access to naturally-occurring food (weight at 40 days of 2,150–2,580 g for 11 fed goslings vs. 1,260–1,880 g for nine controls). In addition, plumage developed earlier in fed (or early-hatched) goslings (ninth primary emerged at 22–26 days old for fed goslings vs. 27 days for early hatched controls and 35 days for late-hatched controls). However, after controlling for hatching date, fed goslings did not have significantly lower mortality than controls.

(1) Lindholm, A., Gauthier, G. & Desrochers, A. (1994) Effects of hatch date and food supply on gosling growth in Arctic-nesting greater snow geese. *The Condor*, 96, 898–908.

15.15.6. Gamebirds

- A controlled study in Tibet (1) found that Tibetan eared pheasants *Crossoptilon harmani* that were fed supplementary food laid significantly larger eggs and clutches than control birds. Nesting success and laying dates were not affected.

A controlled trial at two scrubland sites near Lhasa, Tibet, during 1996 and 1999–2001 (1) found that Tibetan eared pheasants *Crossoptilon harmani* that were fed supplementary food (mainly highland barley provided by Buddhist nuns every day throughout the year) laid significantly larger eggs and had significantly larger clutches than control (unfed) birds, although the differences were small (average of 55 g/egg for 55 eggs from fed birds vs. 53 g/egg for 32 controls; average of 7.7 eggs/clutch for 23 fed birds vs. 7.0 eggs/clutch for 28 controls). Fed birds did not have higher nesting success or lay earlier than controls (fed birds: 96% of 144 eggs fertilised, 94% of these hatched, average first laying date of 6th May for 27 clutches; control birds: 98% of 124 eggs fertilised, 97% of these hatched, average first laying date of 8th March for 35 clutches). No data is provided on fledging success or survival of chicks.

(1) Lu, X. & Zheng, G. M. (2003) Reproductive ecology of Tibetan eared pheasant *Crossoptilon harmani* in scrub environment, with special reference to the effect of food. *Ibis*, 145, 657–666.

15.15.7. Rails and coots

- A small randomised and controlled trial in the USA (1) found that fed American coots *Fulica americana* laid heavier eggs, but not larger clutches than controls.
- However, a randomised, replicated and controlled study in Canada (4) found that clutch size, but not egg size was larger in fed American coot territories. There was also less variation in clutch size between fed territories (3).
- The Canadian study (4) also found that coots laid earlier when fed, whilst a replicated cross-over trial from the UK (2) found three was a shorter interval between common moorhens *Gallinula chloropus* clutches in fed territories, but that fed birds were no more likely to produce second broods.

A small randomised and controlled trial on a lake in Washington State, USA, in 1982 (1), found that American coots *Fulica americana* from three territories given supplementary food (1 kg of dog food provided three times a week in each experimental territory) laid heavier eggs than coots from four control (unfed) territories (30 g/egg for fed birds vs. 28 g/egg for controls), but that there was no consistent effect on clutch size (first laid clutches: 8 eggs/clutch for fed territories vs. 9 eggs/clutch for controls; when including replacement clutched: 8.7 eggs/clutch vs. 9.0 eggs/clutch).

A replicated cross-over trial in a waterfowl park in Cambridgeshire, UK (2), during spring and summer 1986 and 1987 found that common moorhens *Gallinula chloropus* had less time between clutches when provided with supplementary food than when no food was provided (average of 43 days between broods when fed vs. 49.0 days when unfed, nine females tested). However, fed birds were not more likely to produce second broods (84% of 19 fed territories producing second broods vs. 70% of 44 controls). Supplementary food was provided by an 'igloo-shaped feeder' in each fed territory, from five days before the first clutch hatched until the second clutch (if produced) was completed.

A randomised, replicated and controlled study in wetlands in Manitoba, Canada, in 1987–9 (3), found that within-clutch variation in the size of American coot *Fulica americana* eggs was slightly (and significantly) lower in territories where supplementary food was provided (at least one of steam-rolled corn, commercial trout food, commercial rabbit, chicken layer diet or oystershell), compared to in control (unfed) territories (3,219 eggs from 357 clutches measured, standard deviation of 1.2 cm^3 for fed clutches vs. 1.4 cm^3 for controls). Additional results from this study are presented in (4).

A randomised, replicated and controlled study in wetlands in Manitoba, Canada, in 1987–9 and 1991 (4), found that American coots *Fulica americana* laid clutches significantly earlier and had significantly larger clutches when provided with supplementary food (steam-rolled corn and commercial rabbit food), compared with control (unfed) pairs (average of May 5th for first laying date for fed 309 pairs vs. May 7th for 386 unfed pairs). However supplementary food did not affect egg size or laying rate and the authors note that feeding only accounted for 1% and 3% of the variation in laying date and clutch size respectively. No data was presented on the hatching rate or survival of chicks.

(1) Hill, W. L. (1988) The effect of food abundance on the reproductive patterns of coots. *The Condor*, 90, 324–331.
(2) Eden, S. F., Horn, A. G. & Leonard, M. L. (1989) Food provisioning lowers inter-clutch interval in moorhens *Gallinula chloropus*. *Ibis*, 131, 429–432.
(3) Arnold, T. W. (1991) Intraclutch variation in egg size of American coots. *The Condor*, 93, 19–27.
(4) Arnold, T. W. (1994) Effects of supplemental food on egg production in American coots. *The Auk*, 111, 337–350.

15.15.8. Waders

- A small controlled trial from the Netherlands (1) found that Eurasian oystercatchers *Haematopus ostralegus* did not produce larger replacement eggs if provided with supplementary food, and their eggs were, in fact smaller than the first clutch, whereas control females laid larger replacement eggs.

A small controlled trial on a saltmarsh on the island of Schiermonnikoog, The Netherlands (1) found that seven female Eurasian oystercatchers *Haematopus ostralegus*, following the experimental removal of their first clutch, did not produce significantly larger eggs (as part of a replacement clutch) if provided with 50 boiled mussels *Mytilus edulis* (averaging 46 mm long) a day, compared with control (unfed) oystercatchers (average egg volume of 41.2 cm^3 for fed females, n = 7 vs. 43.4 cm^3 for control females, n = 19). In addition, the replacement eggs of fed females were, on average, smaller than original eggs (by 0.9 cm^3), whereas replacement eggs for control females were larger than originals (by 0.2 cm^3).

(1) Jager, T. D., Hulscher, J. A. N. & Kersten, M. (2000) Egg size, egg composition and reproductive success in the oystercatcher *Haematopus ostralegus*. *Ibis*, 142, 603–613.

15.15.9. Ibises

- A study from China (1) found that breeding success of crested ibis *Nipponia nippon* was correlated with the amount of supplementary food provided, although no comparison was made with unfed nests.

A study between 1982 and 2004 in Shaanxi Province, central China (1) found that the breeding success (the percentage of eggs laid fledging chicks) of crested ibis *Nipponia nippon* pairs was correlated with the weight of supplementary food (mainly loach *Misgurnus anguillicaudatus*) provided to supplemented nests in the breeding season. Foraging success, number of eggs and number of young fledged were 'markedly higher' when supplementary food was provided. The weight of food provided ranged from 167–553 kg annually for each nest, with between one and six nests supplied each year. The study does not report the relative breeding success of supplied and unsupplied nests.

(1) Yu, X., Liu, N., Xi, Y. & Lu, B. (2006) Reproductive success of the crested ibis *Nipponia nippon*. *Bird Conservation International*, 16, 325.

15.15.10. Vultures

- Two before-and-after studies from the USA (1) and Greece (3) found that there were population increases in local populations of two vultures (one New World, one Old World) following the provision of food in the area. A study from Israel (2) found that a small, regularly supplied feeding station could provide sufficient food for breeding Egyptian vultures *Neophron percnopterus*.
- A before-and-after study from Italy (4) found that a small population of Egyptian vultures *Neophron percnopterus* declined following the provision of food at a feeding station, and only a single vulture was seen at the feeding station.

A small before-and-after study in California, USA, between February 1971 and May 1973 (1) found that three Californian condors *Gymnogyps californianus* were raised during the study period, compared to only one in the preceding three years. The authors suggest that the only substantial difference between the two time periods was the presence of the feeding station, supplied with approximately one carcass a week, normally of mule deer *Odocoileus hemionus*. The authors note that a longer study would be needed to confirm to role of feeding in reproductive success. This study is also discussed in 'Provide supplementary food to increase adult survival'.

A study in the Negev Desert, Israel, in April–August of 1989 and 1990 (2) found that a large (20–350 kg), irregularly supplied (twice a month) feeding station did not provided sufficient regular food for Egyptian vultures *Neophron percnopterus* feeding young, whereas a feeding station supplied daily with 5–10 kg of chicken did. This study is discussed in more detail in 'Provide supplementary food to increase reproductive success'.

A before-and-after study in a woodland mosaic in north-eastern Greece (3) found that, following the establishment of a feeding station in 1987, the number of European black (cinereous) vultures *Aegypius monachus* overwintering and breeding in the study area increased significantly (overwintering population of 24 in 1984 vs. 59 in 1997; breeding population of 10 pairs in 1984 vs. 21 in 1997). The proportion of pairs successfully fledging young varied considerably but also increased over the study period, as did the number of fledglings produced annually (40–55% fledging success in 1984–1986, average of 47% success and 5 young fledged/year vs. 50–95% and an average of 75% and 12 young/year in 1987–1997). The feeding station was supplied year-round with cattle or horse carcasses every two weeks or so and was protected by a fence to deter mammalian scavengers.

A before-and-after study in southern Italy (4) investigated the effect of a feeding station, active between February 2004 and September 2007, and found that Egyptian vultures *Nephron percnopterus* bred in small numbers until 2003 but only two to three non-breeding adults were seen in the area during 2004–7. Only a single vulture was seen at the station (in 2007), which was regularly used by corvids and raptors, discussed in 'Provide supplementary food to increase reproductive success – Raptors'.

(1) Wilbur, S. R., Carrier, W. D. & Borneman, J. C. (1974) Supplemental feeding program for California condors. *The Journal of Wildlife Management*, 38, 343–346.
(2) Meretsky, V. J. & Mannan, R. W. (1999) Supplemental feeding regimes for Egyptian vultures in the Negev Desert, Israel. *The Journal of Wildlife Management*, 63, 107–115.
(3) Vlachos, C. G., Bakaloudis, D. E. & Holloway, G. J. (1999) Population trends of black vulture *Aegypius monachus* in Dadia Forest, north-eastern Greece following the establishment of a feeding station. *Bird Conservation International*, 9, 113–118.

(4) Gustin, M., Giacoia, V. & Bellini, F. (2009) Establishment of a feeding station near the Laterza LIPU Reserve to provide additional food for three declining necrophagous raptor species in Apulia, Italy. *Conservation Evidence*, 6, 66–70.

15.15.11. Raptors

- A single small before-and-after study in Italy (12) found evidence for a small increase in local kite *Milvus spp.* populations following the installation of a feeding station.
- Four European studies (1,2,4,10) found that kestrels *Falco* spp. and Eurasian sparrowhawks *Accipiter nisus* laid earlier when supplied with supplementary food than control birds. One study (4) found that the earlier feeding began, the earlier average laying date was.
- Three studies from the USA (6,8) and Europe (11) found evidence for higher chick survival or condition when parents were supplied with food, whilst three from Europe found fed birds were more likely to lay (1) or laid larger clutches (2,4) and another (9) found that fed male hen harriers *Circus cyaneus* bred with more females than control birds.
- Four studies from across the world (1,3,7,9) found no evidence that feeding increased breeding frequency (3), clutch size (3,9), laying date (3), eggs size (3) or hatching or fledging success (1,7,9). A study from Mauritius (5) found uncertain effects of feeding on Mauritius kestrel *Falco punctatus* reproduction.
- There was some evidence that the impact of feeding was lower in years with peak numbers of prey species (2).

A replicated and controlled study in mixed conifer forests in southern Scotland between 1971 and 1979 (1), found that 13 pairs of Eurasian sparrowhawks *Accipiter nisus* provided with supplementary food in the pre-laying and laying period had larger clutch sizes, were less likely not to lay and laid earlier than 22 control (unfed) pairs (5.1 eggs/clutch for fed birds, 0% of pairs building nests but not laying, average laying date of 12th May vs. 4.0 eggs/clutch, 27% and an average laying date of 17th May for controls). Food was not provided after clutch completion and fed pairs did not have higher hatching or fledging success (69% hatching success and 62% fledging success for fed birds vs. 54% and 48% for 100 controls). Food provided consisted of either half a pigeon carcass (species not given) or two quail *Coturnix coturnix* carcasses every 1–4 days, was started on the 23rd–30th April and continued until late May at the latest.

A replicated controlled trial near wetlands in the northern Netherlands in 1978–80 (2) found that European kestrel *Falco tinnunculus* pairs that were provided with supplementary food initiated clutches earlier and laid larger clutches than control (unfed) pairs in two out of three years (six fed pairs started laying on 6–17th April, average of 5.0–5.7 eggs/clutch vs. 8–10th May and 4.5–4.6 eggs/clutch for 30 controls). In 1980, the differences were far smaller (23rd April and 6 eggs/clutch for three fed pairs vs. 25th April and 5.6 eggs/clutch for 18 controls), possibly due to it being a peak vole year. Significance levels were not provided. Supplementary feeding consisted of 100–120 g of mouse meat every daily (approximately twice the daily needs of captive kestrels), provided from late January (1978) or early March (1979 and 1980) until the start of incubation in late April or early May.

A replicated controlled trial in subtropical savanna in north-east South Africa in 1989–90 (3) found that Wahlberg's eagle *Aquila wahlbergi* pairs that were provided with supplementary food did not breed more often, have significantly larger clutches or eggs and did not lay earlier than control (unfed) pairs (75% of eight fed pairs breeding vs. 76% of 74 control pairs; average of 1 egg/clutch for both six fed and 56 unfed pairs; average egg volume of 76.7 ml for five fed pairs vs. 76.1 ml for five controls). Feeding consisted of approximately 200 g of meat provided daily from when pairs arrived in a territory until seven days after the first egg was laid (between three and six weeks of feeding).

A randomised, replicated controlled study of kestrels *Falco tinnunculus* in Cuenca, central Spain, in 1985–92 (4) found that food-supplemented breeding pairs laid earlier and had larger clutches than control pairs that had receive no food supplementation, and pairs fed from earlier (28th February, at least 50 days before laying vs. 17th April, 17 days before laying) laid earlier than late-fed pairs (average laying date of 30th April for nine early-fed pairs and May 9th for seven late-fed vs. May 15th for 15 controls; average of 5.1 eggs/clutch for fed pairs vs. 4.3 for controls). The onset of laying was not affected, but the average laying date was earlier. Supplementation affected clutch size independently of laying date. In control pairs, there was a seasonal decline in clutch size, but this decline was not seen with supplemented pairs.

A review of an integrated conservation programme for the endangered Mauritius kestrel *Falco punctatus* from 1973–1994 in montane forest habitat and a captive breeding centre in Black River, Mauritius (5) found that the provision of supplementary food had uncertain effects on breeding productivity. Fed birds produced more eggs, but did not necessarily fledge more chicks and, in some cases, fledged fewer. The authors speculate that Mauritius kestrels are less able to tolerate fur and feathers than other kestrels because young kestrels are fed primarily on Day geckos *Phelsuma* spp. and roughage causes digestive problems. Captive-bred birds and hacked birds were fed mice and day-old chicks.

A randomised, replicated and controlled trial in mixed conifer forests and scrub in New Mexico, USA, in 1992–3 (6), found that northern goshawk *Accipiter gentilis* nestlings from territories supplied with supplementary food had significantly higher survival than those from control (unfed) territories in 1993 but not 1992 and only when chick, rather than nest was the unit of analysis (1993: 90% survival for ten fed chicks vs. 37% survival for eight controls; 80% survival for five fed nests vs. 40% survival for five controls; 1992: 80% survival for 15 fed chicks vs. 100% survival for 16 controls). The authors suggest that this is due to increased attendance by females, as most nestling losses were due to predation and there were no significant differences in nestling size between treatments (average of 590–682 g for fed chicks and 541–676 g for controls). Supplementary food consisted of dead Japanese quail *Cortunix japonica* provided every other day starting the day after hatching and continuing until most control birds left the area. This study also examined differences in adult goshawk weights whilst provisioning, discussed in 'Provide supplementary food to increase adult survival'.

A randomised, replicated, controlled and paired trial in south-east Alaska, USA, in 1994–5 (7), found that bald eagle *Halieetus leucocephalus* pairs provided with supplementary food did not raise significantly more chicks than control (unfed) pairs and fed chicks were not significantly heavier (average of 2 chicks/nest fledged from 18 fed nests, average weight of 4.0–4.4 kg/chick vs. 2 chicks/nest and 4.1–4.2 kg/chick for 18 control nests). The authors note that nest failures and brood reductions were rare following hatching in both fed and control pairs, with most losses being during incubation (22 of 60 nests failed before hatching, three nests failed after hatching). Supplementary food consisted of a pink salmon *Oncorhynchus gorbuscha* provided daily from the date of hatching until hatchlings were seven weeks old, the amount of supplemental food carried to the nests was estimated to provide approximately 50% of the energy requirements of the nestlings.

A randomised, replicated and controlled trial in mixed conifer forests in Utah, USA, in 1996–7 (8), found that northern goshawks *Accipiter gentilis* chicks from territories provided with supplementary food (Japanese quail *Coturnix japonica* provided from close to hatching to chick independence) were significantly heavier (although not larger) than those from control (unfed) territories (average of 778 g for 29 fed chicks vs. 723 g for 22 controls). Nestling survival was significantly higher in fed nests in 1997 (100% survival of 19 fed chicks throughout the study vs. 56% survival of 18 controls) but not 1996 (87% survival of 15 fed chicks throughout the study vs. 89% survival of 18 controls). The authors suggest this difference is due to variations in natural food supply as predation was not a primary mortality factor (although females did stay closer to nests in fed territories, compared to controls). This study also examined differences in adult female weights whilst provisioning, discussed in 'Provide supplementary food to increase reproductive success'.

A randomised, replicated and controlled study in heathland on Orkney Mainland, Scotland, in 1999–2000 (9) found that male hen harriers *Circus cyaneus* provided with supplementary food (chicken *Gallus domesticus* chicks and quarter pieces of European rabbit

Oryctolagus cuniculus or brown hare *Lepus europaeus*) bred with significantly more females than control (unfed) males (100% of 11 fed males mated and 36% mated with more than one female vs. 80% of nine unfed males mated, 11% mated with more than one female). There was no effect of feeding on clutch size or hatching success (average of approximately 5.1 eggs/clutch for 13 fed clutches vs. 4.7 eggs/clutch for four unfed clutches), but productivity still increased. Hooded crows *Corvus cornix* were also removed from all territories, discussed in 'Control avian predators on islands'.

A replicated, randomised, controlled study from February–March in 1999 in 2 mixed experimental and control lesser kestrel *Falco naumanni* colonies, 2 full experimental and 3 full control colonies within an agricultural landscape in La Mancha, Spain (10) found that lesser kestrels provided with food initiated clutch laying earlier than non-supplemented nests or colonies. In mixed colonies, fed pairs laid earlier than unfed ones by 5.8 days on average. There was no significant difference in mean laying date between unfed pairs of mixed colonies and pairs in all-unfed colonies, or between fed pairs of mixed colonies and pairs in all-fed colonies. Additionally, laying date was significantly earlier in all-fed than in all-unfed colonies by 5.7 days on average. The authors suggest, therefore, that laying date is restricted by food availability and is not confounded by individual quality. Extra food consisted of day-old cockerel chicks (35–40 g) placed within each experimental nest.

A replicated before-and-after study in 37 Spanish imperial eagle *Aquila adalberti* territories in Castilla y León, Castilla-La Mancha, Madrid and Extremadura provinces, Spain (11) found that the fledging rate of eagles was significantly higher when territories were supplied with supplementary food, compared to before feeding (average fledging rate of 1.56 young/pair with feeding vs. 0.72 young/pair without, 37 clutches investigated). Siblicide was also lower in fed nests (6% of 50 chicks lost to siblicide in fed nests vs. 45% of 86 in unfed nests) and overall productivity increased in 89% of pairs and declined in just 8%. The increase in fledging rate was even higher in pairs with at least one subadult parent (1.57 young/pair in fed territories vs. 0.54 young/pair in unfed territories) and occurred in both high and low-quality habitats. Food was supplied to territories with more than one chick (and only after the second chick hatched) and consisted of one European rabbit *Oryctolagus cuniculus* provided every two to three days in territories with three chicks or every four days for two chicks.

A small before-and-after study in southern Italy (12) investigated the effect of a feeding station, active between February 2004 and September 2007, and found that no red kites *Milvus milvus* bred between 2004 and 2006, but three pairs bred in 2007. Three pairs of black kites *Milvus migrans* bred in each of 2005–7, before this it had been only an irregular breeder. The station consisted of a 40 × 40 m enclosure surrounded by a 1.8 m high fence. An average of 50 sheep carcasses were provided each year. Between one and 23 red kites and five and 53 black kites were seen every year alongside a large number of non-target corvids. This study also describes the impact of the station on Egyptian vultures *Beophron percnopterus*, described in 'Provide supplementary food to increase reproductive success'.

(1) Newton, I. & Marquiss, M. (1981) Effect of additional food on laying dates and clutch sizes of sparrowhawks. *Ornis Scandinavica*, 12, 224–229.
(2) Dijkstra, C., Vuursteen, L., Daan, S. & Masman, D. (1982) Clutch size and laying date in the kestrel *Falco tinnunculus*: effect of supplementary food. *Ibis*, 124, 210–213.
(3) Simmons, R. E. (1993) Effects of supplementary food on density-reduced breeding in an African eagle: adaptive restraint or ecological constraint? *Ibis*, 135, 394–402.
(4) Aparicio, J. M. (1994) The seasonal decline in clutch size: an experiment with supplementary food in the kestrel, *Falco tinnunculus*. *Oikos*, 71, 451–458.
(5) Jones, C. G., Heck, W., Lewis, R. E., Mungroo, Y., Slade, G. & Cade, T. (1995) The restoration of the Mauritius kestrel Falco punctatus population. *Ibis*, 137, S173–S180.
(6) Ward, J. M. & Kennedy, P. L. (1996) Effects of supplemental food on size and survival of juvenile northern goshawks. *The Auk*, 113, 200–208.
(7) Gende, S. M. & Willson, M. F. (1997) Supplemental feeding experiments of nesting bald eagles in southeastern Alaska. *Journal of Field Ornithology*, 68, 590–601.
(8) Dewey, S. R. & Kennedy, P. L. (2001) Effects of supplemental food on parental-care strategies and juvenile survival of northern goshawks. *The Auk*, 118, 352–365.

(9) Amar, A. & Redpath, S. M. (2002) Determining the cause of the hen harrier decline on the Orkney Islands: an experimental test of two hypotheses. *Animal Conservation*, 5, 21–28.
(10) Aparicio, J. M. & Bonal, R. (2002) Effects of food supplementation and habitat selection on timing of lesser kestrel breeding. *Ecology*, 83, 873–877.
(11) Gonzalez, L. M., Margalida, A., Sánchez, R. & Oria, J. (2006) Supplementary feeding as an effective tool for improving breeding success in the Spanish imperial eagle (*Aquila adalberti*). *Biological Conservation*, 129, 477–486.
(12) Gustin, M., Giacoia, V. & Bellini, F. (2009) Establishment of a feeding station near the Laterza LIPU Reserve to provide additional food for three declining necrophagous raptor species in Apulia, Italy. *Conservation Evidence*, 6, 66–70.

15.15.12. Owls

- Two replicated and controlled trials from Europe (1) and the USA (2) found that owls supplied with supplementary food had higher hatching and fledging rates than control pairs. The European study, but not the American, also found that fed pairs laid earlier and had larger clutches
- The American study (2) also found that owls were no more likely to colonise nest boxes provided with supplementary food.

A replicated, controlled trial in western Finland in spring 1986 (1) found that ten Tengmalm's owl *Aegolius funereus* pairs supplied with supplementary food (50–60 g of dead laboratory mice *Mus musculus* provided daily) during the pre-laying and laying periods started laying earlier, had larger clutches and had higher hatching success than 15 control (unfed) pairs (fed pairs: clutches started on 24th March, 6.8 eggs/clutch, 98% hatching success; control pairs: clutches started on 30th March, 5.9 eggs/clutch, 82% hatching success). Fed pairs also had more fledglings in successful nesting attempts than controls (5.6 fledglings/successful nesting attempt for nine fed pairs vs. 4.0 fledglings/successful nesting attempt for 15 controls). There were no differences in egg dimensions or the weights of parents and fledglings between fed and control nests. The author notes that the study took place in a peak vole year, when naturally-occurring food was extremely plentiful.

A replicated, controlled and paired study at two suburban study sites in Texas, USA, in 1992–5 (2), found that eastern screech owls *Megascops asio* (formerly *Otus asio*) nest in nest boxes provisioned with supplementary food (46% of nests in fed boxes vs. 53% in unfed boxes). First laying dates, clutch sizes and chick survival rates were no different between owls in 13 fed and 16 unfed boxes (fed nest boxes: first eggs laid on 22nd March, 3.0–4.5 eggs/clutch and 77% nestling survival; unfed nest boxes: first eggs laid on 20–22nd March, 3.0–4.5 eggs/clutch and 75% nesting success), but owls in fed boxes had higher fledging success than those in unfed boxes, due to greater hatching success (93% of eggs producing fledglings in 13 fed boxes vs. 74% in 16 unfed boxes). Food provided was either 9 g/day or 26 g/day of dead laboratory mice *Mus musculus* and began approximately 30 days before first laying date and continued through first laying.

(1) Korpimaki, E. (1989) Breeding performance of Tengmalm's Owl *Aegolius funereus*: effects of supplementary feeding in a peak vole year. *Ibis*, 131, 51–56.
(2) Gehlbach, F. R. & Roberts, J. C. (1997) Experimental feeding of suburban eastern screech-owls *Otus asio* has few effects on reproduction apart from non-experimental factors. *Journal of Avian Biology*, 28, 38–46.

15.15.13. Kingfishers

- A controlled study in the USA (1) found that belted kingfisher *Ceryle alcyon* supplied with food had heavier nestlings and were more likely to renest. There was mixed evidence for the effect of feeding on laying date.

A controlled study of breeding pairs of belted kingfisher *Ceryle alcyon* in Colorado, USA (1), found that birds that nested earlier had heavier nestlings and were more likely to renest in the event of nest failure. In 1994 and 1995, food supplementation from early in the prebreeding season (8 March) was associated with earlier nest initiation. However, no supplemented nests occurred earlier than the earliest nests in 1992, in which there had been no feeding stations. A smaller amount of food started later in the season (20 April) in the previous year had been without effect.

(1) Kelly, J. F. & Van Horne, B. (1997) Effects of food supplementation on the timing of nest initiation in belted kingfishers. *Ecology*, 78, 2504–2511.

15.15.14. Pigeons

- A replicated cross-over study in the UK (1) found no differences in reproductive parameters of European turtle doves *Streptopelia turtur* between years when food was supplied and those when it was not.

A replicated cross-over study in 1999–2000 in ten mixed agricultural and natural habitat sites in Norfolk and Suffolk, England (1), found that European turtle dove *Streptopelia turtur* reproductive success, territory size or territory density did not differ between years when supplementary food was provided and control (unfed) years (24 nests studied, daily survival rates of 79–97% for fed nests vs. 85–98 for unfed). However, doves were frequently observed eating the food. The authors argue that the experimental sites were too small (mostly 200–400 ha) to affect the wide-ranging doves.

(1) Browne, S. J. & Aebischer, N. J. (2002) The effect of supplementary feeding on territory size, territory density and breeding success of the turtle dove *Streptopelia turtur*: a field experiment. *Aspects of Applied Biology*, 67, 21–26.

15.15.15. Parrots

- Two studies from New Zealand (1,2) found some evidence that providing supplementary food for kakapos *Strigopus habroptilus* increased the number of breeding attempts made, whilst a third study (3) found that birds provided with specially-formulated pellets appeared to have larger clutches than those fed on nuts.
- One study (3) found no evidence that providing food increased the number of nesting attempts.

A 1998 review of data from three islands in New Zealand between 1990 and 1997 (1), found that providing kakapos *Strigopus habroptilus* with supplementary food (nuts, apples and sweet potatoes provided at feeding stations year-round) may have increased breeding attempts on Little Barrier Island, North Island, with no breeding attempts between 1982 (when 22 kakapos were translocated there) and 1989, but between two and four females nested in four of the next six years, following food provision. Male display behaviour also increased in this period. However, there were no similar increases on two other islands, with one of three females breeding once on Maud Island between 1991 and 1998, despite supplementary feeding from 1991 onwards. Females on Codfish Island only bred once between 1992 and 1998 (with a total of six females nesting), following supplementary feeding starting in 1992 and this coincided with heavy fruiting of rimu *Dacrydium cupressinum* trees. Overall breeding success was low across the study period, owing to high rates of egg fertility, and the starvation or probable predation of many chicks.

A 2001 review of data from Little Barrier Island, North Island, New Zealand, between 1990 and 1999 (2), using largely the same data as (1), found that female kakapos *Strigopus habroptilus* provided with supplementary food (nuts, apples and sweet potatoes provided at feeding stations year-round) bred significantly more frequently than females that did not take supplementary food (11 breeding attempts in 35 'bird years' for fed birds vs. one breeding attempt in 82 'bird years' for unfed birds). The authors note that the mechanism behind the relationship and that the response was extremely variable across other islands. A small analysis in the same study found that four fed females did not produce larger clutches than two unfed females (average of 2 eggs/clutch for both fed and unfed females). Neither of the unfed females raised young (and only one egg hatched), whereas a total of three chicks fledged from fed nests, but samples were too small to determine whether this difference was significant. Failure of both unfed nests was thought to be due to females spending large amounts of time away from the nest during incubation.

A small study on Codfish and Pearl Islands, South Island, New Zealand, during 1997–2005 (3), found that kakapos *Strigopus habroptilus* fed with food pellets (specially formulated to provide nutrients missing from kakapos' diets) laid larger clutches than those provided with either nuts (almonds, walnuts, Brazil nuts and honey-water) or no supplementary food (76% of 29 pellet fed birds having 3–4 eggs/clutch and an average of 2.76 eggs/clutch; 33% and 2.22 for nine nut-fed birds; 33% and 2.00 eggs/clutch for three unfed birds). However, fed birds did not nest more frequently than unfed birds, with nesting apparently related to the fruiting of rimu *Dacrydium cupressinum* trees. Food was either fed *ad libitum* or was 20 g of nuts every three days or 50–150 g of pellets every three days year-round.

(1) Clout, M. N. & Merton, D. V. (1998) Saving the Kakapo: the conservation of the world's most peculiar parrot. *Bird Conservation International*, 8, 281–296.
(2) Elliott, G. P., Merton, D. V. & Jansen, P. W. (2001) Intensive management of a critically endangered species: the kakapo. *Biological Conservation*, 99, 121–133.
(3) Houston, D., Mcinnes, K., Elliott, G., Eason, D., Moorhouse, R. & Cockrem, J. (2007) The use of a nutritional supplement to improve egg production in the endangered kakapo. *Biological Conservation*, 138, 248–255.

15.15.16. Songbirds

- Two studies from the USA (8,10) found evidence for higher population densities of magpies and American blackbirds in areas provided with supplementary food, whilst two studies from the UK (2) and Canada (12) found that population densities did not appear to be affected by feeding.
- Twelve studies from across the world (2,4,7,9,12,13,16,18,22,25,29,33) found that breeding productivity was higher for fed birds than controls. The increases were through higher hatching or fledging rates, or higher chick survival or recruitment rates. One study from the USA (33) found that these increases were only found in dry years.
- Eleven studies from Europe (9,11,15,16,35,37) and the USA (5,6,8,21,29) found that fed birds had no higher, or even lower breeding productivity or chick survival than control birds.
- Nine studies from Europe (3,4,24,27,35) and North America (12,13,21,32) found that the eggs of fed birds were larger or heavier, or that the chicks of fed birds were in better physical condition: being larger, heavier, faster growing, more symmetrical or having a better immune response. In one study (35) this was only true in a heavily polluted site. However, eight studies from across the world (2,7,14–16,21,31,32) found no evidence for better condition or increased size in the eggs or chicks of fed birds.
- Six studies from across the world (4,7,19,22,26,28) found that food-supplemented pairs laid larger clutches than unfed birds, whilst 14 studies from Europe (2,3,11,15,16,27,34,37) and North America (5,6,8,12,20,29) found that fed birds did not lay larger clutches, or even laid smaller ones (5,37).
- Fifteen studies from across the world (1,3,5,7,8,10–12,15,17,19,20,30,34,37) found that birds supplied with supplementary food began nesting or laying earlier than controls, although in two studies this was only true for young females (17) or in one of two habitats (34). In one study (30), a high fat, high protein diet had a greater effect on laying date than a high fat, low protein diet. One study found that fed birds had shorter incubations than controls (9) whilst another (22) found that fed birds re-nested quicker than controls and had shorter second incubations.
- Four studies from the USA (6,29) and Europe (16,27) found that fed birds did not lay any earlier than controls.

- Seven studies from across the world found that fed parent birds showed positive behavioural responses to feeding, such as being more likely to re-nest (7,13), less likely to be parasitised or showing better anti-predator responses (7,31), spending more time incubating (23,29,36) or building larger nests (27).
- Three studies from across the world found neutral or negative responses to feeding, including being more likely to be invaded by conspecifics (5), making no more breeding attempts (17,28) or showing no preference for fed nest boxes compared to controls (28).

A controlled cross-over study in southern Sweden in 1972–3 (1) found that great tits *Parus major* nesting in an 8 ha (1972) or 6 ha (1973) area of oak-hazel woodland supplied with supplementary food began laying eggs significantly earlier than great tits in an adjacent 16 ha or 18 ha control (unfed) area. The difference was significant irrespective of whether all females, one year-old females or older females were examined (average laying date of 1st–5th May for 44 nests in fed areas vs. 6th–11th May for 75 control pairs). Supplementary food consisted of 32 or 33 trays positioned throughout the wood and provided daily with 20–35 g of mealworms from 11th–27th April 1972 and between 10th April and 14th May 1973. On average there were 2 trays/territory in 1972 and 1.5 trays/territory in 1973. The authors note that other species including wood nuthatches *Sitta europea* and chaffinches *Fringilla coelebs* also took food from the trays.

A replicated and controlled study in mixed farmland in north-east Scotland between 1971 and 1973 (2) found that carrion crow *Corvus corone* nestlings in nests provided with supplementary food had significantly higher hatching, survival and fledging rates than those in control (unfed) nests (with human 'predation' included: 79% of 11 fed nests hatching at least one chick, 71% of ten having at least one chick surviving for ten days and 71% fledging at least one chick vs. 61% of 28, 54% of 15 and 43% of 12 for controls). Nestlings from fed nests, however, were no heavier than those from controls, when comparing first-hatched with first-hatched etc. Supplementary food consisted of one domestic hen's egg and five dead hen chicks provided every day from when laying began and increasing to one egg and ten chicks from the seventh day after hatching until fledging. Other experiments in this study found that winter feeding (a hen's egg and five chicks provided between January and April 1973) led to crows laying clutches earlier but did not affect clutch size (average laying date of April 13th and 4.4 eggs/clutch for ten fed territories vs. April 18th and 4.3 eggs/clutch for 21 controls). Further experiments examined the effect of moving supplementary food further from nests, but the author argues that these results are confounded by supplementary food being taken by non-target birds. Finally, additional experiments in the study found that crow nesting density did not increase following the provision of supplementary food and additional nesting sites in breeding territories, discussed in 'Provide artificial nesting sites'.

A controlled trial in coniferous forest in south-west Sweden in 1976–7 (3), found that willow tits *Parus montanus* and crested tits *P. cristatus* in a 1.8 km^2 area provided with supplementary food started laying eggs significantly earlier than individuals in a 5.5 km^2 control (non-supplemented) area (willow tits: average of 2–5 days earlier, 44 pairs tested; crested tits: average 5–8 days earlier, 38 tested). There was no corresponding difference in 1978, when no food was provided. Twelve-day old crested tit nestlings were significantly heavier in the fed area (average of 12.2 g for fed area vs. 11.5 g for control, number of chicks not provided) but there was no difference in willow tit nestlings (average of 10.6 g for fed area vs. 10.7 for control, number of chicks not provided). Clutch sizes did not differ between areas for either species (willow tits: average of 8.1–8.2 eggs/clutch for fed area vs. 7.7–8.2 for controls; crested tits: average of 5.3–5.4 eggs/clutch for fed area vs. 5.0–5.2 for controls). Food was provided at feeding stations located in the approximate centre of each fed territory and contained one feeder of sunflower seeds and three coconut shells containing a mixture of tallow, soy protein, wheat germ, sun-flower seeds, vitamins and minerals. In 1976 feeding started on 20th February and in 1977 on the 20–24th March, in both years it continued until all females had started laying and resumed for the first half of June (when second clutches could be expected to be laid).

A replicated cross-over study in a pine forest site in southern Sweden in 1974–5 (4) found that female black-billed magpies *Pica pica* laid earlier, had larger clutches and laid larger eggs when supplied with supplementary food compared to when no food was provided (average of 3.5 days earlier, an extra 0.56 eggs/clutch and eggs weighing 0.33 g more, ten females studied). Fed pairs were less likely to lose nestlings and had higher fledging success than control (unfed) pairs (88% of 20 fed nests having one surviving nestling and 2.7 fledglings/breeding attempt vs. 48% and 1.3 fledglings/breeding attempt for 32 controls). Supplementary food consisted of 300 g of fish provided in the centre of experimental territories every other day from 15th March until eggs hatched, when it was raised to 600 g. This represents approximately 75% of a pair's daily requirement before hatching and 65–150% after hatching. Food was also provided in seven territories without magpies, but did not attract any new pairs.

A replicated, controlled experiment on marshlands sites in Washington state, USA, in the springs of 1977–8 (5), found that red-winged blackbirds *Agelaius phoeniceus* from territories supplied with supplementary food laid eggs significantly earlier than blackbirds from control (unfed) territories (1977: 17 days earlier in eight territories where feeding began on 31st March, 12 days earlier in five territories fed from 5th April; 1978: 26 days earlier). However, in 1978 clutch size was significantly lower in fed territories (3.4 eggs/clutch, 122 fed clutches vs. 3.6 eggs/clutch for 126 controls), the authors suggest this is due to early-laid clutches being smaller – if clutches laid before April 10th were excluded then clutches were not significantly smaller. Fed territories provided with supplementary food also had higher rates of chick mortality than control territories (numbers not provided) and were invaded by non-territorial red-winged and yellow-headed blackbirds *Xanthocephalus xanthocephalus* at significantly higher rates than controls in 1978 (4.9 intruders/ha in 33 fed territories vs. 1.1/ha for 54 controls), but not in 1977, or when feeder density was reduced in 1978 (0.8 intruders/ha for 17 fed territories vs. 1.3 intruders/ha for 19 controls). Supplementary food consisted of sunflower seeds, cracked corn and puppy food on 30 cm square trays. In 1977 one tray was placed in each of 13 territories on two marshes (controls were on five other marshes); in 1978, trays were placed on a 6 m grid on a single marsh, a tray density approximately ten times that in 1977, reduced to one feeder in each territory several weeks after females arrived.

A small study from January–March in 1979–1980 in Utah, USA (6), found that juvenile black-billed magpies *Pica pica* that established nest sites closer to supplemental food stations experienced higher rates of nestling starvation than nests further away. Very few adult magpies visited food stations. Dominant juvenile males, however, established nests significantly closer to food stations. Of seven nests within 100 m of food stations, six experienced nestling starvation while only one farther than 100 m did. Supplementary feeding did not enhance clutch (6.4 and 6.4 eggs/nest) or brood (5.0 and 4.8 nestlings/nest) size between nests closer to food stations (16 nests) and those further away (13 nests). Similarly, mean clutch initiation was not significantly different between treatments. The authors point out that supplementary feeding can serve as a proximate stimulus to commence nesting on sites that are not inherently of good quality. Food stations were provisioned with beef bones or black-tailed jack rabbit *Lepus californicus* carcasses.

A replicated and controlled study on Mandarte Island, British Colombia, Canada (7), found that song sparrows *Melospiza melodia* provided with supplementary food in 1985 laid earlier, had larger clutches than control (unfed) birds, were more likely to re-nest following breeding failures and were less likely to be parasitised by brown-headed cowbirds *Molothrus ater* (average laying date of 10th April, 3.3 eggs/clutch, 2.8 breeding attempts and 18% of nests parasitised for 15 fed pairs vs. 28th April, 2.8 eggs/clutch, 2.2 breeding attempts and 45% for 34 controls). This led to fed females raising approximately 3.8 young/female to independence compared to 1.0 young/female for controls. However, eggs and nestlings from fed birds were no larger than those from controls nor more likely to hatch or survive to independence (averages of 3.1 g/egg, 16.8 g/nestling at six days old, 78% hatching success and 83% survival to independence vs. 3.0 g/egg, 15.6 g, 73% and 70%) and young from fed territories did not have enhanced reproduction (8.6% of 58 fed young reproducing vs. 12.5% for 40 controls).

Supplementary food consisted of dog food, vitamins, mealworms *Tenerbrio* sp. and millet seed provided continuously in feeders in the centre of each territory from five weeks before laying started (28th February) until the last young had left the nest (24th July). The authors note that 1985 was a year of peak song sparrow density. This study also described the impact of feeding on adult survival, discussed in 'Provide supplementary food to increase adult survival'.

A controlled before-and-after study in Washington, USA, in March–June 1979–82 (8), found that twice as many black-billed magpies *Pica pica* nested, and that they began laying significantly earlier, in a 0.3 ha area of willow *Salix* spp. in 1981 when a dead bullock was cut open to locally increase invertebrate numbers, compared to 1979–80 and 1982 when there was no bullock present (eight nests in 1981, average laying date if 24th March vs. three or four nests and a laying date of 2nd–4th April in other years). There were no corresponding changes in two areas that were not supplied with a carcass (six and nine nests and 1st–6th April for all years studied). There were no differences in clutch size or the number of young fledged from nests with the carcass present. The bullock weighed approximately 450 kg and died between 1st and 7th March 1981 and attracted large numbers of blow flies (Calliphoridae), most of which had left by the 22nd March.

A replicated, paired site study in 1984–1985 in 17 experimental and 17 control pairs of blue tit *Parus caeruleus* nests in Lund, Sweden (9), found that blue tit pairs that were artificially provisioned with food exhibited significantly greater reproductive success. Supplementary feeding significantly shortened the length of the incubation period. The probability that an egg would hatch was higher in provisioned nests than in control nests (98 and 90% of 198 and 199 eggs respectively). However, nestling survival up to 13 days did not differ between the pairs. Each pair was selected so that the last egg in both clutches was laid on the same date and the pairs nested in similar habitats. Four days after the last egg was laid, a container with 7 g of mealworms was placed into the experimental nest, while an empty container was placed into the control nest. Age and wing length did not differ between dyad pairs.

A randomised, replicated and controlled experiment around two lakes in Washington state, USA, in 1981 (10), found that red-winged blackbirds *Agelaius phoeniceus* from territories supplied with supplementary food laid eggs 8–13 days earlier than blackbirds from control (unfed) territories, for first, second, third and fourth nests of the year (20 territories studied). Each lake was divided in half, with one half provided with sunflower seeds from the 4th April. The average number of females in each territory was increased by feeding on one lake (average of 9.0 females/territory on fed territories vs. 6.5 females/territory on controls, eight territories studied) but not the other (average of 5.2 females/territory on fed territories vs. 4.7 females/territory on controls, 12 studied).

A cross-over study in a coppiced oak forest in southern France in 1986–7 (11) found that blue tit *Parus caeruleus* and great tit *P. major* pairs in an area provided with supplementary food started laying clutches significantly earlier than pairs in a control (unfed) area (1986: 9 pairs in a 10 ha fed area laid six days earlier than 26 pairs in a 50 ha control area; 1987: 15 pairs in a 30 ha fed area laid five days earlier than 18 pairs in a 30 ha control area). However there were no significant differences in clutch size or the number of young hatched between treatments (average of 9.2 eggs/clutch for 24 fed clutches vs. 8.3–9.1 eggs/clutch for 44 controls) and, in 1987, significantly fewer chicks fledged from blue tit nests in fed areas than control areas (average of 3.0 chicks/nest for 11 fed pairs vs. 6.3 chicks/nest for 15 controls). There was no significant difference in productivity in 1986 (7.1 chicks/nest for seven from fed areas vs. 7.3 chicks/nest for 23 controls). The authors suggest this could be due to fed clutches hatching before the peak in natural food in 1987.

A study in Alberta, Canada, between 1986 and 1988 (12) found that black-billed magpies *Pica pica* that used supplementary food provided in a single feeder in an urban area laid eggs 6–7 days earlier than magpies that did not use the feeder (seven pairs used the feeder in 1987, six in 1988; 22 pairs did not use the feeder in 1987, 11 in 1988). Nestlings from parents that used feeders were heavier and had higher survival and fledging rates than those from parents that didn't use the feeder, (fed nests: average weight of 184 g, ten nests, 71% survival of seven through a snow storm in 1987, 4.1–4.5 nestlings fledged/nest for 13 nests; unfed nests: 163–168 g for 13, 15% survival of 20, 1.1–1.8 nestlings fledged/nest for 31) although there were no differences in clutch size or breeding density (fed nests: 7.4–7.8 eggs/clutch for ten clutches, 126–134 m between 13 nests; unfed nests: 6.8–6.9 eggs/clutch of 23, 134–153 m between 57 nests). Supplementary food consisted of approximately 1 kg of dog food pellets provided each week from August 1986 until June 1988.

A randomised, replicated and controlled paired study in a Sonoran desert site in Arizona, USA, in the breeding seasons of 1986 and 1987 (13) found that 28 cactus wren *Campylorhynchus brunneicapillus* broods were 10.5–20.0 g heavier from territories provided with supplementary food, compared with control broods. Parents from fed territories were also more likely to have second clutches than controls (12 of 14 fed pairs attempted second broods vs. seven of 14 control pairs). Fed nestlings were also larger and had higher post-fledging survival rates in 1986 but not 1987 (1986: 14 pairs, linear measurements 1.0–2.7 mm larger than controls, 15 fed fledglings surviving for four to six weeks after fledging vs. seven control fledglings; 1987: 14 pairs, linear measurements 0.2–0.6 mm larger than controls, 19 fed fledglings surviving vs. 16 controls). The authors suggest that differences between years were due to higher levels of natural food occurring in 1987. Supplementary food consisted of 35 g of mealworm *Tenebrio molitor* larvae and noctuid moth caterpillars supplied every other day from one or two days after hatching until nestlings fledged (approximately 20 days later) and comprised 75% (1986) and 80% (1987) of food delivered to nests.

A small randomised and controlled paired study in parkland and mixed woodland in southern Scotland in spring and summer 1990 (14) found that great tit *Parus major* nestlings in eight artificially enlarged broods were no larger at 15 days old when provided with supplementary food, compared to nestlings from seven control (enlarged but not fed) broods. Broods were enlarged by the addition of three nestlings (added after hatching). Supplementary food consisted of an average of 2.2 g of minced meat and nutritional supplements fed twice daily to half the experimental brood on days six through 12 after hatching. This represents most of a nestling's daily energetic requirements. This study also investigated the impact on the condition of provisioning females, discussed in 'Provide supplementary food to increase adult survival'.

A replicated and controlled study in grasslands in southern Sweden between 1982 and 1990 (15) found that common starlings *Sturnes vulgaris* supplied with supplementary food began laying significantly earlier than controls (first laying date of 21st April–5th May for fed nests vs. 22nd April–10th May). There were no such differences between nests in the years when supplementary food was not supplied. However there were only occasional differences in egg weight, no differences in clutch size or nestlings weights and fledging rates were actually lower in fed nests in 1990 (4.3 young/nest for fed nests vs. 5.6 for controls). Supplementary food consisted of approximately 100 g of mealworms placed in small feeders either on the outside or inside of nest boxes, supplied to different colonies in 1982 and 1985 and a subset of nests at a third colony in 1990. Feeding began approximately one month before laying started and stopped once all females began laying. Feeding represented more than the daily energetic needs of a pair of starlings.

A replicated, controlled and paired study in a mixed forest in the central Netherlands in 1987 (16) found that pied flycatcher *Ficedula hypoleuca* chicks from pairs provided with supplementary food were significantly more likely to be recruited into the local breeding population than chicks from control (unfed) pairs (7.9% of nestlings from 12 fed clutches recaptured as breeding adults vs. 1.1% from 15 controls). However, fed pairs did not start laying earlier, produce larger clutches, hatch more young, hatch young earlier than control pairs (13 fed pairs started laying on May 18th, average of 6.3 eggs/clutch, 5.5 chicks/clutch and a hatching date of June 6th vs. May 19th, 6.2 eggs/clutch, hatching 5.4 chicks/clutch and June 6th for 16 controls), probably due to the paired nature of the study. In addition, there were no differences between treatments in terms of nestling growth, survival, tarsus length or weight (5.5 chicks/clutch fledging, average weight of 13.9 g, average tarsus length of 17.4 cm for 12 fed pairs vs. 5.5 chicks/clutch fledging, average weight of 13.8 g, average tarsus length of 17.2 cm for 15 controls). Supplementary food consisted of mealworms provided in excess beginning two days after chicks hatched. This study also examines the impact of feeding on adult survival, discussed in 'Provide supplementary food to increase adult survival'.

A replicated and controlled study in alpine meadows and pine woodlands on Honshu, Japan, in May–September 1986–9 (17) found that one year-old alpine accentor *Prunella collaris* females in territories provided with supplementary food began laying clutches significantly earlier than those in control territories (first egg laid on average on 25th–29th June for fed 16 one year-old females vs. 5th–13th July for 17 control one year-olds). There was no such difference in older females (first egg laid on average on 17th–19th June for 18 fed females vs. 21st–23rd June for 19 controls). There were no differences in duration of the copulation period, the number of breeding attempts or the timing of settlement in territories for either one year-old or older females. Supplementary food consisted of 300 g of millet and canary seed provided two to three times a week beginning in May and continuing until September.

A small, controlled before-and-after study on Frégate Island, Seychelles, in 1989–90 (18), found that Seychelles robins *Copsychus sechellarum* had significantly higher reproductive success in 1990 when provided with supplementary food, compared to in 1989, when food was not provided (2.0 nestlings/pair and 1.4 independent young/pair in 1990 vs. 0.8 nestlings/pair and 0.2 young/pair in 1989). There was no corresponding change in four control (unfed) territories. Food consisted of 20–25 freshly killed cockroaches, grated coconuts, boiled rice and fish provided twice a week, and an area of 4 m^2 of soil was broken up in an attempt to facilitate access to natural prey (e.g. worms and beetle larvae). Increases were apparently due to earlier laying dates, more time incubating and nest guarding, higher hatching and provisioning rates, higher chick fledging weight and higher levels of parental care.

A replicated and controlled trial in a dune and scrubland system in Florida, USA, in 1993 (19), found that female Florida scrub jays *Aphelocoma coerulescens* in ten groups provided with supplementary food initiated clutches significantly earlier (on average 16 days earlier) than females from 32 control (unfed) groups. There was no significant difference in clutch sizes between treatments. Non-breeders from fed groups were no less likely to become breeders than those from control groups (fed groups: five of 23 non-breeders established territories; controls: four of 23). Feeding consisted of providing dried dog food, peanuts and mealworms were provided twice daily at feeding stations in the middle of the territories from late January until females finished laying. Food was provided 'in excess'. This study also reported on the effects on adult condition, discussed in 'Provide supplementary food to increase adult survival'.

A replicated and controlled before-and-after trial in scrubland, gullies and caves in southern Spain in 1980–3 (20) found that European jackdaws *Corvus monedula* laid earlier, laid more eggs and had higher breeding success when given supplementary food, compared to either the same colonies in previous years, or to control colonies (average laying date of 25th April, 6.0 eggs/clutch and 48% of all eggs fledging for 13 fed nests vs. 27th April, 5.4 eggs/clutch and 20% of all eggs fledging for 18 experimental colonies in previous years vs. 27th April, 5.4 eggs/clutch and 29% of all eggs fledging for 18 control colonies). Starvation rates were also lower in fed colonies (2.5 nestlings/nest starving in 12 fed nests vs. 3.4 nestlings/nest in 12 controls). Predation rates were lower in fed colonies (0% of 20 fed nests predated vs. 22% of 27 control nests) and the authors argue that this was due to better group defence in fed colonies due to higher breeding densities, although nesting densities were only significantly higher in one colony, comparing 1983 and 1982, other comparisons were non-significant (colony A when not fed: 56%, 63% and 56% of nests occupied in 1980–2 vs. 88% in 1983 when fed; colony B: when not fed: 50%, 67% and 33% of nests occupied in 1980–2 vs. 100% in 1983). Supplementary food consisted of four hens' eggs for each jackdaw pair supplied twice a week and 4–7 kg of bread, continuously supplied.

A randomised, replicated and controlled trial in a forest in Illinois, USA, in 1998 (21), found that 'early-season' house wren *Troglodytes aedon* nestlings (i.e. from clutches laid in mid-May) in nest boxes provided with supplementary food were significantly heavier than nestlings from early-season control broods (fed nestlings approximately 2% heavier than controls, 62 tested). There was no difference in weight between fed and control late-season (laid in late June to early July) broods (53 tested). There were no differences between treatments for tarsus lengths, growth rates or survival until fledgling for either early or late season broods. Food consisted of 30 g of mealworms *Tenebrio molitor* supplied each day within the nest boxes.

A replicated, controlled trial on an island in Lake Rotorua, North Island, New Zealand, between September 1995 and February 2001 (22), found that stitchbirds (hihis) *Notiomystis cincta* in territories provided with supplementary food laid significantly more eggs and had higher fledging and recruitment success than control (unsupplemented) birds (average of 4.4 eggs/clutch for 17 fed nests vs. 3.9 eggs/clutch for 18 controls; 70% fledging success for 22 fed nests vs. 32% for 14 controls and 35% recruitment success for 16 fed nests vs. 13% for 13 controls). In addition, fed females began laying second clutches significantly sooner than controls and incubated second clutches for significantly less time (15.2 days of incubation for eight fed clutches, and a 4.0 day interval before four second clutches vs. 16.8 days incubation for eight controls and a 13.3 day interval for six). There were no significant differences in hatching success or the incubation period of first clutches, between fed and control birds. Supplementary food consisted of either commercial honeyeater food (provided every day) or a solution of sugar or jam (provided every third day) in hummingbird feeders. Territories were considered 'fed' if nests were within 50 m of a feeder. The authors suggest that population viability on Mokoia Island may be dependent on supplementary food.

A randomised, replicated and controlled study in a 10 ha Australian reed *Phragmites australis* wetland in southern Australia, over two breeding seasons (September to January) in 1999–2001 (23), found that Australian reed warbler *Acrocephalus australis* females provided with supplementary food spent longer incubating eggs than control (unfed) females (12 fed females spent approximately 57% of time incubating vs. 51% for ten controls). There was also a significant difference between days when food was provided and those when it was not (on fed days, 59% of time spent incubating vs. 52% of time on unfed days, seven tested). There were no differences between other measures of incubation attendance, such as start and end times of incubation bouts or the average length of bouts. However, broods from fed territories had larger hatching asynchronies (calculated as the difference in size, at one to three days old, between the first- and last-hatched chicks, divided by the mean size of all nestlings) than control pairs (tarsus length asynchrony of approximately 0.15 and weight asynchrony of approximately 0.35 for ten fed broods vs. 0.10 and 0.20 for 19 controls). The effect of feeding on fitness of chicks or parents was not reported. Supplementary food consisted of 30 g of blowfly maggots (over 150% of daily energetic requirements of an adult reed warbler) provided every other day whilst eggs were incubated, but stopped before eggs hatched.

A randomised, replicated and controlled paired study in mixed deciduous forests in the Netherlands in 1999 (24), found that blue tit *Parus caeruleus* nestlings from eleven nest boxes provided with supplementary food had more even leg lengths than nestlings from eleven control (unfed) nest boxes. There were no differences in leg size between treatments and no data were provided on survival or reproductive success. Supplementary food consisted of mealworm *Tenerbio molitor* and wax moth *Galleria mellonella* larvae placed inside the nest boxes and making up one third of the nestlings' total food. Food was provided from day of hatching until the chicks fledged.

A continuation of (26) in 2001 and 2002 (25) found that Florida scrub jay *Aphelocoma coerulescens* chicks from second and third-laid eggs had significantly higher fledging rates from territories provided with high-fat, high-protein supplements compared to those from controls in 2001 but not 2002 (2001: 100% of eleven 2nd and 3rd-hatched chicks fledged from fed nests vs. 53% of seventeen 2nd-hatched and 40% of ten third-hatched chicks from controls; 2002: 63% of eight 2nd-hatched and 71% of seven 3rd-hatched vs. 81% of 16 and 60% of ten for controls). In addition, more third-hatched chicks survived to independence in fed territories in 2001 (50% of four vs. 0% of ten) but not 2002 (43% of seven vs. 30% of ten). Fed chicks also grew flight feathers significantly faster, but there were no other differences in growth (body mass and leg length) or survival between the groups.

A randomised, replicated and controlled cross-over trial in a dune and scrubland system in Florida, USA, in 2000 and 2001 (26), found that that 21 female Florida scrub jays *Aphelocoma coerulescens* provided with high-fat, low-protein (HFLP) supplementary food and 21 females provided with high-fat, high-protein (HFHP) food laid significantly larger clutches than 55 control females in 2000 but not 2001 (2000: approximately 3.7 eggs/clutch for HFLP diets, 3.2 eggs/clutch for HFHP and 2.8 eggs/clutch for controls; 2001: 2.7 eggs/clutch, 3.1 eggs/clutch and 2.8 eggs/clutch respectively). This was due to earlier laying, discussed below. Clutches on the HFHP diet had a tendency to have larger third-laid eggs, but other eggs were not significantly different in size between treatments. Feeding began in early- or mid-January and continued until females began laying, usually six to eight weeks later. Food consisted of food pellets containing 19.3% fat and with a protein content of either 3.5% or 34.5%.

A replicated and controlled paired site study from March–August in 2000–2003 in 20 paired nest box groups (10 placed along wetland edges and 10 in farmlands) in Rutland, England (27) found that tree sparrows *Passer montanus* showed no preference for nest boxes supplied with supplementary food (four fed boxes colonised vs. four unfed). There was no difference in the number of nesting attempts made by birds with or without supplementary food although the mean clutch size was significantly higher in nests closer to supplementary food (5.6 compared to 5.0 eggs/clutch). The authors point out that the small spatial scale of the study (1 km between pairs) may have confounded any effect of supplementary feeding. Nest box groups consisted of 5 nest boxes placed 2–20 m apart; sunflower seeds were randomly provided to one nest box group within each pair.

A replicated, controlled trial in mixed agricultural habitats in southern Spain (28) found that, when provided with supplementary food, male black-billed magpies *Pica pica* built significantly larger nests than controls (average of 5.1 units for 39 nests in fed territories vs. 4.6 units for 123 nests in controls) and females laid significantly larger eggs (average egg volume of 10.0 cm^3 for 36 fed clutches vs. 9.5 cm^3 for 119 controls). However there were no significant differences between groups in terms of laying date or clutch size (average of 25th April for 46 fed clutches vs. 12th April for 162 controls; 6.9 eggs/clutch for 38 fed clutches vs. 6.8 for 128 controls). Supplementary food consisted of 150 g dog food mixed with bread and water supplied every two days from before nest-building began until after the end of the breeding season.

A randomised, replicated and controlled trial at a prairie site in Kansas, USA, in April–August 1997 (29) found that female Bewick's wrens *Thyromanes bewickii* and house wrens *Troglodytes aedon* from nest boxes provided with supplementary food spent longer incubating than females from control (unfed) nest boxes (Bewick's wrens: average incubation bout of 52 min and 82% of time spent on the nest for seven fed females vs. 35 min and 70% for eight controls; house wrens: averages of 30 min and 81% for nine fed females vs. 13 min and 71% for eight controls). There were no differences in date of clutch initiation, clutch size, the average length of time spent away from the nest or hatching success between treatments (Bewick's wrens: average bout away from the nest of 10 min for seven fed females, hatching success of 78% vs. 13 min and 74% for eight controls; house wrens: average of 5 min and 96% for nine fed females vs. 5 min and 82% for eight controls). Supplementary food consisted of 15 g of mealworm *Tenebrio molitor* larvae supplied inside nest boxes every day during incubation, equivalent to almost twice a female wren's daily energetic requirements.

A continuation of (26) between 2000 and 2003 (30), found that breeding groups of Florida scrub jays *Aphelocoma coerulescens* provided with high-fat, low-protein (HFLP, provided in 2000–2 only) supplementary food started laying eggs significantly earlier than control (unfed) territories, whilst territories provided with high-fat, high-protein (HFHP) food laid significantly earlier still (first laying on average of approximately day 16th March for 50 HFHP territories vs. March 20th for 29 HFLP territories vs. March 28th for 115 controls).

A replicated, controlled experiment in Arusha National Park, Tanzania, in October and November 1995 and 1996 (31), found that East African stonechat *Saxicola axillaris* (formerly *S. torquata axillaris*) chicks from ten pairs provided with supplementary food did not grow any faster between hatching and seven days old than chicks from 15 control (unfed) pairs. However, feeding appeared to reduce the impact of common fiscals *Lanius collaris*, a predator of stonechats, the presence of which reduced growth rates in chicks from control pairs (n = 7), but not fed pairs (n = 5). Supplementary food consisted of 30 g of mealworms *Tenebrio molitor* provided every day, from approximately a month before laying began.

A replicated, controlled trial in habitats in southern Spain in 2001 (32), found that black-billed magpie *Pica pica* nestlings provided with supplementary food had significantly higher cell-mediated immune responses (CMI, the change in swelling around insect bites, a measure of immunocompetence) in arid scrub, but not in irrigated farmland and woodland (arid scrub: average CMI of approximately 1.18 mm for fed nestlings vs. 0.85 m for controls; irrigated farmland/woodland: average CMI of 1.20 mm for fed nestlings vs. 1.18 mm for controls; 58 nests studied). In addition, fed nestlings showed significantly lower levels of the ectoparasite *Carnus haemapterus* (Diptera: Carnidae), but only in arid habitats (arid scrub: average infestation intensity of 0.40 arbitrary units for fed nestlings vs. 0.90 arbitrary units for controls; irrigated farmland/woodland: average 1.60 arbitrary units for fed nestlings vs. 1.65 arbitrary units for controls; 28 nests). Feeding had no effect on chick growth, weight or on the prevalence of blood parasites. Supplementary feeding consisted of 0.1 ml of high-calorie, nutrient-rich paste provided every other day for a total of 14 days.

A controlled, replicated study on San Clemente Island, California, USA, between 2000 and 2006 (33) found that pairs of San Clemente loggerhead shrikes *Lanius ludovicianus mearnsi* in territories provided with supplementary food and with rodent control during April–July produced, on average, 2.5 more fledglings each year (55 breeding attempts), compared to control pairs (62 attempts), and 1.4 more fledglings each year compared to territories with rodent control but no feeding (55 attempts). In drier-than-average years, fed pairs also raised more fledglings to independence (40 days old) than other pairs (1.8 more fledglings than control pairs, 0.7 more than rodent control only pairs). There was no effect in wetter-than-average years. Management in December–March did not increase either measure of productivity. This study also investigated the impact of just rodent control, which is discussed in 'Control predators on islands' and the success of captive-bred individuals, in 'Release captive-bred individuals'.

A replicated and controlled study in a range of oak forest habitats on Corsica, France, in the springs of 2006 and 2007 (34) found that blue tits *Parus caeruleus* (also known as *Cyanistes caeruleus*) provided with supplementary food began laying eggs significantly earlier in a holm oak *Quercus ilex* dominated site (females from 30 fed pairs began to lay approximately seven days earlier than controls), but not in two broad-leaved oak *Q. humilis* dominated sites or a second holm oak site. The authors suggest that this difference is due to broad-leaved oak forests having up to ten times more food for blue tits than holm oak forests. Clutch size did not differ between fed and control territories in any of the sites. Supplementary feeding comprised unlimited access to food in feeders within territories, beginning in mid-January and continuing until egg laying began.

A replicated and controlled study in open pine *Pinus sylvestris* forests in south-west Finland in 2005 (35) provided 87 great tit *Parus major* nestlings with three supplementary diets: autumnal moth *Epirrita autumnata* larvae (high in carotenoids – chemicals needed for coloured feathers) and mealworms (diet 1); mealworms and water-dispersed lutein (an important carotenoid, diet 2); mealworms and distilled water (diet 3). Nestlings fed on diets 1 and 2 were larger and heavier than control (unfed) nestlings and nestlings fed on diet 3, but only in areas with high levels of heavy metal pollution (polluted areas: average wing length of approximately 46 mm and body mass of 16.5 g for diet 1 nestlings vs. 46 mm and 15.8 g for diet 2 vs. 44 mm and 15.0 g for diet 3 and 43 mm and 15.0 g for controls; unpolluted areas: 48 mm and 16.5 g for diet 1 vs. 46 mm and 16 g for diet 2 vs. 46 mm and 16.5 g for diet 3 vs. 47 mm and 17 g for controls). In addition, diet 2 nestlings had higher blood lutein levels and correspondingly higher carotenoid chroma (a measure of colour intensity) in breast feathers than other diets and control chicks across both pollution levels (7.5–15.0 μg/ml blood lutein and 0.30–0.35 carotenoid chroma for diet 1 vs. 24.0–28.0 μg/ml and 0.41–0.45 for diet 2 vs. 7.5–16.0 μg/ml and 0.30–0.36 for diet 3 vs. 5.5–17.5 μg/ml and 0.30–0.36 for controls). Supplementary food did not have an effect on fledgling probability. One gram of food was provided every other day from the third day after hatching to the 13th, making approximately 20% of the required food in this time.

A randomised, replicated cross-over experiment in scrubland on South Island, New Zealand in austral spring 2000–1 (36) found that on a day when they were provided with supplementary food, silvereyes *Zosterops lateralis* spent significantly longer incubating and had shorter periods away from the nest, compared to a day when they were not provided with food (adults on fed days spent approximately 94% of time incubating and periods off the nest averaged one minute vs. approximately 84% of time on nests and 3.5 minute periods off the nest, ten nests). However, feeding did not increase the length of individual incubation bouts, or the number of times parents left the nests each hour (longest incubation bouts averaged 39.6 mins when fed vs. 34.1 mins when unfed, ten nests; parents leaving nests an average of 2 times/min when fed vs. 2.3 times/min when unfed, seven nests). Supplementary food consisted of beef fat and sugar mixed and provided in pine cones on one of two experimental days and not provided on the other. No data is provided on the breeding success consequences of feeding.

A replicated, controlled study over 3 breeding seasons in 2006–2008 in 3 treatment blocks (2 experimental and 1 control; 96 nestboxes/block) of broadleaf, deciduous woodland in Worcestershire, UK (37), found that blue tits *Parus caeruleus* (also known as *Cyanistus caeruleus*) and great tits *Parus major* began laying significantly earlier (by averages of two and three days respectively), had reduced clutch size (mean reduction: 0.4 and 0.7 eggs), shortened incubation periods (mean reduction: 0.9 and 0.7 days), lowered hatching success (in blue tits only: mean reduction: 1.4%) and reduced brood size (mean reduction: 0.6 and 0.5 chicks for blue and great tits respectively). Treatment blocks were separated by a 90 m buffer strip. In each year, one treatment block received no supplementary food and two treatment blocks received peanut cake (comprising 50% ground peanuts and 50% beef tallow). Treatments were rotated amongst plots.

(1) Källander, H. (1974) Advancement of laying of great tits by the provision of food. *Ibis*, 116, 365–367.

(2) Yom-Tov, Y. (1974) The effect of food and predation on breeding density and success, clutch size and laying date of the crow (*Corvus corone* L.). *Journal of Animal Ecology*, 43, 479–498.

(3) von Brömssen, A. & Jansson, C. (1980) Effects of food addition to willow tit *Parus montanus* and crested tit *P. cristatus* at the time of breeding. *Ornis Scandinavica*, 11, 173–178.

(4) Hogstedt, G. (1981) Effect of additional food on reproductive success in the magpie (*Pica pica*). *Journal of Animal Ecology*, 50, 219–229.

(5) Ewald, P. W. & Rohwer, S. (1982) Effects of supplemental feeding on timing of breeding, clutch-size and polygyny in red-winged blackbirds *Agelaius phoeniceus*. *Journal of Animal Ecology*, 51, 429–450.

(6) Reese, K. P. & Kadlec, J. A. (1984) Supplemental feeding: possible negative effects on black-billed magpies. *The Journal of Wildlife Management*, 48, 608–610.

(7) Arcese, P. & Smith, J. N. M. (1988) Effects of population density and supplemental food on reproduction in song sparrows. *Journal of Animal Ecology*, 57, 119–136.

(8) Knight, R. L. (1988) Effects of supplemental food on the breeding biology of the black-billed magpie. *The Condor*, 90, 956–958.

(9) Nilsson, J.-Å. & Smith, H. G. (1988) Incubation feeding as a male tactic for early hatching. *Animal Behaviour*, 36, 641–647.

(10) Wimberger, P. H. (1988) Food supplement effects on breeding time and harem size in the red-winged blackbird (*Agelaius phoeniceus*). *The Auk*, 105, 799–802.

(11) Clamens, A. & Isenmann, P. (1989) Effect of supplemental food on the breeding of blue and great tits in Mediterranean habitats. *Ornis Scandinavica*, 20, 36–42.

(12) Dhind, M. S. & Boag, D. A. (1990) The effect of food supplementation on the reproductive success of black-billed magpies *Pica pica*. *Ibis*, 132, 595–602.

(13) Simons, L. S. & Martin, T. E. (1990) Food limitation of avian reproduction: an experiment with the cactus wren. *Ecology*, 71, 869–876.

(14) Johnston, R. D. (1993) The effect of direct supplementary feeding of nestlings on weight loss in female great tits *Parus major*. *Ibis*, 135, 311–314.

(15) Källander, H. & Karlsson, J. (1993) Supplemental food and laying date in the European starling. *The Condor*, 95, 1031–1034.

(16) Verhulst, S. (1994) Supplementary food in the nestling phase affects reproductive success in pied flycatchers (*Ficedula hypoleuca*). *The Auk*, 111, 714–716.

(17) Nakamura, M. (1995) Effects of supplemental feeding and female age on timing of breeding in the alpine accentor *Prunella collaris*. *Ibis*, 137, 56–63.
(18) Komdeur, J. (1996) Breeding of the Seychelles magpie robin *Copsychus sechellarum* and implications for its conservation. *Ibis*, 138, 485–498.
(19) Schoech, S. J. (1996) The effect of supplemental food on body condition and the timing of reproduction in a cooperative breeder, the Florida scrub-jay. *The Condor*, 98, 234–244.
(20) Soler, M. & Soler, J. J. (1996) Effects of experimental food provisioning on reproduction in the jackdaw *Corvus monedula*, a semi-colonial species. *Ibis*, 138, 377–383.
(21) Styrsky, J. D., Dobbs, R. C. & Thompson, C. F. (2000) Food-supplementation does not override the effect of egg mass on fitness-related traits of nestling house wrens. *Journal of Animal Ecology*, 69, 690–702.
(22) Castro, I., Brunton, D. H., Mason, K. M., Ebert, B. & Griffiths, R. (2003) Life history traits and food supplementation affect productivity in a translocated population of the endangered Hihi (stitchbird, *Notiomystis cincta*). *Biological Conservation*, 114, 271–280.
(23) Eikenaar, C., Berg, M. L. & Komdeur, J. (2003) Experimental evidence for the influence of food availability on incubation attendance and hatching asynchrony in the Australian reed warbler. *Journal of Avian Biology*, 34, 419–427.
(24) Grieco, F. (2003) Greater food availability reduces tarsus asymmetry in nestling blue tits. *The Condor*, 105, 599–603.
(25) Reynolds, S. J., Schoech, S. J. & Bowman, R. (2003) Diet quality during pre-laying and nestling periods influences growth and survival of Florida scrub-jay (*Aphelocoma coerulescens*) chicks. *Journal of Zoology*, 261, 217–226.
(26) Reynolds, S. J., Schoech, S. J. & Bowman, R. (2003) Nutritional quality of prebreeding diet influences breeding performance of the Florida scrub-jay. *Oecologia*, 134, 308–316.
(27) Field, R. H. & Anderson, G. Q. (2004) Habitat use by breeding tree sparrows *Passer montanus*. *Ibis*, 146, 60–68.
(28) De Neve, L., Soler, J. J., Soler, M., Pérez-Contreras, T., Martın-Vivaldi, M. & Martınez, J. G. (2004) Effects of a food supplementation experiment on reproductive investment and a post-mating sexually selected trait in magpies *Pica pica*. *Journal of Avian Biology*, 35, 246–251.
(29) Pearse, A. T., Cavitt, J. F. & Cully, J. F. (2004) Effects of food supplementation on female nest attentiveness and incubation mate feeding in two sympatric wren species. *The Wilson Bulletin*, 116, 23–30.
(30) Schoech, S. J., Bowman, R. & Reynolds, S. J. (2004) Food supplementation and possible mechanisms underlying early breeding in the Florida Scrub-Jay (*Aphelocoma coerulescens*). *Hormones and Behavior*, 46, 565–573.
(31) Scheuerlein, A. & Gwinner, E. (2006) Reduced nestling growth of East African stonechats *Saxicola torquata axillaris* in the presence of a predator. *Ibis*, 148, 468–476.
(32) De Neve, L., Soler, J. J., Ruiz-Rodriguez, M., Martin-Galvez, D., Perez-Contreras, T. A. & Soler, M. (2007) Habitat-specific effects of a food supplementation experiment on immunocompetence in Eurasian Magpie *Pica pica* nestlings. *Ibis*, 149, 763–773.
(33) Heath, S. R., Kershner, E. L., Cooper, D. M., Lynn, S., Turner, J. M., Warnock, N., Farabaugh, S., Brock, K. & Garcelon, D. K. (2008) Rodent control and food supplementation increase productivity of endangered San Clemente Loggerhead Shrikes (*Lanius ludovicianus mearnsi*). *Biological Conservation*, 141, 2506–2515.
(34) Bourgault, P., Perret, P. & Lambrechts, M. M. (2009) Food supplementation in distinct Corsican oak habitats and the timing of egg laying by blue tits. *Journal of Field Ornithology*, 80, 127–134.
(35) Eeva, T., Sillanpää, S. & Salminen, J. P. (2009) The effects of diet quality and quantity on plumage colour and growth of great tit *Parus major* nestlings: a food manipulation experiment along a pollution gradient. *Journal of Avian Biology*, 40, 491–499.
(36) Barnett, C. A. & Briskie, J. V. (2010) Silvereyes *Zosterops lateralis* increase incubation attentiveness in response to increased food availability. *Ibis*, 152, 169–172.
(37) Harrison, T. J. E., Smith, J. A., Martin, G. R., Chamberlain, D. E., Bearhop, S., Robb, G. N. & Reynolds, S. J. (2010) Does food supplementation really enhance productivity of breeding birds? *Oecologia*, 164, 311–320.

15.16. Provide supplementary food to allow the rescue of a second chick

- A small controlled study in Spain (1) found that second chicks from lammergeier *Gypaetus barbatus* nests survived longer if nests were provided with food, allowing one chick to be rescued.

Background

Many vultures and birds of prey lay more eggs each year than they can raise, with the final chick hatching several days after the first. This chick will always be weaker and is almost always killed by its siblings or starves to death as it cannot compete for food. It is thought that this chick acts as 'insurance' in case the older chicks die.

In very rare species, it may be beneficial to 'rescue' the younger chick, but doing so can be difficult. It is possible that providing extra food will allow the younger chick to survive for long enough for a rescue attempt to be mounted.

A small controlled study in the Spanish Pyrenees in spring 2004–5 and 2008 (1) found that second-hatched lammergeier (bearded vulture) *Gypaetus barbatus* chicks from two nests provided with supplementary food survived for nine days, as opposed to seven, five and four days for a partially-supplemented nest and two control (unfed) nests, respectively. In a third supplemented nest (in 2008), the second-hatched chick survived for five days and was removed to be hand-reared and incorporated into a captive-breeding programme Supplementary food consisted of 2–3 kg of dead rabbits within 100 m of the nests every two days from the hatching of the first chick until the death (or removal) of the second chick. The partially-supplemented nest was provided with food on just one day.

(1) Margalida, A., Garcia, D., Heredia, R., and Bertran, J. (2009) Video-monitoring helps to optimize the rescue of second-hatched chicks in the endangered bearded vulture *Gypaetus barbatus*. *Bird Conservation International*, 20, 55–61.

15.17. Provide supplementary food to increase adult survival

Background

Well-fed animals are likely to be in better physical condition than those with too little food: having greater muscle mass and larger fat supplies to help them survive lean periods. However, it is worth noting that species that forage in groups can have dominance hierarchies, which alter the relationship between weight and fitness. For example Gentle and Gosler (2001) found that, amongst great tits *Parus major* in Oxfordshire, England, more dominant birds had a lower mass than subdominants, particularly when perceived predation risk was high. Birds with lower masses are better able to take off and therefore escape predators than heavier birds (Krams 2002). However, because of their dominance, they were able to usurp other birds from food resources when hungry. Care should therefore be taken when interpreting results which do not directly examine survival.

Krams, I. (2002) Mass-dependent take-off ability in wintering great tits (*Parus major*): comparison of top-ranked adult males and subordinate juvenile females. *Behavioral Ecology and Sociobiology*, 51, 345–349.

Gentle, L. K. & Gosler, A. G. (2001) Fat reserves and perceived predation risk in the great tit, *Parus major*. *Proceedings of the Royal Society of London. Series B: Biological Sciences*, 268, 487.

15.17.1. Wildfowl

- Two studies from Canada (1) and Northern Ireland (2) found that five species of wildfowl readily consumed supplementary food (grains and seeds).
- Only the Canadian study assessed the physiological effects of feeding, and found that fed birds were heavier and had larger hearts or flight muscles or had more body fat than controls.

A randomised, replicated and controlled trial at Karrak Lake in Northwest Territories, Canada, (1) found that female lesser snow geese *Chen caerulescens caerulescens* and Ross's geese *C. rossii* used supplementary food to different extents during incubation and showed different physiological responses to food. However, both males and females of both species were either heavier, had heavier hearts, more body fat or larger flight muscles when fed, compared to unfed controls. Differences were apparent both after laying and at the end of incubation. Between 250 g and 400 g of cracked and whole corn, durum wheat or shelled rice was provided each day.

A study in a wetland reserve in Northern Ireland (2) found that mallards *Anas platyrhynchos* were attracted to, and ate, 'Wildbird Mix' seeds provided. However, they dominated and excluded other species, so the mix was replaced with white millet seed. This was too small for mallard to eat, and so other species such as wigeon *A. penelope* and teal *A. crecca* were able to feed. The reaction of waders to the same feeding activity is discussed in 'Provide supplementary food to increase adult survival – Waders'.

(1) Gloutney, M. L., Alisauskas, R. T., Hobson, K. A. & Afton, A. D. (1999) Use of supplemental food by breeding Ross's geese and lesser snow geese: evidence for variable anorexia. *The Auk*, 116, 97–108.
(2) McGeehan, A. (2005) Artificial feeding to attract wild birds close to a viewing area at Belfast Lough RSPB Reserve, Antrim, Northern Ireland. *Conservation Evidence*, 2, 28–29.

15.17.2. Waders

- A study in Northern Ireland (1) found that waders fed on millet seed when provided, but were dominated by mallards *Anas platyrhynchos* when larger seeds were provided.

A study at a wetland reserve in Northern Ireland (1) found that black-tailed godwits *Limosa limosa* and northern lapwing *Vanellus vanellus* fed on millet seeds provided near a viewing area. Feeding was only possible, however, after a switch from 'Wildbird Mix' seed – which was eaten by mallards *Anas platyrhynchos* that then dominated other feeders and chased them off. The reaction of wildfowl to feeding is discussed in 'Provide supplementary food to increase adult survival – Wildfowl'.

(1) McGeehan, A. (2005) Artificial feeding to attract wild birds close to a viewing area at Belfast Lough RSPB Reserve, Antrim, Northern Ireland. *Conservation Evidence*, 2, 28–29.

15.17.3. Gamebirds

- Two European studies (1,6) found increased numbers of grey partridge *Perdix perdix* in fed areas, compared to unfed areas. In one study there was no change in the overall population in the study area (1), in the second there was an increase (6).
- One cross-over study from the USA found that northern bobwhites *Colinus virginianus* had higher overwinter survival in fed areas (3), one found lower survival (4) and a literature review found no overall effect of feeding (2).
- A paired sites study from the USA (5) found that fed bobwhites had higher body fat percentages.

Background

Gamebirds are often raised for shooting, with food supplied to support captive-raised birds. However, this section only describes studies on wild and native birds, involving actions that a conservationist might perform.

A small study of feeding as a management option for grey partridges *Perdix perdix* at an arable farm in France (1), found that partridge density was higher in the area with partridge cafeterias, than the area without. In spring 1973 the population on the 424 ha farm was 71 pairs (1 pair/6 ha) and 4 single birds. In spring 1974, a total of 48 pairs (1 pair/4.7 ha) and 4 single birds were recorded in the southern section (224 ha), where 27 partridge cafeterias had been constructed. The northern section (200 ha), with no cafeterias, had 24 pairs (1 pair/8.3 ha). Cafeterias comprised a barrel with a feed mixture (grain and weed seeds) and a sand-bath, sheltered by a leaning roof that collected rainwater in a drinking trough. Small shrubs were planted next to cafeterias to provide shelter. Where possible they were placed one per pair territory.

A 1997 literature review (2) collated data from eight sites in the USA where northern bobwhites *Colinus virginianus* were provided with supplementary food and concluded that feeding neither increases nor decreases bobwhite populations. Feeding ranged from intensive regimes with multiple feeders to more extensive programmes. Fed areas had higher bobwhite densities at three of the eight sites, control areas had higher densities at four and there was no difference at one site. On average, densities were 1.4 birds/ha in fed areas and 1.3 birds/ha in controls.

A controlled study on two 284 ha mixed-prairie sites in Oklahoma, USA, in 1992–6 (3), found that northern bobwhites *Colinus virginianus* had higher winter survival in areas supplied with supplementary food in two out of four winters (19–31% survival of 423 fed birds vs. 3–13% of 396 controls), with lower survival in 1994–5 (11% of 200 vs. 22% of 188) and similar survival in 1995–6 (16% of 200 vs. 23% of 193). Twenty eight percent of seed in birds' crops from the fed area consisted of supplementary food, compared with 6% in birds from the unfed area (783 birds examined). Food consisted of wheat, milo and millet provided in a 55 gallon barrel in the centre of each 8 ha section of the experimental area.

A cross-over experiment on a 796 ha rangeland site in Texas, USA, in the winters of 2000–1 until 2002–3 (4) found that northern bobwhites *Colinus virginianus* had lower winter survival on two sites when they were supplied with food, compared with when no food was supplied (24–57% survival of 89 birds on fed sites vs. 28–72% for 63 controls). Bobwhite home ranges on the fed site were 34–63% of the size of those on controls. Feeding consisted of 40 permanent feeders, supplied twice a week with cracked corn, milo and wheat between October and March each winter. Guards were placed around feeders, but 98% of 152 visits recorded were by non-target species.

A paired sites study in two pairs of 260 ha rangeland sites in Texas, USA, in from spring 1986 until spring 1987 (5) found that northern bobwhites *Colinus virginianus* from sites supplied with supplementary food had significantly higher body fat percentages than bobwhites from control sites (average of 11.8% body fat for 111 bobwhites from fed sites vs. 10.1% for 110 bobwhites from control sites). Supplementary food consisted of 16 feeders supplied with 15 kg of milo (a high-carbohydrate, low fat and low protein supplement) and checked twice weekly. Between 46 and 70% of bobwhites shot on fed sites had milo in their crops. No investigation was made of survival.

A controlled study in 2002–9 on mixed farmland in Hertfordshire, England (6), found that the number of grey partridges *Perdix perdix* increased significantly on an experimental site, where supplementary food was provided (along with several other interventions), but only slightly on a control site without food. This increase was apparent in spring (from fewer than 3 pairs/km^2 in 2002 to 12 pairs/km^2 in 2009, with a high of 18 pairs/km^2 vs. approximately 1 pair/km^2 on the control site in 2002, increasing to approximately 4 pairs/km^2 in 2009) and autumn (from fewer than 10 birds/km^2 in 2002 to approximately 65 birds/km^2 in 2009, with a high of 85 birds/km^2 vs. approximately 4 birds/km^2 on the control site in 2002, increasing to approximately 15 birds/km^2 in 2009). Food consisted of wheat from a hopper, provided from October to March). The experimental site also had predator control present and habitat creation (see 'Control predators not on islands' and 'Plant wild bird seed or cover mixture'). The effects of agri-environment schemes and the provision of set-aside are also discussed in 'Pay farmers to cover the costs of conservation measures' and 'Provide or retain set-aside'.

(1) Westerskov, K. E. (1977) Covey-oriented partridge management in France. *Biological Conservation*, 11, 185–191.
(2) Guthery, F. S. (1997) A philosophy of habitat management for northern bobwhites. *The Journal of Wildlife Management*, 61, 291–301.
(3) Townsend, D. E., Lochmiller, R. L., DeMaso, S. J., Leslie, D. M., Peoples, A. D., Cox, S. A. & Parry, E. S. (1999) Using supplemental food and its influence on survival of northern bobwhite (*Colinus virginianus*). *Wildlife Society Bulletin*, 27, 1074–1081.

(4) Guthery, F. S., Hiller, T. L., Puckett, W. H., Baker, R. A., Smith, S. G. & Rybak, A. R. (2004) Effects of feeders on dispersion and mortality of bobwhites. *Wildlife Society Bulletin*, 32, 1248–1254.
(5) Doerr, T. B. & Silvy, N. J. (2006) Effects of supplemental feeding on physiological condition of northern bobwhite in South Texas. *The Journal of Wildlife Management*, 70, 517–521.
(6) Aebischer, N. J. & Ewald, J. A. (2010) Grey Partridge *Perdix perdix* in the UK: recovery status, set-aside and shooting. *Ibis*, 152, 530–542.

15.17.4. Vultures

- A before-and-after study from Spain (2) found a large increase in griffon vulture *Gyps fulvus* population in the study area following multiple interventions including supplementary feeding.
- Two studies from the USA (1) and Israel (3) found that Californian condors *Gymnogyps californianus* and Egyptian vultures *Neophron percnopterus* fed on many of the carcasses provided for them. The Egyptian vultures were sometimes dominated by larger species at a feeding station supplied twice a month, but not at one supplied every day.

A study in California, USA, between February 1971 and May 1973 (1) found that Californian condors *Gymnogyps californianus* fed on at least 47 of 83 carcasses provided over the study period. Another 27 carcasses may well have been fed on and the remaining nine were taken by black bears *Ursus americanus* before condors could feed. Carcasses were mainly mule deer *Odocoileus hemionus*. This study is also discussed in 'Provide supplementary food to increase reproductive success'.

A before-and-after between 1969 and 1989 in the western Pyrenees, Spain (2), found that the population of griffon vultures *Gyps fulvus* increased from 282 pairs (in 23 colonies) in 1969–75 to 1,097 pairs (46 colonies) in 1989 following the initiation of multiple conservation interventions including the installation of feeding stations between 1969 and 1979. However, the authors note that only two of six feeding stations were used by vultures and food was never apparently a limiting factor for the population in the study area. This study is also discussed in 'Habitat protection', 'Restrict certain pesticides or other agricultural chemicals' and 'Use legislative regulation to protect wild populations'.

A study in the Negev Desert, Israel, in April–August of 1989 and 1990 (3) found that adult Egyptian vultures *Neophron percnopterus* were able to dominate a feeding station supplied daily with 5–10 kg of chicken, but not a station supplied approximately twice a month with large amounts (20–350 kg) of meat. Peak numbers of vultures were higher at the irregularly-stocked station (30–40 vultures present at once vs. 20–30) but they were sometimes excluded by mammals (e.g. striped hyaena *Hyaena hyaena*) or Eurasian griffon vultures *Gypus fulvus*, which did not occur at the regularly-stocked station.

(1) Wilbur, S. R., Carrier, W. D. & Borneman, J. C. (1974) Supplemental feeding program for California condors. *The Journal of Wildlife Management*, 38, 343–346.
(2) Donazar, J. A. (1990) Population trends of the griffon vulture *Gyps fulvus* in northern Spain between 1969 and 1989 in relation to conservation measures. *Biological Conservation*, 53, 83–91.
(3) Meretsky, V. J. & Mannan, R. W. (1999) Supplemental feeding regimes for Egyptian vultures in the Negev Desert, Israel. *The Journal of Wildlife Management*, 63, 107–115.

15.17.5. Raptors

- Two randomised, replicated and controlled studies in the USA (1,2) found that nesting northern goshawks *Accipiter gentilis* were significantly heavier in territories supplied with supplementary food, compared with those from unfed territories.

A randomised, replicated and controlled trial in mixed conifer forests and scrub in New Mexico, USA, in 1992–3 (1), found that northern goshawk *Accipiter gentilis* adults from territories provided with supplementary food during nesting were heavier than adults from control (unfed) territories, but sample sizes were too small for statistical tests (females: average of 1,007 g for six birds from fed territories vs. 975 g for five controls; males: average of 689 g for five fed territories vs. 660 g for two controls). Supplementary food consisted of dead Japanese quail *Cortunix japonica* provided every other day starting the day after hatching and continuing until most control birds left the area. This study also examined differences in chick growth and survival, discussed in 'Provide supplementary food to increase reproductive success'.

A randomised, replicated and controlled trial in mixed conifer forests in Utah, USA, in 1996–7 (2), found that northern goshawk *Accipiter gentilis* females from territories provided with supplementary food (Japanese quail *Coturnix japonica* provided from close to hatching to chick independence) were significantly heavier than those from control (unfed) territories (1,104 g for eight fed females vs. 993 g for nine controls). This study also examined differences in chick growth and survival, discussed in 'Provide supplementary food to increase reproductive success'.

(1) Ward, J. M. & Kennedy, P. L. (1996) Effects of supplemental food on size and survival of juvenile northern goshawks. *The Auk*, 113, 200–208.
(2) Dewey, S. R. & Kennedy, P. L. (2001) Effects of supplemental food on parental-care strategies and juvenile survival of northern goshawks. *The Auk*, 118, 352–365.

15.17.6. Cranes

- A before-and-after study from Japan (1) and a global literature review (2) found that local crane populations increased after the provision of supplementary food.

A before-and-after study of red-crowned cranes *Grus japonensis* in Hokkaido, Japan, in the mid-20th century (1) found that a local population increased from 42 individuals in 1952–4 to 161 in 1960–4. This followed the establishment of an artificial feeding station in 1952, and the author attributes the population rise to supplementary food reducing winter mortality, although no data is provided for the use of the feeding station, any reduction in starvation or increase in reproductive productivity. No details are provided about the supplementary food provided.

A 1998 literature review (2) found that supplementary feeding of cranes appeared to increase local populations of five, and possibly six species of crane in five sites across the world. These were red-crowned cranes *Grus japonensis* in Hokkaido, Japan (1); hooded cranes *G. monachus* and white-naped cranes *G. vipio* wintering at Izumi, Japan; common cranes *G. grus* at Lake Hornborga, Sweden; and demoiselle cranes *Anthropoides virgo* at Khichan in India. It is also possible that winter feeding of whooping cranes *G. americana* in 1993–4 may have encouraged population growth. The author recommends that supplementary feeding is viewed as a potential short-term practice, but that the risks from spreading disease and increased human disturbances may make it unsuitable as a long-term strategy.

(1) Masatomi, H. (1991) Population dynamics of red-crowned cranes in Hokkaido since the 1950s. 297–299 *Proceedings of the 1987 International Crane Workshop* International Crane Foundation, Baraboo, Wisonsin, USA.
(2) Davis, C. (1998) A review of the success of major crane conservation techniques. *Bird Conservation International*, 8, 19–30.

15.17.7. Gulls, terns and skuas

- A randomised, replicated and controlled study in the Antarctic (1) found that female south polar skuas *Catharacta maccormicki* that were fed lost more weight whilst feeding two chicks than unfed birds. There was no difference for birds with single chicks, or male birds.

A randomised, replicated and controlled trial on King George Island, Antarctic Peninsula, in the boreal summer of 2000–1 (1) found that female south polar skuas *Catharacta maccormicki* (also *Stercorarius maccormicki*) that were fed when raising two chicks lost significantly more weight than control (unfed) females (average loss of 7.9% of body weight for fed pairs vs. 4.6% for controls). There was no such effect in male skuas (average loss of 2.1% of body weight for 27 fed males vs. 5.5% for 27 controls) or if females raising single chicks were included in results (loss of 6.9% of body weight for 27 fed pairs vs. 4.5% for 27 controls). Supplementary food consisted of 25–100 g of fish provided to adults every other day, corresponding to approximately 20% of a chick's daily energy needs. This study also includes the impact of feeding on chick growth and survival, see 'Provide supplementary food to increase reproductive success'.

(1) Ritz, M. S. (2006) Sex-specific mass loss in chick-rearing South Polar skuas *Stercorarius maccormicki* – stress induced or adaptive? *Ibis*, 149, 156–165.

15.17.8. Pigeons

- A study of a recently-released pink pigeon *Nesoenas mayeri* population on Mauritius (1) found that fewer than half the birds used supplementary food, and appeared to survive without it.
- However, a later study of the population (2) found that almost all birds were recorded using supplementary feeders.

A study in mixed forests on Mauritius between July 1987 and June 1992 (1) found that, of 42 captive-bred pink pigeons *Nesoenas mayeri* (formerly *Columba mayeri*) released, 16 visited feeders and spent an average of 20.5 mins/day feeding. However, when food provision stopped, only two birds were seen visiting the feeder, suggesting that they could survive without it. Non-native birds also used the feeder but they did not exclude pigeons. Food consisted of mixture of maize, wheat, canary seed, millet, lentils, and occasionally peas and other seeds provided from a hopper outside the release aviary and was provided from June until December 1987. The amount provided was then reduced and the finally stopped, until more pigeons were released in June 1988 and continued until at least 1992. This study is also discussed in 'Use captive breeding to increase or maintain populations', 'Release captive bred individuals', 'Provide supplementary food after release' and 'Predator control on islands'.

A study of restored pink pigeon *Nesoenas mayeri* (formerly *Columba mayeri*) populations in two forest sites on Mauritius, in April to June 2005 (2) found that 99% of the 195 birds studied visited supplementary feeders. More birds appeared to use feeders outside rather than inside release aviaries, although this was not tested statistically. Younger birds used the feeders more frequently older birds but there was no difference in use between nesting and non-nesting birds and feeding stations did not appear to influence the position of breeding territories. Supplementary food consisted of wheat provided from hoppers, protected from rats by being placed on platforms with plastic sheeting around the supporting post.

(1) Jones, C., Swinnerton, K., Taylor, C. & Mungroo, Y. (1992) The release of captive-bred pink pigeons *Columba mayeri* in native forest on Mauritius. A progress report July 1987–June 1992. *Dodo*, 28, 92–125.
(2) Edmunds, K., Bunbury, N., Sawmy, S., Jones, C. G. & Bell, D. J. (2008) Restoring avian island endemics: use of supplementary food by the endangered pink pigeon (*Columba mayeri*). *Emu*, 108, 74–80.

15.17.9. Hummingbirds

- Four studies from the USA (3–6) found that three species of hummingbird showed preferences for higher concentrations of sucrose (measured in weight/volume), consuming more and visiting feeders more frequently.
- A study from the USA (3) found that black-chinned hummingbirds *Archilochus alexanderi* preferentially fed on sugar solutions over artificial sweeteners, and that increasing the viscosity of these solutions (so they appeared more like sugar solutions) did not affect their consumption.
- Two studies from Mexico (*ex situ*, (2)) and Argentina (7) found that four species showed preferences for sucrose over fructose or glucose when equiweight solutions were compared. One (2) found that birds also preferentially fed on sucrose over a sucrose-glucose mix, the other (7) found no preference for sucrose over a glucose-fructose mix.
- A controlled study from the USA (1) found that Anna's hummingbirds *Calypte anna* showed a preference for red-dyed sugar solutions over five other colours, but only if different colours were presented at the same time.

- A replicated study from the USA (4) found that rufous hummingbirds *Selasphorous rufus* preferentially fed on feeders placed higher, over lower ones.

Background

We have separated the literature on feeding hummingbirds from other species because they feed on a food source, nectar, very different from most other birds (although see 'Nectar-feeding songbirds' later in this chapter).

Successfully providing food for nectar-feeding birds is difficult because they can show strong preferences for different sugar solutions, depending on their sugar concentrations, and which sugar is used. Brown *et al.* (2008) argue that care must be taken in interpreting the results of preference studies, because different studies measure solution concentration in different ways. For example, they test the preferences of malachite sunbirds *Nectarinia famosa* (a songbird, not a hummingbird) and found that birds appeared to show a preference for sucrose over hexose, no preference or a preference for hexose over sucrose depending on whether equimolar (i.e. with the same number of sugar molecules/ml), equiweight (the same weight of sugar/ml) or equicalorific (the same energy content/ml) solutions were used.

A controlled study in mountainous pine forests in California, USA, in July 1976 (1) found that Anna's hummingbirds *Calypte anna* fed more on red-dyed sugar solutions than solutions on other colours (17 visits/hour for red solution vs. 13 for yellow, eight for green, four for blue and five for uncoloured). However, birds did not show a preference when solutions were presented one after another. A separate experiment found that, after 40 minutes, approximately 80% of hummingbirds visiting a feeding area went to a feeder containing sugar solution, rather than a control feeder containing water. Solutions consisted of saturated solutions of cane sugar in water and were supplied from a 250 ml feeder with small holes near the base. The feeder had been used for several years before the experiment.

A replicated trial in captivity in Mexico (2) found that three species of hummingbird all showed strong preference for a pure sucrose solution, compared to a 'hexose solution' (1:1 ratio of glucose and sucrose); 'hexose solution' was preferred to glucose by all species; and glucose preferred to fructose. Solutions were equiweight, with 17.1% concentration, by weight. Feeding consisted of providing a pair of sugar solutions in glass feeding tubes (a 42 cm vertical section leading to a lower 7 cm section at a 45° angle, tapering to a 2.5 mm hole from which birds drank) for four hours and measuring the volume consumed. The species tested were cinnamon hummingbird *Amazilia rutila*, broad-billed hummingbird *Cynanthus latirostris* and fork-tailed emerald *Chlorostilbon canivetii*, with six individuals from each species being used in each pairwise comparison.

A replicated study in the summers of 1986–7 in pine scrubland in Arizona, USA (3), found that black-chinned hummingbirds *Archilochus alexanderi* preferentially fed on 40% sucrose solutions (by weight/volume) when offered the choice of 40%, 10% and 20% (taking four times as much 40% solution as the others), but showed no preference when given the choice of 20%, 25% and 30% or 20%, 30% and 40%. Hummingbirds also showed a very strong preference for 20% sucrose solution, compared to saccharin (0.045% concentration), aspartame (0.1%) and Equal® (2.47%) artificial sweeteners (taking between six and 12 times more sugar than sweetener). Increasing the viscosity of the artificial solutions did not increase their consumption. All solutions were provided in ten commercial hummingbird feeders, each with three feeding points, scattered 300 m along a ridge. Experiments ran for 15–30 minutes.

A replicated study in a coniferous forest in Montana, USA, in summer 1995 (4) found that rufous hummingbirds *Selasphorous rufus* preferentially visited feeders containing higher concentrations of sucrose, until 50% sucrose by weight, above which they did not show a preference. They also appeared to preferentially use more highly elevated feeders (approximately 80% of the maximum sucrose consumed was taken from feeders 3 m from the ground vs. a maximum of 55% from feeders hung 2.75 m or less from the ground). Food was supplied at seven stations, each with 12 feeders hung at between 0.25 and 3.00 m from the ground.

A replicated trial in a coniferous forest in Montana, USA, in the summers of 1996 and 1997 (5) found that rufous hummingbirds *Selasphorous rufus* preferentially consumed sucrose at a 50% concentration or above when give a choice of solutions at between 10 and 70%. Solutions were provided from arrays of four 10 ml syringes at 12 feeding stations, 100 m apart. All solutions were measured in weight-volume concentrations (i.e. 40% solution is 40 g sucrose in 100 ml water).

A randomised, replicated trial in a mixed forest and alpine meadow site in Colorado, USA, in May–August 2001 (6) found that female broad-tailed hummingbirds *Selasphorus platycercus* visited feeders more often if they contained higher sucrose concentrations (5.8 visits/hr to feeders with 30% solution vs. 4.2 visits/hr with 20% solution and 0.3 visits/hr with 10%). Sucrose solutions were provided in nine 100 ml commercially-available feeders at least 50 m apart. Feeders were randomly assigned to a sugar concentration at the beginning of the study and monitored for a one hour a day for a total of 17 days. Solutions were measured as weight-volume concentrations.

A randomised replicated and controlled study in temperate forests in southern Argentina (7) found that green-backed firecrowns *Sephanoides sephaniodes* preferentially consumed sucrose solutions over glucose and fructose and preferred all sugar solutions over water. The preference for sucrose was only significant when the total amount of sugar consumed was examined (not the rate of visits, consumption/visit or consumption/hummingbird), with 24% and 68% more sucrose being consumed, compared to fructose and glucose respectively. There were no differences between sucrose and a glucose-fructose mix. Male hummingbirds used feeders more than females and more sugar was consumed at lower temperatures. All sugar solutions were the same concentration (24% by weight) and were provided from five commercial hummingbird feeders placed in three sites throughout the forest for ten days at a time in December 1999, February–May 2000, and January–February 2001.

(1) Wheeler, T. G. (1980) Experiments in feeding behavior of the Anna hummingbird. *The Wilson Bulletin*, 92, 53–62.
(2) Martinez del Rio, C. (1990) Sugar preferences in hummingbirds: the influence of subtle chemical differences on food choice. *The Condor*, 92, 1022–1030.
(3) Stromberg, M. R. & Johnsen, P. B. (1990) Hummingbird sweetness preferences: taste or viscosity? *The Condor*, 92, 606–612.
(4) Blem, C. R., Blem, L. B. & Cosgrove, C. C. (1997) Field studies of rufous hummingbird sucrose preference: does source height affect test results? *Journal of Field Ornithology*, 68, 245–252.
(5) Blem, C. R., Blem, L. B., Felix, J. & Gelder, J. van (2000) Rufous hummingbird sucrose preference: precision of selection varies with concentration. *The Condor*, 102, 235–238.
(6) Camfield, A. F. (2003) Quality of food source affects female visitation and display rates of male broad-tailed hummingbirds. *The Condor*, 105, 603–606.
(7) Chalcoff, V. R., Aizen, M. A. & Galetto, L. (2008) Sugar preferences of the green-backed firecrown hummingbird (*Sephanoides sephaniodes*): a field experiment. *The Auk*, 125, 60–66.

15.17.10. Woodpeckers

- One replicated, controlled study from the USA (1) found that 12 female downy woodpeckers *Picoides pubescens* supplied with supplementary food had higher nutritional statuses than unfed birds.
- However, two analyses of a replicated, controlled study of 378 downy woodpeckers from the USA found that they did not have higher survival rates (2) or nutritional statuses (3) than unfed birds.

A replicated, controlled study in deciduous forests in Ohio, USA, in the winter of 1988–9 (1) found that 12 female downy woodpeckers *Picoides pubescens* grew longer feathers and grew them faster (a proxy for nutritional condition) when supplied with sunflower seeds and suet in excess, compared to six unfed control females. There were no such differences in eight fed and nine control male woodpeckers. The impact on three songbird species is discussed in 'Provide supplementary food to increase adult survival – Songbirds'.

A replicated, controlled study in 54 woodlots and riparian corridors in an agricultural landscape in Ohio, USA, in the winters of 1995–9 (2) found that 378 downy woodpeckers *Picoides pubescens* did not have higher survival rates in either woodlots or riparian strips provided with supplementary food, compared with unfed, control sites. The impact on three songbird species is discussed in 'Provide supplementary food to increase adult survival – Songbirds'. Supplementary food consisted of sunflower seeds and suet provided in excess throughout winter.

Another analysis (3) of the same data as (2) found that downy woodpeckers *Picoides pubescens* did not have higher nutritional statuses (judged by the size of feather growth bars) than woodpeckers in unfed control woodlots. The impact on three songbird species is discussed in 'Provide supplementary food to increase adult survival – Songbirds'.

(1) Grubb, T. C. & Cimprich, D. A. (1990) Supplementary food improves the nutritional condition of wintering woodland birds: evidence from ptilochronology. *Ornis Scandinavica*, 21, 277–281.
(2) Doherty, P. F. & Grubb, T. C. (2002) Survivorship of permanent-resident birds in a fragmented forested landscape. *Ecology*, 83, 844–857.
(3) Doherty, P. F. & Grubb, T. C. (2003) Relationship of nutritional condition of permanent-resident woodland birds with woodlot area, supplemental food, and snow cover. *The Auk*, 120, 331–336.

15.17.11. Songbirds

- Seven studies from Europe (1,19,26,28,31,33) and the USA (17) found higher densities or larger populations in various songbird species in areas close to supplementary food. Six studies from Europe (1,19,28,33), Canada (3) and Japan (16) found that population trends or densities in some species were no different between fed and unfed areas. The American study (18) found that populations appeared to follow food, with populations increasing after feeders were erected and decreasing after they were removed.

- Four studies from Canada (3), Europe (13), Japan (15) and the USA (18) found that birds had higher survival when supplied with supplementary food. However, in two studies this was only apparent in females (13) or in one of two species studied (18). A controlled study in the USA (10) found no evidence that birds were dependent on supplementary food: when food was removed, previously fed birds did not have lower survival than controls.

- A replicated, controlled study from the USA (2) found that song sparrows *Melospiza melodia* had lower survival with feeding stations in their territories.

- Six studies from Europe (5,12) and the USA (7,14,20,21) found that birds supplied with supplementary food were in better physical condition or had larger fat supplies than unfed birds. However, in one replicated, controlled study (5) this was only the case for females; in another two (7,20), only one of three species showed better condition, with one species in one study (20) showing lower condition when fed; a final replicated and controlled study (14) found that differences between treatments were only apparent in the breeding season.

- Two studies investigated the effect of feeding on behaviours: a randomised, replicated and controlled study in the USA (4) found that male Carolina wrens *Thryothorus ludovicianus* spent more time singing when supplied with food; a replicated, controlled study in Sweden (6) found no behavioural differences between wood nuthatches *Sitta europaea* supplied with food, and unfed birds.

- Thirteen studies from the UK (8,23–25,27–31,34), Canada (9) and the USA (11,22) investigated use of feeders. Four studies from the USA (11,22) and the UK (23,24) found high use of supplementary food by several species, with up to 21% of birds' daily energy needs coming from feeders (11). However, another UK study (25) found very low use of food, possibly because the feeder was not positioned close to natural food sources.

- One UK study (29) found that use of feeders peaked in midwinter, although another (34) found that the exact timing of peak use varied between species.
- Two replicated trials from the UK (8) finding that the use of feeders increased with distance to houses and decreased with distance to cover, whilst a replicated Canadian study (9) found that American goldfinches *Carduelis tristis* preferred using bird feeders in high positions. A large-scale replicated study in the UK (27) found that preferences for feeder locations varied between species.
- Three studies from the UK (28,30,31) argue that placing feeders over 1 km apart, and possibly 1.1–1.3 km apart (31) will maximise their use whilst keeping the intervention practical.

A replicated, controlled study in the winters of 1969–71 in three deciduous woods in southern Sweden (1) found that great tit *Parus major* populations increased at two fed sites but declined at an unfed control site in a harsh winter, but increased at all sites in a milder winter. Adult survival over this period was highest in the site fed in both years (37–59% survival of 31 birds), intermediate in the control site (35–47% survival of 22) and lowest at the site fed only in 1969–70 (13–19% survival of seven). More yearlings remained in the two fed sites, compared with the control (14–22% of 85 birds remaining vs. 7–11% of 48). Blue tit *P. caeruleus* populations appeared unaffected by feeding at the same sites. Supplementary food consisted of hoppers filled with sunflower seeds provided from December (1969) or October (1970) until March.

A replicated and controlled study on Mandarte Island, British Colombia, Canada (2), found that song sparrows *Melospiza melodia* provided with supplementary food in 1985 were less likely to survive until the breeding season of 1986 (40–47% survival for fed adults, n = 15 vs. 66–79% for controls, n = 42). The authors note that 1985 was a year of peak song sparrow density and suggest lower adult survival could be due to increased costs of defending feeders from other song sparrows. This study also described the impact of feeding on reproduction, discussed in 'Provide supplementary food to increase reproductive success'.

A controlled study in the winters of 1985–6 and 1986–7 in two small deciduous forest sites in Alberta, Canada (3), found that winter survival of black-capped chickadees *Parus atricapillus* was higher in the 2.6 km^2 area provided with supplementary food, compared to the 1.9 km^2 control (unfed) area (1985–6: 88% of 163 birds in the fed area vs. 80% of 143 controls; 1986–7: 65% of 192 vs. 53% of 137). However, there were no significant differences in the proportion of chickadees acquiring local breeding territories, or in local breeding densities (1986: 54% of fed birds acquiring local breeding territories and 15.3 pairs/km^2 vs. 66% and 17.2 pairs/km^2 for controls; 1987: 66% and 16.1 pairs/km^2 vs. 71% and 14.0 pairs/km^2). Supplementary food consisted of multiple feeding stations, each with between two and four feeders, filled weekly with sunflower seeds.

A replicated, randomised, controlled study from December–January in 1985–1986 in 8 experimental and control (rotated) pairs of Carolina wrens *Thryothorus ludovicianus* in a woodland study site in Tennesse, USA (4), found that male wrens supplemented with food sang significantly more than unsupplemented males. Food supplementation, but not song playback, significantly increased both the song rate and the rate of song-type change (89.4 and 51.2 songs/hour; 1.3 and 0.7 song changes/hour for food supplemented and control males respectively). The authors point out that, because foraging and singing are mutually exclusive behaviours in Carolina wrens, the increase in vocal territorial behaviour associated with food supplementation may reflect a decrease in the time required for foraging. Wren pairs were allocated to food supplementation treatments randomly. Food supplementation consisted of 2 cans filled daily with 100 mealworms/territory. Daily observations began approximately 30 min before sunrise and continued for 4 hours.

A replicated and controlled study in the breeding seasons of 1985–7 in grasslands on Öland, southern Sweden (5), found that female northern wheatears *Oenanthe oenanthe*, but not males, that were provided with supplementary food were significantly heavier than unfed controls (average of 26.9 g for 53 fed females and 24.4 g for 42 fed males vs. 24.3 g and 23.7 g for 48 and 32 unfed controls). However, there was no effect when females were feeding older chicks, (after they were able to regulate their body temperature. A few days after hatching, most food was delivered to chicks, not consumed by adults. Food consisted of 7 g of mealworms provided either during incubation, or for the entire breeding season.

A replicated controlled study in a deciduous wood in southern Sweden in winter 1980 and spring 1981 (6) found that, although wood nuthatches *Sitta europaea* made up 28% of all visits to supplementary feeding stations, there was no difference in the time devoted to different behaviours between five fed pairs and nine control pairs. Supplementary food consisted of 60 kg of sunflower seeds supplied continuously from December 1980 until March 1981 at nine feeders spread evenly across the experimental area. The author suggests that the lack of differences could be due to variations in the intensity of foraging, rather than the time spent foraging.

A replicated, controlled study in ten deciduous forest plots in Ohio, USA, in the winter of 1988–9 (7) found that three species of songbirds grew longer replacement feathers when provided with sunflower seeds and suet in excess, compared to control birds which were not fed. However, most of these differences were not significant, with only tufted titmice *Baeolophus bicolor* (formerly *Parus bicolor*) consistently showing significant differences across ages and sexes. Replacement feathers appeared to grow faster (a proxy for nutritional condition) in titmice and white-breasted nuthatches *Sitta carolinensis* of all age classes and both sexes. Immature and male Carolina chickadees *Parus carolinensis* also showed more rapid growth when fed, the sample size for female chickadees was too small for comparison and there was no significant difference between fed and unfed adult chickadees. When corrected for body size, differences for male titmice and male nuthatches became non-significant. Sample sizes were: 13 male chickadees, five females, 15 adults and 28 immatures; six male titmice, 14 females, 12 adults and 17 immatures; 16 male nuthatches, ten females. The impact of feeding on downy woodpeckers *Picoides pubescens* is discussed in 'Provide supplementary food to increase adult survival – Woodpeckers'.

Two studies in gardens in Cardiff, south Wales, in January–February 1988 and February–April 1989 (8) found that distance to both cover (a dense hedgerow) and housing had significant effects on rate of consumption of supplementary food by five songbird species. At a single site, as distance from cover increased, the proportion of food consumed decreased (32% consumed when the feeder was next to the hedge, 28% at 2.5 m away, 23% at 5 m and 17% at 7.5 m), with a greater impact on house sparrows *Passer domesticus* and blue tits *Parus caeruleus* than on greenfinches *Carduelis chloris*. Overall consumption increased with distance from housing in three other sites (34% of food consumed when three feeders were 10 m from housing, 25% at 7.5 m, 25% at 5 m, and 16% at 2.5 m), however this effect varied between species. Siskins *C. spinus* used all feeders equally; greenfinch use increased with distance from housing and house sparrows used feeders closest to the houses most frequently. Supplementary food consisted of 250 g of peanuts supplied each day.

A replicated study from January–March in 1990 in one suburban garden garden in Ontario, Canada, using four bird tube feeders placed in different locations in Ontario, Canada (9), found that American goldfinches *Carduelis tristis* preferred bird feeders that were placed in higher locations. The number of birds and the amount of food removed were significantly higher for the upper feeder (6.1 m) than the lower feeder (3.4 m) (average 75% birds seen and 67% food removed from high feeder). There was no preference for feeders placed in trees or in the open (both 2.4 m from the ground), although there was a much greater proportion of unexplained variation in the side-by-side experiment than in the height experiment. All tube feeders contained small black oil sunflower seeds. A flock of 15–30 goldfinches used the test feeders daily. Feeders were switched at the end of each trial (3 replicates each).

A controlled study in Wisconsin, USA, over five months in the winter of 1984–5 (10) found that, following the removal of supplementary food, 49 black-capped chickadees *Parus atricapillus* did not have lower survival at a site where they had previously been provided with supplementary food, compared with 35 control chickadees, at a site where they had never been fed (84% monthly survival for previously-fed birds vs. 85% for controls). Food had been provided for 25 winters at the experimental site.

A small study in two deciduous forest sites in Wisconsin, USA, in the winters of 1983–5 (11) found that approximately 83 black-capped chickadees *Parus atricapillus* used two feeders providing sunflower seeds each day throughout winter, and obtained approximately 21% of their daily energy requirements from them. Birds with home territories nearer the feeders used it more than more distant birds, and more bird used the feeders (and fed at a higher rate) in the evening (an average of 39 chickadees using the feeders within two hours of sunset vs. 17 within two hours of sunrise and 36 at midday). Feeders were used the most in autumn and least in spring, with temperature not affecting feeder use. Feeders were monitored for 15 days over the two winters.

A randomised, replicated and controlled paired study in parkland and mixed woodland in southern Scotland in spring and summer 1990 (12) found that great tit *Parus major* females with artificially enlarged broods lost significantly less weight whilst provisioning young when nestlings were provided with supplementary food, compared to females with control (enlarged but not fed) broods (average of 2.2 g lost between day ten of incubation and day 13 of provisioning for experimental females, n = 8 vs. 2.9 g for control females, n = 7). There were no data on survival of adults. Broods were enlarged by the addition of three nestlings (added after hatching). Supplementary food consisted of an average of 2.2 g of minced meat and nutritional supplements fed twice daily to half the experimental brood on days six through 12 after hatching. This represents most of a nestling's daily energetic requirements. This study also investigated the impact on the nestling growth, discussed in 'Provide supplementary food to increase reproductive success'.

A replicated and controlled study in a mixed forest in the central Netherlands in 1987 (13) found that female pied flycatchers *Ficedula hypoleuca* from pairs provided with supplementary food had significantly higher survival rates than those from control pairs (survival in subsequent years of 58% for fed females, n = 12 vs. 27% for controls, n = 60). There was no difference in male survival (survival in subsequent years of 55% for fed males, n = 11 vs. 33% for controls, n = 51) or in adult weights when chicks were seven days old (average of 12.1 g for fed males, n = 11 vs. 12.2 g for control males, n = 13; 12.5 g for fed females, n = 12 vs. 12.5 g for controls, n = 14). Supplementary food consisted of mealworms provided in excess beginning two days after chicks hatched. This study also examines the impact of feeding on reproductive success, discussed in 'Provide supplementary food to increase reproductive success'.

A replicated and controlled trial in a dune and scrubland system in Florida, USA, in 1993 (14), found that breeding Florida scrub jays *Aphelocoma caerulescens* from territories provided with supplementary food had significantly higher body lipid levels (i.e. were in better physical condition) than adults from control (unfed) territories (fed males approximately 6% body fat, n = 9 vs. 3% for controls, n = 18; fed females approximately 4.5%, n = 9 vs. 1.5% for controls, n = 17). There were no differences in non-breeding individuals. Feeding consisted of providing dried dog food, peanuts and mealworms were provided twice daily at feeding stations in the middle of the territories from late January until females finished laying. Food was provided 'in excess'. This study also reported on the effects on reproduction, discussed in 'Provide supplementary food to increase reproductive success'.

A small controlled before-and-after trial in two mixed woodlands on Honshu, Japan in the spring of 1996 (15) found that survival of varied tits *Parus varius* was significantly higher at a 67 ha site supplied with supplementary food, compared to a 46 ha control (unfed) site 3 km away (100% survival over the 51 day study period for 17 tits in the fed area vs. 70% survival for 20 tits from the control area). Many birds changed breeding partners in the control site, whereas pair bonds were maintained in the fed site. Food consisted of 500 g sunflower seeds provided two or three times a week from seven feeders, each at least 150 m from each other. Feeders were visited on average 30 times/hr by three individuals.

A small controlled before-and-after trial (16) in two mixed woodlands on Honshu, Japan (using the same study sites as in (15)), found that neither the local population density nor the home range size of varied tits *Parus varius* increased over a six-week period when supplementary food was provided, compared to a control area where no food was provided. In September 1995 there were 23 tits in the 67 ha experimental site and 17 in the 36 ha control (unfed) site.

By the 12th December (the end of the feeding period), there were 19 tits in the experimental area and 17 in the control area. Food consisted of 500 g sunflower seeds provided two or three times a week from eight feeders, one in each tit territory in the experimental area and all tits in the area used feeders at an average of 5.2 visits/hr.

A randomised, replicated and controlled study in coniferous woods along a road in Maine, USA (17) found that significantly more black-capped chickadees *Parus carolinensis* (also known as *Poecile carolinensis*) were recorded during censuses at four sites fed continuously from late October 1995 to mid-March 1996, compared to at four unfed control sites (average of 5.5 birds/census for fed sites vs. 0.1 birds/census for controls, 18 censuses at each). Sites provided with food from October until January (early-fed sites) had significantly higher chickadee numbers than those fed from January until March (late-fed sites), with both being lower than continuously-fed sites and higher than controls (average of 1.8 birds/census and a maximum of approximately 5 birds/census for early fed vs. average of 0.1 birds/census and maximum of 3.0 birds/census for late-fed, 18 censuses at each). Chickadee numbers declined at early-fed sites when feeders were removed and increased at late-fed when feeders were established. No such patterns were seen at control or continuously-fed sites. Birds took longer to discover feeders at late-fed sites, compared to those supplied from October (all feeders at both continuously- and early-fed sites discovered within 15 days, a maximum of 33 days before the last feeder was discovered in late-fed sites). Feeding consisted of two feeders at each site refilled every week with black oil sunflower seeds.

A replicated, controlled study in 54 woodlots and riparian corridors in an agricultural landscape in Ohio, USA, in the winters of 1995–9 (18) found 315 Carolina chickadees *Parus carolinensis* (also known as *Poecile carolinensis*) had significantly higher survival rates in riparian sites provided with supplementary food, compared to those in unfed control sites (50% survival for fed areas vs. 43% for controls). This effect was not found in 346 white-breasted nuthatches *Sitta carolinensis* or 529 tufted titmice *Baelophus bicolor*. Chickadees in large plots supplied with food also had higher survival than those in unfed large plots. There was no such difference in smaller plots and no significant differences in the other species studied. The impact of feeding on downy woodpeckers *Picoides pubescens* is discussed in 'Provide supplementary food to increase adult survival – Woodpeckers'. Supplementary food consisted of sunflower seeds and suet provided in excess throughout winter.

A replicated, controlled study from May–June in 1992–1998 in one experimental (3 km^2) and four unmanaged arable farms in Leicestershire, England (19) found that the abundance of nationally declining songbird species and species of conservation concern significantly increased through time in the site where supplementary food was provided. Although there was no overall difference in bird abundance, species richness or diversity between the experimental and control sites, numbers of nationally declining species rose by 102% (except for Eurasian skylark *Alauda arvensis* and yellowhammer *Emberiza citrinella*). Nationally stable species rose (non-significantly) by 47% (with 8 species exhibiting net increases, especially greenfinch *Carduelis chloris* 68%, and 4 species exhibiting net decreases). The author concludes that supplementing food (grain provided through winter across the farm), as part of an integrated management package, provides the greatest benefits to species of conservation concern but does not affect species diversity at the farm scale.

Another analysis (20) of the same data as (18) also examined nutritional condition, judged by the size of feather growth bars, and found that 37 Carolina chickadees *Parus carolinensis* (also known as *Poecile carolinensis*) living in large woodlots were in significantly better nutritional condition (judged by the size of feather growth bars) when provided with supplementary food, compared with unfed controls. There were no significant differences in smaller woodlots. White-breasted nuthatches *Sitta carolinensis* had lower growth rates when fed, compared with controls and there was no impact of feeding on 48 tufted titmice *Baelophus bicolor*. The impact of feeding on downy woodpeckers *Picoides pubescens* is discussed in 'Provide supplementary food to increase adult survival – Woodpeckers'.

A replicated, controlled study at two pairs of mixed habitat sites in Kansas, USA, in the winters 2000–1 and 2001–2 (21) found that seven songbird species had higher levels of visible subcutaneous fat in areas supplied with supplementary food, compared with those in control (unfed) areas. Most differences were significant in both years and both pairs of sites, but differences were always small. Body mass relative to size showed a similar trend, but the differences were not significant. Supplementary feeding ran from early December until early March and consisted of four sunflower seed feeders and a 6 × 3 m area beneath them sprinkled with mixed seed. The species studied were dark-eyed junco *Junco hyemalis*, Harris's sparrow *Zonotrichia querula*, song sparrow *Melospiza melodia*, American tree sparrow *Spizella arborea*, northern cardinal *Cardinalis cardinalis*, black-capped chickadee *Parus atricapillus* (also known as *Poecile atricapillus*) and tufted titmouse *Parus bicolor*. The effects of feeding on predation rates are discussed in 'Can supplementary feeding increase predation or parasitism?'.

A replicated study in mountain forests in Tennessee, USA, in 1999–2001 (22) found that 92% of 24 breeding pairs of ovenbirds *Seiurus aurocapilla* and 79% of 38 wood thrush *Hylocichla mustelina* pairs fed on live mealworm *Tenebrio monitor* larvae from feeding stations consisting of moss placed over overhead projector film (clear plastic film) and placed on the ground near nests (6–12 m away from wood thrush nests, 3–6 m from ovenbird nests). Mealworms could burrow into the moss to avoid desiccation but could not escape because of the film. Previous work showed that birds avoided artificial feeders such as bowls and baskets, but removed 70–100% of mealworms within four hours from moss. Food was provided daily and nests were monitored for six (ovenbird) or eight (wood thrush) days.

A study at a farmland site in north-west England between January 2003 and February 2004 (23) found that twite *Carduelis flavirostris* used a supplementary feeding station (established in spring 2002) frequently outside the breeding season, with up to 250 birds being seen at once. However, twite used the station far less during the breeding season, when they relied more on wild seeds. Birds from another feeding station (see (24)) and other breeding colonies up to 20 km away used the feeding station, as well as individuals from a nearby colony of 20–30 birds. Supplementary food consisted of nyjer *Guizotia abyssinca* spread in a thick 2 m × 5 cm line on a 2 m × 2 m patch of bare earth and replenished every week.

A study at a farmland site in north-west England between January 2003 and February 2004 (24) used an identical feeding procedure to (23) at a site 12.6 km away and found that twite *Carduelis flavirostris* used the supplementary feeding station frequently outside the breeding season, with up to 150 birds being seen at once. However, twite used the station far less during the breeding season, when they relied more on wild seeds. A large number of birds from near the feeding station in (23) and colonies up to 20 km away used the feeding station, as well as birds from the two nearby colonies (each approximately 1.5 km away and 20–30 birds).

A study at a farmland site in north-west England between January 2003 and February 2004 (25) used an identical feeding procedure to (23) to establish a feeding station approximately 1 km from a colony of six pairs of twite *Carduelis flavirostris*. This station was only used occasionally and only by one or two birds at a time. The author suggests that the lack of use could have been due to the small size of the colony and the fact that it was not positioned close to natural feeding areas for twite.

A randomised, replicated and controlled study at three farmland sites in England in the winters of 1999–2000 until 2001–2 (26) found that farmland birds showed mixed responses to supplementary food. Chaffinches *Fringilla coelebs*, linnet *Carduelis cannabina* and yellowhammer *Emberiza citrinella* all showed significant short-term increases on at least one plot provided with food (chaffinch densities increased by 80–200% on three of six fed plots; yellowhammer densities increased by 230–400% on four of six; data for linnets not provided). There were no corresponding short-term changes on nearby control plots. Eurasian skylarks *Alauda arvensis* did not show any consistent response to food at any of the sites and there was no longer term impact of feeding on bird densities. Supplementary food consisted of 36 kg/ha of mixed grains broadcast over fields. The authors suggest that the lack of effect of feeding at some sites may be due to a very low natural seed density in the soil, meaning that even with supplementary food, the level of food was too low to attract birds.

The results from two replicated studies found that contextual variables affecting the use of supplementary food by a range of farmland songbirds were not consistent across species or regions (27). The 'BirdAid' programme (run between October and March in the winters of 2000–1 until 2002–3 across the UK) found that all three target species (tree sparrows *Passer montanus*, yellowhammers *Emberiza citrinella* and corn buntings *Miliaria calandra*) used supplementary food, consisting of 25 kg of seeds supplied each week. Tree sparrows and yellowhammers tended to use feeding stations more if they were closer to cover and in mixed landscapes, the opposite was true for corn buntings. The Winter Food for Birds project, run from October 2002 to March 2003 at ten replicates of seven sites across eastern England, found that six of eight target species used supplementary food, consisting of 5 kg each of millet and sunflower seeds supplied each week, sufficiently often for analysis. At the local and landscape scale, only human habitats and woodlands had uniform effects, increasing and decreasing the use for three and four species respectively. All other habitats had different impacts on different species.

A replicated, controlled study covering November–July in 2002–2004 in 10 sites each containing 7 feeding stations (placed at the centre of a 2 × 2 km tetrad) separated at set distances from each other (100 m, 500 m, 1 km, 2 km, 5 km and 10 km) in East Anglia, UK (28) found that supplementary provision of seeds increased local seed-eating bird abundance, especially species of conservation concern. Yellowhammer *Emberiza citrinella* and chaffinches *Fringilla coelebs* used the feeding stations most extensively (93–100% of all stations). Although genuine population trends were difficult to infer from the experimental setup, the authors argue that food provisioning increased the local abundance of several otherwise declining species (yellowhammers, reed buntings *Emberiza schoeniclus*, house sparrows *Passer domesticus* and chaffinches) over two winters. Colour-ring re-sighting and radio-tracking revealed that target granivores move small distances between food resources (500 m–1 km) and the authors suggest placing food resources (over-winter stubbles and wild bird cover crops) at a minimum of 1 km apart in order to be cost effective in reaching the most number of populations. Supplementary seed (10 kg of equally distributed sunflower hearts and millet) was replenished weekly. Avian use of the feeding stations was monitored twice weekly (20 min observation sessions)

A replicated study analysing annual survey data from 458 garden bird feeders in the UK from 1970–2000 (29) found that, of 41 species analysed, 21 increased in occurrence at garden feeders over time. Robins *Erithacus rubecula*, blackbirds *Turdus merula*, blue tits *Parus caeruleus* and greenfinches *Carduelis chloris* occurred at all sites, whilst dunnocks *Prunella modularis*, song thrushes *T. philomelos*, great tits *P. major*, starlings *Sturnus vulgaris*, house sparrows *Passer domesticus* and chaffinch *Fringilla coelebs* occurred at more than 95% of sites. Species were more likely to occur at feeders in years when they had high a country-wide population estimate and peak use tended to occur some time in midwinter.

The Winter Food for Birds project (see (27)) was continued in the winter of 2003–4 and the data from three winters are discussed in (30). For four songbird species (blue tit *Parus caeruleus*, chaffinch *Fringilla coelebs*, great tit *P. major* and robin *Erithacus rubecula*), feeding stations were used more frequently and by more birds if they were more than 500 m from other stations, compared with stations less than 500 m from neighbours. The same pattern was seen (but not significant) in blackbirds *Turdus merula* and house sparrows *Passer domesticus*. Yellowhammers *Emberiza citrinella* and reed buntings *E. schoeniclus*, however, used clustered sites more. There was no significant impact of distance of feeder use by greenfinches *Carduelis chloris*, goldfinches *C. carduelis* or dunnocks *Prunella modularis*. Local populations of all species divided themselves between multiple stations if they were closer than 500 m apart, but used only single stations if they were more widely spaced. The authors use this information to recommend that supplementary food resources provided for conservation purposes should be stations are placed at least 1 km apart to maximise cost-effectiveness (i.e. to ensure the maximum number of birds have access to supplementary food).

A replicated, controlled study from November-March in 2004–2007 in 10 experimental and 10 control tetrads (composed of four 1 km^2 sites) of arable farmland in East Anglia, UK (31) found that provision of seeds during winter significantly increased body mass and breeding population sizes of seed-eating species. Supplementary food was most used in early to mid-winter for generalist species and late winter for specialist species (such as chaffinch *Fringilla coelebs* and yellowhammer *Emberiza citrinella*). Radio tracking and mark-recapture techniques revealed that resource patches (such as wild bird cover crops and over-winter stubbles) should be separated by 1.1–1.3 km to be both cost and conservation effective for priority species (like yellowhammers). The authors suggest that year-round resource delivery could be achieved by placing breeding habitat 2.7–3.6 km from winter food patches. They caution that specific inter-patch distances may vary according to species and habitat but should be based on species of conservation concern. Experimental sites contained one central feeding station provided *ad libitum* with seed (10 kg of equally distributed millet, rape, wheat and sunflower seeds; replenished twice a week) and were fenced (50 cm in height) using 50 mm mesh wire and bamboo canes.

A series of randomised, replicated trials at two sites in England in the winters of 2000–1 and 2001–2 (32) found that five songbird species took supplementary food when provided and preferentially took wheat over oats and oats over barley. Tree sparrows *Passer montanus* and reed buntings *Emberiza schoeniclus* also fed on maize, preferring it to all cereals except wheat, whilst house sparrows *P. domesticus* preferred maize to all cereals. Corn buntings *E. calandra* and yellowhammers *E. citrinella* preferred all cereals to maize. Tree sparrows selected both cereals and oily seeds (sunflower seeds, oilseed rape etc) but avoided rye grass seed. All species preferred cereals to sunflower seeds and none showed any distinction between wheat and a 'weed seed mix'. At one site, food was provided in tubular feeders, in the other it was heaped on the ground. Survival rates of birds were not monitored.

A replicated study using some of the same data as (27) and combining them with data from control areas between 2000 and 2003 did not find robust evidence for supplementary winter feeding increasing breeding abundances of farmland songbirds (33). There were no effects of the Bird Aid programme on target species, although sites used more frequently had increased populations of yellowhammers *Emberiza citrinella* and corn buntings *Miliaria calandra*, but decreased populations of tree sparrows *Passer montanus*. Four of five insect-eating/generalist species declined faster in Winter Food for Birds (WFFB) experimental areas than in controls. There was no such effect for six seed-eating species. Declines in dunnocks *Prunella modularis*, robins *Erithacus rubecula* and yellowhammers *Emberiza citrinella* were lower in WFFB sites provided with more food and centrally-placed WFFB sites, compared to those provided with less food or those around the periphery of WFFB clusters.

Another study, using the same data as (30) and additional data from a second landscape-scale experiment, investigated how use of supplementary food by farmland songbirds varied over the winter months (34). Supplementary food-use peaked in or before January for five generalists and 'human-associated granivores (blackbirds *Turdus merula*, goldfinches *Carduelis carduelis*, greenfinches *C. chloris*, house sparrows *Passer domesticus* and robins *Erithacus rubecula*), whilst yellowhammers *Emberiza citrinella*, reed buntings *E. schoeniclus*, chaffinches *Fringilla coelebs* and dunnocks *Prunella modularis* all used supplementary food most in February or later. Use by great tits *Parus major* and blue tits *P. caeruleus* declined overwinter. The authors suggest food use reflects demand and the first group use food most when temperatures are lowest and daylight hours shortest, whilst the second group (which are heavily dependent on farmland seed) use food most when ambient food sources are at their most scarce and that the third group's pattern of food use reflects the available pool of individuals as mortality occurs through the winter. They caution that these results are likely to be dependent on the mix of farming types across the landscape, with eastern England being dominated by arable fields.

(1) Källander, H. (1981) The effects of provision of food in winter on a population of the great tit *Parus major* and the blue tit *P. caeruleus*. *Ornis Scandinavica*, 12, 244–248.
(2) Arcese, P. & Smith, J. N. M. (1988) Effects of population density and supplemental food on reproduction in song sparrows. *Journal of Animal Ecology*, 57, 119–136.
(3) Desrochers, A., Hannon, S. J. & Nordin, K. E. (1988) Winter survival and territory acquisition in a northern population of black-capped chickadees. *The Auk*, 727–736.

(4) Strain, J. G. & Mumme, R. L. (1988) Effects of food supplementation, song playback, and temperature on vocal territorial behavior of Carolina wrens. *The Auk*, 105, 11–16.
(5) Moreno, J. (1989) Body-mass variation in breeding northern wheatears: a field experiment with supplementary food. *The Condor*, 91, 178–186.
(6) Enoksson, B. (1990) Time budgets of nuthatches *Sitta europaea* with supplementary food. *Ibis*, 132, 575–583.
(7) Grubb, T. C. & Cimprich, D. A. (1990) Supplementary food improves the nutritional condition of wintering woodland birds: evidence from ptilochronology. *Ornis Scandinavica*, 21, 277–281.
(8) Cowie, R. J. & Simons, J. R. (1991) Factors affecting the use of feeders by garden birds: I. The positioning of feeders with respect to cover and housing. *Bird Study*, 38, 145–150.
(9) Dunn, E. H. & Hussell, J. A. T. (1991) Goldfinch preferences for bird feeder location. *Journal of Field Ornithology*, 62, 256–259.
(10) Brittingham, M. C. & Temple, S. A. (1992) Does winter bird feeding promote dependency? *Journal of Field Ornithology*, 63, 190–194.
(11) Brittingham, M. C. & Temple, S. A. (1992) Use of winter bird feeders by black-capped chickadees. *The Journal of Wildlife Management*, 56, 103–110.
(12) Johnston, R. D. (1993) The effect of direct supplementary feeding of nestlings on weight loss in female great tits *Parus major*. *Ibis*, 135, 311–314.
(13) Verhulst, S. (1994) Supplementary food in the nestling phase affects reproductive success in pied flycatchers (*Ficedula hypoleuca*). *The Auk*, 111, 714–716.
(14) Schoech, S. J. (1996) The effect of supplemental food on body condition and the timing of reproduction in a cooperative breeder, the Florida scrub-jay. *The Condor*, 98, 234–244.
(15) Nakamura, M. & Kubota, H. (1998) Food supply in early spring and stability of pair bonds in the varied tit *Parus varius*. *Journal of Avian Biology*, 29, 201–205.
(16) Kubota, H. & Nakamura, M. (2000) Effects of supplemental food on intra- and inter-specific behaviour of the varied tit *Parus varius*. *Ibis*, 142, 312–319.
(17) Wilson, JR, W. E. R. (2001) The effects of supplemental feeding on wintering black-capped chickadees (*Poecile atricapilla*) in central Maine: population and individual responses. *The Wilson Bulletin*, 113, 65–72.
(18) Doherty, P. F. & Grubb, T. C. (2002) Survivorship of permanent-resident birds in a fragmented forested landscape. *Ecology*, 83, 844–857.
(19) Stoate, C. (2002) Multifunctional use of a natural resource on farmland: wild pheasant (*Phasianus colchicus*) management and the conservation of farmland passerines. *Biodiversity and Conservation*, 11, 561–573.
(20) Doherty, P. F. & Grubb, T. C. (2003) Relationship of nutritional condition of permanent-resident woodland birds with woodlot area, supplemental food, and snow cover. *The Auk*, 120, 331–336.
(21) Rogers, C. M. & Heath-Coss, R. (2003) Effect of experimentally altered food abundance on fat reserves of wintering birds. *Journal of Animal Ecology*, 72, 822–830.
(22) Podolsky, A. L., Simons, T. R. & Collazo, J. A. (2004) A method of food supplementation for ground-foraging insectivorous songbirds. *Journal of Field Ornithology*, 75, 296–302.
(23) Raine, A. (2004) Providing supplementary food as a conservation initiative for twite *Carduelis flavirostris* breeding in the South Pennines near Worsthorne, Lancashire, England. *Conservation Evidence*, 1, 23–25.
(24) Raine, A. (2004) Providing supplementary food as a conservation initiative for twite *Carduelis flavirostris* breeding in the South Pennines near Littleborough, West Yorkshire, England. *Conservation Evidence*, 1, 26–28.
(25) Raine, A. (2004) Providing supplementary food as a conservation initiative for twite *Carduelis flavirostris* breeding in the South Pennines near Midgley, West Yorkshire, England. *Conservation Evidence*, 1, 29–30.
(26) Robinson, R. A., Hart, J. D., Holland, J. M. & Parrott, D. (2004) Habitat use by seed-eating birds: a scale-dependent approach. *Ibis*, 146, 87–98.
(27) Siriwardena, G. M. & Stevens, D. K. (2004) Effects of habitat on the use of supplementary food by farmland birds in winter. *Ibis*, 146, 144–154.
(28) Anon (2005) The consequences of spatial scale for agri-environment schemes designed to provide winter food resources for birds.
(29) Chamberlain, D. E., Vickery, J. A., Glue, D. E., Robinson, R. A., Conway, G. J., Woodburn, R. J. W. & Cannon, A. R. (2005) Annual and seasonal trends in the use of garden feeders by birds in winter. *Ibis*, 147, 563–575.
(30) Siriwardena, G. M., Calbrade, N. A., Vickery, J. A. & Sutherland, W. J. (2006) The effect of the spatial distribution of winter seed food resources on their use by farmland birds. *Journal of Applied Ecology*, 43, 628–639.

(31) Anon (2007) Understanding the demographic mechanisms underlying effective deployment of winter prescriptions for farmland bird recovery.
(32) Perkins, A. J., Anderson, G. & Wilson, J. D. (2007) Seed food preferences of seed-eating farmland passerines. *Bird Study*, 54, 46–53.
(33) Siriwardena, G. M., Stevens, D. K., Anderson, G. Q. A., Vickery, J. A., Calbrade, N. A. & Dodd, S. (2007) The effect of supplementary winter seed food on breeding populations of farmland birds: evidence from two large-scale experiments. *Journal of Applied Ecology*, 44, 920–932.
(34) Siriwardena, G. M., Calbrade, N. A. & Vickery, J. A. (2008) Farmland birds and late winter food: does seed supply fail to meet demand? *Ibis*, 150, 585–595.

15.17.12. Nectar-feeding songbirds

- Two studies from Australia (1) and New Zealand (5) found that ten species of honeyeaters and stitchbirds *Notiomystis cincta* readily used feeders supplying sugar solutions, with seasonal variations varying between species (1) and stitchbirds spending more time foraging for insects when food was supplied (5).

- A series of *ex situ* trials using southern African birds (2–4,6) found that most species tested showed a preference for sucrose solutions over glucose or fructose. One study found that sunbirds and sugarbirds only showed such a preference at low (equimolar) concentrations (2). Two more studies (3,4) found that two species showed preferences for sucrose when comparing 20% (by weight) solutions, although a third species did not show this preference (4). All species rejected solutions with xylose (a natural sugar in nectar) added. A final study found that sucrose preferences only became apparent at equicalorific concentrations high enough for birds to subsist on.

Background

As with experiments on hummingbirds, experiments on the preferences of nectar-feeding songbirds can be difficult to interpret and compare between studies. See 'Provide supplementary food to increase adult survival – Hummingbirds' for details.

A replicated study on heathland in New South Wales, Australia in 1986–8 (1) found that ten species of honeyeater were observed using supplementary feeders over 129 hours of observation. The most common were New Holland honeyeaters *Phylidonyris novaehollandiae*, white-cheeked honeyeaters *P. niger*, yellow-faced honeyeaters *Meliphaga chrysops*, white-eared honeyeaters *Meliphaga leucotis*, Little Wattlebirds *Anthochaera chrysoptera*. Species showed seasonal variations in use, but these were not consistent across species. Between eight and 14 feeders were distributed across a 4 ha patch of heath, at least 30 m from the centre of any known honeyeater territory. Feeders consisted of commercial hummingbird feeders modified so that they had a wider drinking hole, or larger plastic bottles with a curved drinking tube. Feeders were filled with 25% by weight sugar solution for four 48 hour periods each month. Feeders were also visited (briefly) by Silvereyes *Zosterops lateralis*, an eastern whipbird *Psophodes olivaceus* and non-birds (*Antechinus* – a small marsupial and honeybees *Apis mellifera*).

A randomised and replicated *ex situ* choice experiment in South Africa (2) found that three species of nectarivorous African songbirds (Gurney's sugarbird *Promerops gurneyi*, malachite sunbirds *Nectarinia famosa* and amethyst sunbird *N. amethystina*) all preferentially consumed sucrose over glucose or fructose when all were offered at low concentrations (0.25 moles/litre), but there were no such preferences when birds were offered higher concentrations (of 0.73 moles/litre). No species preferentially consumed higher concentrations of sucrose when give the choice of high (0.73 moles/litre), medium (0.5 moles/litre) or low (0.25 moles/litre) sucrose solutions. All solutions were provided in 125 ml feeders. A total of five malachite and seven amethyst sunbirds and five sugarbirds were tested.

A randomised, replicated *ex situ* choice experiment in South Africa in 1996 (3) found that 24 pale (Cape) white-eyes *Zosterops pallidus* tested over 12 days preferentially fed on 20% sucrose solution (measured by weight), compared to 20% glucose or fructose solutions. There were no differences between single sugars and a mix of glucose and fructose. Twenty one birds also rejected solutions containing the other sugars if they were mixed with xylose (a sugar isolated from some nectars), with the amount of solution consumed decreasing as xylose concentration increased. Birds appeared to be able to absorb sucrose, glucose and fructose very efficiently (absorbing 97.6–99.9%) but were less able to absorb xylose (averaging 61% absorption). Solutions were provided in 25 ml pipettes with a 5 mm hole in the top.

A randomised and replicated series of *ex situ* choice experiments in the Western Cape, South Africa, in 1995 (4) found that 13 southern (lesser) double-collared sunbirds *Nectarinia chalybea* showed a preference for 20% by weight sucrose solution, compared to 20% fructose or 'hexose' (equal parts fructose and glucose) solutions, both of which were preferred to 20% glucose. In contrast, 13 female Cape sugarbirds *Promerops cafer* did not show a preference for different sugar types. Individuals of both species appeared to avoid solutions with xylose in them, reducing their consumption as the proportion of xylose in the solution increased. Solutions were provided in 25 ml pipettes with a glass bulb on the end with a drinking hole in it, surrounded by red nail varnish to increase visibility.

A controlled before-and-after study on Mokoia Island, Lake Rotoruais, North Island, New Zealand (5), found that stitchbirds (hihi) *Notiomystis cincta* used feeders providing 20% by weight sugar solution frequently and when feeders were present they spent more time foraging for insects and less time foraging for fruit and nectar. However, there were no differences in weight gain or loss or survival rates when feeders were present, compared to when they were not. Feeders were present for 16 days at a time and then removed for 12 days. This alternation continued between January and November 1995. Annual survival in the population was low (38%) and the population appeared likely to decline despite feeding.

A randomised and replicated series of *ex situ* choice experiments in 2006–7 in KwaZulu-Natal, South Africa (6), found that eight malachite sunbirds *Nectarinia famosa* changed preferences for different sugars changed dependent on the concentration being used. At 5% concentration the birds preferentially fed on hexose (equal parts glucose and fructose); at 10%, 15% and 20% they showed no preference and at 25% concentration they preferentially fed on sucrose. Birds also showed a preference for sucrose solutions of higher concentrations, compared to lower-concentration sucrose, although the difference between 25% sucrose and 20% sucrose was not significant. The authors note that 5% sugar solution was not sufficient to maintain the birds' energy levels. Sugar solutions were provided in 20 ml burette tubes, moved periodically to avoid biases.

(1) Armstrong, D. (1992) Use of sugar-water feeders to supplement energy availability to honeyeaters for experimental tests. *Emu*, 92, 170–179.
(2) Downs, C. T. & Perrin, M. R. (1996) Sugar preferences of some southern African nectarivorous birds. *Ibis*, 138, 455–459.
(3) Franke, E., Jackson, S. & Nicolson, S. (1998) Nectar sugar preferences and absorption in a generalist African frugivore, the cape white-eye *Zosterops pallidus*. *Ibis*, 140, 501–506.
(4) Jackson, S., Nicolson, S. W. & Lotz, C. N. (1998) Sugar preferences and 'side bias' in cape sugarbirds and lesser double-collared sunbirds. *The Auk*, 115, 156–165.
(5) Armstrong, D. P. & Perrott, J. K. (2000) An experiment testing whether condition and survival are limited by food supply in a reintroduced hihi population. *Conservation Biology*, 14, 1171–1181.
(6) Brown, M., Downs, C. T. & Johnson, S. D. (2010) Concentration-dependent sugar preferences of the malachite sunbird (*Nectarinia famosa*). *The Auk*, 127, 151–155.

15.18. Can supplementary feeding increase predation or parasitism?

- A replicated, controlled study in the USA (1) found that providing seeds in predictable areas did not increase predation on seven species of songbird.
- A replicated and controlled trial in Spain (2) found higher levels of potentially dangerous gut microflora when fed on livestock carrion, compared to those fed on wild rabbits. A replicated study in Spain (3) found higher levels of predation on artificial nests close to carcasses provided for vultures.

Background

Supplementary feeding could potentially have deleterious effects on both the target population, for example through spreading disease or drawing birds to feeding areas where they can easily be predated, or non-target species, for example by bringing large numbers of predators into the area.

A replicated, controlled study at two pairs of mixed habitat sites in Kansas, USA, in the winters 2000–1 and 2001–2 (1) found that predation rates on seven species of songbird by Cooper's hawk *Accipiter cooperi* and sharp-shinned hawk *A. striatus* were no higher in an area supplied with supplementary food, compared with control (unfed) sites (two attacks in fed sites vs. no attacks in control sites). Supplementary feeding ran from early December until early March and consisted of four sunflower seed feeders and a 6 × 3 m area beneath them sprinkled with mixed seed. The species studied were dark-eyed junco *Junco hyemalis*, Harris's sparrow *Zonotrichia querula*, song sparrow *Melospiza melodia*, American tree sparrow *Spizella arborea*, northern cardinal *Cardinalis cardinalis*, black-capped chickadee *Parus atricapillus* (also known as *Poecile atricapillus*) and tufted titmouse *Parus bicolor*. The effects of feeding on songbird body condition are discussed in 'Provide supplementary food to increase adult survival'.

A replicated, controlled trial in 2004 in Castile and Leon and Madrid, Spain (2), found that red kites *Milvus milvus* that fed on carrion from stabled livestock had higher levels of potentially harmful gut microflora, compared to birds fed largely on wild rabbits (potentially dangerous *Salmonella* serotypes found in 25% of 80 faecal samples from colonies fed with livestock vs. 3% of 33 samples from colonies fed on wild rabbits). More generally, gut flora were more similar between the two colonies supplied with livestock, than either were with wild-fed birds. Livestock consisted mainly of domestic pigs *Sus scrofa*, but also cows, sheep, poultry and domestic rabbits.

A replicated study in dry scrubland on Fuerteventura, Canary Islands, Spain, in spring 1996 (3), found that artificial nests were more likely to be predated when they were within 200 m of a carcase (both naturally-occurring and supplied to a vulture 'restaurant'), compared to more distant nests. A total of 312 nests were laid in 12 lines. Nest predation occurred in 67% of lines, with a maximum of 92% of nests on a line being predated. The restaurant was supplied with approximately 200 kg/week of goat and pig carcasses whilst naturally occurring carcasses consisted of one goat and one yellow-legged gull *Larus michaellis*. Nests imitated either those of lesser short-toed larks *Calandrella refescens* or cream-coloured coursers *Cursorius* cursor and contained two Japanese quail *Cortunix japonica* eggs.

(1) Rogers, C. M. & Heath-Coss, R. (2003) Effect of experimentally altered food abundance on fat reserves of wintering birds. *Journal of Animal Ecology*, 72, 822–830.

(2) Blanco, G., Lemus, J. Ú. S. & Grande, J. (2006) Faecal bacteria associated with different diets of wintering red kites: influence of livestock carcass dumps in microflora alteration and pathogen acquisition. *Journal of Applied Ecology*, 43, 990–998.

(3) Cortés-Avizanda, A., Carrete, M., Serrano, D. & Donázar, J. A. (2009) Carcasses increase the probability of predation of ground-nesting birds: a caveat regarding the conservation value of vulture restaurants. *Animal Conservation*, 12, 85–88.

15.19. Provide supplementary food through the establishment of food populations

- One pre-1950 study in the USA (1) found that waterfowl fed on specially-planted rye grass.
- Three studies from North America (2,4) and Sweden (3) found that attempts to support populations by establishing prey did not succeed. Whooping cranes *Grus americana* in the USA (2) preferentially fed on scattered grains, over planted crops; attempts in Sweden to boost macroinvertebrate numbers (3) were not successful and great horned owls *Bubo virginianus* in Canada did not respond to induced increases in prey populations (4).

Background

Supplementary feeding can be an intensive and potentially expensive intervention. However, it may be possible to achieve the same results by planting crops or establishing prey populations.

A replicated study between over the winters of 1946–1947 in wetlands in Alabama, USA (1), found that waterfowl used rye grass *Lolium multiflorum* plantings as an alternate food source. Waterfowl used the rye grass plantings extensively, often in preference to other green browse, as it added a distinct microhabitat to mud flats exposed during winter draw-down of the reservoir. Rye grass was the only crop able to withstand periodic flooding and silting over the winter period and thus became a reliable food source for waterfowl. Experimental, small-scale plantings began in 1946. Large-scale plantings began in the fall of 1947. Higher mud flats are preferable as planting sites and the rate of seeding is 50 pounds per acre.

A small replicated study over the winters of 1964–1968 in 2 fenced experimental fields (39 ha each) within a coastal wildlife refuge in Texas, USA (2), found that whooping cranes *Grus Americana* preferred supplemented grains and seeds to planted crops during periods of low food availability. On average, 164 and 100 whooping crane use-days were observed for planted crops from October–December and January–April respectively, whereas the average use-days over the same periods for spread grain were 390 and 524 (3607 use-days over the study period in total). Whooping cranes significantly preferred hegari *Sorghum vulgare*, corn and wheat. Whooping cranes preferentially fed along the tidal flats in good weather. However, the crops (especially wheat, corn, legumes, peanuts and peas) were extensively used by sandhill cranes *Grus canadensis*, snow geese *Chen hyperborean* and Canada geese *Branta canadensis* (181 000 goose and 233 000 crane use-days in total from 1964–1967).

A study in a small (1.4 ha) artificial lake in central Sweden between 1978 and 1980 (3) added wheat straw and hay to the lake in an attempt to boost macroinvertebrate biomass and wildfowl numbers. However, the number of wildfowl did not appear to be affected by the addition and there was little change in invertebrate biomass. The authors suggest that high populations of invertebrate predators were responsible for the lack of change.

A replicated, controlled study from 1989–1992 in 3 experimental blocks and 5 control blocks (all 1 km^2) within a forest region in the Yukon, Canada (4), found that artificially increasing the density of prey did not alter the territorial or social structure of great horned owls *Bubo virginianus*. Experimental owls on food-enriched territories did not show a difference in home-range size and patchiness of spatial use compared with control owls. However, the distances of owl locations to treatment blocks were significantly closer to experimental centre-points than expected by chance (on average, 0.6 km closer). At a larger scale, no owls vacated their territories to use experimental plots and no owls switched to a nomadic strategy. The authors speculate that territorial behaviour prevents large aggregations of predators at an intermediate spatial scale. Experimental blocks were provided with commercial rabbit chow added weekly all year; snowshoe hare *Lepus americanus* densities were 2.8–10.3 times higher than in control blocks.

(1) Givens, L. S. & Atkeson, T. Z. (1952) Use of Italian rye grass as a means of attracting waterfowl. *The Journal of Wildlife Management*, 16, 107–108.
(2) Shields, R. H. & Benham, E. L. (1969) Farm crops as food supplements for whooping cranes. *The Journal of Wildlife Management*, 33, 811–817.
(3) Andersson, A. & Danell, K. (1982) Response of freshwater macroinvertebrates to addition of terrestrial plant litter. *Journal of Applied Ecology*, 19, 319–325.
(4) Rohner, C. & Krebs, C. J. (1998) Response of great horned owls to experimental 'hot spots' of snowshoe hare density. *The Auk*, 115, 694–705.

15.20. Use perches to increase foraging success

- Two studies from the USA (1,2) found that raptors and other birds used perches provided, whilst a replicated and controlled study in Sweden (3) found that raptors used clearcuts with perches significantly more than those without.

- However, a controlled study from the USA (1) found that overall bird abundances were not higher in areas provided with perches and a small controlled cross-over trial on an island in the USA (4) found that San Clemente loggerhead shrikes *Lanius ludovicianus mearnsi* did not alter their hunting patterns or increase their success rates following the installation of perches in their territories.

Background

If prey are plentiful but birds have low hunting success then it may be possible to increase population sizes by making hunting more effective, for example by providing perches for birds to use.

A controlled study in June–July 1979 in tallgrass prairie at Konza Prairie Research Natural Area, Kansas, USA (1), found that bird densities in unburned prairie sites were no higher in areas provided with artificial perches than in areas without perches (31 males of all species/ha in both areas). Numbers were higher in a burned area with perches (56 males/ha vs. 27) but not when dickcissel *Spiza americana* and red-winged blackbird *Agelaius phoeniceus* (attracted to a stream in the former area) were excluded. Twenty three perches (1.5 and 2 m long wooden stakes) were added to a 35-ha area of annually burned prairie and 17 to an adjacent 25-ha unburned area. A 12-ha area of burnt and a 39 ha unburnt prairie with no artificial perches served as controls. Eight species used 48% of perches in the burned area, compared with 29% used by four species in the unburned area.

A replicated trial in shrubland on Rhode Island, USA, in winter 1978–9 (2), found that ten raptor species appeared to make frequent use of 14 dead trees erected in 1977, whilst four species used nine man-made perches. In total, raptors were seen using the perches 525 times over 120 days, with most using the perches for resting and American kestrels *Falco sparverius* also using them for hunting and eating.

A replicated, controlled experiment in central Sweden (3) found that raptors used clearcuts with perches significantly more than those without (49 raptor observations in clearcuts with perches vs. 16 in those without). In the 1986 post-breeding season, 11 clearcuts (3.7–19.9 ha) were provided with 6 m high, regularly-spaced, perches (2/ha), and 11 had no artificial perches. Natural perches were virtually absent. Raptor use of the clearcuts was recorded April–May in 1987–1988. In autumn 1987, perches were switched between areas. In total, 33 raptor observations were made in 1987 and 32 in 1988; 85% (55) were common buzzard *Buteo buteo*, 14% (9) common kestrel *Falco tinnunculus* plus one hen harrier *Circus cyaneus*.

A small controlled cross-over trial in shrubland and grassland on San Clemente Island, California, USA (4), found that four pairs of San Clemente loggerhead shrikes *Lanius ludovicianus mearnsi* did not alter their hunting behaviour or success rate following the installation of 15 perches in their territories (approximately 50–75% success with perches vs. 60–65% without). However, some pairs did shift their hunting areas to include perches, suggesting that perches have the potential to increase the area of the island suitable for shrikes.

(1) Knodel-Montz, J. J. (1981) Use of artificial perches on burned and unburned tallgrass prairie. *The Wilson Bulletin*, 93, 547–548.
(2) Reinert, S. E. (1984) Use of introduced perches by raptors: experimental results and management implications. *Raptor Research*, 18, 25–29.
(3) Widén, P. (1994) Habitat quality for raptors: a field experiment. *Journal of Avian Biology*, 25, 219–223.
(4) Lynn, S., Martin, J. A. & Garcelon, D. K. (2006) Can supplemental foraging perches enhance habitat for endangered San Clemente loggerhead shrikes? *Wilson Journal of Ornithology*, 118, 333–340.

15.21. Place feeders close to windows to reduce collisions

- A randomised, replicated and controlled experiment in the USA (1) found that placing bird feeders close to windows reduced the number of collisions with the windows and the number of fatal collisions.

Background
In urban environments, many people feed garden birds in front of their windows, potentially increasing the risk of birds flying into windows. Determining how distance from windows affects collision rates is therefore very important. Other interventions designed to reduce collisions with windows are discussed in 'Threat: Residential and commercial development'.

A randomised, replicated and controlled experiment between October and December 1991 in Pennsylvania, USA (1), found that there were fewest collisions and fatal collisions with windows when platform feeders were placed 1 m away from the window (24% of the 105 collisions, none fatal), compared with when feeders were 5 m (28% of collisions, 33% of fatalities) or 10 m from the window (48% and 67%). Similarly, in a repeat experiment in February 1992, there were fewest collisions and fatalities when feeders were 2 m from the window (23% of 197 collisions, 5% of 21 fatalities) than 3 m (46% and 43%) or 4 m (31% and 52%) away. The proportion of collisions that were fatal increased with distance that feeders were from windows from 0% at 1 m away and 2% at 2 m to 59% at 5 m and 69% at 10 m. Six plate glass, wooden framed windows (1.4 × 1.2 m, 1.2 m off the ground, 55 m apart) and six platform feeders (with the platform level with the base of each window) were used on the edge of deciduous woodland and farmland.

(1) Klem Jr, D., Keck, D. C., Marty, K. L., Ball, A. J., Niciu, E. E. & Platt, C. T. (2004) Effects of window angling, feeder placement, and scavengers on avian mortality at plate glass. *The Wilson Bulletin*, 116, 69–73.

15.22. Provide supplementary water to increase survival or reproductive success

- A controlled study in Morocco (1) found that water supplemented northern bald ibis *Geronticus eremite* pairs had significantly higher reproductive success than those far from water sources.

Background
In arid environments water supply may be as much of limiting factor as food supply and providing drinking water may increase survival, particularly of chicks.

A controlled, multi-year study between 1998 and 2002 in coastal semi-desert steppe in southern Morocco (1) found that northern bald ibis *Geronticus eremite* pairs nesting less than 1 km from supplementary water points had significantly higher reproductive success than pairs nesting more than 5 km from water points (all nests: 1.0–2.2 fledglings/nest for nests close to water vs. 0.37–1.6 fledglings/nest for distant nests; only successful nests: 1.3–2.4 fledglings/nest vs. 1.0–2.3 fledglings/nest). There were no differences in clutch size between treatments, so the authors suggest that differences in productivity were due to failure rates, with 11% of 105 nests close to water failing, compared to 38% of 74 nests far from water. The increase was greatest in years of low natural rainfall but apparent in all years.

(1) Smith, K. W., Aghnaj, A., El Bekkay, M., Oubrou, W., Ribi, M., Armesto, M. J. & Bowden, C. G. R. (2008) The provision of supplementary fresh water improves the breeding success of the globally threatened northern bald ibis *Geronticus eremita*. *Ibis*, 150, 728–734.

15.23. Provide calcium supplements to increase survival or reproductive success

- Eight studies from across the world, including a literature review (1,3,5,6,10–13) from across the world found evidence for positive effects of calcium supplementation on several bird species.

- Positive effects included lower incidence of bone disease (1), higher fledging success (3,5,10,12,13), larger broods (5,10,12), higher quality eggs or chicks (6,10,12,13) and better physical condition of female parents (11). Not all species reacted similarly.
- Six studies including a literature review (2,3,7,9,11–13) did not find any evidence for increased reproductive success in at least one of the species supplied with supplementary calcium.
- One replicated study from Europe (8) found that birds took calcium supplied, and birds at polluted sites took more than those at cleaner sites. The effects on fitness were not monitored.

Background

Calcium is vital for birds to produce eggshell and to produce strong bones, and parents of dependent young therefore feed chicks on calcium-rich foods and precocial chicks (i.e. those that can walk and forage for themselves) seek out calcium-rich foods.

In acidified environments (e.g. those subject to acid rain) the supply of calcium may be limited, leading to low reproductive success. Therefore providing birds with supplementary calcium in a useable form, frequently fragments of bone or shell may increase reproductive output and the subsequent survival of chicks.

A before-and-after study on a northern South African pig farm (1) found that the incidence of osteodystrophy (a bone-deforming disease) in Cape vulture *Gyps coprotheres* chicks declined from an average of 17% in 1974–6 to 2.5% in 1983, following the establishment, in 1977, of a feeding station where carcass skeletons were crushed to provide small bone fragments. A total of 1378 chicks were examined over the study. The authors note that vulture colonies on game reserves not ranches had far lower levels of osteodystrophy (0–1%), probably due to the presence of bone-crushing mammals such as spotted hyenas *Crocuta crocuta*.

A randomised, replicated and controlled trial in woodland patches in Wyoming, USA (2), found that house wrens *Troglodytes aedon* provided with supplementary calcium did not raise more chicks than control (unsupplemented) birds (5.6 chicks/successful nest for 48 supplemented nests vs. 5.4 for 44 controls). In addition, they did not produce larger eggs or chicks (9.0 cm^3/egg and 9.9 g/chick for 53 eggs and 49 chicks from supplemented nests vs. 8.8 cm^3 and 9.9 g for 53 and 41 controls) or have larger clutches (6.7 eggs/clutch for 54 supplemented clutches vs. 6.4 eggs/clutch for 56 controls). Calcium was provided in the form of an equal mix of crushed oyster shells and chicken eggshells in small pots attached to the outside of nest boxes. Control nest boxes also had pots attached, but they were not filled.

A small control trial conducted during the 1996 breeding season on an acid bog in the north-eastern Netherlands (3) found that six of eight black tern *Chlidonias niger* chicks force-fed with supplementary calcium fledged successfully, whereas all 11 unsupplemented chicks died before fledging. The two unsuccessful supplemented chicks died early (within five days), whereas the unsupplemented chicks all showed skeletal deformities and significantly lower rates of weight gain. Commercially-available calcium pills, were given to chicks three times a week, providing a total of approximately 700 mg of calcium over the study period.

A cross-over study in a mixed forest on calcium-poor soils in the Netherlands from 1990–2 (4) found that female great tits *Parus major* supplied with supplementary calcium were more likely to lay eggs (0–3% of supplemented nests empty vs. 10–15% of controls, 622 nests studied) and less likely to desert them (15–25% of supplemented clutches deserted vs. 40–70% of controls, 339 clutches studied). Eggs from supplemented females were less likely to have defective eggshells (15–25% vs. 40–70%, 360 clutches) and a higher proportion of eggs from successful, supplemented nests hatched (80–95% of eggs vs. 55–80% for controls, 204 clutches). There were no significant differences in laying date or clutch size between treatments. This resulted in 5–9 hatchlings/nest for supplemented nests and 2–4 hatchlings/nest for controls. Calcium was provided in the form of 500 mm^2 snail shells or chicken eggshells provided three times a week from early March, in feeding cups on the outside of nestboxes. Birds took supplements mostly during the egg-laying and chick-provisioning stages of reproduction.

A replicated, controlled study from April–June in 1994 in 12 experimental and 15 control plots in mixed oak woodland in Loch Lomond, Scotland (5), found that calcium supplementation did not affect the reproductive success of blue tits (*Parus caeruleus*). Supplementary calcium did not significantly increase egg weight or size (1.12 and 1.14 g/egg; 1,075 and 1,082 mm^3/egg for experimental and control nests respectively). Similarly, there were no differences in shell weight or thickness (both treatments: 0.07 g and 0.038 mm/egg). Clutch size was also similar between pairs (9.3 and 9.5 eggs/nest for experimental and control nests respectively). Hatching success, expressed as the proportion of all eggs in a clutch that hatched, was not significantly different between experimental or control groups (both median = 1). A dry cuttlefish bone, 200 g of oyster grit and 200 g of crushed eggshell were provided in experimental plots.

A replicated, controlled trial in deciduous forests in Estonia in 1995–6 (6) found that pied flycatchers *Ficedula hypoleuca* and great tits *Parus major* used supplementary calcium supplied to their nest boxes, and that, in base-poor habitats, supplemented flycatchers laid larger eggs with thicker shells, compared to controls. There was no difference in tits, or in either species for hatching success or nestling condition. This study used a subset of data from studies described in detail below.

A replicated, controlled study in a 30 km^2 area of base-poor pine forests in Estonia in 1995–7 (7) found that great tits *Parus major* provided with supplementary calcium began laying earlier and had larger clutches than controls (average first laying date of 27th April–6th May for 42 supplemented nests and average clutch volume of 17.6 ml for 36 clutches vs. 29th April–11th May, 84 clutches and 16.6–18.4 ml, 65 clutches). In addition chicks from supplemented nests were larger than controls in 1997, the worst year for reproduction. There were no differences between groups with respect to clutch size (10.6–11.5 eggs/clutch, 40 supplemented clutches vs. 10.3–11.2 eggs/clutch for 81 controls), the size of individual eggs, hatching success or the number of fledglings produced (92–7% success and 7.8–8.9 fledglings/clutch for 29 supplemented clutches vs. 93–5% and 7.0–8.6 fledglings/clutch for 48 controls). Supplementary calcium consisted of snail shell and chicken eggshell supplied constantly in small feeders on the outside of nestboxes, from approximately 2 weeks before the start of nesting. This study uses a subset of the data from (10).

A replicated study at three alpine sites (one heavily polluted in the Czech Republic, two less so in the Slovak Republic and Norway) (8) found that both meadow pipits *Anthus pratensis* and water pipits *A. spinoletta* took calcium-rich supplements placed near their nests, but that both species took supplements more frequently in the heavily polluted site. Snail shells were taken most often by both species, followed by eggshells, bone fragments and plastic snail shells (low in calcium). Quartz (low in calcium) and limestone pieces were taken less frequently than other supplements. A total of 84 meadow pipit nests and 47 water pipit nests were monitored across all sites. The effects of supplementation on reproduction were not monitored.

A replicated, controlled study in 1997 in 15 experimental nests (containing 72 nestlings) and 14 control nests (containing 65 nestlings) in artificial nestboxes in Oklahoma, USA (9) found that purple martin *Progne subis* nestlings were unaffected by calcium supplementation. Calcium-supplemented nestlings did not grow to a larger size or at a faster rate than control nestlings for any of the growth parameters measured in any brood size (on average 53.9 compared 54.6 g/nestling; 4.2 compared to 4.4 g/day respectively). Additionally wing, leg and tail sizes did not differ between calcium-supplemented and control nestlings (on average 29.4 and 29 mm; 15.9 and 15.9; 22.8 and 22.9 mm respectively). Brood sizes were experimentally manipulated to contain either 4 or 5 nestlings. Nestlings in half of the nests of each brood size were fed a 1 ml dose of liquid calcium with a pipette over 3 weeks. Control nestlings received a water placebo.

A replicated, controlled trial in pine and deciduous forests in Estonia in 1995–7 and 1999–2000 (10) continued the study from (6) and found that great tits *Parus major* supplied with supplementary calcium had significantly larger first broods, containing significantly larger chicks and hatching more eggs, than control (unsupplemented) birds (supplemented nests: 10.1–11.6 eggs/clutch, n = 172 nests, 9.2–10.4 hatchlings/nest, n = 101 nests and lower leg lengths of 19.1–19.8 mm, n = 73 chicks vs. 10.2–11.4 eggs/clutch, n = 254 nests, 8.2–10.1 hatchlings/nest, n = 110 nests and 18.8–19.7 mm, n = 97 chicks, for control nests). Supplemented nests also fledged more chicks in 1997, 1999 and 2000 (the only years tested), but this difference was not significant, unless unsuccessful nests were excluded (7.2–9.0 fledglings/nest for 81 supplemented nests vs. 4.2–8.7 fledglings/nest for 115 controls). In 1999 (the only year investigated), second broods were significantly larger and fledged significantly more chicks in supplemented than control nests (8.4–9.1 eggs/clutch, and 5.5–6.6 fledglings/nest for 13 supplemented nests vs. 9.8–9.6 eggs/clutch and 7.9–8.1 fledglings/nest for 15 controls). The authors note that when clutch size was included in models, fledging rates did not differ between treatments, suggesting that calcium provisioning acts mainly to increase clutch size, rather than nest survival. There were no differences between pine and deciduous forests. Calcium was provided in the form of small snail shell and eggshell fragments in a feeder within the nest boxes. Control nest boxes were given empty feeders.

A replicated, controlled trial in a pine forest site in Estonia in 1995–7 and 1999 (11) found that pied flycatchers *Ficedula hypoleuca* provided with supplementary calcium laid larger eggs and their chicks were larger, compared to control (unsupplemented) birds (average size of 1.63 cm^3 for 172 supplemented clutches vs. 159 cm^3 for 178 controls; average leg length of 17.3 mm for 81 supplemented chicks vs. 17.2 mm for 89 controls). In addition, supplemented female flycatchers laying seven or more eggs were significantly heavier than controls. There were no differences in laying date, clutch size, number of fledglings or chick weights between treatments. Calcium was provided in the form of small snail shell and eggshell fragments in a feeder within the nest boxes. Control nest boxes had empty feeders. A subset of the data from this study is discussed in (6).

A 2004 literature review (12) looked at 14 studies of calcium supplementation across a total of seven species. Several studies are discussed in this section. Positive effects on at least one egg-related trait were detected in three species: house wrens *Troglodytes aedon* (clutch size but not egg size), pied flycatchers *Ficedula hypoleuca* (eggshell thickness and egg volume) and great tits *Parus major* (fewer females not nesting, fewer abandoning nests, thicker eggshells and higher hatching success; uncertain evidence for increases in clutch size and advanced laying date (4,7,10)). However, such effects were missing in all traits measured in a study of blue tits *P. caeruleus* in Scotland (egg mass and volume, eggshell thickness, laying date, clutch size, fledging success). Similarly, positive effects on chick traits were detected in four species: cape vultures *Gyps coprotheres* (reduced incidence of bone deformation (1)), black terns *Childonias niger* (higher weight gain (3)), pied flycatchers (higher chick growth rates (11)), great tits (chick growth and fledging success (10)). Such impacts were absent in house wrens (fledgling body mass or number of fledglings, (2)) or purple martins *Progne subis* (growth rate of fledglings, (9)). One study (11) also reported a positive effect on adult female body condition in pied flycatchers but not great tits in Estonia.

A replicated, controlled trial in mixed forest site in Estonia in 2000–1 (13) found that hatching success of great tit *Parus major* chicks did not differ between calcium-supplemented and control nests, but that a higher proportion of supplemented chicks fledged in 2000 (93% of hatched eggs from 14 supplemented nests vs. 78% from 19 controls), but not 2001 (78% of hatched eggs from 25 supplemented nests vs. 70% from 17 controls). At 15 days old, supplemented chicks were no larger or heavier than controls, but showed significantly lower levels of bone calcification activity. This suggests that supplemented nestlings already had fully developed bones by the time measurements were taken. The consequences for fitness and future reproduction may include decreased predation or parasitism in chicks and a lower reproductive costs for parents (because of shorter nestling periods), but these are not investigated in this study. This difference was only apparent in 2000, possibly because 2001 was a very poor year for reproduction, lowering growth rates and fledging success for all broods. Supplementation consisted of chicken eggshell provided in excess in small feeders on the outside of nest boxes.

(1) Richardson, P. R. K., Mundy, P. J. & Plug, I. (1986) Bone crushing carnivores and their significance to osteodystrophy in griffon vulture chicks. *Journal of Zoology*, 210, 23–43.
(2) Johnson, L. S. & Barclay, R. M. (1996) Effects of supplemental calcium on the reproductive output of a small passerine bird, the house wren (*Troglodytes aedon*). *Canadian Journal of Zoology*, 74, 278–282.
(3) Beintema, A. J., Baarspul, T. & De Krijger, J. P. (1997) Calcium deficiency in black terns *Chlidonias niger* nesting on acid bogs. *Ibis*, 139, 396–397.
(4) Graveland, J. & Drent, R. H. (1997) Calcium availability limits breeding success of passerines on poor soils. *Journal of Animal Ecology*, 66, 279–288.
(5) Ramsay, S. L. & Houston, D. C. (1999) Do acid rain and calcium supply limit eggshell formation for blue tits (*Parus caeruleus*) in the UK? *Journal of Zoology*, 247, 121–125.
(6) Tilgar, V., Mänd, R. & Leivits, A. (1999) Effect of calcium availability and habitat quality on reproduction in pied flycatcher *Ficedula hypoleuca* and great tit *Parus major*. *Journal of Avian Biology*, 30, 383–391.
(7) Mänd, R., Tilgar, V. & Leivits, A. (2000) Reproductive response of great tits, *Parus major*, in a naturally base-poor forest habitat to calcium supplementation. *Canadian Journal of Zoology*, 78, 689–695.
(8) Bureš, S. & Weidinger, K. (2001) Do pipits use experimentally supplemented rich sources of calcium more often in an acidified area? *Journal of Avian Biology*, 32, 194–198.
(9) Poulin, R. G. & Brigham, R. M. (2001) Effects of supplemental calcium on the growth rate of an insectivorous bird, the purple martin (*Progne subis*). *Ecoscience*, 8, 151–156.
(10) Tilgar, V., Mänd, R. & Mägi, M. (2002) Calcium shortage as a constraint on reproduction in great tits *Parus major*: a field experiment. *Journal of Avian Biology*, 33, 407–413.
(11) Mänd, R. & Tilgar, V. (2003) Does supplementary calcium reduce the cost of reproduction in the pied flycatcher *Ficedula hypoleuca*? *Ibis*, 145, 67–77.
(12) Reynolds, S. J., Mänd, R. & Tilgar, V. (2004) Calcium supplementation of breeding birds: directions for future research. *Ibis*, 146, 601–614.
(13) Tilgar, V., Mänd, R., Ots, I., Mägi, M., Kilgas, P. & Reynolds, S. J. (2004) Calcium availability affects bone growth in nestlings of free-living great tits (*Parus major*), as detected by plasma alkaline phosphatase. *Journal of Zoology*, 263, 269–274.

Translocations

Translocations, also known as relocations, reintroductions, restocking, or repatriations, involve the intentional release of captive-bred and/or wild-caught birds into the wild in order to re-establish a population that has been lost, or augment a critically small population. This can reduce the risk of inbreeding, or simply increase the range of a species and therefore the maximum possible population. Translocations are also frequently used to move birds to areas that have been cleared of invasive predators, particularly on islands. This intervention is not the same as that of periodically moving next boxes to prevent predators from learning their locations, discussed in 'Invasive and other problematic species'.

Translocations are typically expensive and may risk spreading pathogens to previously unexposed areas.

15.24. Translocate birds to re-establish populations or increase genetic variation

- A review of 239 bird translocation programmes (1) found 63–67% resulted in establishment of self-sustaining populations.

A review of two years (1987 and 1993) of survey data from 239 bird translocations in North America and Australasia (1) showed 63% of 134 translocations in 1987 and 67% of 105 in 1993 resulted in the establishment of self-sustaining populations. Those translocations involving larger numbers of individuals, translocations of native game species (rather than threatened species), and translocations into the core (rather than the periphery) of the species' historical range were most likely to establish self-sustaining populations.

(1) Wolf, C. M., Griffith, B., Reed, C. & Temple, S. A. (1996) Avian and mammalian translocations: update and reanalysis of 1987 survey data. *Conservation Biology*, 10, 1142–1154.

15.24.1. Megapodes

- A replicated study from Indonesia (1) found that up to 78% of maleo *Macrocephalon maleo* eggs hatched after translocation, with higher success if eggs were reburied as they were found. There was only anecdotal evidence that the translocations increased local populations.

Background

Megapodes are a family of birds found in East Asia, Australasia and Oceania that build large nest mounds filled with vegetation which rots to heat and incubate the eggs. The chicks are extremely independent and can run immediately after hatching, with some species even able to fly on the day they hatch. As eggs are not incubated by parents, they can be translocated and re-buried elsewhere.

A replicated study on Sulawesi, Indonesia, in 1972–3 and 1978–9 (1) found that hatching rates for maleo *Macrocephalon maleo* eggs translocated from unprotected beaches to a protected nature reserve ranged from 41% (321 of 789 eggs translocated in 1972–3) to 78% (105 of 134 eggs translocated in 1978–9). Guards at the park reported larger numbers of maleo nesting after the second translocation, although this was not confirmed and it was not certain that any extra birds were the translocated individuals. In the second translocation, eggs were buried in the sand in the same position that they were found in (the top of the egg was marked with a cross before removal), whereas in the first experiment they were placed haphazardly in the sand. In addition, in the second translocation, once one chick from a clutch emerged, the others were dug out. It was found that those placed in the wrong orientation died, those buried too deep hatched but the chicks then died tunnelling to the surface. The highest success rates were with eggs buried approximately 25 cm deep (shallower than the 40–70 cm commonly seen in the wild).

(1) MacKinnon, J. (1981) Methods for the conservation of maleo birds, *Macrocephalon maleo* on the island of Sulawesi, Indonesia. *Biological Conservation*, 20, 183–193.

15.24.2. Petrels and shearwaters

- Two studies from Australia (1,3) and one from New Zealand (2) found that colonies of burrow-nesting Procellariiformes were successfully established on two islands, and in uninhabited areas of another following the translocation and hand-rearing of chicks.

Background

Many species of Procellariiformes (petrels, shearwaters and albatrosses) breed on offshore or oceanic islands and have suffered severe population declines and local extinctions due to introduced mammalian predators. Although many islands have now been cleared of predators, Procellariiformes show very strong homing instincts and return to the area they fledged from to breed. This means that birds are unlikely to recolonise islands, or even other areas of the same island, even if it is safe. Translocations may therefore be a suitable way to increase the range and population size of these species. Using decoys and vocalisations are also possible methods to encourage recolonisation, these are discussed later in this chapter ('Use vocalisations to attract birds to safe areas' and 'Use decoys to attract birds to safe areas').

A replicated, controlled study on Cabbage Tree Island, New South Wales, Australia, in 1995 (1), found that the fledging rate of 30 Gould's petrel *Pterodroma leucoptera* chicks translocated from their burrows to artificial nests nearby and hand-fed was not significantly different from control (unmoved, parent-fed) birds (100% of translocated chicks fledging vs. 29/30 controls). Translocated chicks were also significantly heavier than controls. Gould's petrels only nest in two gullies on the island, 150 m apart, but show such strong philopatry that there is very little interchange between the gullies. As well as increasing inter-breeding, this study acted as a test case, before a possible translocation to another island (see (3)). Chicks were moved before they emerged from burrows (and so could imprint on their surroundings), but as late as possible to minimise the amount of artificial feeding required. This study is also discussed in 'Provide artificial nesting sites', 'Provide supplementary food to increase reproductive success' and 'Artificially incubate and hand-rear birds in captivity'.

A replicated study on Mana Island, North Island, New Zealand (2), found that 49% of 239 common diving petrels *Pelecanoides urinatrix* fledged successfully after being translocated to the island in 1997–9 from two other islands and hand-reared (see 'Artificially incubate and hand-rear birds in captivity') in artificial nests (see 'Provide artificial nesting sites'). The first breeding of translocated birds was recorded in 2000 and by 2003 there were 19 pairs in the colony (which had been empty until 1997). Vocalisations of petrels were also played on the island, see 'Use vocalisations to attract birds to safe areas'.

A before-and-after study in New South Wales, Australia (3), found that a new breeding colony of Gould's petrels *Pterodroma leucoptera leucoptera* was successfully established on Boondelbah Island, following the translocation of 100 nestlings in 1999 (95 fledging) and 100 in 2000 (100 fledging) from Cabbage Tree Island. In 2003–4, 41 petrels (both translocated and unknown birds) were recorded on the island, with 21 breeding pairs producing a total of 24 eggs and 14 fledglings in five years since the translocations. No negative impact was recorded on the productivity of the Cabbage Tree Island population. Nestlings were translocated from Cabbage Tree Island (1.4 km away) when they had reached their maximum weight but before they emerged from burrows (11–28 days old in 1999, 11–22 days old in 2000). They were moved to artificial nests (details in 'Provide artificial nesting sites') and fed fish and squid until they stopped feeding.

(1) Priddel, D. & Carlile, N. (2001) A trial translocation of Gould's petrel (*Pterodroma leucoptera leucoptera*). *Emu*, 101, 79–88.
(2) Miskelly, C. M. & Taylor, G. A. (2004) Establishment of a colony of common diving petrels (*Pelecanoides urinatrix*) by chick transfers and acoustic attraction. *Emu*, 104, 205–211.
(3) Priddel, D., Carlile, N. & Wheeler, R. (2006) Establishment of a new breeding colony of Gould's petrel (*Pterodroma leucoptera leucoptera*) through the creation of artificial nesting habitat and the translocation of nestlings. *Biological Conservation*, 128, 553–563.

15.24.3. Pelicans

- Two reviews of a brown pelican *Pelecanus occidentalis* translocation programme in the USA (1,2) found high survival of translocated nestlings (1) and that the target population grew enormously, to over 16,000 nests (2). The authors note that some of the growth may have been due to immigration from the source populations.

A review of a 1968–76 brown pelican *Pelecanus occidentalis* translocation programme between six colonies in Florida and three coastal sites in Louisiana, USA (1), found that 98% of 778 nestlings (eight to 11 weeks old) moved survived the journey and were successfully released, although all birds released at one site in 1968–9 died, meaning that all subsequent releases were at a single site. The first breeding in Louisiana was recorded in 1971, when the oldest released birds were three years old, and between 1971 and 1976 a total of 221 young fledged successfully. In 1975, an estimated 35–40% of the standing population of 400–450 pelicans died, probably as a consequence of contamination by endrin (an organochloride pesticide). Birds were provided with food twice daily after release.

A 2003 review (2) of the same translocation programme as in (1), found that between 1968 and 1980, a total of 1,276 pelican nestlings were translocated and that the population increased exponentially from 1971 until 1999, with a peak of 16,405 nests across seven colonies in 2001 (a peak of eleven colonies was reached in 2000). Nests produced an average of 1.7 nestlings between 1971 and 2001 (with a peak of 2.1 nestlings/nest in 2001), which, combined with pelicans' long lifespans and a decline in the number of birds in Florida, leads the authors to suggest that the exponential growth of the Louisiana population may have been partly due to immigration from nearby states.

(1) Nesbitt, S. A., Williams Jr, L. E., McNease, L. & Joanen, T. (1978) Brown pelican restocking efforts in Louisiana. *The Wilson Bulletin*, 90, 443–445.
(2) Holm Jr, G. O., Hess Jr, T. J., Justic, D., McNease, L., Linscombe, R. G. & Nesbitt, S. A. (2003) Population recovery of the eastern brown pelican following its extirpation in Louisiana. *The Wilson Bulletin*, 115, 431–437.

15.24.4. Auks

- A replicated study in the USA and Canada (1) found that 20% of 774 translocated Atlantic puffins *Fratercula arctica* remained in or near the release site, with up to 7% breeding.

Background

Like Procellariiformes (petrels, shearwaters and albatrosses, see previous section), auks show strong homing instincts and so are unlikely to recolonise islands or areas where they have become extinct.

A replicated study reviewing a 1973–81 translocation programme for Atlantic puffins *Fratercula arctica* (1) found that less than 0.3% of 774 nestlings moved from Newfoundland, Canada, to Maine, USA, died during the move, with 95% of the remaining 772 birds successfully fledging. Twelve percent (87 birds) were re-sighted at the release site, with a further 8% (60 birds) seen elsewhere in the Gulf of Maine. Translocated birds were first recorded as breeding in 1981, with at least 36 translocated birds and six other birds being confirmed as breeding at the release site in 1985, with 17 more birds breeding on other islands nearby. Seventy eight percent of these breeding attempts (49 attempts) produced chicks surviving for at least 21 days. Nestlings between two and 40 days old were taken from their burrows and moved to the release site within 17 hours. They were then confined to artificial burrows and fed two meals of 50 g of fish and vitamins daily for seven days. They were then allowed to leave the burrows to begin flying. From 1977 onwards, puffin decoys were placed on the island to encourage birds to return, and herring gull *Larus argentatus* and great black-backed gull *L. marinus* were culled and nests destroyed in 1974–5.

(1) Kress, S. W. & Nettleship, D. N. (1988) Re-establishment of Atlantic puffins (*Fratercula arctica*) at a former breeding site in the Gulf of Maine. *Journal of Field Ornithology*, 59, 161–170.

15.24.5. Wildfowl

- Three studies of two duck translocation programmes in New Zealand (3) and Hawaii (4,5) found high post-release survival, breeding and the successful establishment of new populations.
- A replicated study in USA (1) found that none of 391 blue-winged teal *Querquedula discors* stayed in the release site and that there was high mortality after release.
- A replicated, controlled study in the USA (2) found that wing-clipping female wood ducks *Aix sponsa* during translocation prevented them from abandoning their ducklings.

A replicated study of the translocation of 377 flightless young and 14 adult blue-winged teal *Querquedula discors* from Minnesota, USA, to Missouri during 1956–1958 (1) found none had remained to nest at the two release sites by the spring of 1961. Rings were recovered from 2.3% of the released birds at the end of the year of release, suggesting high first year mortality rates. Surviving individuals appeared to migrate from the translocation site once capable of flight.

A replicated, controlled study from 1970–1973 that used three different translocation methods to translocate wild wood duck *Aix sponsa* broods into 32 marsh areas (previously uninhabited by wood ducks) in Maine, USA (2) found that wing-clipping females was successful in preventing duckling abandonment. Two of five females moved in their original boxes successfully cared for their young, the other three abandoned them, as did females from two natural nests that were moved. In the final 25 attempts, the females were wing-clipped and moved with their broods in a release box. Twenty-two of the 25 trials were successful. A release was successful if any duckling from a brood survived to flying age. About 87 female ducklings were transplanted to new areas, of which eight were known to return to nest in the release areas. The most effective technique for releasing the females and ducklings together was a box with a hinged bottom suspended 15–20 cm above the water.

A before-and-after study on Campbell Island, New Zealand, in 2004–5 (3) investigated the success of a joint translocation/reintroduction programme, which transferred 44 wild and 61 captive-bred Campbell Island teal *Anas nesiotis* to the island. Between 75% and 78% of birds survived and breeding occurred. This study is discussed in more detail in 'Release captive-bred individuals'.

A before-and-after study on Midway Atoll, Hawaii, USA (4), found that, following the reintroduction of 42 Laysan ducks (Laysan teal) *Anas laysanensis* in the Octobers of 2004 and 2005, 19 of the 20 birds translocated in 2004 survived their first year. Five of six 2004 females nested in their first year, producing 11 fledgling ducklings by December 2005. Flight feathers of introduced birds were clipped, supplementary feed supplied for the first three months, and individuals monitored with radio telemetry. Although extensive habitat restoration was completed prior to the introductions (including planting native species used as nesting substrates), introduced birds were also observed to use vegetation absent from their original habitat.

Another study (5) of the same 42 Laysan ducks (Laysan teal) *Anas laysanensis* described in (4) found that post-release survival during 2004–2006 was 86% and the population grew to 104 individuals by December 2006, with 17 of 18 founding females attempting to nest. Females translocated as juvenile birds were more likely than those translocated as adults to fledge ducklings successfully, and shorter transport times were observed to lead to reduced loss of condition.

(1) Vaught, R. W. (1964) Results of transplanting flightless young blue-winged teal. *The Journal of Wildlife Management*, 28, 208–212.
(2) Capen, D. E., Crenshaw, W. J. & Coulter, M. W. (1974) Establishing breeding populations of wood ducks by relocating wild broods. *The Journal of Wildlife Management*, 38, 253–256.
(3) McClelland, P. & Gummer, H. (2006) Reintroduction of the critically endangered Campbell Island teal *Anas nesiotis* to Campbell Island, New Zealand. *Conservation Evidence*, 3, 61–63.
(4) Reynolds, M. & Klavitter, J. (2006) Translocation of wild Laysan duck *Anas laysanensis* to establish a population at Midway Atoll National Wildlife Refuge, United States and US Pacific Possession. *Conservation Evidence*, 3, 6–8.
(5) Reynolds, M. H., Seavy, N. E., Vekasy, M. S., Klavitter, J. L. & Laniawe, L. P. (2008) Translocation and early post-release demography of endangered Laysan teal. *Animal Conservation*, 11, 160–168.

15.24.6. Gamebirds

- Three studies from the USA (1,4,5) found that translocation of gamebirds resulted in population establishment (1) or growth (5), or an increase in lekking sites (4).
- Four studies from the USA (3,5–7) found high survival of translocated birds, although one, from Alaska (7) found that translocated birds had high initial mortality, which then fell to levels close to those in resident birds.
- Two studies from the USA (2,4) found high mortality in translocated birds.
- Four studies from the USA (4–7) found breeding rates that were high, or similar to resident birds, amongst translocated birds.

A before-and-after study in Iowa, USA (1), found that a population of 16 eastern wild turkeys *Meleagris gallopavo silvestris* (five males, 11 females) translocated in February 1975 and 1976 from southern Iowa and Missouri, into an area with no resident turkey population, found that the introduced population grew 470% within three breeding seasons, despite slow population growth in the first year (13%). Turkeys had dispersed across an area of 83 km^2 by the 3rd year after release. Birds were captured with rocket or cannon nets, transported in individual wooden crates, held overnight in an unheated building, before being equipped with a radio transmitter and released. Dispersal and nesting success were calculated from observation and radiotelemetry data.

A replicated study of 42 adult and 35 immature ruffed grouse *Bonasa umbellus* translocated from Illinois, USA, to Creek State Forest, Missouri, in autumn 1986 (2) found that 25% of translocated birds survived until May 1987. A maximum of eight of 37 females survived until the breeding season. Mortality rates were highest in the post-release period, with 15 grouse dying within seven days of release. Eighty-six percent of mortalities were attributed to avian or mammalian predation. Birds injured in transit did not have a lower survival rate than those without visible injury at the time of release.

A small controlled study in managed grassland in Illinois, USA (3), found that 20 of 24 greater prairie-chicken *Tympanuchus cupido* eggs, transferred between two sites successfully hatched. This success rate of 83% was significantly higher than the 45% success for 112 eggs not exchanged between populations.

A before-and-after study in Sawtooth Valley, Idaho, USA (4), found that following the translocation of 196 greater sage-grouse *Centrocercus urophasianus* in March and April 1986 and 1987, four of 17 radio-tagged birds (24%) in 1986 and 11 of 27 in 1987 (41%) survived into the summer, with 79% of deaths occurring in the three weeks immediately after release. The number of observed lekking sites increased from one to six by 1987, with one translocated hen nesting in 1986, and seven in 1987. Three of these nests were fertile, producing 14 offspring.

A controlled study between March 1997 and September 1998 in Georgia, USA (5), compared the survival and reproduction rates of 74 translocated and 166 resident northern bobwhite *Colinus virginianus*. No differences were found between the survival, nest production, or nest survival rates of relocated and resident bobwhites using direct observation and radiotelemetry data. Fifty percent of relocated and resident bobwhites died within 123 and 129 days of capture, respectively, with avian predation the greatest cause of mortality (53.3%). Home range size and distances moved from release sites were also all similar for relocated and resident bobwhites. Subsequently, in March and April 2000–2002, 202 wild bobwhites were translocated to three different sites identified as having low population densities relative to surrounding areas. Significant population growth was observed at two of three translocation sites relative to non-translocated areas (108% versus 16.5% and 56.7% versus 12.4%, respectively). The third site showed a non-significant increase in population at the relocation site compared to non-relocation areas.

A replicated study in Strawberry Valley, Utah, USA (6) examined the survival of 141 female greater sage-grouse *Centrocercus urophasianus* introduced into a resident population of 150 during the breeding seasons of 2003–2005. Survival rate was 60% in 2003, with all surviving birds integrated into resident sage-grouse flocks. Across all years, 36% of newly translocated birds, and 73% of females in their second year after translocation attempted nesting. The source populations were tested for presence of infections (in particular *Salmonella pullorum*) prior to translocation. Individuals were captured shortly after sunset, packed in cardboard boxes (30.5 cm × 22.9 cm × 30.5 cm) for 10hrs transit, and equipped with radio-transmitters before release the following morning. The release site was close to an active lekking site, with sagebrush available for immediate cover.

A replicated study reviewing a translocation programme in 2003–6 in the western Aleutian Islands, Alaska, USA (7), found that 15% of 13 newly translocated female Evermann's rock ptarmigans *Lagopus muta evermanni* died within two weeks of release, but that confirmed overwinter mortality was similar for translocated and resident females (30% of ten translocated females known to have died vs. 33% of six resident females). All surviving females nested, laying on average eight days later than 16 resident females (16th June vs. 8th June) and producing significantly fewer eggs (average of 6.8 eggs/clutch vs. 8.3 eggs/clutch). Egg size and nest survival were similar between resident and translocated females, whilst brood survival was higher for translocated females (85% for eight translocated females vs. 25% for 13 residents). Fecundity was also higher for translocated females, but this difference was not significant (0.9 female fledglings/translocated female vs. 0.3 male fledglings/resident female). In total, 75 birds were caught on Attu Island, held for up to 48 hours and fed on melon whilst being moved and then released on Agattu Island immediately upon arrival. One male bird died during transit.

(1) Little, T. W. & Varland, K. L. (1981) Reproduction and dispersal of transplanted wild turkeys in Iowa. *The Journal of Wildlife Management*, 45, 419–427.
(2) Kurzejeski, E. W. & Root, B. G. (1988) Survival of reintroduced ruffed grouse in north Missouri. *The Journal of Wildlife Management*, 52, 248–252.
(3) Westemeier, R. L., Simpson, S. A. & Cooper, D. A. (1991) Successful exchange of prairie-chicken eggs between nests in two remnant populations. *The Wilson Bulletin*, 103, 717–720.
(4) Musil, D. D., Connelly, J. W. & Reese, K. P. (1993) Movements, survival, and reproduction of sage grouse translocated into central Idaho. *The Journal of Wildlife Management*, 57, 85–91.
(5) Terhune, T. M., Sisson, D. C. & Stribling, H. L. (2006) The efficacy of relocating wild northern bobwhites prior to breeding season. *The Journal of Wildlife Management*, 70, 914–921.
(6) Baxter, R. J., Flinders, J. T. & Mitchell, D. L. (2008) Survival, movements, and reproduction of translocated greater sage-grouse in Strawberry Valley, Utah. *Journal of Wildlife Management*, 72, 179–186.
(7) Kaler, R. S., Ebbert, S. E., Braun, C. E. & Sandercock, B. K. (2010) Demography of a reintroduced population of Evermann's rock ptarmigan in the Aleutian Islands. *Wilson Journal of Ornithology*, 122, 1–14.

15.24.7. Rails

- Three reviews of two translocation programmes in the Seychelles (2) and New Zealand (1,3) found high survival amongst translocated rails.
- All studies found that translocated birds bred successfully, although one found that translocated takahe *Porphyrio hochstetteri* had lower reproductive success than birds in the source population (2). The other New Zealand study (3) found no differences in breeding success between recently and formerly translocated takahe.

Background

Thirty-three species of rails and crakes (Rallidae) are globally threatened, of which 13 are flightless (Taylor 1996). These species are unlikely to rapidly colonise new areas or recolonise former ranges, especially as several live on islands. Therefore translocating birds to new suitable habitats is likely to be an important conservation tool.

Taylor, P. B. (1996) Family Rallidae (rails, gallinules and coots) pp. 108–209 in: (eds J. del Hoyo, A. Elliott & J. Sargatal) *Handbook of the birds of the world: Volume 3. Hoatzin to auks*. Lynx Edicions, Barcelona.

A review (1) of adult survival and reproductive success of takahe *Porphyrio hochstetteri* populations established on four offshore islands in New Zealand (by translocating birds from the species' remaining natural range in Fiordland, South Island) found that adult survival was at least as high as in Fiordland (annual survival on islands of 83–100% vs. 73–97% for Fiordland). However, island pairs produced significantly fewer juveniles each year (average of 0.56–0.65 juveniles/pair/year for 43 island breeding attempts vs. 0.85–0.86 juveniles/pair/year for 171 Fiordland breeding attempts), despite laying more eggs (average of 3.4–3.5 eggs/pair/year for 43 island breeding attempts vs. 1.9–2.0 eggs/pair/year for 122 Fiordland breeding attempts), probably due to the milder climate (and hence longer breeding season) and the removal of non-viable eggs on islands. Breeding success was lowest for island pairs in their first year of reproduction (four juveniles from 43 clutches), compared to second or third attempts (11 from 36 and six from 13 clutches, respectively). Island birds were moved during 1984–91 to avoid introduced mammalian predators in Fiordland, whilst the Fiordland population was also intensively managed: single eggs were often removed from two-egg broods and artificially incubated with chicks reared in captivity until around 1 year old, when they were released back into the wild (see 'Artificially incubate and hand-rear birds in captivity' for details).

A review of a translocation programme for white-throated (Aldabra) rails *Dryolimnas cuvieri* (formerly *D. aldabranus*) on Aldabra atoll, Seychelles, in 1999–2001 (2) found that all 18 birds successfully released survived from November 1999 until at least April 2000, with 17 being seen again between December 2000 and April 2001 (the remaining bird was recorded in an inaccessible part of the island). An estimated 15–16 chicks fledged in 1999–2000, with all known mortality (30%) occurring within three weeks of hatching. In 2000–1 at least 20 chicks were fledged, leading to a minimum population of 51 birds, an increase of 183% in 18 months. Birds were transported from Île Malabar during pair formation and transferred to Île Picard within three days. There, birds were held in 30 m^2 enclosures near good quality habitat and provided with shelter, fresh water and food for six or 14 days before release. Food was provided to birds that stayed in or around the cages after release.

An analysis of takahe *Porphyrio hochstetteri* breeding records from four offshore predator-free islands in New Zealand during 1991–2000 (3) found that all eleven translocated birds survived the journey and attempted to breed. There were no differences in age of first breeding attempt between translocated birds of either sex compared with birds born on the islands (2.3–2.6 years old for six male and five female translocated birds vs. 2.1–2.9 for ten male and 11 female resident birds). In addition there were no differences in hatching or fledging rates between pairs containing two, one or no translocated birds (approximately 40% hatching success and 70% fledging success for three pairs of translocated birds vs. 25% and 30% for eight resident pairs and 25% and 30% for five pairs with one resident and one translocated bird). Birds were transferred between Tiritiri Matangi Island, Kapiti Island, Mana Island and Maud Island, and were supplied with supplementary food and kept in pens at the release site for two or three days before release. 'Resident' birds were descended from birds that had been translocated from the New Zealand mainland in the past.

(1) Bunin, J. S., Jamieson, I. G. & Eason, D. (1997) Low reproductive success of the endangered takahe *Porphyrio mantelli* on offshore island refuges in New Zealand. *Ibis*, 139, 144–151.
(2) Wanless, R. M., Cunningham, J., Hockey, P. A., Wanless, J., White, R. W. & Wiseman, R. (2002) The success of a soft-release reintroduction of the flightless Aldabra rail (*Dryolimnas [cuvieri] aldabranus*) on Aldabra Atoll, Seychelles. *Biological Conservation*, 107, 203–210.
(3) Jamieson, I. G. & Wilson, G. C. (2003) Immediate and long-term effects of translocations on breeding success in takahe *Porphyrio hochstetteri*. *Bird Conservation International*, 13, 299–306.

15.24.8. Raptors

- Six studies of three translocation programmes in the UK (1,3,4,6,7) and the USA (5) found that all three successfully established populations of white-tailed eagles *Haliaeetus albicilla* (1,7), red kites *Milvus milvus* (3,4,6) and ospreys *Pandion halieatus* (5). However, the latest review of the programme to reintroduce red kites to England and Scotland (6) reported that one of six populations was very small, with only four pairs, despite 90 birds being released.
- A replicated study in Spain (2) found high survival and establishment of translocated Montagu's harrier *Circus pygargus* fledglings.

Background

Raptor populations across the world have been lost to persecution, pollution (e.g. DDT contamination) and other threats, and have therefore been subject to many reintroduction programmes, using both captive-bred (see 'Release of captive-bred individuals') and wild-bred individuals. These programmes can be controversial as large birds of prey can take livestock, but also have the potential to restore 'flagship' species and boost interest in conservation.

A replicated study on the Isle of Rùm, western Scotland (1), found that, of 14 white-tailed eagles *Haliaeetus albicilla* translocated from Norway in 1975–7, 13 were successfully released. The remaining bird (a male) died of kidney failure after five weeks in captivitiy. Of the released birds, two (15%) were found dead but the others appear to be survive well (at least until publication in 1979). The eagles were collected at five to eight weeks old and kept for two or three months at the release site. After release, food was provided from 'food dumps' until the birds were able to feed themselves. This translocation programme is discussed further below.

A replicated trial (2) found that, of 87 Montagu's harrier *Circus pygargus* fledglings released at a marshland site in south-east Spain between 1988 and 1992, 83% successfully established in the wild (from 66% of 29 birds released in 1992 to 100% of 13 birds released in 1988–9). A further six chicks died during their first flights at the release site. Birds were taken from recovery centres (mainly chicks rescued from agricultural fields) and agricultural fields in Spanish regions with large harrier populations. They were moved to an enclosure at the release site at 20–30 days old and fed there. Five to eight days later the enclosure was opened and the birds could leave. They were then fed until they reached independence (i.e. stopped returning to be fed), an average of 30–37 days after release, depending on the age at release.

A replicated study reviewing a translocation programme for red kites *Milvus milvus* into southern England (3) found that translocated and newly recruited birds fledged a total of 60 young from 35 breeding attempts between 1992 and 1994. A total of 73 birds were translocated between 1989 and 1994 and first breeding was attempted (unsuccessfully) by two pairs in 1991. Five of the breeding attempts were by pairs containing at least one one-year-old bird, of which three were successful. Translocated birds came from Spain (62 birds), Wales (seven birds, taken from the wild as eggs and hatched in captivity) and Sweden (four birds). A further 20 birds from Spain were released in 1994 but their breeding attempts are not analysed here. This translocation programme is also discussed below.

A study (4) reviewed the success, until 1995, of the same red kite *Milvus milvus* translocation programme as (3) as well as translocations to northern Scotland and found that survival and reproductive productivity were higher in England than Scotland. Between 1989–94, 93 juvenile kites (48 males, 45 females) were released in southern England and had an average first-year survival rate of 76%, increasing to 91–2% in second and third years and 100% for fourth and five years after release. There was a slight difference between male and female survival, leading to a gradual change in the sex ratio. During 1989–93, 93 juvenile kites (all from Sweden) were released in northern Scotland and had an average first-year survival rate of 52%, increasing to 67–88% in second and third years and 75–91% for fourth and five years after release. Early breeding attempts in England are described in (3) and by 1995 there were 24 breeding pairs fledging at least 115 young during 1991–5. In Scotland, breeding was first attempted (successfully) in 1992, with 15 breeding pairs in 1995. During 1992–5, 29 clutches were laid, fledging 47 chicks. Survival rates of wild-raised birds in both regions did not differ significantly from released birds during 1992–4. Main causes of mortality were poisoning and collisions in Scotland, with poisoning also being important in England.

A replicated study, reviewing an osprey *Pandion haliaetus* translocation programme in the Twin Cities urban area of Minnesota, USA (5), reported that a total of 143 juvenile ospreys were released by hacking (see 'Release captive-bred individuals') between 1986 and 1995. Breeding was first attempted in the area in 1986, with the first successful nesting in 1988. By the end of 2000, 131 nesting attempts were recorded, with 69% of them successful, producing 194 chicks in total (with an average of 1.6 fledglings/nest or 2.2 fledglings/successful nest). A small number of individuals and sites were responsible for a disproportionate number of chicks, with 85% of successful nest sites being in parks or backyards and 15% in industrial areas. This study is also described in 'Provide artificial nesting sites'.

A 2004 review (6) of the same red kite *Milvus milvus* translocation programmes to the UK as in (3,4) found that the release of 518 subadult birds (between 69 and 103 at each site) between 1989 and 2004 resulted in the establishment of one population of at least 177 breeding pairs (from 93 birds released), four populations of between 16 and 35 breeding pairs and one population of just four pairs (90 birds released, beginning in 2001). Productivity in 2003 was comparable to other parts of their range (1.8–2.0 young/breeding pair) for all populations except the smallest (0.25 young/breeding pair). High mortality rates in the less successful sites are thought to be due to illegal poisoning. Birds were taken from large populations across Europe when 4–6 weeks old, kept in aviaries for a further eight weeks (with minimal human contact) and then released. This paper also discusses the release of captive-bred corncrakes *Crex crex*, discussed in 'Release captive-bred individuals'.

A 2009 review of two white-tailed eagle *Haliaeetus albicilla* translocation programmes in western Scotland (7) found that the release of 82 individuals in 1975–85 and 59 in 1993–8 led to the establishment of a population of 42 territorial pairs in 2007, with the number of territorial adults increasing at 9.7%/year during 1997–2007. Survival rates of released birds were significantly lower than those of wild-bred birds, particularly during the first three years of life (74% survival for one year-old released birds vs. 82% for wild-bred; 94% survival for released birds aged four or more vs. 97% for wild-bred birds; overall probability of surviving until five years old of 37% for released birds vs. 53% for wild-bred). Breeding success of the established population is similar to that of the Norwegian population (in similar environmental conditions) but lower than populations elsewhere in Europe. Overall, breeding success and productivity have increased with time, as reintroduced birds get older (which significantly increases the probability of fledging young) and a higher proportion of the population consists of wild-bred birds (0.61 young fledged/territorial pair in 1993–2000 vs. 0.76 young fledged/territorial pair in 2003–7).

(1) Love, J. A. & Ball, M. E. (1979) White-tailed sea eagle *Haliaeetus albicilla* reintroduction to the Isle of Rhum, Scotland, 1975–1977. *Biological Conservation*, 16, 23–30.
(2) Pomarol, M. (1994) Releasing Montagu's harrier (*Circus pygargus*) by the method of hacking. *Journal of Raptor Research*, 28, 19–22.
(3) Evans, I. M., Cordero, P. J. & Parkin, D. T. (1998) Successful breeding at one year of age by red kites *Milvus milvus* in southern England. *Ibis*, 140, 53–57.
(4) Evans, I. M., Summers, R. W., O'Toole, L., Orr-Ewing, D. C., Evans, R., Snell, N. & Smith, J. (1999) Evaluating the success of translocating red kites *Milvus milvus* to the UK. *Bird Study*, 46, 129–144.
(5) Martell, M. S., Englund, J. V. & Tordoff, H. B. (2002) An urban osprey population established by translocation. *Journal of Raptor Research*, 36, 91–96.
(6) Carter, I. & Newbery, P. (2004) Reintroduction as a tool for population recovery of farmland birds. *Ibis*, 146, 221–229.
(7) Evans, R. J., Wilson, J. D., Amar, A., Douse, A., MacLennan, A., Ratcliffe, N. & Whitfield, D. P. (2009) Growth and demography of a re-introduced population of white-tailed eagles *Haliaeetus albicilla*. *Ibis*, 151, 244–254.

15.24.9. Herons, storks and ibises

- A before-and-after study in the USA (1) found that a colony of black-crowned night herons *Nycticorax nycticorax* was successfully moved, with the new colony producing chicks the year after translocation.

A before-and-after trial at a coastal site in Long Beach, California, USA (1), found that 423 pairs of black-crowned night herons *Nycticorax nycticorax* successfully fledged 1,128 chicks in 2000, following the translocation of the colony beginning in 1999. The former colony was threatened by port development, so 50 mature trees were relocated 2 km away near approximately 70 existing trees. In addition, vocalisations of the original colony were played, decoys placed in the trees and public access stopped. Before the translocation, the old colony held up to 500 pairs of herons.

(1) Crouch, S., Paquette, C. & Vilas, D. (2002) Relocation of a large black-crowned night heron colony in southern California. *Waterbirds*, 25, 474–478.

15.24.10. Owls

- A small study from New Zealand (1) found that translocating two male boobook *Ninox novaeseelandiae novaeseelandiae* allowed the establishment of a small population, when they interbred with the last remaining Norfolk Island boobook *N. n. undulata*.

- A replicated study in the USA (2) found high survival amongst burrowing owls *Athene cunicularia* translocated as juveniles, although no breeding was recorded and all birds left the release site and were not seen again.

A study on Norfolk Island, Australia (1), found that the last remaining Norfolk Island boobook *Ninox novaeseelandiae undulata* (a female) paired with one of two male boobooks *N. n. novaeseelandiae* introduced from New Zealand in 1987 (the other disappeared shortly after release). The pair attempted unsuccessfully to breed in 1988 and 1991–3 but succeeded in 1989 and 1990, and fledged a total of four young. Two of these subsequently bred successfully themselves in 1993 and 1994, increasing the population to eleven birds, all alive in 1995.

A replicated study reviewing a reintroduction programme in Minnesota, USA (2), found that only eight of 105 (8%) burrowing owl *Athene cunicularia* juveniles translocated from South Dakota to Minnesota in 1986–90 were confirmed mortalities, with all other birds seen well past fledging age. However, no birds were seen after leaving the vicinity of the release site and no successful breeding attempts were recorded between 1992 and 1998. Birds were released at prairie sites in the species' former range using 'hacking pens' (see 'Release captive-bred individuals into the wild to restore or augment wild populations') and were fed for 33 days after release. In addition, artificial burrows were used to reduce predation chances and adult owls used as 'parental models'.

(1) Olsen, P. D. (1996) Re-establishment of an endangered subspecies: the Norfolk Island boobook owl *Ninox novaeseelandiae undulata*. *Bird Conservation International*, 6, 63–80.
(2) Martell, M. S., Schladweiler, J. & Cuthbert, F. (2001) Status and attempted reintroduction of burrowing owls in Minnesota, USA. *Journal of Raptor Research*, 35, 331–336.

15.24.11. Woodpeckers

- All five translocation programmes studied were for red-cockaded woodpeckers *Picoides borealis* in the southern USA.
- Six studies of four programmes (1–6) found that >50% of translocated birds remained in their new sites, with two studies of the same programme reporting a large population increase (2,4).
- Birds from four programmes were reported as forming pairs or breeding, although some translocated pairs split up (5) and some translocated nestlings were abandoned (2).
- One study found that translocated nestlings fledged at similar rates to native chicks (7).

Background

Red-cockaded woodpeckers *Picoides borealis* are a vulnerable species from the southeast USA. They are subject to intense management practices including forest manipulations (see 'Natural system modifications') and competitor control ('Invasive and other problematic species'). Red-cockaded woodpeckers live in small family groups, normally with a single breeding pair. These groups occupy small 'clusters' of trees and can become increasingly isolated if populations decline.

Translocating birds between groups and into new areas therefore represents a potential method of maintaining a viable population structure. Pine stands can be specifically managed for woodpeckers to create unoccupied clusters, into which translocated birds can be released.

A small trial at two open pine woodland sites in Texas, USA (1), found that, of five translocated red-cockaded woodpeckers *Picoides borealis*, four remained in their release sites. One pair were moved less than 4 km within the same forest block in February 1991 and the male returned to his original group the next day. The female remained and after a second male (from another forest block) was released the pair nested successfully in both 1991 and 1992. The second pair were translocated from separate sites, released in February 1992 and appeared to remain in the area, using artificial nesting cavities (see 'Provide artificial nesting sites' for details). Birds were captured at their roost cavities, transported to the release site and placed in tree cavities approximately 20 m apart. Wire mesh was placed over the cavity entrances until the birds were released at dawn. Prior to translocation, resin wells at the release site were reopened and potential competitors for cavities (e.g. southern flying squirrels *Glaucomys volans* and red-bellied woodpeckers *Melanerpes carolinus*) were removed (see 'Reduce inter-specific competition for nest sites by removing or excluding competitor species' for details of similar removal programmes).

A small before-and-after study in a loblolly *Pinus taeda* and longleaf *P. palustris* pine forest in South Carolina, USA (2), found that 31% of 16 red-cockaded woodpeckers *Picoides borealis* translocated in 1987–91 died or emigrated from the release site. Adult females settled and bred more successfully than subadults (three of four adult/unknown age females bred vs. one of six subadults) and male translocations appeared less successful than female (one of four males translocated less than 20 km bred). Three nestlings translocated with their parents in 1988 died after being abandoned. By the end of 1991, the local population was six breeding pairs and 15 other birds, compared with one pair and two other birds in 1986. This translocation programme is discussed further below. Competitor species were removed from release sites (see 'Reduce inter-specific competition for nest sites by removing or excluding competitor species') and habitats modified (see 'Threat: Natural system modifications – Forest modifications') throughout the study period.

A series of before-and-after trials in four open pine forests in Texas, USA (3), found that 11 of 19 translocated red-cockaded woodpeckers *Picoides borealis* formed pairs at their release sites. Birds were one-year-old when they were transported to tree clusters containing a single bird of the opposite sex in 1989–92. Success rates did not differ significantly between males (three out of five birds, 60%, establishing) and females (eight out of 14 birds, 57%). This study also discusses installing artificial nesting cavities and managing forests for woodpeckers, see 'Provide artificial nesting sites' and 'Use prescribed burning' for details.

A later review (4) of the same translocation programme as in (2) found that 63% of 49 adult and subadult red-cockaded woodpeckers *Picoides borealis* translocated into a very small population in 1986–95 remained at the release site for at least 30 days and 51% (25 birds) had reproduced by July 1996. Over the same period, the peak woodpecker population increased from 10 to 99 individuals and from one to 19 breeding pairs. Similarly, the total number of fledglings produced each year increased from three in 1985 to 43 in 1996 (average of 2.2 fledglings/breeding pair/year). Birds were translocated to the release site from family groups within the same forest block and from more distant sites (see below). This study also discusses the impact of intensive management of habitats and competitor species in 'Threat: Natural system modifications', 'Provide artificial nesting sites' and 'Reduce inter-specific competition for nest sites by removing or excluding competitor species'.

A before-and-after trial in Texas, USA (5), found that of 17 red-cockaded woodpeckers *Picoides borealis* translocated to an open pine forest site, 12 (71%) established territories: three (18%) at their release sites and the others an average of 2.8 km away. In total, five pairs of subadults and seven individual birds (four male, three female) were released between December 1994 and March 1995. Only one of the pairs released (20%) remained together and five birds (three male, two female) went missing. In total, eight out of nine pairs of woodpeckers newly discovered at the sites in 1995–6 (89%) contained at least one bird translocated from elsewhere. Birds were released into unoccupied 'clusters' (see 'Threat: Natural systems modifications – Forest modifications' for details), with at least three such stands within 1 km of the release site to allow dispersal. Stands were also provided with cavity inserts (see 'Provide artificial nesting sites') and southern flying squirrels *Glaucomys volans* were removed from all sites to reduce competition (see 'Reduce inter-specific competition for nest sites by removing or excluding competitor species').

Another review (6) of the same red-cockaded woodpecker *Picoides borealis* translocation programme as (4) reported that 55% of the 189 nestlings produced between 1986 and 1995 had at least one translocated parent. Two more nestlings were fostered to other birds in the release site, see 'Foster eggs or chicks with wild conspecifics' for details. Above fledging age, age and sex had no impact on translocation success, but long-distance translocations were more likely to succeed, because birds moved shorter distances were more likely to return home (25% success for 12 birds moved less than 7 km; 71% for 21 birds moved 19–23 km; 81% for 16 birds moved 182–483 km). All three groups however were equally likely to breed (25%, 57% and 62% of birds breeding in each group respectively). The presence or not, of a resident male did not significantly alter release success. Birds were released directly into natural or artificial nesting cavities after translocation.

A study in 1997–8 in Louisiana, USA (7), found that red-cockaded woodpeckers *Picoides borealis* translocated through fostering had similar fledging rates to native nestlings. This study is discussed in 'Foster eggs or chicks with wild conspecifics'.

(1) Rudolph, D. C., Conner, R. N., Carrie, D. K. & Schaefer, R. R. (1992) Experimental reintroduction of red-cockaded woodpeckers. *The Auk*, 109, 914–916.
(2) Allen, D. H., Franzreb, K. E. & Escano, R. E. F. (1993) Efficacy of translocation strategies for red-cockaded woodpeckers. *Wildlife Society Bulletin*, 21, 155–159.
(3) Conner, R. N., Rudolph, D. C. & Bonner, L. H. (1995) Red-cockaded woodpecker population trends and management on Texas National Forests. *Journal of Field Ornithology*, 66, 140–151.
(4) Franzreb, K. E. (1997) Success of intensive management of a critically imperiled population of red-cockaded woodpeckers in South Carolina. *Journal of Field Ornithology*, 68, 458–470.
(5) Carrie, N. R., Conner, R. N., Rudolph, D. C. & Carrie, D. K. (1999) Reintroduction and postrelease movements of red-cockaded woodpecker groups in eastern Texas. *The Journal of Wildlife Management*, 63, 824–832.
(6) Franzreb, K. E. (1999) Factors that influence translocation success in the red-cockaded woodpecker. *The Wilson Bulletin*, 111, 38–45.
(7) Wallace, M. T. & Buchholz, R. (2001) Translocation of red-cockaded woodpeckers by reciprocal fostering of nestlings. *The Journal of Wildlife Management*, 65, 327–333.

15.24.12. Parrots

- Three studies of two translocation programmes from the Pacific (3,4) and New Zealand (7) found that populations of parrots were successfully established on islands following translocations, including the colonisation of other islands in the New Zealand study.
- Survival of translocated birds was monitored in five studies of four programmes from across the world (1,2,5–7) and ranged from 41% over 60 days for red-fronted parakeets *Cyanoramphus novaezelandiae* in New Zealand (7) to 98% for kakapos *Strigops habroptila* in New Zealand (1,5).
- Survival for translocated thick-billed parrots *Rhynchopsitta pachyrhyncha* in the USA (2) was higher than for captive-bred birds.
- Despite very high survival, kakapos that were translocated had very low reproductive output in New Zealand (1,5).

Background

Kakapos *Strigops habroptila* are large, flightless parrots that used to be found across New Zealand, but declined catastrophically after the introduction of mammalian predators. The entire population has now been transferred to predator-free islands off the mainland in an attempt to save the species.

A replicated 1994 study on four off-shore islands in New Zealand (1) found that survival of translocated kakapos *Strigops habroptila* was high (63–85% until 1992, see (5) for details). However, reproduction was extremely low, with only two young reared to independence and a third hand-reared in captivity. Between 1974 and 1992, 65 kakapo were translocated from Stewart Island (1,746 km^2, South Island) to Maud Island (300 ha, South Island; 1974–81: nine birds; 1989–91; six birds), Little Barrier Island (3,055 ha; 1982: 22 birds), Codfish Island (1,480 ha; 1987–92: 30 birds) and Mana Island (217 ha; 1992: two males). Translocations occurred because kakapos were suffering extremely high mortality rates due to predation by introduced mammalian predators (particularly cats *Felis catus* and stoats *Mustela erminea*) on Stewart Island. Such predators were removed from target islands prior to translocations.

A replicated study in southeastern Arizona, USA (2), reintroduced 88 thick-billed parrots *Rhynchopsitta pachyrhyncha* into the Chiricahua Mountains between September 1986 and September 1993. Survival two months after release was significantly higher for wild birds caught as adults, compared to parrots caught as juveniles or captive-bred birds, either parent- or hand-reared (6.3% survival for 16 captive-bred, parent-reared birds vs. 0% for four hand-reared birds, 0% for four wild birds caught at juveniles and 43% for 65 wild birds caught as adults). Translocated birds were known to fly more than 110 km away from the release site, with small groups returning each autumn to the Chiricahua Mountains, where at least one pair producing two fledglings. This study is also discussed in 'Release captive-bred individuals' and 'Use holding pens at release sites'.

A before-and-after study in 1992–4 on Hiva, Marquesas Islands, French Polynesia (3), found 14 ultramarine lorikeets *Vini ultramarina* on the island following the translocation of 13 birds from Ua Huka between August 1992 and November 1993. There were also anecdotal reports of juvenile lorikeets being present. This translocation programme is discussed in more detail below.

A second before-and-after study in January 1997 (4) assessed the same population of ultramarine lorikeets *Vini ultramarina* on Fatu Hiva, Marquesas Islands, French Polynesia as discussed in (3), after the translocation of 29 individuals from Ua Huka Island between 1992 and 1994. The introduced population on Fatu Hiva was estimated at 51 individuals. The initial 29 translocated individuals were captured with mist nests, ringed, weighed, and measured before being kept up to seven days in captivity, while further captures took place. The authors identify the absence of black rats *Rattus rattus* and the location of Fatu Hiva within the lorikeet's former range as key considerations in its selection as a release site.

A 1998 review (5) of the kakapo *Strigops habroptilus* translocation programme described in (1) stated that yearly survival rates of 61 kakapo was around 98% following their translocation from Stewart Island to four other offshore islands. In 1997, at least 48 translocated birds (78%) were known to be alive. Productivity of translocated kakapo has been low, with high rates of egg infertility (approximately 40%) and nestling mortality. The population in February 1998 was 57 individuals, approximately 10% fewer than in 1982, when translocations began in earnest.

A small study (6) following the translocation of 14 blue-and-gold macaws *Ara ararauna* from Guyana to their former range in Trinidad found that 11 macaws were observed within 51 km of the release site between three and eight months after release, with several travelling in pairs. One individual did not leave the release site and was recaptured. Translocated macaws were tested for papillomavirus, psittacosis, Newcastle's disease, and avian influenza before transport and provided with supplementary food immediately after release.

A before-and-after study of a 2008–9 translocation programme for red-fronted parakeet *Cyanoramphus novaezelandiae* in North Island, New Zealand (7), found that at least 41% of 32 parakeets translocated to Motuihe Island from Little Barrier Island in 2008 survived at least 60 days after release, with at least two family groups and four juveniles being identified eight months after release. In addition, birds have dispersed to other nearby islands and have bred on them. A further 18 birds were released in 2009, but were not monitored. One bird translocated in 2008 died before release due to trauma and disease, and one female caught in 2009 was very weak and was therefore returned to Little Barrier Island without being released. It is not clear how the other birds died (or if they were just not seen during surveys). Parakeets were caught using mist nets and held in aviaries on Little Barrier Island for up to six days before transport to Motuihe Island by helicopter. Birds were supplied with fruit, water and grains in excess and released immediately upon arrival at a forest fragment on Motuihe.

(1) Lloyd, B. D. & Powlesland, R. G. (1994) The decline of kakapo *Strigops habroptilus* and attempts at conservation by translocation. *Biological Conservation*, 69, 75–85.

(2) Snyder, N. F. R., Koenig, S. E., Koschmann, J., Snyder, H. A. & Johnson, T. B. (1994) Thick-billed parrot releases in Arizona. *The Condor*, 96, 845–862.
(3) Kuehler, C., Lieberman, A., Varney, A., Unitt, P., Sulpice, R. M., Azua, J. & Tehevini, B. (1997) Translocation of ultramarine lories *Vini ultramarina* in the Marquesas Islands: Ua Huka to Fatu Hiva. *Bird Conservation International*, 7, 69–80.
(4) Lieberman, A., Kuehler, C., Varney, A., Unitt, P., Sulpice, R. M., Azua, J. & Tehevini, B. (1997) A note on the 1997 survey of the translocated ultramarine lory *Vini ultramarina* population on Fatu Hiva, Marquesas Islands, French Polynesia. *Bird Conservation International*, 7, 291–292.
(5) Clout, M. N. & Merton, D. V. (1998) Saving the Kakapo: the conservation of the world's most peculiar parrot. *Bird Conservation International*, 8, 281–296.
(6) Oehler, D. A., Boodoo, D., Plair, B., Kuchinski, K., Campbell, M., Lutchmedial, G., Ramsubage, S., Maruska, E. J. & Malowski, S. (2001) Translocation of blue and gold macaw *Ara ararauna* into its historical range on Trinidad. *Bird Conservation International*, 11, 129–141.
(7) Ortiz-Catedral, L. & Brunton, D. H. (2010) Success of translocations of red-fronted parakeets *Cyanoramphus novaezelandiae novaezelandiae* from Little Barrier Island (Hauturu) to Motuihe Island, Auckland, New Zealand. *Conservation Evidence*, 7, 21–26.

15.24.13. Songbirds

- Nine studies from across the world (1,2,4–8,10,11), including a review of 31 translocation attempts in New Zealand (5) found that translocations led to the establishment of songbird populations. The review found that 79% and 100% of translocation programmes for saddlebacks *Philesturnus carunculatus* and New Zealand robins *Petroica australis*, respectively, were successful in establishing populations. Eight of the studies were from islands (2,4–8,10,11), mostly following predator removal.
- Three studies from Zimbabwe (1), New Zealand (9) and the USA (12) report on three translocation programmes that failed to establish populations.
- A methodological paper found that the nesting success of saddlebacks decreased as the latitudinal difference between source area and release site increased (3).

Background

Several species of songbird on islands across the world (e.g. New Zealand, Hawaii and the Seychelles) have been lost from parts of their historic range due to predation by introduced mammalian predators. These predators have now been cleared from some islands, allowing the reintroduction of birds from elsewhere.

A before-and-after study in Zimbabwe from 1975 to 1977 (1) surveyed the populations of yellow-billed oxpeckers *Buphagus africanus* and red-billed oxpeckers *B. erythrorhyncus* in Rhodes Matopos National Park (an area within the former range of both species), following the translocation of 47 yellow-billed and 12 red-billed oxpeckers in 1975. Sightings of *B. africanus* rose from 1975 (3.9 sightings/month) to 1977 (13.7 sightings/month). *B. erythrorhyncus* did not appear to establish in the park, with a single pair seen in the release area in 1977. Translocation mortalities occurred only for those birds released more than 30 hours after capture.

A before-and-after study from September 1988 to January 1993 in the Seychelles (2) found that all 29 Seychelles warblers *Acrocephalus sechellensis* translocated from Cousin Island to each of Aride Island (in September 1988) and Cousine Island (in June 1990) were alive in 1991 and that populations had grown from before translocations to 210 on Aride Island and 53 on Cousine Island. A further census on Aride Island in January 1993 estimated a population of 239 warblers. Prior to translocation, potential introduction sites were identified according to food availability, the absence of feral cats *Felis catus* and black rats *Rattus rattus*, and a sustained commitment to conservation management from the land owners. Translocation was in well-ventilated cardboard cages (15 × 15 × 20 cm) with a stick trellis 1 cm above the floor of each box allowing birds to perch, and the entire process of capture, translocation and release took on average little more than three hours, with no mortalities.

A site comparison study on three islands offshore from South Island, New Zealand, in April–June 2001 (3), found that the nesting success of translocated saddlebacks *Philesturnus carunculatus* declined with increasing difference in latitude from the source population. All birds originally came from Big South Cape island, being translocated to the study islands and others when rats invaded Big South Cape in 1964. Birds on Ulva Island (60 km north of Big South Cape) had 73% nesting success (11 pairs), compared with 32% (16 pairs) for Breaksea Island (190 km north) and 19% success (14 pairs) for Motuara Island (810 km north). Success was calculated using the Mayfield method and differences were largely due to higher egg fertility and hatching success. The authors note that differences in habitat (due to latitude) were unlikely to be the only reason for varying reproductive success, as Breaksea, Ulva and Big South Cape are all similar in habitat, but Breaksea had significantly lower reproductive success.

A before-and-after study on Mokoia Island (135 ha) in Lake Rotorua, North Island, New Zealand (4), found that a population of saddlebacks *Philesturnus carunculatus* (referred to as North Island saddlebacks *P. rufusater*) reintroduced onto the island increased from the 36 birds released in April 1992 to 217 birds in September 1996. The population fell following the attempted eradication of mice but recovered to 200 by September 1997. Reproductive output declined over time as the population grew. Before birds were released, brown rats *Rattus norvegicus* were eradicated from the island.

A replicated study (5) covering a range of time periods assessed the success or failure of 31 translocation attempts of saddlebacks *Philesturnus carunculatus* (24 attempted translocations) and New Zealand robins *Petroica australis* (six attempted) into separate offshore islands around New Zealand, found that both species established successful populations from small founder populations. The average founder population size of robins and saddlebacks was 31 and 34 respectively (ranging from 5–188 individuals). Only five of 24 saddleback populations went extinct or quasi-extinct (population decreased by >50% after 3 years), while none of the 6 robin populations failed. Predation caused 80% of translocation failure. In total, five populations established from fewer than 15 individuals survived and grew. Populations were categorised as extinct if surveys subsequent to translocation failed to record any birds.

A before-and-after study from May 2004 to September 2005 (6) investigated the translocation of 58 adult Seychelles warblers *Acrocephalus sechellensis* in May and June 2004 from Cousin Island to Denis Island, Seychelles. The introduced population grew to 75 individuals by August 2005. Twenty-seven female and 31 male warblers were captured a month before the onset of the breeding season (the time of peak bird weight and condition) and were translocated within 24 hours. Denis Island, which was not part of the warbler's historic range, was selected as the site for introduction as a predator-free island with suitable habitat and food availability. Of the 35 breeding territories vacated on Cousin Island because of the warbler translocations, all but three were occupied within an average of 5.4 days, with the source population rebounding to its pre-translocation level by September 2005.

A before-and-after study in the Cook Islands (7) translocated 30 Rarotonga monarchs *Pomarea dimidiata* (one to two years old) from Rarotonga Island to Atiu Island between August 2001 and August 2003. In June 2004 the monarch population was at least 15 birds, with breeding occurring in 2002 and 2003. Atiu Island was selected as the area for introduction based on the local community's commitment to conservation, the island's size (greater than 500 ha), the apparent absence of black rats *Rattus rattus*, and the presence of suitable areas of habitat. Translocated birds were screened for blood-borne parasites before release, with the risk of infection from other species on Atiu considered low. Birds were translocated in 50 cm × 30 cm × 30 cm plywood boxes, each with a dwelling perch and ventilation holes, within 18 hours of capture.

A before-and-after study from February 2003 to December 2005 investigated the reintroduction of 32 rifleman *Acanthisitta chloris* to Ulva Island, New Zealand in February 2003 (8) and found that at least 20 birds survived and bred by November 2003, and offspring were observed breeding in the second year. The 58 rifleman were captured on Codfish Island and kept in 14 × 4.5 × 2 m aviaries for up to four days before release. Of these, 26 died in captivity and transport, including 14 deaths while in the aviaries. Transport mortality was highest when birds were kept in transfer boxes for long periods of time (6–8 hours), or when transfer boxes were in close proximity and birds attempted to attack one other.

A before-and-after study in New Zealand (9) investigated the translocation of 46 saddlebacks *Philesturnus carunculatus* from Breaksea Island to Erin Island in September 2003 and April 2004 and found that no birds survived until June 2006. Birds carried for an hour before ringing had higher mortality in the first two weeks (20/22 individuals dying) than birds caught within 20 m of the banding hut (9/25 dying). Male saddlebacks with low body condition scores and females with ectoparasites (e.g. ticks, fleas) were most likely to die within 11 days of release. A measure of inbreeding and the total time held in captivity were not good predictors of initial saddleback mortality. Birds were kept in a 2 × 4 × 7 m aviary for up to five days before translocation.

A before-and-after study on Cousin Island, Seychelles (10), found that the population of Seychelles magpie-robins *Copsychus sechellarum* increased from five individuals, translocated from Frégate Island in 1994–5, to 46 individuals in 2006, before declining to 31 birds in 2007. Two males and two females were originally moved, plus a replacement female following the death of one of the original females in early 1995. The birds were kept in individual holding aviaries for two days before translocation. Cousin Island was identified as a suitable introduction site based on its status as a nature reserve, the absence of invasive predators, and the availability of large areas of native forest and food resources.

A before-and-after study on Motuihe Island, New Zealand (11), found that survival of 20 saddlebacks *Philesturnus carunculatus* (formerly North Island saddlebacks *P. rufusater*) translocated from Matangi Island in August 2005, was 70% for the first year, with at least 11 chicks fledged. The reintroduction was one part of a management project for Motuihe Island, including the eradication of brown rats *Rattus norvegicus*, house mice *Mus musculus*, European rabbits *Oryctolagus cuniculus* and feral cats *Felis catus* between 1997 and 2002. Extensive planting of native vegetation also took place from 2003–8. Birds were housed for 1–3 days in an aviary (8 × 5 × 3.5 m) before transport, with no mortalities during captivity and transport. This study is also discussed in 'Provide artificial nesting sites'.

A before-and-after study in Florida, USA, reintroduced 47 brown-headed nuthatches *Sitta pusilla* and 62 eastern bluebirds *Sialia sialis* to Long Pine Key, Everglades National Park, between December 1997 and April 2001 (12). Although 16 eastern bluebirds and 21 nuthatches failed to establish territories and presumably died shortly after release, population growth was initially observed in both species after the translocations were completed. By 2007, however, the populations of both species were either stable or declining, and only at approximately 10% of their predicted carrying-capacity (200 breeding pairs of each species). Nuthatches and bluebirds were captured at either Big Cypress National Preserve or Naples, Florida, transported and then kept in aviaries at the release site for up to three weeks before reintroduction.

(1) Grobler, J. H. (1979) The re-introduction of oxpeckers *Buphagus africanus* and *B. erythrorhyncus* to the Rhodes Matopos National Park, Rhodesia. *Biological Conservation*, 15, 151–158.
(2) Komdeur, I. (1997) Inter-island transfers and population dynamics of Seychelles warblers *Acrocephalus sechellensis*. *Bird Conservation International*, 7, 7–26.
(3) Hooson, S. & Jamieson, I. G. (2004) Variation in breeding success among reintroduced island populations of South Island Saddlebacks *Philesturnus carunculatus carunculatus*. *Ibis*, 146, 417–426.
(4) Armstrong, D. P., Davidson, R. S., Perrott, J. K., Roygard, J. & Buchanan, L. (2005) Density-dependent population growth in a reintroduced population of North Island saddlebacks. *Journal of Animal Ecology*, 74, 160–170.
(5) Taylor, S. S., Jamieson, I. G. & Armstrong, D. P. (2005) Successful island reintroductions of New Zealand robins and saddlebacks with small numbers of founders. *Animal Conservation*, 8, 415–420.
(6) Richardson, D. S., Bristol, R. & Shah, N. J. (2006) Translocation of the Seychelles warbler *Acrocephalus sechellensis* to establish a new population on Denis Island, Seychelles. *Conservation Evidence*, 3, 54–57.
(7) Robertson, H. A., Karika, I. & Saul, E. K. (2006) Translocation of Rarotonga Monarchs *Pomarea dimidiata* within the southern Cook Islands. *Bird Conservation International*, 16, 197–215.

(8) Leech, T. J., Craig, E., Beaven, B., Mitchell, D. K. & Seddon, P. J. (2007) Reintroduction of rifleman *Acanthisitta chloris* to Ulva Island, New Zealand: evaluation of techniques and population persistence. *Oryx*, 41, 369–375.
(9) Taylor, S. S. & Jamieson, I. A. N. G. (2007) Factors affecting the survival of founding individuals in translocated New Zealand saddlebacks *Philesturnus carunculatus*. *Ibis*, 149, 783–791.
(10) López-Sepulcre, A., Doak, N., Norris, K. & Shah, N. J. (2008) Population trends of Seychelles magpie-robins *Copsychus sechellarum* following translocation to Cousin Island, Seychelles. *Conservation Evidence*, 5, 33–37.
(11) Parker, K. A. & Laurence, J. (2008) Translocation of North Island saddleback *Philesturnus rufusater* from Tiritiri Matangi Island to Motuihe Island, New Zealand. *Conservation Evidence*, 5, 47–50.
(12) Lloyd, J. D., Slater, G. L. & Snow, S. (2009) Demography of reintroduced eastern bluebirds and brown-headed nuthatches. *Journal of Wildlife Management*, 73, 955–964.

15.25. Use techniques to increase the survival of species after capture

- A small controlled study from the USA (1) found that providing dark, quiet environments with readily-available food and water increased the survival of small birds after capture and increased the probability that they would accept captivity.
- A study from Hawaii found that keeping birds warm in a 'mock' translocation in Hawaii increased survival (2), although all birds suffered some loss of condition.

Background

Translocations can be a stressful and potentially dangerous procedure, with birds being confined for potentially long periods of time after being captured from the wild. Techniques to maximise survival may therefore be an important part of ensuring the overall success of translocations. In addition, ensuring higher survival can help to reduce the impact of translocations on source populations by requiring fewer individuals to be taken.

A small controlled study over the summers of 1986–1988 tested two transportation methods (prior to reintroduction attempts) from Michigan to Ohio, USA (1) and found that Nashville warblers *Vermivora ruficapilla* were more likely to survive using a modified technique that provided dark, quiet environments, prompter delivery of food and water and reduced handling time. When the standard technique for introducing warblers to captivity was used, 79% of warblers appeared to adapt to the captive environment and five birds died. When the new technique was used, 88% and 96% warblers (1987 and 1988 respectively) adapted, significantly more than when using the standard technique. A total of 188 trips (612 km one-way) were made without fatality.

A small controlled study on Hawaii in December 1996 evaluated the effects of translocation on common amakihi *Hemignathus virens* and Japanese white-eyes *Zosterops japonicus* (2). Birds kept overnight without thermal support had significantly higher mortality rates (4/10 birds in both species) than those provided with thermal support (0/10 common amakihi and 1/10 Japanese white-eyes), and birds that lost the most weight had the highest mortality. Birds were captured, transported by car for four hours and kept in captivity for 48 hours before release. All birds suffered weight loss, and fat and protein store depletion, with all deaths occurring within the first 24 hours following capture, regardless of whether the birds were quarantined and then transported or transported and then quarantined. Bird age, capture weight, or fat score did not affect survival rates.

(1) Bocetti, C. I. (1994) Techniques for prolonged confinement and transport of small insectivorous passerines. *Journal of Field Ornithology*, 65, 232–236.
(2) Work, T. M., Massey, J. G., Johnson, L., Dougill, S. & Banko, P. C. (1999) Survival and physiologic response of common amakihi and Japanese white-eyes during simulated translocation. *The Condor*, 101, 21–27.

15.26. Ensure translocated birds are familiar with each other before release

- Two controlled trials from New Zealand (1,2) found no evidence that translocating birds which were familiar with each other was more likely to succeed than translocating unfamiliar birds.

Background

Translocations are only going to be successful if birds breed after establishing themselves. This may be more likely to occur if birds are familiar with each other before release.

A controlled trial during 1992–3 on two islands (one offshore, one in a lake) in North Island, New Zealand (1), found no evidence that a translocation using North Island robins *Petroica australis longipes* familiar with one another was more likely to succeed than a translocation using unfamiliar birds. Known female survival was similar for the two groups (three females from familiar groups and four from unfamiliar groups alive at the start of the breeding season, a total of 14 females released). All surviving females formed pairs. Aggression between territorial pairs was similar between treatments although it declined with the amount of time pairs had spent in proximity to each other. The lack of treatment effect may therefore have been due to rapid dispersal from the release sites (only 15 birds remained at their release site two weeks after release), meaning that birds did not interact for long with the birds they were released with. A total of 44 birds were moved in four groups: two groups totalling 21 birds were moved as 'intact neighbourhoods' comprising almost all birds from two locations; the remaining 23 birds were in two groups of similar sex ratios and sizes, but with birds from different areas.

A controlled trial during 1992–3 on an island in North Island, New Zealand (2), found no evidence that a translocation using North Island saddlebacks *Philesturnus carunculatus rufusater* familiar with one another was more likely to succeed than a translocation using unfamiliar birds. A total of 36 birds were transferred in two groups of 18, one group (five pairs and eight juveniles) all from a single forest patch and the other (10 unmatched adults and eight juveniles) from multiple patches (all birds came from Tiritiri Matangi, a 135 ha offshore island). Familiar birds formed new pair bonds faster than unfamiliar pairs, although only one translocated pair remained together and overall pairing rates were similar between treatments. Survival (94% vs. 89% over six months for 18 familiar and 18 unfamiliar birds), dispersal (69% dispersal from the release site for 16 familiar birds vs. 59% for 17 unfamiliar birds) and reproductive output (3.1 fledglings/pair for seven familiar pairs vs. 4.0 fledglings/pair for six unfamiliar pairs) were similar between treatments. Birds were kept for two or three days during transport before being released 500 m apart.

(1) Armstrong, D. P. (1995) Effects of familiarity on the outcome of translocations, II. A test using New Zealand robins. *Biological Conservation*, 71, 281–288.
(2) Armstrong, D. P. & Craig, J. L. (1995) Effects of familiarity on the outcome of translocations, I. A test using saddlebacks *Philesturnus carunculatus rufusater*. *Biological Conservation*, 71, 133–141.

15.27. Ensure genetic variation to increase translocation success

- We did not capture any studies on the effects of ensuring genetic variation in translocated birds.

Background

If translocated birds are used to recolonise a part of the species' former range then they will constitute a founder population. If the founders do not have sufficient genetic variation, their breeding may lead to inbreeding depression, potentially jeopardising the long-term viability of the population.

15.28. Translocate nests to avoid disturbance

- Four small trials from the USA (1–4) and a replicated study from Chatham Island, New Zealand (5) found some success in relocating nests whilst they were in use.
- However, one study from the USA (4) found that only 40% of burrowing owls *Athene cunicularia* were moved successfully, another found that American kestrels *Falco sparverius* tolerated movement of their nest, but not repeated disturbance (3) and another found that barn swallow *Hirundo rustica* may follow their nest as it is slowly moved on a car, but may not stay at the new site (2).

Background

Some birds return year after year to the same nest, making moving them difficult: birds are likely to return to the old site the next year. If nesting sites are threatened by development or are unsuitable for some reason then moving birds could be beneficial. However, it is possible that if the nest can be moved whilst it is in use, the birds will return to its eventual site rather than its original position.

A small single-site study from May–August in 1979–1981 reporting a translocation attempt of a pair of golden eagles (*Aquila chrysaetos*) into an undisturbed shrubland area in Wyoming, USA (1) found that temporary nesting platforms can be used to induce relocation into the target area. To encourage the pair to move into the target area, their single nestling was moved sequentially to a series of temporary platforms erected between the nest tree (located in a mining-impacted area) and the target platform at points 175, 715 and 1,375 m from the nest tree. Sticks were added to each temporary platform. Each move was executed only after the nestling's acceptance of the previous platform. A fresh rabbit carcass was placed with the nestling each time it was moved. The nestling fledged from the third temporary platform before the adult pair had fully accepted the temporary platforms. However, in 1981, the adults voluntarily nested on and successfully fledged one nestling from the target platform.

A small replicated study from June–July in 1989 that tested a new method for translocating whole nests on two barn swallow *Hirundo rustica* nests in Ithaca, New York, USA (2), found that nests could be successfully transferred in stages but that only one parent remained with one of the nests. Overall, the 3.5 km journey between the old and new site required 27 and 13 hours of daylight for each nest respectively. All chicks fledged from both nests but only one of four parents remained with the nest post-translocation. Both nests were fastened to a vehicle-roof and moved in stages, waiting at each point until the parents had fed the young. Distance between stops varied from 5–10 m initially to 100–200 m closer to the target site. Nests were moved in 4 m stages at the target site (using a mist-net pole) to the new habitat. The transport box (13 × 20 cm) had one side left open for parents to access the nest.

A single-site study in May 1992 in one field containing an American kestrel *Falco sparverius* breeding pair in an artificial nest box in Ohio, USA (3) found that the breeding pair tolerated initial, but not continued, human disturbance. The nestbox (containing two eggs), initially located in a maple tree, was removed from the tree and placed upright on the ground 3 m away while the tree was felled. The male was found incubating the eggs at this time. The nestbox was subsequently attached to a steel fencepost 10 m away. Although the female kestrel entered the nest initially, the birds appeared to abandon the nest and were later observed copulating near a nest box located 1 km away. When checked, the translocated nest box contained 5 cold eggs. Three eggs were therefore laid after the nest was moved and incubation was initiated while the nest box was on the ground. The author suggests that kestrel nests may be successfully relocated to a short distance if further disturbance is kept to a minimum.

A small study from June–July in 1998 in one field impacted by agriculture in southern Idaho, USA (4) found that burrowing owls *Athene cunicularia* exhibited mixed responses to nest relocation to a nearby natural buffer strip. Relocation distances averaged 153 m from old nests. Overall, two families (five fledglings) accepted the new nests (40%); two families (five fledglings) returned to the vicinity of the old nests one day after relocation (40%); and one family (five fledglings) disappeared from the field (20%). Dates of relocation events did not correlate with relocation outcomes. In 1999, one male and one female returned to the relocated sites and successfully fledged young (20% return rate). However, during 1999, none of the 15 fledglings from the 1998 nests was observed. The buffer strip was 25 m wide on the outskirts of a field zoned for development. All nests were artificial burrow systems.

A replicated study on a beach on Chatham Island, New Zealand (5) found that, of 78 Chatham Island oystercatcher *Haematopus chathamensis* nests gradually moved 1–32 m upshore during 1998–2004, although 11 were subsequently washed away by storm surges or high tides. Nests were moved either by hand (from a shallow nesting scrape to another, man-made scrape close by) or by dragging the artificial nesting platforms that the nests were on (see 'Provide nesting habitat for birds that is safe from extreme weather'). In 2004–5, 26 nests from 33 pairs were washed away by very high storms; no data was available on the number of nests moved in this year. Before nests were moved, all nests on the beach were sometimes washed away. No data were provided on acceptance of movement or breeding success. The beach was also subject to several other conservation interventions: see 'Exclude livestock to reduce trampling or predation', 'Predator control on islands' and 'Remove problematic vegetation'.

(1) Postovit, H. R., Grier, J. W., Lockhart, J. M. & Tate, J. (1982) Directed relocation of a golden eagle nest site. *The Journal of Wildlife Management*, 46, 1045–1048.
(2) Winkler, D. W. & McCarty, J. P. (1990) Method for transplanting nests of barn swallows. *Journal of Field Ornithology*, 61, 426–430.
(3) Carpenter, T. W. (1992) American kestrel completes clutch following movement of its nest box. *Journal of Raptor Research*, 26, 268.
(4) Smith, B. W. & Belthoff, J. R. (2001) Burrowing owls and development: short-distance nest burrow relocation to minimize construction impacts. *Journal of Raptor Research*, 35, 385–391.
(5) Moore, P. & Williams, R. (2005) Storm surge protection of Chatham Island oystercatcher *Haematopus chathamensis* by moving nests, Chatham Islands, New Zealand. *Conservation Evidence*, 2, 50–52.

15.29. Use vocalisations to attract birds to new sites

- Six studies from North America (2,5,7,8), the Galapagos (4) and the Azores (9) found that seabirds were more likely to nest in areas where vocalisations were played (2,9), or were successfully attracted to nest in new areas, following the playing of vocalisations (2,4,5,7,8,10). Four of these studies (5,7,8,10) used several interventions at once. One study found that some calls were more effective than others (2).
- Two studies from the USA (1) and the Galapagos (4) found that birds did not colonise all new areas where vocalisations were played. It is possible that the result from the Galapagos was due to only having a single year's data.
- One controlled study from Hawaii (3) found that albatross were more likely to land in areas where vocalisations were played than in areas without vocalisation playback. A small controlled study from New Zealand (6) found that terns were not more likely to land in areas where vocalisations were played.

Background

As well as physically moving birds to areas (translocations, see separate intervention), it is possible to try and get birds to move of their own accord. Many birds nest colonially for protection and are attracted nesting to areas with conspecifics. Especially for birds that land mainly at night, hearing the calls and other vocalisations of conspecifics may be an important part of guiding them to colonies. Playing recordings of these calls may therefore encourage birds to colonise new areas. A similar intervention is 'Use decoys to attract birds to safe areas'.

A trial between 1983 and 1985 in Alabama, USA (1), found that no little blue herons *Egretta caerulea* or cattle egrets *Bubulcus ibis* were attracted to a 4 ha swamp which had heron calls broadcast from it. Up to 15 herons and egrets of various species were observed perching or roosting at the site, but none nested. The calls were played on a continuous loop, every three minutes during daylight hours over the summer, and decoys were also installed (see 'Use decoys to attract birds to safe areas').

A replicated, controlled study in 1980–3 on four islands in Maine, USA (2), found that significantly more Leach's storm petrels *Oceanodroma leucorhoa* colonised artificial nest chambers when the petrel's 'purr' call was played from speakers near the burrows, compared to when only the 'chuckle' call was played or control burrows, where no vocalisations were played (24% of 40 burrows with just 'purr' calls colonised and 17.5% of 164 burrows with 'purr' and 'chuckle' calls vs. 0% for 20 burrows with 'chuckle' only and 0% for 40 control burrows). Calls were played from 22:00 hours until 04:00 each night from mid-May to mid-August from speakers located in the centre of clusters of burrows. Burrows that were colonised were significantly closer to speakers than expected at random, with over 80% of occupied burrows within 1.5 m of a speaker. Overall, only two chicks fledged successfully from 264 artificial burrows over three years (both in the third year). This study is also discussed in 'Provide artificial nesting sites'.

A controlled study on Kauai, Hawaii, USA, between December 1982 and April 1983 (3) found that Laysan albatrosses *Phoebastria immutabilis* were more likely to land in a study site with albatross decoys on days when albatross vocalisations were also played, than on days when vocalisations were not played or than at a control site with neither decoys nor vocalisations (8.2% of 1,053 flying albatrosses landing when vocalisations were playing vs. 5.2% of 1,300 without vocalisations and 1.8% of 877 at the control site). Albatrosses were also more likely to land close to speakers when vocalisations were playing, compared to when they were turned off (76% of the 97 closest landings when speakers were on vs. 24% when they were off). Each study plot had a speaker surrounded by six decoys. The effect of the decoys in attracting albatrosses is discussed in 'Use decoys to attract birds to safe areas'.

A before-and-after study on two islands in the Galapagos, Ecuador (4), found that playing dark-rumped petrel *Pterodroma phaeopygia phaeopygia* vocalisations from loudspeakers in 1988–90 successfully attracted petrels to an area on Santa Cruz Island provided with artificial burrows, but that playing similar recordings on the predator-free island of Pinta failed to attract any petrels in 1991. The role of artificial burrows in this study is discussed in 'Provide artificial nesting sites', and predator control in 'Control mammalian predators on islands'.

A small trial in 1993–4 at the western end of Lake Ontario, Canada (5), found that the number of Caspian terns *Sterna caspia* nesting on an artificial raft increased from one pair in 1993 to six pairs in 1994. A speaker system played vocalisations from a Caspian tern colony for four hours a day as terns arrived in the area each year. The study is discussed in detail in 'Provide artificial nesting sites'.

A small controlled trial on a shell and sand beach in northern North Island, New Zealand (6), found that New Zealand fairy terns *Sterna nereis davisae* (formerly *S. antillarum*) were no more likely to land in experimental plots on eight days when a tape of fairy tern calls was played, compared with eight days when calls were not played. No data on reproduction were provided. All plots were 120 × 55 m and one of four plots had three decoy models. The experimental plot was rotated each day for a total of 16 days. This study also describes the effect of the decoys on bird behaviour, discussed in 'Use decoys to attract birds to safe areas'.

A before-and-after trial at a coastal site in Long Beach, California, USA (7), reported the successful translocation of a black-crowned night heron *Nycticorax nycticorax* colony using (amongst other interventions) vocalisations of the original colony. This study is discussed in 'Translocate individuals'.

A before-and-after study on two small islands in the Columbia River Estuary, Oregon, USA (8), found that an entire Caspian tern *Sterna caspia* colony (approximately 8,900 pairs) relocated from Rice Island to East Sand Island between 1999 and 2001. Movement was encouraged by using between two and for audio systems, broadcasting the sounds of a Caspian tern colony, as well as several other interventions. This study is discussed in detail in 'Move fish-eating birds to reduce conflict with fishermen'.

A study in 2000–1 on an islet in the Azores, Portugal (9), found that artificial nest chambers occupied by Madeiran storm petrels *Oceanodroma castro* were significantly closer to speakers playing the petrels' 'nest song' every night than unoccupied nest chambers. Forty-seven of 115 chambers were used in early 2000; with 40 of 147 used in September 2000 and 49 in 2001. This study is discussed in more detail in 'Provide artificial nesting sites'.

A before-and-after study on Mana Island, North Island, New Zealand (10), found that a new nesting colony of common diving petrels *Pelecanoides urinatrix* was established by playing vocalisations, providing artificial nesting burrows and translocating chicks. This study is discussed in 'Translocate individuals', 'Provide artificial nests' and 'Artificially incubate and hand-rear birds in captivity'.

(1) Dusi, J. L. (1985) Use of sounds and decoys to attract herons to a colony site. *Colonial Waterbirds*, 8, 178–180.
(2) Podolsky, R. H. & Kress, S. W. (1989) Factors affecting colony formation in Leach's storm petrel. *The Auk*, 106, 332–336.
(3) Podolsky, R. H. (1990) Effectiveness of social stimuli in attracting Laysan albatross to new potential nesting sites. *The Auk*, 107, 119–124.
(4) Cruz, J. & Cruz, F. (1996) Conservation of the dark-rumped petrel *Pterodroma phaeopygia* of the Galapagos Islands, 1982–1991. *Bird Conservation International*, 6, 23–32.
(5) Lampman, K. P., Taylor, M. E. & Blokpoel, H. (1996) Caspian terns (*Sterna caspia*) breed successfully on a nesting raft. *Colonial Waterbirds*, 135–138.
(6) Jeffries, D. S. & Brunton, D. H. (2001) Attracting endangered species to 'safe' habitats: responses of fairy terns to decoys. *Animal Conservation*, 4, 301–305.
(7) Crouch, S., Paquette, C. & Vilas, D. (2002) Relocation of a large black-crowned night heron colony in southern California. *Waterbirds*, 25, 474–478.
(8) Roby, D. D., Collis, K., Lyons, D. E., Craig, D. P., Adkins, J. Y., Myers, A. M. & Suryan, R. M. (2002) Effects of colony relocation on diet and productivity of Caspian terns. *The Journal of Wildlife Management*, 66, 662–673.
(9) Bolton, M., Medeiros, R., Hothersall, B. & Campos, A. (2004) The use of artificial breeding chambers as a conservation measure for cavity-nesting procellariiform seabirds: a case study of the Madeiran storm petrel (*Oceanodroma castro*). *Biological Conservation*, 116, 73–80.
(10) Miskelly, C. M. & Taylor, G. A. (2004) Establishment of a colony of common diving petrels (*Pelecanoides urinatrix*) by chick transfers and acoustic attraction. *Emu*, 104, 205–211.

15.30. Use decoys to attract birds to new sites

- Seven studies (3,5–8,10,11) found that birds bred in areas where decoys (of birds or nests) were used to attract birds. Six of the studies (3,5,6,8,10,11) used several interventions at once. Two studies from the USA (1,2) found that least terns *Sterna antillarum* and herons were not attracted to new areas to breed when decoys were used.
- Five studies from North America (1,4,9,11,12) and France and Spain (7) found that more birds landed near decoys than in control areas.
- The two studies to compare decoy types found that three-dimensional models were better than two-dimensional 'cut-outs' (4) and plastic models of birds were better than rag decoys (12).

Background

As well as physically moving birds to safe areas (translocations, see separate intervention), it is possible to try and get birds to move of their own accord. Many birds nest colonially for protection and are attracted nesting areas with conspecifics. Using realistic decoys may therefore encourage birds to colonise new areas. A similar intervention is 'Use vocalisations to attract birds to safe areas'.

A controlled study in New Jersey, USA (1), found that at one sand and shell island, 81% of 821 least tern *Sterna antillarum* landings between 19th and 25th May 1983 were in a 100 m^2 plot with tern decoys and only 19% in a control plot (without decoys). The first three nests established on the island were within 3 m of a decoy and none of the first 28 nests was in the control plot. At another sand and shell beach, there were only 12 landings in the two study plots between 20th May and 6th June. No terns attempted to nest at the site.

A trial between 1983 and 1985 in Alabama, USA (2), found that no little blue herons *Egretta caerulea* or cattle egrets *Bubulcus ibis* were attracted to a 4 ha swamp which had decoys installed in it. Up to 15 herons and egrets of various species were observed perching or roosting at the site, but none nested. Between eight and 25 decoy herons and egrets were installed each summer and heron calls were broadcast during daylight hours (see 'Use vocalisations to attract birds to safe areas').

A study of an Atlantic puffin *Fratercula arctica* translocation programme in 1973–81 (3) found that a puffin population was established in Maine, USA, after 774 puffin nestlings were translocated from Newfoundland, Canada, and decoys were used to help attract fledged birds back to the release site. This study is discussed in 'Translocate individuals'.

A controlled study in mixed coastal habitats on Kauai, Hawaii, USA, between December 1982 and April 1983 (4) found that Laysan albatrosses *Phoebastria immutabilis* (formerly *Diomedea immutabilis)* were more likely to land in a study site with albatross decoys than at a control site without decoys (5.2% of 1,300 flying albatrosses landing at the experimental site vs. 1.8% of 877 at the control site). In addition, albatrosses landed closer to decoys than would be expected by random. Three-dimensional models of albatross pointing towards the sky attracted more albatrosses to within 3 m than two-dimensional models and paired models attracted more birds than single models. Six decoys were placed (either singly or in pairs) in a 10 m circle at each study plot. This study also describes the effect of playing albatross vocalisations on albatross landings, discussed in 'Use vocalisations to attract birds to safe areas', but does not provide any data on breeding.

A replicated trial in 1990 at Lake Ontario, Canada (5), found that common terns *Sterna hirundo* successfully used four floating wooden rafts, each with six tern decoys on, the same season that they were installed. This study is discussed in detail in 'Provide artificial nesting sites'.

A small trial in 1993–5 at the western end of Lake Ontario, Canada (6), found that the number of Caspian terns *Sterna caspia* nesting on an artificial raft with eight tern decoys on increased from one pair in 1993 to 50 pairs in 1995. This study is discussed in detail in 'Provide artificial nesting sites'.

A review of management at two coastal wetland sites in Bouches-du-Rhône, France and in Andalucia, Spain (7), found that no greater flamingos *Phoenicopterus roseus* used an artificial nesting island created in the French site for the first year. However, following the installation of 500 decoy nests (wicker baskets packed with mud), large numbers of flamingos used the site. When 350 extra decoys were added to another part of the island, flamingos colonised it, in preference to areas without decoys. In Spain, newly created nesting habitat was used, with flamingos showing a preference for areas with artificial nests, depressions and scattered broken eggshell. This study is also discussed in 'Provide artificial nest sites', 'Manage water levels in wetlands' and 'Control predators not on islands'.

A before-and-after study in the upper St. Lawrence River, Canada (8) found that a common tern *Sterna hirundo* colony was re-established when decoys as well as control of ring-billed gulls *Larus delawarensis* were used to try to attract birds. This study is discussed in 'Reduce inter-specific competition for nest sites by removing or excluding competitor species'.

A small controlled trial on a shell and sand beach in northern North Island, New Zealand (9), found that New Zealand fairy terns *Sterna nereis davisae* (formerly *S. antillarum*) were significantly more likely to land in experimental plots with tern decoy models in, compared to control plots, with 80% of all landing episodes were in experimental plots. No data on reproduction were provided. All plots were 120 × 55 m and one of four plots had three decoy models in. The experimental plot was rotated each day for a total of 16 days. This study also describes the effect of using vocalisations to attract birds, discussed in 'Use vocalisations to attract birds to safe areas'.

A before-and-after trial at a coastal site in Long Beach, California, USA (10), reported the successful translocation of a black-crowned night heron *Nycticorax nycticorax* colony using (amongst other interventions) decoys to attract birds. This study is discussed in 'Translocate individuals'.

A before-and-after study on two small islands in the Columbia River Estuary, Oregon, USA (11), found that an entire Caspian tern *Sterna caspia* colony (approximately 8,900 pairs) relocated from Rice Island to East Sand Island between 1999 and 2001. Movement was encouraged by placing 253–415 decoys each year, as well as several other interventions. This study is discussed in detail in 'Move fish-eating birds to reduce conflict with fishermen'.

A randomised, replicated and controlled study on a wetland site in Florida in autumn 1997 (12) found that wading birds (mainly white ibis *Eudocimus albus* and herons, Ardeidae) were more attracted to ponds with white plastic flamingo decoys than to ponds with Texas rag decoys (91 × 91 cm white plastic sheet knotted and on a 122 cm dowel rod) or control ponds with no decoys. Ponds with Tyvek® bag decoys (envelopes stuffed with paper and mounted on a 122 cm dowel rod) were intermediate and not significantly different from the other treatments (approximately 3 birds/pond with flamingos vs. 1.8 birds/pond for Tyvek® bags, 0.6 birds/pond for Texas rag and 1.1 birds/pond for controls). These differences were due to differences in the behaviour of white wading birds – there were no significant differences for dark wading birds. Experiments were conducted over five days, with eight ponds used each day (two for each treatment).

(1) Kotliar, N. B. & Burger, J. (1984) The use of decoys to attract least terns (*Sterna antillarum*) to abandoned colony sites in New Jersey. *Colonial Waterbirds*, 7, 134–138.
(2) Dusi, J. L. (1985) Use of sounds and decoys to attract herons to a colony site. *Colonial Waterbirds*, 8, 178–180.
(3) Kress, S. W. & Nettleship, D. N. (1988) Re-establishment of Atlantic puffins (*Fratercula arctica*) at a former breeding site in the Gulf of Maine. *Journal of Field Ornithology*, 59, 161–170.
(4) Podolsky, R. H. (1990) Effectiveness of social stimuli in attracting Laysan albatross to new potential nesting sites. *The Auk*, 107, 119–124.
(5) Dunlop, C. L., Blokpoel, H. & Jarvie, S. (1991) Nesting rafts as a management tool for a declining common tern (*Sterna hirundo*) colony. *Colonial Waterbirds*, 14, 116–120.
(6) Lampman, K. P., Taylor, M. E. & Blokpoel, H. (1996) Caspian terns (*Sterna caspia*) breed successfully on a nesting raft. *Colonial Waterbirds*, 135–138.
(7) Martos, M. R. & Johnson, A. R. (1996) Management of nesting sites for greater flamingos. *Colonial Waterbirds*, 167–183.
(8) Blokpoel, H. & Tessier, G. D. (1997) Successful restoration of the Ice Island common tern colony requires on-going control of ring-billed gulls. *Colonial Waterbirds*, 20, 98–101.
(9) Jeffries, D. S. & Brunton, D. H. (2001) Attracting endangered species to 'safe' habitats: responses of fairy terns to decoys. *Animal Conservation*, 4, 301–305.
(10) Crouch, S., Paquette, C. & Vilas, D. (2002) Relocation of a large black-crowned night heron colony in southern California. *Waterbirds*, 25, 474–478.
(11) Roby, D. D., Collis, K., Lyons, D. E., Craig, D. P., Adkins, J. Y., Myers, A. M. & Suryan, R. M. (2002) Effects of colony relocation on diet and productivity of Caspian terns. *The Journal of Wildlife Management*, 66, 662–673.
(12) Crozier, G. E. & Gawlik, D. E. (2003) The use of decoys as a research tool for attracting wading birds. *Journal of Field Ornithology*, 74, 53–58.

15.31. Alter habitat to encourage birds to leave an area

- A single before-and-after study in the USA (1) found that an entire Caspian tern *Sterna caspia* population moved following (amongst other interventions) the alteration of nesting habitat at the old colony site.

Background

Many ground-nesting seabirds require little or no vegetation to be able to nest. Therefore, if conservationists are trying to move a colony, for a number of reasons, they can make the habitat at the old site unsuitable by encouraging vegetation to grow there.

Many species of seabird disperse from colonies outside the breeding season, allowing work to be carried out without disturbing birds.

A before-and-after study on two small islands in the Columbia River Estuary, Oregon, USA (1), found that an entire Caspian tern *Sterna caspia* colony (approximately 8,900 pairs) relocated from Rice Island to East Sand Island between 1999 and 2001, following habitat alteration on Rice Island. Wheat *Triticum aestivum* was planted and silt fencing erected to encourage vegetation growth, whilst suitable habitat was prepared on East Sand Island (see 'Habitat creation – intertidal habitats') along with several other interventions (see 'Use decoys to attract birds to new nesting areas', 'Use vocalisations to attract birds to new nesting areas' and 'Control avian predators on islands'). The impact on conflict reduction (the purpose of the translocation) is discussed in 'Move fish-eating birds to reduce conflict with fishermen'.

(1) Roby, D. D., Collis, K., Lyons, D. E., Craig, D. P., Adkins, J. Y., Myers, A. M., and Suryan, R. M. (2002) Effects of colony relocation on diet and productivity of Caspian terns. *The Journal of Wildlife Management*, 66, 662–673.

16. Captive breeding, rearing and releases (*ex situ* conservation)

Key messages – captive breeding

Use captive breeding to increase or maintain populations

We reviewed 21 studies describing 18 captive-breeding programmes (of tinamous, seabirds, rails, cranes, bustards, storks, ibises, raptors, pigeons and songbirds), all describing at least some successful breeding in captivity. Twelve studies from nine programmes describe extensive breeding or the establishment of an *ex situ* population through captive breeding. Five describe small-scale programmes with one or two pairs of birds. Two studies describe some failures or low levels of success. Two studies on Mauritius kestrels and northern bald ibises found that captive-bred birds had lower reproductive success than birds hatched from 'harvested' wild eggs or wild birds.

Can captive breeding have deleterious effects on individual fitness?

We captured no studies investigating the effects of captive-breeding on fitness. Three studies using wild and captive populations or museum specimens found physiological or genetic changes in populations that had been bred in captivity. One found that changes were more likely to be caused by extremely low population levels than by captivity.

Use artificial insemination in captive breeding

A replicated study from Saudi Arabia found that artificial insemination could increase fertility in houbara bustards. A study of the same programme and a review found that repeated inseminations increased fertility, with the review arguing that artificial insemination had the potential to be a useful technique. Two studies from the USA found that artificially-inseminated raptors had either zero fertility, or approximately 50%.

Freeze semen for use in artificial insemination

Two small trials from the USA found that using thawed frozen semen for artificial insemination resulted in low fertility rates. A small trial from the USA found that a cryprotectant increased fertility rates achieved using frozen semen.

Wash contaminated semen and use it for artificial insemination

A replicated, controlled study from Spain found that washed, contaminated semen could be used to successfully inseminate raptors.

Artificially incubate and hand-rear birds in captivity

We captured 38 studies examining the success or failure of the artificial incubation or hand-rearing of birds. Thirty-three of these found that birds were successfully fledged or showed similar behaviour or fitness to non-artificially-reared birds. Five found that birds were poorly adapted, died either during rearing or shortly after release or had low fitness compared with other birds and one study found that eggs removed at a very young age were unlikely to be successfully raised. In addition, a study found that hand-reared partridges were potentially more vulnerable to ground predators than other birds. Five studies reported on the effectiveness of different methods such as increasing the food given or the temperature of incubation. One found that only intensively-reared ibises appeared to form social bonds similar to those in wild birds. A study on houbara bustards found that removing eggs for artificial incubation led to females laying large replacement clutches.

Use puppets to increase the success of hand-rearing

Three studies from the USA and Saudi Arabia found that crows and bustards raised using puppets did not have higher survival, dispersal or growth than chicks hand-reared conventionally.

Key messages – release of captive-bred individuals

Release captive-bred individuals into the wild to restore or augment wild populations

We captured 60 studies of 61 release programmes, with 19 programmes resulting in the increase or establishment of wild populations (one with additional predator control and one with uncertainty over the viability of the population). Twenty studies found that released birds had high survival, bred or showed normal behaviours. Two studies found that wild populations were not increased by released birds, and 21 found low survival of released birds (<50% each year). It should be noted that some successful programmes found very low survival for some releases, especially in the first year of releases

Use appropriate populations to source released populations

Two studies from Europe found that birds from populations near release sites adapted better and in one case had higher reproductive productivity than those from more distant populations.

Use holding pens at release sites

Three of four studies from North America and Saudi Arabia found that birds released into holding pens were more likely to form pairs or had higher survival than birds released into the open. One study found that parrots released into pens had lower survival than those released without preparation. A review of northern bald ibis releases found that holding pens could be used to prevent birds from migrating from the release site and so increase survival.

Clip birds' wings on release

Two of four studies found that bustards and geese had lower survival when released into holding pens with clipped wings compared to birds released without clipped wings. One study found no differences in survival for clipped or unclipped northern bald ibis. One study found that adult geese released with clipped wings survived better than geese released before they were able to fly.

Release birds in groups

A study from New Zealand found that released stilts were more likely to move long distances after release if they were released in larger groups.

Release chicks and adults in 'coveys'

Two out of three studies found that geese and partridges released in coveys had higher survival than young birds released on their own or adults released in pairs. A study from Saudi Arabia found that bustard chicks had low survival when released in coveys with flightless females.

Release birds as adults or sub-adults, not juveniles

Three out of nine studies from across the world found that birds released as sub-adults had higher survival than those released as juveniles. Two studies found lower survival of wing-clipped sub-adult geese and bustards, compared with juveniles and one study found lower survival of all birds released as sub-adults, compared to those released as juveniles. Three studies found no differences in survival for birds released at different ages, although one found higher reproduction in birds released at greater ages.

Use 'anti-predator training' to improve survival after release

Both studies captured found higher survival for birds given predator training before release, compared with un-trained birds. One found that using a live fox, but not a model, for training increased survival in bustards, but that several birds were injured during training.

Use 'flying training' before release

A study from the Dominican Republic found that parrots had higher first-year survival if they were given pre-release flying training.

Provide supplementary food during and after release

All three studies captured found that released birds used supplementary food provided. One study from Australia found that malleefowl had higher survival when provided with food and a study from Peru found that supplementary food could be used to increase the foraging ranges of Andean condors after release.

Use microlites to help birds migrate

A study from Europe found that northern bald ibises followed a microlite south in the winter but failed to make the return journey the next year.

Captive breeding

16.1. Use captive breeding to increase or maintain populations

Background

Captive breeding or *ex situ* conservation involves establishing and maintaining populations of wild species in captivity whilst attempting to keep the fundamental nature of the species the same (i.e. avoiding domestication). It is frequently used as a technique when wild populations become very small indeed, very fragmented or are declining very rapidly. Captive breeding potentially offers a way to maintain a population of the species whilst the threats to it in the wild are reduced or removed; or of increasing reproductive output and outbreeding beyond what would be possible in the wild.

Large-scale captive breeding programmes typically use specially-designed breeding centres, but zoos and wildlife parks can also be used. Whilst the differences between these may impact on the success of captive breeding, we have included both in the following section, as the techniques used should be applicable to both.

16.1.1. Tinamous

- A replicated study from Costa Rica over three years (1) found that great tinamous *Tinamus major* successfully bred in captivity, with similar reproductive success to wild birds.

A study in a breeding centre in Costa Rica between 2003 and 2005 (1) found that 42 great tinamous, *Tinamus major*, (28 female, 14 male) successfully bred in four breeding enclosures and laid a comparable number of eggs to wild birds (672 eggs laid by 24 females over three years in the centre vs. 3–6 eggs laid 3–4 times each breeding season in the wild), although the breeding season was a month longer than in the wild. Captive-laid eggs were similar in size to wild eggs. Older females laid more eggs and fertility in two enclosures appeared to increase with the number of years that birds spent together (50–61% infertility in the first year together vs. 16% in the second year). An enclosure with a male: female ratio of 4:1 had higher fertility in its first year than an enclosure with a ratio of 1:5 (13% infertility vs. 59%, number of eggs not provided). There was egg predation in two enclosures, which tinamous shared with chestnut-mandibled toucans, *Ramphastos swainsonii*, and great curassows, *Crax rubra*, but not in enclosures shared with only songbirds or no other species. The eggs were removed from cages and artificially incubated, but the effectiveness of this was not reported.

(1) Fournier, L., Fournier, R. & Janik, D. (2007) Techniques for the captive breeding of *Tinamus major fuscipennis* ZooAve, (Tinamiformes, Tinamidae), Costa Rica. *Zeledonia*, 11, 20–25.

16.1.2. Seabirds

- A study from Spain over five years (1) found that a single pair of Audouin's gulls, *Larus audouinii*, successfully bred in captivity.

A small study in a captive-breeding centre in eastern Spain (1) found that a pair of captive Audouin's gulls *Larus audouinii* successfully bred in captivity each year between 1996 and 2000. This followed a pair of wild gulls nesting inside the centre but outside the cages. Comparisons with the wild pair are made in 'Artificially incubate and hand-rear birds in captivity'.

(1) Martínez-Abraín, A., Viedma, C., Ramón, N., and Oro, D. (2001) A note on the potential role of philopatry and conspecific attraction as conservation tools in Audouin's gull *Larus audouinii*. *Bird Conservation International*, 11, 143–147.

16.1.3. Rails

- A study from an island in Australia (1) found that Lord Howe Island woodhens *Tricholimnas sylvestris* successfully bred in captivity, with 66 chicks being produced over four years from three pairs of adults.

A study on Lord Howe Island, Australia, between 1980 and 1983 (*1*) found that Lord Howe Island woodhens, *Tricholimnas sylvestris*, successfully reproduced in a purpose-built breeding centre. In 1980, the only three healthy adult pairs of woodhen were taken into captivity and they produced 66 chicks which were reared in captivity. This study also describes the impact of predator control (see 'Control mammalian predators on islands') and release of captive-bred individuals (see 'Release captive-bred individuals').

(1) Miller, B. & Mullette, K. J. (1985) Rehabilitation of an endangered Australian bird – the Lord Howe Island woodhen *Tricholimnas sylvestris* (Sclater). *Biological Conservation*, 34, 55–95.

16.1.4. Cranes

- A study from Canada over 32 years (1) found that whooping cranes *Grus americana* successfully bred in captivity eight years after the first eggs were removed from the wild. The authors note that young 'downy' chicks suffered high mortality in captivity.

A study between 1967 and 1991 (1) found that the removal of 355 whooping crane, *Grus Americana*, eggs from a wild population in north-west Territories and Alberta, Canada, to start a captive population, did not negatively affect the wild population, which increased from 48 to 146 birds during the study period, with no nests being abandoned. The captive population had high hatching success (78–100% for 50 eggs shipped to Patuxent Wildlife Research Centre in 1967–74, and 77% for 166 eggs shipped to Grays Lake National Wildlife Refuge in 1975–88) but 'downy young' suffered 68% mortality, mainly due to disease and anatomical abnormalities. However, cranes first bred in captivity in 1975 (in Patuxent), with five females laying 19 eggs in 1989 (nine hatching). This study is also discussed in 'Release captive-bred individuals', 'Foster eggs or chicks with wild conspecifics' and 'Foster eggs or chicks with wild non-conspecifics (cross-fostering)'.

(1) Kuyt, E. (1996) Reproductive manipulation in the whooping crane *Grus americana*. *Bird Conservation International*, 6, 3–10.

16.1.5. Bustards

- We captured four studies (*1–4*) of a houbara bustard, *Chlamydotis undulata, macqueenii* captive breeding programme in Saudi Arabia.
- The project successfully raised chicks in captivity, with 285 chicks hatched in the 7th year of the project (*1*) after 232 birds were used to start the captive population. Captive birds bred earlier (*2*) and appeared to lay more eggs (*4*) than wild birds. Forty-six percent of captive eggs hatched and 43% of chicks survived to ten years old (*3*), although no comparison was made with wild birds.

A review of a houbara bustard, *Chlamydotis undulata macqueenii*, captive breeding programme in Saudi Arabia, starting in 1986 (*1*) found that the captive population first bred in 1989, producing 17 chicks. In 1992, 138 chicks hatched, establishing a self-sustaining captive population. In 1993, 285 chicks hatched from 75 females. However, 18% of females never became accustomed to captivity and did not lay eggs. The captive population began as 103 chicks from Pakistan and 129 from the African subspecies *C. u. undulata*, all collected between 1986 and 1988. This study is also discussed in 'Release captive-bred individuals', 'Use artificial insemination in captive breeding' and 'Artificially incubate and hand-rear birds in captivity'.

A review (2) of the same project as in (*1*) found that captive houbara bustards, *Chlamydotis undulata macqueenii*, first reproduced at an earlier age than wild birds, with 2% of females laying at one year old; 23% at two years; 62% at three years and 82% (i.e. all birds that became accustomed to captivity and so would be expected to lay) at four years old. This study is also discussed in 'Use artificial insemination in captive breeding', 'Release captive-bred individuals', 'Release birds in 'coveys', 'Use holding pens at site of release' and 'Use holding pens at site of release and clip birds' wings'.

Another review (3) of the same project found that, between 1992 and 1999, 46% of 2,917 captive-laid eggs successfully hatched. Survival of 1,135 hatchlings to six months old was 75%, to three years, 69% and to ten years, 43%. Hatching failures were mainly caused by infertility and death during incubation, whilst mortality was caused mainly by trauma and disease. This study also describes the impact of artificial insemination and incubation, discussed in 'Artificially incubate and hand-rear birds in captivity'.

A review (4) of the same project found that the number of eggs laid by females ranged from approximately two (for five-year-old first-time breeders) to 8.5 (for four-year-old females that had bred before). The number of eggs laid increased with breeding experience and, although no comparisons were made to productivity in wild bustards in this study, other studies suggest that wild females normally lay between one and four eggs.

(1) Seddon, P. J., Jalme, M. S., Van Heezik, Y., Paillat, P., Gaucher, P. & Combreau, O. (1995) Restoration of houbara bustard populations in Saudi Arabia: developments and future directions, *Oryx*, 29, 136–142.
(2) Jaime, M. S., Combreau, O., Seddon, P. J., Paillat, P., Gaudier, P. & Heezik, Y. (1996) Restoration of *Chlamydotis undulata macqueenii* (houbara bustard) populations in Saudi Arabia: A progress report, *Restoration Ecology*, 4, 81–87.
(3) van Heezik, Y. & Ostrowski, S. (2001) Conservation breeding for reintroductions: assessing survival in a captive flock of houbara bustards, *Animal Conservation*, 4, 195–201.
(4) van Heezik, Y., Jalme, M. S., Hemon, S. & Seddon, P. (2002) Temperature and egg-laying experience influence breeding performance of captive female houbara bustards, *Journal of Avian Biology*, 33, 63–70.

16.1.6. Storks and ibises

- We captured a small study and a review (1,2) describing the captive breeding of storks (Ciconiidae) and a study and a review (3,4) describing the breeding of northern bald ibis, *Geronticus eremita*.
- Both studies on storks were from the USA. The small study found that a pair bred (1); the review found that only seven of 19 species had been successfully bred in captivity (2).
- A review of bald ibis conservation (4) found that 1,150 birds had been produced in captivity from 150 founders over 20 years. However, some projects had failed, and a study from Turkey (3) found that captive birds had lower productivity than wild birds.

A small study at the Audubon Park Zoo, New Orleans, USA, in 1983 (1) found that a pair of Abdim's storks, *Ciconia abdimii*, successfully bred in captivity. However, they did not re-nest following the removal of two eggs for artificial incubation. This study is also discussed in 'Artificially incubate and hand-rear birds in captivity'.

A 1987 review of the captive-breeding of storks (2) found that only seven species had been bred in captivity, and many of these only on a few occasions. These seven were: wood stork, *Mycteria americana*; yellow-billed stork, *M. ibis*; painted stork, *M. leucocephala*; black stork, *Ciconia nigra*; Abdim's stork, *C. abdimii*; white stork, *C. ciconia*; and marabou stork, *Leptoptilos crumeniferus*.

A study in south-east Turkey in 1977–88 (3) found that efforts to breed northern bald ibis (waldrapp) *Geronticus eremita* in captivity were not very successful. From a captive population of between 11 and 45 individuals (with 41 taken from the wild over the study period), a maximum of 19 young/year were produced, with a total of 90 over the study period. This was equivalent to 1.45 young/nest, lower than in wild birds, possibly due to high levels of pre-fledging mortality due overcrowding in the cages. Forty-six young were healthy and 67 individuals were released. Post-release survival is discussed in 'Release captive-bred individuals' and this study is also described in 'Provide artificial nesting sites'.

A 2007 review of northern bald ibis (waldrapp), *Geronticus eremita*, conservation (4) found that three lineages of birds in North America, Japan and Europe, comprising a total of

1,150 birds were produced from an original 150 birds taken from the wild in Morocco in 1988. About 800 additional birds are also thought to be present in captivity but are not registered. However, a programme at a group of zoos, headed by Munich Zoo, failed to establish a productive captive colony for three years. The authors note that ibises frequently swallow small objects they find including nails and wire, which has led to many birds dying from a punctured gut. This study is also discussed in 'Release captive-bred individuals into the wild to restore or augment wild populations', 'Artificially incubate and hand-rear birds in captivity', 'Use holding pens at release sites', 'Release birds as adults or sub-adults, not juveniles', 'Clip birds' wings on release', 'Use microlites to help birds migrate' and 'Foster birds with non-conspecifics'.

(1) Farnell, G. & Shannon, P. W. (1987) The breeding of Abdim's storks at the Audubon Park Zoo. *Colonial Waterbirds*, 10, 251–254.
(2) Johnson, R. E., Coulter, M. C., Luthin, C. S., King, C. E. & Valenzuela, A. J. (1987) Storks: status, conservation and captive breeding. *Colonial Waterbirds*, 10, 236–241.
(3) Akçakaya, R. (1990) Bald ibis *Geronticus eremita* population in Turkey: an evaluation of the captive breeding project for reintroduction. *Biological Conservation*, 51, 225–237.
(4) Bowden, C. G. R., Boehm, C., Jordan, M. J. R. & Smith, K. W. (2007) Why is reintroduction of northern bald ibis *Geronticus eremita* so complicated? An overview of recent progress and potential. 27–35 in: M. M. Lamont (eds) *The Proceedings of the IV International Symposium on Breeding Birds in Captivity; 2007 Sept 12–Sept 16* Toronto, Ontario, Canada.

16.1.7. Raptors

- Three small studies (1–3) and a review (4) from around the world all found that raptors bred successfully in captivity.
- Two small studies on *Accipiter* spp. (1,2) found that wild-caught birds bred in captivity after a few years, with one pair producing 15 young over four years (2), whilst a study on bald eagle, *Haliaeetus leucocephalus*, captive breeding (3) found low fertility in captive-bred eggs, but that birds still produced chicks after a year or so together.
- A review of Mauritius kestrel, *Falco punctatus*, captive breeding (4) found that 139 independent young were raised over 12 years from 30 eggs and chicks taken from the wild (of which 13 survived).
- An update of the same programme (5) found that hand-reared Mauritius kestrels were less successful if they came from captive-bred eggs, compared to wild 'harvested' eggs.

A small study reports that a pair of black sparrowhawks, *Accipiter melanoleucus*, successfully bred in an aviary in Paris, France, in 1979 (1). The birds were caught in the wild (in Gabon) at four years (female) or three months (male) old in 1971 and 1976 respectively. The pair built a nest in 1977, attempted to breed in 1978 but laid only infertile eggs. In 1979 they laid three eggs, all of which hatched, although only a single chick fledged.

A small study from Canberra, Australia (2), found that a pair of wild-caught brown goshawks, *Accipiter fasciatus*, failed to breed in captivity in 1975 (two years after being caught). However, following transfer to a smaller outdoor cage (from an indoor room with no windows) and the falconry training of the male, the pair did breed successfully each year from 1976–9, including a second brood in 1978. A total of 17 eggs were laid and 15 chicks survived and were released, discussed in 'Release captive bred individuals'.

A small study at a wildlife research centre in Maryland, USA (3), found that, between 1976 and 1980, between one and five pairs of captive bald eagles, *Haliaeetus leucocephalus*, incubated 31 eggs. Of these, 15 were fertile (48%) and 14 hatched (45% of all, 93% of fertile eggs). The one failure was possibly due to a wild bird disturbing the pair and causing them to cease incubation. The main cause of infertility appeared to be lack of matings between pairs, with each pair producing infertile clutches the first year they were paired. Clutches averaged 2.5 eggs for first clutches and 1.9 for second. This study is also discussed in 'Use artificial insemination in captive breeding', 'Artificially incubate and hand-rear birds in captivity', 'Foster birds with wild conspecifics' and 'Release captive-bred individuals'.

A 1993 review (4) details the captive-breeding programme for the Mauritius kestrel, *Falco punctatus*, on Mauritius. Between 1981 and 1986, 28 fertile eggs and two young were removed from the wild, resulting in 13 healthy captive birds, which began breeding in 1984. By 1986–7, more than 30 birds had been reared and by 1993 a total of 618 eggs had been laid, of which 253 were fertile, 164 hatched and 139 produced independent young. The rate of egg fertility in captivity (41%) was lower than that of wild eggs (74% of 265 eggs). Before the release of captive-bred individuals, the wild population had grown from five individuals in 1973 to 31 in 1986. This study is also discussed in 'Artificially incubate and hand-rear birds in captivity', 'Foster chicks or eggs with wild conspecifics' and 'Release captive-bred individuals'.

A 1995 update (5) of the same conservation programme studied in (4), found that hand-rearing Mauritius kestrel, *Falco punctatus*, eggs from birds in captivity was less successful than rearing wild-laid eggs. This study is discussed in more detail in 'Artificially incubate and hand-rear birds in captivity'.

(1) Brosset, A. (1981) Breeding the black sparrowhawk *Accipiter melanoleucus* in captivity. *Raptor Research*, 15, 58–64.
(2) Olsen, J. & Olsen, P. (1981) Natural breeding of *Accipiter fasciatus* in captivity. *Raptor Research*, 15, 53–57.
(3) Wiemeyer, S. N. (1981) Captive propagation of bald eagles at Patuxent Wildlife Research Center and introductions into the wild, 1976–80. *Raptor Research*, 15, 68–82.
(4) Cade, T. J. & Jones, C. G. (1993) Progress in restoration of the Mauritius kestrel. *Conservation Biology*, 7, 169–175.
(5) Jones, C. G., Heck, W., Lewis, R. E., Mungroo, Y., Slade, G. & Cade, T. (1995) The restoration of the Mauritius kestrel *Falco punctatus* population. *Ibis*, 137, S173–S180.

16.1.8. Pigeons

- A review of a captive-breeding programme on Mauritius and in the UK (1) found that 42 pink pigeons, *Nesoenas mayeri*, were successfully bred in captivity.

A review of a captive-breeding programme on Mauritius (1) found that at least 40 pink pigeons, *Nesoenas mayeri* (formerly *Columba mayeri*), were successfully bred in captivity, fostered under African collared doves, *Streptopelia roseogrisea*, (formerly *S. risoria*) and released. A further two birds were raised at the Jersey Wildlife Preservation Trust (Jersey, UK) and released. This study is also discussed in 'Provide supplementary food to increase adult survival', 'Release captive bred individuals', 'Provide supplementary food after release' and 'Predator control on islands'.

(1) Jones, C., Swinnerton, K., Taylor, C. & Mungroo, Y. (1992) The release of captive-bred pink pigeons *Columba mayeri* in native forest on Mauritius. A progress report July 1987–June 1992. *Dodo*, 28, 92–125.

16.1.9. Songbirds

- A replicated study from Australia (2) and two small studies from the USA (1,3) found that three species of songbird were successfully bred in captivity.
- Four out of five pairs of wild-bred, hand-reared puaiohi, *Myadestes palmeri*, formed pairs and laid a total of 39 eggs in 1998 (3) and a breeding population of helmeted honeyeaters, *Lichenostomus melanops cassidix*, was successfully established through a breeding programme (2).
- Only one pair of loggerhead shrikes, *Lanius ludovicianus*, formed pairs from eight wild birds caught (1) and their first clutch died.

A small study in New York, USA (1), reported that a pair of wild-caught loggerhead shrikes *Lanius ludovicianus* (from a total of eight birds caught and hand-reared from the age of 8–9 days in 1970) formed a pair-bond and laid seven eggs, all of which hatched. However, none of the young survived. A replacement clutch of three eggs was laid, of which two survived and were hand-reared (discussed in 'Artificially incubate and hand-rear birds in captivity')

A replicated study reviewing a helmeted honeyeater, *Lichenostomus melanops cassidix*, captive breeding programme from 1989–1991 in Victoria, Australia (2), found that in total, 25 honeyeaters were successfully reared in captivity to establish a founding breeding population. The cost of removing eggs from wild populations and using foster parents to hatch and rear them was calculated as a quarter of the cost of removing nestlings. Thirteen birds died, three when nestlings were less than ten days old, with most deaths associated with respiratory diseases. Birds were kept in a complex of aviaries configured to simulate natural communities. The temperature in the hand-rearing area was set at 28°C initially and gradually reduced to ambient temperature by the time chicks were 40 days old. Mature birds were given an artificial diet supplemented with live insects to satisfy both the birds' nutritional requirements and allow them to learn foraging techniques.

A small study at a breeding centre on Kauai, Hawaii, USA (3), found that four out of five pairs of hand-reared puaiohi, *Myadestes palmeri*, (a critically endangered thrush) successfully formed pairs and laid a total of 39 eggs in 1998. Birds were taken as eggs from wild nests and artificially incubated (see 'Artificially incubate and hand-rear birds in captivity'). Releases are discussed in 'Release captive-bred individuals'.

(1) Cade, T. J. (1992) Hand-reared loggerhead shrikes breed in captivity. *The Condor*, 94, 1027–1029.
(2) Smales, I., Miller, M., Middleton, D. & Franklin, D. (1992) Establishment of a captive-breeding programme for the helmeted honeyeater *Lichenostomus melanops cassidix*. *International Zoo Yearbook*, 31, 57–63.
(3) Kuehler, C., Lieberman, A., Oesterle, P., Powers, T., Kuhn, M., Kuhn, J., Nelson, J., Snetsinger, T., Herrmann, C., Harrity, P., Tweed, E., Fancy, S., Woodworth, B. & Telfer, T. (2000) Development of restoration techniques for Hawaiian thrushes: Collection of wild eggs, artificial incubation, hand-rearing, captive-breeding, and re-introduction to the wild. *Zoo Biology*, 19, 263–277.

16.2. Can captive breeding have deleterious effects on individual fitness?

- Three studies of wild populations (*1*), wild and captive populations (*3*) and museum specimens (*2*), one replicated, found evidence for potentially deleterious physiological or genetic changes due to captive breeding. These studies did not investigate fitness.
- A study of a wild Mauritius kestrel, *Falco punctatus*, (*1*) population derived totally from captive individuals found high inbreeding and a loss of genetic diversity, but this was caused more by the very low population size (four wild birds) than by captivity *per se*.
- The museum-based study found reduced relative brain volume in captive wildfowl, compared with wild birds (*2*), whilst a comparison of wild and captive populations of white-headed ducks *Oxyura leucocephala* (*3*) found lower genetic diversity in captive populations.

Background

One potential problem with captive breeding is that it limits the effective population of a species to the number of captive individuals that are available to interbreed. This could lead to inbreeding depression and a loss of fitness. Similarly, the different selection pressures on individuals in captivity may alter the physiology of captive populations, as has occurred in domesticated species.

It is important to note that low genetic diversity, inbreeding and changes due to captivity do not necessarily lead to reduced fitness. Care should therefore be taken in interpreting the results from the studies below.

A study of a population of Mauritius kestrels *Falco punctatus* in eastern Mauritius (*1*) found that the population had grown from 12 individuals in 1987 to a minimum of 154 by 2002. Over this time, the degree of inbreeding increased by 2.6% each generation and by 2002, 25% of pairs were either closely or moderately related. Over this period 1.6% of genetic diversity was lost each generation. Effects on reproduction or survival were not monitored.

A replicated study of brain volume in 21 species of wildfowl from museum collections (2) found that brain volume was lower in captive populations for 16 of the species, with an average decrease of 4.7% (range of 1–33%). Relative brain volume (brain volume in relation to other body measurements) was also lower in captive populations for 20 of the species, with an average reduction of 7.7% (range of 2–30%). A total of 268 skeletons were examined, at least one member of each sex was examined for each species. The effect of these decreases on behaviour, survival or reproduction is not known.

A controlled 2008 study of genetic diversity in white-headed ducks *Oxyura leucocephala* (3) found that two captive-bred populations had significantly lower genetic diversity than wild birds from Greece and Spain. A total of 38 captive-bred birds were tested, (27 from a Spanish collection and 11 from a UK collection) and compared with 70 wild birds collected between 1993 and 2003 (63 from Spain, seven from Greece). The captive Spanish birds descended from eight wild birds caught in Spain in 1982, the UK birds from three wild birds caught in Pakistan in 1968. Both microsatellite and mitochondrial DNA were less diverse in captive populations. Effects on reproduction or survival were not monitored.

(1) Ewing, S. R., Nager, R. G., Nicoll, M. A. C., Aumjaud, A., Jones, C. G. & Keller, L. F. (2008) Inbreeding and loss of genetic variation in a reintroduced population of Mauritius kestrel. *Conservation Biology*, 22, 395–404.
(2) Guay, P.-J. & Iwaniuk, A. N. (2008) Captive breeding reduces brain volume in waterfowl (Anseriformes). *The Condor*, 110, 276–284.
(3) Muñoz-Fuentes, V., Green, A. J. & Sorenson, M. D. (2008) Comparing the genetics of wild and captive populations of white-headed ducks *Oxyura leucocephala*: consequences for recovery programmes. *Ibis*, 150, 807–815.

16.3. Use artificial insemination in captive breeding

- A review of artificial insemination (5) argued that it could be a useful tool to conservationists, but that there were challenges to its use. Deep and repeated inseminations increased fertility.
- Two trials from the USA (1,2) found that artificial insemination of raptors achieved approximately 50% fertility (1) or 0% (2).
- A review of a houbara bustard *Chlamydotis undulata macqueenii* captive breeding programme in Saudi Arabia found that artificial insemination increased fertility (3), whilst another review (4) found that the highest fertility levels were achieved with inseminations of at least 10 million spermatozoa every 4–5 days.

Background

Artificial insemination involves artificially introducing sperm into a female's reproductive tract to fertilise her eggs. It may have advantages over natural insemination through mating if, for example, captive birds will not form pair bonds, or if males in some pairs are infertile. The ability to transport frozen sperm between geographically isolated populations can also substantially increase the genetic diversity of species. However, the methods used may be very species-specific and research may be needed before it can be adopted for any one species.

A replicated *ex situ* study in the USA in 1973–5 (1) found that five prairie falcons *Falco mexicanus* that were artificially inseminated laid a total of 37 eggs (although none in 1973), of which 19 were fertile and 15 hatched. Twelve chicks eventually fledged. Sperm was taken from a single male and either used immediately or refrigerated and used within ten hours.

A small study in a breeding centre in Maryland, USA, in 1980 (2) found that an artificially inseminated, wild-bred, female bald eagle *Haliaeetus leucocephalus* laid two eggs, but both were infertile. The sperm was taken from one wild-bred male and one captive-bred male and mixed and the female was inseminated within an hour. The authors suggested that repeated inseminations would have been preferable, but that the disturbance caused by capturing and inseminating the female risked disturbing other breeding birds nearby. This study is also discussed in 'Use captive breeding to increase or maintain populations', 'Release captive-bred individuals' and 'Foster birds with wild conspecifics'.

A review of a houbara bustard *Chlamydotis undulata macqueenii* captive breeding programme in Saudi Arabia (3) found that artificial insemination increased fertility rates from 50% to 85% between 1989 and 1993. At least 440 chicks were hatched during the study period. This study is also discussed in 'Use captive breeding to increase or maintain populations', Artificially incubate and hand-rear birds in captivity' and 'Release captive-bred individuals'.

Another review (4) of the same houbara bustard *Chlamydotis undulata macqueenii* programme as (3) found that artificial insemination achieved the highest rates of fertility with inseminations of more than 10 million spermatozoa every 4–5 days. This study is also discussed in 'Use captive breeding to increase or maintain populations', 'Release captive-bred individuals', 'Use holding pens at site of release', 'Release birds in 'coveys' and 'Use holding pens at site of release and clip birds' wings'.

A review in 2009 (5) of the challenges of artificial insemination argued that it has the potential to be an important component of bird conservation and population restoration. Extracting good quality sperm is a major barrier, with the 'abdominal massage technique' being the most practical and widely applicable method (although it is prone to urine contamination). Trained birds are more likely to accept insemination. The authors found that fertility is up to 400% higher when sperm are deposited into the vagina compared to the cloaca. However, this constraint can be overcome by boosting sperm volume and insemination frequency (2–3 times/week), which has resulted in 80% fertility in crane species using cloacal insemination. Multiple, deep inseminations improves fertility and can help to overcome poor semen quality. Even when eggs are mostly fertile, artificial insemination can increase fertility by an additional 5–10%.

(1) Boyd, L. L., Boyd, N. S. & Dobler, F. C. (1977) Reproduction of prairie falcons by artificial insemination. *The Journal of Wildlife Management*, 41, 266–271.
(2) Wiemeyer, S. N. (1981) Captive propagation of bald eagles at Patuxent Wildlife Research Center and introductions into the wild, 1976–80. *Raptor Research*, 15, 68–82.
(3) Seddon, P. J., Jalme, M. S., Van Heezik, Y., Paillat, P., Gaucher, P. & Combreau, O. (1995) Restoration of houbara bustard populations in Saudi Arabia: developments and future directions. *Oryx*, 29, 136–142.
(4) Jaime, M. S., Combreau, O., Seddon, P. J., Paillat, P., Gaudier, P. & Heezik, Y. (1996) Restoration of *Chlamydotis undulata macqueenii* (houbara bustard) populations in Saudi Arabia: A progress report. *Restoration Ecology*, 4, 81–87.
(5) Blanco, J. M., Wildt, D. E., Höfle, U., Voelker, W. & Donoghue, A. M. (2009) Implementing artificial insemination as an effective tool for *ex situ* conservation of endangered avian species. *Theriogenology*, 71, 200–213.

16.4. Freeze semen for use in artificial insemination

- A small controlled trial in the USA (2) found that using frozen semen for artificial insemination resulted in lower fertility in falcons, and a second small trial from the USA (1) found that an American kestrel *Falco sparverius* had only 33% fertility when inseminated with frozen semen.
- A small trial from the USA (3) found that fertility rates were highest when semen contained 10% dimethyl sulphoxide (DMSO, a cryoprotectant), compared to semen containing 6% or 8% DMSO.

Background

Spermatozoa only survive for a short time and so if artificial insemination is not going to happen immediately, the semen must be frozen. This in itself can damage cells, so a cryoprotectant such as dimethyl sulphoxide or glycerol must be added to protect the cells. Some cyroprotectants (e.g. glycerol) must then be removed by dialysis before insemination.

A small *ex situ* study (1) found that a single American kestrel *Falco sparverius* inseminated with frozen peregrine falcon *F. peregrinus* semen (defrosted and dialysed to remove glycerol added before freezing) laid six eggs of which two were fertile and one hatched.

A small, controlled *ex situ* study in New York, USA, in 1986 (2) found that four female prairie falcons *Falco mexicanus* and one American kestrel *F. sparverius* had significantly lower fertility when inseminated with frozen sperm (defrosted and dialysed to remove glycerol added before freezing), compared to two prairie falcons and one kestrel inseminated with fresh semen (25% of 28 eggs for frozen vs. 94% of 17 eggs for fresh semen). Only two prairie falcons and the kestrel produced any fertile eggs from frozen semen. The birds inseminated with fresh sperm were a subset of the larger group and were inseminated with semen from a peregrine falcon *F. peregrinus* and a peregrine-gyrfalcon *F. rusticolus* hybrid. Frozen semen preparation took approximately 90 minutes after thawing, during which it was kept at 0–4°C. Insemination used 80 μl (frozen) or 20–70 μl (fresh) semen.

A small *ex situ* study in 1983–5 (3) found that six American kestrels *Falco sparverius* inseminated with previously-frozen semen produced 14 infertile eggs only when the semen contained 4% dimethyl sulphoxide (DMSO). This compared with 35% of 17 eggs being fertile when 6% DMSO was used; 40% of ten eggs for 8% DMSO and 57% of seven eggs for 10% DMSO. Sperm mobility in semen containing 10% DMSO was lower (44%) than in semen containing 8% or 6% DMSO (61% and 62% respectively). Semen was taken from 15 male kestrels and frozen for between one and 14 months prior to thawing and insemination.

(1) Parks, J. E., Heck, W. R. & Hardaswick, V. (1986) Cryopreservation of peregrine falcon semen and post-thaw dialysis to remove glycerol. *Raptor Research*, 20, 15–20.
(2) Parks, J. E. & Hardaswick, V. (1987) Fertility and hatchability of falcon eggs after insemination with frozen peregrine falcon semen. *Journal of Raptor Research*, 21, 70–72.
(3) Gee, G. F., Morrell, C. A., Franson, J. C. & Pattee, O. H. (1993) Cryopreservation of American kestrel semen with dimethylsulfoxide. *Journal of Raptor Research*, 27, 21–25.

16.5. Wash contaminated semen and use it for artificial insemination

- A single replicated controlled study in Spain (1) found that semen contaminated with urine could be successfully washed to increase its pH and produced three raptor nestlings.

Background

Collecting semen can be difficult and expensive, and if it is contaminated it is likely to be cheaper to wash it, than to collect more.

A replicated controlled study from February–April in a captive breeding programme in Spain (1) found that urine-contaminated sperm can be used to artificially inseminate raptor females after washing the sperm with an alkalinised diluent. Urine contamination of ejaculate samples was high in all 4 species (25 individuals) analysed (37% for golden eagle *Aquila chrysaetos*, 43% for Spanish imperial eagle *Aquila adalberti*, 29% for Bonelli's eagle *A. fasciatus* (also *Hieraaetus fasciatus*) and 48% for peregrine falcon *Falco peregrinus*) and significantly reduced semen pH (6.5–6.9 compared to 7.2–7.6). However, sperm motility was significantly higher in sperm washed with an alkalinised diluent (compared to a neutral diluent). An intramagnal insemination technique of washed semen produced 1 golden eagle and 2 peregrine falcon nestlings (11 and 16% of clutch size respectively). Each sperm sample was divided into two equal amounts and washed with either neutral (pH 7.0) or alkalinised (pH 8.0) diluent (Lake's formula with 300 mg/100 ml of citric acid added) before being incubated for 30 min (21°C).

(1) Blanco, J. M., Gee, G. F., Wildt, D. E. & Donoghue, A. M. (2002) Producing progeny from endangered birds of prey: treatment of urine-contaminated semen and a novel intramagnal insemination approach. *Journal of Zoo and Wildlife Medicine*, 33, 1–7.

16.6. Artificially incubate and hand-rear birds in captivity

Background

Artificial incubation involves removing eggs from incubating parents and using an incubator to hatch them. Techniques can be extremely complex, with precision humidity and temperature control and turning of the eggs to ensure correct development Hand-rearing can be used with chicks from artificially-incubated eggs or with chicks removed from parents after hatching and involves manually feeding chicks until independence. Both techniques can be used to encourage parents to produce more offspring, or when naturally-raised chicks and eggs have low survival.

16.6.1. Seabirds

- Five studies from across the world (1–5) found evidence for the success of hand-rearing seabirds.
- One small study in Spain (2) found that one of five hand-reared Audouin's gulls *Larus audouinii* successfully bred in the wild.
- Four studies (1,2,4,5) found that various petrel species (Procellariiformes) successfully fledged after hand-rearing. One controlled study found that fledging rates of hand-reared birds was similar to parent-reared birds (3), although a study on a single bird (1) found that the chick fledged at a lower weight and later than parent-reared chicks.

A small study on Bermuda in 1971 (1) reported on the successful hand-rearing of a Bermuda petrel chick *Pterodroma cahow*. The chick was abandoned by one parent, causing its development to slow and it was not ready to fledge when it reached the normal age for departure (84 days). It was therefore hand-fed on blended squid and shrimp using a squeezable pipette. The chick reached a lower weight than most petrels, probably due to stunted growth before being hand-reared. It was released successfully but at a greater age than parent-reared birds fledged.

A small study at a captive-breeding centre in eastern Spain (2) found that, of five captive-bred, hand-reared Audouin's gulls *Larus audouinii* released in 1992, one bird returned to the centre and successfully bred every year from 1995–2000. A second bird, released in 1995, returned in 1998 but did not breed. The released bird and its mate moved the location of their nest each year, each time nesting close to a captive pair also breeding. There were no significant differences between clutch size or hatching success of the released and captive pairs (an average of 2.6 eggs/clutch and 53% hatching success for released birds vs. 2.4 eggs/clutch and 67%). This study also describes the captive-breeding efforts, discussed in 'Use captive breeding to increase or maintain populations'.

A replicated, controlled study on Cabbage Tree Island, New South Wales, Australia, in 1995 (3), found that the fledging rate of 30 hand-reared Gould's petrels *Pterodroma leucoptera* moved from their burrows to artificial nests nearby was not significantly different from control (unmoved, parent-fed) birds or from chicks provided with supplementary food (100% of hand-reared chicks fledging vs. 29/30 fed chicks and 29/30 controls). Hand-reared chicks were also significantly heavier than controls, but lighter than supplementary-fed chicks. Hand-rearing consisted of approximately 25 g of food every three days. This study is also discussed in 'Provide artificial nesting sites', 'Translocate individuals' and 'Provide supplementary food to increase reproductive success'.

A replicated study in November–December 1997–9 (4) found that 118 of 239 common diving petrel *Pelecanoides urinatrix* nestlings (49%) were successfully hand-reared after being translocated to Mana Island, New Zealand, from other offshore islands. Chicks were between four and eight weeks old when caught and fed a krill-based paste (also containing calcium and other supplements) with a 12 ml syringe either once (in 1997) or twice (1998–9) a day until fledged. Fledging rates were higher in 1997 (58% of 90 chicks) than 1998 (40% of 100) or 1999 (53% of 49), but these differences were not investigated statistically. Information on translocation success and other interventions are discussed in 'Use vocalisations to attract birds to safe areas', 'Provide artificial nesting sites' and 'Translocate individuals'. This study describes a technique usually used in captivity being used in the wild.

A small study on Bermuda in summer 1997 (5) found that an abandoned Bermuda petrel *Pterodoma cahow* chick was successfully hand-reared from approximately three months old until fledging, 20 days later. The chick was fed on 60–90 cm^3 of blended squid and shrimp in a 2:1 ratio, nutrient tablets and warm water. The chick was allowed outside to exercise its wing muscles for a week before eventual fledging.

(1) Wingate, D. B. (1972) First successful handrearing of an abandoned Bermuda Petrel Chick. *Ibis*, 114, 97–101.
(2) Martínez-Abraín, A., Viedma, C., Ramón, N. & Oro, D. (2001) A note on the potential role of philopatry and conspecific attraction as conservation tools in Audouin's gull *Larus audouinii*. *Bird Conservation International*, 11, 143–147.
(3) Priddel, D. & Carlile, N. (2001) A trial translocation of Gould's petrel (*Pterodroma leucoptera leucoptera*). *Emu*, 101, 79–88.
(4) Miskelly, C. M. & Taylor, G. A. (2004) Establishment of a colony of common diving petrels (*Pelecanoides urinatrix*) by chick transfers and acoustic attraction. *Emu*, 104, 205–211.
(5) Raine, A. F. & Abernethy, K. E. (2006) The hand-rearing of an abandoned Bermuda petrel *Pterodroma cahow* chick from Nonsuch Island, Bermuda. *Conservation Evidence*, 3, 4–5.

16.6.2. Penguins

- Two replicated and controlled studies from South Africa (1,2) found that hand-reared and released African penguins *Spheniscus demersus* had similar survival and breeding success as birds which were not orphaned and hand-reared.

A replicated study in the Western Cape, South Africa, in 1994–9 (1), found that orphaned African penguins *Spheniscus demersus* that were hand-reared and released had similar survival and breeding probabilities as naturally-fledged chicks (11% of 437 hand-reared chicks seen at colonies, 1% breeding, 2% found dead vs. 9%, 1% and 1% of 399 naturally-fledged chicks). Of 507 chicks that were hand-reared, 437 (86%) were successfully released into the wild. Survival of rehabilitated adults is discussed in 'Clean birds following oil spills' in 'Threat: Pollution'.

A controlled and replicated study on Robben and Dassen Islands, South Africa (2), found that the survival to breeding age and breeding success of African penguins *Spheniscus demersus* during 2001–6 were similar for birds that were orphaned in the *Treasure* oil spill in 2000 and hand-reared, compared to birds that were not orphaned and hand-reared (1.6 chicks fledged/pair for 24 pairs with at least one hand-reared bird vs. 1.1 chicks fledged/pair for 227 pairs without hand-reared birds). The authors note that the sample size of hand-reared pairs was too small for statistical tests to determine significance.

(1) Whittington, P. A. (2003) Post-release survival of rehabilitated African penguins. 8–17 in: D. C. Nel, P. A. Whittington (eds) *Rehabilitation of oiled African Penguins: a conservation success story* BirdLife South Africa and Avian Demography Unit, Cape Town.
(2) Barham, P. J., Underhill, L. G., Crawford, R. J. M., Altwegg, R., Mario Leshoro, T., Bolton, D. A., Dyer, B. M. & Upfold, L. (2008) The efficacy of hand-rearing penguin chicks: evidence from African Penguins (*Spheniscus demersus*) orphaned in the Treasure oil spill in 2000. *Bird Conservation International*, 18, 144–152.

16.6.3. Wildfowl

- Two replicated studies in Canada (1) and India (2) found high success rates for hand-rearing buffleheads *Bucephala albeola* and bar-headed geese *Anser indicus* in captivity. Eggs were artificially incubated (1) or incubated under foster parents (2).
- A replicated, controlled study in England (3) found that Hawaiian geese (nene) *Branta sandvicensis* chicks showed less well-adapted behaviours if they were raised without parental contact, compared to chicks raised by parents.

A replicated study in 1964–1965 in two sites in British Columbia, Canada (1) found that bufflehead *Bucephala albeola* eggs can safely be transported and hatched within a portable incubator. In 1964, 11 of 12 eggs transported in the incubator hatched. In 1965, nine of 14 eggs hatched. Egg collections each year were treated similarly (all eggs partially incubated on the nests) and were in transit for 10 hours (across 300 miles) before they were set under bantam hens for final incubation. The temperature varied only 0.3°C and was constant most of the time. The custom-designed incubator consisted of three main elements: the plywood housing, the electrical heating element, and the thermo-switch. A small glass window was installed in the centre of the upper door to observe the eggs and temperature without opening the door.

A replicated study in Kashmir, India (2), found that 25 of 30 bar-headed geese *Anser indicus* eggs transported from the UK to India were successfully raised to independence. One egg failed to hatch and four goslings died (two from choking and two from the disease afflotoxicosin). Eggs were incubated under foster parents and turned twice a day. Chicks were fed on 'chick crumbs' at first and then a mix of maize, wheat, soybean, fishmeal and supplements. Eleven goslings were released into a wetland reserve in Kashmir.

A replicated, controlled *ex situ* study in western England in 1990 (3) found that Hawaiian geese (nene) *Branta sandvicensis* raised by parents were more dominant, more vigilant, more wary of a potential predator (a domestic dog *Canis familiaris*) and integrated into the adult flock sooner than goslings raised with limited or no contact with adults. Gosling growth rate and final body size were not affected by rearing regime. A total of 42 goslings were tested: 12 were reared in an outdoor pen adjacent to a pen with a pair of adult geese, allowing interaction from 16–30 days after hatching; eleven chicks reared in a large outdoor pen with no adults present; nine chicks were raised in three groups, in sight of adults for the first 14 days after hatching; ten chicks in four groups were reared by parents in either a large pen with 90 other geese, or a small pen with just the family present.

(1) Crouch, D. E. & Crouch, L. S. (1966) A portable avian egg incubator. *The Journal of Wildlife Management*, 30, 187–189.
(2) Qadri, S. S. (1987) Attempted introduction of bar-headed goose through a new habitat. *Environmental Conservation*, 14, 264–265.
(3) Marshall, A. P. & Black, J. M. (1992) The effect of rearing experience on subsequent behavioural traits in Hawaiian geese *Branta sandvicensis*: implications for the recovery programme. *Bird Conservation International*, 2, 131–147.

16.6.4. Gamebirds

- A single, replicated study (1) in Finland found that hand-reared grey partridges *Perdix perdix* did not take off to fly as effectively as wild-caught birds, potentially making them more vulnerable to predation from ground predators.

A replicated *ex situ* study in 1993 and 1994 in Finland (1) found that hand-reared grey partridges *Perdix perdix* took flight with a shallower take off angle (average of 31° tested on 12 birds) and climbed more slowly (climbing rate of 1.8 m/s for 11 birds) than wild-caught birds (average 44° for 19 birds tested and 2.7 m/s for 18 birds), potentially making them more vulnerable to predation from ground predators.

(1) Putaala, A., Oksa, J., Rintamaki, H. & Hissa, R. (1997) Effects of hand-rearing and radio transmitters on flight of gray partridge. *The Journal of Wildlife Management*, 61, 1345–1351.

16.6.5. Rails

- A controlled before-and-after study from New Zealand (1) found that post-release survival of hand-reared takahe *Porphyrio hochstetteri (formerly P. mantelli)* was as high as wild-reared birds and that six of ten released females raised chicks.

A controlled before-and-after study between 1991 and 1994 in Fiordland National Park, South Island, New Zealand (1), found that post-release survival of hand-reared takahe

Porphyrio hochstetteri (*formerly P. mantelli*) was at least as high as that of wild-reared birds, despite differences between movements and habitat selection (50–100% survival between one and two years for 31 hand-reared birds vs. 0–100% of 12 wild birds; ranges come from different release years). In addition, survival of wild birds between six months and one year old (before hand-reared birds were released) was lower than for captive birds (100% survival for 62 captive-reared birds vs. 25–100% survival for 24 wild birds), especially in years with particularly low temperatures. Causes of death were varied and often unknown. Survival rates did not differ between male and female hand-reared birds, but females were significantly more likely to form pairs than males (89% of nine females forming pairs vs. 25% of eight males). Six of ten captive-reared females nested in the summer after release and two raised a chick to at least six months old. The authors note that low reproductive success in first-time breeders is common. Eggs were taken from the wild and hand-reared using a glove puppet, whilst being played calls of brooding adults (see 'Use puppets to increase the survival or growth of hand-reared chicks' for more studies of this intervention). They were moved to outdoor pens at three months old, to larger pens at six months and finally released at approximately one year old.

(1) Maxwell, J. M. & Jamieson, I. G. (1997) Survival and recruitment of captive-reared and wild-reared takahe in Fiordland, New Zealand. *Conservation Biology*, 11, 683–691.

16.6.6. Cranes

- A replicated and controlled study (2) and a small study (1), both from the USA, found that hand-reared birds showed normal reproductive behaviour (1) and higher survival than parent-reared birds (2).

In order to test hand-rearing techniques for use with whooping cranes *Grus americana*, a small study in the USA in 1992–3 (1) investigated the behaviour of hand-reared male greater sandhill cranes *G. canadensis tabida* after release and found that they exhibited normal reproductive behaviour. All six paired with females in 1992 (none nested); four pairs nested in 1993, with one nest flooding but the others producing one or two eggs each. The hand-reared males incubated the eggs and three hatched (the remaining nest with two eggs was abandoned following the researchers' visit), although none of the chicks survived more than a week. The authors conclude that reproductive behaviour is not affected by hand-rearing, which consisted of 'isolation rearing' – with the birds not given any access to humans, but instead reared by puppets heads (to avoid imprinting on human carers, see 'Use puppets to increase the survival or growth of hand-reared chicks' for studies on this intervention).

A replicated, controlled study in a breeding centre in Mississippi, USA, between 1989 and 1996 (2) found that first-year survival of captive-bred, hand-reared Mississippi sandhill cranes *Grus canadensis pulla* was higher than for captive-bred, parent-reared birds (approximately 85% survival for 56 hand-reared birds vs. 77% for 76 parent-reared birds). Hand-reared birds were reared with mounts of brooding adults and heat lamps, and were taught to feed by costumed humans with mounts of crane (see 'Use puppets to increase the survival or growth of hand-reared chicks' for more information). Details of the releases are discussed in 'Release captive-bred individuals'.

(1) Duan, W. & Bookhout, T. A. (1997) Breeding behavior of isolation-reared sandhill cranes. *Journal of Field Ornithology*, 68, 200–207.
(2) Ellis, D. H., Gee, G. F., Hereford, S. G., Olsen, G. H., Chisolm, T. D., Nicolich, J. M., Sullivan, K. A., Thomas, N. J., Nagendran, M. & Hatfield, J. S. (2000) Post-release survival of hand-reared and parent-reared Mississippi sandhill cranes. *The Condor*, 102, 104–112.

16.6.7. Bustards

- A review of a houbara bustard *Chlamydotis undulata macqueenii* captive breeding programme in Saudi Arabia (2) found that there was no difference in survival between artificially and parentally incubated eggs.

- A second review of the same programme found that removing eggs from clutches as they were laid increased the number laid by females (1).

A review of a houbara bustard *Chlamydotis undulata macqueenii* captive breeding programme in Saudi Arabia (1) between 1989 and 1993 found that removing eggs from females to artificially incubate them increased the number of eggs produced from one to four eggs/year for wild birds to nearly nine eggs/female. This study is also discussed in 'Use captive breeding to increase or maintain populations', 'Use artificial insemination in captive breeding' and 'Release captive-bred individuals'.

A review (2) of the same programme as (1) found that, between 1992 and 1999, there was no significant difference in survival between artificially incubated eggs and those hatched by parental incubation, once breeding experience of mothers, year of lay and the cohort of birds that the mother came from were taken into account. A total of 1,012 eggs were studied. This study is also discussed in 'Use captive breeding to increase or maintain populations'.

(1) Seddon, P. J., Jalme, M. S., Van Heezik, Y., Paillat, P., Gaucher, P. & Combreau, O. (1995) Restoration of houbara bustard populations in Saudi Arabia: developments and future directions. *Oryx*, 29, 136–142.
(2) van Heezik, Y. & Ostrowski, S. (2001) Conservation breeding for reintroductions: assessing survival in a captive flock of houbara bustards. *Animal Conservation*, 4, 195–201.

16.6.8. Waders

- Three out of four replicated and controlled studies from the USA (1,2) and New Zealand (3,4) found that artificially incubated and/or hand-reared waders had higher hatching and fledging success than controls.
- One study from New Zealand (4) found that hatching success of black stilt *Himantopus novaezelandiae* was lower for artificially-incubated eggs.

A replicated, controlled trial on coastal habitats in California, USA, in 1986 (1), found that 79% of 28 artificially incubated Kentish (snowy) plover *Charadrius alexandrinus* eggs, taken from the wild, hatched (excluding 16 that were thought to be dead or infertile). This compared with 92% of 185 parent-incubated eggs, or 43% including those destroyed by humans or natural causes. Fledging rates were 79% and 38% for hand-reared and wild-reared chicks respectively. In 1987, eight hand-reared birds' breeding attempts were monitored, alongside 16 wild-reared birds. Hand-reared birds laid eggs later than wild-reared, although this difference was not significant. Hand-reared females tended to nest in less productive areas than wild-reared birds and produced fewer young (two of eight hand-reared birds produced young vs. six of sixteen wild-reared), although this difference was not significant. Artificial incubation consisted of a 37.6°C incubator with 80–85% humidity, with chicks then fed on tubifex worms, supplements, krill and crickets. Birds were released at between 41 and 72 days old.

A replicated and controlled experiment on two islands in Lake Michigan, USA, in 1989 (2) found that artificially incubated killdeer *Charadrius vociferus* eggs had significantly higher hatching and fledging success than either wild-reared or cross-fostered (raised by spotted sandpipers *Actitis macularia*) eggs and chicks (82% hatching success, 2.3 fledglings/pair and 78% fledging success for captive-reared chicks from six broods vs. 47%, 0.8 fledglings/pair and 48% for cross-fostered chicks from 16 broods, and 54%, 0.6 fledglings/pair and 27% for parent-reared chicks for 24 broods). Eggs were removed during the first week of incubation where possible and incubated at 39°C, transferred to a 35°C box and then released into outdoor pens. Killdeer chicks are precocial and so fed themselves from tubifex worms, mealworms, earthworms and cat food spread in water over sand (to simulate natural feeding conditions). They also fed on insects attracted to heat lamps. Chicks were released into the wild at 35 days-old and no behavioural differences were observed after release. No parent-reared chicks were seen in following years, whereas one captive-reared chick was seen the next year.

A replicated controlled study in South Island, New Zealand (3), investigated the survival of black stilts *Himantopus novaezelandiae*, fostered by both conspecifics and black-winged stilts *H. himantopus*. The eggs were removed from nests as soon as possible after laying. They were then incubated under a 37.2°C dry bulb and 28.9°C wet bulb until 2–3 days before hatching, when the wet bulb was increased to 32.2°C. Hatching eggs were then returned to the wild. Fledging success of managed nests was at least ten times that reported from unmanaged nests (13–27 chicks fledging in the population each year, a 20–42% fledging rate vs. 2% reported in another study for unmanaged nests).

In New Zealand, captive breeding is a major conservation intervention for the black stilt *Himantopus novaezelandiae*. From 1981 to 2003, 1,879 eggs were collected from wild and captive pairs, artificially incubated and most chicks hand-reared until release. Analysis was undertaken to access factors that might influence rearing success (4). Hatching success was 78% for captive-laid and 91% for wild-laid eggs. Most egg mortality occurred early on and around hatching, but timing of death was similar regardless of whether captive or wild, hybrid or pure black stilt, or when eggs were laid. Heavier hatchlings, and chicks from wild parents, had higher initial survival. Chick survival at 10 months of age was 82% regardless of egg origin. Survival of chicks subjected to major health interventions was 69% after 4 months. Survival of birds subjected to minor health interventions was as healthy chicks (82%).

(1) Page, G. W., Quinn, P. L. & Warriner, J. C. (1989) Comparison of the breeding of hand- and wild-reared snowy plovers. *Conservation Biology*, 3, 198–201.
(2) Powell, A. N. & Cuthbert, F. J. (1993) Augmenting small populations of plovers: an assessment of cross-fostering and captive-rearing. *Conservation Biology*, 7, 160–168.
(3) Reed, C. E. M., Ron J. Nilsson & Murray, D. P. (1993) Cross-fostering New Zealand's black stilt. *The Journal of Wildlife Management*, 57, 608–611.
(4) van Heezik, Y., Lei, P., Maloney, R. & Sancha, E. (2005) Captive breeding for reintroduction: influence of management practices and biological factors on survival of captive kaki (black stilt). *Zoo Biology*, 24, 459–474.

16.6.9. Storks and ibises

- A small study in the USA (1) describes the successful artificial incubation and hand-rearing of two Abdim's stork *Ciconia abdimii* chicks, whilst a review of northern bald ibis *Geronticus eremita* conservation (2) found that only very intensive rearing of a small number of chicks appeared to allow strong bonds to form between chicks – thought to be important for the successful release of birds into the wild.

A small study the Audubon Park Zoo, New Orleans, USA, in 1983 (1) found that a pair of Abdim's storks *Ciconia abdimii* successfully bred in captivity (see 'Use captive breeding to increase or maintain populations'), producing two eggs which were artificially incubated and hand-reared. The two chicks successfully integrated with the captive population and displayed normal behaviours. The eggs were incubated in a forced-air incubator at 36.9°C, moved to a 34°C brooder after hatching, with the temperature gradually reduced to 26°C by the time chicks were four weeks old. Hand-rearing consisted of seven feeds a day until four weeks old, when they were fed three times a day and then once a day from seven weeks old. Food consisted of commercial bird-of-prey food, fish, insects and yoghurt.

A 2007 review of northern bald ibis (waldrapp) *Geronticus eremita* conservation (2) found that intensive hand-rearing of ibis chicks by a small number of human foster-parents appeared to lead to the formation of strong bonds between chicks which appear important in successful releases of the species. This study is also discussed in 'Use captive breeding to increase or maintain populations', 'Release captive-bred individuals into the wild to restore or augment wild populations', 'Use holding pens at release sites', 'Release birds as adults or sub-adults, not juveniles', 'Clip birds' wings on release', 'Use microlites to help birds migrate' and 'Foster birds with non-conspecifics'.

(1) Farnell, G. & Shannon, P. W. (1987) The breeding of Abdim's storks at the Audubon Park Zoo. *Colonial Waterbirds*, 10, 251–254.

(2) Bowden, C. G. R., Boehm, C., Jordan, M. J. R. & Smith, K. W. (2007) Why is reintroduction of northern bald ibis *Geronticus eremita* so complicated? An overview of recent progress and potential. 27–35 in: M. M. Lamont (eds) *The Proceedings of the IV International Symposium on Breeding Birds in Captivity; 2007 Sept 12–Sept 16* Toronto, Ontario, Canada.

16.6.10. Vultures

- A study in Peru (1) found that hand-reared Andean condors *Vultur gryphus* had similar survival to parent-reared birds after release into the wild.

A study using Andean condors *Vultur gryphus* in arid mountains in Peru in 1980–1 to develop release techniques for Californian condors *Gymnogyps californianus* (1) found that there was no difference in survival between hand-reared birds released at natural fledging age (approximately six months old) and parent-reared birds released at between one and three years old (three of five hand-reared birds alive 18 months after release vs. four of six parent-reared birds). All mortalities occurred in the first six months after release. Hand-reared birds were fed using puppets heads (to avoid imprinting on human carers, see 'Use puppets to increase the survival or growth of hand-reared chicks' for studies on this intervention). Puppet-reared birds were kept in aviaries at the release site for five months before release, parent-reared birds were kept for seven weeks. After release, parent-reared birds integrated into wild populations faster than puppet-reared birds, and their foraging area increased to approximately 1,300 km^2 after 170 days, puppet-reared birds took approximately 320 days to increase foraging area to this size. The authors suggest that they were able to manipulate the foraging behaviour (discussed in 'Provide supplementary food after release') of puppet-reared birds more effectively than parent-reared birds. This study is also discussed in 'Release captive-bred individuals'.

(1) Wallace, M. P., and Temple, S. A. (1987) Releasing captive-reared Andean condors to the wild. *The Journal of Wildlife Management*, 51, 541–550.

16.6.11. Raptors

- Six studies from across the world (1,2,4,5,7,10) found high success rates for artificial incubation (1,2,4,7) and hand-rearing (2,5,7,10) of raptors.
- A replicated and controlled study from France (9) found that artificially incubated raptor eggs had significantly lower hatching success than parent-incubated eggs. This study found that fledging success for hand-reared chicks was similar to wild chicks, whilst a replicated and controlled study from Canada (3) found that hand-reared chicks had slower growth and attained a lower weight than parent-reared birds.
- A replicated study from Mauritius (8) found that hand-rearing of wild eggs had higher success than hand-rearing captive-bred chicks.
- Three studies that provided methodological comparisons found that American kestrel *Falco sparverius* eggs were more likely to hatch at 38.5°C, compared to 36°C or 40°C (1), that peregrine falcon *F. peregrinus* eggs should be incubated over 37°C (4) and that falcon chicks gained far more weight when saline was added to their diet (6).

A replicated study in a breeding centre in New York, USA, in spring 1970 (1) found that artificially incubated American kestrels *Falco sparverius* eggs were more likely to hatch when incubated at 38.5°C (100% of 11 eggs hatching), than at 36°C (34% of 13) or 40°C (25% of 12). Whether the eggs were cooled to 21°C twice daily or not did not affect hatching success (61% of 18 cooled eggs hatching vs. 44% of 18 non-cooled eggs). Sixteen of the 19 hatched chicks were raised to fledging on a diet of minced meat. Eggs had been naturally incubated for 2–26 days before being taken from the wild.

A replicated study in a breeding centre in Maryland, USA, in 1978–80 (2) found that, of 16 bald eagle *Haliaeetus leucocephalus* eggs removed from captive breeding pairs (in a total of 11 clutches) and artificially incubated, 11 (69%) hatched. All 11 chicks were successfully raised until they were transferred to foster nests (see 'Foster birds with wild conspecifics'). They were incubated at 56% humidity under one 37.6°C ('dry') bulb and one 30°C ('wet') bulb and turned every two hours. Once hatched, the chicks were fed on chopped meat and fish and provided with vitamin and calcium carbonate supplements. A further five eggs were removed from eagle nests with poor reproductive histories and artificially incubated. Three of these hatched and two of the eaglets survived hand-rearing to be fostered by more successful wild pairs. This programme is also discussed in 'Use artificial insemination in captive breeding', 'Use captive breeding to increase or maintain populations' and 'Release captive-bred individuals'.

A replicated and controlled *ex situ* study in a research centre in Quebec, Canada (3), found that 25 hand-reared American kestrels *Falco sparverius* grew more slowly than 19 parent-reared birds, also in captivity. Parent-reared birds also achieved greater body sizes than hand-reared birds (predicted weight of hand-reared birds of 119–130 g vs. 133–138 for parent-reared). Hand-reared birds were fed until sated four times a day, whilst parent birds were provided with food in excess. Survival and reproductive output were not measured in this study.

A replicated study in a breeding facility in Colorado, USA (4), found that for peregrine falcon *Falco peregrinus*, artificial incubation led to the hatching of 83% of approximately 300 captive-laid eggs and over 90% of 100 wild-obtained eggs incubated between 1978 and 1980. Eggs were incubated at between 37.2°C and 37.8°C and 60% humidity and were turned every 30 minutes. Eggs were 'treated individually', with weight loss being calculated and, if losing weight too rapidly, eggs were partially coated with paraffin. If losing weight slower than expected, shells were sanded very carefully above the air cell. Hatching success was approximately 20% higher if eggs received five days of natural incubation before being placed in incubators. Symptoms of low incubation temperatures (physical deformities and abnormalities) were found if eggs were incubated at below 37°C.

A small study in wetlands in the Doñana National Park, southern Spain, in summer 1984 (5), found that an orphaned fledgling Spanish imperial eagle *Aquila adalberti* (91 days old) increased in weight by 1,250 g in nine days in captivity (growing from 2,300 g to 3,550 g). The chick was then successfully released and 'adopted' by a foster pair, discussed in 'Foster eggs or chicks with wild conspecifics'.

A small study in a breeding centre in New York, USA, in 1977 and 1980 (6) found that hand-reared chicks from five pairs of peregrine falcons *Falco peregrinus* and one pair of gyrfalcons *F. rusticolus* were twice as heavy at ten days old in 1980, when saline solution was added to their food, compared with 1977, when no saline was added. This was a significant difference, despite the small sample size. The authors note that food supply was not strictly controlled and no monitoring of chick health and survival was performed after day ten, therefore it is not certain whether the saline caused the increase in weight. Chicks were hand-raised on ground common quail *Coturnix coturnix* with or without 0.9% saline added. The amount of saline was not measured, but it increased the water content of the food to 73% from 68%.

A 1993 replicated study in Mauritius (7) found that, of 265 Mauritius kestrel *Falco punctatus* eggs removed from wild nests and artificially incubated, 195 (74%) were fertile (higher than captive-bred eggs), 156 (80% of fertile eggs) hatched and 147 (94% of hatched eggs) were successfully hand-reared or fostered to other birds. This study is also discussed in 'Use captive breeding to increase or maintain populations', 'Foster chicks or eggs with wild conspecifics' and 'Release captive-bred individuals'.

A 1995 update (8) of the same conservation programme studied in (7), found that hand-rearing young Mauritius kestrels *Falco punctatus* hatched from harvested eggs was significantly more successful than rearing eggs laid in captivity (96% and 80% hatching rate respectively). Of 292 fertile eggs, 83% were hatched artificially. Some chicks were retained for the captive breeding programme while the rest were released in areas outside the range of the original population (see 'Release captive-bred individuals into the wild to restore or augment wild populations').

A replicated controlled study in cereal fields in western France in 1995–6 (9) found that a programme to rescue Montagu's harrier *Circus pygargus* eggs and chicks from nests in fields about to be harvested resulted in the release of 129 birds into the wild. The hatching success of 54 artificially incubated eggs was 62%: significantly lower than that of 322 naturally-incubated, wild eggs (hatching success not provided). However, fledging success of 33 hand-reared chicks was 64%, comparable to that of 313 wild chicks (wild fledging rates not provided). Among released birds, those that spent longer in captivity had shorter periods of dependence on the food provided and were in better condition. Captive-reared birds were re-observed more frequently following release (16–21% of 129 released birds re-observed vs. 8–9% of 181 naturally-fledged young), although the authors warn that this could be an artefact of lower dispersal in captive-reared birds. Eggs and chicks up to 15 days old were removed from at risk nests and released in groups of between one and eleven at two sites. Food (chicks and mice) was provided in excess during rearing and then at the release sites.

A replicated study in pine forests in Slovakia in summers between 1993 and 2000 (10) found that golden eagle *Aquila chrysaetos* chicks removed from nests with two chicks, hand-reared or fostered in captivity and then fostered by wild conspecifics were successfully raised 74% of the time (of 35 fostering attempts). This study is discussed in detail in 'Foster eggs or chicks with wild conspecifics'.

(1) Snelling, J. C. (1972) Artificial incubation of sparrowhawk eggs. *The Journal of Wildlife Management*, 36, 1299–1304.

(2) Wiemeyer, S. N. (1981) Captive propagation of bald eagles at Patuxent Wildlife Research Center and introductions into the wild, 1976–80. *Raptor Research*, 15, 68–82.

(3) Bird, D. M. & Clark, R. G. (1983) Growth of body components in parent- and hand-reared captive kestrels. *Raptor Research*, 17, 77–84.

(4) Burnham, W. (1983) Artificial incubation of falcon eggs. *The Journal of Wildlife Management*, 47, 158–168.

(5) Gonzalez, J. L., Heredia, B., Gonzalez, L. M. & Alonso, N. (1986) Adoption of a juvenile by breeding Spanish imperial eagles during the postfledging period. *Raptor Research*, 20, 77–78.

(6) Oliphant, L. W. (1988) Effect of saline added to food on weight gain of hand-raised falcons. *Journal of Raptor Research*, 22, 119–120.

(7) Cade, T. J. & Jones, C. G. (1993) Progress in restoration of the Mauritius kestrel. *Conservation Biology*, 7, 169–175.

(8) Jones, C. G., Heck, W., Lewis, R. E., Mungroo, Y., Slade, G. & Cade, T. (1995) The restoration of the Mauritius kestrel *Falco punctatus* population. *Ibis*, 137, S173–S180.

(9) Amar, A., Arroyo, B. E. & Bretagnolle, V. (2000) Post-fledging dependence and dispersal in hacked and wild Montagu's Harriers *Circus pygargus*. *Ibis*, 142, 21–28.

(10) Kornan, M., Majda, M., Macek, M. & Kornan, J. (2003) An unusual case of adoption of a golden eagle (*Aquila chrysaetos*) chick in the Mala Fatra mountains, northwestern Slovakia. *Journal of Raptor Research*, 37, 259–260.

16.6.12. Parrots

- Two studies from South America (2,4) describe the successful hand-rearing of parrot chicks, with ten of 12 yellow-shouldered amazons *Amazona barbadensis* surviving for a year after release and blue-fronted amazons *Amazona aestiva* fledging at higher weights than wild birds.
- A review of the kakapo *Strigops habroptilus* management programme found that chicks could be successfully raised and released, but that eggs incubated from a young age had low success.
- A study from the USA (1) found that all hand-reared thick-billed parrots *Rhynchopsitta pachyrhyncha* died within a month of release: significantly lower survival than for wild-caught birds also translocated to the release site.

A study in south-east Arizona, USA (1), released seven hand-reared thick-billed parrots *Rhynchopsitta pachyrhyncha* into the Chiricahua Mountains between September 1986 and September 1993 as part of a wider release and translocation programme. None was alive a month after release. This was significantly lower survival than wild-caught birds translocated as part of the same programme. This study is discussed in more detail in 'Release captive-bred individuals', 'Translocate Individuals' and 'Use holding pens at release sites'.

A small study in 1990–1 on Margarita Island, Venezuela (2), found that all 14 yellow-shouldered amazon *Amazona barbadensis* chicks hand-reared at a release centre fledged successfully. Two of the birds were killed by predators before release, but the remaining 12 were released and at least ten survived for at least a year (see 'Release captive-bred individuals' for details). Hand-rearing involved feeding birds three times a day using a syringe with commercial parrot food and fruits. From 55 days old, chicks were provided with chunks of fruit and whole fruits. This study is also discussed in 'Foster eggs or chicks with wild conspecifics' and 'Use education programmes and local engagement to help reduce pressures on species'.

A 2001 review of data from the kakapo *Strigops habroptilus* management programme in New Zealand (3) found that 22 chicks were removed from nests between 1992 and 1999, with nine surviving (41%) and being released into the wild. These nine represent 60% of the 15 birds fledged between 1990 and 2001. However, the authors note that eggs artificially incubated from an age of less than ten days old had low success rates. None of the birds raised had reached breeding age by the time of the review, so their breeding behaviour and success was unknown.

A replicated, controlled study at a rehabilitation centre in southern Brazil in 1997–9 (4) found that hand-reared blue-fronted amazons *Amazona aestiva* grew for longer than wild birds (with an average growth period of 64 days for 124 captive-reared birds vs. 56 days for 86 wild nestlings). This meant that fledging weights of captive-reared birds were significantly higher than wild birds (average weights of 401 g for 34 captive males and 362 g to 28 females vs. 364 g for nine wild males and 343 g for 12 wild females). Birds were initially fed on a mix of mashed fruit, vegetables, corn flour and puppy food three times per day, changing to fruit, vegetables, seeds and grain as they grew.

(1) Snyder, N. F. R., Koenig, S. E., Koschmann, J., Snyder, H. A. & Johnson, T. B. (1994) Thick-billed parrot releases in Arizona. *The Condor*, 96, 845–862.
(2) Sanz, V. & Grajal, A. (1998) Successful reintroduction of captive-raised yellow-shouldered amazon parrots on Margarita Island, Venezuela. *Conservation Biology*, 12, 430–441.
(3) Elliott, G. P., Merton, D. V. & Jansen, P. W. (2001) Intensive management of a critically endangered species: the kakapo. *Biological Conservation*, 99, 121–133.
(4) Seixas, G. H. F. & Mourao, G. (2003) Growth of nestlings of the blue-fronted amazon (*Amazona aestiva*) raised in the wild or in captivity. *Ornitología Neotropical*, 14, 295–305.

16.6.13. Songbirds

- Four studies from the USA (1–4) found high rates of success for artificial incubation (2–4) and hand-rearing (1,2,4) of songbirds.
- The one study to compare techniques (3) found that crow chicks fed more food had higher growth rates, but that these rates never matched those of wild birds.

A small study at a captive-breeding centre in New York, USA (1), found that eight loggerhead shrike *Lanius ludovicianus* chicks removed from the wild and hand-reared from 8–9 days old survived. Two bred (see 'Use captive breeding to increase or maintain populations') and two of their chicks were removed, hand-reared and released. Both survived for several weeks until they disappeared.

A replicated study on Hawaii, USA, in 1993–4 (2) found that a total of 12 Hawaiian crow (alala) *Corvus hawaiiensis* nestlings were successfully artificially incubated and hand-reared from 17 eggs taken from wild nests. Three eggs were infertile, one failed to hatch due to 'embryonic malpositioning' and one of the 13 successful hatchlings died as a result of a yolk sac infection. This gave a total of 93% hatchability and 92% survival to 30 days. The release of some of these birds is discussed in 'Release captive-bred individuals'.

A replicated *ex situ* study in Idaho, USA (3), found that hand-reared corvid (ravens *Corvus corax*, American crows *C. brachyrynchos* and black-billed magpies *Pica pica*) chicks fed large amounts of food regularly grew faster and were healthier than those fed less food, or less regularly (e.g. ravens fed 40% of their weight every hour grew the fastest, those fed 15% of their weight grew the slowest; crows fed unlimited amounts every hour grew fastest, those fed every two hours grew the slowest; magpies fed unlimited amounts every 30 minutes grew the fastest). However, no feeding regime allowed chicks to grow as fast as wild-raised birds. Crows and magpies were most affected by feeding frequency, whereas ravens were more influenced by the initial amount of food offered. Survival three months after release was positively related to feeding frequency. Eggs were collected from the wild after 0–18 days of incubation and incubated at 37.5°C (crows and ravens) or 38.0°C (magpies) and at 31–69% humidity. Overall, 72% of 430 eggs hatched. After hatching, chicks were transferred to heated aquaria and fed with minced meat, egg, insects and dog food. This study is also discussed in 'Use puppets to increase the survival or growth of hand-reared chicks'.

Two replicated studies from 1995–9 at breeding centres in Hawaii, USA (4), found that hatching rates for artificially incubated wild-laid eggs were 93% for 29 omao *Myadestes obscurus* (a thrush) and 91% for 43 puaiohi *M. palmeri* (eggs consisted of both captive-laid and wild). Two out of five parent-incubated puaiohi eggs also hatched, whilst an additional 15 eggs were not viable. Both species were hand-reared, with 93% of omao and 92% of puaiohi chicks surviving until 30 days old (producing a total of 38 fledged chicks during 1996–8). The two parent-hatched chicks also survived. Eggs were incubated under a 37.2°C dry bulb and a 26.7–31.1°C wet bulb. Chicks were fed a high protein diet of insects, egg and fruit, initially provided every hour but with decreasing frequency as chicks grew. This study is also discussed in 'Use captive breeding to increase or maintain populations' and 'Release captive-bred individuals'.

(1) Cade, T. J. (1992) Hand-reared loggerhead shrikes breed in captivity. *The Condor*, 94, 1027–1029.
(2) Kuehler, C., Harrity, P., Lieberman, A. & Kuhn, M. (1995) Reintroduction of hand-reared alala *Corvus hawaiiensis* in Hawaii. *Oryx*, 29, 261–266.
(3) Whitmore, K. D. & Marzluff, J. M. (1998) Hand-rearing corvids for reintroduction: importance of feeding regime, nestling growth, and dominance. *The Journal of Wildlife Management*, 62, 1460–1479.
(4) Kuehler, C., Lieberman, A., Oesterle, P., Powers, T., Kuhn, M., Kuhn, J., Nelson, J., Snetsinger, T., Herrmann, C., Harrity, P., Tweed, E., Fancy, S., Woodworth, B. & Telfer, T. (2000) Development of restoration techniques for Hawaiian thrushes: Collection of wild eggs, artificial incubation, hand-rearing, captive-breeding, and re-introduction to the wild. *Zoo Biology*, 19, 263–277.

16.7. Use puppets to increase the success of hand-rearing

- Three replicated studies from the USA (1,3) and Saudi Arabia (2) found that corvids (1,3) and bustards (2) raised using puppets did not have higher survival (2), dispersal (3) or growth (1) than conventionally hand-reared chicks.

Background

A potential problem with hand-rearing chicks is that they will 'imprint' on the humans raising them and be unable to adjust to life in the wild or breed with conspecifics. It has therefore been suggested that chicks should be fed with as little human contact as possible and that puppets, designed to look like parent birds, be used to feed them.

A replicated *ex situ* study in Idaho, USA (1), found that the growth of raven *Corvus corax* chicks did not vary between 30 individuals fed with a puppet and 82 fed by keepers. Post-release survival and reproduction were not compared. This study is also discussed in 'Artificially incubate and hand-rear birds in captivity'.

A randomised, replicated and controlled study in Idaho, USA, between 1993 and 1995 (3), found that 25 raven *Corvus corax* chicks (used as surrogates for Hawaiian crows *C. hawaiiensis* and Mariana crows *C. kubaryi*) hand-raised using puppets did not behave differently towards other ravens before or after release, or differ in dispersal from the release site, compared to 49 chicks raised without puppets. Puppet-rearing appeared to increase post-release survival, but the whereabouts of 49% of released birds were unknown, adding considerable uncertainty to this conclusion. Puppet-raised birds were more fearful of keepers following release, which could be beneficial for some species. Puppet-reared birds were separated from each other at 7–10 days old (before their eyes opened).

A replicated trial in Saudi Arabia in 1995 (2) found that hand-reared houbara bustards *Chlamydotis undulata macqueenii* raised with a puppet to minimise human contact were not significantly more likely to survive following release at a desert site, than control (reared with human contact) birds (42% of 12 puppet-reared birds alive the year after release vs. 27% of 12 controls). This study also is also discussed in 'Use "anti-predator training" to improve survival after release'.

(1) Whitmore, K. D. & Marzluff, J. M. (1998) Hand-rearing corvids for reintroduction: importance of feeding regime, nestling growth, and dominance. *The Journal of Wildlife Management*, 62, 1460–1479.

(2) van Heezik, Y., Seddon, P. J. & Maloney, R. F. (1999) Helping reintroduced houbara bustards avoid predation: effective anti-predator training and the predictive value of pre-release behaviour. *Animal Conservation*, 2, 155–163.

(3) Valutis, L. L. & Marzluff, J. M. (1999) The appropriateness of puppet-rearing birds for reintroduction. *Conservation Biology*, 13, 584–591.

Release of captive-bred individuals

16.8. Release captive-bred individuals into the wild to restore or augment wild populations

Background

Captive breeding is normally used to provide individuals which can then be released into the wild to either restore a population in part of the species' former range, or to augment an existing population.

Release techniques vary considerably, from 'hard releases' involving the simple release of individuals into the wild to 'soft releases' which involve a variety of adaptation and acclimatisation techniques before release or post-release feeding and care. The following section includes studies describing the overall effects of release projects. Studies that compare specific release techniques are described elsewhere ('Use holding pens at release sites', 'Use 'anti-predator training' to improve survival after release' etc).

16.8.1. Wildfowl

- Two studies of reintroduction programmes of ducks in New Zealand (5,6) found high survival of released birds and population establishment, with one (6) describing successful breeding. One study (5) describes higher success in the second year of the release programme, potentially because there was then a population present in the wild and more intensive predator control.
- A before-and-after study from Alaska (4) found low survival of released cackling geese *Branta hutchinsii*, but that the population recovered from 1,000 to 6,000 birds after releases and the control of mammalian predators.
- A review of a reintroduction programme from Hawaii (2) found that the release of 2,150 Hawaiian geese (nene) *Branta sandvicensis* had not resulted in the establishment of a self-sustaining population, although some birds bred.
- Two studies from Canada (1,3) found very low return rates for released ducks with one finding no evidence for survival of released birds over two years (1), although there was some evidence that breeding success was higher for released birds than wild ones (3).

A replicated study in parkland in central Saskatchewan, Canada, in 1971–3 (1) found that only six of 93 female canvasbacks *Aythya valisineria* returned to the vicinity of the release site a year later, with only three nesting (two on the year after release and one the following year). No birds were seen more than a year after their release. The survival or return rates of 83 males released over the two years is not discussed. Differences in survival between birds released at different ages are discussed in 'Release birds as adults or sub-adults, not juveniles'.

A 1997 review of the Hawaiian goose (nene) *Branta sandvicensis* reintroduction programme (2) concluded that the release of 2,150 captive-bred birds on Hawaii and Maui, USA, starting in 1949 had not resulted in a self-sustaining wild population. Estimated mortality rates ranged from 0–87% annually, although were generally low until droughts in 1973–86, when 1,200 released geese died. Mortality rates were lower in the lowest-altitude release site (at <1,300 m a.s.l.), with only three years between 1976 and 1983 having mortality rates over 15%. By contrast, the few geese released in uplands that survived the droughts did so by migrating away from their release site. Over the study period there were 515 nests recorded, with 37% raising at least one gosling. Overall there were 473 goslings raised (0.92 goslings/nest), with the highest rates in lowland sites. Birds were all released into temporary enclosure, with differences in release techniques discussed in 'Clip birds' wings on release' and 'Release birds as adults or sub-adults, not juveniles'.

A replicated, controlled study in wetlands in south-west Manitoba, Canada, between 1992 and 1995 (3) found that only 2–3% of 1,766 released captive-bred female mallards *Anas platyrhynchos* were re-sited close to the release site (36 females positively identified and 19 more possibly identified). Annual rates ranged from nearly 10% of 1992 releases to only 1% of 1994 releases being seen again. If these numbers are adjusted for an average mortality of 60% for juvenile females, return rates were still only 6–9%. Comparisons of reproductive success were difficult due to small sample sizes (12 captive-reared females and 30 wild females were monitored): with 60% of wild females and either 71% (1993) or 0% (1994) of captive-reared females nesting. Nest success was 80% for five captive-reared females; it ranged from 11% (1993) to 67% (1994) for wild females, dependent on whether the majority nested on the ground or on artificial structures. Releases involved acclimatising groups of ducklings for 6–10 hours in open-topped cages before allowing them out. Supplementary food was also provided for three weeks after release (see 'Provide supplementary after release' for more information on this release technique).

A before-and-after study from the Aleutian Islands, Alaska, USA on the cackling goose *Branta hutchinsii* recovery programme (4) found that the goose population increased from fewer than 1,000 birds in the 1970s to over 6,000 by 1991, following the release of captive-bred birds and the eradication of Arctic foxes *Alopex lagopus* from breeding islands (see 'Predator control on islands'). The authors note, however, that the release of captive-bred geese was not very successful overall. Release techniques are discussed in 'Release birds as adults or sub-adults, not juveniles' and 'Clip birds' wings on release'.

A replicated study at a reintroduction programme in the north of North Island, New Zealand (5) found that 60 captive-bred brown teal *Anas chlorotis* released in 2003 had an annual survival rate of just 45%, but the survival rate of 40 individuals released in 2004 was 85%. This difference was probably due to a more intensive control of feral cats *Felis catus* (a major cause of mortality in 2003) between releases (see 'Control predators not on islands'). The site also had on-going control of stoats *Mustela erminea* and ferrets *M. putorius*. The authors suggest that higher survival may also have been due to the presence of an established teal population in 2004 but not 2003. Teal were not kept in aviaries at the site before release, but were provided with supplementary food (see 'Provide supplementary food after release' and 'Use holding pens at release sites' for details on these techniques).

A reintroduction programme on Campbell Island, New Zealand, in 2004–5 (6) found that at least 78% (2004) and 75% (2005) of 105 Campbell Island teal *Anas nesiotis* survived reintroduction or translocation. The birds also bred in 2006, with at least two nests and four young being produced. Forty-four of the released birds were wild-caught (see 'Translocate individuals') and 61 captive bred. All birds were kept individually or in pairs for 2–10 days in small holding pens on Campbell Island and provided with food before being released into the wild.

(1) Sugden, L. G. (1976) Experimental release of canvasbacks on breeding habitat. *The Journal of Wildlife Management*, 40, 716–720.
(2) Black, J. M., Marshall, A. P., Gilburn, A., Nelson Santos, Hoshide, H., Medeiros, J., Mello, J., Hodges, C. N. & Katahira, L. (1997) Survival, movements, and breeding of released Hawaiian geese: an assessment of the reintroduction program. *The Journal of Wildlife Management*, 61, 1161–1173.
(3) Yerkes, T. & Bluhm, C. (1998) Return rates and reproductive output of captive-reared female mallards. *The Journal of Wildlife Management*, 62, 192–198.
(4) USFWS (2001) Aleutian Canada goose road to recovery.
(5) O'Connor, S. (2005) Captive breeding and release of brown teal *Anas chlorotis* into the Moehau Kiwi Sanctuary, Coromandel, New Zealand. *Conservation Evidence*, 2, 72–73.
(6) McClelland, P. & Gummer, H. (2006) Reintroduction of the critically endangered Campbell Island teal *Anas nesiotis* to Campbell Island, New Zealand. *Conservation Evidence*, 3, 61–63.

16.8.2. Gamebirds

- One of five studies from across the world (2) found that releasing gamebirds established a population or bolstered an existing population, although the authors argued that the population of 30–40 western capercaillie *Tetrao urogallus* (from nearly 400 released) was unlikely to be self-sustaining.
- A review of a reintroduction programme in Pakistan (1) found some breeding success in released cheer pheasants *Catreus wallichii*, but that habitat change at the release site then excluded released birds.
- Three studies from Europe (3,5) and the USA (4) found that released birds had low survival, low reproductive success and had no impact on the wild population.

A review of a 1978–89 reintroduction programme for cheer pheasants *Catreus wallichii* in northern Pakistan (1) found that post-release survival was low between 1978 and 1985 (see 'Use "anti-predator training" to improve survival after release' for details), but that 10–15% of 305 birds released in 1986 and 1988 survived at least one year after release. Breeding was recorded in 1987 (one pair nested, nine eggs laid, seven hatched but all chicks died within six weeks) and 1989 (three females nested, with eight chicks surviving to at least eight weeks and leaving the release pen). However, a survey in 1990 suggested that the release area (a national park) was becoming too overgrown to support cheer pheasants, and that rotational burning (similar to traditional agriculture) may be necessary to maintain the population in the area. Releases were conducted at a medium elevation site (700m) in 1978–81, and two higher elevation sites (approximately 1,000 m) from 1983 and 1988 respectively. Birds were released into open-topped release pens, with 54 birds released between 1978 and 1981; 279 released in 1983–6 and 305 in 1988–9.

A before-and-after study in western Germany between 1980 and 1992 (2) reports that a western capercaillie *Tetrao urogallus* reintroduction programme succeeded in establishing a population of 30–40 individuals by spring 1992, with breeding recorded in 1986. Before the releases, the species had died out in the region probably by 1974. A total of 393 birds (226 males, 167 females) were released during the study period, with 200 in the last four years. A large number of released birds were predated, but some released males lived for five to seven years following release. Before release, chicks were hatched in incubators, raised in 'post-hatch' cages until four months old and then moved to large outdoor pens for a month before release. The authors argue that the wild population size is too small to be self-sustaining in the long-term.

A controlled, replicated study (1991–1996) in mixed arable land in central Finland found that, due mainly to poor survival and low reproductive success, releasing hand-reared female grey partridges *Perdix perdix* contributed little to boosting the local wild population (3). Hand-reared females had lower survival during the breeding period than wild females (19% vs. 69%) and wild partridges produced more fledglings than released ones (2.09/female vs. 0.05/female). There was no significant difference in spring dispersal (3.1 km wild; 2.3 km hand-reared), nesting chronology, clutch size (wild average 20.5 eggs vs. hand-reared 19.3 eggs), or nest predation (main cause of mortality in both sets of birds) between wild and hand-reared birds.

A review of a reintroduction programme in two prairie sites in Texas, USA, in 1996–7 (4) found that two-week survival rates of 119 released, captive-bred and hand-reared Attwater's prairie chickens *Tympanuchus cupido attwateri* (an endangered subspecies of the greater prairie chicken) were 51–82%. The date of release (July–October), release habitat (prairie or soybean 'food plot') or type of radio-transmitter used to track birds did not affect six-month survival rates, whereas time spent in cages prior to release did (discussed in 'Use holding pens at release sites'). Movements and range-sizes were similar for released and wild birds but there was no known recruitment into the population from released birds. Mortality was mainly from predation, whilst known nesting failures appeared to be due to invasive red fire ants *Solenopsis invicta*.

A replicated, controlled study in two arable farmland areas in Angus, Scotland, found that, due to poor survival and low reproductive success, releasing commercially reared grey partridges *Perdix perdix* did not contribute to bolstering declining wild populations (5). Studies were conducted at one site in autumn 1997 to summer 2001 and a second in autumn 2001 to summer 2004. Eight-week old commercially-reared partridges were placed in release pens in September each year and released 2–3 weeks later. Released birds (520) were monitored by spring and autumn counts, night-time surveys and radio-telemetry. Some wild female partridges were radio-tagged for comparison. Survival of captive-reared birds from autumn to the following spring was low (averaging 10%). Breeding-season survival of released females averaged 30% and for wild females 44%. The major cause of mortality was predation (69% of losses). Of the few reared birds that survived to breed, none fledged chicks in their first breeding season. Only one released female survived to breed in her second year, but this individual raised 14 young.

(1) Garson, P. J., Young, L. & Kraul, R. (1992) Ecology and conservation of the cheer pheasant *Catreus wallichii*: studies in the wild and the progress of a reintroduction project. *Biological Conservation*, 59, 25–35.
(2) Spittler, H. (1994) Wiedereinbürgerungsversuch mit Auerwild (*Tetrao urogallus* L.) im Hochsauerland. *Zeitschrift für Jagdwissenschaft*, 40, 185–199.
(3) Putaala, A. & Hissa, R. (1998) Breeding dispersal and demography of wild and hand-reared grey partridges *Perdix perdix* in Finland. *Wildlife Biology*, 4, 137–145.
(4) Lockwood, M. A., Clifton P. Griffin, Morrow, M. E., Randel, C. J. & Silvy, N. J. (2005) Survival, movements, and reproduction of released captive-reared Attwater's prairie-chicken. *The Journal of Wildlife Management*, 69, 1251–1258.
(5) Parish, D. M. B. & Sotherton, N. W. (2007) The fate of released captive-reared grey partridges *Perdix perdix*: implications for reintroduction programmes. *Wildlife Biology*, 13, 140–149.

16.8.3. Rails

- One replicated study from Australia (1) found that released Lord Howe Island woodhens *Tricholimnas sylvestris* successfully bred in the wild, re-establishing a wild population.
- A replicated study from the UK (3) found high survival of released corncrake *Crex crex* in the first summer (although no data were available on overwinter survival or breeding).
- A replicated study in New Zealand (2) found very low survival of North Island weka *Gallirallus australis greyi* following release, mainly due to predation by invasive mammals.

A replicated study on Lord Howe Island, Australia, in 1981–3 (1) found that captive-bred Lord Howe Island woodhens *Tricholimnas sylvestris* survived for up to two years in the wild (released after a pig *Sus scrofa* and goat *Capra hircus* control programme had been running for several years, see 'Control mammalian predators on islands'). In addition, 19 wild-bred young were reared successfully. Before captive breeding, there were only three pairs known in the wild, which were transferred to captivity. In total, 57 birds were released over three years. This study also describes the captive-breeding efforts, discussed in 'Use captive breeding to increase or maintain populations'.

A replicated study on North Island, New Zealand, in 1992–3 (2) found that, of 17 North Island weka *Gallirallus australis greyi* released between October and March at a mixed habitat site, only one bird was alive more than seven months after release. Most individuals were killed by predators (mainly by domestic dogs *Canis familiaris*) and 12 individuals (71%) survived less than 50 days. Weka had small home ranges (average of 2.7 ha) and dispersed an average of only 1.3 km during the study.

A 2004 review of a corncrake *Crex crex* release programme in a wet grassland site in eastern England (3) found that only six chicks could be released into the wild in 2002 (due to predation in captivity), and that none was seen in the area the following year. From 140 eggs laid in 2003, 52 chicks were released during summer. Survival was apparently high, but data on overwinter survival and subsequent reproduction were not available. Captive birds were kept in a flock in autumn and winter and then paired off in the spring. Eggs were removed before hatching and incubated artificially. Once hatched, they were hand-fed until they could feed themselves and then released into a pen at the release site when ten days old before being released at 28 days old. This paper also discusses the translocation of red kites *Milvus milvus* to the UK, discussed in 'Translocate individuals'.

(1) Miller, B. & Mullette, K. J. (1985) Rehabilitation of an endangered Australian bird – the Lord Howe Island woodhen *Tricholimnas sylvestris* (Sclater). *Biological Conservation*, 34, 55–95.
(2) Bramley, G. N. & Veltman, C. J. (1998) Failure of translocated, captive-bred North Island weka *Gallirallus australis greyi* to establish a new population. *Bird Conservation International*, 8, 195–204.
(3) Carter, I. & Newbery, P. (2004) Reintroduction as a tool for population recovery of farmland birds. *Ibis*, 146, 221–229.

16.8.4. Cranes

- Four studies of five release programmes from the USA and Russia (1,4–6), from a total of eight programmes, found that released cranes had high survival or bred in the wild. Two studies from two release programmes in the USA (3,4) found low survival of captive-bred eggs fostered to wild birds, compared with wild eggs, or a failure to increase the wild flock size.
- A worldwide review (4) found that releases of migratory species only tended to be successful if birds were released into existing flocks, with higher success for non-migratory populations.
- One study from the USA (2) found that birds released as sub-adults had higher survival than birds cross-fostered to wild birds.
- One study from the USA (1) found that 73% of all mortalities occurred in the first year after release.

A replicated study describing the success of releasing captive-bred Mississippi sandhill cranes *Grus canadensis pulla* onto a wet pine savanna site in Mississippi, USA (1) found that, of 40 birds released between 1979 and 1985, 46% were alive at the end of the study (between one and six years after release). Of the 22 mortalities, 16 (73%) occurred during the first year after release, with three during each of the second and third years. Predation and human-caused mortality were the main causes. Birds were bred in captivity and parent-raised before being rendered temporarily flightless with wing brails and moved to acclimatisation pens. They were between four months and one year old at release.

A replicated study as part of the planning for a whooping crane *Grus americana* reintroduction programme, a study in Florida, USA, in 1986–7 (2) found that greater sandhill cranes *Grus canadensis tabida* released as sub-adults in a 'soft release' programme had higher survival than birds fostered to Florida sandhill cranes *G. c. pratensis* (56% of 27 birds surviving for one year vs. 39% survival for 34 fostered birds, discussed in 'Foster eggs or chicks with wild non-conspecifics (cross-fostering)'). The nine to ten month-old cranes were prevented from flying and kept in an open-topped 1.5 ha enclosure for four to six weeks until they were released. Food was provided until the birds no longer returned to the enclosure. Greater sandhill cranes are migratory, whilst Florida sandhill cranes are not. Migratory movements of released birds were larger than a control group of Florida sandhill cranes, but not significantly so.

A replicated study in Idaho, USA, between 1975 and 1991 (3) found that 215 wild-sourced whooping crane *Grus americana* eggs that were cross-fostered into sandhill crane *G. canadensis* nests had higher hatching success than 73 captive-bred whooping crane eggs, fostered at the same time (77% hatching success for wild-sourced eggs vs. 60% for captive-bred). This study is also discussed in 'Use captive breeding to increase or maintain populations', 'Foster eggs or chicks with wild conspecifics' and 'Foster eggs or chicks with wild non-conspecifics (cross-fostering)'.

A 1998 review (4) found that crane *Grus* spp. reintroduction programmes have had mixed success, with reintroductions of migratory species generally failing when birds were not released into existing flocks. Reintroductions of Siberian cranes *G. leucogeranus* has not increased wild flock size, with no reintroduced birds being seen after migration and high mortality during rearing, high poaching levels and few wild birds to guide migration. Releases of semi-wild red-crowned cranes *G. japonensis* and white-naped cranes *G. vipio* in south-east Russia found that four of ten released birds migrated successfully, and at least two pairs nested, one successfully. At least 84% of 38 greater sandhill cranes *G. canadensis tabida* survived for a year after release in Michigan, USA, 74% returned after migration and four males nested. Non-migratory releases generally had higher success: first-year survival of non-migratory whooping cranes *G. americana* has increased from approximately 34% (1993–4) to 71% (1996), although the population remains very small and may rely on continued releases. Captive-reared Mississippi sandhill cranes *G. c. pulla* had an overall first-year survival of 70%, an adult survival over 91% following release and in 1992 represented 80% of the wild population. In 1996 there were 13 nesting pairs (the most recorded), with 60% of known pairs having at least one captive-reared individuals. The population, however, remains dependent on releases. The author argues that post-release monitoring is vital to identify causes of mortality.

A replicated, controlled study in a breeding centre in Mississippi, USA, between 1989 and 1996 (5) found that first year survival of captive-bred Mississippi sandhill cranes *Grus canadensis pulla* was high, with approximately 80% of 132 birds surviving. Birds were released either in mixed flocks (both hand-reared and parent-reared birds) or non-mixed flocks (with just one rearing type). Survival rates over four years were highest for hand-reared birds in mixed flocks (approximately 95% survival for 17 birds), followed by parent-reared birds in mixed flocks (89% of 31 birds), hand-reared in non-mixed flocks (78% of 39 birds) and were lowest in parent-reared, non-mixed flocks (56% of 45). By the end of the study, however, differences between parent and hand-reared birds were no longer statistically significant, although mixed flock birds still had higher survival. Birds were kept in 'cohorts' for four to five weeks, before being moved to the release site and kept for a month in uncovered pens before wing brails (which prevent flying) were removed in December. Details of hand-rearing are found in 'Artificially incubate and hand-rear birds in captivity'.

A replicated study of a whooping crane *Grus americana* reintroduction programme in 2001–5 in wetlands in Florida, USA (6), found that winter-releases of this migratory bird proved effective. Average first-year survival of 71 winter-released juvenile birds was 87%, and was higher in later years as techniques improved. Birds were reared by humans wearing costumes (to avoid imprinting on human carers, see 'Use puppets to increase the survival or growth of hand-reared chicks' for studies on this intervention) and guided to the release site by an ultralight aircraft. Once there they were kept in holding pens by costumed caretakers. When the habitat prevented this, the juveniles were vulnerable to bobcat *Lynx rufus* predation, but this problem was overcome by vegetation clearance. Winter releases of this type were advantageous because the intensive care reduced predation by bobcats, juveniles were kept away from harassment by adult birds and juveniles did not lose their fear of humans through contact with tame sandhill cranes *G. canadensis*. Once released, juveniles showed ordinary migratory and summer behaviour.

(1) Zwank, P. J. & Wilson, C. D. (1987) Survival of captive, parent-reared Mississippi sandhill cranes released on a refuge. *Conservation Biology*, 1, 165–168.

(2) Nesbitt, S. A. & Carpenter, J. W. (1993) Survival and movements of greater sandhill cranes experimentally released in Florida. *The Journal of Wildlife Management*, 57, 673–679.

(3) Kuyt, E. (1996) Reproductive manipulation in the whooping crane *Grus americana*. *Bird Conservation International*, 6, 3–10.
(4) Davis, C. (1998) A review of the success of major crane conservation techniques. *Bird Conservation International*, 8, 19–30.
(5) Ellis, D. H., Gee, G. F., Hereford, S. G., Olsen, G. H., Chisolm, T. D., Nicolich, J. M., Sullivan, K. A., Thomas, N. J., Nagendran, M. & Hatfield, J. S. (2000) Post-release survival of hand-reared and parent-reared Mississippi sandhill cranes. *The Condor*, 102, 104–112.
(6) Urbanek, R. P., Fondow, L. E. A., Zimorski, S. E., Wellington, M. A. & Nipper, M. A. (2010) Winter release and management of reintroduced migratory whooping cranes *Grus americana*. *Bird Conservation International*, 20, 43–54.

16.8.5. Bustards

- Three reviews (1–3) of a release programme for houbara bustard *Chlamydotis undulata macqueenii* in Saudi Arabia and a replicated trial as part of the same programme (4) found low initial survival of released birds (1), but the establishment of a breeding population (3) and an overall success rate of 41% (4).
- The programme tested many different release techniques, discussed elsewhere, with releases being most successful if sub-adults were released, able to fly, into a large exclosure (2).

A review of a captive breeding programme in south-west Saudi Arabia (1) reported that, of six houbara bustards *Chlamydotis undulata macqueenii* released in three separate trials in 1991–2, only two survived, of which one was recaptured. This programme is discussed in more detail in 'Use captive breeding to increase or maintain populations', 'Use artificial insemination in captive breeding' and 'Artificially incubate and hand-rear birds in captivity'.

A review (2) of the same release programme as in (1) found that houbara bustard *Chlamydotis undulata macqueenii* releases were most successful when sub-adult birds were released into a large (4 km^2) fenced enclosure, compared with releases of birds without an enclosure, releases of chicks or releases of sub-adult birds with clipped wings. Four adults and sub-adults were released without an enclosure in 1991 and all were killed within three days by foxes. This study is also discussed in 'Use captive breeding to increase or maintain populations' and 'Use artificial insemination in captive breeding'. The other release techniques are discussed in 'Release birds in "coveys", "Use holding pens at site of release" and "Use holding pens at site of release and clip birds" wings'.

A review (3) of the same release programme as in (1) reports a houbara bustard *Chlamydotis undulata macqueenii* population established in south-west Saudi Arabia through releases of captive-bred birds in 1993–4 (discussed in (4)), bred for the first time in 1995. Of 22 females and 13 males in the population, three females and one male were confirmed as breeding. One yearling female raised three chicks to fledging (at 38–42 days old) whilst two other females laid a total of three eggs, all of which were infertile. The authors suggest that infertility may be caused by a low density of males and inexperienced females. One infertile brood was replaced with fertile eggs, discussed in 'Foster eggs or chicks with wild conspecifics'.

A replicated trial (4) as part of the same release programme as in (1,3) found that 35 of 85 houbara bustards *Chlamydotis undulata macqueenii* (41%) released at a desert steppe site in south-west Saudi Arabia were successfully introduced. This study investigates how different release techniques affected survival, discussed in 'Release birds as sub-adults or adults, not juveniles' and 'Use holding pens at release sites'. The effect of predator removal is discussed in 'Control predators not on islands'.

(1) Seddon, P. J., Jalme, M. S., Van Heezik, Y., Paillat, P., Gaucher, P. & Combreau, O. (1995) Restoration of houbara bustard populations in Saudi Arabia: developments and future directions. *Oryx*, 29, 136–142.
(2) Jaime, M. S., Combreau, O., Seddon, P. J., Paillat, P., Gaudier, P. & Heezik, Y. (1996) Restoration of *Chlamydotis undulata macqueenii* (houbara bustard) populations in Saudi Arabia: A progress report. *Restoration Ecology*, 4, 81–87.
(3) Gelinaud, G., Combreau, O. & Seddon, P. J. (1997) First breeding by captive-bred houbara bustards introduced in central Saudi Arabia. *Journal of Arid Environments*, 35, 527–534.

(4) Combreau, O. & Smith, T. R. (1998) Release techniques and predation in the introduction of houbara bustards in Saudi Arabia. *Biological Conservation*, 84, 147–155.

16.8.6. Waders

- A review of black stilt *Himantopus novaezelandiae* releases in New Zealand (1) found that birds had low survival (13–20%) and many moved away from their release sites so, in consequence, that they could not be managed and were unlikely to interact with stilt populations in the wild.

A review of critically endangered black stilt (kaki) *Himantopus novaezelandiae* releases in riverine habitats in South Island, New Zealand, between 1993 and 2005 (1) found that 13–20% of 464 birds released were alive two years after release. However, 32% of birds that reached breeding age did not remain at their release site and 15% moved to an area where they could no longer be managed and were unlikely to reproduce successfully. The authors argue that this second category of birds is 'effectively dead' as they no longer contribute to the wild population. Birds were released into populations that needed supplementation; therefore movements away from the release site could also be detrimental. Eggs were taken from wild and captive-bred birds and artificially incubated. Birds were not held at the release site before release, but food was provided at release site for between six weeks and two months. This study is also discussed in 'Release birds as adults or sub-adults, not juveniles' and 'Release birds in groups'.

(1) van Heezik, Y., Maloney, R. F. & Seddon, P. J. (2009) Movements of translocated captive-bred and released critically endangered kaki (black stilts) *Himantopus novaezelandiae* and the value of long-term post-release monitoring. *Oryx*, 43, 639–647.

16.8.7. Storks and ibises

- A replicated study (1) and a review (2) of northern bald ibis *Geronticus eremita* release programmes in Europe and the Middle East found that only one of four had resulted in a wild population being established or supported, with many birds dying or dispersing, rather than forming stable colonies.

A replicated study in south-east Turkey in 1977–88 (1) found that northern bald ibis *Geronticus eremita* bred or kept in captivity did not increase the Turkish population. From 1981–88, 67 individuals were released and 12 (18%) migrated with the wild population (note: this excludes 1984, 1986 and 1987, for which data were not available). There was high winter mortality among birds that did not migrate and also high mortality on migration. The wild population in the study area declined over the study period, with five pairs in 1986, seven in 1987 and only four birds returning in 1988. This study is also described in 'Use captive breeding to increase or maintain populations' and 'Provide artificial nesting sites'.

A 2007 review of northern bald ibis (waldrapp) *Geronticus eremita* conservation (2) found varying success in release programmes, dependent on the techniques used. Trials in Israel using a variety of techniques (described in 'Release birds as adults or sub-adults, not juveniles' and 'Clip birds' wings on release') found that all 56 birds released became emaciated and disorientated and formed poor social bonds. Similarly, the release of 73 birds in Spain between 2004 and 2006 has not resulted in the formation of a stable colony. However, the release of 43 birds in Austria has led to the establishment of a colony in the wild. This study is also discussed in 'Use captive breeding to increase or maintain populations', 'Artificially incubate and hand-rear birds in captivity', 'Use holding pens at release sites', 'Release birds as adults or sub-adults, not juveniles', 'Clip birds' wings on release', 'Use microlites to help birds migrate' and 'Foster birds with non-conspecifics'.

(1) Akçakaya, R. (1990) Bald ibis *Geronticus eremita* population in Turkey: an evaluation of the captive breeding project for reintroduction. *Biological Conservation*, 51, 225–237.
(2) Bowden, C. G. R., Boehm, C., Jordan, M. J. R. & Smith, K. W. (2007) Why is reintroduction of northern bald ibis *Geronticus eremita* so complicated? An overview of recent progress and potential. 27–35 in: M. M. Lamont (eds) *The Proceedings of the IV International Symposium on Breeding Birds in Captivity; 2007 Sept 12–Sept 16* Toronto, Ontario, Canada.

16.8.8. Vultures

- Four studies (2–5) of two release programmes found that release programmes led to large population increases in Andean condors *Vultur gryphus* in Colombia (2) and griffon vultures *Gyps fulvus* in France (3–5).
- A small study in Peru (1) found high survival of released Andean condors *Vultur gryphus* over 18 months, with all fatalities occurring in the first six months after release.

A small study using Andean condors *Vultur gryphus* to develop release techniques for Californian condors *Gymnogyps californianus* (1) found that 7 of 11 (64%) Andean condors released in arid mountains in northern Peru in 1980–1 survived for at least 18 months after release. All mortalities occurred in the first six months after release. This study is discussed in more detail in 'Artificially incubate and hand-rear birds in captivity' and 'Provide supplementary food after release'.

A replicated study of a reintroduction programme for Andean condors *Vultur gryphus* found that 19 of the 22 birds released at Páramo sites in the Colombian Andes between 1989 and 1992 were alive in 1993 (2). This represents an increase of almost 100% on the previous Colombian population, estimated at ten pairs. The three birds that died did not appear to be affected by human activity. Releases were at three sites, using birds bred in zoos in the USA, with acclimatisation periods of between 17 and 103 days. After releases, food was provided irregularly at multiple sites (to encourage birds to search for carrion). Birds ranged over nearly 200 km^2 over two years, which is less than ranges for reintroduced condors in arid habitats in Peru.

A replicated study over ten years of a griffon vulture *Gyps fulvus* reintroduction programme in river gorges in Aveyron, southern France (3) found that 39 adult and 20 immature (less than four years old) birds released between 1980 and 1986 had high survival rates. Survival rates of the wild-bred offspring of the released birds were also high: 86% for the first three years of life and 99% from then on. Of the 18 marked vultures recovered dead between 1981 and 1991, 12 (67%) had died as a result of electrocution. Reproduction of the released birds is discussed in (4), the effect of age at release is discussed in 'Release birds as adult or sub-adults, not juveniles' and the education programme that accompanied the releases is discussed in 'Use education programmes and local engagement to help reduce pressures on species'.

A replicated study over 12 years (4) of the same programme as in (3) showed that the number of nesting pairs of griffon vultures *Gyps fulvus* in the release site in southern France increased steadily from three to 33 (fledging a total of 95 young) over 11 breeding seasons following the release of 59 captive-bred birds during 1981–1986. The majority of wild-born and young-released birds nested first at four years old. The effect of the age birds were released at is discussed in 'Release birds as adults or sub-adults, not juveniles'.

Further analysis of the reintroduction programme discussed in (3,4) found that the breeding population of griffon vultures *Gyps fulvus* in the release area in southern France increased from approximately 13 pairs in 1987 to 130 breeding pairs in 2005, following the release of 61 captive-bred birds (5). Eleven young adult (between five and nine years old) releases fed their chicks at a lower rate than eight wild-bred birds, but in all age classes there was no difference in feeding rates or dominance of released, compared with wild-bred birds.

(1) Wallace, M. P. & Temple, S. A. (1987) Releasing captive-reared Andean condors to the wild. *The Journal of Wildlife Management*, 51, 541–550.

(2) Lieberman, A., Rodriguez, J. V., Paez, J. M. & Wiley, J. (1993) The reintroduction of the Andean condor into Colombia, South America: 1989–1991. *Oryx*, 27, 83–90.

(3) Sarrazin, F., Bagnolini, C., Pinna, J. L., Danchin, E. & Clobert, J. (1994) High survival estimates of griffon vultures (*Gyps fulvus fulvus*) in a reintroduced population. *The Auk*, 111, 853–862.

(4) Sarrazin, F., Bagnolini, C., Pinna, J. L. & Danchin, E. (1996) Breeding biology during establishment of a reintroduced griffon vulture *Gyps fulvus* population. *Ibis*, 138, 315–325.

(5) Bose, M. & Sarrazin, F. C. O. (2007) Competitive behaviour and feeding rate in a reintroduced population of griffon vultures *Gyps fulvus*. *Ibis*, 149, 490–501.

16.8.9. Raptors

- Five studies (6,7,9,13,14) of three release programmes from across the world found the establishment or increase of wild populations of falcons *Falco* spp.
- Five studies from the USA (2,4,5,8,10) found high survival of released raptors (with between one and 204 birds released), whilst two (11,12) found that released birds behaved normally and hunted successfully.
- One study from Australia (1) found that a wedge-tailed eagle *Aquila audax* had to be taken back into captivity after acting aggressively towards humans, whilst another Australian study (3) found that only one of 15 brown goshawks *Accipiter fasciatus* released was recovered, although the authors do not draw conclusions about survival rates from this.

Background

Raptors can be released through a process called 'hacking' which was developed as a falconry method to train young birds to hunt. It involves placing broods (once they can feed themselves and regulate their body temperature) in an artificial nest site where they can be fed without imprinting on humans. The birds can learn about their environment and gradually learn to fly and hunt, with food provided until it is no longer taken (Jones 2004).

Jones, C. G. (2004) Conservation management of endangered birds pp. 269–303 in eds. W. J. Sutherland, I. Newton & R. Green: *Bird ecology and conservation a handbook of techniques*, Oxford University Press, Oxford.

A small, pre-1980, study in a national park in the Australian Capital Territory, Australia (1), found that a female wedge-tailed eagle *Aquila audax* released into a 5,500 ha nature reserve successfully adapted to release and began hunting European rabbits *Oryctolagus cuniculus*. However, the bird had to be recaptured after it attacked people entering her hunting area, two months after release. The eagle came from Melbourne Zoo and was fearful of humans both before release and after recapture.

A replicated study from the eastern USA between 1975 and 1979 (2) found that 72% of 204 captive-bred peregrine falcons *Falco peregrinus* that were hacked in artificial and natural sites survived to independence, with three groups of releases being 'adopted' by wild adults. Success was higher for birds released at artificial sites (i.e. from a tower), compared to natural sites (i.e. from cliffs), mainly because of high rates of predation by great horned owls *Bubo virginianus* at cliffs. Most birds stayed in the release area and first year survival appears comparable with wild birds. In 1979, three pairs consisting of released birds were known.

A replicated study of bald eagle *Haliaeetus leucocephalus* reintroductions from a breeding centre in Maryland, USA (4), found that all eleven captive-bred, parent-reared birds hacked at two sites in New York and Georgia, USA, successfully reached independence. This study is also discussed in 'Use captive breeding to increase or maintain populations', 'Use artificial insemination in captive breeding' and 'Foster birds with wild conspecifics'.

A replicated study from Canberra, Australia (3), found that, only 1 of 15 captive-bred brown goshawk *Accipiter fasciatus* chicks released into a suburban habitat between 1976 and 1979 was recovered: a male hit by a car 960 km away and nine months after release. The authors note that all young were very secretive after release. Young were hacked by being fed for between two weeks and two months after release. This study is also discussed in 'Use captive breeding to increase or maintain populations'.

A study from wetlands in Kentucky and Tennessee, USA (5), describes the successful release, through hacking, of a captive-bred, juvenile bald eagle *Haliaeetus leucocephalus* in summer 1981. The eagle was fed in an enclosure until 14.5 weeks old and began flying immediately after it was released. The eagle remained close to the release site for 39 days, hunted successfully (with a 50% success rate) and appeared to behave normally until it dispersed from the study area.

A replicated 1993 study (6) found that 77% of 164 captive-bred and raised Mauritius kestrels *Falco punctatus* released into the wild in tropical forests in southern Mauritius between 1986 and 1992 survived until independence. Release involved hacking on an offshore island for several weeks before being released on the mainland. Before the release of captive-bred individuals, the wild population had grown from five individuals in 1973 to 31 in 1986. Following fostering (see 'Foster chicks or eggs with wild conspecifics') and releases, the wild population reached at least 30 breeding pairs in 1991–2. This study is also discussed in 'Use captive breeding to increase or maintain populations', 'Artificially incubate and hand-rear birds in captivity'.

A 1995 update (7) of the same conservation programme studied in (6), found that hacking released captive-bred Mauritius kestrels *Falco punctatus* nestlings significantly contributed to the recovery of the natural population. A total of 331 birds were released into various sites from 1984–1985 and 1993–1994 of which 78% became independent and 61% survived their first winter. Of 208 fledglings hacked (25–34 day old nestlings were put into small groups in a nest box and food was provided while hunting skills were honed), 79% became independent. Most moved out of the release area between 85–100 days; similar to that of natural parent-raised birds. However, only 38% of released first-year females successfully fledged young whereas older females averaged 2 fledglings/nest (from 64% of nests). The remainder of the released captive-bred young were fostered (see 'Foster eggs or chicks with wild conspecifics'). At the end of the 1993–1994 breeding season, the natural population had recovered to 222–286 birds (containing at least 56 breeding pairs and 40–70 non-breeding birds), from a low of four wild individuals.

A replicated study in 1993–4 (8) found that four-week survival rates of captive-bred aplomado falcons *Falco femoralis* hacked at a wetland site in southern Texas, USA, ranged from 58% (five known mortalities from 12 birds released in 1994) to 85% (four known mortalities from 26 birds released in 1993). Predation by great horned owls *Bubo virginianus* and coyotes were the main causes of mortality. Released birds had larger range sizes than predicted, which the authors suggest is due to birds having expanded ranges before pairing up. Birds were transported to the release site when four weeks old and fed there before being released at 37 days old. Food was then provided until birds no longer returned to feed.

A replicated study of a captive-release programme in eastern Germany (9) found that the release of 201 captive-reared peregrine falcons *Falco peregrinus* between 1990 and 2000 led to a population of at least 22 adult peregrines in the study area in 2000. This population of peregrines was unique in that it nested in trees, and imprinting techniques meant that all but 11 released birds nested in trees. The remaining 11 used buildings or cliffs, and none of their offspring reverted to tree-nesting.

A study in the summers of 1999–2000 at a river cliff site in Iowa, USA (10), found that two week survival of 38 (21 in 1999, 17 in 2000) juvenile peregrine falcons *Falco peregrinus* released through hacking was between 74% and 89%, with overall weekly survival estimated at 98.8%. Movement away from the release site was higher in 2000, possibly due to the large numbers of great horned owls *Bubo virginianus* seen in the area (although no mortalities were due to owl predation).

Observations in Panama of two captive-bred harpy eagles *Harpia harpyja*, indicated that after release, prey diversity collected and predation rates were broadly consistent with that of wild birds (11). In 1998 in captive breeding facilities in USA, two harpy eagles were hatched and reared using puppets (to avoid imprinting on human carers, see 'Use puppets to increase the survival or growth of hand-reared chicks' for studies on this intervention), then placed in an enclosure with an adult female eagle. Near fledging (161–165 days old) they were transferred to an aviary at a release site in Panama, where they were habituated for 4–5 weeks prior to release. They were provided with supplementary food until they ceased to visit (11 months). Both birds were recaptured and relocated to a nearby safer site, the male re-released on 16 June and the female on 10 October, 1999. The eagles were monitored during June 1999 to August 2000. Both made captures of wild prey with apparent ease, despite lack of human training or parental guidance.

A replicated study in Texas, USA, between 1993 and 2002 (12), found that all of the 154 northern aplomado falcons *Falco femoralis septentrionalis* studied displayed hunting behaviour without having been taught it. Birds were hacked from 22 sites in groups of between two and eight birds, taken to the hacking site at 30 days old, released at 38–41 days old and provided with food for a further six weeks. Males began hunting earlier (19 days after release for 78 birds vs. 24 days after release for 76 birds), but made their first kills later (35 days after release for 19 kills vs. 32 days after release for 19 kills by females). Group hunting was also observed.

A 2004 review (13) of the same Mauritius kestrel *Falco punctatus* release programme as in (6), between 1987 and 2001 found that survival estimates for adult kestrels were similar, irrespective of whether they were 'hacked' as fledglings (and provided with supplementary food until independence), or fostered to wild breeding pairs or wild-bred (80% for 42 fostered birds; 80% for 46 hacked birds and 75% for 284 wild-bred birds). Survival estimates for juvenile kestrels were far more variable, but did not appear to differ between treatments (36–72% for hacked, 23–100% for fostered and 31–80% for wild-bred). A total of 40 breeding pairs were monitored in 2000–1. Overall, the wild kestrel population across Mauritius reached an estimated 500 to 800 individuals in 2000, compared to five individuals in 1973.

A review of a reintroduction programme for northern aplomado falcons *Falco femoralis septentrionalis* in coastal plains in Texas, USA (14), found that the release of captive-bred falcons since 1993 had led to the establishment, by 2002–4, of 38 breeding pairs in the two study areas. During 2001–3, 141 captive-bred falcons reached independence in the study area, and 75 chicks fledged. Of these, 43 (19 released and 24 wild-bred) were seen after fledging at least once. Of 18 birds recruited into the breeding population (i.e. forming breeding pairs), only three (17%) were captive-bred and released and 15 (83%) were wild-bred. Captive-bred birds were 'hacked' during release. This involved providing cohorts of 2–8 birds with food for 21 days after release. Those birds seen after 21 days were said to have reached independence.

(1) Olsen, J. & Olsen, P. (1980) Some considerations for future raptor rehabilitation. *Raptor Research*, 14, 10–12.

(2) Barclay, J. H. (1980) Release of captive-produced peregrine falcons in the eastern United States, 1975–1979.

(3) Olsen, J. & Olsen, P. (1981) Natural breeding of *Accipiter fasciatus* in captivity. *Raptor Research*, 15, 53–57.

(4) Wiemeyer, S. N. (1981) Captive propagation of bald eagles at Patuxent Wildlife Research Center and introductions into the wild, 1976–80. *Raptor Research*, 15, 68–82.

(5) Altman, R. L. (1983) Post-release flight and foraging behavior of a bald eagle hacked in western Kentucky. *Raptor Research*, 17, 37–42.

(6) Cade, T. J. & Jones, C. G. (1993) Progress in restoration of the Mauritius kestrel. *Conservation Biology*, 7, 169–175.

(7) Jones, C. G., Heck, W., Lewis, R. E., Mungroo, Y., Slade, G. & Cade, T. (1995) The restoration of the Mauritius kestrel *Falco punctatus* population. *Ibis*, 137, S173–S180.

(8) Perez, C. J. & Zwank, P. J. (1996) Survival, movements and habitat use of aplomado falcons released in southern Texas. *Journal of Raptor Research*, 30, 175–182.

(9) Kirmse, W. (2001) Wiedereinbürgerung baumbrütender Wanderfalken (*Falco peregrinus*) in Mitteleuropa. *Zeitschrift für Jagdwissenschaft*, 47, 165–177.

(10) Powell, L. A., Calvert, D. J., Barry, I. M. & Washburn, L. (2002) Post-fledging survival and dispersal of peregrine falcons during a restoration project. *Journal of Raptor Research*, 36, 176–182.

(11) Touchton, J. M., Hsu, Y. & Palleroni, A. (2002) Foraging ecology of reintroduced captive-bred subadult harpy eagles (*Harpia harpyja*) on Barro Colorado Island, Panama. *Ornitología Neotropical*, 13, 365–379.

(12) Brown, J. L., Heinrich, W. R., Jenny, J. P. & Mutch, B. D. (2004) Development of hunting behaviour in hacked Aplomado falcons. *Journal of Raptor Research*, 38, 148–152.

(13) Nicoll, M. A. C., Jones, C. G. & Norris, K. (2004) Comparison of survival rates of captive-reared and wild-bred Mauritius kestrels (*Falco punctatus*) in a re-introduced population. *Biological Conservation*, 118, 539–548.

(14) Brown, J. L., Collopy, M. W., Gott, E. J., Juergens, P. W., Montoya, A. B. & Hunt, W. G. (2006) Wild-reared Aplomado falcons survive and recruit at higher rates than hacked falcons in a common environment. *Biological Conservation*, 131, 453–458.

16.8.10. Owls

- A study in the USA (1) found that a barn owl *Tyto alba* population was established following the release of 157 birds in the area over three years.
- A replicated, controlled study in Canada (2) found that released burrowing owls *Athene cunicularia* had similar reproductive output, but higher mortality than wild birds, and no released birds returned after migration, although return rates for released birds' offspring were no different from wild birds.

A study in 1985 and 1986 (1) found three barn owl *Tyto alba* nests each year in a riverine marsh site in Missouri, USA, with at least 11 eggs being laid and a minimum of seven chicks fledging (complete data are not included). The site was the location of a reintroduction programme in 1983–5 which released 157 owls, and at least two and probably more of the parents at the nests found were captive-bred.

A replicated, controlled study in mixed grasslands in Saskatchewan, Canada, in the springs of 1997–2000 and 2002 (2), found that 12 of 26 pairs of burrowing owls *Athene cunicularia* released together stayed together for the first breeding season, with eight pairs fledging a total of 43 young. Six birds paired with wild owls, raising 31 young in their first years. Reproductive output did not differ between wild and captive pairs, but mortality of released owls was significantly higher than wild birds (19% of 52 released birds dying vs. 4% of 780 wild birds). Five birds (10%) failed to migrate and no released birds returned after migration. Only one offspring from released birds returned to the area the following year, but this was not significantly different from return rates for the offspring of wild birds. This study also describes fostering and release techniques (see 'Foster chicks with wild conspecifics' and 'Use holding pens at release sites').

(1) Henke, R. J. & Crawford, W. C. (1987) Common barn-owls from captive propagation found nesting in the wild. *Journal of Raptor Research*, 21, 74.
(2) Poulin, R. G., Danielle Todd, L., Wellicome, T. I. & Brigham, R. M. (2006) Assessing the feasibility of release techniques for captive-bred burrowing owls. *Journal of Raptor Research*, 40, 142–150.

16.8.11. Pigeons

- A single review of a captive-release programme in Mauritius (1) found that that released pink pigeons *Nesoenas mayeri* had a first year survival of 36%.

A review of a pink pigeon *Nesoenas mayeri* (formerly *Columba mayeri*) release programme in mixed forest habitats at Black River Gorges in southern Mauritius between 1987 and 1992 (1) found that 36% of 42 pigeons were known to be alive one year after release. This study is also discussed in 'Provide supplementary food to increase adult survival', 'Use captive breeding to increase or maintain populations', 'Provide supplementary food after release' and 'Predator control on islands'.

(1) Jones, C., Swinnerton, K., Taylor, C. & Mungroo, Y. (1992) The release of captive-bred pink pigeons *Columba mayeri* in native forest on Mauritius. A progress report July 1987–June 1992. *Dodo*, 28, 92–125.

16.8.12. Parrots

- A before-and-after study from Venezuela (2) found that the local population of yellow-shouldered amazons *Amazona barbadensis* increased significantly following the release of captive-bred birds, along with other interventions.
- A replicated study in Costa Rica and Peru (4) found high survival and some breeding of scarlet macaw *Ara macao* after release.
- Three replicated studies in the USA (1), Dominican Republic (3) and Puerto Rico (5) found low survival in released birds (4–41% in the first year after release), although the Puerto Rican study also found that released birds bred successfully.

A replicated study in 1986–93 in pine forests in south-eastern Arizona, USA (1), found that captive-bred thick-billed parrots *Rhynchopsitta pachyrhyncha* released into the wild were significantly less likely to survive for two months after survival than translocated birds caught from the wild (4% survival for 23 captive-bred birds vs. 41% for 69 wild birds). This study is also discussed in 'Translocate individuals' and 'Use holding pens at release sites'.

A 1998 before-and-after study, reviewing a yellow-shouldered amazon *Amazona barbadensis* release programme in semi-dry tropical shrubland on Margarita Island, Venezuela (2), found that the population on the island increased from 750 to approximately 1,900 individuals between 1989 and 1996. Conservation measures are also discussed in: 'Release captive-bred individuals', 'Artificially incubate or hand-rear birds in captivity', 'Foster eggs or chicks with wild conspecifics and 'Use education programmes and local engagement to help reduce pressures on species'. For releases, birds were kept in large outdoor aviaries at the release site and released when either 18 or 30 months old. Food was placed outside aviaries twice daily for 15 days after release and once daily for another 15 days. At least ten of 12 birds released survived for at least a year and integrated into wild groups five days to nine months after release. At least three birds scouted nest holes and one nested and fledged two chicks. The programme is estimated to have cost US$2,800 for each bird.

A replicated study in the Dominican Republic in 1997–8 (3) found that survival rate estimates of captive-reared Hispaniolan parrots *Amazona ventralis* released in a subtropical forest site were only 30–35% for 24 parrots released in 1997 (with seven birds alive 53 weeks after release, 12 definitely dead and five with unknown fates) and 29% for 25 birds released in 1998 (with ten birds definitely dead). In 1997, five birds died within five days of release, however all birds released in 1998 survived at least ten weeks. Mortality in 1998 may have been affected by Hurricane Georges hitting the release site in September 1998. Birds were held in training cages at the release site for a quarantine period of at least 40 days before release. This study is also discussed in 'Use "flying training" before release'.

A replicated study (4) at three scarlet macaw *Ara macao* release centres in Costa Rica and Peru found that annual post-release survival of 71 captive-bred birds and former pets was 89% (77% first-year survival and 96% after). First-year survival ranged from 60% to 90% and survival was higher for birds released in larger groups and in areas with birds already present. Pairs formed at all three sites, with at least four chicks fledged at the Peruvian site. Birds began to breed at four to seven years old. Birds were not raised in isolation from humans and did not show fear of humans after release. Five former pets released all survived for at least two years, but they appeared to socialise less with other released macaws. At the two Costa Rican sites, birds were kept in aviaries at the release sites for at least six months, there was little pre-release training at the Peruvian site.

A replicated study of the release of 34 captive-bred Puerto Rican parrots *Amazona vittata* in a subtropical rainforest in north-east Puerto Rico, in 2000–2 (5), found that first-year survival was estimated at 41% (ten confirmed alive, 13 confirmed dead and 11 unaccounted for). Three released and one wild bird attempted to breed in 2004: one attempt (by a pair of birds released in 2002) failed, but the other (with a male released in 2001 and a wild female) successfully fledged two chicks. Seven mortalities (54%) were due to avian predation. Birds were held for four months in large aviaries close to the release site before being moved to acclimatisation cages at the release site one month before release. Birds were given flight and predator aversion training.

(1) Snyder, N. F. R., Koenig, S. E., Koschmann, J., Snyder, H. A. & Johnson, T. B. (1994) Thick-billed parrot releases in Arizona. *The Condor*, 96, 845–862.

(2) Sanz, V. & Grajal, A. (1998) Successful reintroduction of captive-raised yellow-shouldered amazon parrots on Margarita Island, Venezuela. *Conservation Biology*, 12, 430–441.

(3) Collazo, J. A., White Jr, T. H., Vilella, F. J. & Guerrero, S. A. (2003) Survival of captive-reared Hispaniolan parrots released in Parque Nacional del Este, Dominican Republic. *The Condor*, 105, 198–207.

(4) Brightsmith, D., Hilburn, J., Del Campo, A., Boyd, J., Frisius, M., Frisius, R., Janik, D. & Guillen, F. (2005) The use of hand-raised psittacines for reintroduction: a case study of scarlet macaws (*Ara macao*) in Peru and Costa Rica. *Biological Conservation*, 121, 465–472.
(5) White Jr, T. H., Collazo, J. A. & Vilella, F. J. (2005) Survival of captive-reared Puerto Rican parrots released in the Caribbean National Forest. *The Condor*, 107, 424–432.

16.8.13. Songbirds

- A before-and-after study in Mauritius (6) describes the establishment of a population of Mauritius fody *Foudia rubra* following the release of captive-bred individuals.
- Four studies (1–4) of three release programmes on Hawaii found high survival of all three species released (Hawaiian crows *Corvus hawaiiensis* and two thrushes: omao *Myadestes obscurus* and puaiohi *M. palmeri*), with the two thrushes successfully breeding. The authors in one (3) note that many of the released puaiohi dispersed from the release site, meaning that repopulating specific areas may require multiple releases.
- A replicated, controlled study from the USA (5) found that San Clemente loggerhead shrike *Lanius ludovicianus mearnsi* pairs with captive-bred females had lower reproductive success than pairs where both parents were wild-bred.

A replicated study on Hawaii, USA, in 1993–4 (1) found that at least ten of 12 Hawaiian crows (alala) *Corvus hawaiiensis* released into the wild survived for at least one month (with three bird surviving at least a year). The status of the other two birds was unknown. Eight of the released birds (including both with unknown statuses) were hand-reared from wild eggs (see 'Artificially incubate and hand-rear birds in captivity' for details), the remaining four were captive-bred birds. Birds were transferred to small cages at the release site when 46–63 days old and then into a larger aviary when 62–96 days old. Birds were then slowly released, with the timing dependent on their ability to fly and find food. Supplementary food was provided for several months after release and non-native predators (mongoose *Herpestes auropunctatus* and black rats *Rattus rattus*) were trapped from around the aviary whilst releases were on-going (see 'Invasive and other problematic species' for more studies of invasive species control).

A replicated study in 1995–6 on Hawaii, USA (2), found that 80% of 25 captive-bred omao *Myadestes obscurus* (a thrush) survived for at least 30 days after being released, with at least two chicks being raised. The same study found that 14 (six male, eight female) captive-bred puaiohi *Myadestes palmeri* (a critically endangered thrush) released at a marshland site on Kaua'i, Hawaii, USA, in 1999 successfully fledged at least seven chicks (from six pairs). Both species were 'hacked' by being kept in predator-proof cages at the release site for 6–14 days before release. Food was provided for 17 days after release and predators (feral cats and rats) were poisoned and trapped for 2.5 months before the first puaiohi releases. Details of survival are provided in (3). This study is also discussed in 'Use captive breeding to increase or maintain populations' and 'Artificially incubate and hand-rear birds in captivity'.

A replicated study (3) reviewing the same programme as in (2) found that all 14 captive-bred puaiohi *Myadestes palmeri* released survived for at least 56 days after release. Six of the birds (43%) established breeding territories and two of the remaining females formed pairs with local males. The authors note that repopulating specific areas may require multiple releases because of the 57% dispersal out of the release area.

A continuation of the programme described in (3), found that 91% of 21 female and 13 male puaiohi *Myadestes palmeri* released between 1999 and 2001 survived to independence (defined as 30 days after release) (4). Seventy-five percent of 12 birds monitored for longer survived the next 50 days. All 12 birds (ten female, two male) monitored during the breeding season had active nests, with 31 nests being built over two years by the ten females and 28 becoming active. The fate of 24 nests was known, with 42% fledging at least one young and 38% being predated (probably by rats). Clutch size (average of 2 eggs/nest, 16 nests), daily survival rates (97%) and fledglings/successful nest (1.4 fledglings/nest, ten nests) were similar for released and wild birds, although fewer fledglings/active nest were produced (0.58 fledglings/nest vs. 1.1 fledglings/nest). Release techniques were the same as in (3), but food was provided for up to 30 days.

A controlled, replicated study on San Clemente Island, California, USA, between 2000 and 2006 (5) found that pairs of San Clemente loggerhead shrikes *Lanius ludovicianus mearnsi* with captive-bred females produced fewer fledglings and reared fewer chicks to independence than pairs with wild-bred females (2.6 fledglings/pair and 1.9 independent young/pair for 65 breeding attempts with captive-bred females vs. 3.5 fledglings/pair and 2.6 independent young/pair for 107 attempts with wild-bred females). The same pattern was seen with the origin of the male in a pair, but this was not a significant effect (2.6 fledglings/pair and 1.9 independent young/pair for 54 breeding attempts with captive-bred males vs. 3.6 fledglings/pair and 2.6 independent young/pair for 118 attempts with wild-bred females). Other interventions used are discussed in 'Control predators on islands' and 'Provide supplementary food to increase reproductive success'.

A before-and-after study on Ile aux Aigrettes, Mauritius (6), reports that the release of 93 captive-bred Mauritius fodies *Foudia rubra* in the breeding seasons of 2003–4, 2004–5 and 2005–6 has led to the establishment of a population of 142 individuals and 47 breeding pairs by December 2008. Survival to one year was between 33% (2003–4) and 75% (2005–6), with increases possibly due to the presence of established birds in later years. The first successful breeding was during 2004–5, when five chicks from two females fledged. This increased to 40 from 19 in 2005–6 and 47 from 38 in 2006–7. First-year survival for wild-bred birds was 60–88%. Birds were kept in large aviaries at the release site for at least seven days before release (birds that had not been put in large aviaries before were first placed in small cages within aviaries) and fed a diet of fruit, commercial insectivore food and eggs. Adults were released in groups of one or two (after 30 days in the aviaries), whereas juveniles were released in groups of two to nine birds. Food was provided continuously at the release site.

(1) Kuehler, C., Harrity, P., Lieberman, A. & Kuhn, M. (1995) Reintroduction of hand-reared alala *Corvus hawaiiensis* in Hawaii. *Oryx*, 29, 261–266.

(2) Kuehler, C., Lieberman, A., Oesterle, P., Powers, T., Kuhn, M., Kuhn, J., Nelson, J., Snetsinger, T., Herrmann, C., Harrity, P., Tweed, E., Fancy, S., Woodworth, B. & Telfer, T. (2000) Development of restoration techniques for Hawaiian thrushes: Collection of wild eggs, artificial incubation, hand-rearing, captive-breeding, and re-introduction to the wild. *Zoo Biology*, 19, 263–277.

(3) Tweed, E. J., Foster, J. T., Woodworth, B. L., Oesterle, P., Kuehler, C., Lieberman, A. A., Powers, A. T., Whitaker, K., Monahan, W. B., Kellerman, J. & Telfer, T. (2003) Survival, dispersal, and home-range establishment of reintroduced captive-bred puaiohi, *Myadestes palmeri*. *Biological Conservation*, 111, 1–9.

(4) Tweed, E. J., Foster, J. T., Woodworth, B. L., Monahan, W. B., Kellerman, J. L. & Lieberman, A. (2006) Breeding biology and success of a reintroduced population of the critically endangered puaiohi (*Myadestes palmeri*). *The Auk*, 123, 753–763.

(5) Heath, S. R., Kershner, E. L., Cooper, D. M., Lynn, S., Turner, J. M., Warnock, N., Farabaugh, S., Brock, K. & Garcelon, D. K. (2008) Rodent control and food supplementation increase productivity of endangered San Clemente Loggerhead Shrikes (*Lanius ludovicianus mearnsi*). *Biological Conservation*, 141, 2506–2515.

(6) Cristinacce, A., Handschuh, M., Switzer, R. A., Cole, R. E., Tatayah, V., Jones, C. G. & Bell, D. (2009) The release and establishment of Mauritius fodies *Foudia rubra* on Ile aux Aigrettes, Mauritius. *Conservation Evidence*, 6, 1–5.

16.9. Use appropriate populations to source released populations

- A replicated study from Sweden (1) and a small study from France (2) found that birds sourced from populations distant from where they were released were less successful than birds from the area.

- In Sweden, released white storks *Ciconia ciconia* from North Africa produced fewer than half the chicks as those that naturally re-colonised, whilst both studies found that storks and little bustards *Tetrax tetrax* were less likely to migrate than birds originating in the release area.

Background

Some bird species follow instinctive migration patterns, which vary between populations (Berthold *et al.* 1984), and populations are likely to be adapted for local environmental conditions. If individuals from one population are released into another area, this could result in maladaptive behaviours, with birds migrating in the wrong direction, or breeding at the wrong time. Sourcing individuals for release from the area that they will be released into may, therefore, improve success rates.

Berthold, P. (1984) The endogenous control of bird migration: a survey of experimental evidence. *Bird Study*, 31, 19–27.

A replicated study in southern Sweden in 1989–2005 (1) found white storks *Ciconia ciconia* that naturally re-colonised the region in 1989 from the nearest remaining population (in north-east Europe) and their direct descendants fledged over twice as many chicks as birds descended from a reintroduced population which originated in north Africa (average of 1.9 fledglings/pair for birds descended from wild birds vs. 0.9 fledglings/pair for birds descended just from reintroduced birds). In addition, birds with wild ancestry were significantly more likely to migrate than birds only descended from captive individuals (11 of 18 storks confirmed as migrating had some wild ancestry, as did eight of ten storks that probably migrated. A total of 101 storks in the population had some wild ancestry, compared to 189 descended solely from captive storks). The original reintroduction was of 15 birds from a breeding centre in Switzerland, of which eight bred, leading to 470 descendants between 1980 and 2005. Approximately 82% of the current Swedish population is descended from four captive birds. A total of 12 native birds re-colonised, with 14% of the total population being descended from four of these.

A small study in southern France and Spain in 1997–2007 (2) found that six little bustards *Tetrax tetrax* originating in Spain but hand-reared and released in France did not migrate to Spain (with the possible exception of one bird that could not be tracked). By contrast, 13 out of 21 wild adults from France (62%) and six of eight hand-reared French chicks (75%) migrated. The authors conclude that hand-rearing does not affect migration probability, but that genetic origin appears to.

(1) Olsson, O. (2007) Genetic origin and success of reintroduced white storks. *Conservation Biology*, 21, 1196–1206.
(2) Villers, A., Millon, A., Jiguet, F., Lett, J. M., Attie, C., Morales, M. B. & Bretagnolle, V. (2010) Migration of wild and captive-bred little bustards *Tetrax tetrax*: releasing birds from Spain threatens attempts to conserve declining French populations. *Ibis*, 152, 254–261.

16.10. Use holding pens at release sites

- Three replicated and one small study from three release programmes in Saudi Arabia (2,3), the USA (4) and Canada (5) found that released birds had higher survival (2–4) or were more likely to pair up (5) if kept at release sites in holding pens before release.
- A replicated study in the USA (1) found lower survival for thick-billed parrots *Rhynchopsitta pachyrhyncha* released in holding pens, compared to birds released without preparation.
- A review of northern bald ibis *Geronticus eremita* conservation (6) found that holding pens successfully prevented most birds from migrating (which resulted in 100% mortality), although some 200 birds 'escaped' over 25 years.

Background

Captive-bred individuals may well take time to adjust to life in the wild, without regular food and potentially in an unfamiliar habitat. It may therefore increase the chances of birds surviving if they are kept at the release site for some time before being released, either enclosed in pens or with access to the habitat outside.

This section includes only those studies that make a comparison between releases using holding pens and those without. Many of the studies in 'Release captive-bred individuals' describe releases that use holding pens, but do not test the effects of using them.

A replicated study in south-eastern Arizona, USA (1), found that survival in thick-billed parrots *Rhynchopsitta pachyrhyncha* translocated and released into the Chiricahua Mountains between September 1986 and September 1993 was higher for birds that were released without preparation into the wild, compared to those that were soft-released. This difference held both for all birds (26% of 69 soft release birds alive after two months vs. 48% of 23 'standard release' birds) and for wild-caught birds (37% of 49 soft release birds vs. 63% of 16 standard releases). Soft release involved preconditioning to the local environment and supplying birds with local food while in captivity. This study is discussed in more detail in 'Release captive-bred individuals' and 'Translocate individuals'.

A replicated study, reviewing a houbara bustard *Chlamydotis undulata macqueenii* release programme in south-west Saudi Arabia between 1991 and 1993 (2) found that releases were most successful when subadult birds were released into a large (4 km^2) fenced enclosure, compared with releases of birds without an enclosure. All four birds released without an enclosure were killed by foxes within three days, whilst one of 25 birds released into the enclosure survived for at least seven months and ten other established themselves in territories in the release site. Twelve of the dead birds were predated and two contracted pox. Other release techniques and descriptions of the captive-breeding programme are discussed in 'Use captive breeding to increase or maintain populations', 'Use artificial insemination in captive breeding', 'Release captive-bred individuals', 'Release birds in "coveys" and "Use holding pens at site of release and clip birds" wings'.

A replicated, controlled study in 1991–4 in desert steppe in Saudi Arabia (3) found that the survival of released houbara bustards *Chlamydotis undulata macqueenii* (released as part of the same programme as in (2)) was significantly higher when birds were released into a release pen, compared to direct release into the reserve (48% survival for 59 sub-adults released in 1992–4 vs. 0% survival for birds released into the reserve or the pen without a 'transition period' in 1991). The effect of clipping sub-adults' wings before release is discussed in 'Use holding pens and clip birds' feathers', with further details of the programme in 'Release captive bred individuals', 'Release birds as sub-adults or adults, not juveniles' and 'Control predators not on islands'.

A replicated study reviewing a reintroduction programme in two prairie sites in Texas, USA, in 1996–7 (4) found that six month survival rates of released, captive-bred and hand-reared Attwater's prairie chickens *Tympanuchus cupido attwater* (an endangered subspecies of the greater prairie chicken) were higher for birds that spent 14 days in an 'acclimatisation pen' (47% of 97 birds alive after six months) compared to birds that spent only three days in the pens (19% of 31 birds alive). Further analysis revealed that this difference was due to increased survival in the first two weeks after release. This study is discussed further in 'Release captive-bred individuals'.

A small study at a mixed grassland site in Saskatchewan, Canada, in the springs of 1997–2000 and 2002 (5), found that released pairs of burrowing owls *Athene cunicularia* were more likely to form pairs when they were held in release enclosures for five days, or until they began laying, compared with birds only held for three days before the enclosure was removed (all eight pairs remained together when enclosures were kept for five days or until laying vs. 22% of 18 pairs remaining together when enclosures were kept for just three days). This study is also discussed in 'Release captive bred individuals' and 'Foster chicks with wild conspecifics'.

A 2007 review of northern bald ibis (waldrapp) *Geronticus eremita* conservation (6) found that the seasonal use of holding pens, designed to prevent birds from migrating from a release site in Turkey, kept the majority of released individuals in the area, but over 25 years some 200 birds avoided capture and were 'lost' from the colony. Previous work had found that migrating birds were unlikely to survive or return. These birds migrated but never returned the following spring. This study is also discussed in 'Release captive-bred individuals into the wild to restore or augment wild populations', 'Use captive breeding to increase or maintain populations', 'Artificially incubate and hand-rear birds in captivity', 'Release birds as adults or sub-adults, not juveniles', 'Clip birds' wings on release', 'Use microlites to help birds migrate' and 'Foster birds with non-conspecifics'.

(1) Snyder, N. F. R., Koenig, S. E., Koschmann, J., Snyder, H. A. & Johnson, T. B. (1994) Thick-billed parrot releases in Arizona. *The Condor*, 96, 845–862.
(2) Jaime, M. S., Combreau, O., Seddon, P. J., Paillat, P., Gaudier, P. & Heezik, Y. (1996) Restoration of *Chlamydotis undulata macqueenii* (houbara bustard) populations in Saudi Arabia: A progress report. *Restoration Ecology*, 4, 81–87.
(3) Combreau, O. & Smith, T. R. (1998) Release techniques and predation in the introduction of houbara bustards in Saudi Arabia. *Biological Conservation*, 84, 147–155.
(4) Lockwood, M. A., Clifton P. Griffin, Morrow, M. E., Randel, C. J. & Silvy, N. J. (2005) Survival, movements, and reproduction of released captive-reared Attwater's prairie-chicken. *The Journal of Wildlife Management*, 69, 1251–1258.
(5) Poulin, R. G., Danielle Todd, L., Wellicome, T. I. & Brigham, R. M. (2006) Assessing the feasibility of release techniques for captive-bred burrowing owls. *Journal of Raptor Research*, 40, 142–150.
(6) Bowden, C. G. R., Boehm, C., Jordan, M. J. R. & Smith, K. W. (2007) Why is reintroduction of northern bald ibis *Geronticus eremita* so complicated? An overview of recent progress and potential. 27–35 in: M. M. Lamont (eds) *The Proceedings of the IV International Symposium on Breeding Birds in Captivity; 2007 Sept 12–Sept 16* Toronto, Ontario, Canada.

16.11. Clip birds' wings on release

- Two studies from Saudi Arabia and Hawaii found that bustards (1) and geese (2) had lower survival when released in temporary exclosures with clipped wings, compared to birds released with unclipped wings.
- A review of cackling goose *Branta hutchinsii* conservation (3) found that wing-clipped or moulting wild adult geese proved a better strategy than releasing young geese.
- A review of northern bald ibis (waldrapp) *Geronticus eremita* conservation (4) found no differences in survival between birds released with clipped and unclipped wings in Israel.

Background

Holding pens are designed to allow birds to adjust to their environment before being released, but if they leave before they are ready, the pens may have no effect. Therefore preventing birds from leaving by clipping their wings may increase the time they spend in the pens and their survival.

A similar intervention is the use of 'wing brails' to prevent birds from flying. Brails are bands attached to wings to prevent birds extending them and so preventing birds from flying. We captured no evidence of the effects of wing brails, but several studies describe their use in the release of cranes Grus spp. These are described in 'Release captive-bred individuals'.

A review of a houbara bustard *Chlamydotis undulata macqueenii* release programme in south-west Saudi Arabia between 1991 and 1993 (1) found that three to five month sub-adult birds released in a large (4 km^2) fenced enclosure (designed to reduce predation by mammalian predators) with clipped wings had significantly lower survival (2 of 13 birds surviving to join wild birds), compared to two month-old birds released with unclipped wings (one of 25 birds released survived for at least seven months, ten other established territories in the release site). Six of the wing-clipped birds were killed by avian predators within the enclosure, one died of pox and three were killed by mammalian predators after they left the enclosure. Twelve of the unclipped birds were also predated. Other release techniques and descriptions of the captive-breeding programme are discussed in 'Use captive breeding to increase or maintain populations', 'Use artificial insemination in captive breeding', 'Release captive-bred individuals', 'Release birds in 'coveys' and 'Use holding pens at site of release'.

A 1997 review of the Hawaiian goose (nene) *Branta sandvicensis* reintroduction programme (2) concluded that birds released into temporary exclosures with their wings clipped survived less well than those released into the wild before fledging. This study is discussed in more detail in 'Release captive-bred individuals into the wild to restore or augment wild populations'.

A before-and-after study from the Aleutian Islands, Alaska, USA on the cackling goose *Branta hutchinsii* recovery programme (3) found that releasing wing-clipped or moulting wild adult geese proved a better strategy than releasing young geese (captive or wild, see 'Release birds as adults or sub-adults, not juveniles'). The authors note that the release of captive-bred geese was not very successful overall. This study also investigates the effect of Arctic fox *Alopex lagopus* control on breeding islands (see 'Predator control on islands').

A 2007 review of northern bald ibis (waldrapp) *Geronticus eremita* conservation in Israel (4) found no differences in survival between 16 birds released with clipped wings and 40 birds released without clipping. All 56 birds released became emaciated and disorientated and formed poor social bonds. This study is also discussed in 'Release captive-bred individuals into the wild to restore or augment wild populations', 'Use captive breeding to increase or maintain populations', 'Artificially incubate and hand-rear birds in captivity', 'Use holding pens at release sites, 'Release birds as adults or sub-adults, not juveniles', 'Use microlites to help birds migrate' and 'Foster birds with non-conspecifics'.

(1) Jaime, M. S., Combreau, O., Seddon, P. J., Paillat, P., Gaudier, P. & Heezik, Y. (1996) Restoration of *Chlamydotis undulata macqueenii* (houbara bustard) populations in Saudi Arabia: A progress report. *Restoration Ecology*, 4, 81–87.
(2) Black, J. M., Marshall, A. P., Gilburn, A., Nelson Santos, Hoshide, H., Medeiros, J., Mello, J., Hodges, C. N. & Katahira, L. (1997) Survival, movements, and breeding of released Hawaiian geese: an assessment of the reintroduction program. *The Journal of Wildlife Management*, 61, 1161–1173.
(3) USFWS (2001) Aleutian Canada goose road to recovery.
(4) Bowden, C. G. R., Boehm, C., Jordan, M. J. R. & Smith, K. W. (2007) Why is reintroduction of northern bald ibis *Geronticus eremita* so complicated? An overview of recent progress and potential. 27–35 in: M. M. Lamont (eds) *The Proceedings of the IV International Symposium on Breeding Birds in Captivity; 2007 Sept 12–Sept 16* Toronto, Ontario, Canada.

16.12. Release birds in groups

- A replicated study from New Zealand (1) found that released black stilts *Himantopus novaezelandiae* were more likely to move long distances after release if they were released in larger groups.

A replicated study of critically endangered black stilt (kaki) *Himantopus novaezelandiae* releases in South Island, New Zealand, between 1993 and 2005 (1) found that birds were more likely to move long distances from the release site when released in large groups, compared to birds released in smaller numbers. This study is discussed in more detail in 'Release captive-bred individuals' and 'Release birds as adults or sub-adults, not juveniles'.

(1) van Heezik, Y., Maloney, R. F. & Seddon, P. J. (2009) Movements of translocated captive-bred and released critically endangered kaki (black stilts) *Himantopus novaezelandiae* and the value of long-term post-release monitoring. *Oryx*, 43, 639–647.

16.13. Release chicks and adults in 'coveys'

- A replicated study in Saudi Arabia (1) found that houbara bustard *Chlamydotis undulata macqueenii* survival was low when chicks were released in coveys with flightless females.

- A review of cackling goose *Branta hutchinsii* conservation (2) and a replicated study in England (3) found that geese and grey partridge *Perdix perdix* releases were more successful for birds released in coveys than for young birds released on their own or adults released in pairs.

Background

Coveys are 'family groups' of birds (normally gamebirds): broods of chicks with an accompanying adult. Groups of birds may be safer from predators than individual chicks and the presence of parents or foster parents may also help survival.

A replicated study reviewing a houbara bustard *Chlamydotis undulata macqueenii* release programme in central Saudi Arabia between 1991 and 1993 (1) found that releasing coveys of chicks into a large (4 km^2) enclosure designed to exclude mammalian predators had low levels of success. Eight coveys with a total of 15 chicks were released along with eight females (rendered flightless). Of these only five birds integrated into wild flocks with five leaving the enclosure and being predated by mammalian predators; two were killed inside the enclosure and three died of disease. Other release techniques and descriptions of the captive-breeding programme are discussed in 'Use captive breeding to increase or maintain populations', 'Use artificial insemination in captive breeding', 'Release captive-bred individuals', 'Use holding pens at site of release' and 'Use holding pens at site of release and clip birds' wings'.

A before-and-after study from the Aleutian Islands, Alaska, USA on the cackling goose *Branta hutchinsii* recovery programme (2) found that releasing geese in family groups proved a better strategy than releasing young geese (captive or wild, see 'Release birds as adults or sub-adults, not juveniles'). The authors note that the release of captive-bred geese was not very successful overall. This study also investigates the effect of Arctic fox *Alopex lugopus* control on breeding islands (see 'Predator control on islands').

A replicated study on four farms in Gloucestershire and Oxfordshire, England, in 2007 (3) found that grey partridge *Perdix perdix* released in coveys in autumn had significantly higher survival (78% survival of 92 monitored birds over 13 days) than adult birds released in pairs in spring (42% survival for 70 birds). The authors suggest that the differences were due to different habitat use: spring-released birds spent a lot of time in fields and field margins, whilst autumn birds spent more time in game cover crops. Use of field margins was negatively associated with survival, whilst use of crops was positively associated.

(1) Jaime, M. S., Combreau, O., Seddon, P. J., Paillat, P., Gaudier, P. & Heezik, Y. (1996) Restoration of *Chlamydotis undulata macqueenii* (houbara bustard) populations in Saudi Arabia: A progress report. *Restoration Ecology*, 4, 81–87.
(2) USFWS (2001) Aleutian Canada goose road to recovery.
(3) Rantanen, E. M., Buner, F., Riordan, P., Sotherton, N. & Macdonald, D. W. (2010) Habitat preferences and survival in wildlife reintroductions: an ecological trap in reintroduced grey partridges. *Journal of Applied Ecology*, 47, 1357–1364.

16.14. Release birds as adults or sub-adults, not juveniles

- Three replicated studies (3,6,7) found that malleefowl *Leipoa ocellata*, houbara bustards *Chlamydotis undulata macqueenii* and cackling geese *Branta hutchinsii* released as sub-adults, not juveniles had higher survival rates.
- A replicated study from New Zealand (9) found lower survival for black stilts *Himantopus novaezelandiae* released as sub-adults, compared with juveniles.
- Two replicated studies from Hawaii (5) and Saudi Arabia (6) found lower survival for Hawaiian geese *Branta sandvicensis* and bustards released as wing-clipped sub-adults, compared with birds released as juveniles.

- Three replicated studies (1,2,8) found no differences in survival between ducks (1), vultures (2) and ibises (8) released at different ages, but a second study of the vulture release programme (4) found that birds released when more than three years old had lower reproductive success than birds released at an earlier stage.

Background

Most captive-bred birds are released as juveniles as adults may have become too conditioned to life in captivity. Under some circumstances, however, adults or sub-adults may survive better.

A replicated study in parkland in central Saskatchewan, Canada, in 1971–3 (1) found there was no difference in return rate for female canvasbacks *Aythya valisineria* released as flightless young or when a year old (three of 43 birds released when flightless vs. three of 50 yearlings), although most year-old birds formed brief pairs when released. Overall survival and return rates are discussed in 'Release captive-bred individuals into the wild to restore or augment wild populations'.

A ten-year study of a griffon vulture *Gyps fulvus* reintroduction programme in river gorges in Aveyron, southern France (2) found that annual survival rates were similar or higher for birds released as adults (74% in the first year after release and 98% from the second year onwards, 39 birds released), compared to those released as immature (75% during the first two years after release, 20 birds released). This study is also discussed in 'Release captive-bred individuals' and 'Use education programmes and local engagement to help reduce pressures on species'.

A replicated study in mallee scrub in New South Wales, Australia, in 1987–90 (3) found that survival of released malleefowl *Leipoa ocellata* was significantly higher for birds released as sub-adults (14–28 months old) than for birds released as juveniles (3–5 months old) (four sub-adults, of 12 released – 33% survived for 36 days; three for at least 428 days and two for at least 787 days vs. all 24 juveniles were dead within 104 days with at least 83% dead within 36 days). The difference in survival was only evident more than eight days after release, with 50% and 42% of sub-adult and juvenile birds surviving the first eight days respectively. At least 21 juveniles (87%) and seven sub-adults (58%) were killed by predators (mainly foxes *Vulpes vulpes*). Eggs were collected from the wild and artificially incubated and the chicks fed on seeds, mealworms and vegetables before release.

A replicated study over 12-years (4) of the same programme as in (2), found that the nesting success of griffon vultures *Gyps fulvus* released at the age of three or more (0.42 fledglings/pair for 103 nesting attempts) was significantly lower than that of younger releases and wild-bred birds (0.82 young/pair for 11 attempts). This difference was partially due to lower hatching success for older released birds (55% hatching success for 79 eggs), compared with younger releases and wild-bred birds (75% for 11 eggs). The overall success of the programme is discussed in 'Release captive-bred individuals into the wild to restore or augment wild populations'.

A 1997 review of the Hawaiian goose (nene) *Branta sandvicensis* reintroduction programme (5) concluded that birds released into temporary exclosures before fledging survived better than older birds released in exclosures with their wings clipped. This study is discussed in more detail in 'Release captive-bred individuals into the wild to restore or augment wild populations'.

A replicated study at a desert steppe site in 1992–4 in south-west Saudi Arabia (6) found that houbara bustards *Chlamydotis undulata macqueenii* released as sub-adults had higher survival than those released as chicks, except when released with their wings clipped (48% survival for 59 unclipped sub-adults released in 1992–4 vs. 36% of 14 chicks released in 1993 and 17% of 12 sub-adults with clipped wings released in 1992). Details of the chick releases are discussed in 'Release birds in "coveys"'. This study is also discussed in 'Release captive-bred individuals' and 'Use holding pens at release sites'. The effect of predator removal is discussed in 'Control predators not on islands'.

A before-and-after study from the Aleutian Islands, Alaska, USA on the cackling goose *Branta hutchinsii* recovery programme (7) found that releasing young geese was less successful than other strategies (see 'Clip birds' wings on release' and 'Release birds in 'coveys'). However, the authors note that the release of captive-bred geese was not very successful overall. This study also investigates the effect of Arctic fox *Alopex lugopus* control on breeding islands (see 'Predator control on islands').

A 2007 review of northern bald ibis (waldrapp) *Geronticus eremite* conservation (8) found no differences in survival between birds released in Israel as breeding adults, juveniles or fledglings. All 56 birds released became emaciated and disorientated and formed poor social bonds. This study is also discussed in 'Release captive-bred individuals into the wild to restore or augment wild populations', 'Use captive breeding to increase or maintain populations', 'Artificially incubate and hand-rear birds in captivity', 'Use holding pens at release sites, 'Clip birds' wings on release', 'Use microlites to help birds migrate' and 'Foster birds with non-conspecifics'.

A review of black stilt (kaki) *Himantopus novaezelandiae* releases in South Island, New Zealand, between 1993 and 2005 (9) found that 20% of 150 birds released as juveniles (60–90 days old) and 13% of those released as sub-adults (nine months old) were alive two years after release (with 25 juvenile releases and 52 sub-adults not yet at breeding age). Neither group was more likely to be seen at the release site. Eggs came from both wild and captive birds and were artificially-incubated until hatching. This study is discussed in more detail in 'Release captive-bred individuals' and 'Release birds in groups'.

(1) Sugden, L. G. (1976) Experimental release of canvasbacks on breeding habitat. *The Journal of Wildlife Management*, 40, 716–720.

(2) Sarrazin, F., Bagnolini, C., Pinna, J. L., Danchin, E. & Clobert, J. (1994) High survival estimates of griffon vultures (*Gyps fulvus fulvus*) in a reintroduced population. *The Auk*, 111, 853–862.

(3) Priddel, D. & Wheeler, R. (1996) Effect of age at release on the susceptibility of captive-reared malleefowl *Leipoa ocellata* to predation by the introduced fox *Vulpes vulpes*. *Emu*, 96, 32–41.

(4) Sarrazin, F., Bagnolini, C., Pinna, J. L. & Danchin, E. (1996) Breeding biology during establishment of a reintroduced griffon vulture *Gyps fulvus* population. *Ibis*, 138, 315–325.

(5) Black, J. M., Marshall, A. P., Gilburn, A., Nelson Santos, Hoshide, H., Medeiros, J., Mello, J., Hodges, C. N. & Katahira, L. (1997) Survival, movements, and breeding of released Hawaiian geese: an assessment of the reintroduction program. *The Journal of Wildlife Management*, 61, 1161–1173.

(6) Combreau, O. & Smith, T. R. (1998) Release techniques and predation in the introduction of houbara bustards in Saudi Arabia. *Biological Conservation*, 84, 147–155.

(7) USFWS (2001) Aleutian Canada goose road to recovery.

(8) Bowden, C. G. R., Boehm, C., Jordan, M. J. R. & Smith, K. W. (2007) Why is reintroduction of northern bald ibis *Geronticus eremita* so complicated? An overview of recent progress and potential. 27–35 in: M. M. Lamont (eds) *The Proceedings of the IV International Symposium on Breeding Birds in Captivity; 2007 Sept 12–Sept 16* Toronto, Ontario, Canada.

(9) van Heezik, Y., Maloney, R. F. & Seddon, P. J. (2009) Movements of translocated captive-bred and released critically endangered kaki (black stilts) *Himantopus novaezelandiae* and the value of long-term post-release monitoring. *Oryx*, 43, 639–647.

16.15. Use 'anti-predator training' to improve survival after release

- A review from Pakistan (1) and a small trial from Saudi Arabia (2) found that pheasants (1) and bustards (2) had higher survival after release, when given pre-release predator training, compared to birds without training, many of which were predated.
- The Saudi Arabian study found that introducing a model fox (as opposed to a live predator) to cages did not increase post-release survival. Introducing a live fox to the cage increased post-release survival more than other techniques used.

Background

If individuals have been raised in captivity, it is unlikely that they will have encountered predators and, in consequence, may have lost some of their innate fear leaving them vulnerable after release. Using predators under controlled conditions can therefore teach them to avoid dangerous species and potentially increase post-release survival.

A review of a 1978–89 reintroduction programme for cheer pheasants *Catreus wallichii* in northern Pakistan (1) found that post-release survival was low between 1978 and 1985, with all 17 birds released in 1981 predated by foxes *Vulpes vulpes*. This was thought to be because birds nested on the ground and were relatively fearless due to rearing techniques. From 1982 onwards, birds were flushed into trees at dusk by workers in their release enclosures and appeared to survive better, with 10–15% of 305 birds released in 1986 and 1988–9 surviving for at least one year. This programme is discussed in more detail in 'Release captive-bred individuals'.

A small trial in Saudi Arabia in 1995–7 (2) found that captive-bred houbara bustards *Chlamydotis undulata macqueenii* released at a desert site were significantly more likely to avoid predation if they were exposed to a live fox *Vulpes vulpes* in their cages before release, compared to control birds (raised without contact with predators) (44% of 18 predator-trained birds killed by predators vs. 83% of 18 controls). However, exposing birds to a model fox which 'lunged' at the birds had no impact on post-release survival (53% of 15 model-trained birds killed by predators vs. 36% of 11 controls). Survival of birds exposed to a live fox was also higher than for birds reared with puppets to minimise human contact (42% of 12 puppet-reared birds predated, see 'Use puppets to increase the survival or growth of hand-reared chicks' for details). Exposure to the fox consisted of introducing a hand-reared fox into the birds' cage on a leash for between 40 seconds and 15 minutes (depending on how aggressive the fox was). There were 12 training sessions, with the fox muzzled for eight of them. Three birds were seriously injured during training, one of which later died (birds were then moved to larger cages and the fox more tightly muzzled which prevented any further injuries).

(1) Garson, P. J., Young, L. & Kraul, R. (1992) Ecology and conservation of the cheer pheasant *Catreus wallichii*: studies in the wild and the progress of a reintroduction project. *Biological Conservation*, 59, 25–35.
(2) van Heezik, Y., Seddon, P. J. & Maloney, R. F. (1999) Helping reintroduced houbara bustards avoid predation: effective anti-predator training and the predictive value of pre-release behaviour. *Animal Conservation*, 2, 155–163.

16.16. Use 'flying training' before release

- A replicated study from the Dominican Republic (1) found that captive-reared Hispaniolan parrots *Amazona ventralis* had higher initial survival if they were given pre-release predator training, although this difference was not present a year after release.

Background

Birds in captivity will rarely fly as far or as often as those in the wild, potentially leading to weaker flight muscles.

A replicated study in the Dominican Republic in 1997–8 (1) found that 25 captive-reared Hispaniolan parrots *Amazona ventralis* released in 1998 in a subtropical forest site after intensive pre-release flying 'training' had flight muscles in significantly better condition than 24 birds released in 1997 after less intensive training. Intensive training consisted of exercise (making birds fly around their aviary) three or four times a week, as opposed to twice a week, and every day in the final week before release. Survival over the first five weeks was much higher in 1998 than 1997 (no deaths in 1998 vs. five deaths in 1997), but overall first-year survival estimates were similar in both years (29% survival in 1998 vs. 30–35% in 1997). This study is discussed in more detail in 'Release captive-bred individuals into the wild to restore or augment wild populations'.

(1) Collazo, J. A., White Jr, T. H., Vilella, F. J. & Guerrero, S. A. (2003) Survival of captive-reared Hispaniolan parrots released in Parque Nacional del Este, Dominican Republic. *The Condor*, 105, 198–207.

16.17. Provide supplementary food during and after release

- Three studies from (1–3) found that malleefowl *Leipoa ocellata* (2), Andean condors *Vultur gryphus* (1) and pink pigeons *Nesoenas mayeri* (3) used supplementary food when it was provided after release.
- A replicated, controlled study from Australia (2) found that malleefowl had higher survival when supplied with supplementary food.
- A study in Peru (1) found that supplementary food could be used to increase the foraging range of condors after release, or to guide them back to suitable feeding areas.

Background

Supplementary feeding is a technique frequently used to support bird populations or increase reproductive success (see separate section). Newly released birds may be unable to find enough food when they are first released and therefore supplementary food may offer a way of ensuring they survive until they have learned to forage successfully.

A study using Andean condors *Vultur gryphus* in arid mountains in Peru in 1980–1 to develop release techniques for Californian condors *Gymnogyps californianus* (1) found that both parent- and puppet-reared birds foraged on carcasses provided in the vicinity of the release site. In addition, moving where carcasses were placed and increasing the distance from the release site appeared to help increase the size of the foraging area used by birds. It also allowed researchers to guide birds back to good feeding areas when they were at risk of starvation in bad weather. Carcasses were moved by 50–75 m each day initially, and then by distances of up to 1.5 km as birds began to search more widely. This study is also discussed in 'Artificially incubate and hand-rear birds in captivity' and 'Release captive-bred individuals'.

A replicated controlled trial in New South Wales, Australia, in 1987 (2) found that significantly more malleefowl *Leipoa ocellata* chicks survived the first 30 days after release when provided with supplementary food, compared to control (unfed) birds, birds provided only with water or birds supplied with water but kept in an enclosure with 15 rabbits *Oryctolagus cuniculus* (89% of nine fed chicks survived for 30 days vs. all 20 other chicks surviving for less than 20 days, with 85% dying within eight days of release). The one fed bird that died survived for six days before being drenched in heavy rain and dying. Of the other releases, six were killed by raptors, five died of starvation, five died of chilling following heavy rain, two died of unknown causes, one died of a cloacal blockage, and one was removed after it fractured its leg. All were found to have little or no food in the crop or gizzard, suggesting that food shortage was a contributing factor in all their deaths. Eggs were taken from wild nesting mounds, artificially incubated and released into 1 ha enclosures with electric fences to keep out mammalian predators. Food supplied consisted of 24 kg of seed mix in both feeders and spread on the ground and replenished at least once a week.

A replicated study in mixed forests in Mauritius in 1987–91 (3) found that 61% of 44 released captive pink pigeons *Nesoenas mayeri* (formerly *Columba mayeri*) continued to use a supplementary feeding station at the release site one month after release. This study is also discussed in 'Provide supplementary food to increase adult survival', 'Use captive breeding to increase or maintain populations', 'Release captive bred individuals' and 'Predator control on islands.'

(1) Wallace, M. P. & Temple, S. A. (1987) Releasing captive-reared Andean condors to the wild. *The Journal of Wildlife Management*, 51, 541–550.
(2) Priddel, D. & Wheeler, R. (1990) Survival of malleefowl *Leipoa ocellata* chicks in the absence of ground-dwelling predators. *Emu*, 90, 81–87.
(3) Jones, C., Swinnerton, K., Taylor, C. & Mungroo, Y. (1992) The release of captive-bred pink pigeons *Columba mayeri* in native forest on Mauritius. A progress report July 1987–June 1992. *Dodo*, 28, 92–125.

16.18. Use microlites to help birds migrate

- A review of northern bald ibis Geronticus eremita conservation (1) found that a group of birds followed a microlite from Austria to Italy but none made the return journey.

A 2007 review of northern bald ibis (waldrapp) *Geronticus eremita* conservation (1) found that a group of ibis successfully followed a microlite from Austria to Italy in 2004. However, by 2006, no birds had successfully returned, although several had made northwards journeys of up to 300 km. This study also discusses several other *ex situ* conservation interventions, discussed in the relevant sections.

(1) Bowden, C. G. R., Boehm, C., Jordan, M. J. R. & Smith, K. W. (2007) Why is reintroduction of northern bald ibis *Geronticus eremita* so complicated? An overview of recent progress and potential. 27–35 in: M. M. Lamont (eds) *The Proceedings of the IV International Symposium on Breeding Birds in Captivity; 2007 Sept 12–Sept 16* Toronto, Ontario, Canada.

Index